Flora of Central India (Indore Division)

Editors

Dr. V.B. Diwanji
Dr. C.M. Solanki
Dr. Swati Samvatsar Pandit
Dr. B.S. Annigeri
Dr. Arjun P. Tiwari
Dr. Mujaffar Shaikh
Dr. Seemavati Sisodiya
Dr. Kiran Surage

SCIENTIFIC PUBLISHERS

Published by

SCIENTIFIC PUBLISHERS (INDIA)
5 A, New Pali Road
P.O. Box 91
Jodhpur - 342 001 INDIA
E-mail: info@scientificpub.com
Website: http://www.scientificpubonline .com

Print 2026

ISBN: 978-93-89832-32-7
eISBN: 978-93-89832-34-1

Printed in India

FOREWORD

Numerous scientific texts have been credited with the knowledge of a large number of plants, viz. The Rigveda, The Atharvaveda, besides their repeated mention in the Sacred Ramayana, Srimad Bhagwat and Bhagwat Gita which teach apart from humanity, not only the aesthetic and economic values of the plants, but also our total dependence on plants as a whole. The role of Charak Samhita in this regard is invaluable.

For the biologist of today, particularly the plant taxonomist, the foremost priority that stares, is the proper management and conservation of the present-day global biodiversity. This is possible only through consistent exploration and systematic documentation of the various biotic communities as well as their components.

The advice of the late Fr. Santapau (1956, 1958) holds great relevance in this context. He had rightly suggested that the Universities and colleges should undertake extensive explorations of the plant wealth in their areas.

The volume under discussion pertains to the floristic wealth of eight districts of Indore Division of the State of Madhya Pradesh. All these districts have been systematically explored by the college teachers for their Ph.D. dissertations.

The Flora is represented by 1,432 species of wild and naturalized plants coming within 641 genera and 120 families. The system of Bentham and Hooker has been followed with modifications in the light of The Plant List, version 1.1, Sept., 2013. This monumental work has been brought out by the Royal Botanic Garden, Kew, England, The Missouri Botanical Garden U.S.A. and in collaboration with a very large number of experts drawn from various countries. The nomenclature is based upon this List. There are easy to follow keys for identification of families, genera and species. Basionyms and synonyms are provided as required for understanding the position of a species within the family, in various floras. Local/vernacular names are provided, where available. This is followed by brief description, phenology and distribution.

To the best of my knowledge, this is perhaps the only full-fledged flora, using inputs from The Plant List. Nowadays many research papers are being published, based upon this list.

A unique feature of the flora is a systematic listing of venerated plants, species dedicated to the 27 Nakshatras (Constellations), plants of the 12 Zodiac signs, species under which the Jain Teerthankars obtained salvation etc.

Phytogeographically the flora is classified into 15 provinces, the Indian component being 34% or just 1/3 of the total flora, the rest being contributed by the remaining 14 provinces to a lesser or greater degree.

Another appreciable feature is the list of plants confined to a single district, starting from Alirajpur and ending in Khargone. This automatically singles out the rare and endangered species. Chapter VI is dedicated to conservation strategies and also provides a list of Protected Areas in the state.

The volume includes several thousand names, citation of authors, books, journals etc., and certainly therefore, some anomalies might have crept in. The editor will be glad to receive critical comments from those who happen to notice them.

Prof. G.S. Paliwal,
D.Sc. Ph.D. FLS (London),
Delhi 110 034.

PREFACE

Plant taxonomic researches in India started some time in the 16 th century. Since then, a large number of Europeans started taking interest in the Indian plants, obviously attracted by the rich tropical flora. Naturally the Indian plains did not receive that kind of attention. During those early days network of roads did not exist and most of the collection tours were undertaken using bullock-carts.

The earliest record of exploration is that of V. Jacquemont (1832), pertaining to our study area of Indore and West Nimar, followed by an array of workers like, N. Vicary (1833), Ritchie (1838), Edgeworth (1848), J.F. Duthie (1888- 1891) etc. All were not botanists, some were collectors, administrators, foresters, surgeons etc.

The landmark publication of The Flora of India, by J.D. Hooker 1872-1897, gave a real impetus to botanical collections. From 1948 onwards, the interest taken by the Forest Dept., Universities and colleges, broughout the botanical wealth of this state, through various publications.

The Central Indian State of Madhya Pradesh came into existence in Nov. 1956 and the Central Circle of Botanical Survey of India was established in 1972 in Allahabad.

My own Ph.D. students and later their Ph.D. students undertook extensive floristic explorations of the area and within a span of several years, all the 8 districts of Indore Division were thoroughly worked out floristically.

The advantage of working out limited areas is very clear, when we see that 12,000 + sq. cms. of the total area has yielded 1432 species and what is more noteworthy is the addition of more than 6 species to the Indian flora and 50 + species new to the State.

The flora is in a single volume, dealing largely with the plants of Dry-Deciduous forest, but would be equally useful in areas surrounding this state, which have an identical forest type.

This inventory, it is hoped would provide data for studies in various fields of Botany as well as plant based industries.

ACKNOWLEDGEMENTS

I wish to place on record my deep sense of gratitude to Prof. G.S. Paliwal who gave valuable advice to me from time to time in the progress of this venture. I am grateful to him for writing the foreword.

I am equally thankful to Prof. S.R. Yadav, Prof. Emeritus, Shivaji University Kolhapur, for his guidance and suggestions. He willingly gave his consent to use his excellent photographs of grasses. I appreciate his readiness to help.

As stated in the preface, this is entirely the work of my Ph.D. students and that of Prof. C.M. Solanki's Ph.D. students. Dr. Solanki workd out the flora of Indore District, Dr. Swati Samvatsar completed flora of Jhabua and Dhar Districts for her Ph.D. dissertation and later published a book on the flora of these districts. Dr. Mujaffar Shaikh did the flora of 2 districts of East Nimar, Dr. S. Sisodiya, flora of Barwani and K. Surage Grasses of 4 districts of Nimar. This the basis of the present work. I am thankful to all of them. I also thank Dr. Arjun Prasad Tiwari for providing photographs of plants and in nomenclatural problems.

To Dr. B.S. Annigeri, I am specially thankful for writing the chapter on Conservation Strategies and Protected Area. His expertise as a Chief Conservator of Forest was useful in various ways.

I sincerely wish to thank the assistance rendered by the Forest Department throughout the progress of this book. Sometimes it was accommodation, at others, help from the forest guard.

I am glad to acknowledge the help rendered by my son, Col. Nikhil, in typing the MSS. I am not computer savvy and his suggestions in the technical problems I faced, were very useful.

We thank the organization, Maps of India, for maps of the 8 districts. This has enhanced the beauty of the book.

It is a pleasure to thank my daughter Archana, daughter in law Apeksha, my grand children, Sooravi, Tanishka, Akshada and Abeer, the members of my family, for encouragement throughout the progress of this work.

Authors

ABBREVIATIONS

auct. = author

auct.non. = not of authors

auct.plur. = of many authors

ca. circa. = about, more or less

DAVV= Devi Ahilya Vishva Vidyalaya, Indore

e.g.= for *example*

et.al.= *et al,iorum*; and other authors

excl. = *exclusus;* excluding

excl.syn. = *exclusus synomynon*; excluding synonym/s

f = *filius*; son

Fl.-Fr. = flowering-fruiting period

l.c = *loco citato;* in the place previously quoted

nom.illeg.= *nomen* illegitimum; illegitimate name

nom.nud. = *nomen nudum*; name published without description/without reference to previous publication

n.s. = new series

p.p. = *pro parte*; partly, in part

quoad syn = *quaod synonym*; as far as synonym is concerned

viz. = *videlicet*; namely

op.cit. = *opere citato*; in the work cited

sensu lato = in a broad sense

sensu stricto = in a limited, narrow sense

ser. = series

subsp = sub species

spp. = species, singular or plural

syn = *synonymon*; synonym

var. = *varietas.*

GLOSSARY

Abrupt= changing suddenly

Acaulis= stemless (apparently stem not visible)

Achene= dry, indehiscent, 1-seeded fruit

Acicular= needle-shaped

Acuminate= apex of leaf long drawn like a tail

Acute= sharp, ending in a point

Adenophorus= gland bearing

Adnate= union between two different organs (e.g. petals and stamens)

Aestivalis= pertaining to summer

Albidus, albus= white

Alatus= winged

Antiquorum= of the ancients

Appressed, adpressed= closely pressed against, not united

Arachnoid= Cobweby, like net of spider

Arenarius= of sandy places

Areolatus= pitted

Aril= an appendage of a seed growing out of hilum or funiculus

Aristate= bearing a stiff, bristle-like awn or seta

Articulate= jointed

Arundinaceous= reed-like

Asper= rough, as with hairs

Atrovirens= dark green

Augustus= notable, majestic

Auricomus= golden haired

Auricle= an ear-shaped part or appendage

Azureus= sky blue

Baccate= berry-like, soft, fleshy

Barbarus= foreign

Barbatus= barbed, bristles or awns provided with terminal or lateral hooks

Bifid= two cleft, as apex of leaf or petal

Bifurcate= forked, 2-branched, as Y- shaped hairs

Bilabiate= 2-lipped, as calyx or corolla

Borealis= northern

Bract= much reduced leaf bearing a flower in its axis/l

Bulbiferus= bulb bearing

Canus= ash-coloured

Caesius= bluish-grey

Caeruleus= sky blue

Caespitosus= tufted as in bunch

Callosus= thick-skinned, hard

Campestris= related to fields or plains

Canescens= grey-pubescent

Carinatus= keeled

Carnosus= fleshy

Caudatus= tailed

Cauline= pertaining to stem

Cernuus= drooping, nodding

Caespitose= matted, growing in tufts

Chartaceous= papery or tissue-like in texture

Chloranthus= green-flowered

Chrysophyllus= golden-leaved

Cinereus= ash –coloured

Circumscissile= opening or dehiscing by a line around the fruit or anther, coming off as a lid

Cirrhosus= tendrilled

Cladophyll= a flat, leaf-like stem, arising in the axil of a reduced bract, as in *Asparagus*

Clavatus= club-shaped, long body thickened at apex

Claw= narrow, petiole-like base of petals as in Cruciferae

Coarctatus= crowded together
Coccigera= bearing berries
Coccus= one seeded, fleshy part of a fruit
Cochlearis= spoon-shaped
Collinus= pertaining to a hill
Communis= growing in common or community
Commutatus= changing
Comosus= with long hairs
Concolor= coloured similarly
Concinnus= neat, well made, elegant
Confertus= crowded
Conglomeratus= crowded together
Connate= union between similar parts, like gamosepalous
Connivent= coming together, but not fused, as anthers in *Solanum*
Coriaceous= of leathery textured, as leaves in many *Ficus* species
Crassipes= thick footed or stalked
Crenate= shallowly round-toothed, obtuse toothed
Crinitus= with long hairs, as in *Abelmoschus crinitus*
Cruentus= bloody, red
Cucculatus= hooded
Cuneate= wedge=shaped, triangular, with the narrow end at the point of attachment
Cuspidatus= with a sharp, stiff point
Cyananthus= blue flowered
Cymbiformis= boat-shaped
Debilis= weak, frail
Decipiens= deceptive
Decurrent= running down the stem
Deltoid= triangular
Demersus= under water
Dendroid= tree-like
Denudatus= naked
Didymus= in pairs
Digitate= members arising from one point, as palmate leaf
Dimorphic= occurring in two forms
Diurnus= of the day, as flowering in day
Downy= covered with soft, short , weak hairs
Dulcis= sweet, as the aril in *Pithecellobium dulce*
Eburneus= ivory white
Echinatus= prickly, bristly
Effusus= loose spreading
Elatior= taller
Emarginate= with a shallow notch at the apex
Ensiformis= sword-shaped
Ephemeral= short lived
Eriocarpus= wooly-fruited
Exaltatus= very tall
Exotic= from another country
Falcatus= sickle-shaped
Fastigiate= branches erect and close together
Ferox= very thorny
Ferrugineous= rust-coloured
Fetid (foetid) = having a bad smell
Ficoides= fig-like
Filiform= thread-like
Fimbriate= fringed
Fistulosus= hollow-cylindrical
Flavus= yellow
Flabellate= spreading like a hand-fan
Flagelliformis= whip-like
Flexuosus= zigzag, wavy
Floccose= woolly
Floribundus= flowering freely
Foveolatus= pitted
Fragilis= fragile, weak
Frondosus= leafy
Frumentaceous= related to grain
Frutescens= shrubby, bushy
Fugacious= falling very early
Fulgens= shining
Furcate= forked
Fusiform= spindle-shaped
Gangeticus= of the Ganges
Gemmifera= bearing buds
Geniculate= bent like a knee

Gibbous= swollen on one side

Glabrous= not hairy

Gladiatus= sword-like

Glaucus= covered with a bloom or whitish substance that rubs off

Glochid= minute barbed spine or bristle, often in tufts

Glomerate= clustered

Glutinous= sticky, like gum

Gracilis= graceful, slender

Grandiflorus= large flowered

Gratissimus= pleasing

Graveolens= heavy scented

Grossus= thick, large

Guttatus= spotted or speckled

Halophilus= salt-loving

Hastate= arrow-shaped with the lateral lobes at right angles to the stem

Heterogamous= with two or more kinds of flowers

Hirsute= with coarse hairs

Hoary= covered with a close whitish pubescence

Humilis= dwarf. Low growing

Hyaline=transluscent like a candle

Indumentum= hairy or pubescent with heavy covering

Induratus= hardy

Insignis= distinguished, remarkable

Insularis= belonging to an island

Jubatus= crested

Keeled= ridged like the bottom of a boat

Labiate= lipped

Lacerate= irregularly torn or cleft

Laciniate= cut into narrow, pointed lobes

Lacuna= gap, cavity

Laetus= bright

Laevigatus= smooth

Lanate= woolly, with long, intertwined hairs

Lanceolate= shaped like a lance

Lasianthus= woolly flowered

Latifolius= broad-leaved

Lateritus= brick-red

Lax= loose

Lenticular= lens- shaped, biconvex

Lepidophyllus= scaly-leaved

Lepis= scale

Lepto= thin

Leuco= white

Lignosus= woody

Linear= long and narrow, sides nearly parallel

Littoralis= of the sea shore

Loculicidal= fruit dehiscing from the back of locule

Lucidus= bright, shining, clear

Luteus= yellow

Lyrate= pinnatfid with a large terminal lobe

Maculatus= spotted

Maritimus= of the sea

Melano= black

Micans= shining like mica

Miliaceous= related to millet

Mollis= soft-hairy, soft

Moniliform= beaded

Monspeliensis= of France

Montanus= pertaining to mountains

Moschatus= scented like musk (kasturi)

Mucronate= terminated by a sharp point

Multicaulis= many-stemmed

Muralis= of walls or growing on walls

Muricate= rough due to raised points

Mutabilis= colour changing (*Hibiscus mutabilis*-flower changes colour from white to pink up to evening)

Nannus= small, dwarf

Natans= floating

Nervosus= nerved

Niger= black

Nitidus= shining

Niveus= snowy

Nobilis= famous, noble

Nocturnus= of night

Nodiflorus= with flowers at nodes
Nodosus= jointed, with nodes
Nutans= nodding
Nucifera= nut-bearing
Nudicaulis= naked-stemmed
Obtuse= blunt, rounded
Occidentalis = Western
Odoratus= fragrant, good smelling
Officinalis= medicinal
Oleifera= oil-bearing
Oleraceous= plant used in cooking
Olitorius= pertaining to vegetable garden or gardens
Operculate= with a cap or lid
Orbiculate= circular or disc-shaped
Orchioides= like an orchid
Orientalis= Eastern
Ortho= straight
Patens, patulus= spreading
Pallidus= pale
Palustris= marsh-loving
Panduratus= violin-shaped
Pannosus= ragged, tattered
Papillate= bearing pimple-like protuberences
Parvi= small
Pauci= few in number
Pavoninus= peacock-like
Pectinate= comb-like, with close, narrow divisions
Pedate= palmately divided with side lobes divided again
Pellucid= clear, transparent
Peltate= attachment of stalk inside the margin
Peltophorus= shield-bearing
Pendulous= hanging
Peigrinus= exotic, foreign
Personate= two-lipped corolla where lips are closed
Pictus= painted
Pilifera= bearing soft hairs
Pilose= shaggy with soft hairs
Pitted= having little depressions or cavities
Platy= broad
Plicate= folded like a hand fan
Plumose= feathery
Polygamous= bearing uni and bisexual flowers on the same plant
Praemorsus= bitten at the end
Pratensis= of the meadows
Princeps= princely, first
Procumbent= lying flat or trailing but not rooting
Pumilus= small, dwarf
Pubescent= covered with short, soft hairs, downy
Pulcher, pulchellus, pulcherimus= beautiful, pretty
Pulvinatus= cushion-like
Pungens= sharp ponted
Punctatus, punctate= dotted with glands
Purpurascens= purplish, becoming purple
Pycnanthus= densely flowered
Pygmaeus= short
Pyrene= nutlet, particularly the nutlet in a drupe as in Verbenaceae
Quiquefolius= five-leaved
Rhachilla, rachilla= much reduced axis in grasses and sedges that bears the florets
Radicans= rooting
Radiatus= spreading from a common center like rays
Ramosus= branched
Rariflorus= scattered-flowered
Repens, repent, reptans= creeping
Regalis= regal, royal
Religiosus= sacred
Reniformis= kidney-shaped
Repand= weakly sinuate
Retrorse= bent or turned over back or downward
Retuse= notched slightly at the usually obtuse apex
Revolute= rolled backwards
Rhomboidal= shaped like a rhomboid
Rigidus= stiff, rigid

Ringens= open, two lipped corolla where lips are open

Riparius= of river beds

Rivalis= pertaining to streams

Rivularis= stream-loving

Rostratus= beaked or beak-like projection

Rotundifolius= rounded-leaved

Ruber= red

Rubiginosus= rusty

Rupestris= rock-loving

Rupicola= growing on cliffs or ridges

Saccharum= pertaining to sugar

Saccate= having a pouch, bag-shaped

Sagittate= like head of an arrow

Salverform= narrow tube suddenly expanding above

Sanctus= holy, sacred

Sanguineus= blood red

Sapindus= pleasing to taste

Sapientum= of the wise men

Saponarius= soapy

Sarmentosus, sarmentose= bearing runners

Scabrous= rough, not smooth

Scape= leafless peduncle arising from the ground

Scapose= bearing flowers on a scape

Scarious, scariosus= thin and not green (pertaining to leaves, bracts or tepals

Scoparius= broom-like

Scorpioides= like a scorpion

Sepium= of hedges or fences

Secundus, secund= one-sided (as in helicoid cyme)

Senilis= old, white-haired

Sericeous= silky

Serrate= toothed like a saw

Setigerus= bristly

Setose= full of bristles

Sigmoid= S-shaped

Simplex= simple, unbranched

Sinensis= Chinese, of China

Sinuate= wavy

Sinus= space between the lobes of a leaf and its petiole

Spathulate= spoon-shaped

Speciosus= showy, good-looking

Spicigera= spike bearing

Spine= sharp pointed projection, being a modification of leaf

Spiralis= spiral

Splendens= splendid

Spurious= false

Spur= tubular, sac-like elongation of a sepal or petal

Squamose= full of scales, these scales are coarse

Stans= standing erect, upright

Stellate= star-like

Stenophyllus= narrow-leaved

Sterilis= sterile, infertile

Strigose= hispid

Stolen= branch bending to the ground and bearing roots

Striate= with fine, longitudinal lines or ridges

Strictus, strict= upright, erect, often rigid

Suaveolens= sweet- scented

Subulate= tapering from base to apex

Suffruticose= somewhat shrubby

Sulcate= grooved or furrowed lengthwise

Supinus= prostrate

Sylvaticus= forest-loving

Sylvestris= of forest or woods, growing wild

Tenax= strong

Tenuis= slender, thin

Terete= round in cross section

Ternate= in threes

Torulose= somewhat contracted at intervals

Tessellatus= laid off in squares or dice-like pattern

Throat= the opening of a gamopetalous corolla or perianth

Tinctorius= belonging to dyes

Tomentose= densely woolly

Tortuosus= much twisted

Trichome= hair or a bristle
Tristis= bitter, dull, sad
Trivialis= common, ordinary, found everywhere
Truncate= cut off at one end
Tuberculatus = having tubercles
Tuberosus= tuberous
Tumid= swollen, inflated
Tunicate= with concentric scaly leaves as in onion
Turgid= swollen from fullness
Turbinate= top-shaped, inversely conical
Urens= burning
Uliginosus= of wet or marshy places
Uncinate= hooked at point
Undulatus= wavy, undulate
Urceolate= cup-shaped
Utricle= bladdery, a small bladder
Vaginate= sheathed
Velutinus= velvety
Ventricose= 1-sided swelling or inflation
Vernus= of spring
Verrucosus= warty
Versicolor= variously coloured, as one colour passing into another or changing in colour
Vesicle= small bladdery cavity filled with air or water
Vestitus= covered, clothed as with hairs or pubescence
Virens= green
Villous= soft hairy
Violaceous= violet
Viridis= green
Virgate= long, straight and slender
Viscid= sticky, glandular
Volubilis= twining
Voluble= twining
Xanthinus= yellow
Zeylanicus= of Ceylon (Ceylanicus)

CONTENTS

Plate 1

Fig. 1. *Dillenia pentagyna*

Fig. 2. *Miliusa tomentosa*

Fig. 3. *Cissampelos pareira*

Fig. 4. *Capparis sepiaria*

Fig. 5. *Cochlospermum religiosum*

Fig. 6. *Abelmoschus crinitus*

Fig. 7. *Azanza lampas*

Fig. 8. *Grewia hirsuta*

Plate 2

Fig. 1. *Grewia rothii*

Fig. 2. *Helicteres isora*

Fig. 3. *Hibiscus caesius*

Fig. 4. *Kavalama urens*

Fig. 5. *Sterculia villosa*

Fig. 6. *Soymida febrifuga*

Fig. 7. *Celastrus paniculatus*

Fig. 8. *Ampelocissus latifolia*

Plate 3

Fig. 1. *Leea asiatica*

Fig. 2. *Lannea coromandelica*

Fig. 3. *Semecarpus anacardium*

Fig. 4. *Abrus precatorius*

Fig. 5. *Acacia catechu*

Fig. 6. *Acacia leucophloea*

Fig. 7. *Albizia lebbeck*

Fig. 8. *Alysicarpus monilifer*

Plate 4

Fig. 1. *Bauhinia racemosa*

Fig. 2. *Bauhinia vahlii*

Fig. 3. *Caesalpinia bonduc*

Fig. 4. *Crotalaria spectabilis*

Fig. 5. *Erythrina suberosa*

Fig. 6. *Flemingia strobilifera*

Fig. 7. *Indigofera cordifolia*

Fig. 8. *Indigofera linnaei*

Plate 5

Fig. 1. *Mucuna pruriens*

Fig. 2. *Tephrosia purpurea*

Fig. 3. *Uraria picta*

Fig. 4. *Anogeissus latifolia*

Fig. 5. *Careya arborea*

Fig. 6. *Lagerstroemia parviflora*

Fig. 7. *Citrullus colocynthis*

Fig. 8. *Trichosanthes cucumerina*

Plate 6

Fig. 1. *Trichosanthes tricuspidata*

Fig. 2. *Ceriscoides turgida*

Fig. 3. *Gardenia latifolia*

Fig. 4. *Ixora pavetta*

Fig. 5. *Tamilnadia uliginosa*

Fig. 6. *Plumbago zeylanica*

Fig. 7. *Anagallis arvensis*

Fig. 8. *Embelia basaal*

Plate 7

Fig. 1. *Dregea volubilis*

Fig. 2. *Canscora diffusa*

Fig. 3. *Nymphoides indica*

Fig. 4. *Cordia macleodii*

Fig. 5. *Porana paniculata*

Fig. 6. *Oroxylum indicum*

Fig. 7. *Stereospermum chelonoides*

Fig. 8. *Martynia annua*

Plate 8

Fig. 1. *Barleria montana*

Fig. 2. *Eranthemum purpurascens*

Fig. 3. *Lavandula bipinnata*

Fig. 4. *Leucas aspera*

Fig. 5. *Rotheca serrata*

Fig. 6. *Aristolochia indica*

Fig. 7. *Viscum articulatum*

Fig. 8. *Ficus microcarpa*

Plate 9

Fig. 1. *Ficus racemosa*

Fig. 2. *Flacourtia indica*

Fig. 3. *Vanda tessellata*

Fig. 4. *Cheilocostus speciosus*

Fig. 5. *Tacca leontopetaloides*

Fig. 6. *Gloriosa superba*

Fig. 7. *Commelina benghalensis*

Fig. 8. *Sagittaria guyanensis*

Plate 10

Fig. 1. *Potamogeton nodosus*

Fig. 2. *Andropogon pumilus*

Fig. 3. *Apluda mutica*

Fig. 4. *Cymbopogon martini*

Fig. 5. *Eleusine indica*

Fig. 6. *Heteropogon contortus*

Fig. 7. *Hygroryza aristata*

Fig. 8. *Ischaemum afrum*

Plate 11

Fig. 1. *Ischaemum rugosum*

Fig. 2. *Melanocenchris jacquemontii*

Fig. 3. *Parahyparrhenia bellariensis*

Fig. 4. *Paspalum vaginatum*

Fig. 5. *Pennisetum pedicellatum*

Fig. 6. *Sacciolepis interrupta*

Fig. 7. *Themeda quadrivalvis*

Fig. 8. *Thysanolaena latifolia*

CHAPTER - I

INTRODUCTION

The study area comprises of 8 districts of MP, viz., Alirajpur, Barwani, Burhanpur, Dhar, Indore, Jhabua, Khandwa and Khargone.

Location: The geographical location of the study area is, Longitude 2 1°4'26.4" to 23°14'27. 6'' N and 74°2'6'' to 77°13' 8.4'' E., correspondingly in western part of MP. The Eastern side touches Dahod district of Gujarat, south is Nandurbar and Jalgaon districts of Maharashtra, North and east side is surrounded by Ratlam, Ujjain, Dewas and Harda districts of MP.

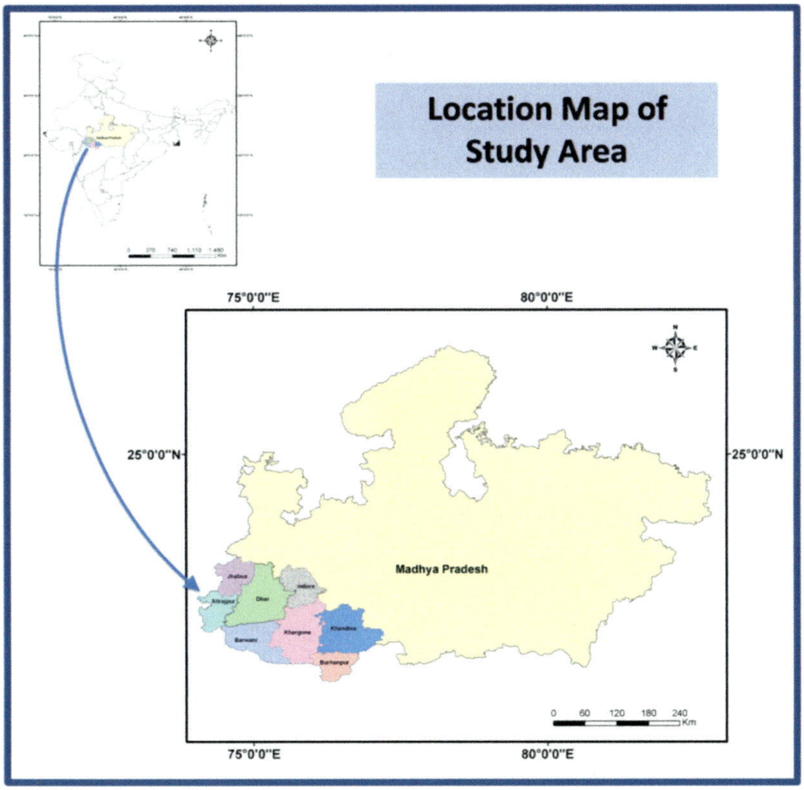

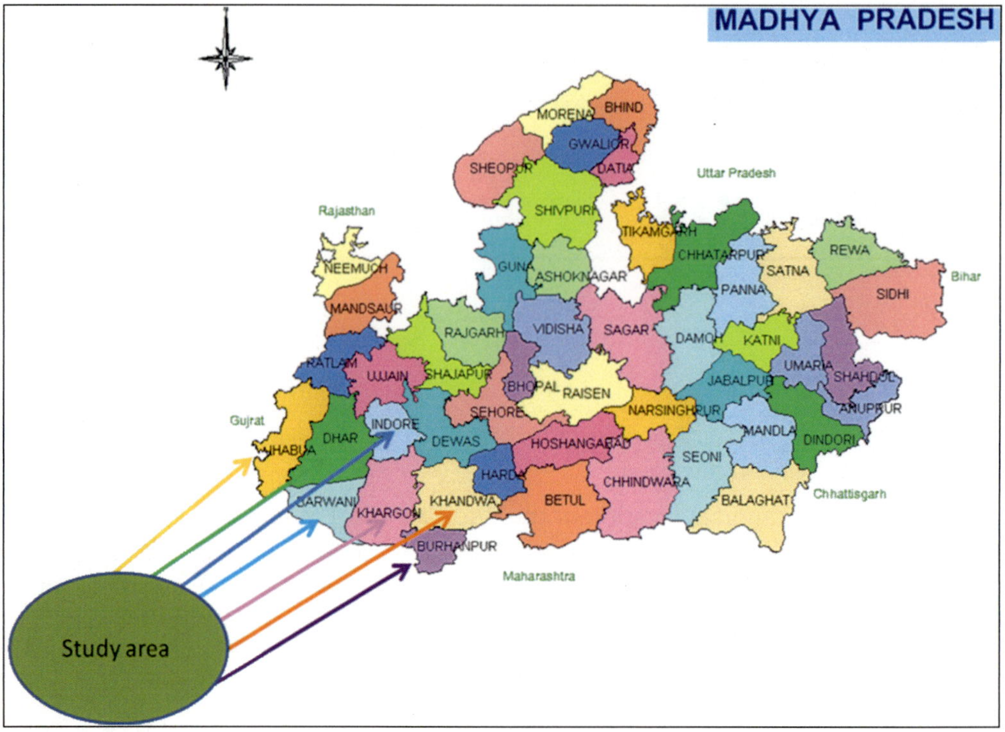
MADHYA PRADESH
MORENA
BHIND
GWALIOR
SHEOPUR
DATIA
Uttar Pradesh
SHIVPURI
Rajasthan
TIKAMGARH
CHHATARPUR
REWA
NEEMUCH
GUNA
ASHOKNAGAR
SATNA
SIDHI
Bihar
PANNA
MANDSAUR
RAJGARH
VIDISHA
SAGAR
DAMOH
KATNI
RATLAM
UJJAIN
SHAJAPUR
UMARIA
SHAHDOL
JABALPUR
BHOPAL
RAISEN
SEHORE
NARSINGHPUR
ANUPPUR
Gujrat
INDORE
JHABUA
DHAR
DEWAS
HOSHANGABAD
MANDLA
DINDORI
SEONI
HARDA
BARWANI
KHARGON
KHANDWA
BETUL
CHHINDWARA
BALAGHAT
Chhattisgarh
BURHANPUR
Maharashtra
Study area

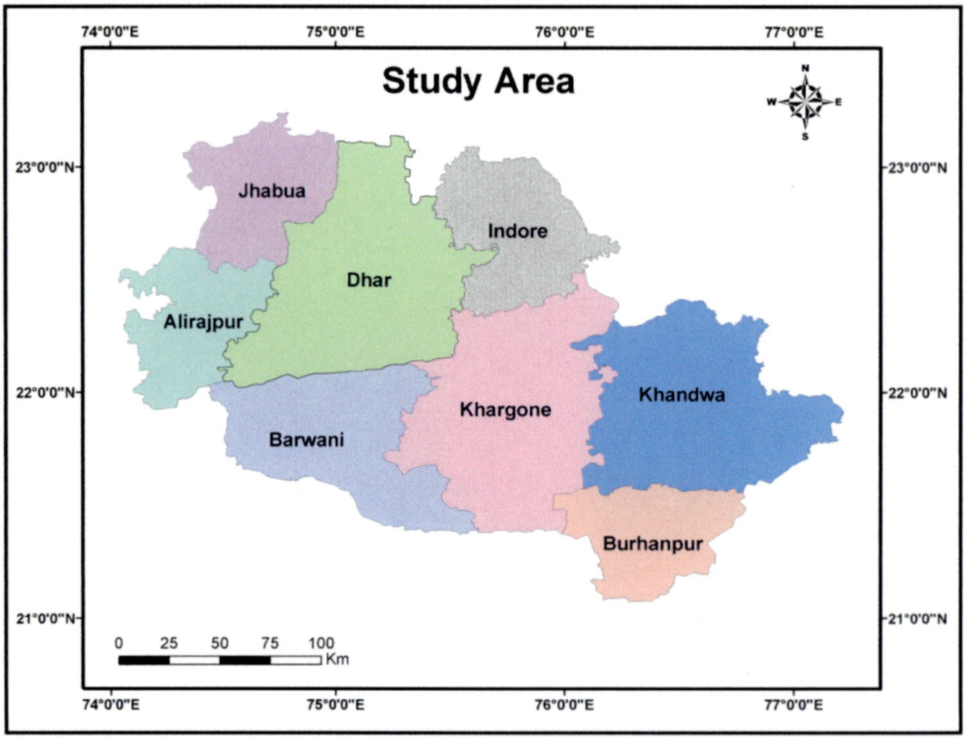
Study Area
74°0'0"E
75°0'0"E
76°0'0"E
77°0'0"E
23°0'0"N
22°0'0"N
21°0'0"N
N
W
E
S
Jhabua
Indore
Dhar
Alirajpur
Khandwa
Khargone
Barwani
Burhanpur
0 25 50 75 100
Km

North of river Narmada, consists of Alirajpur, Dhar, Indore and Jhabua, while the second group, south of Narmada consists of Barwani, Burhanpur, Khandwa and Khargone. These 4 districts constitute Nimar region, Burhanpur and Khandwa forming East Nimar and Barwani and Khargone forming West Nimar. In east Nimar two forest tracts of Chandgarh and Selani lie north of Narmada and in west Nimar the southern part of Satpura touches northern part of Vindhyan ranges. We are dealing with all aspects of the flora, except taxonomical enumeration, under these two headings. This area can be conveniently divided into 2 groups.

PEOPLE

Alirajpur, Jhabua, Barwani and Dhar districts are predominantly tribal. In all, 59% population belongs to the tribal community as would be seen by the Table 1

TABLE 1

District Name	Number of villages	Total Households	Total Population of Village	Total Scheduled Castes Population of Village	Total Scheduled Tribes Population of Village
Alirajpur	543	113129	671925	23503	627835
Barwani	716	204644	1181812	64344	925650
Burhanpur	263	97842	497560	41672	222132
Dhar	1535	339173	1772572	109481	1142263
Indore	629	159400	848988	164465	134702
Jhabua	813	175934	933065	11397	861498
Khandwa	711	216182	1050625	122463	440789
Khargone	1417	309315	1574190	177708	700953
Total	**6627**	**1615619**	**8530737**	**715033**	**5055822**

TABLE 2. Land Use Pattern (area in sq. km.)

District	Geog. Area	Forest Area	Area under Non-Agricultural Uses	Barren & Un-cultivable Land Area	Permanent Pastures and Other Grazing Land Area	Land Under Miscellaneous Tree Crops etc. Area	Culturable Waste Land Area	Fallow Land other than Current Fallow Area	Current Fallow Area	Net Area Sown
Alirajpur	3182	675	233	456	55	36	96	38	58	1617
Barwani	5427	914	277	573	119	40	80	48	50	2417
Burhanpur	3427	1309	123	88	110	27	16	26	25	1118
Dhar	8153	686	480	667	526	97	148	114	66	5074

District	Geog. Area	Forest Area	Area under Non-Agricultural Uses	Barren & Un-cultivable Land Area	Permanent Pastures and Other Grazing Land Area	Land Under Miscellaneous Tree Crops etc. Area	Cultu-rable Waste Land Area	Fallow Land other than Current Fallow Area	Current Fallow Area	Net Area Sown
Indore	3898	679	134	911	111	11	77	105	55	2385
Jhabua	3600	229	295	358	53	18	220	46	33	1773
Khandwa	7352	2032	424	124	346	39	141	185	113	3010
Khargone	8025	1309	401	197	603	76	196	161	76	4006
Total	**43064**	**7833**	**2370**	**2557**	**1926**	**348**	**977**	**725**	**480**	**21403**

TABLE 3. STATE OF THE FOREST (area in sq. km.)

	Geog. Area	Forest Area	Very dense forest	Moderately Dense forest	Open Forest	Total	% of forest ------- -- Area to Geog. area
MADHYA PRADESH	308250	77414	6563	34571	36280	77414	25.11
STUDY AREA	43064	7853	207	3181	4465	7853	18.23
District							
Alirajpur	3182	675	00	212	463	675	21.21
Barwani	5427	914	00	168	746	914	16.84
Burhanpur	3427	1309	58	643	608	1309	38.19
Dhar	8153	686	00	119	567	686	08.41
Indore	3898	679	00	349	330	679	17.41
Jhabua	3600	229	00	33	196	229	06.36
Khandwa	7352	2032	148	1161	723	2032	27.63
Khargone	8025	1309	01	476	832	1309	16.31

DRAINAGE

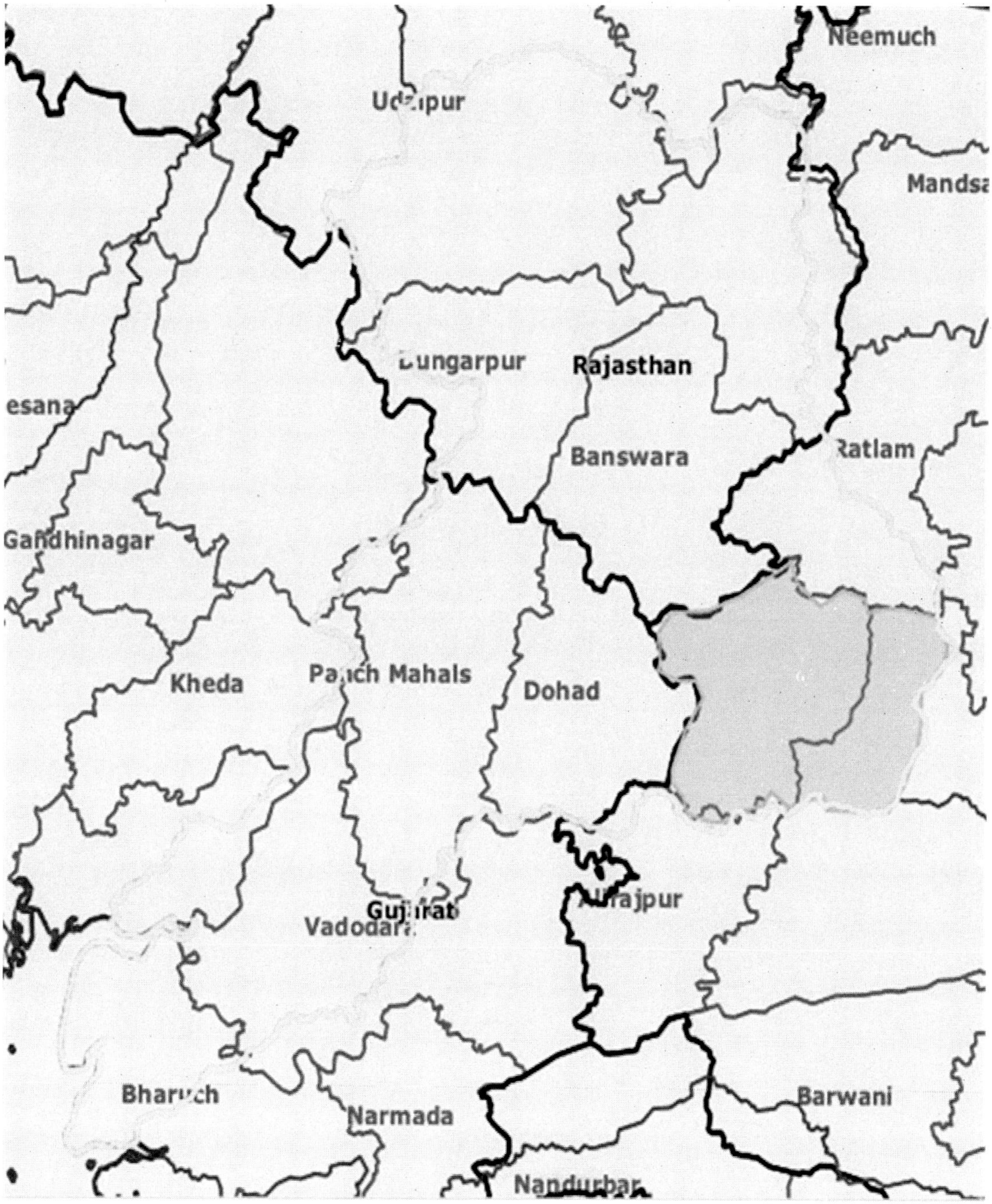

Upper Mahi : Jhabua and Parts of Dhar

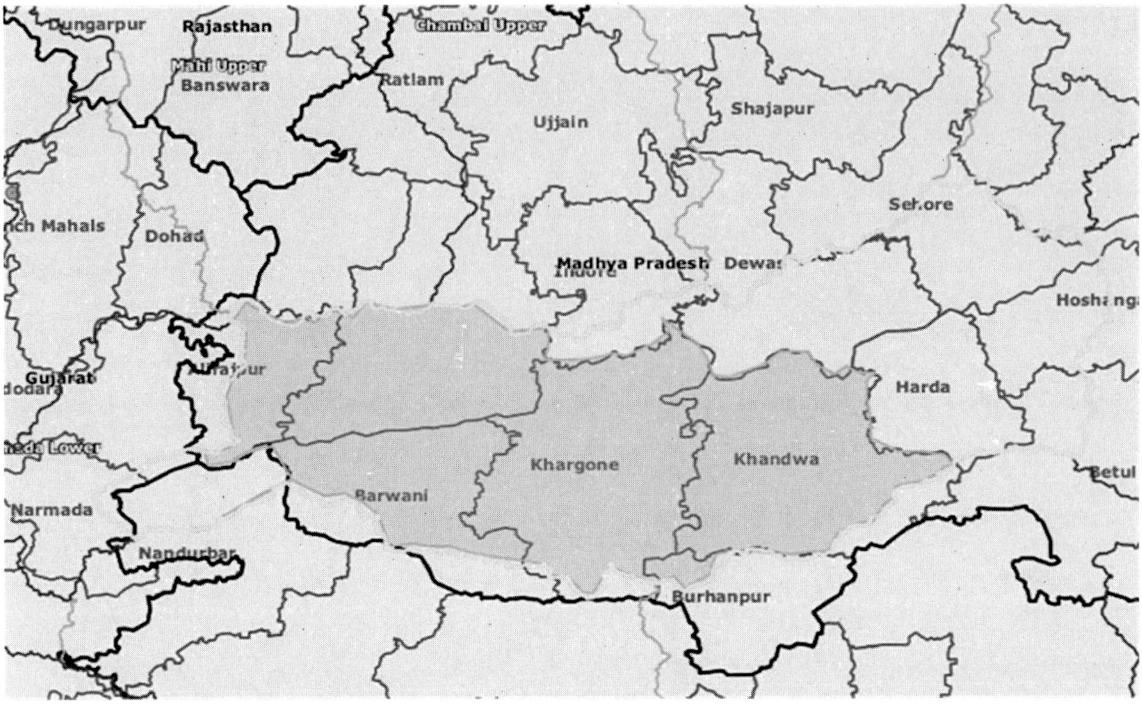

Middle Narmada : Major parts of Barwani, Khargone, Khadwa, Alirajpur Dhar and tracer part of Burhanpur

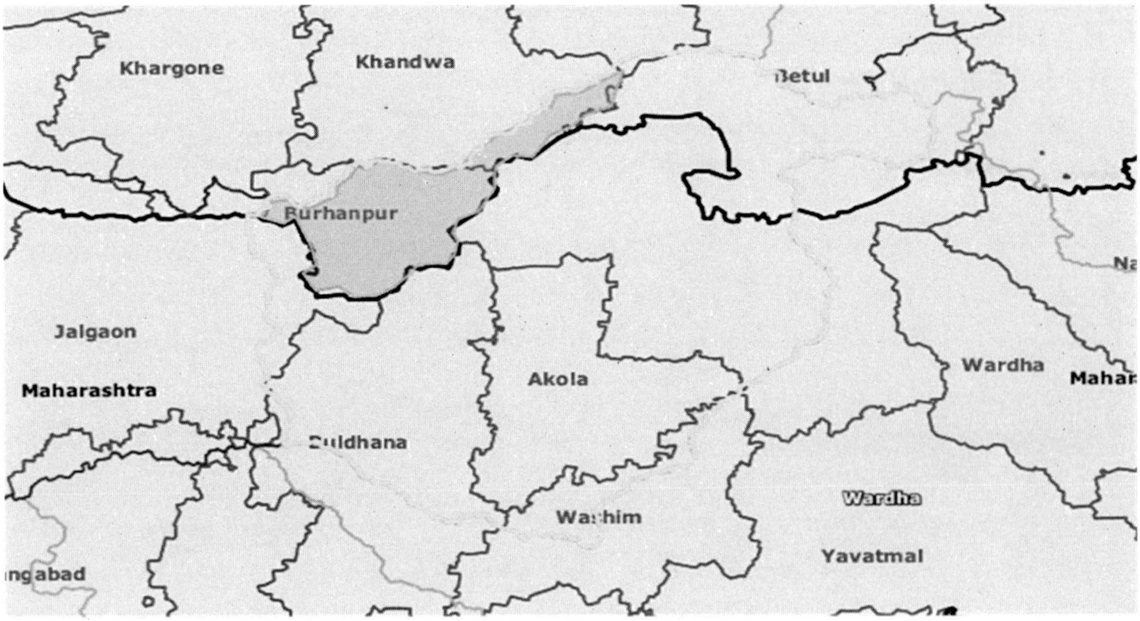

Upper Tapi : Major parts of Burhanpur

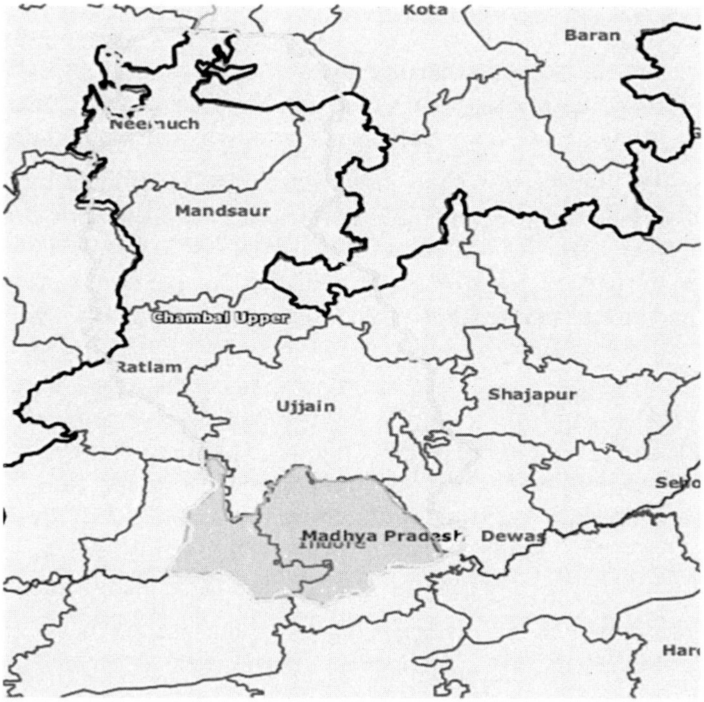

Upper Chambal : Indore and Parts of Dhar

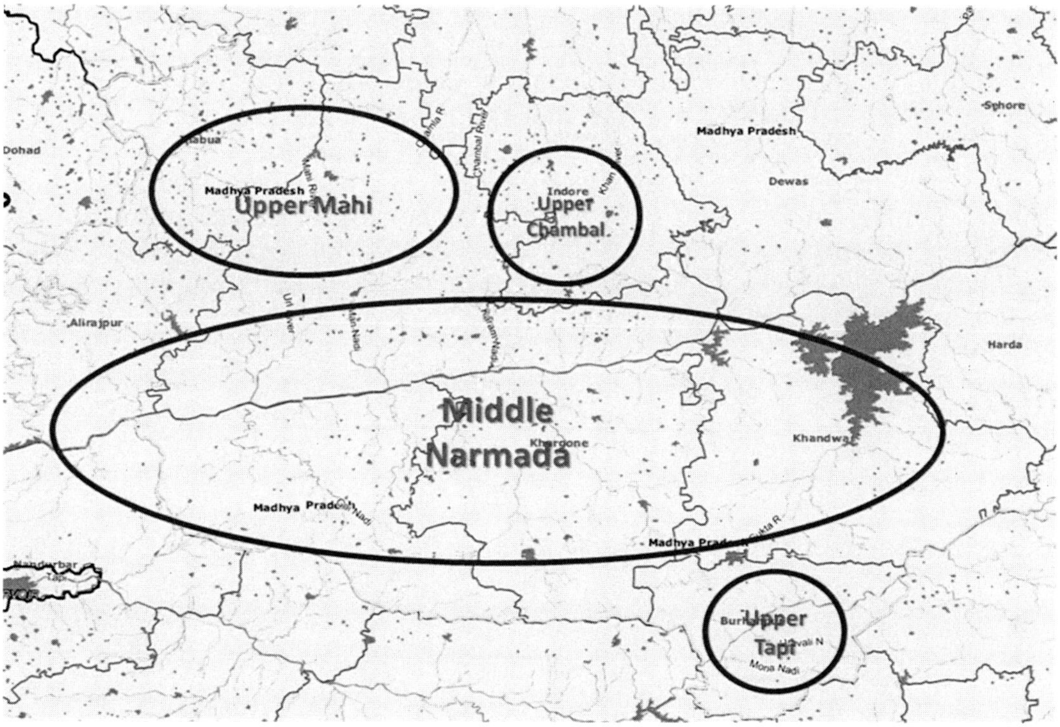

HISTORY OF BOTANICAL EXPLORATIONS

Sengupta (1977) has already summarized the history of Botanical explorations in MP and later Khanna and Mudgal (2002) have provided further data on the same subject. Instead of repeating, we are presenting below contributions of various workers pertaining to the study area.

The most important milestone in this respect is the publication of the Flora of MP by the Botanical Survey of India. The first vol., by Verma *et al.,* appeared in 1993, second vol., by Mudgal *et al.,* in 1997 and the third vol., by Singh *et al.,* in 2001. These publications covered Chhattisgarh also. In 2001 itself, Khanna *et al.,* published supplement to the flora of M.P. In the meantime, college and university teachers were also active and presented their investigations either in the form of Ph. D. dissertations (most of them unpublished), or research papers communicated to various journals. So far as the study area is concerned, I am fortunate to have an easy access to the work of my Ph. D. students and subsequently their students in turn. Solanki (1983) worked out the flora of Indore district, making the publication of Rao and Sastry (1964) as the base, because the authors had stated that they worked out the flora in "three fairly long periods (14 to 22 December, 1961: 18 to 26 October, 1962 and 17 to 26 April, 1963)", a total period of 28 days. They have reported a total 566 species within 373 genera and 93 families. Their list includes common cultivated plants also. That is a good achievement, just in 28 days.

Solanki (1983) has reported 951 species. This is quite natural if we consider that Solanki required 3 years to complete the work. Samvatsar earlier worked for her Ph. D. dissertation on Jhabua district, later extended her work to Dhar district and after receiving the degree, published her work (1996), reporting 1159 species within 553 genera and 128 families. Shaikh Mujaffar worked out the flora of Eastern Nimar (districts Burhanpur and Khandwa) for his Ph.D. 2011, unpublished) and has described 1110 species. Sinha and Shukla (2007 and 2009) published synoptic Flora of Khargone district and as the title indicates, it is a listing of plants only. S. Mujaffar *et al.* (2012), reported **Amorphophallus konkanensis** Hett *et al.,* as a new record for M.P., in the same year, he published **Salvadora alii** as a new record for the flora of India. Sisodiya (2012, unpublished), worked out the flora of Barwani district. S. Mujaffar *et al.* (2014) discovered a new species in the genus **Tinospora** and named it **T. maqsoodiana.** K. Surage (2015) worked out the grasses of Nimar (all 4 districts) for her Ph. D. dissertation (unpublished), reporting 242 grass species. Tiwari, A.P., S. Gavade and Mujaffar *et al.* (2016) published lectotypification of **Eriocaulon duthiei.** S. Mujaffar (2017), discovered **Chlorophytum diwanjii** as a new species in that genus.

CHAPTER - II

GENERAL CHARACTERISTICS OF THE VEGETATION

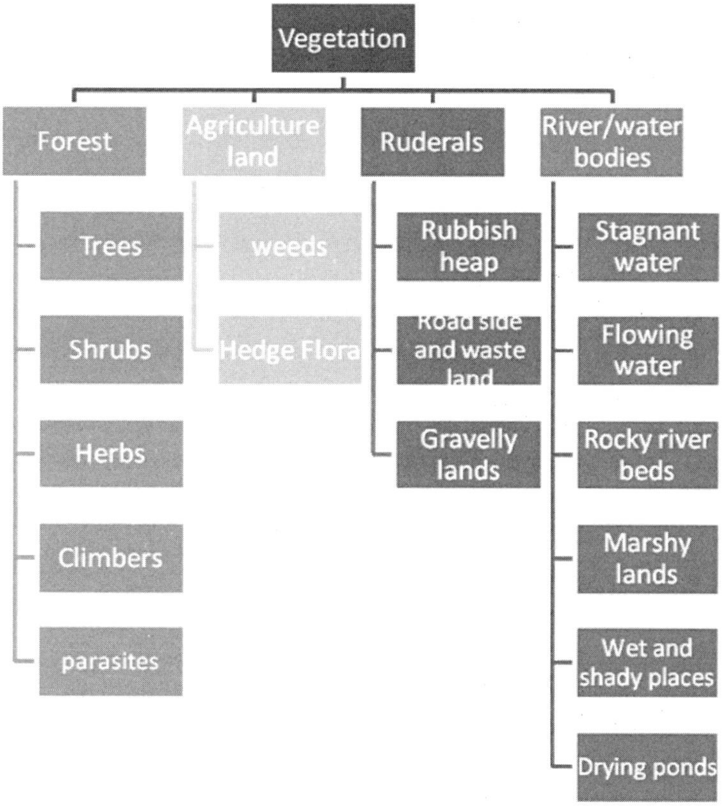

1. FORESTS

Forest cover on the whole is meagre. The names of places having rich pockets of forest are given below.

Alirajpur : Aamkhut-Kattiwada, Andharkach, Chhaktala, Mathwad, Shakarja and Udaigarh

Barwani : Pansemal, Sendhwa, Bokarata, Pati, Julwania, Katbeli, Chirmiriya

Burhanpur: Sukta reservoir, Dhulkot, Shekhpura, Dedtalai, Tukaithad, Khaknar, Jambupani, Asirgarh, Nepanagar

Dhar : Dharampuri bet, Manavar, Mandu plateau and valleys below

Indore : Patalpani-Kalakund, Choral, Gautampura, Janapao hills, Naharjhabua, Manpur, Tillor khurd, Mendh, Khurda-Khurdi

Jhabua : Meghnagar, Ranapur, Dholya bad, Jhaknawada

Khandwa : Khalwa, Awliya, Punasa, Omkareshwar, Mortakka, Kalibhit east and west, Indira sagar reservoir, Sunder dev

Khargone : Bhagwanpura, Jhirnia, Katkut, Satak reservoir, Deo nala, Pipalzopa, Sirvel

FOREST TYPES

The category of the forest is Dry deciduous, Vb type (Champion and Seth, 1968), but not uniformly. Along with the dominants (Teak) and co-dominants like *Albizia procera, Ailanthus excelsa, Lagerstroemia parviflora, Mitragyna parvifolia* etc., some noteworthy and rare elements are : *Acacia catechu, Aegle marmelos, Alangium salvifolium, Anogeissus latifolia, A. pendula, Boswellia serrata, Bridelia retusa, Butea monosperma, Cochlospermum religiosum, Dendrocalamus strictus, Desmodium oojeinense, Diospyros melanoxylon, Ficus* spp., *Firmiana colorata, Haldina cordifolia, Hardwickia binata, Kavalama urens, Kydia calycina, Lannea coromandelica, Madhuca longifolia* var. *latifolia, Mallotus philippensis, Mitragyna parvifolia. Albizia amara,* while less common in the forest, is typical of scrub and road-side. This species is very common in Dhar district, compared to other areas. From a distance, it can be easily mistaken for an *Acacia. Borassus flabellifer* is more common in Alirajpur and rarely encountered in other districts. District-wise list of rare plants is given separately in the chapter on discussion.

In valleys and gorges, where there is more humidity and comparatively less temperature, we have wet deciduous forest. Where the ground is open, there is a savannah type of forest, dominated, for example by *Gymnosporia montana*.

The herbaceous ground cover comprises mainly of: *Acalypha indica, Acanthospermum hispidum, Ageratum conyzoides, Alysicarpus monilifer, A. rugosus, Anisomeles indica, Biophytum sensitivum, Bidens biternata, Capillipedium filiculme, Crotalaria* spp., *Curculigo orchioides, Cyanthillium cinereum, Desmodium neomexicanum, Dyschoriste erecta, Euphorbia* spp., *Indigofera cassioides, Justicia* spp., *Leucas* spp., *Ocimum americanum, Oldenlandia corymbosa, Orthosiphon pallidus, Parasopubia delphinifolia, Phyllanthus* spp., *Rungia repens, Tridax procumbens, Xanthium strumarium*, etc. There is a very huge, steep bare rock in the Mathwar range, which is completely covered by *Combretum ovalifolium*. Among twiners and climbers are: *Celastrus paniculatus, Cissampelos pariera, Cissus* spp., *Cryptolepis dubia, Cyphostemma auriculatum, Dioscorea* spp., *Hemidesmus indicus, lpomoea* spp., *Pergularia daemia* etc.

2. OPEN GRASS-LAND

Poaceae, which is one of the 6 families of any given area, is also well represented here by 239 species. According to a number of Agrostologists, including Bor (1960), India does not have a true grass land. What we have is the secondary sere in ecological terms. It has originated due to tremendous pressure of grazing and other anthropomorphic factors.

Open grass land is always a mixture of grasses and many non-grass elements. Some such elements are: *Achyranthes aspera, Alysicarpus pubescens* var. *pubescens, Blepharis maderaspatanus, Boerhavia diffusa, Bulbostylis barbata, Celosia argentea, Cyperus* spp. *Desmodium* spp. *Euphorbia* spp. *Hemigraphis urens, Hypoxis aurea, Indigofera* spp. *Leucas cephalotus, L. lavandulifolia, Mollugo stricta, Neanotis* spp. *Orthosiphon pallidus, Scleria tessellata, Zornia gibbosa* etc.

Among grasses we have *Dichanthium annulatum, D. caricosum, Heteropogon contortus, Iseilema anthephoroides, Sehima nervosum, Themeda quadrivalvis, T. triandra* as dominant species, while occasionally *Capillipedium filiculme* and *Pseudanthistiria heteroclita* form pure patches. Since this habitat is common throughout the area, names of places are omitted.

3. AGRICULTURE

3. 1. HEDGE FLORA

Hedges serve a dual purpose: that of demarcating boundaries between fields and houses and protecting the interior from grazing or browsing animals. Common plants used for a live hedge are: *Acacia farnesiana, Caesalpinia bonduc, C. decapetala, Delonix elata, Dodonea viscosa, Erythrina variegata, Euphorbia neriifolia, E. tirucalli, E. umbellata, Ipomoea carnea, Jatropha curcas, Lantana* × *aculeata, Lawsonia inermis, Opuntia elatior, Pithecellobium dulce, Volkameria inermis, Ziziphus* spp. etc. These hedges support quite a number of climbers and twiners of families like Asclepiadaceae, Convolvulaceae, Cucurbitaceae, Menispermaceae etc. The herbaceous or shrubby ground cover is composed mainly of *Apluda mutica, Boerhavia chinensis, Dicliptera verticillata,* some herbaceous *Euphorbias, Parkinsonia aculeata, Peristrophe bicalyculata, Plumbago zeylanica, Xanthium strumarium* etc. Occasionally climbers like *Gloriosa superba* are met with. We have also noted that *Delonix elata, Capparis decidua, Salvadora persica* are more commonly used as hedge plants in Dhar and Khargone, while in Jhabua, *Ziziphus* spp. is more commonly used.

Parahyparrhenia khannae, a rare grass within the area (but abundant at its place of occurrence) was collected from this habitat from Khujava near Dharampuri.

3.2 WEEDS

In the whole area, it was noticed that plants are virtually allowed a 'free hand' and as such, must be affecting the yield to a great extent. Many of these weed elements are Neotropical in origin and abundantly found within fields. Some of the common plants of this type are: *Ageratum conyzoides, Amaranthus polygonoides* and other *Amaranthus* spp., *Celosia argentea, Cleome viscosa, Cocculus hirsutus, Convolvulus arvensis, Corchorus olitorius, Euphorbia hirta, E. heyneana, Fumaria indica, Ipomoea obscura, Indigofera* spp., *Justicia simplex, Mollugo pentaphylla, Ocimum americanum, Neanotis montholonii, Polygala erioptera* etc. In addition to these, there are a number of Cyperaceae and Poaceae, which constitute a major component in the category. It is pertinent to note that *Celosia argentea* is so abundant in the fields that one wonders, whether it is planted along with other crops.

The following are alien weeds, categorized region-wise for better appreciation (Source: Naik, 1998).

Tropical American: *Chloris virgata, Corchorus aestuans, Euphorbia prostrata, Iseilema laxum, Martynia annua, Nicotiana plumbaginifolia, Sida cordata.*

South American: *Acanthospermum hispidum, Alternanthera paronychioides, A. pungens, Croton bonplandianus, Gomphrena serrata, Hyptis suaveolens.*

Mexican: *Argemone mexicana.*

Eurasian: *Anagallis arvensis, Lathyrus aphaca, Melilotus albus, Setaria pumila, Sonchus arvensis.*

Afro-Asian: *Cleome monophylla, Cymbopogon martini, Dichanthium foveolatum, Digera muricata, Emilia sonchifolia, Laggera aurita, Merremia emarginata, Pupalia lappacea, Vigna trilobata.*

Austro-Asian: *Crotalaria medicaginea, Indigofera glandulosa, I. linnaei, I. trifoliata, I. trita, Ottelia alismoides, Salvia plebeia, Scleria tessellata, Sporobolus diandrus.*

Eurasian and African: *Medicago polymorpha, Sonchus oleraceous.*

Temperate: *Convolvulus arvensis, Potamogeton crispus, P. perfoliatus, Stuckenia pectinata, Veronica anagallis-aquatica.*

Paleotropical: *Abelmoschus ficulneus, Acalypha ciliata, Achyranthes aspera, Canscora diffusa, Cleome viscosa, Commelina forsskalii, Corchorus fascicularis, C. trilocularis, Cyperus alopecuroides, C. difformis, C. iria, Datura innoxia, Eragrostis tremula, Glinus oppositifolius, Heliotropium ovalifolium, Hibiscus panduriformis, H. vitifolius, Ipomoea aquatica, I. eriocarpa, I. pes-tigridis, Indigofera cordifolia, I. hirsuta, I. linifolia, Kyllinga tenuifolia, Ludwigia perennis, Persicaria barbata, Physalis angulata, Pycreus flavidus, Saccharum spontaneum, Solanum virginianum, Spermacoce articularis* and *S. pusilla.*

Pantropical: *Abrus precatorius, Ageratum conyzoides, Amaranthus spinosus, A. viridis, Biophytum sensitivum, Blainvillea acmella, Boerhavia diffusa, Cardiospermum halicacabum, Celosia argentea, Chamaecrista pumila, Cleome gynandra, Corchorus capsularis, C. olitorius, C. tridens, Cressa cretica, Cyanthillium cinereum, Cyperus compressus, C. michelianus* subsp. *pygmaeus, Dactyloctenium aegyptium, Desmodium triflorum, Eclipta prostrata, Elephantopus scaber, Euphorbia hirta, Fimbristylis quinquangularis, Hackelochloa granularis, Merremia aegyptia, Mucuna pruriens, Polycarpaea corymbosa, Rhynchosia minima, Rottboellia cochinchinensis, Sesbania bispinosa, Senna obtusifolia, S. occidentalis, S. tora, Sida cordifolia, S. spinosa, Tephrosia purpurea, Teramnus labialis, Triumfetta rhomboidea, Tribulus terrestris, Zornia gibbosa.*

Cosmopolitan: *Bacopa monnieri, Ceratophyllum demersum, Chenopodium murale, Cynodon dactylon, Cyperus rotundus, Digitaria ciliaris, Oxalis corniculata, Portulaca oleracea, Vallisneria spiralis.*

4 RUDERALS

4. 1 RUBBISH HEAPS

These are anthropogenic in origin and are naturally found near human habitation and corners of fields. This habitat shows plants like: *Abutilon indicum* var. *indicum* and var. *guineense,*

Amaranthus spp. especially *A. spinosus, Argemone mexicana, Calotropis gigantea, Chrozophora rottlerii, Chloris virgata, Datura stramonium, Ipomoea carnea, Launaea asplenifolia, Martynia annua, Opuntia elatoir, Phyllanthus fraternus, Setaria intermedia, Solanum virginianum* etc.

4.2 ROAD-SIDES AND WASTE LANDS

The affected area has 12037 ha. waste and fallow land. This category shows plants like *Acacia leucophloea, Albizia amara, Azadirachta indica, Balanites aegyptiaca, Borassus flabellifer, Butea monosperma, Delonix elata, Phoenix sylvestris, Prosopis chilensis, Senna auriculata, Ziziphus* spp. etc. In addition, we have *Acanthospermum hispidium, Achyranthes aspera, Alternanthera bettzickiana, Amaranthus* spp., *Baliospermum solanifolium, Boerhavia diffusa, Calotropis procera, Martynia annua, Parthenium hysterophorus, Senna occidentalis, S. tora* and some Poaceae constitute the herbaceous elements. We have observed only one plant of *Acacia nilotica* subsp. *cupressiformis* near Katnera in Dhar district as a road-side tree.

4.3 GRAVELLY LAND AND WALL FLORA

These are clubbed together as the habitat is almost similar. This is exposed, dry, stony or gravelly substratum, supporting some plants like: *Acalypha indica, Achyranthes aspera, Aerva lanata, Ageratum conyzoides, Amaranthus viridis, Andropogon pumilus, Anisochilus carnosus, Apluda mutica, Aristida adscensionis, A. redacta, A. setacea, Blumea membranacea, Boerhavia diffusa, Cenchrus* spp., *Chrysopogon fulvus, Commelina benghalensis, Convolvulus prostratus, Corchorus depressus, Cyanotis fasciculata, Cyanthillium cinereum, Dichanthium annulatum, Eragrostiella bifaria, Eragrostis ciliaris, Euphorbia hirta, Evolvulus alsinoides, Ficus religiosa, Heteropogon contortus, Lantana × aculeata, Lepidagathis trinervis, Lindenbergia indica, Melanocenchris jacquemontii, Microchloa indica, Nanorrhinum ramosissimum, Oplismenus burmannii, Oropetium thomaeum, O. villosulum, Peristrophe bicalyculata, Schoenfeldia gracilis, Tetrapogon tenellus, Tridax procumbens, Tripogon jacquemontii, Tragus mongolorum, Zornia gibbosa* etc., during or immediately after monsoon.

PHYTO-RESOURCES - Man has been using plants for various purposes from times immemorial. A complete list of all the plants used by us is too lengthy to be given here. Common, easily available species have been listed under many headings.

1. Timber-wood (for home building, construction, agricultural implements, furniture etc.) *Acacia nilotica* subsp. *indica* (Babool), *Ailanthus excelsa* (Maharukh), *Anogeissus latifolia* (Dhawda), *Azadirachta indica* (Neem), *Hardwickia binata* (Anjan), *Mangifera indica* (Aam), *Prosopis juliflora* (Vilayati babool), *Syzygium cumini* (Jamun), *Tectona grandis* (Sagwan), *Terminalia tomentosa* (Saja, Sajad). Other trees occasionally used, when the above are not available are: *Acacia leucophloea* (Reunja), *Alangium salvifolium* (Ankol, Alirajpur, especially), *Cleistanthus collinus* (Garari), *Garuga pinnata* (Kakad, Kekad), *Lagerstroemia parviflora* (Lendia), *Mitragyna parvifolia* (Kalam), *Ziziphus jujuba* (Ber).

2. Ornamentals in gardens: This is not a complete list, as it would be too lengthy. *Acalypha hispida, A. wilkesiana, Agave americana, A. cantala, Aglaonema commutataum, Alcea rosea* (Hollyhock), *Allamanda cathartica, Alocasia macrorrhiza, Anthurium andreaenum, Antigonon leptopus, Antirrhinum majus, Argyreia nervosa, Asclepias curassavica, Asparagus gonocladus, A. setaceous, Aster amellus, Barleria cristata* (this species occurs wild also), *Bassia* *indica* (*Kochia indica*-old), *Beaumontia grandiflora, Bougainvillea spectabilis, Bryophyllum pinnatum* (*Kalanchoe pinnata* –old), *Caladium bicolor, Calathea ornata, Calendula officinalis, Callistemon rigidus, Callistephus chinensis, Canna indica* (Kardali), *Caryota urens, Casuarina equisetifolia, Celosia argentea, Cestrum nocturnum* (Rat rani), *Chrysanthemum indicum, Clerodendrum splendens, Clitoria ternatea, Cocos nucifera* (Nariyal), *Codiaeum variegatum* (Croton), *Combretum indicum, Cordia sebestena, Cordyline terminalis, Coreopsis tinctoria, Cosmos bipinnatus, C. sulphureus, Crinum asiaticum* (Sudarshan), *Crossandra infundibuliformis, Cyperus involucratus, Dianthus chinensis, Dahlia pinnata, Dieffenbeckia picta, Dolichandra unguis-cati, Dracaena elliptica, Epiprenum aureum* (Money plant), *Eranthemum pulchellum, Euphorbia milii, E. pulcherrima* (Peela ya Lal patta), *Ferandoa adenophyllum* (*Haplophragma adenophyllum*-old), *Ficus benjamina, Fourcroya gigantea, Gaillardia aristata, G. pulchella, Gladiolus gandavensis, Gomphrena globosa* (Button rose), *Grevillea robusta* (Silver oak), *Hamelia patens, Helianthus annuus* (Bara Sooraj mukhi), *H. pauciflorus* (Chhota Sooraj mukhi), *Heliconia rostrata, Hibiscus rosa-sinensis* (Gudhal), *H. schizopetalus, Hiptage benghalensis* (Madhavi lata), *Homalocladium platycladum, Iberis amara* (Wall flower), *Ipomoea cairica* (Railway creeper), *I. quamoclit* (Ganesh bel), *Ixora coccinia, I. pavetta, Jasminum multiflorum* (Kunda), *J. grandiflorum* (Chameli), *J. sambac* (Mogra), *Jatropha integerrima, Lachenalia ensifolia* (Tube rose, Nishigandha *Polyanthes tuberosa* old.), *Lagerstroemia indica* (Sawni), *L. speciosa* (Queen of forest), *Livistona chinensis, Malva sylvestris, Malvaviscus arboreus* var. *mexicana, Maranta cristata, Mentha spicata, Mirabilis jalapa, Murraya paniculata* (Kamini), *Nerium oleander, Nyctanthes arbor-tritis* (this species is abundant in the hills, known locally as Sirali and used extensively for making baskets, mats, storage for grain etc.), *Pancratium zeylanicum, Pedilanthes tithymaloides, Pentas lanceolata, Petunia hybrid, Petrea volubilis, Phlox drummondii, Plumeria rubra,Portulaca grandiflora, Pyrostegia venusta, Rhoeo discolor, Rosa canina, R.* × *damascena, Russelia equisetiformis, Salvia coccinia, S. fulgens, S. leucantha, Solidago canadensis,Tabebuia aurea, T. rosea, Tabernaemontana divaricata, Tagetes erecta, T. patula, Tecoma stans, Thunbergia alata, T. grandiflora, Tradescantia spathacea* var. *zebrina, Vallaris solanacea, Verbena bipinnatifida, Zephyranthes candida* etc.

3. Trees or shrubs commonly planted along roadsides: *Acacia longifolia, Aegle marmelos, Albizia amara, A. lebbeck, A. procera, Alstonia scholaris, Azadirachta indica, Bauhinia malabarica, B. purpurea, B. tomentosa, B. variegata, Cascabela thevetia, Cassia fistula, Chloroxylon swietenia, Delonix elata, D. regia, Erythrina variegata, Eucalyptus camaldulensis, E. tereticornis, Ficus arnottiana, F. benghalensis, F. religiosa, F. virens, Gliricidia sepium, Gmelina arborea, Grevillea robusta, Hamelia patens, Haldina cordifoilia, Ixora pavetta, Jacaranda mimosifolia, Kigelia africana, Lagerstroemia indica, L. parviflora, Limonia acidissima, Madhuca longifolia* var. *latifolia, Melia azedarach, Mitragyna parvifolia, Morinda coreia, Nerium oleander, Parkia biglandulosa, Peltophorum pterocarpum, Plumeria rubra, Polyalthia longifolia, Pterygota alata, Pterospermum acerifolium, Schleichera oleosa, Senna siamea, S. surattensis, Soymida febrifuga, Spathodea campanulata, Sterculia villosa, Thespesia populnea, T. populneoides, Toona ciliata* etc. Some species like *Hamelia patens* are confined to gardens.

4. Plants in religion, culture, festivals (all the plants listed below do not occur in the study area)

Nakshtras

1. Ashwini-*Strychnos nux-vomica* (Kuchila) **2. Bharani**-*Phyllanthus emblica* (Aaonla)**3. Krittica**-*Ficus racemosa* (Gular) **4. Rohini**-*Syzygium cumini* (Jamun) **5. Mrigashirsha**- *Acacia catechu* (Khair)**6. Ardra**- *Dalbergia sissoo* (Sissoo, Sheesham) **7. Punarvasu**-*Dendrocalamus strictus* (Bans, Vansh) **8. Pushya**-*Ficus religiosa* (Peepal, Ashwattha) **9.** * **Aashlesha**- *Mammea suriga* (Nag kesar) **10. Magha**-*Ficus benghalensis* (Bargad) **11. Poorva phalguni**-*Butea monosperma* (Palash, Tesu) **12. Uttara phalguni**-*Ficus virens* (Plaksha, Pakar) **13. Hasta**-*Sapindus emarginatus* (Aritha, Ritha) **14. Chitra**-*Aegle marmelos* (Bael, Bilwa patra) **15. Swati**-*Terminalia arjuna* (Arjun) **16. Vishakha**-*Flacourtia indica* (Bilanga, Kanju, Katai) **17. Anuradha**-*Mimusops elengi* (Morsali, Bakul) **18. Jyeshtha**-**Pinus roxburghii* (Chir) **19. Mool**-**Shorea robusta* (Sal) **20. Poorva ashadha**-*Salix tetrasperma* (Banjari, Bod, Nained) **21. Uttara ashadha**-**Artocarpus heterophyllus* (Kathal) **22. Shravan**-*Calotropis procera* (Aak, Aakda) **23. Dhanishtha**- *Prosopis cineraria* (Shami) **24. Shatabhisha**-*Neolamarckia cadamba* (Kadamb) **25. Poorva bhadrapada**-*Mangifera indica* (Aam) **26. Uttara bhadrapada**-*Azadirachta indica* (Neem) **27. Rewati**-*Madhuca longifolia* var. *latifolia* (Mahua).

Sacred plants of the 9 planets: Surya-Aak, Aakda-*Calotropis procera,* **Chandra**- Palash, Tesu-*Butea monosperma*, **Mangal**-Khair-*Acacia catechu*, **Budha**-Apamarga, Latjira-*Achyranthes aspera,* **Brihaspati**-Peepal-*Ficus religiosa,* **Shukra**-Gular-*Ficus racemosa,* **Shani**-Shami-*Prosopis cineraria*, **Rahu**-Doob-*Cynodon dactylon*, **Ketu**-Kusha, Darbha-*Desmostachya bipinnata*

Sacred plants of the Rashis: Zodiac signs

Mesha- Aries -Rakta chandan (**Pterocarpus santalinus*), **Gular** (*Ficus racemosa*)

Vrishabha- Taurus-Saptaparni (*Alstonia scholaris*)

Mithun- Gemini-Kathal- (**Artocarpus heterophyllus*)

Karka- Cancer- Palash, Tesu, Dhak (*Butea monosperma*)

Simha- Leo- Shami- (*Prosopis cineraria*)

Kanya- Virgo- Aam (*Mangifera indica*)

Tula- Libra- Morsali (*Mimusops elengi*)

Vrishchik- Scorpious- Kattha- (*Acacia catechu*)

Dhanu- Sagittarius- Peepal (*Ficus religiosa*)

Makar- Capricornus- Sheesham (*Dalbergia latifolia*)

Kumbha- Aquarius- Seeta ashok (**Saraca asoca*)

Meen-Pisces- Bargad (*Ficus benghalensis*)

There are some more trees of rashis, but all of them are not indigenous to our country. It is logical that persons belonging to these should worship and if possible, plant these trees. This is true for other trees in these categories also.

Plants associated with deities or auspicious days

1. *Acacia ferruginea* - Bara khair, Safed khair-Flowers offered to **Lord Shiva** and to the sage **Vashishtha**

2. *Aegle marmelos* -Bael, Bilwa patra-Favourite plant of **Lord Shiva**
3. *Albizia lebbeck*-Kala siris- Tree of Venus and pious to **Buddhists**
4. *Azadirachta indica*- Neem-Young twigs eaten on Hindu New Year, **Gudi padwa** to remain healthy for the whole year
5. *Bauhinia purpurea*- Koliar, Peddare- flowers used in worship of **Sun**
6. *Bauhinia racemosa*-Asta- Leaves offered on the occasion of **Dussera** festival as best wishes for the whole year to friends and family
7. *Bombax ceiba*-Semal, Shalmali-sacred tree for worshipping **Lord Bramha**, **Indra** and **Goddess Laxmi**
8. *Boswellia serrata*-Salai, Loban-used in worship (the resin, loban is burnt during worship)
9. *Butea monosperma*- Palash, Tesu-twigs used as samidha in **Yagyna, thread ceremony**, worship of **Lord Vishnu, Narsimha**, the tree is dedicated to planet Moon
10. **Elaeocarpus sphaericus*-Rudraksh-the rosary made from fruits is worn while worshipping **Lord Shiva,** supposed to be originated from the tears of Shiva and also for control of high blood pressure
11. *Erythrina variegata*-Mandar-in worship of **Shiva** and **Vishnu**
12. *Ficus benghalensis*-Bargad, Bad- the three main gods (Trinity), are situated in: bark-**Vishnu**, branches-**Shiva** and roots-**Bramha**. Married women worship the tree for long life to their husband, on full moon day in the month of Jyeshtha (June or July)
13. *Ficus racemosa*-Gular- Favourite abode of **Lord Dattatreya**
14. *Ficus religiosa*- Peepal, Ashwattha-represents **Lord Krishna**
15. *Limonia acidissima*- Kaith, Kabith-the Gods **Ganesh** and **Vishnu** are supposed to reside in this tree
16. *Mallotus philippensis*-Kamala, Sinduri- Newly married girls worship the tree on every **Tuesday** of **Shravan** month, for the longevity of life of their husband
17. **Mammea suriga*-Nag kesar- used in worshp on **Vaikunth Chaturdashi** day
18. *Mangifera indica*-Aam-This tree is believed to be an incarnation of **Prajapati**, flowers are offered to **Shiva** on Shivaratri day
19. *Manilkara hexandra*-Khirni- roots used in **Talisman** for obtaining blessings of **Moon**
20. **Michelia champaka*- peela champa, sone champa-flowers in worship of **Vishnu**, leaves in **Shiva** and **Parvati**
22. *Mimusops elengi*-Morsali, bakul, -worship of **Vishnu** by flowers fallen on the ground, not plucked from the tree
23. *Nyctanthes arbor-tristis*-Harsingar, Parijat-tree brought from heaven by **Krishna**, for his wife **Satyabhama**
24. *Phyllanthus emblica*-Aaonla, Amalki-leaves in worship of : **Satyanarayana** pooja, **Shiva** on **Shani pradosh**, **Shiva** and **Gauri** on Nitya **Som vrata**

25. *Prosopis cineraria*- Shami- leaves in worship of **Vishnu**, tree is abode of deity **Vasu**, the Pandwas had hidden their weapons in the hollow of this tree when they were secretly abiding their time, assuming different names, profession, wardrobe etc.

26. **Pterocarpus santalinus*- Rakta chandan-piece of root worn in **Talisman** to appease planet **Rahu**'s wrath

27. *Santalum album*-Chandan-used in worship of **Mahadeva** and **Lakshmi**, root piece in **Talisman** to please planet **Rahu**

28. **Saraca asoka*-Ashok- in worship of **Vishnu, Buddha, Indra**, tree of **Cupid**-love god (**Kamdeo**)

29. *Schleichera oleosa*-Kusum-tree is supposed to be originated from **Vishnu**

30. *Syzygium cumini*-Jamun-tree dedicated to **God Megh** (clouds), fruits are a favourite of **Lord Krishna,** because they are dark in colour

This is just a part of the pious plants, the list is almost unending. We are giving below a list of plants, universally used in worship.

Cocos nucifera- **Nariyal**, *Areca catechu*-**Supari**, *Piper betel*-Betal, **Paan leaves,** *Nelumbium speciosum*-**Kamal**, *Nymphaea nouchali* and *N. pubescens*-**Kamal**, Musa × paradisiacal-**kela**-false stem, leaves and fruits, *Mangifera indica*-**Aam**-leaves in Kalash of pooja.

Jain religion: These are plants under which the Jain Tirthankars obtained enlightenment (Salvation) after meditation

Name of Tirthankar Vernacular name of plant Botanical name

1. **Rishabha Nath** Bargad *Ficus benghalensis*
2. **Ajit Nath** Sapta parni *Alstonia scholaris*
3. **Sambhar Nath** Sal **Shorea robusta*
4. **Abhinandan** Cheed **Pinus roxburghii*
5. **Sumati Nath** Priyangu (Daya) *Callicarpa macrophylla*
6. **Padma Nath** Priyangu (Daya) *Callicarpa macrophylla*
7. **Suparshwa Nath** Kala siris *Albizia lebbeck*
8. **Chandraprabha Nath** Nag kesar **Mammea suriga*
9. **Pushpadant** Baheda *Terminalia bellirica*
10. **Sheetal Nath** Bael, Bilwa patra *Aegle marmelos*
11. **Shreyans Nath** Tendu **Diospyros malabarica*
12. **Vasu poojya** Kadamba *Neolamarckia cadamba*
13. **Vimal Nath** Jamun *Syzygium cumini*
14. **Anant Nath** Peepal *Ficus religiosa*
15. **Dharma Nath** Kaitha, Kabeeth *Limonia acidissima*
16. **Shanti Nath** Toon *Toona ciliata*

17. **Kunthu Nath** Chatpati, Tilwan **Wendlandia exserta*

18. **Araha Nath** Aam *Mangifera indica*

19. **Malli Nath** Ashok **Saraca asoka*

20. **Munisuvrata Nath** Peela champa **Michelia champaca*

21. **Nami Nath** Morsali *Mimuspos elengi*

22. **Nemi Nath** Bans *Dendrocalamus spp.*

23. **Parshwa Nath** Deodar **Cedrus deodara*

24. **Mahavir/Vardhaman** Sal **Shorea robusta*

Panchvati: Group of 5 trees- 1. **Aaonla**-*Phyllanthus emblica,* 2. **Bargad**-*Ficus benghalensis* 3. **Bael**-*Aegle marmelos* 4. **Peepal** - *Ficus religiosa* 5. **Ashok**-**Saraca asoka*. There are more combinations, but only one is given here.

Species marked with * do not occur in this area.

Even a cursory glance at these examples shows that many are common. However, the most significant aspect is, our rishis, sages (Pious people) etc., were far sited and aware of the fact that our people are very religious minded and if they are to preserve our plant wealth, it must be connected with religion. The same is true regarding the clan (gotra) trees of tribal people, e. g., Bhils, who belong to Jamunia clan, will not harm Jamun tree (but it is also true that he will not hesitate to cut a tree belonging to Aawalia clan (aaonla) tree).

Plants associated with fairs

Name of fair Papaver somniferum MP Mandsaur

6. Edible plants (Used after cooking) - *Dioscorea bulbifera* (Dang kanda, Ratalu), *D. hispida* (Baichandi), *D. pentaphylla* (Bara kand, Suwar kand), *Nymphaea nouchali* and *N. pubescens* (Kamal)

1. Leaves and young shoots (after cooking) - Species of *Amaranthus* (Chavlai), *Ipomoea aquatica* (Kalmi sag), *Oxalis corniculata* (Khatti bhaji), *Portulaca oleracea* (Bari kulpha, sold in the market also)

2. Flowers (after cooking) - *Annona squamosa* (Sitaphal), *Cordia dichotoma* (Gondi, Lasoda), *Diospyros melanoxylon* (Tendu, Temru), *Ficus hispida* (Bhui gular, young figs), *F. racemosa* (Gular, young figs), *Syzygium cumini* (Jamun).

3. Fruits - *Annona squamosa* (Sitaphal), *Buchanania lanzan* (Achar, Char), *Carissa carandas C. spinarum* (Karonda), *Diospyros melanoxylon* (Tendu, Temru), *Grewia asiatica* (Phalsa), *Manilkara hexandra* (Khirni, most of it comes from Mandu plateau, Dhar), *Syzygium cumini* (Jamun), *S. salicifolium* (Kat jamun), *Ziziphus jujuba* (Ber), *Z. nummularia* (Jhar beri) .

4. Seeds (either roasted or cooked) - *Indigofera glandulosa* (Barbata, Jhugin), *Kavalama urens* (Kadhai, Kullu), *Vigna* spp. (Jangli moong), *Pithecellobium dulce* (Jangle jalebi, ripe seed stalk-funiculus).

5. Medicinal plants - Disease-wise The list is too long to be given here. Only a few plants are enumerated here. Authorities for binomials not given.

Skin Diseases

Name of Plant	Vernacular name	Part used	Administration	Used for
Dalbergia sissoo	Sissoo	Leaves	Internal	Ringworm
Dioscorea bulbifera	Bela	Root	External	Irritation, Burning sensation
Leea macrophylla	Mothi dhani	Stem	External	Irritation
Stereospermum chelonoides	Padar	Stem bark	External	Irritation
Terminalia arjuna	Arjun, Kahu	Stem bark	External	Irritation
Tinospora sinensis	Giloy, Guduchi	Aerial roots	Internal	Irritation
Xanthium strumarium	Adha sisi	Seed oil	External	Irritation
Cuts, wound healing, fractures				
Cissus quadrangularis	Harjod	Stem	External, Internal	Fracture
Curculigo orchioides	Kali musli	Root	External (paste)	Wound healing
Eclipta prostrata	Bhangra, Maka	Whole plant	External (paste)	Cuts and Wounds
Emilia sonchifolia	Sadamandi	Whole plant	External (paste)	Cuts and Wounds
Soymida febrifuga	Rohan	Stem bark	Poultice	Fracture
Tribulus terrestris	Chhota gokhru	Leaf paste	Poultice	Wound healing
Uraria picta	Prista parni, Dabia	Root and leaves extract	Internal	Wound healing
Fever				
Aristolochia bracteolata	Kidamar	Root extract	Internal	
Boerhavia repens	Safed punarnava	Root extract	Internal	
Caesalpinia bonduc	Gatar, Fever nut	Embryo paste	Internal	Fever due to various reasons
Calotropis procera	Aak, Aakda	Root bark extract	Internal	
Celosia argentea	Safed murga, Kuyada jad (root)	Root extract	Internal	
Enicostema axillare	Kadvi nai	Whole plant extract	Internal	
Eranthemum roseum	Dasha mooli	Root extract	Internal	
Leptadenia reticulata	Dodi	Leaf extract	Internal	
Marsdenia tenacissima	Dudhia bela, Kadvi dudhi	Root extract	Internal	

Name of Plant	Vernacular name	Part used	Administration	Used for
Ocimum basilicum	Marwah, Babui tulsi	Leaf extract	Internal	
Phyllanthus virgatus	Aamli	Extract of root and leaf	Internal	
Gynic problems				
Asparagus racemosus	Khargosh ki ghas	Root extract	Internal	
Biophytum sensitivum	Lajwanti	Whole plant extract	Internal	
Bridelia retusa	Ekdania, Kasai	Stem bark extract	Internal	
Butea monosperma	Palash, Khakra, Tesu	Stem bark extract	Internal	Gum called Kamar kas given in back ache for ladies after delivery
Corchorus fascicularis	Mal-rajan	Whole plant extract	Internal	
Curculigo orchioides	Kali musli	Root extract	Internal	
Diplocyclos palmatus	Shivlingi	Root extract	Internal	
Pueraria tuberosa	Vidari kand	Root extract	Internal	
Soymida febrifuga	Rohan	Root extract	Internal	
Mental weakness, memory				
Aloe vera	Gawar phata	Leaf sap extract	Internal	
Bacopa monnieri	Nir bramhi	Plant extract	Internal	
Celastrus paniculatus	Mal kangni	Seed oil	One drop on puffed sugar (Patasa)	
Epilepsy				
Calotropis procera	Aak, Aakda	Root bark	Extract, internal	
Measles				
Ailanthus exselsa	Maharukh, Mahanim	Stem bark extract	Internal	
Leucoderma				
Plumbago zeylanica	Chitrak	Leaves and roots paste	External	
Leprosy				
Cryptolepis dubia	Kala patha	Root extract	Internal	

Name of Plant	Vernacular name	Part used	Administration	Used for
Martynia annua	Bagh nakhi	Seed paste	External	
Aches and Pains				
Achyranthes aspera	Lat jeera	Head, Stomach	Whole plant paste	External
Calotropis procera	Aak, Aakda	Head	Root extract	Internal
Commelina benghalensis	Kan kauwa	Tooth	Whole plan paste	External
Cassine glauca	Jamrasi	Ear, Head	Leaves and stem bark extract	Internal
Tinospora sinensis	Giloy, Guduchi	Stomach	Aerial root extract	Internal
Triumfetta rotundifolia	Gadar lapti	Stomach	Root extract	Internal
Urena lobata, U. sinuata	Bachita	Stomach	Root extract	Internal
Dysentary & Diarrhoea				
Asparagus racemosus	Khargosh ki ghas	Dysentary, Flatulence	Root extract	Internal
Celosia argentea	Safed murga	Vomiting & Dysentary	Root extract	Internal
Cissampelos pariera	Pahad bel	Vomiting & Dysentary	Root extract	Internal
Clitoria annua Terminalia bellirica	Luijad & Baheda	Dysentary	Leaves of luijad and stem bark of Baheda extract	Internal
Corchorus fascicularis	Mal rajan	Indigestion	Root extract	Internal
Cheilocactus speciosus	Keo kand	Heart burn (acidity)	Root extract	Internal
Eranthemum roseum	Dash mooli	Dysentary	Root extract	Internal
Desmodium gangeticum	Salparni	Dysentary & Vomiting	Root extract	Internal
Hygrophila auriculata	Tal makhana	Dysentary	Seed extract	Internal
Jatropha curcas	Ratan jyote	Dysentary & Vomiting	Root extract	Internal
Solanum virginianum	Bhui ringni	Dysentary	Root extract	Internal
Rheumatism				
Alangium salvifolium	Ankol		Leaf paste	External
Asparagus racemosus	Khargosh ki ghas		Root paste	External

Name of Plant	Vernacular name	Part used	Administration	Used for
Boswellia serrata	Salai		Gum-resin paste	External
Bridelia retusa	Kasai		Stem bark paste	External
Cardiospermum halicacabum	Kan phuti		Leaf paste	External
Cassia fistula	Amaltas		Leaf paste	External
Celastrus paniculatus	Mal kangni		Seed oil	External
Cleome gynandra	Safed hurhur		Leaf paste	External
Cuscuta reflexa	Amar bel		Stem paste	External
Euphorbia neriifolia	Patton ki send		Stem paste	External
Hemidesmus indicus	Anant mool		Leaf paste	External
Holoptelea integrifolia	Bandar roti		Leaf paste	External
Jatropha curcas	Ratan jyote		Seed oil	External
Kydia calycina	Baranga		Leaf paste	External
Lantana × aculeata	Rai muniya		Leaf paste	External
Madhuca longifolia var. *latifolia*	Mahua		Seed oil	External
Momordica charantia	Karela		Fruit paste	External
Nyctanthes arbor-tristis	Sirali		Leaf extract	Internal
Pergularia daemia	Sadovani		Leaf paste	External
Plumbago zeylanica	Chitrak		Root extract	Internal
Solanum viginianum	Bhui ringni		Extract whole plant	Internal
Soymida febrifuga	Rohan		Paste stem bark	External
Tephrosia purpurea	Sar punkha		Root extract	Internal
Tinospora sinensis	Giloy, Guduchi		Fruit extract	Internal
Ventilago denticulata	Keoti		Stem bark paste & Sesame oil	External
Vitex negundo	Samhalu		Leaf paste	External
Withania somnifera	Asgandha		Plant paste	External
Jaundice				
Calotropis procera	Aak, Aakda		Root extract	Internal
Cocculus hirsutus	Jal jamni		Fruit extract	Internal

Name of Plant	Vernacular name	Part used	Administration	Used for
Cheilocostus speciosus	Keo kand		Rhizome paste & extract	Former external, latter internal
Dalbergia volubilis	Dhoban bel		Leaf extract	External & internal
Luffa acutangula	Kadvi torai		Leaf & fruit paste	External
Morinda coreia	Aal		Root extract	Internal
Ocimum basilicum	Marwah		Extract whole plant	Internal
Phyllanthus fraternus	Bhui aawli		Extract whole plant	Internal
Terminalia arjuna	Arjun, Kahu		Extract stem bark	Internal

Note:- The data given above is based upon Adibasis (mainly Bhils, Bhilalas), village elders. We have not verified any of these claims. Great caution should be taken in application, both external and internal.

WILD GERM PLASM OF CULTIVATED PLANTS

S. no	Name of Species	English-Vern. Name	Name of wild relative species
	Cereals & Millets		
1	**Oryza sativa** L.	Rice/ Chaval	**Oryza rufipogon** Griff. *Jangli Dhan*
2	**Echinochloa frumentacea** Link		**Echinochloa crus-galli** (L.) P. Beauv.
3	**Eleusine coracana** (L.) Gaertn.	Finger millet/ Mandua	**Eleusine indica** (L.) Gaertn. *Pandhar*
4	**Sorghum bicolor** (L.) Moench	Jawar	**Sorghum halepense** (L.) Pers. *Boru*
5	**Panicum miliaceum** L.	Common millet/ Morha	**Panicum paludosum** Roxb.
6	**Paspalum scrobiculatum** L.	Kodo	**Paspalum notatum** Fluegge *Joker band*
7	**Pennisetum glaucum** (L.) R. Br.	Pearl millet/ Bajra	**Pennisetum pedicellatum** Trin. *Dinanath*
8	**Amaranthus cruentus** L.	Rajgira	**Amaranthus hybridus** L.
9	**Setaria italica** (L.) P. Beauv.	Italian millet	**Setaria verticillata** (L.) P. Beauv. *Chiktya*
	Pulses		
10	**Cajanus cajan** (L.) Millsp.	Pigeon pea/ Tuwar	**Cajanus scarabaeoides** (L.) Thouars
11	**Vigna radiata** R. Wilczek	Green gram/ Moong	**Vigna radiata** (L.) R. Wilczek var. **sublobata** (Roxb.) Verdc.
	Fruits		
12	**Musa × paradisiaca** L.	Banana / Kela	**Musa acuminata** Colla, **M. balbisiana** Colla

S. no	Name of Species	English-Vern. Name	Name of wild relative species
13	**Citrullus lanatus** (Thunb.) Matsum. & Nakai	Water melon/ Tarbuz	**Citrullus colocynthis** (L.) Schrad. *Indrayan*
14	**Physalis peruviana** L.	Rasbhary/ Chirpoti	**Physalis angulata** L. *Ban phutka*
	Vegetables		
15	**Momordica cochinchinensis** (Lour.) Spreng.	Bhat karda	**Momordica dioica** Roxb. ex Willd. *Kakoda*
16	**Abelmoschus esculentus** (L.) Moench	Lady's fingers/ Bhindi	**Abelmoschus manihot** subsp. **tetraphyllus** (Roxb. ex Hornem) Borss Walk.
17	**Cucumis melo** L.	Musk melon/ Kharbuz	**Cucumis prophetarum** L.
18	**Solanum melongena** L.	Egg plant / Baingan	**Solanum incanum** L. *Jangli bhata*
	Fibers		
19	**Corchorus capsularis** L.	Jute / Patsan	**Corchorus aestuans** L., **C. trilocularis** L.
	Spices and condiments		
20	**Curcuma longa** L.	Turmeric / Haldi	**Curcuma pseudomontana** J. Graham
21	**Zingiber officinale** L.	Ginger / Adrak	**Zingiber cernuum** Dalz. *Jangli adrak*
	Sugar		
22	**Saccharum officinarum** L.	Sugar / Shakar, Chini	**Saccharum spontaneum** L. *Kans*
	Medicinal plants		
23	**Catharanthus roseus** (L.) G. Don	Periwinkle/ Sada bahar	**Catharanthus pusillus** (Murray) G. Don
	Narcotic plant		
24	**Nicotiana tabacum** L.	Tobaco /Tamakhu	**Nicotiana rustica** L. *Kalkatia tamakhu*

Source:- S. Ray & J. Sainkhedia, Phytotaxonomy 16: 96-98. 2017.

Country cigars (Biris) *Diospyros melanoxylon*, sometimes leaves of *Bauhinia vahlii*, where available.

Brooms- *Aristida adscencionis*, *Eragrostis japonica* (Ghadela, Ghodila), *Phoenix sylvestris* (Leaves folded from the midrib, so the halves of lamina face each other), *Thysanolaena latifolia* (Phul buhari).

Silk cotton for stuffing pillows - *Bombax ceiba*.

Shampoo- *Sapindus emarginatus* (used for washing silk clothes also).

Dyes- *Acacia catechu, Butea monosperma, Lawsonia inermis, Woodfordia fruticosa, Wrightia tinctoria*

Fibres- *Abutilon* spp., *Helicteres isora* (Maror phalli), *Marsdenia tenacissima* (Dudhia bela, Kadwi dudhi), *Typha angustifolia* (Er, Ghao bajra)

Fish poison- *Casearia graveolens* (Girchi), *Catunaregam spinosa* (Mendal), *Cleistanthus collinus* (Garari)

Fuel- *Acacia nilotica* subsp. *indica, Balanites aegyptiaca* (Hingot), *Lagerstroemia parviflora, Lannea coromandelica* (Moyan), *Miliusa tomentosa* (Kari, Om), *Prosopis juliflora* (Vilayati babool), *Ventilago denticulata* (Keoti). Actually trees available in plenty, or when the above mentioned trees are scarce

Green manure- *Desmodium gangeticum* (Salparni), *D. triflorum* (Jangli methi), *Indigofera glandulosa* (Barbada), *Peristrophe bicalyculata* (Atrilal), *Teramnus labialis* (Mash parni)

Insect repellent- *Azadirachta indica* (Neem), *Blumea eriantha* (Jangli kapoor-used in Alirajpur), *Cymbopogon martini* (Rosha ghas, Tikhadi)

Non edible oils (used in paints, varnishes, medicine etc.) *Buchanania lanzan* (Achar, Char), *Casearia graveolens* (Girchi), *Celastrus paniculatus* (Mal kangni), *Kavalama urens* (Kullu), *Madhuca longifolia* var. *latifolia* (Mahua), *Pongamia pinnata* (Karanj), *Ventilago denticulata* (Keoti)

Paper pulp- *Boswellia serrata* - Salai (critically endangered, Nepa nagar), *Dendrocalamus strictus* (Bans), the following plants have a potential capacity for use in making paper. Everything depends upon abundant and assured supply : *Acacia nilotica* subsp. *indica, Arundo donax, Bombax ceiba, Butea monosperma, Chrysopogon zizanoides, Cordia dichotoma, Crotalaria juncea, Cyperus exaltatus, Dendrocalamus strictus, Desmodium gangeticum, D. velutinum, Ficus benghalensis, F. religiosa, Helicteres isora, Hibiscus cannabinus, Imperata cylindrica, Lannea coromandelica, Lantana × camara, Linum usitatissimum, Melilotus albus, Ricinus communis, Saccharum spontaneum, Sehima nervosum.*

CHAPTER III

VINDHYAN AND SATPUDA RANGES

I. VINDHYAN REGION-NORTH OF NARMADA RIVER

1. Malwa Plateau: Most part of the area is covered by this plateau viz., Dhar, Badnavar and Sardarpur tehsils of Dhar dist., Jhabua, Meghnagar, Thandla and Petlawad tehsils of Jhabua dist. and Alirajpur, Bhabra and Sondwa tehsils of Alirajpur dist. This area is confined to Upper Mahi-Anas hill rocks and Chambal-Chhoti Kali Sindh basin of Western Malwa. Average elevation above msl is 500 m, ranging between 300-600 m. Generally the elevation is low in western part of Jhabua. This part of Malwa plateau proceeds southwards and merges with the Vindhyan Scarp.

2. Vindhyan Scarp: It constitutes southern boundary of Malwa plateau and northern boundary of Narmada basin. Southern part of Dhar tehsil including the Mandu hills, northern part of Kukshi tehsil, Alirajpur and Bhabra tehsils of Alirajpur dist. In Indore dist., the Vindhyan Scarp forms water shed. Several streams originate here to meet Chambal river in the north and Narmada River in the south. Here the Vindhyan ranges spread in west to south-west and east to north-east direction. Actually this is Mhow tehsil of Indore dist., where forest of Indore dist. is found. Rest of Indore is plain area.

The highest peaks in this area are: Magarba (751 mts.) and Neelkanth (702 mts.) in Dhar dist., Singarcholi (906 mts.), Nahar khodra (903 mts.), Janapao (903 mts.) in Indore dist. Rest of the area is plain.

3. Narmada valley: South-east part of Manavar and Kukshi tehsils of Dhar district, Alirajpur and Jobat tehsils of Alirajpur district constitute this valley.

GEOLOGY

Following formations are met with.

	Age	Formations
	Recent	Alluvium
	Pleistocene	Laterite
Cretaceous		
	Eocene	Deccan trap
	Upper cretaceous	Lametta
	Lower cretaceous	Bagh beds

Late pre Cambrian	Vindhyan system
Archean	Bijawar series
	Granite and Gneissen

SOIL: The whole area is occupied by black cotton soil or Regur. Texture of soil is loamy and clayey. There is a belt of red soil in north –west area of Jhabua.

RIVER SYSTEM: The 2 main rivers in Alirjpur, Dhar and Jhabua districts are the Narmada and Chambal. Narmada along with its tributaries drains into the Arabian Sea. Chambal follows a northerly course and meets the river Yamuna along with its tributaries. Some other important rivers in these 3 districts are: Mahi, Karam, Man, Anas, Uri, Wagh, Hathni etc.

Indore district has 5 rivers, all originating in the district. They are: 1. Chambal or Charmavati of ancient times, originates from the western slope of Janapao hill. It flows in a northerly direction, forming the western boundary of the district near Ghatabillod, Gautampura town and station. After receiving the rivers from Malwa plateau, it joins Yamuna River in the north. 2. River Kshipra originates from Kakri-Bardi hill, about 11 km south-east of Indore. It forms eastern boundary of the district and joins river Chambal in Ujjain. 3. Kanh river is a tributary of Kshipra. It originates near village Umaria. After flowing through Indore city, it also joins Chambal in Ujjain. 4. Gambhir river. It originates from the eastern slope of Janapao hills, flows through Mhow tehsil and joins Kshipra in Ujjain. These 4 rivers meet Yamuna in the north. We have seen that Kshipra, Kanh and Gambhir already join Chambal in Ujjain. 5. River Choral- This river originates near Chhoti Jam hill in the North West direction. At Kundia village it leaves Indore district and joins Narmada near Barwaha, forming part of the Narmada drainage system.

CLIMATE: It is monsoon climate throughout. We have 3 well defined seasons. Monsoon- from mid June to Sept. end, winter from October to February and summer from March to June. The rain fall varies from place to place. Jhabua receives the least rainfall, while Kattiwada in Alirajpur distrct receives the highest.

II. SATPURA REGION

EASTERN NIMAR- DISTRICTS BURHANPUR AND KHANDWA

TOPOGRAPHY

These two districts lie between Narmada and Tapti river basin. Both these rivers flow from east to west and ultimately join the Arabian Sea. There are 4 natural divisions. 1. Narmada valley 2. Tapti valley. 3. Satpura ranges and 4. Hatti range

1. **Narmada valley** forms northern boundary of East Nimar, and as mentioned earlier, only Chandgarh and Selani lie north of Narmada (in the Vindhyan region). Eastern part of Narmada valley is drained by Chhoti Tawa River. At Punasa, there is Indira Sagar reservoir. Fertile valleys of Abna and Sukta rivers lie in the southern part of the region. This region is devoid of hills and forests.
2. **Tapti valley** lies between two parallel ranges of Satpura covering the southern part of this region. In the eastern part are Manjrod and Nathwa, covered by forest. This valley opens west of Samardeo hills towards Burhanpur.

3. **Satpura ranges** - This is a complex system separating Narmada and Tapti valleys. Here, Satpura is bifircated into parallel ridges on either side of Tapti valley. The ridge on the northern side penetrates south-eastern part of Harsud and also marks a boundary between Khandwa and Burhanpur districts. In Burhanpur, Asirgarh peak rises to a height of 667 mts. The hills west to Asirgarh are called Rajpipla hills. Amba reserve forest of east Nimar is located here. Towards east of Asirgarh are the Kalibhit ranges. Some notable peaks here are: Kalibhit 678 mts, Kali pahar 754 mts and Phepri 651 mts.
4. **Hatti range** - This is bifurcated part of Satpura system, lying south of Tapti valley. In the extreme east of this range, there are some high peaks, including the highest in east Nimar-Pipar dol, 917 mts, Bain garh 828 mts, Sitoli 755 mts and Bhingar 755 mts. This entire range is covered by thick forest.

RIVER AND DRAINAGE

Both Narmada and Tapti rivers flow from east to west. The major portion of the area drains towards north into Narmada through Chhota tawa and Kaveri rivers along with a large number of streams.

River Narmada enters East Nimar just after its confluence with Machak river at Panghat. After passing a narrow gorge at Dhairi, a point opposite Punasa, the river widens at Mandhata. Its extent in east Nimar is about 86 Kms.

Main tributaries of Narmada are - Chhota Tawa, Abna, Sukta and Kaveri.

Tapti river-Originates in Betul district of MP (Multai) and enters East Nimar near Golai of Harsud tehsil. Burhanpur and Jainabad are situated on either side of this river. It enters Maharashtra after 14 kms. Its tributaries are-Moha, Utaoli, Umraoti and Damini.

GEOLOGY: Major part of East Nimar is Deccan trap, except a small part along Narmada in north and alluvial tract round about Burhanpur. The rocks on either side of Narmada are granite and Bijawar rocks belonging to the Archean formation of late Cambrian. Lameta rocks of Upper Cretaceous formation are found at Punasa, which is the northern part of East Nimar.

SOIL: The composition of soil varies from place to place. Alluvium deposits are found mainly along water courses. Soil types are Regur, Pandhar, Man and Thari. The last is enriched with silt and occurs only in Burhanpur tehsil. Pandhar is light grey containing organic matter. Man is yellow sandy soil with less fertility. A very large region is occupied by Regur or black cotton soil.

CLIMATE: East Nimar has typical Tropical climate. Winter is mild, persisting from October to February. January is the coldest month. Summer is from Mach to June. May is the hottest month, characterized by hot winds (*Loo*). July and August are monsoon months, although September witnesses some rains. Winter rains or Maotha, as it is locally called, rarely occurs nowadays.

RAINFALL: Average rainfall is 701 mm. In 2007, the highest rainfall of 1019 mm was recorded.

TEMPERATURE: May is the hottest, the temperature being in the range of 43 to 45 degrees centigrade. January is coldest, the temperature being 10.5 degrees centigrade.

HUMIDITY: Relative humidity varies from 29% in April to 82.7% in August.

WESTERN NIMAR- BARWANI AND KHARGONE DISTRICTS

TOPOGRAPHY: Two natural regions are recognized. 1. Satpura ranges and 2. Narmada valley

1. **Satpura ranges**: these ranges cover most of the south-western region. The highest peak is Tasdin-Valli which is about 1050 mts high above msl. Most of the forest is situated in this region. Sendhwa, Pansemal, Newali and Pati tehsils come under this Satpura range.
2. **Narmada valley**: This valley lies between Khargone and Dhar districts in the northern region. Major portion of Barwani district comes under this. It has a width of about 8-16 kms. Most of the area is agricultural.

GEOLOGY: The plains are covered by the Deccan Trap rocks of different basaltic lava flows. The formations are:

Quaternary	Recent	Alluvial
Tertiary	Cretaceous to Eocene	Deccan trap
Mesozoic	Jurassic –Triassic	Upper Gondwana
Paleozoic	Upper Lower Gondwana,	Upper Vindhyas
Proterozoic	Pre Cambrian	Lower Vindhyas
Archean	Granite and Gneisis	Kudappa

SOIL

1. Black cotton soil- This is found in major part of West Nimar. Generally this type is found in plains, lower slopes and valleys
2. Red soil-Shallow, gravelly soil of poor quality, termed 'kankar'. This is fertile in nature but lacks organic matter and minerals. It is generally found on tops of hills and upper slopes
3. Brown soil-This is clay-loam soil, light grayish, medium brown to dark brown, commonly found near river and stream banks and hill slopes. It is rich in organic matter
4. Alluvial-This is light sandy and clay soil in the beds of rivers and streams. This type supports better quality of teak forests

CLIMATE- There are 3 seasons. Monsoon, from mid June to nearly end of September, winter-from October to February and summer-from March to early June. July and August are the wettest months. Average rain fall is around 800 mm. Maotha or winter rains are uncertain.

TEMPERATURE-In summer, the temperature goes up to 48 degrees centigrade. Hot winds are common. In winter, the temperature goes down to 7-8 degrees centigrade.

HUMIDITY- It is 76% during rainy months, gradually going down with the rise in temperature.

VEGETATION- Forests in west Nimar can be classified into 7 groups

1. Dry teak forests, 2. Dry, mixed deciduous forests, 3. Open scrub forests, 4. Boswellia forests, 5. Hardwickia forests, 6. Butea forests and 7. Dry savanna forests. It is worthwhile to note that district Barwani is mostly devoid of hills and there is very little forest. Therefore, the account of forest being presented, largely applies to district Khargone.

1. **Dry teak forests**- This is found in Barwah, Sanawad, Mandleshwar and Bhikangaon range. These are mainly concentrated on hill slopes and valleys. The soil is clay-loam and rich in moisture. Some species found here are: *Anogeissus latifolia, Bauhinia racemosa, Bridelia retusa, Dalbergia latifolia, Diospyros melanoxylon, Grewia tiliifolia, Holoptelea integrifolia, Lagerstroemia parviflora* and *Tectona grandis.*

2. **Dry mixed deciduous forests**- This category consists of a large number of species. The top canopy has elements like: *Anogeissus latifolia, Boswellia serrata, Diospyros melanoxylon, Hardwickia binata, Lannea coromandelica, Madhuca longifolia* var. *latifolia, Terminalia tomentosa, Tectona grandis.* Other species which are also encountered are: *Acacia* spp., *Aegle marmelos, Albizia procera, Bauhinia malabarica, Bombax ceiba, Buchanania lanzan, Butea monosperma, Cassia fistula, Diospyros melanoxylon, Dolichandrone falcata, Flacourtia indica, Gmelina arborea, Haldina cordifolia, Holoptelea integrifolia, Lagerstroemia parviflora, Terminalia bellirica, T. tomentosa, Wrightia arborea* etc.

The shrubby elements which are normally found are: *Azanza lampas, Carissa carandas, Colebrookia oppositifolia, Grewia hirsuta, Gymnosporia montana, Helicteres isora, Vitex negundo, Woodfordia fruticosa* etc.

The twiners and climbers which are more common are *Ampelocissus latifolia, Canavalia gladiata, Celastrus paniculatus, Cissampelos pariera, Cocculus hirsutus, Mucuna pruriens, Tinospora sinensis* etc.

The ground cover is composed of *Acanthospermum hispidum, Alysicarpus* spp., *Blainvillea acmella, Crotalaria albida, C. hirsuta, Desmodium triflorum, Euphorbia* spp., *Indigofera* spp., *Phyllanthus* spp. and a number of Poaceae.

3. **Open scrub forests** - This is a transitional zone between open mixed forests and dry grasslands localized at Khargone, Bistan, Bhikangaon, Barwah and Barwani regions and characterized by growth of grasses with widely scattered trees; dominant species in this category are *Acacia catechu, A. leucophloea, Butea monosperma, Dolichandrone falcata, Gymnosporia montana, Helicteres isora, Holoptelea integrifolia, Lagerstroemia parviflora, Lannea coromandelica, Vitex negundo, Woodfordia fruticosa* and *Ziziphus* spp.

Common climbers are: *Ampelocissus latifolia, Cissampelos pariera, Dregea volubilis, Tinospora sinensis* and among members of grasses are: *Apluda mutica, Aristida adscensionis, Brachiaria ramosa, Cenchrus ciliaris, C. setiger, Chloris virgata, Chrysopogon aciculatus, Dichanthium annulatum, D. foveolatum, Dactyloctenium aegyptium, Digitaria setigera, Eragrostis* spp., *Heteropogon contortus, Iseilema prostratum, Rottboellia cochinchinensis, Setaria* spp. and *Themeda quadrivalvis.*

Depending upon dominance of a particular species, the forest vegetation can be further divided.

4. **Boswellia forests** - Here this species forms an over wood of stunted trees and shrubs of dry deciduous forest. It is confined to southern parts of Satpura hill tracts and northern parts of

Vindhyan scrub. Dominant tree species are: *Boswellia serrata, Hardwickia binata, Kavalama urens, Lannea coromandelica, Tectona grandis, Terminalia tomentosa* etc. *Acacia catechu, Anogeissus latifolia, Bauhinia malabarica* and *Nyctanthes arbor-tristis* are sparsely present.

5. **Hardwickia forests**- In these forests the % of *Hardwickia* is more than 70. They occur in the southern parts of Satpura hills: Barwani, Khargone, Bistan, Bhikangaon and Sendhwa ranges. The common tree species apart from *Hardwickia binata* are: *Acacia catechu, Albizia lebbeck, Anogeissus latifolia, Boswellia serrata, Lagerstroemia parviflora, Lannea coromandelica, Tectona grandis* etc.

6. **Butea forests**-It occurs in the northern part of Satpura, along Narmada river. *Butea monosperma* is 70% here. Other species are: *Acacia leucophloea, Prosopis juliflora* (now naturalized), *Ziziphus mauritiana* and *Z. nummularia.*

7. **Dry savanna forests**-This is characterized by tree species occurring individually or in small groups with heavy grasses. They harbor many thorny shrubs like *Gymnosporia montana.*

RIVERINE VEGETATION

This type of vegetation occurs in low lying, water-logged areas, along river banks, streams, ponds etc. Some of the common big or small trees are: *Acacia nilotica* subsp. *indica, Alangium salvifolium, Bridelia retusa, Hymenodictyon orixense, Mitragyna parvifolia, Pongamia pinnata, Syzygium salicifolium, Terminalia arjuna* etc. Shrubby elements of this area are: *Ficus hispida* (assumes tree form in favourable habitat), *Tamarix dioica, Typha angustifolia, Vitex negundo, Woodfordia fruticosa* along with herbs like - *Alternanthera sessilis, Ammannia* spp., *Bacopa monnieri, Coix aquatica, C. lacryma-jobi, Cyathocline purpurea, Cyperus* spp., *Fimbristylis* spp. etc.

AQUATIC VEGETATION

A complete list is not given here (The habitat is not much different from the above; many plants are common).

1. Free floating plants: *Eichhornia crassipes, Lemna* spp., *Pistia stratiotes, Spirodela polyrrhiza, Wolffia arrhiza* etc.
2. Plants rooted in mud with floating or submerged parts: *Hydrilla verticillata, Najas graminea, Nymphaea nouchali, Nymphoides hydrophylla, Potamogeton nodosus, P. perfoliatus, Sagittaria trifolia, Stuckenia pectinata, Typha angustifolia* etc.
3. Plants of marshy habitat : *Aeschynomene indica, Ammannia* spp., *Alternanthera sessilis, Bacopa monnieri, Caesulia axillaris, Cleome chelidonii, Coix lacryma-jobi, Cyperus* spp., *Hygrophila auriculata, Paspalum geminatum, Persicaria glabra, Phyla nodiflora, Schoenoplectiella lateriflora* etc.

Vegetation of drying ponds puddles etc. - After the rainy season is over, this habitat supports plants like: *Alternanthera sessilis, Glinus lotoides, Gnaphalium polycaulon, Grangea maderaspatana, Jamesbrittenia dissecta, Laphangium affine, L. luteoalbum, Polygonum plebium* etc. With the onset of summer, the soil starts becoming dry, but before it is fully devoid of any moisture, the habitat supports plants like: *Bergia ammannioides, Coldenia procumbens, Eclipta prostrata, Hoppea dichotoma* etc.

CHAPTER IV

METHODOLOGY

The study area is composed of 8 districts. These were explored floristically, by various college teachers for their (unpublished) Ph. D. dissertations, following the procedures laid down by Jain and Rao (1977). Regular field trips during all seasons were undertaken for a period of 3 years. Later on, Samvatsar published the flora of Dhar, Jhabua and Alirajpur (1996). For identification, various published floras were consulted. Dried herbarium specimens were poisoned and mounted on sheets of recommended size. The herbariums are either with the research scholars or in the institutions where the investigation was carried out. A copy is also deposited in the library of DAVV, as per regulations.

This is the foundation on which the present study is based.

The system of Bentham and Hooker (1862-1883) has been followed with some changes in the light of The Plant List.

The second and most important aspect is the use of The Plant List, Sept. 2013, version 1. 1. This list was prepared by the Royal Botanic Gardens, Kew, England, The Missouri Botanical Garden, N. America and with inputs from experts drawn from all over the world. There are numerous changes in the traditional concept and contents of the families. The basic principle behind these changes is the close phylogenetic relationship between the affected taxa. A few examples would clearly drive home this point.

In Malvaceae, the familes Bombacaceae, Sterculiaceae and Tiliaceae are merged, following the characters of order Malvales and moreover, differences like monoadelphous stamens with 1-celled anthers, echinulate pollen (Malvaceae), palmate compound leaves and smooth pollen in Bombacaceae, monoadelphous stamens, 2-celled anthers, staminodes usually present- Sterculiaceae, free stamens and absence of staminodes - Tiliaceae are minor characters to keep them as separate families. Monocarpellary ovary with marginal placentation combines all 3 separate families, Fabaceae, Caesalpiniaceae and Mimosaceae in one family, Leguminosae. Presence of gynostegium alone was separating Asclepiadaceae from Apocynaceae. Verbenaceae and Lamiaceae were separated on the basis of fleshy and dry fruit, now they are separated from each other on the basis of inflorescence. Members of Verbenaceae having racemose inflorescence are in Verbenaceae, but those having cymose inflorescence (*Clerodendrum, Gmelina*) are placed in Lamiaceae. There are many more examples like this. The exhaustive keys provided will be helpful.

Citations are according to the Plant List and the pattern is chronological. Accepted name followed by basionym (if any), and other synonyms in a chronological order.

Authorities for binomials are given in taxonomic enumeration. At other places where botanical names appear, authorities for them are not given. In sections like, exotic genera, monotypic genera, dominent genera, etc., authorities for them are given.

LIST OF FAMILIES ACCORDING TO THE PLANT LIST

		World	MP	SA
1.	Ranunculaceae	65/2377	4/9/	2/3
2.	Dilleniaceae	11/ 219	1/3	1/1
3.	Annonaceae	128/ 2106	2/5	2/3
4.	Menispermaceae	68/448	4/6	3/5
5.	Nelumbonaceae	1/2	1/1	1/1
6.	Nymphaeaceae	8/70	1/2	1/2
7.	Papaveraceae	41/920	2/3	2/3
8.	Brassicaceae	372/ 4060	12/19	5/7
9.	Cleomaceae	12/257	1/8	1/8
10.	Capparaceae	34/449	4/10	4/8
11.	Violaceae	25/ 806	2/2	1/1
12.	Bixaceae	4/23	1/1	1/1
13.	Polygalaceae	27/1163	2/16	1/6
14.	Caryophyllaceae	91/ 2456	4/6	3/4
15.	Portulacaceae	10/258	1/3	1/2
16.	Talinaceae	2/28	1/1	1/1
17.	Tamaricaceae	3/79	1/ 3	1/2
18.	Elatinaceae	2/57	2/3	1/2
19.	Malvaceae	245/ 4465	25/77	23/63
20.	Linaceae	16/213	2/2	1/1
21.	Malpighiaceae	81/1301	2/2	2/2
22.	Zygophyllaceae	26/211	3/3	2/2
23.	Oxalidaceae	8/601	2/10	2/5
24.	Balsaminaceae	2/488	1/5	1/2
25.	Rutaceae	158/ 1730	5/6	4/4
26.	Simaroubaceae	19/121	1/1	1/1
27.	Burseraceae	18/649	3/3	3/3
28.	Meliaceae	52/669	4/4	4/4
29.	Olacaceae	27/149	1/2	1/1

		World	MP	SA
30.	Celastraceae	87/1168	4/7	4/4
31.	Rhamnaceae	53/839	4/12	2/7
32.	Vitaceae	16/985	7/22	5/13
33.	Sapindaceae	138/1751	3/5	3/5
34.	Anacardiaceae	77/701	6/ 7	4/4
35.	Moringaceae	1/13	1/1	1/1
36.	Leguminosae	946/24505	89/301	64/189
37.	Rosaceae	133/1966	2/2	1/1
38.	Vahliaceae	1/3	1/1	1/1
39.	Crassulaceae	48/1312	1/3	1/1
40.	Combretaceae	17/480	4/15	3/ 8
41.	Myrtaceae	144/ 5774	1/4	1/2
42.	Lecythidaceae	24/319	2/3	1/1
43.	Lythraceae	31/540	6/17	6/14
44.	Onagraceae	43/720	1/5	1/5
45.	Passifloraceae	27/694	1/2	1/1
46.	Cucurbitaceae	134/965	13/25	11/20
47.	Cactaceae	176/2233	3/6	2/2
48.	Aizoaceae	146/2271	2/4	2/3
49.	Molluginaceae	10/59	2/6	2/4
50.	Apiaceae	418/3257	13/21	6/6
51.	Cornaceae	10/115	1/1	1/1
52.	Rubiaceae	609/13673	20/42	15/27
53.	Compositae	1911/ 32913	64/118	47/73
54.	Campanulaceae	88/2385	3/9	3/4
55.	Plumbaginaceae	24/635	2/2	2/2
56.	Primulaceae	68/ 2788	5/6	3/4
57.	Sapotaceae	60/1343	2/2	2/2
58.	Ebenaceae	3/751	1/6	1/3
59.	Oleaceae	25/688	3/13	3/5
60.	Salvadoraceae	3/8	1/3	1/2
61.	Apocynaceae	410/5556	28/41	23/29
62.	Loganiaceae	15/351	3/3	3/3
63.	Gentianaceae	96/1682	6/12	5/8
64.	Menyanthaceae	7/55	1/2	1/2
65.	Boraginaceae	155/2686	8/27	8/21

		World	MP	SA
66.	Convolvulaceae	67/1296	12/16	11/39
67.	Solanaceae	115/2678	6/16	6/14
68.	Linderniaceae	16/164	3/21	2/9
69.	Plantaginaceae	120/1614	12/23	11/14
70.	Phrymaceae	10/199	3/3	3/3
71.	Scrophulariaceae	76/1576	4/6	3/3
72.	Orobanchaceae	89/1613	6/13	6/10
73.	Lentibulariaceae	4/312	1/11	1/5
74.	Gesneriaceae	164/3122	4/4	1/1
75.	Bignoniaceae	86/852	5/6	1/5
76.	Martyniaceae	5/21	1/1	1/1
77.	Acanthaceae	242/3941	23/72	21/50
78.	Verbenaceae	34/1035	6/7	6/7
79.	Lamiaceae	245/7886	30/78	24/54
80.	Nyctaginaceae	34/450	4/8	1/6
81.	Amaranthaceae	178/2052	14/30	12/28
82.	Polygonaceae	59/1384	4/15	3/9
83.	Aristolochiaceae	8/624	1/3	1/2
84.	Piperaceae	13/2658	2/2	1/1
85.	Lauraceae	66/2978	2/3	1/1
86.	Loranthaceae	79/886	2/4	2/2
87.	Santalaceae	42/992	3/3	2/2
88.	Euphorbiaceae	228/6547	11/42	9/38
89.	Phyllanthaceae	58/2099	9/26	5/15
90.	Putranjivaceae	3/216	1/1	1/1
91.	Urticaceae	54/1465	8/9	6/6
92.	Ulmaceae	8/64	1/1	1/1
93.	Cannabaceae	8/102	2/4	1/3
94.	Moraceae	40/1217	4/25	1/13
95.	Salicaceae	54/1269	3/5	3/4
96.	Ceratophyllaceae	1/4	1/2	1/1
MONOCOTYLEDONS				
97.	Hydrocharitaceae	16/133	5/7	5/7
98.	Orchidaceae	899/27801	25/49	11/17
99.	Costaceae	7/139	1/1	1/1
100.	Zingiberaceae	52/1587	7/18	2/6

		World	MP	SA
101.	Musaceae	2/78	2/2	2/2
102.	Amaryllidaceae	80/2258	2/4	2/3
103.	Hypoxidaceae	10/154	2/3	2/2
104.	Dioscoreaceae	6/653	2/13	2/8
105.	Asparagaceae	128/2929	4/9	4/9
106.	Colchicaceae	18/284	3/4	2/3
107.	Xanthorrhoeaceae	34/1236	2/2	2/2
108.	Smilacaceae	2/261	1/1	1/1
109.	Pontederiaceae	6/33	3/3	2/3
110.	Commelinaceae	41/728	4/24	3/16
111.	Arecaceae	185/2522	3/4	3/4
112.	Typhaceae	2/65	1/1	1/1
113.	Acoraceae	1/2	1/1	1/1
114.	Araceae	117/3368	13/25	11/19
115.	Alismataceae	17/120	4/6	3/4
116.	Aponogetonaceae	2/58	1/2	1/1
117.	Potamogetonaceae	6/186	2/5	3/5
118.	Eriocaulaceae	11/1206	1/14	1/8
119.	Cyperaceae	110/5784	14/117	14/80
120.	Poaceae	759/11554	94/262	89/239

CHAPTER V

TAXONOMIC ENUMERATION

ARTIFICIAL KEY TO FAMILIES OF DICOTYLEDONS

Characters pertain to study area

There are 3 pairs of figures opposite a family. The first pair indicates the number of genera and species in that family in the world; the second pair, the same in MP and the third pair, the same in study area. Similarly, there are 3 figures opposite every genus. The first figure shows the number of species in that genus in the world, the second figure shows the same in MP and the third figure shows the same in the study area. Source: World-The Plant List, version 1.1; MP: The Flora of MP including the supplement to that flora and the third is the study area: SA

1. Perianth usually 2-seriate, of sepals and petals
  2. Petals free; stamens not epipetalous
    3. Sepals free from the ovary
      4. Petals and stamens indefinite; carpels free and sunk in the receptacle or not; flowers very large, above 10 cms across
        5. Aquatic plants
          6. Sepals 4; carpels united, not in sunken pits; ovules many; fruit baccate; leaves floating on surface of water;petiole not peltate .... NYMPHAEACEAE
          6. Sepals 5; carpels free, sunk in the receptacle; ovules 1-2; fruit achenial.. NELUMBONACEAE
        5. Terrestrial plants; flowers comparatively small
          7. Carpels free; flowers bisexual
            8. Climbing perennials or herbs; leaves compound, opposite or simple, alternate; petals present or absent; fruit an etario of achenes .... RANUNCULACEAE
            8. Trees; fruit indehiscent
              9. Sepals deciduous, not enclosing the fruit; flowers usually 3-merous; anthers extrorse .... ANNONACEAE
              9. Sepals persistent, fleshy, tightly enclosing the fruit; anthers introrse; leaves very large, lateral veins numerous, obliquely parallel, ending in serratures .... DILLENIACEAE
          7. Carpels united; fruit drupaceous; leaves alternate; scandent shrubs

10. Flowers bisexual; fertile stamens 3, rest staminodes; petals densely hairy on the inner side .... OLACACEAE

10. Flowers unisexual, normally 3-merous; stamens 4-6, staminodes absent .... MENISPERMACEAE

4. Petals and stamens definite (indefinite in **Argemone**); flowers bisexual; carpels united

11. Sepals 2 or 3; ovary unilocular; herbs

12. Placentation parietal; flowers actinomorphic or zygomorphic; stamens 6 or more; style reduced; latex present or absent .... PAPAVERACEAE

12. Placentation free-central; sepals 2

13. Stamens 8-12; capsule circumscissile; herbs .... PORTULACACEAE

13. Stamens more than 20; capsule dehiscing by valves .... TALINACEAE

11. Sepals 4 or 5; herbs

14. Petals 3; filaments united; sepals persistent; fruit capsular; seeds arillate .... POLYGALACEAE

14. Petals 4; filaments free or united; sepals 4, deciduous; seeds exarillate

15. Stamens 6, tertradynamous; corolla cruciform; fruit siliqua or silicula .... BRASSICACEAE

15. Stamens not tetradynamous when 6; corolla not cruciform; fruit not siliqua

16. Leaves opposite, fleshy; stamens 8, free; hypogynous scales present; fruit follicular .... CRASSULACEAE

16. Leaves alternate; stamens 6-20; fruits baccate or capsular

17. Gynophore always present; fruit baccate; leaves simple; shrubs, twiners or scramblers .... CAPPARACEAE

17. Gynophore normally absent; fruit capsular; leaves simple or compound; herbs .... CLEOMACEAE

11. Sepals 5; petals 5 (except **Commiphora**)

18. Ovary unilocular

19. Leaves reduced, scale-like; flowers uni or bisexual; stamens 5 or 10; shrubs inhabiting rocky river beds .... TAMARICACEAE

19. Leaves well developed

20. Placentation parietal; leaves alternate

21. Flowers zygomorphic; fruit capsular

22. Lowermost petal spurred; leaves simple, stipulate; anthers crested, united; staminodes absent; seeds not winged .... VIOLACEAE

22. Lower petal not spurred; leaves 2-3 pinnate, exstipulate; anthers not crested, free; staminodes present; seeds winged; trees .... MORINGACEAE

21. Flowers actinomorphic; fruit capsular or baccate; leaves palmately veined

23. Tendril climbers; corona and andro-gynophore present; fruit baccate, enclosed in fimbriate lobes of epicalyx (our species); seeds not cottony .... PASSIFLORACEAE

23. Trees; corona and andro-gynophore absent; fruit capsular; capsule large, 5-valved; seeds cottony .... BIXACEAE

20. Placentation free central; leaves opposite; fruit capsular; herbs .... CARYOPHYLLACEAE

18. Ovary 2 or more locular

24. Disc absent

25. Sepals imbricate; fruit capsular

26. Leaves opposite, stipulate; flowers minute; marshy herbs .... ELATINACEAE

26. Leaves alternate, exstipulate; herbs of drier habitat .... LINACEAE

25. Sepals valvate; leaves stipulate, venation usually palmate; petals with twisted aestivation; fruit various; mucilage present .... MALVACEAE

24. Disc present

27. Disc enlarged, adnate to calyx base or lining it

28. Flowers polygamous, uni or bisexual

29. Fruit drupaceous; ovary 1-celled; stamens 1-10; trees .... ANACARDIACEAE

29. Fruit drupaceous or samaroid; ovary 1-3 celled; stamens 8; climbers, shrubs or trees .... SAPINDACEAE

28. Flowers bisexual

30. Stamens opposite petals; petals valvate

31. Inflorescence axillary; fruit drupaceous or samaroid; woody twiners, trees or shrubs .... RHAMNACEAE

31. Inflorescence leaf-opposed; fruit baccate; tendrils present or absent; leaves simple or compound .... VITACEAE

30. Stamens alternate with petals; petals imbricate

32. Stamens in 2 whorls, all fertile; leaves opposite; fruit samaroid; twiners (our area) .... MALPIGHIACEAE

32. Stamens in 1 whorl; leaves alternate or sub-opposite; fruit drupaceous or capsular; seeds arillate or exarillate; shrubs, trees or climbers .... CELASTRACEAE

27. Disc ring-like (annular), adnate to stamens

33. Stipules present or absent; leaves opposite, compound, leaflets 2 to many, opposite or alternate; flowers bisexual, polygamous or unisexual; fruit drupaceous or schizocarpic .... ZYGOPHYLLACEAE

33. Stipules absent; leaves alternate

34. Leaves gland-dotted, compound ovary deeply lobed, seated on a stalk; stamens twice as many as petals or many; fruit a woody or soft berry .... RUTACEAE

34. Leaves not gland-dotted; ovary entire (not lobed)

35. Flowers zygomorphic; posterior sepals spurred at base; leaves simple; fruit capsular, dehiscing elastically; fleshy herbs .... BALSAMINACEAE

35. Flowers actinomorphic; sepals not as above; leaves compound

36. Herbs; leaves palmate or pinnate compound; fruit capsular .... OXALIDACEAE
36. Trees; leaves pinnate compound
37. Filaments united into a tube (except **Toona**); leaves uni or bipinnate compound; fruit capsular or drupaceous .... MELIACEAE
37. Filaments free
38. Flowers unisexual or polygamous; cells of ovary 1-ovuled; fruit samaroid .... SIMAROUBACEAE
38. Flowers bisexual; cells of ovary 1 or more ovuled; fruit drupaceous or capsular .... BURSERACEAE
3. Flowers peri or epigynous
39. Carpel 1, if more, then free; leaves stipulate
40. Carpel 1, placentation marginal; fruit mostly a legume, but not always; flowers regular or irregular, stamens definite or indefinite, free, mono, di or polyadelphous; plants showing all types of habit .... LEGUMINOSAE
40. Carpels many, free (our area); fruit an etario of achenes .... ROSACEAE
39. Carpels more than 1, syncarpous; ovary superior or inferior
41. Inflorescence a simple or compound umbel, subtended by an involucre/s of bracts or sometimes involucre absent; ovary inferior; fruit a cremocarp; leaves simple or compound; aromatic herbs .... APIACEAE
41. Inflorescence not an umbel
42. Ovary free from the hypanthium
43. Petals present (absent in some **Ammannia** species); style 1; stamens inflexed in bud .... LYTHRACEAE
43. Petals absent; styles equal to number of carpels; stamens erect in bud
44. Perianth free; seeds estrophiolate; flowers in terminal or leaf-opposed, lax cymes .... MOLLUGINACEAE
44. Perianth united; seeds strophiolate; flowers in axillary fascicles .... AIZOACEAE
42. Ovary partly or completely fused with the hypanthium, inferior
45. Succulent plants; leaves reduced to spines; stamens indefinite; fruit baccate .... CACTACEAE
45. Non succulent plants; leaves well developed; stamens definite or indefinite
46. Ovary unilocular
47. Flowers unisexual; tendril climbers; fruit a pepo (rind hard) or a berry; sap watery .... CUCURBITACEAE
47. Flowers bisexual
48. Ovary unilocular; ovules pendulous .... COMBRETACEAE
48. Ovary 2-many locular; ovules erect
49. Stamens indefinite
50. Leaves opposite, gland-dotted; flowers small .... MYRTACEAE

50. Leaves alternate, not gland-dotted; flowers large .... LECYTHIDACEAE

49. Stamens definite

51. Trees; sepals, petals and stamens 5-10 .... CORNACEAE

51. Herbs, glandular; sepals, petals and stamens 4-8

52. Glandular, pubescent herbs; sepals, petals and stamens 5; styles 2 .... VAHLIACEAE

52. Eglandular herbs of marshy habitat; sepals, petals and stamens 4-8 .... ONAGRACEAE

2. Petals united; stamens mostly epipetalous

53. Ovary inferior

54. Inflorescence capitulum; anthers united, filaments free; calyx modified as pappus or greatly reduced; fruit a cypsela .... COMPOSITAE

54. Inflorescence not a capitulum; anthers free (except **Lobelia**); calyx well developed; fruit not cypsela

55. Leaves opposite, stipulate(except **Rubia**); flowers regular; fruit baccate or capsular; seeds winged or not; herbs, shrubs or trees .... RUBIACEAE

55. Leaves alternate, exstipulate; flowers regular or irregular; anthers free or united; herbs inhabiting moist ground .... CAMPANULACEAE

53. Ovary superior

56. Flowers regular

57. Ovary unilocular; placentation free central or parietal

58. Leaves alternate or opposite; stamens opposite corolla lobes

59. Style branches 5; capsule circumscissile, ultimately dehiscent by 5 valves from below upwards; calyx glandular or not .... PLUMBAGINACEAE

59. Style unbranched; leaves alternate or opposite; fruit indehiscent or circumscissile .... PRIMULACEAE

58. Leaves opposite; stamens alternate with corolla lobes

60. Petals and stamens 4; leaves with a pungent smell; trees .... SALVADORACEAE

60. Petals and stamens 5; leaves without pungent smell; herbs .... GENTIANACEAE

57. Ovary 2-many locular; placentation axile; leaves simple

61. Flowers unisexual; stamens free from the corolla; ovules always twice the number of styles; seeds arillate .... EBENACEAE

61. Flowers bisexual; stamens epipetalous

62. Carpels 3 or more; petals and sometimes sepals in 2 series; stamens equal to or twice the number of petals, alternating normally with staminodes; latex present; trees .... SAPOTACEAE

62. Carpels 2

63. Leaves opposite

64. Stamens 2; fruit capsular or baccate; seeds winged or not .... OLEACEAE

64. Stamens 4 or more

65. Ovaries free, styles and stigmas uniting to form a gynostegium or not; corolla twisted; corona present or absent; flowers 5-merous; fruit follicular or baccate; latex present .... APOCYNACEAE

65. Ovaries and styles united; stigma 2-lobed; corolla imbricate; flowers 4-merous; fruit capsular or baccate; latex absent .... LOGANIACEAE

63. Leaves alternate

66. Aquatic herbs; ovary unilocular ....MENYANTHACEAE

66. Terrestrial plants; ovary 2 or more locular

67. Cells of ovary many ovuled; placenta swollen; fruit baccate or capsular; plants armed or not .... SOLANACEAE

67. Cells of ovary 1 or 2 ovuled; inflorescence cymose

68. Style terminal; fruit capsular or indehiscent; ovary not lobed; twining autotrophs or total parasites ..........................CONVOLVULACEAE

68. Style gynobasic; ovary 4-lobed; fruit of 4 nutlets; herbs or trees .... BORAGINACEAE

56. Flowers irregular; stamens didynamous, when 4

69. Ovary unilocular

70. Placentation parietal; fertile stamens 2; anthers 2-celled

71. Calyx divided less than half way down; anthers connivent or coherent in pairs; ovary superior or half inferior; capsule many-seeded ...............GESNERIACEAE

71. Calyx divided nearly to the base; anthers free; ovary superior; drupe with 2, sharp, curved claws ....MARTYNIACEAE

70. Placentation basal; stamens 2; anthers 1-celled; capsule globose; seeds many; aquatic or marshy plants .... LENTIBULARIACEAE

69. Ovary 2-4 locular

72. Seeds numerous per chamber

73. Seeds not winged; herbs

74. Anther cells confluent at apex; stamens 2 or more; corolla 2-lipped or unequally 5-lobed .... PHRYMACEAE

74. Anther cells not confluent at apex

75. Anther cells divaricate; perfect stamens 2; capsule dehiscing septicidally; staminodes present ....LINDERNIACEAE

75. Anther cells not divaricate; perfect stamens 2 or 4; staminodes absent

76 . Flowers solitary, axillary (flowers racemose and stamens 2 in **Veronica**)...PLANTAGINACEAE

76. Flowers in terminal racemes or spikes

77. Flowers in terminal racemes . ... SCROPHULARIACEAE

77. Flowers in terminal spikes; partial or total parasites or plants with parasitic tendency .... OROBANCHACEAE

73. Seeds winged; trees; stamens 4 (5 in **Oroxylum**); leaves opposite, compound..BIGNONIACEAE

72. Seeds normally 4 (more than 4 in some **Acanthaceae**)

78. Ovules superposed; leaves with cystoliths; bracts and bracteoles prominent; stamens 2 or 4. .... ACANTHACEAE

78. Ovules laterally placed; leaves without cystoliths

79. Inflorescence racemose .... VERBENACEAE

79. Inflorescence cymose . ... LAMIACEAE

1. Perianth of 1 whorl (sometimes both whorls are present or altogether absent)

80. Flowers bisexual; perianth of atleast 1 whorl present

81. Ovary inferior

82. Ovary 4-6 celled; ovules many per chamber; perianth 3-lobed, large, irregular; stamens 6-12, adnate to stylar column; fruit capsular; twiners .... ARISTOLOCHIACEAE

82. Ovary 1-celled; ovules 1-4 per chamber; perianth not showy, regular; partial or total parasites

83. Ovule 1, basal, erect; petals present; partial parasites .... LORANTHACEAE

83. Ovules 2-4, pendulous; sepals present; initial root parasites .... SANTALACEAE

81. Ovary superior, mostly 1-celled

84. Anthers dehiscing by valves; stamens more than perianth lobes, accompanied by glands; staminodes present; ovule 1; stipules absent .... LAURACEAE

84. Anthers not dehiscing by valves

85. Stipules present; perianth petaloid, pentamerous or trimerous; nut biconvex or trigonous .... POLYGONACEAE

85. Stipules absent; perianth petaloid or not

86. Perianth petaloid, tubular, its base fusing with ovary and forming an anthocarp .... NYCTAGINACEAE

86. Perianth not petaloid, lobes scaly, dry or herbaceous, basally united, not forming an anthocarp; fruit nut or utricle (in **Celosia** it is an indehiscent capsule) .... AMARANTHACEAE

80. Flowers mostly unisexual, when bisexual, perianth absent

87. Submerged, dichotomously branched herbs; leaves in whorls, dichotomously branched in filiform segments .... CERATOPHYLLACEAE

87. Terrestrial plants; stipules almost always present, when absent, flowers bisexual and perianth absent

88. Ovary normally 3-celled; ovules 1 or 2 per chamber; perianth usually single, rarely double or altogether absent; latex present

89. Ovule 1 per cell .... EUPHORBIACEAE

89. Ovules 2 per cell

90. Disc and pistillode absent; stamens 3 (our area), alternating with tepals; tepals 4, free; fruit drupaceous .... PUTRANJIVACEAE

90. Disc present; pistillode present or absent; stamens 2-5, free or united, opposite tepals; fruit drupaceous or capsular . ... PHYLLANTHACEAE

88. Ovary 1-celled, 1 ovuled

91. Perianth usually absent

92. Trees; fruit capsular or baccate; flowers uni or bisexual; perianth 4-5 merous or absent; seeds arillate or not, sometimes with penicillate hairs; shrubs .... SALICACEAE
92. Herbs; fruit drupe-like; perianth absent; flowers bisexual; peltate bract always present below each flower; stamens 2-6; stigmas 1-4; seed 1 ....PIPERACEAE

91. Perianth usually present; flowers unisexual
93. Anthers neither inflexed in bud nor reversed; ovule pendulous
94. Leaves palmately veined, margin serrate; ovary sessile; fruit drupaceous; herbs or evergreen trees ....CANNABACEAE
94. Leaves pinnately veined, margin entire; ovary stalked; fruit samara; deciduous trees...ULMACEAE
93. Anthers inflexed in bud and reversed
95. Style 1; stamens 3-5; ovule basal, erect; leaves alternate or opposite; herbs, shrubs or trees with watery sap and sometimes with stinging hairs ....URTICACEAE
95. Style branches 2; stamen 1 or 2; ovule pendulous; leaves alternate, rarely opposite; flowers in hypanthodium; latex milky .... MORACEAE

RANUNCULACEAE 65/2377/; 4/9; 2/3

Key to Genera

1. Petals absent; leaves opposite, compound; achenes feathery; woody twiners ....CLEMATIS
1. Petals present; leaves simple, alternate; achenes not feathery; herbs ....RANUNCULUS

CLEMATIS L. W. 373; MP 4; SA 2

Key to Species

1. Flowers more than 2.5 cm across ....**C. heynei**
1. Flowers less than 2.5 cm across ....**C. gouriana**

Clematis gouriana Roxb. ex DC., Syst. Nat. 1: 138. 1817; Hook.f. & Thoms. in Hook.f., Fl. Brit. India 1: 4. 1872; M.A. Rau in B.D. Sharma *et al.,* Fl. India 1: 64. 1993; Mukherjee in Verma *et al.,* Fl. Madhya Pradesh 1: 120.1993; Samvatsar, Fl. West. Trib. Madhya Pradesh 37. 1996. *Belkun*

Shrubby climbers with pinnately trifoliate to 5-foliate, opposite, 4-5 × 2-2.5 cm; leaves petioles twining; sepals white, normally 4; connective not produced beyond the anther lobes; carpels 5-6; styles as many, feathery, persistent; fruit etario of achenes.

Fl.-Fr.: Oct. -Dec.

Common in hilly areas on forest trees and shrubs, cultivated in Indore garden.

Clematis heynei M.A. Rau in Fl. India 1: 67. 1993; *Clematis triloba* Heyne ex Roth, Nov. Pl. Sp. 251. 1821; *nom. illegit*; Hook.f. & Thoms. in Hook.f., Fl. Brit. Ind. 1: 3. 1872; Mukherjee in Verma *et al.,* Fl. Madhya Pradesh 1: 121. 1993.

Extensive climbers; stems and branches sulcate; leaves simple or pinnately 3-foliolate; panicles many-flowered; flowers white, 2-2.5 cm across; bracts leafy; carpels many; achenes compressed, in ovoid heads, villous with long feathery styles.

Fl.-Fr.: Oct.-Dec.

As undergrowth in mixed deciduous forest, **Burhanpur** only, Badal khora, Jambupani.

RANUNCULUS L. W. 413; MP 1; SA 1

Ranunculus scleratus L., Sp. Pl. 551. 1753; Hook.f. & Thoms. in Hook.f., Fl. Brit. India 1: 19. 1872; M.A. Rau in B.D. Sharma *et al.*, Fl. India 1: 128. 1993; Mukherjee in Verma *et al.,* Fl. Madhya Pradesh 1: 123. 1993. *Jal dhania*

Herbs; radical leaves reniform, 2.5-5 cm across, 3-partite at base; cauline leaves obovate, unlobed or bluntly 3-5-toothed or shortly 3-partite; flowers several, bright yellow, diffusely racemose; sepals elliptic, *ca* 3 × 1.5 mm, caducous; petals yellow, as long as sepals or shorter; achenes small, obovoid, *ca* 0.75 mm long, compressed.

Fl.-Fr.: Feb.– March

Burhanpur, Indore, Khandwa.

DILLENIACEAE 11/219/; 1/3/; 1/1

DILLENIA L. W. 58; MP 3; SA 1

Dillenia pentagyna Roxb., Pl. Coromandel 1: 21. 1795; Hook.f. & Thoms. in Hook.f., Fl. Brit. India 1: 38. 1872; Majumdar in B.D. Sharma *et al.,* Fl. India 1: 156. 1993; Mukherjee in Verma *et al.,* Fl. Madhya Pradesh 1: 125. 1993; Samvatsar, Fl. West. Trib. Madhya Pradesh 40.1996. *Shalya karni* **(Plate 1, Fig. 1)**

Deciduous trees, 10-15 m tall; leaves oblong-lanceolate or oblong-obovate, 30-60 × 20-26 cm, obtuse at apex, obtuse or cuneate at base, nearly entire to dentate; petioles 1.5-5 cm long, with persistent wing-like stipules; flowers 2-7, yellow, fragrant, 2.5-3 cm across; calyx fleshy; fruits yellow, orange or red, drooping, sub-globose, *ca* 1.5 × 1.5 cm, indehiscent, fleshy; seeds black, ovoid, *ca* 5 × 3.5 mm, glabrous.

Fl.-Fr.: March-May

Alirajpur and Khargone

ANNONACEAE 128/2106/; 2/5; 2/3

Key to Genera

1. Ripe carpels fused, forming a fleshy fruit; anthers concealed by the overtopping connective . ... ANNONA
1. Ripe carpels free, forming an etario of berries; anthers not concealed by the connective . ... MILIUSA

ANNONA L. W. 166; MP 2; SA 1

Annona squamosa L., Sp. Pl. 537. 1753; Hook.f. & Thoms. Hook.f., Fl. Brit. India 1: 78. 1872; D. Mitra in B.D. Sharma *et al.,* Fl. India 1: 207. 1993; Mukherjee in D.M. Verma *et al.,* Fl. Madhya Pradesh 1: 128. 1993; Samvatsar, Fl. West. Trib. Madhya Pradesh 37. 1996. *Sitaphal*

Shrubs or small trees; leaves elliptic-oblong, 8-14 × 2.5-5 cm, obtuse or slightly acute at apex, acute to cuneate at base, glabrous; flowers axillary, solitary or a few together, leaf opposed; sepals 3, broadly ovate; petals dimorphic, 3+3 in two whorls; outer ones greenish-white or yellow, reddish at base inside; inner rudimentary or absent; fruits ovoid-globose, 8-9 cm in diam., tuberculate with rounded tips; seeds many, black, shiny.

Fl.-Fr.: April-Nov.

Naturalised throughout the area.

MILIUSA Leschn. ex DC. W. 14; MP 2; SA 2

Key to Species

1. Leaves 12-30 cm long, young ones covered on both sides with dense, silky hairs; peduncles 0.6-2.5 cm long; pedicels 3.3-8.8 cm long; sepals covered with dense, silky hairs; carpels densely woolly; ovules 2 per chamber .... **M. velutina**

1. Leaves 10-15 cm long, membranous; peduncles 1-1.5 cm long; pedicels 3-6 cm long; ovules 4-6 per chamber .... **M. tomentosa**

Miliusa tomentosa (Roxb.) Sinclair, Gard. Bull. Singapore 14:378. 1955; *Uvaria tomentosa* Roxb., Pl. Coromandel 1: 31. 1795; *Saccopetalum tomentosum* Hook.f. & Thoms. in Fl. Ind. 1:152. 1855; Hook.f., Fl. Brit. India 1: 88. 1872; D. Mitra in B.D. Sharma *et al.,* Fl. India 1: 221. 199;. Mukherjee in D.M. Verma *et al.,* Fl. Madhya Pradesh 1: 129. 1993; Samvatsar, Fl. West. Trib. Madhya Pradesh 37. 1996. *Kari* **(Plate 1, Fig. 2)**

Deciduous trees, up to 20 m high; leaves ovate or oblong, 10-15 × 8-9 cm, obtuse or acute at apex, rounded or subacute at base, tomentose; flowers greenish- yellow; sepals 3, linear-lanceolate, tomentose; petals 6, in 2 whorls of 3 each, outer sepaloid, linear-lanceolate, 4-6 mm long; inner ones oblong-ovate; ripe carpels many, purple, subglobose, 2-3 cm in diam., tapering to base, succulent, tomentose; seeds 4-5.

Fl.-Fr.: May-July

Common in forest, in all districts.

Miliusa velutina (A. DC.) Hook.f. & Thoms. Fl. Ind. 151. 1855 & in Hook.f., Fl. Brit. India 1: 87.1872; *Guatteria velutina* A. DC., Mem. Soc. Phys. Geneve 5:218. 1852; *Uvaria velutina* Dunal, Monogr. Anonac. 91. 1817; D. Mitra in B.D. Sharma *et al.,* Fl. India 1: 222. 1993; Mukherjee in D.M. Verma *et al.,* Fl. Madhya Pradesh 1: 129. 1993.

Large trees; leaves ovate or oblong-elliptic, 12-30 × 5-12 cm, acute to acuminate at apex, obliquely cordate at base, velvety; flowers in leaf-opposed cymes, pale yellow or greenish; sepals 3, ovate, 2-4 mm long, acute, velvety in bud, densely tomentose afterwards; petals 6 in 2 whorls of 3 each; outer whorl sepaloid, inner dark brown, broadly ovate, densely tomentose outside, glabrous inside; ripe carpels many, bluish purple, ovoid or ellipsoid, 1.5-2.5 cm in diam., tomentose; seeds 1-2.

Fl.-Fr.: March – Oct.

Rare, Barwani, Khargone.

MENISPERMACEAE 68/448; 4/5; SA 3/5

Key to Genera

1. Sepals 4; stamens 4, connate; carpel one .... CISSAMPELOS
1. Sepals 6; stamens 6, free; carpels 3-6
2. Leaves obtuse or sub-cordate at base; seeds horse-shoe shaped . ... COCCULUS
2. Leaves deeply cordate at base; seeds oblong or globose . ... TINOSPORA

CISSAMPELOS L. W. 21; MP 1; SA 1

Cissampelos pareira L., Sp. Pl. 1031 1753; Hook.f. & Thoms. in Hook.f., Fl. Brit. India 1: 103. 1872; *C. hirsuta* Buch. -Ham. ex DC. Syst. Nat. 1: 535. 1817; *Cissampelos pareira* L. var. *hirsuta* (Buch.-Ham. ex DC.) Foreman in Kew Bull. 22: 356. 1968; Gangop. in B.D. Sharma *et al.,* Fl. India 1: 317.1993; Mukherjee in D.M. Verma *et al.,* Fl. Madhya Pradesh 1: 132. 1993; Samvatsar, Fl. West. Trib. Madhya Pradesh 38. 1996. *Pahad bel* **(Plate 1, Fig. 3)**

Slender climbers; leaves reniform, orbicular or cordate, peltate, 4-5.5 × 4-6 cm, base subcordate, apex apiculate or retuse, glabrescent above, pubescent below; flowers greenish-white, minute, unisexual, males in panicles, females in pendulous racemes, in axils of conspicuous bracts; drupes red, ovoid, tomentose.

Fl.-Fr.: July - Dec.

Common in all districts.

COCCULUS A. DC. W. 9; MP 2; SA 2

Key to Species

1. Leaves 1.5-5 cm long, oblong-lanceolate to ovate or trapezoid; female flowers 1-2; drupe black at maturity .... **C. pendulus**
1. Leaves 2-5 cm long, ovate-oblong, villous on both surfaces; female flowers 2-3; drupes reddish-black at maturity ... . **C. hirsutus**

Cocculus hirsutus (L.)W. Theob, Burmah ed 3, 2:657. 1860; *Menispermum hirsutum* L., Sp. Pl. 341. 1753; *Menispermum villosum* Lam., Encycl. 4: 97. 1797; *Cocculus villosus* (Lam.) DC., Syst. Nat. 1: 525. 1817; Hook.f. & Thoms. in Hook.f., Fl. Brit. India 1: 101. 1872; Gangop. in B.D. Sharma *et al.,* Fl. India 1: 318. 1993; Mukherjee in D.M. Verma *et al., Fl.* Madhya Pradesh 1: 133. 1993; Samvatsar, Fl. West. Trib. Madhya Pradesh 38. 1996. *Jal jamni*

Climbers; leaves variable, ovate, ovate-oblong or sagittate, 2-4.5 × 1-2.8 cm, pubescent above, velvety tomentose beneath, base cordate or rounded, apex obtuse, mucronate; flowers greenish-yellow, unisexual, minute; male flowers in slender panicles; female flowers in fascicles or racemes; drupes reddish-purple, globose.

Fl.-Fr.: Oct. - May

Throughout the area.

Cocculus pendulus (J. R. Forst. & G. Forst.) Diels, Pflanzenr. IV, 94: 237 1910; *Epibaterium pendulum* J. R. Forst. & G. Forst., Char. Gen. Pl. 54 1775; *Menispermum loeba* Delile, Descr. Egypte, Hist. Nat. 284 1813; *Cocculus loeba* (Delile) DC., Syst. Nat. 284. 1813; Hook.f. & Thoms. in Hook.f., Fl. Brit. Ind. 1: 102. 1872; Gangop. in B.D. Sharma *et al.,* Fl. India 1: 320.1993; Mukherjee in D.M. Verma *et al.,* Fl. Madhya Pradesh 1: 133. 1993.

Twining monoecious shrubs, with pubescent stem and branches; leaves petiolate, narrowly elliptic, acute at base, sub-glabrous; flowers greenish-yellow, male in axillary fascicles, females sub-solitary; drupes ovoid, ridged, compressed; stylar mark basal.

Fl.-Fr.: Oct. -Dec.

In drier habitats, **Burhanpur** only, Khaknar, Jambupani.

TINOSPORA Miers. W. 13; MP 1; SA 2

Key to Species

1. Young shoots papilllose-glandular; petiole 7-11 cm long; lamina deltoid; basal lobes of sinus cm broad; endocarp surface papillose to smooth .... **T. maqsoodiana**
1. Young shoots glabrous; petiole 5-7 cm long; lamina deeply cordate, basal lobes of sinus 7-9 cm broad .... **T. sinensis**

Tinospora maqsoodiana Mujaffar, S. Moinuddin & S. Mustakim, Indian Forester 140(5):528-530.2014. *Gul vel*

Woody, deciduous twiners; stem glandular-papillose; aerial roots very long, abounding in watery sap; inflorescence *ca* 15 cm long; staminodes 6; drupes 1-3; drupelets orange-red at maturity; seeds oblong, reniform dorsally, deeply grooved ventrally, surface glabrous.

Fl.-Fr.: Oct. -Dec.

Khandwa only, Kalibhit.

Tinospora sinensis (Lour.) Merr. in Sunyatsenia 1(4): 193. 1934; *Campylus sinensis* Lour., Fl. Cochinch. 113. 1790; *Menispermum cordifolium* Willd., Sp. Pl. 4: 826. 1806; *Tinospora cordifolia* (Willd.) Miers in Ann. Mag. Nat. Hist., ser. 2 7: 35, 38 1851; Hook.f., Fl. Brit. India 1: 97. 1872; A. Pramanik in B.D. Sharma *et al.,* Fl. India 1: 347. 1993; Mukherjee in D.M. Verma *et al.,* Fl. Madhya Pradesh 1:136. 1993; Samvatsar, Fl. West. Trib. Madhya Pradesh 38. 1996. *Giloy*

Succulent climbers; bark rough, corky; leaves broadly ovate, 3-10 × 4.5-8 cm, base cordate, apex acute or acuminate, glabrous; flowers in slender, pseudo-racemes on leafless branchlets; male flowers fascicled, yellow; female flowers solitary; petals spathulate; drupes red, globose, sessile.

Fl.-Fr.: Aug. -Feb.

Common everywhere.

NELUMBONACEAE 1/2; 1/1; 1/1

NELUMBO Adans. W. 2; MP 1; SA1

Nelumbo nucifera Gaertn., Fruct. Sem. Pl. 1: 73. 1788; *Nelumbium speciosum* Willd., Sp. Pl. 2: 1258. 1799; Hook.f., & Thoms. in Hook.f., Fl. Brit. India 1: 116. 1872; D. Mitra in B.D. Sharma *et al.,* Fl. India 1: 441. 1993; Mukherjee in Verma *et al.,* Fl. Madhya Pradesh 1: 138. 1993; Samvatsar, Fl. West. Trib. Madhya Pradesh 40.1996. *Kamal*

Perennial, rhizomatous, aquatic herbs; leaves orbicular, peltate, 32-60 cm across; flowers white to pink, 10-25 cm across, solitary, erect or cernuous, scape sheathing, as long as the petioles; sepals 5, caducous; petals many seriate; carpels many, free, embedded in an obconical receptacle.

Fl.-Fr.: Aug. - Dec.

Occasional in ponds, Dhar, Indore and Khargone.

NYMPHAEACEAE 8/70; 1/2; 1/2

NYMPHAEA L. W. 44; MP 2; SA2

Key to Species

1. Connectives produced beyond anther lobes; flowers white or blue; leaves glabrous beneath, obscurely dentate .... **N. nouchali**
1. Connectives not produced beyond anther lobes; flowers white to pink; leaves pubescent beneath .... **N. pubescens**

Nymphaea nouchali Burm.f., Fl. Ind. 120.1768; *N. stellata* Willd., Sp. Pl. 2. 1153. 1799; Hook.f. & Thoms. in Hook.f., Fl. Brit. India 1: 114. 1872; D. Mitra in B.D. Sharma *et al.,* Fl. India 1: 430.1993; Mukherjee in D.M. Verma *et al.,* Fl. Madhya Pradesh 1:139. 1993; Samvatsar, Fl. West. Trib. Madhya Pradesh 40.1996. *Neel kamal*

Aquatic, rhizomatous herbs; leaves floating, long petioled, peltate, orbicular, 6-25 × 5-24 cm, green above, pale- purple beneath, irregularly sinuate-dentate with obtuse teeth or subentire; flowers solitary, fragrant, 5-10 cm across; nutlets green, spongy, globose or ovoid, 1.8-3.5 cm across; seeds numerous, ellipsoid.

Fl.-Fr.: May-Nov.

Not common, Dhar, Khandwa.

Nymphaea pubescens Willd., Sp. Pl. 2: 1154. 1799; Hook.f. & Thoms. in Hook.f., Fl. Brit. India 1: 114. 1872; D. Mitra in B.D. Sharma *et al.,* Fl. India 1: 431. 1993; Mukherjee in Verma *et al.,* Fl. Madhya Pradesh 1: 139. 1993; Samvatsar, Fl. West. Trib. Madhya Pradesh 40.1996.

Aquatic, rhizomatous herbs; leaves long petioled, orbicular, up to 20 cm across, sagittate and purplish when young, mature leaves green above, densely pubescent beneath; flowers carmine red; fruits not developed in our area *Kanval*

Fl.-Fr.: July-Oct.

Common in ponds during rains.

PAPAVERACEAE 41/920; 3/5; 2/3

Key to Genera

1. Sepals 2; flowers zygomprphic; petals 4, 2-large, 2 small; stamens 6 in two bundles of 3 each .... FUMARIA

1. Sepals 2 or 3; flowers actinimorphic; petals equal; stamens indefinite or 8-12, free .... ARGEMONE

ARGEMONE L. W. 32; MP 2; SA 2

Key to Species

1. Flowers bright yellow, up to 2.5 × 1.5 cm; stigma lobes more or less appressed ....**A. mexicana**

1. Flowers white to yellowish, up to 3.5 × 2 cm; stigma lobes spreading .... **A. ochroleuca**

Argemone mexicana L., Sp. Pl. 508. 1753; Hook.f. & Thoms. in Hook.f., Fl. Brit. India 1: 117. 1872; Debnath & M.P. Nayar in B.D. Sharma *et al.,* Fl. India 2: 2. 1993; Mukherjee in Verma *et al.,* Fl. Madhya Pradesh 1: 141. 1993; Samvatsar, Fl. West. Trib. Madhya Pradesh 41. 1996. *Pili kateri*

Annual herbs with yellow latex; leaves sessile, elliptic-obovate, 3-22 × 2-8 cm, pinnatifid, sinuate-lobulate, segments spiny along margins; flowers bright yellow, 4-7 cm across, sessile, subtended by 2-3 foliaceous bracts; capsules ellipsoid to oblong, 1-4 cm long; seeds many, blackish brown to deep brown, 1.6-2 mm long, deeply reticulate, suborbiculate.

Fl.-Fr.: Feb. -May

Found almost throughout the area in wasteland.

Argemone ochroleuca Sweet, Brit. Fl. Gard. 3: t. 242. 1828; N. C. Nair in Rec. Bot. Surv. India 21(1): 7-8. 1978; Debnath & M.P. Nayar in B.D. Sharma *et al.,* Fl. India 2: 5. 1993; Mukherjee in Verma *et al.,* Fl. Madhya Pradesh 1: 141. 1993.

Large herbs; leaves oblanceolate, sinuate to pinnatifid, basal leaves deeply lobed, lobes oblong, glaucous; flowers whitish or pale lemon- yellow, 2.5-3.5 cm across, sessile, subtended by 2-3 foliaceous bracts; capsules ovoid or lanceolate-ovoid, spiny on valves, the largest spine up to 10 mm long; seeds black, 1.5-2 mm across, finely reticulate.

Fl.-Fr.: Feb. -June

Not common, Burhanpur and Khargone.

FUMARIA L. W. 57; MP 1; SA 1

Fumaria indica (Hausskn.) Pugsley, J. Linn. Soc., Bot. 44: 313. 1919; *F. vaillantii* Loisel., var. *indica* Hausskn., in Flora 56: 443. 1873; *F. parviflora* Lam. subsp. *vaillantii* (Loisel.) Hook.f., Fl. Brit. India 1: 128. 1872; J.L. Ellis & N.P. Balakr., in B.D. Sharma *et al.,* Fl. India 2: 84. 1993; Mukherjee in Verma *et al.,* Fl. Madhya Pradesh 1: 142. 1993; Samvatsar, Fl. West. Trib. Madhya Pradesh 41. 1996. *Pit papra*

Annual herbs, up to 60 cm tall; leaves 2-3 pinnatisect, 2-5 cm long, segments linear-oblanceolate, apiculate; flowers purplish-red, spurred, in terminal or leaf opposed, bracteate racemes; sepals inciso-dentate; nuts subglobose, 2-3 mm in diameter, rugose when dry.

Fl.-Fr.: Jan. -Apr.

Common winter weed.

BRASSICACEAE 372/4060; 12/19; 5/7

Key to Genera

1. Fruit silicula; stem clothed with bifid hairs; fruit ob-triangular; flowers white ....CAPSELLA
1. Fruit siliqua; stem glabrous or nearly so
   2. Siliqua compressed
      3. Valves of fruit winged ....LEPIDIUM
      3. Valves of fruit not winged ....CARDAMINE
   2. Siliqua not compressed; fruits up to 15 mm long
      4. Fruit 10-15 mm long, cylindrical; sepals and petals 2-2.5 mm long . ...RORIPPA
      4. Fruit 3-5 mm across, globose; sepals and petals less than 2 mm long .... COCHLEARIA

CAPSELLA Medik. W 9; MP 1; SA 1

Capsella bursa-pastoris (L.) Medik., Pfl. –Gatt. 85. 1792; *Thlaspi bursa-pastoris* L., Sp. Pl. 647. 1753; Hook.f., & T. Anderson in Hook.f., Fl. Brit. India 1: 159. 1872; Bhaumik in B.D. Sharma *et al.,* Fl. India 2: 189. 1993; Samvatsar, Fl. West. Trib. Madhya Pradesh 41. 1996; Khanna *et al.,* Suppl. Fl. Madhya Pradesh 6.2001.

Erect herbs, up to 40 cm high; basal leaves rosulate, oblong-lanceolate, 2-14 × 0.5-3.5 cm, pinnati-partite; cauline leaves ovate-lanceolate to linear, auricled, 1-7 × 0.2-2 cm, clasping at base, entire or denticulate; racemes up to 30 cm long in fruit; fruits obcordate to ob-tirangular, 5-9 × 3.5-6 mm, with wide apical notch, apical lobes subacute; seeds reddish- brown to black, ellipsoid.

Fl.-Fr.: March-Oct.

Alirajpur, Indore.

CARDAMINE L. W. 233; MP 3; SA 2

Key to Species

1. Plants usually with a simple stem; basal leaves densely rosulate .... **C. hirsuta**
1. Plants with several branches; basal leaves not rosulate .... **C. flexuosa**

Cardamine flexuosa With., Arr. Brit. Pl. ed. 3 3:578. 1796; *C. scutata* subsp. *flexuosa* (With.) H. Hara, J. Fac. Sci. Univ. Tokyo, Sect. 3, Bot. 6:59. 1952; Samvatsar, Fl. West. Trib. Madhya Pradesh 42. 1996; Khanna *et al.,* Suppl. Fl. Madhya Pradesh 6.2001.

Annual, erect herbs; basal leaves 7-15, scarcely rosulate, orbicular, sinuate-dentate to lobed; cauline leaves 4-10, ovate to lanceolate, pinnate, 2.5-5 cm long; racemes 2-12 cm long; flowers few to many, *ca* 3 mm across; pods elliptic-oblong, obtuse, 1.2-2.5 cm long; seeds yellow, ovoid, compressed, *ca* 1 mm long.

Fl.-Fr.: Jan. -Aug.

Alirajpur, Dhar, Jhabua.

Cardamine hirsuta L., Sp. Pl. 2: 655. 1753; Bhaumik in B.D. Sharma *et al.,* Fl. India 2: 192. 1993; *Cardamine hirsuta* var. *sylvatica* (Link) Hook.f. & T. Anderson, Fl. Brit. India 1: 138 1872; Khanna *et al.,* Suppl. Fl. Madhya Pradesh 7.2001.

Annual herbs; stems diffuse or much-branched, glabrous or hairy; leaves very variable, glabrous or hairy, 3-25 cm long; basal leaves rosulate; petiole ciliate; cauline leaves petioled or subsessile; leaflets lanceolate, narrowly oblong or obovate; inflorescence racemose or subcorymbose, 10-20-flowered, up to 20 cm long; flowers white, 3-5 mm across; pod cylindric, somewhat compressed, 15-25 cm long; seeds light brown, 6-12 on either side of replum.

Fl.-Fr.: March-May

Nursery weed, Burhanpur, Indore.

COCHLEARIA L. W. 27; MP 1; SA 1

Cochlearia cochlearioides (Roth) Santapau & Maheshw. in J. Bombay Nat. Hist. Soc. 54: 804. 1957; *Alyssum cochlearioides* Roth, Nov. Pl. Sp. 322. 1821; *Cochlearia flava* Roxb. ex Hook.f. & T. Anderson in Hook.f., Fl. Brit. India 1: 145. 1872; Bhaumik in B.D. Sharma *et al.,* Fl. India 2: 191. 1993; Mukherjee in Verma *et al.,* Fl. Madhya Pradesh 1: 145. 1993; Samvatsar, Fl. West. Trib. Madhya Pradesh 42. 1996.

Annual, erect, diffusely branched glabrous herbs; leaves lanceolate, pinnatifid, up to 12 cm long, auricled at base; racemes numerous; flowers small, yellow; fruits globose, smooth, valves membranous, nerveless; seeds numerous, rugose.

Fl.-Fr.: Aug. – Oct.

River banks and damp places, common.

LEPIDIUM L. W. 234; MP 1; SA1

Lepidium didymum L., Mant. Pl. 92. 1767; *Coronopus didymus* (L.) Sm., Fl. Brit. 2: 691. 1800; Bhaumik in B.D. Sharma *et al.,* Fl. India 2: 192. 1993; Mukherjee in Verma *et al.,* Fl. Madhya Pradesh 1: 147. 1993; Samvatsar, Fl. West. Trib. Madhya Pradesh 42. 1996.

Diffuse herbs; stems branched, glabrous or hairy; leaves 6-10 jugate, 5-10 cm long, pinnatifid-pinnatipartite; ultimate segments 3-5 jugate, lanceolate; racemes 30-60-flowered, 6-8 cm long; flowers minute, about 1 mm across; pods of two indehiscent lobes; valves globose, septum very narrow; seed 1 in each lobe, reniform, reddish-brown.

Fl.-Fr.: Nov – June

Weed of cultivated fields, not common.

RORIPPA L. W. 9; MP 3; SA 2

Key to Species

1. Petals absent; leaves serrate-dentate, upper petiolate; siliqua 2-3 cm long .... **R. dubia**
1. Petals present; leaves irregularly pinnatifid, sessile; siliqua 1-1.8 cm long................................. **R. indica**

Rorippa dubia (Pers.) H. Hara in J. Jap. Bot. 30:196. 1955*; Sisymbrium dubium* Pers., Syn. Pl. 2: 199 1806*; Nasturtium indicum auct. non* (L.) DC. 1821; Hook.f. & T. Anderson in Hook.f., Fl. Brit. India. 1:132. 1872; Hajra & Chowdhery in B.D. Sharma *et al.,* Fl. India 2: 129. 1993; Samvatsar, Fl. West. Trib. Madhya Pradesh 43. 1996; Khanna *et al.,* Suppl. Fl. Madhya Pradesh 7.2001.

Erect herbs; lower leaves deeply lyrate-pinnatifid to pinnatipartite; upper ones ovate-lanceolate to lanceolate-oblong, entire, lyrate-pinnatifid; racemes terminal, 7-30 cm long, many-flowered; flowers bright yellow, *ca* 2.5 mm across; fruits 10-20 × *ca* 1.5 mm, slightly curved, beaked, on 4-8 mm long stalks; seeds red-brown, biseriate, minute, rugose, spherical.

Fl.-Fr.: April – June

Alirajpur, Burhanpur, Dhar, Khandwa.

Rorippa indica (L.) Hiern. Cat. Afr. Pl. 1: xxvi. 1896; *Sisymbrium indicum* Hiern, Mant. Pl. 93. 1767; *Nasturtium indicum* (L.) DC., Syst. Nat. 2: 199. 1821; Hook.f. & T. Anderson in Hook.f., Fl. Brit. India. 1: 134. 1872; Hajra & Chowdhery in B.D. Sharma *et al.,* Fl. India 2: 129. 1993. Mukherjee in Verma *et al.,* Fl. Madhya Pradesh 1: 149. 1993; Samvatsar, Fl. West. Trib. Madhya Pradesh 43. 1996.

Erect herbs; lower leaves deeply lyrate-pinnatifid to pinnatipartite; upper ones ovate-lanceolate to lanceolate-oblong, entire, lyrate-pinnatifid; racemes terminal, 7-30 cm long, many-flowered; flowers bright yellow, *ca* 2.5 mm across; fruits 10-20 × *ca* 1.5 mm, slightly curved, beaked, on 4-8 mm long stalks; seed red-brown, biseriate, minute, rugose, spherical.

Fl.-Fr.: April-June

Alirajpur, Burhanpur, Dhar, Indore, Jhabua, Khandwa.

CLEOMACEAE 12/257; 1/9; 1/8

CLEOME L. W. 206; MP 9; SA 8

Key to Species

1. Leaves simple
   2. Stamens 6; seeds granulate or transversely rugose
      3. Plants scabrid; bracts sessile; seeds granulate .... **C. scaposa**
      3. Plants hairy; bracts stalked; seeds transversely rugose .... **C. monophylla**
   2. Stamens 12; seeds smooth .... **C. simplicifolia**
1. Leaves compound
   4. Leaves 3-foliolate; flowers pink-purple
      5. Ovary raised on a gynophore; bracts trifoliolate; stamens 6; seeds with longitudinal striations and prominent transverse ridges .... **C. rutidosperma** var. **burmannii**
      5. Ovary sessile; bracts unifoliolate; stamens many; seeds tubercled .... **C. felina**
   4. Leaves 5 or more foliolate
      6. Leaves 5-foliolate; flowers white; stamens and carpels borne on gynandrophore stamens 6........ **C. gynandra**
      6. Leaves 5 or more foliolate; flowers purple or yellow; stamens and carpels sessile; stamens 10 or more
         7. Flowers purple; ovary and capsules glabrous; plants glabrous or with few scattered glandular hairs .... **C. chelidonii**
         7. Flowers yellow; ovary and capsules densely villous; plants viscous with stalked glands........ **C. viscosa**

Cleome chelidonii L.f., Suppl. Pl. 300.1782; Hook.f. & Thoms. in Hook.f., Fl. Brit India 1: 170.1872; Raghavan in B.D. Sharma *et al.,* Fl. India 2: 306. 1993; Mukherjee in Verma *et al.,* Fl. Madhya Pradesh 1: 154. 1993; Samvatsar, Fl. West. Trib. Madhya Pradesh 43. 1996.

Strigose, suffrutiscent herbs; leaflets 5-9, 1-7 × 0.5-1 cm, obovate, lower petiolate, upper gradually becoming sessile; flowers in leafy racemes; stamens numerous, anthers purplish-black; capsules slender, cylindric, tapering at both ends; seeds yellowish-brown, echinate.

Fl.-Fr.: Aug.-Oct.

Not common, weed of moist places, ponds, Burhanpur, Dhar, Indore, Khandwa, Khargone.

Cleome felina L.f., Suppl. Pl. 300.1781; Hook.f. & Thoms. in Hook.f., Fl. Brit. India 1: 170.1872; Raghavan in B.D. Sharma *et al.,* Fl. India 2: 308. 1993; Mukherjee in Verma *et al.,* Fl. Madhya Pradesh 1: 154. 1993.

Annual, prostrate herbs; stems and leaves entirely covered with stiff, appressed scale- like hairs; leaves tri-foliolate; leaflets obovate 0.7-1.7 × 0.5-1 cm; flowers pink, axillary solitary; capsules linear, glabrous; seeds reniform, yellowish- brown.

Fl.-Fr.: May-Nov.

Confined to **Indore**, in rocky river beds, Kalakund.

Cleome gynandra L., Sp. Pl. 671. 1753; *Gynandropsis pentaphylla* (L.) DC., Prodr. 1: 238. 1824; Hook.f., & Thoms. in Hook.f., Fl. Brit. India 1: 171. 1872; Raghavan in B.D. Sharma *et al.,* Fl. India 2: 309. 1993; Mukherjee in Verma *et al.,* Fl. Madhya Pradesh 1: 154. 1993; Samvatsar, Fl. West. Trib. Madhya Pradesh 43. 1996. *Safed hur hur*

Erect, glandular-hairy herbs; leaflets 5,4-6.5 × 1.5-3.5 cm, obovate to oblanceolate, denticulate to subentire; flowers white; stamens 6, equal in length; capsules 2.5-8.0 × 0.3-0.4 cm, glandular-pubescent, obliquely striated, many seeded; seeds reniform, with superficial concentric ribs and irregular cross-ribs, black-brown.

Fl.-Fr.: June-Nov.

Common in waste places, road-side refuse dumps.

Cleome monophylla L., Sp. Pl. 672. 1753; Hook.f., & Thoms. in Hook.f., Fl. Brit. India 1: 168. 1872; Raghavan in B.D. Sharma *et al.,* Fl. India 2: 312. 1993; Mukherjee in Verma *et al.,* Fl. Madhya Pradesh 1: 156. 1993; Samvatsar, Fl. West. Trib. Madhya Pradesh 44. 1996.

Glandular hairy herbs; leaves simple, 3-5.5 × 0.4-1 cm, lower long-petioled, upper sub-sessile, shorter, ovate to lanceolate, subcordate; flowers pale purple in leafy racemes; seeds transversely rugose.

Fl.-Fr.: May-Oct.

Common, open grass-lands, moist soil.

Cleome rutidosperma var. **burmannii** (Wight & Arn.) Siddiqui & S. N. Dixit in Pl. Sci. (Lucknow) 6: 68. 1974; *Cleome burmannii* Wight & Arn. Prodr. Fl. Ind. Orient. 22. 1834; Hook.f., & Thoms. in Hook.f., Fl. Brit. India 1: 170.1872; Raghavan in B.D. Sharma *et al.,* Fl. India 2: 304. 1993.

Annual herbs up to 50 cm high; stems clothed with scattered soft,minute prickles; leaves compound; leaflets 3-5, ovate to elliptic-lanceolate; flowers white, axillary, solitary; capsules linear, terete, torulose, glabrous, 3.5-4.5 cm long; seeds reddish-brown, with concentric ribs.

Fl.-Fr.: May-June

Rare, in **Burhanpur**, along railway tracks, Mandwa, Chandani.

Cleome scaposa DC., Prodr. 1: 239.1824; *Cleome gracilis* Edgew. in J. Asiat. Soc. Bengal 16: 1212 1847; Hook.f. & Thoms. in Hook.f., Fl. Brit. India 1: 168. 1872; Raghavan in B.D. Sharma *et al.,* Fl. India 2: 314. 1993; Samvatsar, Fl. West. Trib. Madhya Pradesh 44. 1996; Khanna *et al.,* Suppl. Fl. Madhya Pradesh 9.2001.

Glandular hairy, strongly aromatic herbs; stems much-branched, setose-scabrous; leaves elliptic-ovate to orbicular, 1-2.5 × 0.5-1.5 cm; flowers yellow, in lax, slender, elongated racemes; capsules linear-falcate, glabrous or sparsely glandular; seeds brownish-black, reniform, granulate.

Fl.-Fr.: Aug. – Dec.

Rare, in open rocky slopes, **Alirajpur** Bakhat garh.

Cleome simplicifolia Hook.f. & Thoms. in Hook.f., Fl. Brit. India 1: 169. 1872; *Polanisia simplicifolia* Cambess. Voy Inde 4:20.1841; Raghavan in B.D. Sharma *et al.,* Fl. India 2:

314. 1993; Mukherjee in Verma *et al.,* Fl. Madhya Pradesh 1: 156. 1993; Samvatsar, Fl. West. Trib. Madhya Pradesh 44. 1996.

Annual, 20-60 cm high herbs, clothed with white, stiff, glandular hairs; leaves obovate-oblong or lanceolate, 3-8 × 0.5-2.5 cm, strigose; flowers violet, axillary, solitary, passing into racemes; ovary sessile; capsule 1.5-5.0 cm long, many ribbed, beaked, slightly torulose; seeds yellowish-brown, smooth.

Fl.-Fr.: July-Oct.

Common in open rocky places, also in grass-lands, all over the area.

Cleome viscosa L., Sp. Pl. 672. 1753; Hook.f., & Thoms. in Hook.f., Fl. Brit. India 1: 170.1872; Raghavan in B.D. Sharma *et al.,* Fl. India 2: 317. 1993; Mukherjee in Verma *et al.,* Fl. Madhya Pradesh 1: 156. 1993; Samvatsar, Fl. West. Trib. Madhya Pradesh 44. 1996. *Pili hur hur.*

Annual herbs, clothed with viscid-glandular hairs; leaflets 3-7, 1-4 × 0.5-1.5 cm, elliptic, oblong or obovate; flowers yellow, axillary, solitary, passing upwards into lax, bracteate corymbose-racemes; capsule cylindric, 4-5 cm long, viscid, many-ribbed, beaked; seeds subglobose, pitted in the centre, transversely ribbed, brown.

Fl.-Fr.: June-Sept.

Common in waste land, refuse dumps, in all districts.

CAPPARACEAE 34/449; 4/10; 4/8

Key to Genera

1. Flowers hypogynous; sepals 2-seriate; leaves simple
   2. Unarmed, erect shrubs; stamens 4-6, inserted on the gynophores; berry thin, cylindric; funnel-shaped nectary present at the base of gynophore; petals clawed .... CADABA
   2. Armed shrubs, trees or scramblers; stamens indefinite, inserted at the base of gynophores; berry globose; petals not clawed .... CAPPARIS
1. Flowers perigynous; sepals uniseriate; leaves simple or palmately compound; stamens many
   3. Leaves palmately 3-foliolate; sepals free; flowers polygamous; petals clawed; berry globose; trees .... CRATEVA
   3. Leaves simple; sepals united; flowers bisexual; petals not clawed; berry moniliform; woody twiners .... MAERUA

CADABA Forssk. W. 29; MP 1; SA 1

Cadaba fruticosa (L.) Druce in Bot. Exch. Club Soc. Brit. Isles 3: 415. 1914; *Cleome fruticosa* L., Sp. Pl. 671. 1753; *Cadaba indica* Lam., Encycl. 1: 544. 1785; Hook.f. & Thoms. in Hook.f., Fl. Brit. India 1: 172. 1872; Raghavan in B.D. Sharma *et al.,* Fl. India 2: 250.1993; Samvatsar, Fl. West. Trib. Madhya Pradesh 46. 1996; Khanna *et al.,* Suppl. Fl. Madhya Pradesh 8. 2001.

An unarmed, scandent shrub; leaves elliptic-oblong, 0.9-4 × 0.4-1.5 cm, obtuse or mucronate at apex, rounded at base, entire; flowers white, in terminal racemes; berries olive-green, 1.5-2 cm long, irregularly torulose; pericarp fleshy, thick, densely glandular pubescent; seeds many.

Fl.-Fr.: Nov. –June.

Alirajpur, Barwani, Burhanpur, Dhar, Jhabua, Khargone.

CAPPARIS L. W. 181; MP 6; SA 4

Key to Species

1. Leaves caducous; flowers orange-red or scarlet .... **C. decidua**
1. Leaves well developed; flowers greenish-white, white or pink
    2. Trees; flowers white, in racemes; stipular spines minute; fruit 1.5-2cm, purple when ripe .... **C. grandis**
    2. Shrubs or scramblers; flowers 2-3, fascicled or in cymose umbels; stipules always present
        3. Flowers large, white or pink, in fascicles of 2-3; berry berry 5-7 cm long, red, when ripe; scramblers .... **C. zeylanica**
        3. Flowers small, greenish-white, in cymose umbels; berry small, to 1 cm, black when ripe; shrubs .... **C. sepiaria**

Capparis decidua (Forssk.) Edgew. in J. Proc. Linn. Soc., Bot. 6: 184. 1862; *Sodada decidua* Forssk., Fl. Aegypt. –Arab. 81. 1775; *Capparis aphylla* Roth, Nov. Sp. Pl. 238. 1821; Hook.f., & Thoms. in Hook.f., Fl. Brit. India 1: 174. 1872; Raghavan in B.D. Sharma *et al.,* Fl. India 2: 265. 1993; Mukherjee in Verma *et al.,* Fl. Madhya Pradesh 1: 151. 1993; Samvatsar, Fl. West. Trib. Madhya Pradesh 46. 1996. *Karil.*

Densely branched shrubs or small trees, 2-6 m tall, with leafless (except the young shoots), green, spiny branches; leaves when present small, linear, caducous; stipular spines straight; flowers orange-red or scarlet, in many flowered corymbs; fruits 1.0-1.5 cm across, globose or ovoid, bright red when ripe.

Fl.-Fr.: March–July

Burhanpur, Dhar, Indore, occasional.

Capparis grandis L.f., Suppl. Pl. 263. 1781; Hook.f. & Thoms. in Hook.f. Fl. Brit. India 1: 176 . 1872; Mukherjee in Verma *et al.,* Fl. Madhya Pradesh 1: 152. 1993; Samvatsar, Fl. West. Trib. Madhya Pradesh 46. 1996. *Pachar*

Small to medium sized trees; branches pubescent; stipules rarely present; leaves 4-7 cm long, pubescent, lanceolate to obovate; flowers white, in terminal racemes; fruits globose to subglobose, smooth, 2-6 seeded.

Fl.-Fr.: March-June

Rare, along road sides, Barwani, Dhar, Indore, Khandwa.

Capparis sepiaria L.,Syst. Nat. ed. 10 2: 1071. 1759; Hook.f., & Thoms. in Hook.f. Fl. Brit. India 1: 177. 1872; Jacobs in Blumea 12: 489. 1965; Mukherjee in Verma *et al.,* Fl. Madhya Pradesh 1: 152. 1993; Samvatsar, Fl. West. Trib. Madhya Pradesh 46. 1996 **(Plate I, Fig. 4)**

Woody, 1-2 m high, straggling shrubs; leaves 3-5 × 2-3 cm, ovate, elliptic, entire, retuse; stipular spines hooked; umbels axillary and terminal; berries 1 cm across, globose, smooth, black when ripe.

Fl.-Fr.: Feb. -May

Common in hedges throughout the area and in scrub forest.

Capparis zeylanica L., Sp. Pl. ed. 2 720.1762; *C. horrida* L.f., Suppl. Pl. 264. 1782; Hook.f. & Thoms. in Hook.f., Fl. Brit. India 1: 178. 1872; Raghavan in B.D. Sharma *et al.,* Fl. India 2: 298. 1993; Mukherjee in Verma *et al.,* Fl. Madhya Pradesh 1: 152. 1993; Samvatsar, Fl. West. Trib. Madhya Pradesh 46, 1996. *Ardanda*

Straggling shrubs; leaves 3-5 × 2-3 cm ovate or oblong, entire; stipular spines hooked; petals densely hairy inside; berries subglobose, 4-8 × 3-4 cm, reddish-brown, with woody pericarp; seeds subglobose.

Fl.-Fr.: Nov. -April

Common among bushes and shrubs in wastelands, throughout the area.

CRATEVA L. W. 15; MP 2; SA 2

Key to Species

1. Leaflets ovate-lanceolate, tapering at apex; berry ovoid, tuberculate .... **C. religiosa**
1. Leaflets ovate to obovate abruptly acuminate at apex; berry globose, smooth ....**C. adansonii** subsp. **odora**

Crateva adansonii DC. subsp. **odora** (Buch. -Ham.) Jacobs in Blumea 12: 198. 1964; *C. odora* Buch. -Ham. in Trans. Linn. Soc. London 15: 118. 1827; *C. religiosa* var. *roxburghii* (R. Br.) Hook.f. & Thoms. in Hook.f., Fl. Brit. India 1: 172. 1872; Raghavan in B.D. Sharma *et al.,* Fl. India 2: 322. 1993; Mukherjee in Verma *et al.,* Fl. Madhya Pradesh 1: 157. 1993; Samvatsar, Fl. West. Trib. Madhya Pradesh 46. 1996. *Barna*

Trees, 5-10 m high; leaves 3-foliolate; leaflets ovate to elliptic-lanceolate, 8-10 × 5-6 cm, entire; petioles glandular at tip; flowers fragrant, whitish, in densely flowered corymbs; berries orange yellow, globose, *ca* 3.5 cm across; seeds brownish, broadly ellipsoid, 5-6 mm long.

Fl.-Fr.: April – Nov.

Rare, Burhanpur and Dhar.

Crateva religiosa G. Forst., Pl. Esc. 45. 1786; Hook.f. & Thoms. in Hook.f., Fl. Brit. India 1: 172. 1872; Raghavan in B.D. Sharma *et al.,* Fl. India 2: 325. 1993; Mukherjee in Verma *et al.,* Fl. Madhya Pradesh 1: 158. 1993; Samvatsar, Fl. West. Trib. Madhya Pradesh 46. 1996. *Barna*

Trees, 10-15 m high; young shoots brown with lenticels; leaves 3-foliolate; leaflets lanceolate or elliptic, 8-20 × 2-6 cm, acuminate at apex, cuneate at base; flowers white or creamy; berries subglobose, pericarp tuberculate, 3-3.5 cm across; seeds broadly ellipsoid, 12-18 mm long, dorsally keeled.

Fl.-Fr.: March–Oct.

Rare in deciduous and mixed forests, Alirajpur, Dhar.

(Both species occur as just 1 or 2 trees in the area of their occurrence)

MAERUA Forssk. W. 53; MP 1; SA 1

Maerua oblongifolia (Forssk.) A. Rich., Tent. Fl. Abyss. 1:32. 1847; *Capparis oblongifolia* Forssk. Fl. Aegypt. -Arab. 99. 1775; *Maerua arenaria* (DC.) Hook.f. & Thoms. in Fl. Brit. India 1:171. 1872; Samvatsar, Fl. West. Trib. Madhya Pradesh 48. 1996.

Scandent shrubs; leaves linear to elliptic, obtuse, 2.5-4.5 × 1.3 cm; flowers greenish-white, in terminal and lateral racemes; bracts yellow, ovate-acuminate; berry fleshy, elongated, irregularly moniliform; seeds many, bluish-black.

Fl.-Fr.: Oct. -Feb.

Fairly common over hedges in villages, in all districts.

VIOLACEAE 25/806; 2/; SA 1/1

HYBANTHUS Jacq. W. 70; MP 1; SA 1

Hybanthus enneaspermus (L.) F. Muell., Fragm. 10: 81. 1867; *Viola enneasperma* L., Sp. Pl. 937. 1753; Hook.f. & Thoms. in Hook.f., Fl. Brit. India 1: 185. 1872; S.P. Banerjee & Pramanik in B.D. Sharma *et al.,* Fl. India 2: 343. f. 64. 1993; Mukherjee in Verma *et al.,* Fl. Madhya Pradesh 1: 159. 1993; Samvatsar, Fl. West. Trib. Madhya Pradesh 48. 1996. *Ratan purus* [*Ionidium suffruticosum* Ging., is synonym of *Hybanthus linearifolius* (Vahl) Urb.]

A small suffrutiscent, perennial herb; leaves 2-5 × 0.2-0.4 cm linear-lanceolate, subsessile; stipules subulate, gland-tipped; flowers dark pink, axillary, solitary; sepals lanceolate, acute, keeled; petals unequal, lower petal spurred; capsule subglobose; seeds striated.

Fl.-Fr.: Oct. –Jan.

Common forest under growth, especially in moist soil, in all districts.

BIXACEAE 4/23; 1/1; 1/1

COCHLOSPERMUM Kunth ex DC. W. 13; MP 1; SA 1

Cochlospermum religiosum (L.) Alston, Handb. Fl. Ceyl. 6 (Suppl.): 14. 1931; *Bombax religiosum* L., Sp. Pl. 512. 1753; *Cochlospermum gossypium* DC., 1824 *nom illeg*; Hook.f. & Thoms. in Hook.f. Fl. Brit. India 1: 190.1872; N.P. Balakr. in B.D. Sharma *et al.,* Fl. India 2: 383. 1993; Mukherjee in Verma *et al.,* Fl. Madhya Pradesh 1: 162. 1993; Samvatsar, Fl. West. Trib. Madhya Pradesh 49. 1996. *Galgal* **(Plate 1, Fig. 5)**

Deciduous trees, up to 8 m tall, with crooked branches; leaves orbicular, 6-18 cm across, 3-5-lobed, cordate at base, pubescent beneath, lobes undulate, acuminate at apex; flowers yellow, up to 7 cm across, in terminal racemes; capsules obovoid, 5-10 × 2-7 cm, valves 5, striate-ribbed; seeds brown, reniform, 5-6 mm across, woolly.

Fl.-Fr.: Dec. –June

Alirajpur, Barwani, Burhanpur, Dhar, Indore, Khargone but nowhere common, only 1-3 trees seen at places.

POLYGALACEAE 27/1163; 2/16; 1/6

POLYGALA L. W. 623; MP 15; SA 6

Key to Species

1. Wing sepals symmetric, petaloid; flowers purple .... **P. persicariifolia**
1. Wing sepals asymmetric, green; flowers yellow
    2. Strophiole with 2-appendages and a short tooth........ **P. chinensis**
    2. Strophiole with 3-appendages, tooth absent
        3. Racemes longer than leaves; seeds hairy .... **P. elongata**
        3. Racemes shorter than leaves
            4. Leaves up to 1.5 cm long; seeds densely pilose .... **P. erioptera**
            4. Leaves up to 3.5 cm long
                5. Seeds pubescent .... **P. jacobii**
                5. Seeds villous .... **P. arvensis**

Polygala arvensis Willd., Sp. Pl. 3(2): 876. 1802; A.W. Benn. in Hook.f. & Thoms. in Fl. Brit. India 1: 204. 1872; *P. chinensis* var. *linarifolia* sensu Mukerjee in Bull. Bot. Soc. Bengal 12: 40.1958; Banerjee in B.D. Sharma *et al.,* Fl. India 2: 460.1993; Mukherjee in Verma *et al.,* Fl. Madhya Pradesh 1: 168. 1993; Samvatsar, Fl. West. Trib. Madhya Pradesh 50.1996. *Meradu*

Annual, procumbent or erect herbs; leaves orbicular, obovate, elliptic or elliptic-oblong, 1-3 cm long; flowers usually yellow, solitary or in lateral racemes; capsules ovate-oblong, *ca* 3.5 mm long; seeds black, oblong-ellipsoid, *ca* 2.5 mm long, pilose; caruncle hood-shaped, 3-fid almost to base with short appendages.

Fl.-Fr.: Throughout the year.

Common in plains and grassy fields, all over the area.

Polygala chinensis L., Sp. Pl. 704. 1753; *P. glomerata* Lour., Fl. Cochinch. 2: 426. 1790; Bennett in Hook.f., Fl. Brit. India 1: 206. 1872; Banerjee in B.D. Sharma *et al.,* Fl. India 2: 464. 1993; Mukherjee in Verma *et al.,* Fl. Madhya Pradesh 1: 169. 1993; Samvatsar, Fl. West. Trib. Madhya Pradesh 50.1996.

Annual, erect herbs; leaves 0.6-4 × 0.4-2 cm long, oblong or obovate, acute; flowers in axillary few flowered racemes, keel petal crested with filiform appendages; capsule sub-orbicular, apex emarginate; seeds with 2 oblong appendages and a short tooth.

Fl.-Fr.: July – Dec.

Common in grass lands, hill slopes, commonly occurs along with *P. arvensis*. Willd.

Polygala elongata Klein ex Willd., Sp. Pl. 3: 879. 1802; Bennett in Hook.f., Fl. Brit. Ind. 1: 203. 1872, *p.p;* Banerjee in B.D. Sharma *et al.,* Fl. India 2: 466. 1993; Mukherjee in Verma *et al.,* Fl. Madhya Pradesh 1: 169. 1993.

Annual herbs; leaves oblong or elliptic-oblong, shortly petioled; flowers yellow, in lateral, leaf-opposed or terminal racemes; capsules rhomboid, margined, obliquely emarginate at apex; seeds oblong-obovate, densely hairy; caruncle hood-shaped with 3-toothed appendages.

Fl.-Fr.: Aug. - Feb.

An undergrowth in mixed forests, Burhanpur, Khandwa.

Polygala erioptera DC., Prodr. 1: 326. 1824; Bennett in Hook.f., Fl. Brit. India 1: 203. 1872; Banerjee in B.D. Sharma *et al.,* Fl. India 2: 467. 1993; Mukherjee in Verma *et al.,* Fl. Madhya Pradesh 1: 170.1993; Samvatsar, Fl. West. Trib. Madhya Pradesh 50.1996.

Erect, 10-30 cm high herbs; leaves linear to elliptic-oblong, 1.4-0.1 × 0.3-1.0 c,. entire; flowers yellow, in axillary or extra-axillary racemes, shorter than leaves; wings nigro-punctate; keel of petals crested with filiform appendnges; capsule oblong, obliquely emarginate at apex, margined; seeds oblong, black, densely pilose; strophiole with 3 appendages.

Fl.-Fr.: Sept. – Nov.

Common everywhere during rains.

Polygala jacobii Chandrabose in Bull. Bot. Surv. India 9:288. 1968; *P. chinensis sensu* P. Mudaliar & Rao, Handb. South Indian Weeds 85. t. 32. 1955, non L. 1753; *P. dunnii* Panigrahi in Kew Bull. 29:655. 1974; Mukherjee in Verma *et al.,* in Fl. Madhya Pradesh 1:170.1993.

Erect, perennial herbs; leaves short petioled, oblong-obovate, oblanceolate; flowers yellow, solitary, axillary or in short racemes; wing sepals obliquely ovate-oblong; lateral petals as long as the keel, notched at apex; seeds blackish-brown, obovoid, silky hairy; caruncle hood-shaped with 3, broad, membranous winged appendages, up to 2/3 length of the seed.

Fl.-Fr.: Sept. -March

Indore only, Khurda Khurdi.

Polygala persicariifolia DC., Prodr. 1: 326. 1824; Bennett in Hook.f., Fl. Brit. India 1: 202. 1872, '*persicariaefolia*'; Banerjee in B.D. Sharma *et al.,* Fl. India 2: 480.1993; Mukherjee in Verma *et al.,* Fl. Madhya Pradesh 1: 171. 1993; Samvatsar, Fl. West. Trib. Madhya Pradesh 50.1996.

Erect, annual herbs; leaves subsessile, linear- elliptic to lanceolate, margins ciliolate; flowers in terminal and extra axillary racemes; corolla shorter than wings; capsule notched, ciliate; seeds villous.

Fl.-Fr.: Aug. -Sept.

Common, as forest under-growth and along water streams, in all districts.

CARYOPHYLLACEAE 91/2456/; 5/7; 3/4

Key to Genera

1. Erect plants; stamens 5 .... POLYCARPAEA
1. Prostrate plants
   2. Stamens 3-5; leaves 0.5-1.5 cm long, obovate or linear-oblong, venation pinnate; flowers light pink .... POLYCARPON
   2. Stamens 2-3; leaves 1-3.5 cm long, almost orbicular, broader than long, base deeply cordate, venation palmate; flowers white; petals deeply divided .... DRYMARIA

DRYMARIA Willd. ex Roem. & Schult. W. 55; MP 1; SA 1

Drymaria cordata (L.) Willd. ex Schult. Syst. Veg. 5:406. 1819; *Holosteum cordatum* L., Sp. Pl. 88. 1753; Edgew. & Hook.f. in Hook.f., Fl. Brit. India 1: 244. 1874; Majumdar in B.D. Sharma *et al.,* Fl. India 2: 533. 1993; Mukherjee in Verma *et al.,* Fl. Madhya Pradesh 1: 173. 1993.

Annual, prostrate or decumbent herb; leaves rounded or orbicular-ovate, 0.6-2.5 × 0.4-2 cm, apiculate, 3-5-nerved from the base; flowers greenish- white, in lax, glandular puberulous, repeatedly forked, axillary or terminal cymes; capsules trigonous, 1.5-2.5 mm long; seeds reniform, compressed, 1.5-2 mm across, tuberculate.

Fl.-Fr.: Sept. – Dec.

Grows in moist shady places, **Burhanpur** only, Mirachica garden.

POLYCARPAEA Lam. W. 19; MP2; SA 2

Key to Species

1. Internodes long, densely white pubescent; leaves 1-5 cm long; stipules scarious; flowers reddish-brown; inflorescence cymose; sepals brown; seeds purplish- brown .... **P. aurea**
1. Internodes not as above; leaves 0.5-2 cm long; stipules fimbriate; flowers light pink,white afterwards; inflorescence cymose, paniculate, silvery white; sepals white; seeds yellowish-brown .... **P. corymbosa**

Polycarpaea aurea Wight & Arn. in Ann. Nat. Hist. 3: 91. 1839; Edgew. & Hook.f., Fl. Brit. Ind. 1: 245. 1874; Majumdar in B.D. Sharma *et al.,* Fl. India 2: 533. 1993; Mukherjee in Verma *et al.,* Fl. Madhya Pradesh 1: 174. 1993.

Much-branched shrubs; leaves subulate with inrolled margins, axillary leaves a few or absent; stipules hardly fimbriate; internodes densely white-tomentose; flowers very small; sepals highly coloured, much exceeding the petals and capsules; capsules oblong-ovoid, 3-valved; seeds numerous, glabrous, somewhat compressed, reniform, yellowish-brown.

Fl.-Fr.: Aug. -Feb.

Growing in rock crevices, Burhanpur and Khandwa.

Polycarpaea corymbosa (L.) Lam., Encycl. 2: 129. 1797; *Achyranthes corymbosa* L., Sp. Pl. 205. 1753; Edgew. & Hook.f. in Hook.f., Fl. Brit. Ind. 1: 245. 1874; Majumdar in B.D. Sharma *et al.,* Fl. India 2: 549. 1993; Mukherjee in Verma *et al.,* Fl. Madhya Pradesh 1: 174. 1993; Samvatsar, Fl. West. Trib. Madhya Pradesh 50.1996.

Erect herbs with woody rootstock; leaves opposite, linear, flat, sometimes pseudo-whorled from the axillary leaf fascicles; flowers small, silvery, in terminal, panicled, dense, dichotomous cymes; capsules oblong-ovoid, 3-valved, brown; seeds numerous, glabrous, somewhat compressed, reniform, yellowish-brown.

Fl.-Fr.: Aug.-Feb.

Common in sandy soil, all districts.

POLYCARPON L. W. 9; MP 1; SA1

Polycarpon prostratum (Forssk.) Asch. & Schweinf. in Oesterr. Bot. Z. 39: 128. 1889; *Alsine prostrata* Forrsk., Fl. Aegypt. -Aarab. 207. 1775; Majumdar in B.D. Sharma *et al.,* Fl. India 2: 549. 1993; Mukherjee in Verma *et al.,* Fl. Madhya Pradesh 1: 176. 1993; Samvatsar, Fl. West. Trib. Madhya Pradesh 51. 1996. *Surete*

Prostrate, dichotomously much branched herbs; leaves small, pseudo-whorled, narrow, oblong, oblanceolate or spathulate; flowers in crowded cymes, with silvery bracts or sessile in the forks of branches; capsule ovoid; seeds pale brownish, minutely tuberculate.

Fl.-Fr.: April - June

Growing in moist sandy soil, all over the area.

PORTULACACEAE 10/258; 1/3; 1/2

PORTULACA L. W. 109; MP 3; SA 2

Key to Species

1. Nodal appendages absent; leaves alternate or sub-opposite, 0.8-2.5 cm long; petals 5 .... **P. oleracea**
1. Nodal appendages present, hairy; leaves opposite, 0.2-0.6 cm long; petals 4 .... **P. quadrifida**

Portulaca oleracea L., Sp. Pl. 445. 1753; Dyer in Hook.f., Fl. Brit. India 1: 246. 1874; M.K.V. Rao in B.D. Sharma *et al.,* Fl. India 3: 4. 1993; Mukherjee in Verma *et al.,* Fl. Madhya Pradesh 1: 178. 1993; Samvatsar, Fl. West. Trib. Madhya Pradesh 51. 1996. *Bari kulpha*

Prostrate or diffuse, succulent herbs; stem swollen at the nodes; leaves obvate, sessile; flowers yellow, terminal, solitary or fascicled; capsule ovate-oblong or conical; seeds many, brown, concentrically granulate.

Fl.-Fr.: Throughout the year

Very common, throughout the area, along water streams in fields, open sewage lines.

Portulaca quadrifida L., Mant. Pl. 73. 1767; Dyer in Hook.f., Fl. Brit. India 1:247. 1874; M.K. Rao in B.D. Sharma *et al.,* Fl. India 3:6. 1993; Mukherjee in Verma *et al.,* Fl. Madhya Pradesh 1:179. 1993; Samvatsar, Fl. West. Trib. Madhya Pradesh 52. 1996. *Chhoti kulpha*

Small diffuse annuals; stem filiform, rooting at nodes; leaves opposite, fleshy, ovate, on the thickened nodes; sepals united at base; capsule conical; seeds minutely tubercled.

Fl.-Fr.: Aug. -Nov.

Common in moist and shady places in fields throughout the area, not outside fields.

TALINACEAE 2/28; MP 1/1; SA 1/1

TALINUM L. W. 16; MP 1; SA 1

Talinum portulacifolium (Forssk.) Asch. & Schweinf. in Bull. Herb. Boissier 4 (App. 2): 172. 1896; *Orygia portulacifolia* Forssk., Fl. Aegypt.–Arab. 103. 1775; *Talinum cuneifolium* Willd., Sp. Pl. 2: 864. 1799; Dyer in Hook.f., Fl. Brit. India 1: 247. 1874; M.K.V. Rao in B.D. Sharma *et al.,* Fl. India 3: 9. 1993; Samvatsar, Fl. West. Trib. Madhya Pradesh 52. 1996; Khanna *et al.,* Suppl. Fl. Madhya Pradesh 13.2001.

A perennial, suffrutiscent plant, to 1 m tall, root tuberous; leaves 6-8 × 3-4 cm alternate, exstipulate, base cuneate, flowers 1.5 – 2 cm across, in terminal panicles; peduncles *ca* 5 cm long; bracts linear; sepals 2, 6 × 3mm, 3-nerved; petals 5, pink, 1 × 0.5 cm; stamens many; ovary unilocular; ovules indefinite on free central placenta, style 3-fid; capsule 3-valved; seeds subreniform with regular concentric striations.

Fl.-Fr.: Sept. -Dec.

Rare, Narmada river bed, **Dhar** only (Dharampuri bet).

TAMARICACEAE 3/79; 1/3; 1/2

TAMARIX L. W. 57; MP 3; SA 2

Key to Species

1. Flowers unisexual; stamens 5; stigmas clavate .... **T. dioica**
1. Flowers bisexual; stamens 10; stigmas ellipsoid .... **T. ericoides**

Tamarix dioica Roxb. ex Roth, Nov. Pl. Sp. 185. 1821; Dyer in Hook.f., Fl. Brit. India 1: 249. 1874; B.V. Shetty & R.P. Pandey in B.D. Sharma *et al.,* Fl. India 3: 24. 1993; Mukherjee in Verma *et al.,* Fl. Madhya Pradesh 1: 180.1993; Samvatsar, Fl. West. Trib. Madhya Pradesh 52. 1996. *Saran*

Shrubs; leaves vaginate, free part triangular-ovate, 1-3 mm long, acuminate at apex; flowers pink or pinkish red, unisexual, in simple or compound racemes; bracts ovate-lanceolate, 2-3 mm long, keeled; capsules 4-5 mm long; seeds *ca* 0.5 mm long, with *ca* 3 mm long coma.

Fl.-Fr.: June – Jan.

Common in rocky river beds all over the area.

Tamarix ericoides Rottler & Willd., Ges. Naturf. Freunde Berlin Neue Schrifften 4: 214, pl. 4. 1803; Dyer in Hook.f., Fl. Brit. India 1: 249. 1874; B.V. Shetty & R.P. Pandey in B.D.

Sharma *et al.,* Fl. India 3: 25. 1993; Mukherjee in Verma *et al.,* Fl. Madhya Pradesh 1: 180. 1993; Samvatsar, Fl. West. Trib. Madhya Pradesh 52. 1996.

Undershrubs or shrubs; leaves vaginate in lower part, broadly triangular or triangular-ovate, 2-5 mm long, acute or acuminate at apex, keeled; flowers pink or pale pink, in 8-15 cm long racemes; bracts ovate-lanceolate, 3-5 mm long, acuminate at apex; capsules *ca* 1.2 cm long; seeds *ca* 1 mm long, with *ca* 6 mm long coma.

Fl.-Fr.: Nov. – March

Same habitat, as above.

ELATINACEAE 2/57; 2/3; 1/2

BERGIA L. W. 29; MP 2; SA 2

Key to Species

1. Flowers crowded in dense axillary clusters; stamens 5; stipules subulate .... **B. ammannioides**
1. Flowers solitary or in fascicles of 2-8; stamens 10; stipules linear .... **B. suffruticosa**

Bergia ammannioides Roxb. ex Roth, Nov. Pl. Sp. 219. 1821; Dyer in Hook.f., Fl. Brit. India 1: 251. 1874; J. Bhattacharya in B.D. Sharma *et al.,* Fl. India 3: 33. 1993; Anand Kumar in Verma *et al.,* Fl. Madhya Pradesh 1: 181. 1993; Samvatsar, Fl. West. Trib. Madhya Pradesh 54. 1996.

Diffusely branched, stiff annuals; stem and branches glandular hairy; leaves sessile, 0.6 -0.8 cm long, oblong oblanceolate, serrulate; flowers red, minute, in axillary clusters; sepals, petals and stamens 5; capsule sub-globose; seeds numerous, ovoid, polished.

Fl.-Fr.: Nov. -Feb.

Near water courses in fields, margins of ponds, other moist and shady habitats throughout the area.

Bergia suffruticosa (Del.) Fenzl, in Denkschr. Bayer. Bot. Ges. Regensberg 3:183. 1841; *Lancretia suffruticosa* Del. Descr. Egypte, Hist. Nat. 213. 1813; *Bergia odorata* Edgew. in Journ. Asiat. Soc. Bengal 7:765. 1838; Dyer in Hook.f. Fl. Brit. India 1:254. 1874; J. Bhattacharya in B.D. Sharma *et al.,* Fl. India 3: 36. 1993; Samvatsar, Fl. West. Trib. Madhya Pradesh 54. 1996.

A small, decumbent or spreading glandular shrub with papery, deciduous, rust coloured bark; leaves 0.4-2 cm. long, sessile, elliptic or ovate with crenate-serrate margin, glandular hairy on both sides; flowers white, 4 mm across, solitary or in fascicles; stamens 10, 5 alternately shorter or subequal; capsule 5-celled; seeds shining, brown, foveolate.

Fl.-Fr.: Oct. -Dec.

Occasional, in wet and marshy places, Burhanpur, Dhar.

MALVACEAE 245/4465; 25/77; 23/63

Key to Genera

1 . Anthers 1-celled; stamens indefinite, mono or polyadelphous; leaves simple, palmately lobed or palmate compound

2. Pollen echinate; stamens monoadelphous; leaves simple or palmately lobed

3. Fruit capsular

4. Style one, furrowed; capsule 3-5 valved; epicalyx 8-lobed; flowers yellow, in 3-flowered axillary or terminal peduncles; leaves palmately 3-lobed; shrubs with peltate scales .... AZANZA

4. Styles more than one

5. Epicalyx present

6. Styles and carpels 3; flowers white, unisexual or polygamous; petals clawed; epicalyx segments spathaceous, accrescent in fruit, persistent; leaves 7-nerved at base .... KYDIA

6. Styles and carpels 5 or more

7. Calyx spathaceous, toothed .... ABELMOSCHUS

7. Calyx not spathaceous, deeply lobed .... HIBISCUS

5. Epicalyx absent; ovules 2 or more per chamber; flowers yellow .... ABUTILON

3. Fruit schizocarpic

8. Epicalyx absent

9. Style branches equal to number of carpels; flowers not in heads; plants not bristly .... SIDA

9. Style branches double the number of carpels; flowers in heads, surrounded by foliaceous bracts; plants bristly .... MALACHRA

8. Epicalyx present

10. Epicalyx members 3; styles equal to number of carpels; flowers yellow; mericarps hooked at apex .... MALVASTRUM

10. Epicalyx members 5 or more; style branches twice the number of carpels; flowers pink

11. Mericarps glochidiate, not winged; epicalyx segments 5; leaves with gland on the midrib near the base; plants not viscous .... URENA

11. Mericarps not glochidiate, winged; epicalyx segments 10-16, filiform; leaves without a gland; plants viscous .... PAVONIA

2. Pollen smooth; stamens polyadelphous; leaves palmately compound

12. Fruit indehiscent, 20- 35 cm long; seeds embedded in farinaceous pulp; flowers white, hanging from long pedicels; girth of stem 4 meters or more .... ADANSONIA

12. Fruit dehiscent, capsular; seeds cottony; flowers red; girth of stem not so large .... BOMBAX

1. Anthers 2-celled

13. Fruit capsular, drupaceous, follicular or indehiscent

14. Fruit capsular

15. Flowers pink; petals persistent, not clawed; styles 5; capsule 5-valved; herbs .... MELOCHIA

15. Flowers yellow; capsule 2-10 valved

16. Capsule 2-valved; corolla persistent; petals clawed; staminodes absent; epicalyx deciduous; style not lobed; seed 1 at maturity; herbs .... WALTHERIA
16. Capsule 5 valved; corolla deciduous; petals not clawed; staminodes present; epicalyx persistent and accrescent; seeds many; style 5-lobed ....MELHANIA
17. Capsule 3-5 valved; seeds not winged; herbs .... CORCHORUS
17. Capsule 10-valved, tuberculate; seeds winged; shrubs ....ERIOLAENA
13. Fruit indehiscent, follicular or drupaceous
18. Fruit echinate; herbs or shrubs ....TRIUMFETTA
18. Fruit drupaceous or follicular
19. Fruit drupaceous; flowers yellow (mostly); petals with a gland on the inner side, facing the ovary; trees or shrubs ....GREWIA
19. Fruit follicular
20. Petals present, clawed; follicles twisted, smooth; flowers crimson; corolla 2-lipped, zygomorphic; shrubs ....HELICTERES
20. Petals absent
21. Flowers orange-red; follicles papery; seeds 2 per follicle, attached on the margin; leaves palmately lobed .... FIRMIANA
21. Flowers yellow or pink; follicles thick, spreading; seeds many
22. Follicles densely bristly, deep red, when young with stinging hairs; leaves palmately lobed, not compound; bark smooth, white, exfoliating in flakes ....KAVALAMA
22. Follicles not bristly, smooth; leaves palmately lobed or palmate compound ....STERCULIA

ABELMOSCHUS Medik. W. 10; MP 5; SA 4

Key to Species

1. Epicalyx bracteoles persistent ....**A. manihot**
1. Epicalyx bracteoles caducous or deciduous
2. Bracteoles 4-6, setaceous; flowers in leafless racemes, white or with a pink centre ....**A. ficulneus**
2. Bracteoles 8-12, linear; flowers yellow with a dark purple centre, either axillary or in racemes
3. Capsule enclosed in calyx tube; bracteoles with fimbriate margins ....**A. crinitus**
3. Capsule exserted from the calyx tube; bracteoles linear, margin entire ....**A. moschatus**

Abelmoschus crinitus Wall., Pl. Asiat. Rar. 1: 39, t. 44. 1830; *Hibiscus crinitus* (Wall.) G. Don, Gen. Hist. 1:480.1831; *H. cancellatus* Roxb. ex G. Don, 1831, *nom. illeg*; Maast. in Hook.f., Fl. Brit. India 1: 342. 1874; T.K. Paul in B.D. Sharma *et al.,* Fl. India 3: 302. 1993; Verma in Verma *et al.,* Fl. Madhya Pradesh 1: 188. 1993. **(Plate 1, Fig. 6)**

Annual, erect, herbs, with tuber- like tap roots; stems hirsute due to shiny, simple and stellate hairs; leaves rounded or obtusely 5-angled, 5-8 cm across, deeply cordate at base, crenate, hirsute on both the surfaces; flowers solitary or in terminal racemes by reduction of upper leaves; epicalyx-segments linear, 10-16, 2-5 cm long, ciliate, stellately hairy; capsules ovoid or globose,

2-4 × 2-3 cm, shortly acuminate or rounded, hirsute; seeds globose to reniform, 3-5 mm long, rusty tomentose, covered by concentric rings.

Fl.-Fr.: July – Dec.

Burhanpur, Indore, Khandwa, not common.

Abelmoschus ficulneus (L.) Wight & Arn. ex Wight, Cat. Ind. Pl. 14. 1833; *Hibiscus ficulneus* L., Sp. Pl. 2:695. 1753; Wight & Arn., Prodr. Fl. Ind. Orient. 53. 1834; Mast. in Hook.f., Fl. Brit. India 1: 340.1874; T.K. Paul in B.D. Sharma *et al.,* Fl. India 3: 304. 1993; Verma in Verma *et al.,* Fl. Madhya Pradesh 1: 190.1993; Samvatsar, Fl. West. Trib. Madhya Pradesh 55. 1996. *Ran bhindi*

Undershrub; leaves suborbicular, 2-15 cm across, cordate, palmately 3-5-lobed; petioles 1.5-15 cm long; flowers axillary, solitary or in terminal racemes; pedicels 1-1.5 cm long; epicalyx-segments 5-6, usually falling before anthesis; corolla white turning pink, mauve, with dark purple centre; capsules ovoid, 5-angled, tomentose, valves inside with long white hairs; seeds black, globose.

Fl.-Fr.: Sept. – Nov.

Very common in all districts.

Abelmoschus manihot (L.) Medik. subsp. **tetraphyllus** (Roxb. ex Hornem.) Borss. Waalk., Blumea 14: 97 1966; *Hibiscus tetraphyllus* Roxb. ex Hornem. Hort. Bot. Hafn. 661. 1815; *Abelmoschus tetraphyllus* (Roxb. ex Hornem.) Wall., Numer. List 87. 1925. 1829; T.K. Paul in B.D. Sharma *et al.,* Fl. India 3: 306. 1993; Verma in Verma *et al.,* Fl. Madhya Pradesh 1: 190.1993; Samvatsar, Fl. West. Trib. Madhya Pradesh 55. 1996. *Jangli bhindi*

An annual herbs, up to 60-90 cm tall; stems minutely prickly; leaves broadly ovate-suborbicular, 5-18 × 4-15 cm, 3-7-lobed, base cordate with 7 palmate veins; flowers yellow with deep purple centre, axillary, solitary or in terminal racemes; bracteoles 4; capsules ovoid-oblong, 3-4 cm long, acuminate, covered with prickly hairs; seeds dark brown, subreniform, pubescent.

Fl.-Fr.: Aug. – Nov.

Common weed of fields, also in hedges along village roads.

Abelmoschus moschatus subsp. **tuberosus** (Span.) Borssum in Blumea 14: 93, 1966; *Hibiscus longifolius* var. *tuberosus* Span. in Linnaea 15: 170.1841; *Abelmoschus rugosus* Wall. ex Wight & Arn., Prodr. Fl. Ind. Orient. 1: 43 1834; *Hibiscus rugosus* Roxb. ex Mast. in Fl. Brit. India 1: 342 1874; Verma in Verma *et al.,* Fl. Madhya Pradesh 1: 191. 1993; Samvatsar, Fl. West. Trib. Madhya Pradesh 55. 1996.

Stellately pubescent herbs with fusiform root-stock; stem bristly at apex, hairs spreading; leaves 8-13 × 9-19 cm, orbicular, 3-7 lobed; flower buds up to 2 cm long; epicalyx segments linear-filiform, 1.5 cm long, prominently setaceous; petals yellow, turning pink, with purple centre; seeds usually tomentose.

Fl.-Fr.: Sept. – Oct.

Occasional weed in *Hibiscus esculentus* field as well as along village roads, Alirajpur, Barwani, Dhar, Khandwa, Khargone.

ABUTILON Miller W. 216; MP 5; SA 4

Key to Species

1. Flowers orange with a purple centre, solitary, axillary; capsule segments 20-30 .... **A. hirtum**
1. Flowers uniformly yellow; seeds more than 20 per segment of capsule
   2. Segments of capsule 10; capsule cylindric, less than 1 cm long, beaked .... **A. ramosum**
   2. Segments of capsule 15-30
      3. Each segment 3-seeded; flowers in pseudo-racemes; corolla thrice longer than calyx.. **A. pannosum**
      3. Each segment 2-seeded (normally); flowers solitary, axillary .... **A. indicum**

Abutilon hirtum (Lam.) Sweet, Hort. Brit., ed. 1.53. 1826; *Sida hirta* Lam., Encycl. 1: 7. 1783; *Abutilon graveolens* (Roxb. ex Hornem) Wight & Arn. ex Wight, Cat. Ind. Pl. 13. 1833; Mast. in Hook.f., Fl. Brit. India 1: 327. 1874; T.K. Paul in B.D. Sharma *et al.,* Fl. India 3: 264. 1993; Verma in Verma *et al.,* Fl. Madhya Pradesh 1: 192. 1993. *Bar kanghi*

Viscous undershrubs, up to 2 m high; leaves orbicular to broadly ovate, 1.5-6 × 1-5 cm, acute or slightly acuminate at apex, cordate at base; flowers axillary, solitary; corolla orange-yellow with a purple centre; capsule segments globular, indented at apex; seeds reniform, *ca* 2.5 mm across, punctate by minute stellate hairs.

Fl.-Fr.: Sept. – Nov.

Rare, Barwani, Burhanpur, Indore, Khargone.

Abutilon indicum (L.) Sweet, Hort. Brit., 54. 1826; *Sida indica* L., Cent. Pl. II 2: 26. 1756; Mast. in Hook.f., Fl. Brit India 1: 326. 1874; T.K. Paul in B.D. Sharma *et al.,* Fl. India 3: 266. 1993; Verma in Verma *et al.,* Fl. Madhya Pradesh 1: 192. 1993; Samvatsar, Fl. West. Trib. Madhya Pradesh 56. 1996.

Key to varieties

1a. Calyx shorter than fruit, lobes spreading .... var. **indicum**
1b. Calyx as long as fruit, lobes appressed .... var. **guineense**

var. **indicum**

Herbs or undershrubs, annual or perennial, densely tomentose, 1-1.5 m tall; leaves ovate, 9-14.5 × 7.5-11 cm, acuminate at apex, cordate at base, crenate-dentate, pubescent on both the surfaces; flowers axillary, solitary; calyx campanulate, lobes 5, ovate, apiculate; corolla yellow-orange, lobes obovate, 1.5-2.5 cm long; capsule segments dorsally and ventrally stellate hairy, 2-3–seeded; seeds brownish black, reniform, 2-3 mm across, minutely stellate- hairy or glabrescent. *Kanghi*

Fl.-Fr.: Aug. – March

Common everywhere.

var. **guineense** (Schumach.) K.M. Feng, Acta Bot. Yunnan 4:28 1982; *Sida guineensis* Schumach., Beskr. Guin. Pl. 307. 1827; *Abutilon asiaticum* (L.) Sweet, Hort. Brit., 53. 1826; Mast. in Hook.f., Fl. Brit. India 1: 326. 1874; T.K. Paul in B.D. Sharma *et al.,* Fl. India 3:

267. 1993; Borss. Waalk., Blumea 14:175 1966 (as subsp. *guineense*); Samvatsar, Fl. West. Trib. Madhya Pradesh 56. 1996. *Kanghi*

A perennial undershrub; leaves 5-10 cm long, ovate, cordate, acute or acuminate, serrate, sparsely hairy and scabrid above, velvety below; pedicels stout, longer than the petioles, jointed near the apex; capsule depressed, sub-globose, 2.5 cm in diameter.

Fl.-Fr.: Oct. – Nov.

Common in all districts

Abutilon pannosum (G. Forst.) Schltdl. Bot. Zeitung (Berlin) 9:828. 1851; *Sida pannosa* G. Forst., Commentat. Soc. Regiae Sci. Gott. 9: 62 1789; *Abutilon muticum* (Delile ex DC.) Sweet, Hort. Brit. ed. 2:65. 1830; Mast. in Hook.f., Fl. Brit. Ind. 1: 327. 1874; T.K. Paul in B.D. Sharma *et al.,* Fl. India 3: 268. 1993; Verma in Verma *et al.,* Fl. Madhya Pradesh 1: 193. 1993; Samvatsar, Fl. West. Trib. Madhya Pradesh 56. 1996. *Kasili*

Suffrutiscent, silky tomentose herbs; leaves broadly orbicular, rounded to acute at apex; stipules lanceolate, recurved; flowers yellow, solitary, axillary or often in terminal racemes; calyx lobes triangular, acuminate; petals obliquely obovate; staminal column with short neck, stelletely pubescent at base; capsule globose, depressed at apex, densely velvety; seeds sparsely hairy at hilum.

Fl.-Fr.: Oct. - Dec.

In wastelands and along cultivated fields, Barwani, Burhanpur, Indore, Khargone.

Abutilon ramosum (Cav.) Guill. & Perr., Fl. Seneg. Tent. 68. 1831; *Sida ramosa* Cav., Diss. 1: 28. 1785; Mast. in Hook.f., Fl. Brit. India 1: 328. 1874; T.K. Paul in B.D. Sharma *et al.,* Fl. India 3: 271. 1993; Khanna *et al.,* Suppl. Fl. Madhya Pradesh 16.2001.

Shrubs; stems densely stellate and simple hairy; leaves ovate, 6.5-12 × 5.5-11 cm, acuminate at apex, cordate at base, sometimes 3-angled, crenate-serrate, palmately 5-7 nerved at base; flowers yellow, axillary or terminal, solitary or paired; peduncle tomentose, divided above dichotomously as in cymes; capsules dehiscent, *ca* 1 cm across, segments 8-10, each with two long villose awns; seeds brownish- black, reniform, *ca* 2 mm across, glabrescent.

Fl.-Fr.: Aug. – March

Burhanpur and Khandwa.

ADANSONIA L. W. 8; MP 1; SA 1

Adansonia digitata L., Sp. Pl. 1190.1753; Mast. in Hook.f., Fl. Brit. India 1: 348. 1874; M.P. Nayar & M.C. Biswas in B.D. Sharma *et al.,* Fl. India 3: 404. 1993; Hanfi in Verma *et al.,* Fl. Madhya Pradesh 1: 211. 1993; Samvatsar, Fl. West. Trib. Madhya Pradesh 62. 1996. *Gorakh Imli*

Large trees; trunk abruptly narrowing into thick branches; leaves digitate, 3-9-foliolate; leaflets obovate-oblong, *ca* 12 × 6 cm, acuminate-caudate at apex, cuneate at base; flowers solitary, hanging from long pedicels, white or cream-coloured; fruits oblong-ovoid, 15-30 cm long, pendulous, woody, indehiscent; seeds numerous, reniform, embedded in soft, sour pulp.

Fl.-Fr.: May – Oct.

Not wild anywhere, abundant on Mandu plateau (Dhar), a few trees in Asirgarh (Burhanpur) and one or 2 trees in Indore.

AZANZA Alefeld W. 2; MP 2; SA 1

Azanza lampas (Cav.) Alef. in Bot. Zeitung (Berlin) 19: 297 1861; *A. zollingeri* Alef., Bot. Zeitung (Berlin) 19:298. 1861; Mast. in Hook.f., Fl. Brit. India 1: 345. 1874; T.K. Paul in B.D. Sharma *et al.,* Fl. India 3: 350.1993; Verma in Verma *et al.,* Fl. Madhya Pradesh 1: 208. 1993; Samvatsar, Fl. West. Trib. Madhya Pradesh 61. 1996. [*Thespesia lampas* (Cav.) Dalz. & Gibs., is a separate species with basionym, *Hibiscus lampas* Cav., treated as synonyms of *Azanza lampas* in most of the floras.] *Ban kapas* **(Plate 1, Fig. 7)**

Shrubs, 0.5-3 m high; leaves orbicular, 6-12 cm across, 3-5-lobed with deltoid segments, cordate at base, upper leaves smaller and narrower, angled or entire; flowers axillary, solitary or in few-flowered racemes; calyx cupular, *ca* 8 mm across, with distant subulate lobes; corolla large and showy, yellow with a dark purple centre; capsules ovoid, 2-3 cm long, dehiscing into 5 valves; seeds 8-14 per cell, black, obovoid, *ca* 4 mm long, hairy.

Fl.-Fr.: Sept. – July.

Common forest undergrowth.

BOMBAX L. W. 9; MP 2; SA 1

Bombax ceiba L., Sp. Pl. 511. 1753; *B. malabaricum* DC., Prodr. 1: 479. 1824; Mast. in Hook.f., Fl. Brit. India 1: 349. 1874; M.P. Nayar & M.C. Biswas in B.D. Sharma *et al.,* Fl. India 3: 398. 1993; Hanfi in Verma *et al.,* Fl. Madhya Pradesh 1: 212. 1993; Samvatsar, Fl. West. Trib. Madhya Pradesh 62. 1996. *Semal*

Large trees; trunk with bulbous-based, woody prickles when young; leaves digitate, 5-7-foliolate; leaflets lanceolate or elliptic, 8-18 × 5-8 cm, acuminate at apex, tapering at base; flowers bright red, solitary or in clusters near the ends of branches, appearing on leafless branches and when the new leaves appear; capsules oblong-ovate, 10-15 cm long, 5-valved; seeds numerous, pyriform, bearing creamy white, silky fibres. The spines are absorbed in the trunk due to increase in the girth as a result of secondary growth.

Fl.-Fr.: Feb. – May

Common all over the area.

CORCHORUS L. W. 77; MP 7; SA 7

Key to Species

1. Capsules globose, without a beak .... **C. capsularis**
1. Capsules elongated, with a beak
    2. Capsules ending in 3-radiating beaks
        3. Capsules 6-angled, 3-winged .... **C. aestuans**
        3. Capsules 3-angled, not winged .... **C. tridens**
    2. Capsules ending in a solitary beak

4. Capsules short
   5. Capsules 3-valved, pubescent; leaves 2-4 cm long; annuals ....**C. fascicularis**
   5. Capsules 4-valved, glabrous; leaves 0.6-2 cm long; perennials .... **C. depressus**
4. Capsules elongated
   6. Capsules cylindric, glabrous ....**C. olitorius**
   6. Capsules 3-angled, aculeate, scabrous .... **C. trilocularis**

Corchorus aestuans L., Syst. Nat. ed. 10 2: 1079. 1759; *C. acutangulus* Lam., Encycl. 2: 104. 1786; Mast. in Hook.f., Fl. Brit. India 1: 398. 1874; P. Daniel & Chandrab. in B.D. Sharma *et al.,* Fl. India 3: 485. 1993; Hanfi in Verma *et al.,* Fl. Madhya Pradesh 1: 224. 1993 Samvatsar, Fl. West. Trib. Madhya Pradesh 65. 1996.

Annual, glabrous herbs; leaves 3-9 × 1.5-5 cm, ovate or ovate-lanceolate, serrate, the lower serratures produced into filiform appendages on two sides; flower yellow, 2-3 together in leaf opposed, fascicles; capsule 2-3.5 cm long, with 6-10 serrate, broad winged angles; seeds brownish-black, angular;

Fl.-Fr.: Aug. -Oct.

Common weed in wastelands and cultivated fields, throughout the area.

Corchorus capsularis L., Sp. Pl. 529. 1753; Mast. in Hook.f., Fl. Brit. India 1: 397. 1874; P. Daniel & Chandrab. in B.D. Sharma *et al.,* Fl. India 3: 485. 1993; Hanfi in Verma *et al.,* Fl. Madhya Pradesh 1: 224. 1993; Samvatsar, Fl. West. Trib. Madhya Pradesh 65. 1996. *Jute*

Annual, glabrous herbs; leaves 5-10 × 1.5-3 cm, linear-lanceolate to ovate-lanceolate, acuminate, serrate, lower serratures produced into filiform appendages; flowers yellow, in cymes; capsule 1.25-2 cm across, muricate, 5-valved; seeds smooth, brown, wedge-shaped.

Fl.-Fr.: Aug. -Nov.

Occassional in wet places, waterlogged soil, Burhanpur, Dhar, Khandwa.

Corchorus depressus (L.) Stocks, Proc. Linn. Soc. London 1:367. 1848; *Antichorus depressus* L., Mant. Pl. 64. 1767; *Corchorus antichorus* Raeusch., Nomencl. Bot. ed. 3: 158. 1797; Mast. in Hook.f., Fl. Brit. India 1: 398. 1874; P. Daniel & Chandrab. in B.D. Sharma *et al.,* Fl. India 3: 486. 1993; Samvatsar, Fl. West. Trib. Madhya Pradesh 65. 1996; Khanna *et al.,* Suppl. Fl. Madhya Pradesh 24.2001.

Much branched perennials; leaves 0.6-2 cm × 0.4-0.7 cm, oblong, glabrous; flowers yellow, many, in leaf-opposed fascicles; capsule 1-1.5 cm long, cylindric, beaked, glabrous, 4-valved; seeds trigonous, black.

Fl.-Fr.: Aug. -Jan.

Rare, Burhanpur and Dhar.

Corchorus fascicularis Lam., Encycl. 2: 104. 1786. Mast. in Hook.f., Fl. Brit. India 1: 398. 1874. P. Daniel & Chandrab. in B.D. Sharma *et al.,* Fl. India 3: 486; 1993; Hanfi in Verma *et al.,* Fl. Madhya Pradesh 1: 225. 1993; Samvatsar, Fl. West. Trib. Madhya Pradesh 65. 1996.

Woody herbs; leaves 2-5 × 0.5-1.5 cm, elliptic-oblong or lanceolate, serrate; flowers yellow, in fascicles of usually 2-5 floweres; capsule 0.8-1.5 × 0.2 cm, pubescent; seeds wedge-shaped, smooth, dark brown.

Fl.-Fr.: Sept. -Nov.

Common as a weed of fields and in drying moist ground, throughout the area.

Corchorus olitorius L., Sp. Pl. 529. 1753. Mast. in Hook.f., Fl. Brit. India 1: 397. 1874; P. Daniel & Chandrab. in B.D. Sharma *et al.,* Fl. India 3: 487. 1993; Hanfi in Verma *et al.,* Fl. Madhya Pradesh 1: 225. 1993; Samvatsar, Fl. West. Trib. Madhya Pradesh 65. 1996.

Annual herbs; leaves 3-10 × 2-5 cm, ovate-lanceolate, serrate, lower serratures produced in to filiform appendages; flowers yellow, leaf-opposed, solitary or 2-together, on short pedicels; capsule cylindric, 3-7 cm long; beak shortly 5-dentate; valves transversely compressed between the seeds; seeds 3-gonous, black.

Fl.-Fr.: Aug-Oct.

Common weed in wastelands and cultivated fields.

Corchorus tridens L., Mant. Pl. 566. 1771; Mast. in Hook.f., Fl. Brit. India 1: 398. 1874; P. Daniel & Chandrab. in B.D. Sharma *et al.,* Fl. India 3: 488. 1993; Hanfi in Verma *et al.,* Fl. Madhya Pradesh 1: 225. 1993; Samvatsar, Fl. West. Trib. Madhya Pradesh 65. 1996.

Erect or suberect, annual herbs, with terete or angular branches; leaves 7-12 × 2-3.5 cm, ovate-lanceolate, serrate, basal serratures produced in to filiform appendages; flowers yellow in leaf opposed 1-4 flowered, corymbose, more or less sessile cymes; capsule 2-3 cm long 3-valved; seeds truncate at both ends, brownish-black.

Fl.-Fr.: Aug. -Dec.

Occasional in fields and as forest undergrowth, Alirajpur, Dhar, Indore.

Corchorus trilocularis L., Mant. Pl. 1: 77. 1767; Mast. in Hook.f., Fl. Brit. India 1: 397. 1874; Hanfi in Verma *et al.,* Fl. Madhya Pradesh 1: 226. 1993; P. Daniel & Chandrab. in B.D. Sharma *et al.,* Fl. India 3: 488. 1993; Samvatsar, Fl. West. Trib. Madhya Pradesh 66 1996.

Erect, annual herbs; leaves 3-10 × 1.5-4 cm, elliptic-oblong, serrate, with basal serratures produced into filiform appendages; flowers yellow, 1-3 together in leaf-opposed fascicles; capsules up to 8 cm long, acute or shortly beaked, scabrous or aculeate; seeds lead black, smooth, trigonous, truncate.

Fl.-Fr.: July-Nov.

Frequent, weed in cultivated fields and wastelands, throughout the area.

ERIOLAENA A. DC. W. 10; MP 2; SA 2

Key to Species

1. Capsule 10-valved; petiole 5 cm long .... **E. hookeriana**
1. Capsule 5-6 valved; petiole 1.5-3 cm long .... **E. candollei**

Eriolaena candollei Wall., Pl. Asiat. Rar. 1: 57.1830; Mast. in Hook.f., Fl. Brit. India 1: 370. 1874; Malick in B.D. Sharma *et al.,* Fl. India 3: 415. 1993; Verma *et al.,* Fl. Madhya Pradesh 1: 215. 1993.

Large shrubs; leaves broadly ovate, 12-17 × 7-12 cm, acuminate at apex, cordate at base; flowers in pedunculate, corymbose cymes; petals yellow, oblong, notched at apex; capsules ovoid, 4-5 cm long, beaked, 10-lobed; seeds many.

Fl.-Fr.: March – Oct.

Barwani, Indore, Khargone.

Eriolaena hookeriana Wight & Arn., Prodr. Fl. Ind. Orient. 70.1834; Mast. in Hook.f., Fl. Brit. India 1: 370.1874; Malick in B.D. Sharma *et al.,* Fl. India 3: 415. 1993; Samvatsar, Fl. West. Trib. Madhya Pradesh 63. 1996. *Bhont*

Shrubs or small, 2-3 m high trees; leaves broadly ovate, 5-14 × 4-12 cm, acuminate at apex, cordate at base; flowers in axillary, pedunculate cymes; involucral bracts multisect; sepals linear-lanceolate, 1-2 cm long; petals golden yellow, obovate, 3-4 cm long; capsules ovoid, 8-10-valved, tuberculate.

Fl.-Fr.: June – Feb.

Alirajpur, Burhanpur, Dhar, as forest undergrowth.

FIRMIANA Marsili W. 16; MP 1; SA 1

Firmiana colorata (Roxb.) R. Br., Pterocymbium 235. 1844; *Sterculia colorata* Roxb., Pl. Coromandel 1:26. 1795; Mast. in Hook.f. Fl. Brit. India 1:359. 1874; Hanfi in Verma *et al.,* Fl. Madhya Pradesh 1:216. 1993; Samvatsar, Fl. West. Trib. Madhya Pradesh 64. 1996. *Kuwarin*

Deciduous trees of medium height; leaves crowded at the ends of branches, 15-20 cm long, palmately 3-5 lobed, deeply cordate at base, tomentose below; stipules lanceolate; flowers in terminal panicles, unisexual, up to 2 cm long, clothed with red, velvety tomentum; follicles 2.5-3.5 cm long, leaf-like with 2 seeds attached near the base, on the margin.

Fl.-Fr.: March-June

Rare, Alirajpur and Indore.

GREWIA L. W. 321; MP 16; SA 10

Key to Species

1. Flowers polygamous
    2. Stamens 20 or less .... **G. helicterifolia**
    2. Stamen 40 or more .... **G. hirsuta**
1. Flowers bisexual
    3. Petals bifid or notched at the apex; gland of petal up to ½ its length
        4. Stipules foliaceous .... **G. tiliifolia**
        4. Stipules not foliaceous

5. Leaves glabrous or nearly so .... **G. tenax**
5. Leaves stellately hairy or tomentose, atleast on lower surface
6. Leaves ovate or broadly suborbicular .... **G. asiatica**
6. Leaves not as above
7. Leaf margin subentire to minutely serrate .... **G. damine**
7. Leaf margin serrate to crenate-serrate
8. Leaves 3-nerved .... **G. flavescens**
8. Leaves 5-6 nerved .... **G. eriocarpa**
3. Petals entire or slightly emarginate at apex; leaves hoary tomentose beneath
9. Petals about ¼ th length of the sepals .... **G. abtulifolia**
9. Petals about ½ the length of sepals .... **G. rothii**

Grewia abutilifolia Vent. ex Juss. in Ann. Mus. Natl. Hist. Nat. 4: 92. 1804; *G. aspera* Roxb. Fl. Ind. ed. 1832 2:591. 1832; Mast. in Hook.f., Fl. Brit. India 1: 390.1874; P. Daniel & Chandrab. in B.D. Sharma *et al.,* Fl. India 3: 493. 1993; Hanfi in Verma *et al.,* Fl. Madhya Pradesh 1: 225. 1993; Samvatsar, Fl. West. Trib. Madhya Pradesh 65. 1996. *Kali gangeru*

Shrubs; young parts stellate-pubescent; leaves broadly ovate, 4-12 × 4-9 cm; flowers in axillary, umbellate cymes; peduncles 1-3 together; pedicels *ca* 5 mm long; sepals oblong, *ca* 8 mm long; petals yellow, oblong, *ca* 3 mm long; drupes subglobose, 1-1.2 cm across, obscurely 4-lobed.

Fl.-Fr.: Oct. -May

Not common, Alirajpur, Barwani, Indore, Khandwa, Khargone.

Grewia asiatica L. Mant. Pl. 1:122. 1767; *G. subinaequalis* DC., Prodr. 1:511. 1824; Mast. in Hook.f., Fl. Brit. India 1: 386. 1874; P. Daniel & Chandrab., in B.D. Sharma *et al.,* Fl. India 3:494 1993; Hanfi in Verma *et al.,* Fl. Madhya Pradesh 1: 232. 1993; Samvatsar, Fl. West. Trib. Madhya Pradesh 67. 1996.

Shrubs or small trees, 3-6 m high; leaves ovate to broadly ovate or suborbicular, 5-17 × 5-11.5 cm; flowers in axillary, pedunculate, umbellate cymes; sepals oblong or oblanceolate, up to 1 cm long, pubescent; petals yellow or reddish- yellow, oblong, *ca* 6 mm long; drupes dark purple, subglobose, 8-12 mm across.

Fl.-Fr.: Feb. – Aug.

Cultivated and also as an escape, in Alirajpur and Dhar.

Grewia damine Gaertn., Fruct. Sem. Pl. 2: 113. 1790; *G. salvifolia* Heyne ex Roth, Nov. Pl. Sp. 239. 1821; Mast. in Hook.f., Fl. Brit. India 1: 386. 1874; P. Daniel & Chandrab. in B.D. Sharma *et al.,* Fl. India 3: 496. 1993; Hanfi in Verma *et al.,* Fl. Madhya Pradesh 1: 228. 1993; Samvatsar, Fl. West. Trib. Madhya Pradesh 66. 1996.

Scandent shrubs to small trees; leaves 4-7 × 1-2.5 cm, ovate- oblong or lanceolate, rounded and oblique at the base, 3-nerved, serrate or subentire, pubescent on lower surfaces; flowers yellow, in small umbels, gland of petal hairy; drupes sparsely pubescent, reddish-yellow when ripe.

Fl.-Fr.: June-Aug.

Occasional at fringe of forest, Alirajpur, Barwani, Dhar, Indore, Khandwa.

Grewia eriocarpa A. Juss. in Ann. Mus. Natl. Hist. Nat. 4: 93. 1804; *G. asiatica* L. var. *vestita* (Wall. ex Brandis) Mast. in Hook.f., Fl. Brit. India 1: 387. 1874; P. Daniel & Chandrab. in B.D. Sharma *et al.,* Fl. India 3: 497. 1993; Hanfi in Verma *et al.,* Fl. Madhya Pradesh 1: 228. 1993; Samvatsar, Fl. West. Trib. Madhya Pradesh 66. 1996.

Small trees. leaves ovate or ovate-lanceolate, 6-12 × 3.5-5 cm. flowers in densely crowded, axillary cymes. sepals linear-oblong, up to 1 cm long; petals yellow, oblong, *ca* 3 mm long. drupes globose, 5-8 mm across, obscurely 2-4-lobed.

Fl.-Fr.: Feb. – May

Dhar only, Mandu plateau.

Grewia flavescens Juss., Ann. Mus. Natl. Hist. Nat. 4: 91. 1804; *G. pilosa* Lam. ex Poir., Encycl. Suppl. 3:43. 1813; Mast. in Hook.f., Fl. Brit. India 1: 388. 1874; P. Daniel & Chandrab. in B.D. Sharma *et al.,* Fl. India 3: 498. 1993; Hanfi in Verma *et al.,* Fl. Madhya Pradesh 1: 228. 1993.

Shrubs; leaves 3-nerved, 7-9 × 4-5 cm, ovate, elliptic or obvate, serrate, stellately pubescent on both surfaces; stipules linear; flowers yellow in axillary or extra axillary 2-3 flowered cymes, buds oblong, constricted in the middle; petals notched; drupe 15 × 6 mm two lobed, hirsute, stones 4.

Fl.-Fr.: March-June

Alirajpur, Burhanpur, Dhar.

Grewia helicterifolia Wall. ex G. Don, Gen. Hist. 1: 548. 1831; Mast. in Hook.f., Fl. Brit. India 1: 391. 1874; *G. hirsuta* Vahl var. *helicterifolia* (Wall. ex G. Don) Haines, Forest Fl. Chota Nagpur 196. 1910; P. Daniel & Chandrab. in B.D. Sharma *et al.,* Fl. India 3: 500.1993; Hanfi in Verma *et al.,* Fl. Madhya Pradesh 1: 230.1993. *Bono sulai*

Undershrubs or shrubs; young branches grey tomentose; leaves lanceolate, linear-lanceolate or lanceolate-elliptic, 6-11 × 1.5-3.5 cm, acute or acuminate at apex, usually rounded at base, sometimes oblique, serrulate, hispid above, densely pubescent beneath, 3-nerved from base; petioles 3-5 mm long; stipules subulate; flowers white, polygamous, in axillary, umbellate cymes; drupes orange-red, reddish- brown when ripe, obscurely 4-lobed.

Fl.-Fr.: Sept. – Feb.

Indore only, Hatod.

Grewia hirsuta Vahl, Symb. Bot. 1: 84. 1790; Mast. in Hook.f., Fl. Brit. India 1: 391. 1874; P. Daniel & Chandrab. in B.D. Sharma *et al.,* Fl. India 13: 501. 1993; Hanfi in Verma *et al.,* Fl. Madhya Pradesh 1: 230.1993; Samvatsar, Fl. West. Trib. Madhya Pradesh 66.1996. *Gursukri* **(Plate 1, Fig. 8)**

Tomentose shrubs; leaves 4-15 × 1-3.5 cm, obliquely elliptic- ovate, acute, serrate, upper surface glabrescent or with few stellate hairs, lower surface densely villous; stipules hairy, subulate, longer than the petioles; flowers white, polygamous, in axillary umbellate cymes, gland of petal with stiff hairs on upper margins, ciliate on lateral margins; drupe fleshy, globose, pilose.

Fl.-Fr.: June-Dec.

Occasional, along forest edges, all districts.

Grewia rothii DC., Prodr. 1: 509. 1824; Mast. in Hook.f., Fl. Brit. India 1: 385. 1874; P. Daniel & Chandrab. in B.D. Sharma *et al.,* Fl. India 3: 507. 1993; Hanfi in Verma *et al.,* Fl. Madhya Pradesh 1: 231.1993; Samvatsar, Fl. West. Trib. Madhya Pradesh 66. 1996. *Bhansuli* [*Grewia excelsa* Vahl is synonym of *G. arborea* (Forssk.) Lam.] **(Plate 2, Fig. 1)**

Shrubs or small trees; leaves ovate-oblong or lanceolate, 6-9 × 1.5-2.5 cm, acute or acuminate at apex, rounded or unequally subcordate at base, serrulate, densely tomentose beneath; flowers in axillary, clustered cymes; sepals elliptic-oblong, *ca* 5 mm long, tomentose; petals greenish, oblong or ovate-lanceolate, *ca* 3 mm long; drupes globose, white tomentose.

Fl.-Fr.: May – Oct.

On hilly slopes, Barwani, Dhar, Khargone.

Grewia tenax (Forssk.) Fiori in Agric. Colon. 5 (Suppl.): 23. 1912; *Chadara tenax* Forssk., Fl. Aegypt. -Arab. 105. 1775; *Grewia populifolia* Vahl, Symb. Bot. 1: 33. 1790; Mast. in Hook.f., Fl. Brit. India 1: 385. 1874; P. Daniel & Chandrab. in B.D. Sharma *et al.,* Fl. India 3: 511. 1993; Hanfi in Verma *et al.,* Fl. Madhya Pradesh 1: 232. 1993. *Gangeran*

Shrubs, *ca* 2 m high; leaves ovate, broadly elliptic or suborbicular, 1-4 × 1-3 cm; flowers solitary, leaf-opposed; pedicels *ca* 6 mm long; sepals linear-oblong, *ca* 1.5 cm long, tomentose outside; petals white, linear-oblong, *ca* 6 mm long; drupes orange-red, 2-partite, each half didymous, glabrescent.

Fl.-Fr.: May – Jan.

Confined to **Burhanpur** Dhulkot, Hansapura.

Grewia tiliifolia Vahl, Symb. Bot. 1: 35. 1790; Mast. in Hook.f., Fl. Brit. India 1: 386. 1874, '*tiliaefolia*'; P. Daniel & Chandrab. in B.D. Sharma *et al.,* Fl. India 3: 511. 1993; Hanfi in Verma *et al.,* Fl. Madhya Pradesh 1: 232. 1993; Samvatsar, Fl. West. Trib. Madhya Pradesh 67. 1996. *Dhaman*

A small tree with pale brown bark; leaves 8-12 × 0.5-0.8 cm, ovate, acuminate, crenate dentate, often lobate at the apex, hairy or subglabrous above and tomentose beneath; cordate and oblique at base; stipules sagittate; flowers yellowish-brown, small, umbellate; petals about half as long as the sepals, gland about one third the length of the petal; drupe about 7 mm across, deeply bilobed.

Fl.-Fr.: May-Sept.

Common, in all districts.

HELICTERES L. W. 56; MP 1; SA 1

Helicteres isora L., Sp. Pl. 963. 1753; Mast. in Hook.f., Fl. Brit. India 1: 365. 1874, *incl.* vars; Malick in B.D. Sharma *et al.,* Fl. India 3: 426. 1993; Hanfi in Verma *et al.,* Fl. Madhya Pradesh 1:216. 1993; Samvatsar, Fl. West. Trib. Madhya Pradesh 63. 1996. *Maror phali* **(Plate 2, Fig. 2)**

Small trees, 3-6 m high; leaves variable, elliptic-ovate, obovate or suborbicular, 8-16 × 5-10 cm; flowers axillary, solitary or fascicled; calyx narrowly campanulate, *ca* 1.5 cm long, gibbous; petals red, reflexed, clawed at base, 2 lower shorter; follicle cylindrical, 5-8 cm long, spirally twisted; seeds many, wrinkled.

Fl.-Fr.: Aug. – Feb.

Common forest undergrowth.

HIBISCUS L. W. 241; MP 12; SA 6

Key to Species

1. Epicalyx absent; flowers white .... **H. lobatus**
1. Epicalyx present
    2. Flowers small, white or pink; seeds cottony .... **H. micranthus**
    2. Flowers large, yellow, with a purple centre; seeds not cottony
        3. Lobes of calyx spathulate
            4. Epicalyx without appendages; capsule shorter than calyx .... **H. panduriformis**
            4. Epicalyx with appendages arising from the spathulate tip; capsule equal to calyx in length .... **H. beddomei**
        3. Lobes of calyx not spathulate
            5. Epicalyx shorter than calyx; capsule winged .... **H. vitifolius**
            5. Epicalyx longer than calyx, spreading like rays; capsule not winged .... **H. caesius**

Hibiscus beddomei Rakshit & Kundu, Sci. & Cult. 27. 193. 1961; Paul in Sharma *et al.,* Fl. India 3: 324. 1993; Samvatsar, Fl. West. Trib. Madhya Pradesh 57. 1996; Khanna *et al.,* Suppl. Fl. Madhya Pradesh 18. 2001.

Erect herbs, up to 2 m high; branches covered with dense rigid, stellate hairs; leaves ovate, 3-angled or slightly 3-lobed, 5-7 cm across, cordate at base, acute at apex, serrate; flowers yellow, axillary, solitary or in terminal racemes; epicalyx with linear appendages arising from the base of the spathulate tip; capsules ovoid, beaked, densely hairy.

Fl.-Fr.: Sept. – Dec.

Very rare, **Alirajpur**, Phulmal.

Hibiscus caesius Garcke, Bot. Zeitung (Berlin) 7: 850.1849; *H. pentaphyllus* F. Muell. Frag. 2:13. 1860; *H. gibsonii* Stocks ex Harv., Fl. Cap. 2: 587. 1862; Mast. in Hook.f., Fl. Brit. India 1: 339. 1874; T.K. Paul in B.D. Sharma *et al.,* Fl. India 3: 334. 1993; Samvatsar, Fl. West. Trib. Madhya Pradesh 58. 1996; Khanna *et al.,* Suppl. Fl. Madhya Pradesh 18. 2001. **(Plate 2, Fig. 3)**

Undershrubs, 1-2 m high; branches bristly with minute prickles; leaves deeply palmately divided, 3-5-lobed, lobes oblong-lanceolate, 2-7 × 1-3 cm; flowers yellow, axillary, solitary; pedicels jointed towards apex; epicalyx-segments up to 10, linear, hairy, persistent; capsules ovoid, enclosed within calyx, prickly hairy outside, glabrous inside; seeds reniform, *ca* 3 mm long, minutely pilose.

Fl.-Fr.: Sept. – Dec.

Not common, along village roads, hill slopes, Barwani, Burhanpur, Dhar, Indore, Khargone.

Hibiscus lobatus (Murray) Kuntze, Revis. Gen. Pl. 3(2):19. 1898; *Solandra lobata* Murray, Comment. Soc. Regiae Sci. Gott. 6:20.1785; *Hibiscus solandra* L'Hér., Stirp. Nov. 1: 103. 1789; Mast. in Hook.f., Fl. Brit. India 1: 336. 1874; T.K. Paul in B.D. Sharma *et al.,* Fl. India 3: 336. 1993; Verma in Verma *et al.,* Fl. Madhya Pradesh 1: 197. 1993; Samvatsar, Fl. West. Trib. Madhya Pradesh 58. 1996.

Erect herbs; leaves 2-12 × 1-8 cm, entire to 3-5 lobed orbicular ovate, obtuse or acute, serrate; flowers white, usually solitary or racemed by reduction of leaves; calyx deeply 5-lobed, hairy; capsule ovoid, beaked, pilose at the top; seeds black, tubercled.

Fl.-Fr.: Sept. -Nov.

Common as forest undergrowth, along water streams, throughout the area.

Hibiscus micranthus L.f., Suppl. Pl. 308.1782; *Hibiscus ovalifolius* (Forssk.) Vahl, Symb. 1: 50. 1790; *Urena ovalifolia* Forssk., Fl. Aegypt. -Arab. 124.1775; Mast. in Hook.f., Fl. Brit. India 1: 335.1874; T.K. Paul in B.D. Sharma *et al.,* Fl. India 3: 330.1993; Verma in Verma *et al.,* Fl. Madhya Pradesh 1:197.1993; Samvatsar, Fl. West. Trib. Madhya Pradesh, 58, 1996.

Small erect shrubs, scabrid with scattered, stellate bristles; leaves simple 1.5-3.5 cm, ovate or oblong, acute or obtuse, serrate; flowers whitish-pink, axillary, solitary and in terminal racemes; capsule globose, pubescent, 5-valved; seeds reniform black, with white silky hairs.

Fl.-Fr.: Aug. -Nov.

Common, along road sides in interior parts of the forest, in all districts.

Hibiscus panduriformis Burm.f., Fl. Indica 151. 1768; *Abelmoschus panduriformis* Hassk., Tijdschr,Ned. -Indie 1:134. 1843; Mast. in Hook.f., Fl. Brit. India 1: 338. 1874; T.K. Paul in B.D. Sharma *et al.,* Fl. India 3: 339. 1993; Verma in Verma in Verma *et al.,* Fl. Madhya Pradesh 1: 198. 1993; Samvatsar, Fl. West. Trib. Madhya Pradesh 58. 1996.

Densely stellate or simple tomentose, erect under shrubs; leaves 3.5-8 × 2-7, cordate, lobed, acute, irregularly toothed, silky hairy on both surfaces; flowers yellow, with purple centre, solitary, axillary, or racemed by abortion of leaves; stipules filiform, 3-fid; epicalyx of 10 segments; sepals hispid, 3-nerved; capsule ovoid, very hairy; seeds reniform, densely hairy.

Fl.-Fr.: Oct. - Dec.

Common on hedges of fields, bunds etc., throughout the area.

Hibiscus vitifolius L., Sp. Pl. 696. 1753; *Abelmoschus vitifolius* (L.) Hassk., Tijdschr. Natuurl. Gesch. Physiol. 10:134. 1843; Mast. in Hook.f., Fl. Brit. Ind. 1: 338. 1874; *Fioria vitifolia* (L.) Mattei, Boll. Reale Orto Bot. Palermo, n.s., 2: 72. 1917; Borss. in Blumea 14: 82. 1966; T.K. Paul in B.D. Sharma *et al.,* Fl. India 3:310 1993; Verma in Verma *et al.,* Fl. Madhya Pradesh 1: 199. 1993; Samvatsar, Fl. West. Trib. Madhya Pradesh 58. 1996. *Ban kapas*

Annual slender climbing or trailing herbs; leaves ovate, cordate, 3-10 cm long, palmately 3-7 lobed; flowers axillary, solitary or clustered, drooping, 5-6 cm across, yellow with a purple centre; capsules suborbicular, apiculate, *ca* 15 mm across, 5-winged, wings reticulately nerved; seeds reniform, minutely tubercled.

Fl.-Fr.: Oct. – March

Hill slopes, Alirajpur, Barwani, Dhar, Incfdore, Khargone.

KAVALAMA Raf. Monotypic

Kavalama urens (Roxb.) Raf., Sylva Tellur. 72. 1838; *Sterculia urens* Roxb. Pl. Coromandel 1:25. 1795; Mast. Hook.f. Fl. Brit. India 1:355. 1874; Verma in Verma *et al.,* Fl. Madhya Pradesh 1:222. 1993; Samvatsar in Fl. West. Trib. Madhya Pradesh 64. 1996. *Kadhai* **(Plate 2, Fig. 4)**

Large deciduous trees; bark off-white, peeling in papery fragments; leaves crowded at the ends of branches, palmately 5-lobed, 20-35 × 20-45 cm, on long, thick petioles, base cordate with 7 palmate nerves; flowers greenish-yellow, polygamous, in viscidly hairy terminal panicles; follicles 4-6, 3-6 × 2-3 cm, ovoid-oblong, bristly, pink when young; seeds black, minutely wrinkled.

Fl.-Fr.: Dec. -March

Very common in all districts, especially on hill slopes.

KYDIA Roxb. W. 2; MP 1; SA 1

Kydia calycina Roxb. Pl. Coromandel 3:11. 1811; Mast. in Hook.f. Fl. Brit. India 1:348. 1874; Verma in Verma *et al.,* Fl. Madhya Pradesh 1:200.1993; Samvatsar in Fl. West. Trib. Madhya Pradesh 58. 1996. *Baranga*

Medium sized trees with soft wood; leaves 5-15 × 7-17 cm, suborbicular, sinuate, angled with large gland on the nerves beneath,base 7-nerved; flowers white, in axillary and terminal, tomentose panicles; epicalyx segments 4-6, accrescent; petals clawed, emarginate, fimbriate along margins; capsules globose, tomentose with a slightly umbonate top; seeds reniform, black.

Fl.-Fr.: Aug. -Dec.

Common forest species encountered in all districts.

MALACHRA L. W. 8; MP 1; SA 1

Malachra capitata (L.) L. Syst. Nat. ed. 12 2:458. 1767; *Sida capitata* L. Sp. Pl. 685. 1753; Mast. in Hook.f., Fl. Brit. India 1:329.1874; Borss. in Blumea 14:146. 1966; Paul, in Sharma *et al.,* Fl. India 3:367. f. 104. 1993; Verma in Verma *et al.,* Fl. Madhya Pradesh 1:202 1993.

Annual herbs or undershrubs, about 2 m tall; stem covered with simple and stellate prickly hairs; leaves 5-10 cm across, ovate-orbicular, margin serrate, angular to shallowly lobed, base cordate; heads axillary, of 3-7 yellow flowers, encircled by 3-4 leafy bracts; stipules filiform; schizocarps 5-6 mm across; mericarps 5, each *ca* 3 × 2 mm, trigonous; seeds trigonous, 2-2.5 mm long.

Fl.-Fr.: Aug. -Dec.

Burhanpur, Indore, Khandwa.

MALVASTRUM A. Gray W. 27; MP 1; SA 1

Malvastrum coromandelianum (L.) Garcke in Bonplandia 5:298. 1857; *Malva coromandelianum* L. Sp. Pl. 687. 1753; *Malvastrum tricuspidatum* (R. Br.) A. Gray, Pl. Wright 1:16. 1852; Mast. in Hook.f. Fl. Brit. India 1: 321. 1874; Borss., in Blumea 14:152. 1966; Paul in Sharma *et al.,* Fl. India 3:277. f. 75. 1993; Verma in Verma *et al.,* Fl. Madhya Pradesh 1:203. 1993; Samvatsar in Fl. West. Trib. Madhya Pradesh 59. 1996.

Annual hairy herbs or undershrubs (becoming perennial when moisture available); leaves 4-7 cm long, ovate-oblong, margin serrate or dentate; epicalyx of 3, free bracteoles; flowers yellow,1.5 cm across, solitary, axillary; schizocarps globular, shortly awned at the tip and with 2 dorsal awns in the middle; seeds reniform.

Fl.-Fr.: Oct. -Jan.

Common in all districts, along irrigation canals of fields, waste places, old walls.

MELHANIA Forssk. W. 44; MP 2; SA 1

Melhania futteyporensis Munro ex Mast. in Hook.f. Fl. Brit. India 1:373. 1874; Hanfi in Verma *et al.,* Fl. Madhya Pradesh 1:219. 1993.

An undershrub, up to 70 cm tall; leaves 5-6 nerved at base, lanceolate to ovate, normally unlobed, occasionally 3-lobed, 2-9 × 1-6 cm, velvety on both surfaces, crenate-serrate; stipules filiform; flowers in 2-4 flowered cymes, 1.5-2 cm across; pedicels elongating in fruit; bracts ovate to cuspidate; capsules 6-15 × 3-8 mm, obtuse, villous; seeds black, angular, muricate.

Fl.-Fr.: Aug. -Dec.

Rare, **Indore** only, Patal pani and Tinchha falls

MELOCHIA L. W. 57; MP 1; SA 1

Melochia corchorifolia L., Sp. Pl. 675. 1753; Mast. in Hook.f. Fl. Brit. India 1:1874; Malick in Sharma *et al.,* Fl. India 3:441. 1993; Hanfi in Verma *et al.,* Fl. Madhya Pradesh 1:219 1993; Samvatsar in Fl. West. Trib. Madhya Pradesh 63. 1996.

Herbs up to 1 m tall; branches stellately tomentose; leaves 5 × 3 cm, ovate, elliptic-lanceolate, margin serrate, base 5-nerved; flowers pink,1 cm across in densely crowded terminal clusters; bracts and bracteoles hispid along margins; petals clawed, persistent; filaments of stamens connate; capsules depressed-globose, 5 mm across; seeds angled, mottled black and brown.

Fl.-Fr.: May-Sept.

Alirajpur, Barwani, Indore, Khandwa and Khargone, in waste land, along banks of canals.

PAVONIA Cav. W. 250; MP 2; SA 2

Key to Species

1. Leaves 2-3 × 3-3.5 cm, deeply 3-lobed, petiole 3-4 cm long; epicalyx segments 8-10, 7 mm long. mericarps 4 × 3 mm .... **P. zeylanica**

1. Leaves 5×5 cm, orbicular, 3-5 angled, petiole 1-3 cm long. epicalyx segments 10-12, 7-8 mm long. mericarps 3 × 2 mm .... **P. odorata**

Pavonia odorata Willd., Sp. Pl. 3: 837. 1822; Mast. in Hook.f., Fl. Brit. India 1: 331.1874; T.K. Paul in B.D. Sharma *et al.,* Fl. India 3: 373.1993; Verma in Verma *et al.,* Fl. Madhya Pradesh 1: 204.1993; Samvatsar, Fl. West. Trib. Madhya Pradesh 58. 1996. *Sugandha bala*

Annual herbs, up to 40 cm high; stems, petioles and pedicels covered with simple glandular hairs; leaves broadly ovate or suborbicular, 2-10 × 1.5-4 cm, obscurely 3-5-lobed; flowers axillary, solitary; epicalyx-lobes 10-12, free, linear; corolla pink with deep striations; mericarps obovate, wingless, glabrous; seeds brownish black, reniform, *ca* 2 mm long, minutely papillose.

Fl.-Fr.: Aug. – Feb.

Alirajpur only.

Pavonia zeylanica (L.) Cav., Diss. 3: 134. 1787; *Hibiscus zeylanicus* L., Sp. Pl. 697. 1753; Mast. in Hook.f., Fl. Brit. India 1: 331. 1874; T.K. Paul in B.D. Sharma *et al.,* Fl. India 3: 377. 1993; Verma in Verma *et al.,* Fl. Madhya Pradesh 1: 205. 1993.

Herbs or undershrubs; stems, petioles and pedicels pubescent with stellate and glandular simple hairs; leaves orbicular or obovoid-rounded, 1-4 × 0.8-3.5 cm, deeply 3-5-lobed, mid lobe longest, acute or rounded at apex, cordate at base; flowers axillary, solitary; epicalyx-segments 8-12, free, linear, 8-12 mm long; corolla rotate, light pinkish; mericarps oblong, *ca* 4 mm long, trigonous, narrowly winged, glabrous; seeds brown, reniform, minutely pubescent.

Fl.-Fr.: Aug.– Dec.

Barwani, Burhanpur, Khandwa.

SIDA L. W. 156; MP 7; SA 7

Key to Species

1. Petiole with small, spine-like tubercles; styles and mericarps 5; leaf base cordate or subcordate .... **S. spinosa**

1. Petiole without spiny tubercles; styles and mericarps 7-14

   2. Procumbent herbs or subshrubs; stem with both stellate and simple pilose hairs; mericarps awnless .... **S. cordata**

   2. Erect herbs or subshrubs

      3. Calyx with long, simple hairs; plants sticky with simple and glandular hairs; mericarps smooth, not dehiscent .... **S. mysorensis**

      3. Calyx stellately pubescent or glabrous; plants not sticky; mericarps sometimes dehiscent, smooth or grooved , vertically or transversely, reticulately wrinkled

         4. Awns of mericarps *ca* 2 mm long, conspicuous, retrorsely barbed; tube of filaments with long hairs .... **S. cordifolia**

         4. Awns of mericarps absent or less than 2 mm long

            5. Flowers white; leaves downy on both surfaces .... **S. ovata**

            5. Flowers yellow

               6. Leaves distichous; stipules of two types, one filiform, 1-veined and the other linear- lanceolate, two or more veined; calyx and mericarps glabrous .... **S. acuta**

               6. Leaves spiral; stipules uniform; calyx hairy; mericarps awned .... **S. rhombifolia**

Sida acuta Burm.f., Fl. Ind. 147. 1768; T.K. Paul in B.D. Sharma *et al.* Fl. India 3: 281.1993; Verma in Verma *et al.,* Fl. Madhya Pradesh 1 : 206.1993; Samvatsar, Fl. West. Trib. Madhya Pradesh 60. 1996. *Maha bala*

Herbs or undershrubs; stems pubescent with minute stellate hairs; leaves lanceolate to linear or ovate-oblong, 2-6 × 2-3 cm, subacute at apex, rounded at base; flowers axillary, solitary; corolla yellow, lobes obliquely obovate, 7-8 mm long; mericarps 6-10, 2-awned, reticulately rugose; seeds dark brown, *ca* 2 mm long, trigonous.

Fl.-Fr.: Aug. – Nov.

In wastelands and along roadsides, throughout the area.

Sida cordata (Burm.f.) Borss. Waalk., Blumea 14: 182. 1966; *Melochia cordata* Burm.f., Fl. Ind. 143. 1768; *Sida veronicifolia* Lam., Encycl. 1:5. 1783; T.K. Paul in B.D. Sharma *et al.,* Fl. India 3: 283. 1993; Verma in Verma *et al.,* Fl. Madhya Pradesh 1: 206. 1993; Samvatsar, Fl. West. Trib. Madhya Pradesh 60.1996.

Perennial, trailing, 20-70 cm tall herbs; leaves ovate to orbicular, 1-6 × 1-5 cm, acuminate at apex, cordate at base, crenate-serrate, hairy on both the surfaces; flowers axillary, solitary or in pairs; corolla yellow; lobes obliquely obovate, 6-7 mm long; schizocarps globose, 3.5-4 mm across; mericarps 5, somewhat tetrahedral, glabrous, smooth, 1-seeded; seeds black, trigonous.

Fl.-Fr.: Aug. -Dec.

Common in wastelands and along roadsides, throughout the area.

Sida cordifolia L., Sp. Pl. 684. 1753; Mast. in Hook.f., Fl. Brit. India 1: 324. 187; T.K. Paul in B.D. Sharma *et al.,* Fl. India 3: 285. 1993; Verma in Verma *et al.,* Fl. Madhya Pradesh 1: 207. 1993; Samvatsar, Fl. West. Trib. Madhya Pradesh 60.1996. *Bala*

Perennial herbs or undershrubs; stems, petioles and pedicels densely pubescent mixed with simple and stellate hairs. leaves ovate, 3.5-6 × 2-4 cm, obtuse at apex, shallowly cordate at base. flowers axillary, solitary or in fascicles; corolla yellow; lobes obliquely obovate, 7-8 mm long. mericarps 10, *ca* 3 mm long, 3-sided, retrorsely hairy; seeds *ca* 2 mm long, trigonous, pubescent around hilum.

Fl. & Fr. : Sept. – Nov.

Very common in wastelands and along roadsides, throughout the area.

Sida mysorensis Wight & Arn., Prodr. 59. 1834; Mast. in Hook.f., Fl. Brit. Ind. 1: 322. 1874. Verma in Verma *et al.,* Fl. Madhya Pradesh 1: 207. 1993; Samvatsar, Fl. West. Trib. Madhya Pradesh 60.1996.

Annual herbs or undershrubs; stems, petioles and pedicels covered by minute stellate hairs; leaves ovate-oblong, or sometimes orbicular, cordate, acuminate, 2-9 cm long; flowers axillary, solitary, or racemed or panicled; corolla yellow, 10-12 mm long; mericarps 5, tetrahedral, *ca* 2.5 mm long, hairy at the top, awnless; seeds ovoid, *ca* 2 mm long.

Fl.-Fr.: Oct. -Jan.

In wastelands, Alirajpur, Burhanpur, Dhar, Indore, Khandwa.

Sida ovata Forssk., Fl. Aegypt. -Arab. 124. 1775; *S. grewioides* Guill. & Perr., Fl. Seneg. Tent. 1: 71.1831; Mast. in Hook.f., Fl. Brit. India 1: 323. 1874; T.K. Paul in B.D. Sharma *et al.,* Fl. India 3: 288. 1993; Verma in Verma *et al.,* Fl. Madhya Pradesh 1 : 207. 1993; Samvatsar, Fl. West. Trib. Madhya Pradesh 60.1996.

Erect undershrubs, up to 30 cm high; leaves ovate-oblong, 1.5-5 × 1-4 cm, obtuse at apex, rounded at base, crenate, densely stellate-velutinous on both the surfaces; flowers axillary, solitary or sometimes paired; corolla yellow or yellowish white, lobes obliquely truncate, *ca* 1 × 0.5 cm; mericarps more or less indehiscent, *ca* 5 mm across; mericarps 8, 3-4 mm across; seeds blackish, round-reniform, *ca* 2 mm across.

Fl.-Fr.: July – Feb.

Common in wastelands in throughout the area.

Sida rhombifolia L., Sp. Pl. 684.1753; Mast. in Hook.f., Fl. Brit. India 1: 323.1874; T.K. Paul in B.D. Sharma *et al.,* Fl. India 3: 289.1993; Verma in Verma *et al.,* Fl. Madhya Pradesh 1: 207.1993; Samvatsar, Fl. West. Trib. Madhya Pradesh 60.1996.

Key to Subspecies

1. Calyx stellate pubescent; pedicels jointed above the middle; tube of filaments glabrous; mericarps 7-10 .... subsp. **rhombifolia**
1. Calyx stellate tomentose; pedicels jointed below the middle; tube of filaments hairy; mericarps 6-8 .... subsp. **alnifolia**

subsp. **alnifolia** L., Sp. Pl. 684. 1753; *S. alnifolia* L., Sp. Pl. 684. 1753; *S. rhombifolia* L. subsp. *retusa* (L.) Borss. in Blumea 14: 198. 1966; T.K. Paul in B.D. Sharma *et al.,* in Fl. India 3 : 289. 1993.

Erect herbs or shrubs, up to 1 m tall; stem stellate tomentose; leaves ovate to elliptic-lanceolate, *ca* 5x4 cm; flowers axillary, solitary; corolla orange-yellow, *ca* 8x6 mm; mericarps 6-8, rugose with a pair of short, stellate hairs; seeds 2 mm long.

Fl.-Fr.:-Aug. -March

A weed in grassy slopes and roadsides, Burhanpur, Dhar, Indore, Jhabua.

subsp. **rhombifolia**

Herbs or undershrubs, 40-90 cm tall; stems, petioles and pedicels covered with minute stellate hairs; leaves ovate to oblong, often rhomboid, 2-6 × 0.8-4.5 cm, serrate, entire towards base; flowers axillary or clustered at the ends of branches; corolla yellowish, lobes oblique, glabrous; mericarps 7-10, pubescent with 2 short awns; seeds black, trigonous, *ca* 2 mm long.

Fl.-Fr.: Sept. -Dec.

In wastelands and on outskirts of forests, all districts.

Sida spinosa L. Sp. Pl. 683. 1753; Mast. in Hook.f., Fl. Brit. India1 4:323. 1874; Borss. Blumea 14:191. 1966; T.K. Paul in Sharma *et al.,* Fl. India 3:292. 1993; Verma in Verma *et al.,* Fl. Madhya Pradesh 1:206. 1993; Samvatsar, Fl. West. Trib. Madhya Pradesh 61. 1996.

Diffuse, suffruticose herbs, up to 70 cm tall; leaves 2-6 × 1,5-3 cm, elliptic to ovate, cuneate at base, dentate; flowers pale-white; mericarps 5, trigonous, strongly reticulately veined, apex with 2 short horns; seeds smooth.

Fl.-Fr.: Oct. -Dec.

Common weed of fields and gardens, in all districts.

STERCULIA L. W. 91; MP 2; SA 1

Sterculia villosa Roxb., Fl. Ind. ed. 1832 3:153-154. 1832; Mast. in Hook.f. Fl. Brit. India 1: 355. 1874; Malick in B.D. Sharma *et al.,* Fl. India 3: 472. 1993; Verma in Verma *et al.,* Fl. Madhya Pradesh 1: 222. 1993; Samvatsar, Fl. West. Trib. Madhya Pradesh 64. 1996. *Udal* **(Plate 2, Fig. 5)**

Trees, 10-12 m high; leaves palmately 5-7-lobed, cordate at base, lobes oblong or ovate-oblong, acuminate at apex, entire or 3-fid; flowers unisexual or polygamous, pinkish yellow, in terminal, drooping, rusty pubescent panicles; calyx campanulate, stellate-hairy outside, pinkish inside; petals absent; follicles 3-5, oblong, spreading, rusty villous outside, red inside; seeds black, oblong.

Fl.-Fr.: March – Aug.

Rare, Burhanpur and Dhar.

TRIUMFETTA L. W. 102; MP 5; SA 5

Key to Species

1. Spines on capsules exceeding their diameter; leaves ovate to acuminate, not lobed
   2. Sepals tomentose; spines on capsules strigose; hairs on leaf blade unbranched .... **T. bogotensis**
   2. Sepals glabrous; spines on capsules glabrous; hairs on leaf blade stellate .... **T. annua**
1. Spines on capsules not exceeding their diameter; leaves rhomboid or roundish, lobed or not
   3. Leaves not lobed, roundish, 2-5 cm across...**T. rotundifolia**
   3. Leaves palmately lobed
      4. Spines on the capsules hispid ciliate on one side .... **T. pentandra**
      4. Spines on the capsules glabrous, except at the base ....**T. rhomboidea**

Triumfetta annua L., Mant. Pl. 73. 1767; Mast. in Hook.f., Fl. Brit. India 1: 396. 1874; P. Daniel & Chandrab. in B.D. Sharma *et al.,* Fl. India 3: 518. 1993. Hanfi in Verma *et al.,* Fl. Madhya Pradesh 1: 233. 1993.

Erect herbs; stems glabrous except a line of hairs on one side; leaves ovate-lanceolate, 6-10 × 2-5 cm, acute-acuminate at apex, cuneate at base; flowers 3, in leaf-opposed, pedunculate cymes; petals yellow-orange, spathulate, *ca* 4 mm long, obtuse; fruits globose, *ca* 6 mm across, glabrous; spines uncinate, glabrous.

Fl.-Fr.: Aug. – Feb.

Along village roadsides, **Khandwa** only, Aawlia.

Triumfetta bogotensis DC., Prodr. 1: 506. 1824; *Triumfetta pilosa* Roth, 1821, *nom. illeg*; Mast. in Hook.f., Fl. Brit. India 1: 394. 1874; P. Daniel & Chandrab., in B.D. Sharma *et al.,* Fl. India 3: 519. 1993; Hanfi in Verma *et al.,* Fl. Madhya Pradesh 1: 234. 1993.

Erect herbs; leaves ovate-oblong or ovate-lanceolate, 6-10 × 2-5 cm, acute-acuminate at apex, rounded or subcordate at base; flowers in leaf-opposed, shortly pedunculate, many-flowered cymes; petals yellow, oblong-spathulate, *ca* 7 mm long; fruits subglobose, tomentose; spines uncinate, hairy except apical portion.

Fl.-Fr.: Aug. – Dec.

Rare, Burhanpur, Dhar, Khandwa.

Triumfetta pentandra A. Rich., Fl. Seneg. Tent. 1:93. 1831; *T. neglecta* Wight & Arn. Prodr. Fl. Ind. Orient. 1:75. 1834; Mast. in Hook.f., Fl. Brit. India 1: 396. 1874; P. Daniel & Chandrab., in B.D. Sharma *et al.,* Fl. India 3: 519. 1993; Hanfi in Verma *et al.,* Fl. Madhya Pradesh 1: 234. 1993; Samvatsar, Fl. West. Trib. Madhya Pradesh 67. 1996.

Herbs or undershrubs; leaves elliptic, rhomboid or rhomboid-ovate, 3-9 × 1.5-9, 5 cm, acuminate at apex, cuneate to subcordate at base, irregularly serrate, often shallowly 3-lobed; flowers in leaf-opposed, shortly pedunculate cymose clusters arranged in interrupted racemes; petals yellow, spathulate, *ca* 2 mm long, obtuse; fruits ovoid-oblong, tomentose; spines uncinate, with a line of hairs.

*Fl.-Fr.:*Sept. – April

Common along forest roads, in all districts.

Triumfetta rhomboidea Jacq., Enum. Syst. Pl. 22. 1760; Mast. in Hook.f., Fl. Brit. India 1: 395. 1874; P. Daniel & Chandrab. in B.D. Sharma *et al.,* Fl. India 3: 520.1993; Hanfi in Verma *et al.,* Fl. Madhya Pradesh 1: 234. 1993; Samvatsar, Fl. West. Trib. Madhya Pradesh 67. 1996. *Gadar lapti*

Annual, erect herbs; leaves variable, rhomboid or ovate-elliptic, 3-9 × 2.5-8 cm, acute-acuminate at apex, rounded at base, 3-lobed; flowers in axillary or terminal, few-flowered cymes; petals yellow, obovate-oblong, *ca* 5 mm long; fruits subglobose, *ca* 4 mm across, white-tomentose; spines uncinate, glabrous.

Fl.-Fr.: Aug. – Dec.

Common species of grasslands, all districts.

Triumfetta rotundifolia Lam., Encycl. 3: 421. 1792; Mast. in Hook.f., Fl. Brit. India 1: 395. 1874; P. Daniel & Chandrab. in B.D. Sharma *et al.,* Fl. India 3: 521. 1993; Hanfi in Verma *et al.,* Fl. Madhya Pradesh 1: 236. 1993; Samvatsar, Fl. West. Trib. Madhya Pradesh 67. 1996.

Undershrubs; young branches densely tomentose; leaves suborbicular, *ca* 4 cm across, acute at apex, obtuse at base, densely, stellate-hairy beneath; flowers in leaf-opposed cymes arranged in interrupted racemes; petals yellow, oblong-spathulate, *ca* 5 mm long, ciliate at base; fruits globose or subglobose, pubescent; spines uncinate, pubescent.

*Fl.-Fr.:*Oct. -Dec.

Common along forest and village roads, all districts.

URENA L. W. 10; MP 2; SA2

Key to Species

1. Leaves not divided beyond the middle; epicalyx cupular in fruit .... **U. lobata**
1. Leaves divided beyond the middle; epicalyx spreading in fruit .... **U. sinuata**

Urena lobata L., Sp. Pl. 692. 1753; Mast. in Hook.f., Fl. Brit. India 1: 329. 1874; T.K. Paul in B.D. Sharma *et al.,* Fl. India 3: 380.1993; Verma in Verma *et al.,* Fl. Madhya Pradesh 1: 209. 1993; Samvatsar, Fl. West. Trib. Madhya Pradesh 61. 1996. *Bachita*

Undershrubs; leaves highly variable; lower suborbicular, angular or shallowly palmately-lobed, 5-10 cm across, upper ones oblong lanceolate, shallowly cordate, serrate, densely stellate hairy on both the surfaces, midrib beneath or sometimes also the side nerves with a linear nectary at base; flowers axillary, solitary or 2-3 in clusters; corolla pink with a purple centre, rotate, 2-3 cm across; mericarps 4-5 mm long, reticulately veined with retrorse sharp hooks at top; seeds reniform.

Fl.-Fr.: July-Nov.

Common along streams, all districts.

Urena sinuata L., Sp. Pl. 692. 1753; Mast. in Hook.f., Fl. Brit. India 1: 329. 1874; *U. lobata* subsp. *sinuata* (L.) Borss. in Blumea 14: 142. 1966; T.K. Paul in B.D. Sharma *et al.,* Fl. India 3: 382. 1993; Verma in Verma *et al.,* Fl. Madhya Pradesh 1: 209. 1993; Samvatsar, Fl. West. Trib. Madhya Pradesh 61. 1996. *Bachita*

Undershrubs; leaves palmately-lobed or palmatifid, 3-11 cm across; flowers axillary, solitary or 2-3 in clusters; epicalyx spreading or reflexed in fruits, segments linear to lanceolate; corolla pink with a purple centre, rotate, 2-3 cm across; mericarps 4-5 mm long, reticulately veined with retrorse, sharp hooks at top; seeds reniform.

*Fl.-Fr.:*Aug. – Dec.

Common along banks of streams in forest, in all districts.

WALTHERIA L. W. 14; MP 1; SA 1

Waltheria indica L. Sp. Pl. 673. 1753; *W. americana* L. Sp. Pl. 673. 1753; Mast. in Hook.f., Fl. Brit. India 1:374. 1874; Hanfi in Verma *et al.,* Fl. Madhya Pradesh 1:223. 1993; Samvatsar, Fl. West. Trib. Madhya Pradesh 64. 1996.

Perennials, up to 90 cm tall (favourable habitat); leaves 2-2.7 × 0.8-4 cm ovate to ovate-oblong, crenate-serrate; flowers yellow in axillary and terminal clusters, about 1 cm across; sepals red-tinged, strongly 3-nerved; capsules enclosed in the calyx; seed one, obovoid, black or brown-black.

Fl.-Fr.: Sept. -Feb.

Common in wet soil, waste land, in all districts.

LINACEAE 16/213; 2/2; 1/1

LINUM L. W. 141; MP 2; SA 1

Linum mysurense B. Heyne ex Benth. in Edward's Bot. Reg. 16: t. 1326. 1830; Hook.f., in Fl. Brit. India 1:411. 1874; Srivastava in Fl. Madhya Pradesh 1:237. 1993.

An erect, glabrous, 15-35 cm tall, annual herb, branches in the upper part giving a corymbose appearance; leaves sessile, faintly 3-nerved, 1-1.2 cm long, linear, apex acute; flowers yellow, in panicled corymbs; stamens 5, connate at base, alternating with minute staminodes; capsules glabrous, globose, surrounded by persistent calyx; seeds 10, compressed.

Fl.-Fr.: Sept. -Nov.

Rare, on hill slopes, **Indore** only, Janapao hills.

MALPIGHIACEAE 81/1301; 2/2; 2/2

Key to Genera

1. Samara 3-winged, one wing longer than remaining 2; petals fimbriate .... HIPTAGE
1. Samara winged all around, seed enclosed between dorsal and ventral flaps of wings; petals not fimbriate .... ASPIDOPTERYS

ASPIDOPTERYS A. Juss. W. 25; MP 2; SA 1

Aspidopetrys cordata (B. Heyne ex Wall.) A. Juss. in Ann. Sci. Nat. Bot. II, 13:267. 1840; *Hiraea cordata* Heyne ex Wall., Pl. As. Rar. 1: 13. 1829; Hook.f., Fl. Brit. Ind. l: 421. 1874; R. C. Srivastava in Verma *et al.,* Fl. Madhya Pradesh 1: 239. 1993; Samvatsar, Fl. West. Trib. Madhya Pradesh 68. 1996; R. C. Srivastava in Hajra *et al.,* Fl. India 4: 5. 1997.

Twiners; young branches white silky tomentose; leaves orbicular or ovate, 5-14 × 4-13 cm, cordate at base, glabrous above, white silky tomentose beneath; panicles tomentose, 9-20 cm long; flowers white or pale yellow, *ca* 1 cm across; samaras deep brown, suborbicular, 2-3 × 1.5-2 cm, distinctly wavy at margins.

Fl.-Fr.: Sept. -Dec.

Found only in **Alirajpur**, Aamkhut.

HIPTAGE Gaertn. W. 13; MP 1; SA 1

Hiptage benghalensis (L.) Kurz in J. Asiat. Soc. Bengal, Pt. 2, Nat. Hist. 43:136. 1874; *Banisteria benghalensis* L., Sp. Pl. 427. 1753; *Hiptage madablota* Gaertn., Fruct. Sem. Pl. 2: 169. 1790; Hook.f., Fl. Brit. India 1: 418. 1874; R. C. Srivastava in Verma *et al.,* Fl. Madhya Pradesh 1: 240.1993 & in Hajra *et al.,* Fl. India 4: 14. 1997. *Madhvi lata*

Large climbers; leaves oblong or ovate-lanceolate, 9-21 × 4-9 cm, acute or acuminate at apex, cuneate at base, glabrous; racemes 10-20 cm long, erect, softly tomentose; flowers fragrant, 2-3 cm across; petals white with yellow center, suborbicular to obovate, *ca* 1 cm long, hairy outside, fimbriate on margins; samaras 4-6 × 1-1.5 cm, acute, obtuse or rounded at apex.

Fl.-Fr.: Jan. -April

Planted as ornamental (Indore), wild in Barwani, Burhanpur, Khargone.

ZYGOPHYLLACEAE 26/211; 3/3; 2/2

Key to Genera

1. Stipules present; leaves opposite, pinnately compound; flowers bisexual, yellow; fruit schizocarpic; mericarps 5, tuberculate, with 2 long and 2 short, downwardly directed spines .... TRIBULUS

1. Stipules absent; leaves alternate, palmately 2-foliolate; flowers polygamous, greenish-yellow; fruit drupaceous .... BALANITES

BALANITES Delile W. 10; MP 1; SA 1

Balanites aegyptiaca (L.) Delile, Descr. Egypte, Hist. Nat. 221. 1813; *Ximenia aegyptiaca* L., Sp. Pl. 1194. 1753; Bennett in Hook.f., Fl. Brit. India 1: 522. 1875; Khanna in Verma *et al.,* Fl. Madhya Pradesh 1: 262. 1993; Samvatsar, Fl. West. Trib. Madhya Pradesh 71. 1996; P. Singh & V. Singh in Hajra *et al.,* Fl. India 4: 40.1997. *Hingot.*

Small trees; thorns *ca* 6 cm long, simple or forked; leaflets 2, elliptic-oblong or oblanceolate, 2-5 × 1-3 cm, acute or acuminate at apex, rounded or cuneate at base; flowers pale greenish-yellow, fragrant, 4-12, in axillary fascicled cymes; drupes yellow when ripe, ovoid, subspherical, 5-grooved, 3-7 × 2-4.5 cm, with foetid smelling pulp; seed solitary, stony with fibrous testa.

Fl.-Fr.: Throughout the year.

Common along road sides of villages, all districts.

TRIBULUS L. W. 12; MP 1; SA 1

Tribulus terrestris L., Sp. Pl. 387. 1753; Edgew. & Hook.f., Fl. Brit. India 1: 423.1874; Ram Lal in Verma *et al.,* Fl. Madhya Pradesh 1: 242. 1993; Samvatsar, Fl. West. Trib. Madhya Pradesh 68. 1996; P. Singh & V. Singh in Hajra *et al.,* Fl. India 4: 55.1997. *Chhota gokhru*

Prostrate herbs; leaves opposite, 3-7 cm long, abruptly pinnate; leaflets 8-12 pairs, ovate to elliptic, 4-11 × 2-5 mm, acute or obtuse at apex, oblique at base; flowers yellow, solitary, axillary or leaf-opposed, *ca* 1 cm across; schizocarps 10-15 mm across, excluding spines; mericarps 4-8 × 5-7 mm, dorsally tuberculate; spines 4 or 2.

*Fl.-Fr.:*During and after rains.

Along village roads, common, in all districts.

OXALIDACEAE 8/601; 2/10; 2/5

Key to Genera

1. Leaves pinnate compound; leaflet pairs more than 3 .... BIOPHYTUM

1. Leaves palmate compound; leaflets 3 .... OXALIS

BIOPHYTUM DC. W. 81; MP 4; SA 3

Key to Species

1. Leaflets 10-20 pairs; sepals hardly exceeding the capsule; pedicels equalling or exceeding the flowers .... **B. reinwardtii**
1. Leaflets 5-15 pairs; sepals much exceeding the capsule; pedicels shorter than the flowers
    2. Leaflets 5-7 pairs; seeds with simple tubercles .... **B. umbraculum**
    2. Leaflets 10-15 pairs; seeds with transverse, tubercled rings .... **B. sensitivum**

Biophytum reinwardtii (Zucc.) Klotzsch, Naturw. Reise Mossambique 85. 1861; *Oxalis reinwardtii* Zucc., Abh. Math.-Phys. Cl. Konigl. Bayer Akad. Wiss. 1:274. 1831; *B. sensitivum* var. *reinwardtii* (Zucc.) Guillaumin, Bull. Mus. Natl. Hist. Nat. 15:127. 1909; Edgew. & Hook.f. in Hook.f., Fl. Brit. India 1: 437. 1874; Ram Lal in Verma *et al.,* Fl. Madhya Pradesh 1: 245. 1993; Manna in Hajra *et al.,* Fl. India 4: 236. 1997.

Annual herbs; leaves pinnate; leaflets 12-17 pairs, terminal pair largest, obovate, 7-15 × 4-7 mm, rounded at apex, equal at base, glabrous; flowers yellow, in pseudo-umbels, clustered or on 5-12 cm long peduncles; capsules globose, glabrous; seeds with spiral warts.

*Fl.-Fr.:*Oct. – Dec.

Weeds in wet areas, Burhanpur and Khandwa.

Biophytum sensitivum (L.) DC., Prodr. 1: 690.1824; *Oxalis sensitiva* L., Sp. Pl. 434. 1753; Edgew. & Hook.f. in Hook.f., Fl. Brit. India 1: 436. 1874; Ram Lal in Verma *et al.,* Fl. Madhya Pradesh 1: 245. 1993; Samvatsar, Fl. West. Trib. Madhya Pradesh 68. 1996; Manna in Hajra *et al.,* Fl. India 4: 238. 1997. *Lajalu*

Erect, unbranched annual herbs; leaves 6-10 cm long sensitive, leaflets opposite, sessile, oblong, terminal longest; flowers yellow, in terminal umbels; capsule ellipsoid, apiculate, slightly exceeding the sepals; seeds ovoid, transversely striate.

Fl.-Fr.: Feb-Sept.

Common weed of wet and moist places, forest and fields, throughout the area.

Biophytum umbraculum Welw. Apont. 590.1859; (as *B. petersianum* Klotz.); *B. apodiscias* (Turcz.) Edgew. & Hook.f. in Hook.f., Fl. Brit India 1:437. 1874; Ram Lal in Verma *et al.,* Fl. Madhya Pradesh 1: 245. 1993.

Annual, erect, 2-5 cm tall herbs; leaves crowded, 2.5-4 cm; leaflets 5-7 pairs, reniformly obovoid, widening towards the tip, coriaceous; lowest pair smallest, upper pairs gradually becoming larger; pedicels very short; flowers *ca* 5 mm across; sepals exceeding the capsule; seeds with simple tubercles.

Fl.-Fr.: July-Nov.

Rare, **Indore** only, as forest undergrowth, Sitala Mata falls area.

OXALIS L. W. 504; MP 4; SA 2

Key to Species

1. Flowers yellow; umbels 2-8 flowered ...... **O. corniculata**
1. Flowers bright pink; umbels 8-15 flowered .... **O. debilis**

Oxalis corniculata L., Sp. Pl. 435. 1753; Edgew. & Hook.f., Fl. Brit India 1:436. 1874; Ram Lal in Verma *et al.,* Fl. Madhya Pradesh 1:246. 1993; Samvatsar, Fl. West. Trib. Madhya Pradesh 69. 1996. *Amrul sak*

Procumbent, appressed hairy herbs; leaves digitately 3-foliolate; leaflets 1-1.7 × 1-1.5 cm, subsessile, cuneate or acute at base, appressed hairy; flowers yellow, 2-8 in umbellate clusters; capsule oblong, pubescent; narrowed to the apex; seeds transversely ribbed.

*Fl.-Fr.:*Oct. -May

Common weed throughout the area in moist places, gardens etc.

Oxalis debilis Kunth, Nov. Gen. Sp. 5:236.1822; Almeida, Fl. Maharashtra 1:174.1996. *Khatti buti*

Perennial, bulbous herbs; leaflets obcordate, deeply notched at apex; flowers violet with purple streaks, 2 to many in subumbellete corymbs; capsules 5-angled, beaked; seeds rugose.

*Fl.-Fr.:*Oct. -March

Barwani, Burhanpur, Khargone as a weed in gardens.

BALSAMINACEAE 2/488; 1/5; 1/2

IMPATIENS L. W. 487; MP 45; SA 2

Key to Species

1. Leaves up to 2 cm long; flowers solitary or fascicled; capsules 1.2 cm long .... **I. dasysperma**
1. Leaves 8-14 cm long; flowers 1-3; capsules 1-1.5 cm long .... **I. balsamina**

Impatiens balsamina L., Sp. Pl. 938. 1753; Hook.f., Fl. Brit. India 1: 453. 1874; Ram Lal in Verma *et al.,* Fl. Madhya Pradesh 1: 248. 1993; Samvatsar, Fl. West. Trib. Madhya Pradesh 69. 1996; Vivek. *et al.,* in Hajra *et al.,* Fl. India 4: 123. 1997. *Gultevdi*

Annual herbs; leaves alternate, elliptic-lanceolate, 3-15 × 1-3 cm, acute at both ends, serrate; petioles *ca* 10 cm long, glandular; flowers purple or white, solitary or in fascicles of 2-3; capsules *ca* 2 cm long, tomentose or woolly ellipsoid; seeds globose, reticulate.

*Fl.-Fr.:*Aug. -Oct.

Very common during rains along road sides of villages, all districts.

Impatiens dasysperma Wight, Mad. J. Lit. Sci. 5:7. 1837. *I. lucida* B. Heyne ex Hook.f., Fl. Brit India 1:451. 1874; Mudaliar *et al.,* Fl. Maharashtra 1:487.2000 (as *I. lucida*).

Annual, slender,10-20 cm tall herbs; leaves 1.3-2 cm long, subopposite or whorled, ovate or ovate-lanceolate, crenate serrate, apex acuminate; flowers pink, up to 1 cm across, solitary or fascicled; spur slender; capsule *ca* 1.2 cm long, ellipsoid, turgid in the middle, narrowed at both ends.

Fl.-Fr.: Aug. -Sept.

Khandwa only, as forest undergrowth in moist ground, at Kalibhit.

RUTACEAE 158/1730; 5/6; 4/4

Key to Genera

1. Unarmed shrubs or trees; stamens 10; leaves imparipinnate; flowers in axillary and terminal panicles; capsule 3-valved; seeds winged ....CHLOROXYLON

1. Armed trees; stamens 8-many; fruit baccate

 2. Leaves 3-foliolate; flowers in axillary panicles; stamens more than 20; berry 5-15 cm across, rough with thick rind .... AEGLE

 2. Leaves 3-9 foliolate; flowers polygamous or bisexual; stamens 8-12

  3. Flowers red, polygamous, in panicles; ovules many per locule; fruit 5-8 cm across, pericarp woody; seeds embedded in pulp................................ LIMONIA

  3. Flowers white, in axillary, sub-umbellate racemes; ovule 1 per locule; fruit 1.5 cm across; pericarp not woody .... HESPERETHUSA

AEGLE Correa W. 2; MP 1; SA 1

Aegle marmelos (L.) Corrêa in Trans. Linn. Soc. London 5: 223. 1800; *Crateva marmelos* L., Sp. Pl. 444. 1753; Hook.f., Fl. Brit. India 1: 516. 1875; Ram Lal in Verma *et al.,* Fl. Madhya Pradesh 1: 252. 1993; Samvatsar, Fl. West. Trib. Madhya Pradesh 69. 1996; K. N. Nair & M.P. Nayar in Hajra *et al.,* Fl. India 4: 264. 1997. *Bael*

Armed trees; leaves pinnately 3 foliolate; leaflets ovate-elliptic, terminal ones *ca* 13 × 6.5 cm, lateral ones *ca* 7 × 4 cm, acuminate at apex, oblique at base, glandular-punctate; inflorescence few flowered, *ca* 5 cm long, axillary and terminal; petals 4-5, greenish- white; fruits subglobose, 5-10 cm across many seeded; pericarp hard, woody; seeds oblong, embedded in a sweet, thick, mucilaginous pulp.

Fl.-Fr.: March – Dec.

Common on hill slopes and also cultivated.

CHLOROXYLON DC. Monotypic

Chloroxylon swietenia DC., Prodr. 1: 625. 1834; Hiern in Hook.f., Fl. Brit. Ind. 1: 569. 1875; Ram Lal in Verma *et al.,* Fl. Madhya Pradesh 1: 254. 1993; K. N. Nair & M.P. Nayar in Hajra *et al.,* Fl. India 4: 355. 1997. *Bhirra*

Trees; leaves paripinnate; leaflets 20-40, 3 × 1 cm, opposite, subopposite or alternate, glabrous, pale and dotted on lower surface, oblong-lanceolate, obtuse with oblique base; flowers white; peduncles up to 10 cm long; capsule ovoid or oblong, 3 × 1.5 cm long, dark brown; seeds compressed, winged above.

Fl.-Fr.: May-Aug.

Alirajpur, Burhanpur, Khandwa, Khargaone.

HESPERETHUSA M. Roem. Monotypic

Hesperethusa crenulata (Roxb.) M. Roem., Fam. Nat. Syn. Monogr. 1: 38. 1846; *Limonia crenulata* Roxb., Pl. Coromandel 1:60, pl. 86. 1795; *Naringi crenulata* (Roxb.) Nicolson in Sald. & Nicols., Fl. Hassan 387. 1976; Ram Lal in Verma *et al.,* Fl. Madhya Pradesh 1: 259. 1993; Samvatsar, Fl. West. Trib. Madhya Pradesh 70.1996; K. N. Nair & M.P. Nayar in Hajra *et al.,* Fl. India 4: 302. 1997. *Bhel sai*

Spiny trees; leaves 5-7-foliolate; petiole and rachis auriculate; leaflets elliptic-obovate, 2-5 cm long, crenate, obtuse, usually notched; racemes subumbellate, pubescent, often leafy; flowers 4-merous, white; berry globose, *ca* 1.5 cm across, bluish-black.

Fl.-Fr.: Sept-Dec.

Rare, **Alirajpur,** Mogarba.

LIMONIA L. Monotypic

Limonia acidissima Groff, Lingnaam Agric. Rev. 2(L):22. 1924; *Schinus limonia* L. Sp. Pl. 389. 1753; *Feronia elephantum* Corrêa in Trans. Linn. Soc. 5: 225. 1800; Hook.f., Fl. Brit. India 1: 516. 1875; *Feronia limonia* (L.) Swingle in J. Washington Acad. Sci. 4: 328. 1914; Ram Lal in Verma *et al.,* Fl. Madhya Pradesh 1: 257. 1993; Samvatsar, Fl. West. Trib. Madhya Pradesh 69. 1996; K. N. Nair & M.P. Nayar in Hajra *et al.,* Fl. India 4: 294. 1997. *Kabith*

A moderate sized, spiny tree; leaves alternate, imparipinnate; leaflets 3-9, opposite, 2-3.5 × 1-1.8 cm, petiole winged; flowers red, small, in lateral or terminal pubescent panicles; fruit globose, hard; pericarp woody, rough, grey coloured; pulp edible, many seeded.

Fl.-Fr.: March – Dec.

Occasional along village roads, hills all dists.

SIMAROUBACEAE 19/121; 1/1; 1/1

AILANTHUS Desf. W. 7; MP 1; SA 1

Ailanthus excelsa Roxb., Pl. Coromandel 1: t. 23. 1795; A. W. Benn. in Hook.f., Fl. Brit. India 1: 518. 1875; Khanna in Verma *et al.,* Fl. Madhya Pradesh 1:262. 1993; Samvatsar, Fl. West. Trib. Madhya Pradesh 70; 1996; Basak in Hajra *et al.,* Fl. India 4: 410.1997. *Maha rukh*

Deciduous trees, with large leaf scars; leaves pinnate, 20-40 cm, long, rarely *ca* 90 cm or more; petioles tomentose; leaflets 8-14 pairs, alternate or opposite, variable in shape, 4-16 × 3-8 cm, acute at apex, unequal at base, coarsely irregularly dentate; flowers greenish- yellow, large, in much branched panicles; samaras linear or oblong to lanceolate, 4-7.5 × 1-1.5 cm, strongly reticulate.

Fl.-Fr.: Dec. -July

Common roadside tree in villages, all districts.

BURSERACEAE 18/649; 3/3; 3/3

Key to Genera

1. Flowers tetramerous; leaflets 3-foliolate ....COMMIPHORA
1. Flowers pentamerous; leaflets many foliolate
   2. Fruits 3-angled, capsular; seeds flat and winged .... BOSWELLIA
   2. Fruits globose, drupaceous; seeds not as above ....GARUGA

BOSWELLIA Roxb. ex ColebR.W. 28; MP 1; SA 1

Boswellia serrata Roxb. ex Colebr. in Asiat. Res. 9: 379. 1807; A. W. Benn. in Hook.f., Fl. Brit. India 1: 528. 1875; Srivastava in Verma *et al.,* Fl. Madhya Pradesh 1: 265. 1993; Samvatsar, Fl. West. Trib. Madhya Pradesh 71. 1996; Chithra & A.N. Henry in Hajra *et al.,* Fl. India 4: 432. 1997. *Salai*

Trees, 10-15 m high; bark papery, ash-coloured; leaves pubescent,imparipinnate, crowded at the ends of branches, 15-20 cm long; leaflets 8-20, variable in shape and size, 4-9 × 1.5-2 cm, acute or mucronate at apex, oblique, cuneate or rounded at base, coarsely serrate; panicles axillary, *ca* 20 cm long; flowers pink or white, small, fragrant; capsules green, 3-gonous, 1.2-2.4 × 1-1.4 cm, smooth.

Fl. - Fr.: Feb. -April

In all districts, but more common in Burhanpur.

COMMIPHORA Jacq. W. 208; MP 1; SA 1

Commiphora wightii (Arn.) Bhandari in Bull. Bot. Surv. India 6: 327. 1965; *Balsamodendron wightii* Arn. in Ann. Nat. Hist. 3: 86. 1839; *Balsamodendron mukul* Hook. ex Stocks in Hook., J. Bot. Kew Gard. Misc. 1: 259. t. 8. 1849; A. W. Benn. in Hook.f., Fl. Brit. India 1: 529. 1875; *Commiphora roxburghii* (Stocks) Engl., Pflanzenfam. 3(4):253. 1890; Samvatsar, Fl. West. Trib. Madhya Pradesh 71. 1996; Chithra & A.N. Henry in Hajra *et al.,* Fl. India 4: 445. 1997; Khanna *et al.,* Suppl. Fl. Madhya Pradesh 33.2001. *Gugal*

A much branched shrub or small tree with silvery, paper-like bark; branches knotty and crooked, usually ending in sharp spines; leaves 1-3 foliolate, subsessile, rhomboid-ovate, serrate; flowers brownish-red in fascicles of 2-3; drupe ovoid, acute, readily splitting into two, red when ripe.

Fl.-Fr.: May-Dec.

Rare in forest, normally at the edges and clearings, as hedge, in Alirajpur, Burhanpur, Dhar, Indore, Khandwa.

GARUGA Roxb. W. 4; MP 1; SA 1

Garuga pinnata Roxb., Pl. Coromandel 3: 5. 1811; A. W. Benn. in Hook.f., Fl. Brit. India 1: 528. 1875; Srivastava in Verma *et al.,* Fl. Madhya Pradesh 1: 266. 1993; Samvatsar, Fl. West. Trib. Madhya Pradesh 72. 1996; Chithra & A.N. Henry in Hajra *et al.,* Fl. India 4: 448. 1997. *Kakad*

Deciduous trees; leaves odd pinnate, leaflets imparipinnate, pairs 6-10, 5-14 × 2-5 cm, opposite, oblique, crenate; flowers yellow, in panicles; drupes black, fleshy, edible; seeds with membranous wings.

Fl.-Fr.: Feb. -June

Common in forest, everywhere.

MELIACEAE 52/669; 4/4; 4/4

Key to Genera

1. Fruits capsular; ovary 5-celled; seeds winged
    2. Stamens 10, monoadelphous, staminal tube 10-lobed .... SOYMIDA
    2. Stamens 5, free .... TOONA
1. Fruit drupaceous; ovary 3-5 celled; seeds not winged
    3. Leaves unipinnate; flowers white; disc absent; staminal tube laciniate at apex; ovary 3-celled; drupe 1-seeded .... AZADIRACHTA
    3. Leaves bi or tripinnate; flowers lilac; disc present; staminal tube 20-toothed; ovary 5-celled; drupe 4-seeded .... MELIA

AZADIRACHTA A. Juss. W. 2; MP 1; SA 1

Azadirachta indica A. Juss. in Mém. Mus. Hist. Nat. 19: 221. 1830; *Melia azadirachta* L., Sp. Pl. 385. 1753; Hiern in Hook.f., Fl. Brit. India 1: 544. 1875; Srivastava in Verma *et al.,* Fl. Madhya Pradesh 1: 267. 1993; Samvatsar, Fl. West. Trib. Madhya Pradesh 72. 1996; S. K. Jain & S. S.R. Bennet in Hajra *et al.,* Fl. India 4: 478. 1997. *Neem*

A large tree with straight trunk; leaves uni-pinnate, leaflets 9-15, 4-8 × 2-4 cm sub-opposite, obliquely lanceolate, acuminate, serrate, glabrous on both sides; flowers white to whitish-yellow, fragrant, in axillary panicles; drupes lemon yellow on maturity, succulent, 1-seeded, ovoid-oblong.

Fl.-Fr.: Feb. -Sept.

Common tree in villages, urban areas, naturalized, in all districts.

MELIA L. W. 2; MP 2; SA 1

Melia azedarach L., Sp. Pl. 384. 1753; Hiern in Hook.f., Fl. Brit. India 1: 544. 1875; Srivastava in Verma *et al.,* Fl. Madhya Pradesh 1: 268. 1993; Samvatsar, Fl. West. Trib. Madhya Pradesh 73. 1996; S. S. Jain & S. S.R. Bennett in Hajra *et al.,* Fl. India 4: 494. 1997. *Bakain*

A moderate sized tree; leaves bipinnate; leaflets 1.5 × 4-1.2 cm, ovate or lanceolate irregularly serrate, glabrous, dark green; flowers fragrant, lilac, in panicles; staminal tube purple; drupe ellipsoid globose, 4-seeded, yellow when ripe.

Fl.-Fr.: March – June

Commonly planted along roadsides, not wild in the area.

SOYMIDA A. Juss. Monotypic

Soymida febrifuga (Roxb.) A. Juss in Mém. Mus. Hist. Nat. 19: 251. 1830; *Swietenia febrifuga* Roxb., Pl. Coromandel 1: 1795 & Fl. Ind. 2: 398. 1832; Hiern in Hook.f., Fl. Brit. India 1: 567. 1875; Srivastava in Verma *et al.,* Fl. Madhya Pradesh 1: 269. 1993; Samvatsar, Fl. West. Trib. Madhya Pradesh 73. 1996; S. S. Jain & S. S.R. Bennett in Hajra *et al.,* Fl. India 4: 501. 1997. *Rohan* **(Plate 2, Fig. 6)**

A tall tree; leaves crowded towards the ends of the branches; leaflets 3-6 pairs, 7-13 × 1-12 cm, opposite, elliptic-oblong, coriaceous peninerved; flowers greenish-white in large terminal or axillary panicles; capsule obovoid, woody, reddish-purple,5-valved.

Fl.-Fr.: Jan. -July

Rare, Barwani, Burhanpur, Khandwa, Khargone.

TOONA (Endl.) M. Roem. W. 5; MP 1; SA 1

Toona ciliata M. Roem., Fam. Nat. Syn. Monogr. 1: 139. 1846; *Cedrela toona* Roxb. ex Rottler, Neue Schriften Ges. Naturf. Freunde Berlin 2:148. 1803; Hiern in Hook.f., Fl. Brit. India 1: 568. 1875; Srivastava in Verma *et al.,* Fl. Madhya Pradesh 1: 270.1993; Samvatsar, Fl. West. Trib. Madhya Pradesh 73. 1996; S. S. Jain & S. S.R. Bennett in Hajra *et al.,* Fl. India 4: 504. 1997. *Toon*

Deciduous, large trees; leaves usually paripinnate, 30-60 cm long; leaflets 3-10 pairs, obliquely ovate or lanceolate, 6-10 × 2-3 cm, acuminate at apex, entire or undulate on margin; panicles shortly pedunculate, 10-45 cm long, drooping, pubescent; flowers greenish or creamish-yellow, shortly pedicellate; capsules ovate-oblong, 1.5-3 × 0.8-1.5 cm, blackish when dry; seeds winged at both ends.

Fl.-Fr. : : March – May

Alirajpur, Indore, doubtfully wild.

OLACACEAE 27/149; 4/7; 1/1

OLAX L. W. 21; MP 2; SA 1

Olax scandens Roxb., Pl. Coromandel 2: 2. 1800; Mast. in Hook.f., Fl. Brit. India 1: 575. 1875; A. K. Tripathi in Verma *et al.;* Fl. Madhya Pradesh 1: 271. 1993; Samvatsar, Fl. West. Trib. Madhya Pradesh 73. 1996.

Large climbing prickly shrubs; leaves distichous, elliptic or ovate-oblong, *ca* 9 × 3.2 cm, obtuse at apex, glabrous above, glabrous or puberulous beneath; flowers white, in short, axillary racemes; bracts ovate-oblong, *ca* 2 mm long, caduceus; calyx cupular, ciliate; petals usually 5, rarely 6, linear-oblong, acute, connate at base, cleft, *ca* 8 mm long; stamens 3, staminodes 2-cleft at apex; ovary ovoid, glabrous, style linear, stigma 3-lobed; drupes yellow, globose or ovoid, 5-6 mm across, enclosed by accrescent calyx.

Fl.-Fr.: April - Oct.

Alirajpur, Barwani, Khargone.

CELASTRACEAE 87/1168; 4/7; 4/4

Key to Genera

1. Leaves opposite or sub-opposite; flowers bisexual in axillary, compound dichasia; fruit drupaceous; seeds exarillate .... CASSINE
1. Leaves alternate; flowers bisexual or polygamous, axillary or terminal; fruit capsular; seeds arillate
   2. Climbing shrubs; flowers polygamous, paniculate; ovary free from the disc .... CELASTRUS
   2. Armed shrubs; ovary confluent with the disc
      3. Axillary thorns bearing leaves and flowers; leaves up to 7 cm long........................GYMNOSPORIA
      3. Axillary thorns without leaves or flowers; leaves 7-10 cm long .... MAYTENUS

CASSINE L. W. 64; MP 1; SA 1

Cassine glauca (Rottb.) Kuntze, Rev. Gen. Pl. 1:114. 1891; *Mangifera glauca* Rottb. in Nye Saml. Kongel. Danske Vidensk. Selsk. Skr. 2: 534 1783; *Celastrus glaucus* Vahl, Symb. Bot. 2:42. 1791; *Elaeodendron glaucum* (Rottb.) Pers., Syn. Pl. 1:241. 1805; *Cassine albens* (Retz.) Kosterm. in Gard. Bull. Singapore 39(2): 178. 1986; *Schrebera albens* Retz., Observ. Bot. 6: 25. 1791; Khanna in Verma *et al.,* Fl. Madhya Pradesh 1: 273. 1993; Samvatsar, Fl. West. Trib. Madhya Pradesh 73. 1996; Ramam. in N.P. Singh *et al.,* Fl. India 5: 80.2000. *Jamrasi*

A 5-8 m tall tree; leaves 5-12 × 2-6 cm, orbicular-elliptic, glabrous, acute or acuminate, often twisted at the apex, crenate or serrate; flowers greenish-yellow, small, numerous in axillary, dichotomously branched cymes; drupes 1-1.5 cm long, obovoid, 1-seeded, tipped with persistent style.

Fl.-Fr.: Sept. -Nov.

Frequent on the hilly slopes, throughout the area.

CELASTRUS L. W. 38; MP 1; SA 1

Celastrus paniculatus Willd., Sp. Pl. 1: 1125. 1798; M.A. Lawson in Hook.f., Fl. Brit. India 1: 617. 1875; Khanna in Verma *et al.,* Fl. Madhya Pradesh 1: 274, 1993; Samvatsar, Fl. West. Trib. Madhya Pradesh 74. 1996; Ramam. in N.P. Singh *et al.,* Fl. India 5: 87.2000. *Malkangni* **(Plate 2, Fig. 7)**

Deciduous, unarmed, shrubby twiners; leaves 5-10 × 3-5 cm, alternate, ovate-elliptic, crenulate; flowers yellowish or greenish-white, unisexual, in terminal pyramidal panicles; sepals fimbriate at tip; apsules subglobose, 3-valved, transversely wrinkled, tipped by persistent styles; seeds 1-6, crimson, arillate.

Fl.-Fr.: Nov. -Feb.

Common in the forests and on hill slopes, throughout the area.

GYMNOSPORIA (Wight & Arn.) Benth. & Hook.f. W. 102; MP 3; SA 1

Gymnosporia montana (Roth) Benth., Fl. Austral. 1:400.1863; *Celastrus montanus* Roth, Syst. Veg. 5:427. 1819; Khanna in Verma *et al.,* Fl. Madhya Pradesh 1:278. 1993; Samvatsar, Fl. West. Trib. Madhya Pradesh 74. 1996. *Baikal*

A tall, armed shrub or small tree; leaves 2.5-5 × 1.5-2.5 cm, elliptic or obovate, rounded at the apex; flowers white, in axillary cymes; capsule purple or nearly black when ripe; seeds glabrous, brownish-red, rugose.

Fl.-Fr.: Sept. -Dec.

Common on road-sides, and very common in open forest, throughout the area.

MAYTENUS Molina W. 160; MP 2; SA 1

Maytenus rothiana Lobr. –Callen, Adansonia n.s., 15: 223. 1975; *Celastrus rothiana* Wight & Arn., Prodr. Fl. Pen. Ind. Orient. 1:159. 1834; *Gymnosporia rothiana* M.A. Lawson in Hook.f., Fl. Brit. Ind. 1: 620.1875; Khanna in Verma *et al.,* Fl. Madhya Pradesh 1: 277. 1993. *Baikal prajati*

Armed large shrubs or small trees; leaves coriaceous, elliptic or obovate, 7.5-10 × 3-5- 5 cm, apex rounded or acuminate, base cuneate, glabrous, crenulate; flowers greenish- white in short peduncled or subsessile cymes or in fascicles; fruit bright red, 3-lobed; seeds brownish orange, shining, ellipsoid with a basal aril.

Fl.-Fr.: April-July

Rare, **Burhanpur,** Asirgad and Chandni.

RHAMNACEAE 53/839; 4/12; 2/7

Key to Genera

1. Unarmed lianas; style enlarged in a linear, oblong wing .... VENTILAGO
1. Armed (sometimes unarmed) shrubs, trees or stragglers; style not enlarged .... ZIZIPHUS

VENTILAGO Gaertn. W. 9; MP 2; SA 2

Key to Species

1. Calyx tube covering more than half the nut; nut and its wing yellow, pubescent; disc villous .... **V. denticulata**
1. Calyx tube covering only base of nut; nut and its wing grey, nearly glabrous; disc glabrous .... **V. maderaspatana**

Ventilago denticulata Willd. in Neue Schriften Ges. Naturf. Freunde Berlin 3:417. 1801; Khanna in Verma *et al.,* Fl. Madhya Pradesh 1: 283. 1993; Samvatsar, Fl. West. Trib. Madhya Pradesh 75. 1996; Bhandari & Bhansali in N.P. Singh *et al.,* Fl. India 5: 217.2000. *Keoti*

Sarmentose shrubs or lianas with grey bark, having bright red furrows; leaves 3-12 × 2.6-6 cm, elliptic-oblong, subacute, crenate-serrate; flowers creamy- white in paniculate spikes; fruit 3-5 cm long (with wing), broad, rounded at the apex, 1-nerved.

Fl.-Fr.: Aug. -June

Common in all districts.

Ventilago maderaspatana Gaertn., Fruct. Sem. Pl. 1: 223, pl. 49, f. 2. 1788; M.A. Lawson in Hook.f., Fl. Brit. India 1: 631. 1875; Samvatsar, Fl. West. Trib. Madhya Pradesh 75. 1996; Bhandari & Bhansali in N.P. Singh *et al.,* Fl. India 5: 220.2000.

A scandent shrub; young branches and panicles pubescent; leaves 5-10 × 3-5 cm, oblong-lanceolate, sub-acute, crenate pubescent beneath; flowers greenish in terminal panicles; fruit about 1.0-1.25 cm across, yellow with persistent calyx in the form of disc at the base; wings broad, linear-oblong, 1-nerved.

Fl.-Fr. : Nov. -Jan.

Burhanpur, Dhar, Khandwa.

ZIZIPHUS Miller W. 53; MP 8; SA 5

Key to Species

1. Petals absent; styles 2 .... **Z. rugosa**
1. Petals present; styles 2 or 3
    2. Plants straggling; leaves very oblique at base; disc pitted .... **Z. oenopolia**
    2. Plants not as above
        3. Styles 3; flowers in pedunculate cymes .... **Z. xylopyrus**
        3. Styles 2; flowers in sessile fascicles or short peduncled cymes
            4. Small trees; drupes more than 2 cm across .... **Z. jujuba**
            4. Diffuse shrubs with zigzag branches; drupe not exceeding 1.5 cm across; leaf apex rounded, tomentose on upper surface .... **Z. nummularia**

Ziziphus jujuba Mill., Gard. Dict. ed. 8. no. 1. 1768; *Rhamnus jujuba* L. Sp. Pl. 1:194. 1753; *Z. sativa* Gaertn., Fruct. Sem. Pl. 1: 202. 1788; *Ziziphus vulgaris* Lam., Encycl. 3:316. 1789; *Z. mauritiana* Lam., Encycl. 3:319. 1789; M.A. Lawson in Hook.f., Fl. Brit. India 1: 633. 1875; Khanna in Verma *et al.,* Fl. Madhya Pradesh 1: 295. 1993; Samvatsar, Fl. West. Trib. Madhya Pradesh 75. 1996; Bhandari & Bhansali in N.P. Singh *et al.,* Fl. India 5: 232.2000. *Ber*

Armed, glabrous, 6-9 m tall trees; leaves obliquely ovate-lanceolate, 1.3-5 × 0.6-2 cm, obtusely acute at apex, rounded at base, crenulate, glabrous, 3-nerved at base; inflorescence an axillary fascicle; flowers *ca* 5 mm across; drupes dark red or black, ovoid-oblong, *ca* 1.7 cm across; seeds 2, black, compressed.

Fl.-Fr.: June – Jan.

Common in forests, all over the area.

Ziziphus nummularia (Burm.f.) Wight & Arn., Cat. Ind. Pl. 31. 1833; *Rhamnus nummularia* Burm.f., Fl. Indica 61. 1768; *Ziziphus rotundifolia* Lam., Encycl. 3:319. 1789; M.A. Lawson in Hook.f., Fl. Brit. India 1: 633. 1875; Khanna in Verma *et al.,* Fl. Madhya Pradesh 1: 285. 1993; Samvatsar, Fl. West. Trib. Madhya Pradesh 75. 1996; Bhandari & Bhansali in N.P. Singh *et al.,* Fl. India 5: 235. 2000. *Jharberi*

A much branched bushy shrub; leaves 1.5-2 cm, orbicular or oblong, crenate, pubescent above and silky tomentose beneath, with 3 veins at the rounded base; stipular spines tomentose, one straight, other hooked; flowers greenish-white, in small axillary cymes; drupe globose, scarlet red when ripe.

Fl.-Fr.: July-Jan.

Common in fields, wastelands throughout the area.

Ziziphus oenopolia (L.) Mill., Gard. Dict. (ed. 8) no. 3. 1768; *Rhamnus oenopolia* L., Sp. Pl. 194. 1753; M.A. Lawson in Hook.f., Fl. Brit. India 1: 634. 1875; Khanna in Verma *et al.,* Fl. Madhya Pradesh 1: 286. 1993. Samvatsar, Fl. West. Trib. Madhya Pradesh 76. 1996; Bhandari & Bhansali in N.P. Singh *et al.,* Fl. India 5: 236.2000. *Makor*

Rambling shrubs with rusty-tomentose, spinous branches; leaves 2.5-4.5 × 1-2 cm, ovate-lanceolate, oblique at base, basally 3-5 veined, rusty tomentose on both surfaces; flowers greenish, in axillary cymes; drupes purple.

Fl.-Fr.: Aug. -Nov.

Common in hedges, throughout the area.

Ziziphus rugosa Lam., Encycl. 3: 319. 1789; M.A. Lawson in Hook.f., Fl. Brit. India 1: 636. 1875; Khanna in Verma *et al.,* Fl. Madhya Pradesh 1: 286. 1993; Samvatsar, Fl. West. Trib. Madhya Pradesh 76. 1996; Bhandari & Bhansali in N.P. Singh *et al.,* Fl. India 5: 240.2000. *Choorna*

Rambling, armed shrubs; leaves 5-10 × 3.2-5 cm, broadly ovate or elliptic, obtuse, crenate-serrate, 3-5 veined, glabrous above, tomentose beneath; flowers small, pale, in densely grey tomentose panicles; drupe globose, white when ripe.

Fl.-Fr.: Nov. -March

Not common, on slopes of hills, Alirajpur, Dhar, Indore.

Ziziphus xylopyrus (Retz.) Willd., Sp. Pl. 1: 1104. 1798; *Rhamnus xylopyrus* Retz., Observe. Bot., 2: 11. 1781; M.A. Lawson in Hook.f., Fl. Brit. India 1: 634; 1875; Khanna in Verma *et al.,* Fl. Madhya Pradesh 1: 286. 1993; Samvatsar, Fl. West. Trib. Madhya Pradesh 76. 1996; Bhandari & Bhansali in N.P. Singh *et al.,* Fl. India 5: 243.2000. *Ghatbor*

An armed straggling shrub or small tree; leaves 4-7 × 2-4.5 cm,oblong,suborbicular, denticulate, acute, 3-nerved, base oblique, glabrous above, brownish-tomentose beneath; stipular spines 2, one straight, another curved; flowers greenish-white, in paniculate cymes; peduncles and pedicels tomentose; drupe woody, hard, oblong, black when ripe, 2-3 seeded.

Fl.-Fr.: May-Aug.

Common on hill slopes in forest, in all districts. Fruit not edible.

VITACEAE 16/985; 7/22; 5/13

Key to Genera

1. Erect herbs; ovary 3-6 celled, cells 1-ovuled .... LEEA
1. Tendril climbers; ovary 2-celled, cells 2-ovuled
   2. Flowers pentamerous; inflorescence tendril bearing; style short conical .... AMPELOCISSUS
   2. Flowers tetramerous; inflorescence not tendril bearing
      3. Leaves digitately 3-5 foliolate; disc cup-shaped .... CAYRATIA
      3. Leaves simple or 3-5 foliolate
         4. Inflorescence leaf-opposed; corolla not contracted in the middle; disc entire, with a ring of 4 lobes; stigma entire, discoid .... CISSUS
         4. Inflorescence pseudoaxillary or pseudoterminal; corolla always contracted in the middle; disc of 4, free glands; stigma bifid .... CYPHOSTEMMA

AMPELOCISSUS Planch. W. 24; MP 3; SA 2

Key to Species

1. Seeds obcordate, 2-furrowed, pitted on the back; leaves tomentose; disc not furrowed .... **A. tomentosa**
1. Seeds plano-convex, transversely grooved on the back; leaves glabrous; disc furrowed ....**A. latifolia**

Ampelocissus latifolia (Roxb.) Planch., Vigne Amer. Vitic. Eur. 8: 374. 1884; *Vitis latifolia* Roxb., Hort. Bengal. 18. 1814; M.A. Lawson in Hook.f., Fl. Brit. India 1: 652. 1875; Khanna in Verma *et al.,* Fl. Madhya Pradesh 1: 288. 1993; Samvatsar, Fl. West. Trib. Madhya Pradesh 77. 1996; B.V. Shetty & P. Singh in N.P. Singh *et al.,* Fl. India 5: 256.2000. **(Plate 2, Fig. 8)**

Extensive climbers with hollow, striated branches; leaves orbicular to cordate, 3-7 angled, 10-15 cm in diameter, dentate, many-nerved at the base; flowers reddish-brown in leaf-opposed cymes; peduncles with a forked tendril just below the cymes; petals and anthers free; berries globose, black, glabrous, 2-celled; seeds ellipsoid.

Fl.-Fr.: May-Sept.

Common in forests at the foot of hills, in all districts.

Ampelocissus tomentosa (Roth) Planch., Vigne Amer. Vitic. Eur. 8: 375. 1884; *Vitis tomentosa* Roth in Roem. & Schult, Syst. Veg. 5: 318. 1819; M.A. Lawson in Hook.f., Fl. Brit. India 1: 650.1875; Khanna in Verma *et al.,* Fl. Madhya Pradesh 1: 290.1993; Samvatsar, Fl. West. Trib. Madhya Pradesh 77. 1996; B.V. Shetty & P. Singh in N.P. Singh *et al.,* Fl. India 5: 261.2000. *Dhotto*

A stout tendrillar shrub; tendrils woody, 3-5 forked; leaves simple, 10-25 cm long, palmately 3-5 lobed, woolly on both sides, petiole 4-8 cm long; flowers pentamerous, sessile, in shortly peduncled, woolly cymes; berry subglobose, 2-3 seeded.

Fl.-Fr.: Aug. -Nov.

Rare, Alirajpur, Dhar, Jhabua, Khandwa.

CAYRATIA A. L. Juss. W. 67; MP 2; SA 1

Cayratis trifolia (L.) Domin in Biblioth. Bot. 89:370.1927; *Vitis trifolia* L., Sp. Pl. 1: 203. 1753; *Cissus carnosa* Lam. Encycl. 1:31. 1783; *Vitis carnosa* Wall. ex M.A. Lawson in Hook.f., Fl. Brit. India 1:654. 1875; Khanna in Verma *et al.,* Fl. Madhya Pradesh1:291. 1993; Samvatsar, Fl. West. Trib. Madhya Pradesh 77. 1996. *Amal bel*

Woody climbers; branches grooved; leaves digitately 3-foliolate; leaflets 4-6 × 2.5-3 cm, ovate-elliptic, subcordate at base,serrate-dentate; tendrils branched; flowers greenish-white in corymbose cymes; berries depressed, globose, glabrous, 2-4 seeded, initially red, turning black at maturity.

Fl.-Fr.: July-Oct.

Very common in all districts. The plant has a deep, tuberous root.

CISSUS L. W. 324; MP 6; SA 5

Key to Species

1. Stem 4-winged, fleshy; tendrils simple; flowers white; berry red at maturity .... **C. quadrangularis**
1. Stem not winged
    2. Leaf margin 5-12 toothed; tendrils bifurcate; seed surface with sparse ridges; berry normally 1-seeded .... **C. repens**
    2. Leaf margin 15-44 toothed
        3. Tendril racemosely branched; seed surface smooth .... **C. repanda**
        3. Tendril 2-3 branched; seed surface with conspicuous ridges; berries purplish-blue; leaf base deeply cordate
            4. Flowers yellow; stem densely pubescent; leaves 5-lobed, apex acute......................... **C. vitiginea**
            4. Flowers white or faint yellow; stem pubescent; leaves not lobed, apex abruptly acuminate .... **C. adnata**

Cissus adnata Roxb., Fl. Ind. 1: 405. 1820; *Vitis adnata* (Roxb.) Wall., Prodr. Fl. Ind. Orient. 1: 126. 1834; M.A. Lawson in Hook.f., Fl. Brit. India 1: 650.1875; Khanna *et al.,* Suppl. Fl. Madhya Pradesh 38. 2001.

Climbers; leaves broadly ovate, sometimes 3-5-lobed, 6-15 × 3-12 cm, cordate at base, acuminate at apex, serrate-dentate, tomentose beneath; flowers red, in terminal, dichotomous cymes; berries ovoid *ca* 8 mm long, beaked.

Fl.-Fr.: Dec. –March

Khandwa and Khargaone.

Cissus quadrangularis L., Syst. Nat. ed. 12 2:124. 1767; *C. quadrangula* L., Mant. Pl. 1:39. 1767; *Vitis quadrangularis* (L.) Wall. ex Wight, Cat. Ind. Pl. 26. 1833; *Vitis quadrangularis* Wall. ex Lawson in Hook. f, Fl. Brit. India 1: 645. 1875; Samvatsar, Fl. West. Trib. Madhya Pradesh 78. 1996; Khanna *et al.,* Suppl. Fl. Madhya Pradesh 38. 2001. *Har jod*

Herbaceous, fleshy climber with stout root-stock; stem fleshy, glabrous, quadrangular; leaves broadly ovate or reniform, sometimes lobed, denticulate, cordate, short lived; tendrils long, simple; flowers white, in cymes with spreading umbellate branches; calyx cup- shaped, truncate; petals ovate oblong, acute; berry obovoid or globose, red when ripe.

Fl.-Fr.: July-Sept.

Occasional in forest but sometimes cultivated because of its medicinal value.

Cissus repanda (Wight & Arn.) Vahl, Symb. Bot. 3: 18. 1794; *Vitis repanda* (Vahl) Wight & Arn., Prodr. Fl. Ind. Orient. 1: 125. 1834; M.A. Lawson in Hook.f., Fl. Brit. India 1: 648. 1875; Khanna in Verma *et al.,* Fl. Madhya Pradesh 1: 291. 1993; Samvatsar, Fl. West. Trib. Madhya Pradesh 78. 1996; B.V. Shetty & P. Singh in N.P. Singh *et al.,* Fl. India 5: 290.2000; *Vitis repanda* (Vahl) Wight & Arn., Prodr. Fl. Ind. Orient. 125. 1834. M.A. Lawson in Hook.f., Fl. Brit. India 1: 648. 1875. *Pani bel*

A large woody climber; stem minutely muricate; tendrils with flattened disc attaching to the support; leaves 6-18 × 5-15 cm, broadly ovate, cordate, acuminate, pubescent; petiole striate and flattened below the middle; flowers pink, in leaf opposed, compound cymes; berry usually 1-seeded, dark purple when ripe; seeds obovoid, smooth.

Fl.-Fr.: Feb. -May

Common as forest undergrowth, Alirajpur, Dhar, Indore, Khandwa, Khargone.

Cissus repens Lam., Encycl. 1: 31. 1783; *Cissus glauca* Roxb. Fl. Ind. 1: 406, 1820; *Vitis glauca* Wight & Arn., 126. 1834; *V. repens* (Lam.) Wight & Arn., Prodr. Fl. Ind. Orient. 1:125-126. 1834; Samvatsar, Fl. West. Trib. Madhya Pradesh 78. 1996; Khanna *et al.,* Suppl. Fl. Madhya Pradesh 38. 2001.

A glabrous, slender, scandent, herb; branches subterete, or angled; tendrils forked; leaves 8-15 × 3-9 cm, broadly ovate, margin spinous toothed, deeply cordate, glabrous; flowers in leaf opposed, compound cymes; berries subglobose to pyriform, normally 1-seeded, black when ripe.

Fl.-Fr.: Jan. -April

Not common, forest undergrowth, Alirajpur, Dhar, Khandwa.

Cissus vitiginea L., Sp. Pl. 117. 1753; *Vitis vitiginea* (L.) W. Theob., Burmah ed. 3, 2:564. 1883; *Vitis linnaei* Wall. ex Wight & Arn. Prodr. 126. 1834; M.A. Lawson in Hook.f., Fl. Brit. India 1: 649. 1875; Khanna *et al.,* Suppl. Fl. Madhya Pradesh 38. 2001.

Climbing shrubs; branches hoary tomentose; leaves simple, ovate, palmately 3-5-anglular, acuminate at apex, cordate or truncate at base, pubescent; flowers pale yellow, in umbelliform, compound cymes; berries ovoid-pyriform, apiculate.

Fl.-Fr.: July–Feb.

Rare, **Burhanpur**, in hedges, also at Asirgad.

CYPHOSTEMMA (Planch.) Alston W. 253; MP 2; SA 2

Key to Species

1. Extensive climbers; leaves 5-foliolate; stipules large, ear-shaped; plants glabrous .... **C. auriculatum**

1. Prostrate plants, spreading on walls or ground; leaves 3-foliolate; stipules small; plants glandular .... **C. setosum**

Cyphostemma auriculatum (Roxb.) P. Singh & B.V. Shetty in Taxon 35: 596. 1986; *Cissus auriculata* Roxb., Fl. Ind. 1: 411. 1820; *Vitis auriculata* (Roxb.) Wall. Numer. List. 6031. 1831; Wall. ex M.A. Lawson in Hook.f., Fl. Brit. India 1: 658. 1875; Khanna in Verma *et al.,* Fl. Madhya Pradesh 1: 291. 1993; Samvatsar, Fl. West. Trib. Madhya Pradesh 78. 1996; B.V. Shetty & P. Singh in N.P. Singh *et al.,* Fl. India 5: 298. 2000. *Jangli angoor*

Climbing shrubs; young branches pubescent; leaves digitately 5-foliolate, obovate-rhomboid, 9-12 × 5-8 cm, acute at apex, cuneate at base, serrate, lateral leaflets 6-11 × 4-6.2 cm; inflorescence axillary or pseudo-terminal, divaricate, much branched cyme; flowers *ca* 2.5 mm long; berries red, globose, *ca* 1 cm across, glabrous, 1-seeded; seeds subglobose, *ca* 6 mm across.

Fl.-Fr.: July – Nov.

Common as forest undergrowth, in all districts.

Cymphostemma setosum (Roxb.) Alston, Handb. Fl. Ceylon 6(Suppl.): 53. 1931; *Cissus setosa* Roxb., Fl. Ind. 1: 410.1820; *Vitis setosa* (Roxb.) Wall. ex Wight & Arn., Cat. Ind. Pl. 26. 183; Lawson in Hook.f., Fl. Brit. India 1: 654. 1875; Samvatsar, Fl. West. Trib. Madhya Pradesh 78. 1996; Khanna *et al.,* Suppl. Fl. Madhya Pradesh 40.2001.

Glandular-pubescent, prostrate succulent, striate, tendrillar herbs; tendrils leaf-opposed, forked; leaves trifoliate, leaflets sub-succulent, obovate-oblong, 5-9 × 3-5 cm; flowers in dichasial cymes; berry 0.6-1 cm across, red when ripe.

Fl.-Fr.: Sept. -Dec.

Alirajpur, Burhanpur, Dhar, Indore, Khandwa, Khargaone.

LEEA D. Royen W. 36; MP 4; SA 3

Key to Species

1. Leaves simple, unifoliate or unipinnate, very large .... **L. macrophylla**

1. Leaves bi or tripinnate compound

2. Leaves unipinnate, relatively small; stem, rachis and peduncles crispate or winged .... **L. asiatica**

2. Leaves bi or tripinnate; stem rachis and peduncles not crispate or winged .... **L. indica**

Leea asiatica (L). Ridsdale, Bot. Hist. Hort. Malab. 189. 1980; *Phytolacca asiatica* L., Suppl. Pl. 1:441. 1753; *L. crispa* L. Mant. Pl. 1:124. 1767; M.A. Lawson in Hook.f., Fl. Brit. India 1: 665. 1875; Khanna in Verma *et al.,* Fl. Madhya Pradesh 1: 293. 1993; Samvatsar, Fl. West. Trib. Madhya Pradesh 78. 1996; B.D. Naithani in N.P. Singh *et al.,* Fl. India 5: 330.2000. *Kurum* **(Plate 3, Fig. 1)**

An erect shrub, stems somewhat woody, jointed, swollen above the joints; leaves usually simply pinnate, leaflets oblong, ovate or shortly acuminate, sharply serrate, rounded at base; flowers greenish-white, in small terminal cymes; calyx lobes triangular, gland tipped; petals oblong; berry depressed-globular, usually 2-6 lobed, black when ripe.

Fl.-Fr.: Aug. -Oct.

Common in forest, all districts.

Leea indica (Burm.f.) Merr., Philipp. J. Sci. 14(2): 245. 1919; *Staphylea indica* Burm.f., Fl. Indica 75, pl. 24, f. 2. 1768; *Leea sambucina* Willd., 1798 (unresolved); M.A. Lawson in Hook.f., Fl. Brit. India 1: 666. 1875; Khanna in Verma *et al.,* Fl. Madhya Pradesh 1: 293. 1993; Samvatsar, Fl. West. Trib. Madhya Pradesh 79. 1996; B.D. Naithani in N.P. Singh *et al.,* Fl. India 5: 337.2000. *Karkani*

Large shrubs; leaflets up to 18 × 5 cm, ovate or oblong- lanceolate, subacuminate, truncate at base, glabrous; stipules fleshy, sheathing; flowers greenish, in leaf axils or corymbose cymes; berry about 6 mm across, dorsally compressed, 1-4 lobed.

Fl.-Fr.: Aug. -Oct.

Alirajpur, Barwani, Burhanpur, Dhar.

Leea macrophylla Roxb. ex Hornem., Hort. Bot. Hafn. 1: 231. 1813; *L. aspera* Wall. ex G. Don, Gen. Hist. 1:713. 1831; M.A. Lawson in Hook.f., Fl. Brit. India 1: 664. 1875; Khanna in Verma *et al.,* Fl. Madhya Pradesh 1: 295. 1993; Samvatsar, Fl. West. Trib. Madhya Pradesh 79. 1996; B.D. Naithani in N.P. Singh *et al.,* Fl. India 5: 339.2000. *Hathi pan*

Perennial herbs; leaves simple 25-70 × 15-45 cm, ovate, cordate at base, acute, pubescent beneath, serrate; flowers greenish-white, in mealy pubescent corymbs; calyx lobes triangular-ovate; petals oblong; berry black, 3-6 celled, glabrous or nearly so, depressed globose; seeds usually 6.

Fl.-Fr.: July-Sept.

Occurs as forest undergrowth, in all districts.

SAPINDACEAE 138/1751; 4/5; 3/5

Key to Genera

1. Tendril bearing herbaceous annuals .... CARDIOSPERMUM
1. Tendrils absent; trees or shrubs
    2. Petals present; seeds exarillate .... SAPINDUS
    2. Petals absent; seeds arillate .... SCHLEICHERA

CARDIOSPERMUM L.W. 17; MP 1; SA 2

Key to Species

1. Flowers less than 2 cm long; capsule pyriform .... **C. halicacabum**
1. Flowers 2 cm long; capsule turbinate (top-shaped) .... **C. microcarpum**

Cardiospermum halicacabum L, Sp. Pl. 366. 1753; Hiern in Hook.f., Fl. Brit. India 1: 670.1875; Anand Kumar & Ram Lal inVerma *et al.,* Fl. Madhya Pradesh 1: 296, 1993; Samvatsar, Fl. West. Trib. Madhya Pradesh 79 1996; P.C. Pant in N.P. Singh *et al.,* Fl. India 5: 356.2000. *Kanphuti*

Annual, climbing herbs; leaves biternate, deltoid or ovate, 3-8 cm long; leaflets ovate or ovate-lanceolate, 4-4.5 × 1-2 cm, acute at apex, rounded at base; flowers white, in axillary cymes; capsules pyriform, 1.2-2 cm across, trigonous, truncate at top, with winged angles; seeds black, globose, 4-5 mm across, with white aril.

Fl.-Fr.: Almost throughout the year.

Climbing on bushes and hedges along the road side and forest-clearings, in all districts.

Cardiospermum microcarpum Kunth, Nov. Gen. Sp. 5: 104.1821; *Cardiospermum halicacabum* var. *microcarpum* (Kunth) Blume in Rumphia 3:185. 1847; *C. canescens* Wall. Pl. Asiat. Rar. 1:14. 1830; *C. microspermum* E. Mey, Zwei Pflanzengeogr. Dokum. 170. 1843; Hiern in Hook.f., Fl. Brit. India 1:670.1875; Almeida, Fl. Maharashtra 1:276. 1996. *Kanphuta*

Tendrillar herbs; leaves exstipulate, alternate bipinnate; leaflets inciso-dentate; flowers zygomorphic, 2 cm long, in few-flowered axillary cymes; cymes subtended by a pair of tendrils; capsule obversely deltoid; seeds *ca* 4 mm in across.

Fl.-Fr.: Aug. -Jan.

Rare, **Burhanpur**, Jainabad and Shahpur.

SAPINDUS L. W. 12; MP 3; SA 2

Key to Species

1. Leaves broadly oblong or elliptic, obtuse or emarginate ....**S. emarginatus**
1. Leaves obliquely ovate-lanceolate, acuminate .... **S. trifoliatus**

Sapindus emarginatus Vahl, Symb. Bot. 3: 54. 1794; Hiern in Hook.f., Fl. Brit. India 1: 682. 1875; Anand Kumar & Ram Lal in Verma *et al.,* Fl. Madhya Pradesh1: 300.1993; P.C. Pant in N.P. Singh *et al.,* Fl. India 5: 381.2000. *Ritha*

Trees, 5-15 m high; young parts tomentose; leaves alternate, 10-30 cm long; leaflets opposite or alternate, ovate-oblong or elliptic-oblanceolate, 3-11 × 2-6 cm, obtuse to emarginate at apex; flowers white, in terminal panicles; drupes ovoid, *ca* 1.5 cm across, smooth; seeds 5-7 mm across.

Fl.-Fr.: Oct. – Jan.

Confined to villages, not encountered in forest.

Sapindus trifoliatus L., Sp. Pl. 36 (1):586. 1753; Hiern in Hook.f., Fl. Brit. India 1: 682. 1875; P.C. Pant in N.P. Singh *et al.,* Fl. India 5: 382.2000. *Ritha*

Trees, 8-12 m high; leaves pinnate; leaflets elliptic-lanceolate, 5-14 × 3-5 cm, acuminate at apex, oblique at base; flowers white, in terminal, up to 15 cm long, rusty pubescent, panicles; drupes 1.5-2 cm long, 2-3-lobed, ferrugineous-velvety, wrinkled when ripe; seeds black, 6-8 mm long.

Fl.-Fr.: Sept. -March

Same as above species, but rare.

SCHLEICHERA Willd. W. 2; MP 1; SA 1

Schleichera oleosa (Lour.) Merr., Interpr. Rumph. Herb. Amboin. 337. 1917; *Pistacia oleosa* Lour., Fl. Cochinch. 615. 1790; *Schleichera trijuga* Willd., Sp. Pl. 4: 1096. 1806; *Schleichera oleosa* (Lour.) Oken, Allg. Naturgesch. 3(2): 1341. 1841; Hiern in Hook.f., Fl.

Brit. India 1: 681. 1875; Anand Kumar & Ram Lal in Verma *et al.,* Fl. Madhya Pradesh1: 302. 1993; Samvatsar, Fl. West. Trib. Madhya Pradesh 80.1996; P.C. Pant in N.P. Singh *et al.,* Fl. India 5: 384.2000. *Kusum*

Trees, 10-15 m high; leaves pinnate, bright red when young; leaflets oblong-lanceolate, 6-20 × 2-8 cm, coriaceous, truncate at base; flowers greenish- white, in axillary, 5-10 cm long racemes; drupes ovoid, 1-2 cm across, with persistent style; seeds 1-2, brown, compressed, enclosed in pulpy aril.

Fl.-Fr.: March – Aug.

Common in dry deciduous forests, often along the river bank, in all districts.

ANACARDIACEAE 77/701; 5/6; 4/4

Key to Genera

1. Flowers tetramerous; leaves pinnately compound; sepals, petals and carpels 4; stamens 6 .... LANNEA
1. Flowers pentamerous; leaves simple; flowers bisexual or polygamous
  2. Styles 3, terminal; carpels 3; drupe resting on a fleshy receptacle; flowers unisexual or polygamous .... SEMECARPUS
  2. Style 1, terminal or lateral; carpels 5 or 1
    3. Carpels 5; stamens 10; flowers bisexual .... BUCHANANIA
    3. Carpel 1; stamen 1; flowers polygamous; drupe large .... MANGIFERA

BUCHANANIA Spreng. W. 27; MP 2; SA 1

Buchanania lanzan Spreng., Schrad., J. Bot. 1800(2):234. 1800; *B. latifolia* Roxb., Fl. Ind. 2: 385. 1824; Hook.f., Fl. Brit. India 2: 23. 1876; Anand Kumar in Verma *et al.,* Fl. Madhya Pradesh 1: 306. 1993; Samvatsar, Fl. West. Trib. Madhya Pradesh 81 1996. *Char* (*Buchanania cochinchinensis* (Lour.) M.R. Almeida becomes *nom illeg.* as Lour's basionym *Toluifera cochinchinensis* is common for this species as well as for *Glycosmis cochinchinensis*, this is tentative till the matter is resolved).

Deciduous trees; leaves oblong or elliptic-oblong, 4-18 × 2-10 cm, obtuse at apex, rounded at base, coriaceous, tomentose beneath; flowers greenish- white, sessile, in 6-20 cm long, terminal panicles; drupes black, lenticular, 5-8 mm across; stone woody.

Fl. & Fr. : Jan – May

Occasional, all dists.

LANNEA A. Rich. W. 27; MP 1; SA 1

Lannea coromandelica (Houtt.) Merr., J. Arnold Arbor. 19(4): 353. 1938; *Dialium coromandelicum* Houtt., Nat. Hist. 2(2): 39, pl. 5,f. 2 1774; *Odina wodier* Roxb., Fl. Ind., ed. 1832 2: 293-294. 1832; Hook.f., Fl. Brit. India 2: 29. 1876; Anand Kumar in Verma *et al.,* Fl. Madhya Pradesh 1: 306. 1993; Samvatsar, Fl. West. Trib. Madhya Pradesh 81. 1996; D. Chandra & Mukh. in N.P. Singh *et al.,* Fl. India 5: 463.2000. *Moyan.* **(Plate 3, Fig. 2)**

Large trees, with ash-coloured bark; leaflets 7-11, ovate-oblong, 4.5-13 × 2.5-6 cm, acuminate, entire, without submarginal veins. flowers yellowish-green in drooping racemes borne at the ends of branches; males drooping; females erect; petals imbricate; drupes ovoid or sub-reniform, 1-seeded.

Fl.-Fr.: Feb. – May

Common in all districts.

MANGIFERA L. W. 10; MP 1 SA1

Mangifera indica L., Sp. Pl. 200.1753; Hook.f., Fl. Brit. India 2: 13. 1876; Anand Kumar in Verma *et al.,* Fl. Madhya Pradesh 1: 307. 1993; Samvatsar, Fl. West. Trib. Madhya Pradesh 81. 1996; D. Chandra & Mukh. in N.P. Singh *et al.,* Fl. India 5: 466.2000. *Aam*

Large evergreen trees; leaves 8-30 × 2-8 cm, oblong-lanceolate, acuminate, crowded at the ends of branches; flowers yellowish-green, polygamous, in terminal panicles; disc lobed; drupes ovoid, with sweet pulp when mature; seeds compressed.

Fl.-Fr.: Feb. – July

Common, cultivated and wild in deep forest in Khandwa and Khargone.

SEMECARPUS L.f. W. 5; MP 1; SA 1

Semecarpus anacardium L.f., Suppl. Pl. 182. 1782; Hook.f., Fl. Brit. India 2: 30.1876; Anand Kumar in Verma *et al.,* Fl. Madhya Pradesh 1: 309. 1993; D. Chandra & Mukh. in N.P. Singh *et al.,* Fl. India 5: 500.2000. *Bhilawa* **(Plate 3, Fig. 3)**

Trees; leaves obovate, 8-35 × 4.3-19 cm, broadly rounded at apex, cuneate or rounded at base, pubescent beneath; flowers greenish- yellow, in terminal, up to 25 cm long panicles; drupes black, obliquely ovoid or oblong, 1.5-3 cm long.

Fl. & Fr. : March – Jan.

Alirajpur, Badwani, Dhar, Indore, Khandwa, Khargone, but nowhere common, only 1-2 trees are seen.

MORINGACEAE 1/13; 1/2; 1/1

MORINGA Adans. W. 13; MP 2; SA 1

Moringa concanensis Nimmo, J. Grah. Cat. Bomb. Pl. 43. 1839; Hook.f., Fl. Brit. India 2:45. 1876; Anand Kumar in Verma *et al.,* Fl. Madhya Pradesh 1:310.1993; Samvatsar, Fl. West. Trib. Madhya Pradesh 82. 1996. *Kadwa surjana*

Soft wooded, medium sized trees, up to 12 m tall; leaves bipinnate, about 45 cm long, pinnules 9-15, elliptic or suborbicular, 1.5-3 cm long; bracts caducous; flowers in lax panicles, yellow,up to 1.5 cm long, streaked with red; stamens 5; staminodes 4-5; capsules acutely triquetrous, 30-50 × 1.5-2 cm, slightly constricted between the seeds; seeds 3-winged, bitter.

Fl.-Fr.: Nov. -April

Common in forests, village roads.

LEGUMINOSAE 946/24505; 71/258; 63/189

Key to Genera

1. Corolla papilionaceous; aestivation descending imbricate
   2. Stamens monoadelphous
      3. Leaflets stipellate; plants twining
         4. All anthers fertile; pod not hooked
            5. Leaflets 3
               6. Leaves pinnately tri-foliolate; keel petals united; pods not clustered
                  7. Nodes of raceme tumid;
                     8. Pod thick, fleshy, smooth, 4- angled, ribbed on the upper suture .... CANAVALIA
                     8. Pod membranous, flat, constricted between the seeds and covered with bristly brown hairs .... PUERARIA
                  7. Nodes of raceme not tumid; pod septate between the seeds, densely hairy, hairs not bristly; whole plant densely hairy .... NEONOTONIA
               6. Leaves palmately tri-foliolate; keel petals almost free; pods clustered in 3; prostrate herbs .... ROTHIA
            5. Leaflets 5-7 pairs; nodes of raceme densely silky, not tumid; pod, flat, indehiscent, velvety- brown .... MILLETIA
         4. Alternate anthers sterile; pod dehiscent, prominently hooked at apex; leaflets 3 ...... TERAMNUS
      3. Leaflets exstipellate
         9. Anthers dimorphic; stamens (10)
            10. Pod indehiscent, jointed, echinate, joints folded one on other; leaflets 2, gland-dotted; herbs .... ZORNIA
            10. Pod dehiscent, turgid, glabrous or hairy; leaves simple or 3-foliolate; stipulate or not........ CROTALARIA
         9. Anthers uniform; stamens (9)
            11. Leaflets even pinnate; seeds black and red; pod dehiscent; shrubby twiners .... ABRUS
            11. Leaflets odd pinnate; seeds uniform in colour; pod indehiscent, samaroid (not samara) or dehiscent next season; trees
               12. Leaflets alternate; pod samaroid .... DALBERGIA
               12. Leaflets opposite; pod woody, dehiscent next season .... PONGAMIA
   2. Stamens diadelphous, either 1+ (9) or (5) + (5)
      13. Stamens (5) + (5); pods indehiscent
         14. Leaflets alternate, odd; pods flat, samaroid; trees .... DALBERGIA
         14. Leaflets opposite; pods jointed; herbs
            15. Leaflets even in number; pods twisted, enclosed within the bristly calyx .... SMITHIA
            15. Leaflets odd in number; pods straight, elongated; exserted from the glabrous calyx .... AESCHYNOMENE
      13. Stamens 1+ (9)
         16. Rachis ending in a tendril

17. Pod compressed; staminal tube oblique at mouth .... VICIA
17. Pod turgid; staminal tube truncate at mouth ....LATHYRUS
16. Rachis not ending in a tendril
18. Pods indehiscent; seeds 1 or 2
19. Leaflets 3 or more (sometimes one)
20. Leaflets 5-7, alternate; fruit a samara; trees ....PTEROCARPUS
20. Leaflets 1- 3, opposite; fruit not a samara
21. Leaflets 3
22. Pod 1-seeded; leaflets not gland-dotted; calyx not accrescent; trees
23. Flowers large, red, not paniculate; trees or lianas .... BUTEA
23. Flowers very small, dull white, paniculate; lianas .... SPATHOLOBUS
22. Pod many seeded; leaves gland-dotted; calyx persistent and accrescent; twiners .... PARACALYX
21. Leaflets 1-3, gland-dotted; bracts foliaceous; seeds 2 .... FLEMINGIA
19. Leaflets 1 or 3; seed 1
24. Leaflets gland-dotted; seed adhering to pericarp .... CULLEN
24. Leaflets not gland-dotted; seed free from the pericarp; pod boat-shaped; leaf orbicular.. ELEIOTIS
25. Pods jointed; seeds more than two
26. Joints of pod echinate on the margins .... TAVERNIERA
26. Joints of pod not echinate
27. Pods twisted, enclosed in calyx; joints face to face ....URARIA
27. Pods not twisted, exserted from the calyx; herbs or trees
28. Leaves 1-foliolate
29. Unarmed herbs; joints of pods turgid; sepals scarious; pod less than 2.5 cm long ....ALYSICARPUS
29. Armed shrubs; joints of pod moniliform; sepals not scarious; pod 1.5-2.5 cm long .... ALHAGI
28. Leaves 1 and 3 foliolate intermixed; both sutures of pod indented ....DESMODIASTRUM
30. Leaves 1 or 3 foliolate; joints of pod flat; sepals herbaceous; herbs, shrubs or trees .... DESMODIUM
30. Leaves 1-foliolate; petiole broadly winged .... TADEHAGI
31. Leaves 3-foliolate
32. Bracts of 2 types, primary bracts large, foliaceous, orbicular, enclosing 5-8 flowers; segments of pod 1-2 .... PHYLLODIUM
32. Bracts not as above; upper suture of pod straight, lower indented; seeds strophiolate; pod hooked .......CODARIOCALYX
25. Pods not jointed; leaflets 3
33. Armed trees; petals very unequal; pod torulose ....ERYTHRINA

33. Un-armed herbs; petals sub-equal; pod not torulose
    34. Pods exserted from calyx
        35. Pods spirally coiled; their dorsal sutures hooked .... MEDICAGO
        35. Pods elongated, their dorsal sutures smooth .... TRIGONELLA
    34. Pods not exserted from the calyx; leaves toothed; lateral veins ending in teeth
        36. Leaves pinnately 3-foliolate; flowers in elongated racemes .... MELILOTUS
        36. Leaves palmately 3-foliolate; flowers in dense spikes .... TRIFOLIUM

18. Pods dehiscent
    37. Twining or trailing plants
        38. Petals very unequal; pods 'S'shaped, covered with irritating bristles; anthers alternately long and short .... MUCUNA
        38. Petals sub-equal; pods not curved, bristles absent; anthers uniform
            40. Pods septate between the seeds; seeds strophiolate; leaves gland-dotted; stipels absent; flowers yellow .... CAJANUS
            40. Pods not septate; seeds estrophiolate; leaves not gland-dotted
                41. Stipels present; flowers bluish-purple in 6-12 cm long raceme; pod straight; standard petal not appendaged .... GALACTIA
                41. Stipels absent; flower normally 1, axillary; pod falcately curved; standard petal with 2 appendages on either side .... MACROTYLOMA
                    31. Leaves 5 or more foliolate
                        42. Leaves 5-foliolate (our area); flowers 2, axillary, white; pod straight, glabrous .... CLITORIA
                        42. Leaflets 5-8 pairs; stipellate; pod torulose dehiscing next season ....ENDOSAMARA
    37. Erect herbs or shrubs; leaflets one to many; flowers in axillary clusters or racemes
        43. Anthers apiculate
            44. Standard petal with hairs on the back; keel petals not beaked; pod 1-many seeded, torulose or not; septa between seeds pithy; endocarp often spotted .... INDIGOFERA
            44. Standard petal glabrous; keel petals beaked; pod bilaterally flattened; septa between seeds membranous; endocarp never spotted ....INDIGASTRUM
        43. Anthers obtuse
            45. Style bearded in the upper part or near apex; leaflets 3............................ . VIGNA
            45. Style glabrous throughout; leaflets1-many
                46. Racemes leaf-opposed or terminal; lateral veins of leaflets very strong; leaves simple or compound .... TEPHROSIA

46. Racemes axillary; leaves compound with many leaflets .... SESBANIA

1. Corolla not papilionaceous; aestivation ascending imbricate or valvate

47. Aestivation ascending imbricate

48. Leaves simple, shallowly or deeply lobed; fertile stamens 3-10 .... BAUHINIA

48. Leaves compound;

49. Leaves unipinnate compound;

50. Petals absent; pod samaroid in the lower part, indehiscent; seed 1, near the tip; leaflets one pair...HARDWICKIA

50. Petals present; pod not samaroid

51. Petals 3, 2 reduced to scales; stamens (3); pod indehiscent .... TAMARINDUS

51. Petals 5

52. Petiole and rachis without glands; stamens 10, filaments of 3 abaxial (lower) stamens sigmoidly curved, usually longer than the anthers; pod cylindric, indehiscent .... CASSIA

52. Petiole and rachis with glands; stamens 10 or 5 fertile; filaments straight; pod dehiscent or not

53. Bracteoles absent; stamens 10 all fertile or 3 staminodial; pod either indehiscent or if dehiscent by both sutures, then valves not coiling or valves breaking into 1-seeded joints .... SENNA

53. Bracteoles present; stamens 5 or 10, all fertile; pods elastically dehiscent, valves coiling.. CHAMAECRISTA

49. Leaves bipinnate compound

54. Sepals imbricate; stragglers studded densely with spines; stipules simple or foliaceous; pod prickly or not, 1-6 seeded ....CAESALPINIA

54. Sepals valvate

55. Armed, armature due to stipular spines and primary rachis; secondary rachis flat, green,ribbon-like; pinnules minute; pod torulose ....PARKINSONIA

55. Un-armed; rachis and pinnules not as above; pod not torulose .... DELONIX

47. Aestivation valvate

56. Stamens definite, 5-10

57. Anthers gland crested; flowers pentamerous

58. Flowers in globose heads; plants not prickly; peduncles with 2 large bracts ....NEPTUNIA

58. Flowers in elongated spikes; plants prickly

59. Pods turgid, straight, septate (septa obscure due to mesocarp); flowers uniform in colour ....PROSOPIS

59. Pods not turgid, finally twisted, aseptate; flowers dimorphic in colour, upper yellow, lower pink .... DICHROSTACHYS

60. Anthers not gland crested; flowers tetramerous; pods slightly curved with segmented valves and persistent sutures .... MIMOSA

56. Stamens indefinite; flowers pentamerous

60. Stamens free; spines stipular and or spread all over ....ACACIA

60. Stamens monoadelphous; spines if present, of stipular origin

61. Spines absent; pods straight, flat; pinnae and leaflets in several pairs; funicle filiform .... ALBIZIA

61. Spines present, stipular and primary rachis also modified into a spine; pods coiled; pinnae and leaflets one pair; funicle fleshy .... PITHECELLOBIUM

ABRUS Adans. W. 19; MP 2; SA 1

Abrus precatorius L., Syst. Nat. ed. 12.2: 472. 1767; Baker in Hook.f., Fl. Brit. India 2: 175. 1876; Sanjappa, Legumes of India 74. 1991; Singh in Verma *et al.*, Fl. Madhya Pradesh 1: 317. 1993; Samvatsar, Fl. West. Trib. Madhya Pradesh 84. 1996. *Gumchi* **(Plate 3, Fig. 4)**

Twining shrubs; leaflets 10-20 pairs, oblong or obovate-oblong, 1-5.5 × 0.5-1 cm, entire, apiculate, appressed hairy; flowers pink or white, fascicled on the swollen nodes of axillary racemes; pods oblong, turgid, 2-5 × 1.5 cm, 3-5 seeded, appressed hairy; seeds subglobose, scarlet, shining, with black spot at the hilum.

Fl.-Fr.: Aug. - March

Common among clumps of trees and shrubs in wastelands and open forests.

ACACIA Miller W. 1393; MP 21; SA 15

Key to Species

1. Pods cylindric, turgid; flowers in globose heads
    2. Leaflets 5-10 pairs; stipular spines alternately long, straight and short, curved; pod cicinately coiled; seeds uniseriate; peduncle bracteate near the base; heads white; trees having umbrella-shaped canopy .... **A. planifrons**
    2. Leaflets 10-20 pairs; stipular spines uniform; pod not coiled; seeds biseriate; peduncle bracteate near the apex; heads bright yellow; shrubs .... **A. farnesiana**
1. Pods flat
    3. Plants prickly all over; scramblers
        4. Flowers pedicellate; pods dark, chocolate-brown; leaflets overlapping .... **A. pennata**
        4. Flowers sessile; pod not as above; leaflets not overlapping; stem 5-angled
            5. Pinnae 5-6 pairs; leaflets 15-20 pairs .... **A. caesia**
            5. Pinnae 12-20 pairs; leaflets 30-45 pairs .... **A. torta**
    3. Plants armed with stipular spines only
        6. Flowers in globose heads
            7. Flowers in terminal panicles, fragrant, whitish .... **A. leucophloea**
            7. Flowers in axillary heads, yellow
                8. Pod moniliform, persistently silvery canescent .... **A. nilotica**
                8. Pod not moniliform, nearly glabrous; leaves glabrous
                    9. Flowers fragrant; pod transversely or reticulately veined .... **A. jacquemontii**
                    9. Flowers foul smelling; pod longitudinally veined .... **A. eburnea**
        6. Flowers in elongated spikes
            10. Stipular spines 3-nate .... **A. senegal**

10. Stipules paired
    11. Pinnae up to 4 pairs ....**A. lenticularis**
    11. Pinnae more than 4 pairs
        12. Pod beaked at apex; sepals and petals villous ....**A. catechu**
        12. Pod not beaked; sepals and petals not villous
            13. Corolla pale yellow ....**A. ferruginea**
            13. Corolla white
                14. Petals equal to sepals or slightly longer................................ ....**A. polyacantha**
                14. Petals thrice longer than sepals... ....**A. chundra**

Acacia caesia (L.) Willd., Sp. Pl. 4(2): 1090.1806; *Mimosa caesia* L., Sp. Pl. 1: 522. 1753, *p.p; Acacia intsia* (L.) Willd. var. *caesia* (L.) Wight & Arn. ex Baker in Hook.f., Fl. Brit. India 2: 297. 1878; Sanjappa, Legumes of India 37. 1991; Samvatsar, Fl. West. Trib. Madhya Pradesh 119. 1996; Khanna *et al.,* Suppl. Fl. Madhya Pradesh 1: 82.2001.

Extensively large, climbing shrubs; branchlets, petioles and peduncles pubescent and armed; leaves 2-pinnate; rachis 7.5-30 cm long, raised gland above base and 2 to 3 smaller ones between upper most pinnae; pinnae 3-6 pairs, up to 5 cm long; leaflets 10-35 pairs, linear or oblong, acute at apex, oblique at base, pubescent; flowers creamy white, in globose heads, in terminal panicles; pods flat, 10-15 cm long, dehiscent, slightly curved, rusty or brown tomentose when young, glabrous on maturity, 5-10-seeded.

Fl.-Fr.: Aug. – March

Rare, Burhanpur, Dhar, Khandwa.

Acacia catechu (L.f.) Willd., Sp. Pl. 4(2): 1079. 1806; *Mimosa catechu* L.f., Suppl. Pl. 439. 1782; Baker in Hook.f., Fl. Brit. India 2: 295. 1878; Sanjappa, Legumes of India 38. 1991; Singh in Verma *et al.,* Fl. Madhya Pradesh 1: 431. 1993; Samvatsar, Fl. West. Trib. Madhya Pradesh 119. 1996. *Khair* **(Plate 3, Fig. 5)**

Trees, 3-6 m tall; bark dark grey; young branchlets pubescent, prickles brown or grey; leaves 9-16 cm long, with a large gland above the middle of the petiole and several glands between the pinnae; pinnae 10-30 pairs; leaflets 20-50 pairs subessile, linear-oblong, subacute; flowers pale yellow, mildly fragrant; pods 5-8.5 × 1.5 cm, flat with a triangular beak at the apex, shining, brown.

Fl.-Fr.: May-Sept.

Common along forest paths, open grasslands roadsides, throughout the area.

Acacia chundra (Roxb. ex Rottler) Willd. Sp. Pl. 4(2):1078. 1806; *Mimosa chundra* Roxb. ex Rottler in Ges. Naturf. Freunde Berlin Neue Schriften 4: 207. 1803; Baker in Hook.f., Fl. Brit. India 2: 295. 1878; Sanjappa, Legumes of India 38. 1991; Singh in Verma *et al.,* Fl. Madhya Pradesh 1: 432. 1993; Samvatsar, Fl. West. Trib. Madhya Pradesh 119. 1996. *Lal khair*

Medium sized, totally glabrous trees, with rough bark; spines up to 5 mm long; pinnae 10-20 pairs; rachis with small glands between the upper pinnae; petiolar gland at the base of first pinna;

leaflets 20-40 pairs, ligulate; spikes white, drooping; calyx, corolla and leaves glabrous; pods strap-shaped, straight, up to 6 × 1 cm, subacute at both ends.

Fl.-Fr.: July-Oct.

Common in forests, all over the area.

Acacia eburnea (L.f.) Willd., Sp. Pl. 4(2): 1081. 1806; *Mimosa eburnea* L.f. Suppl. Pl. 437, 1781; Baker in Hook.f., Fl. Brit. India 2: 293, 1878; Sanjappa, Legumes of India 39. 1991; Singh in Verma *et al.,* Fl. Madhya Pradesh 1: 432. 1993; Samvatsar, Fl. West. Trib. Madhya Pradesh 119. 1996. *Mormati*

Small trees; leaves 2-3 cm long; stipular spines straight, up to 2 cm long; pinnae 2-7 pairs, 1-3 cm long; leaflets 5-8 pairs, linear subotuse; heads yellow, pedunculate; flowers fetid; pods stalked, flat 6-13 × 0.6-1 cm, glabrous.

Fl.-Fr.: Nov. -Feb.

Not common, Alirajpur, Dhar, Indore, Khandwa.

Acacia farnesiana (L.) Willd., Sp. Pl. 4(2): 1083-1084. 1806; *Mimosa farnesiana* L., Sp. Pl. 521. 1753; Baker in Hook.f., Fl. Brit. India 2: 292. 1878; Sanjappa, Legumes of India 39. 1991; Singh in Verma *et al.,* Fl. Madhya Pradesh 1: 432. 1993; Samvatsar, Fl. West. Trib. Madhya Pradesh 120.1996. *Deo babool*

Small trees or large shrubs; pinnae 4-8 pairs, 2-3 cm long leaflets 12-20 pairs, linear-oblong; spines stipular; rachis eglandular; petiolar gland near the middle; heads yellow,fragrant, solitary or fascicled; pods hard, 3.6-8 × 1.2 cm with pulpy mesocarp; seeds broadly oblong or suborbicular, biseriate.

Fl.-Fr.: Aug. -March

Alirajpur, Badwani, Burhanpur, Indore, Khandwa, Khargone.

Acacia ferruginea DC., Prodr. 2: 458. 1825; *Mimosa ferruginea* Roxb., Fl. Ind. 2: 561. 1832; Baker in Hook.f., Fl. Brit. India 2: 295. 1878; Sanjappa, Legumes of India 39. 1991; Singh in Verma *et al.,* Fl. Madhya Pradesh 1: 433. 1993; Samvatsar, Fl. West. Trib. Madhya Pradesh 120.1996. *Safed khair*

Medium-sized trees, 6-10 m high; leaf-rachis with one gland towards the base and one between the upper most pinnae; pinnae 2-6 pairs, 5-8 cm long; leaflets 10-20 pairs, ligulate, oblong, 7-10 × 2-3 mm, obtuse at apex, oblique at base; stipular spines dark brown, paired, hooked, sometimes suppressed; flowers yellow, in 7-10 cm long, dense spikes; pods dark brown, 8-12 × 1.8-3 cm, glabrous, stipitate, 4-6-seeded.

Fl.-Fr.: March – Jan.

Badwani, Dhar, Indore, Jhabua, Khargone.

Acacia jacquemontii Benth. in Hook., London J. Bot. 1: 499. 1842; Baker in Hook.f., Fl. Brit. India 2: 293. 1878; Singh in Verma *et al.,* Fl. Madhya Pradesh 1: 433. 1993; Samvatsar, Fl. West. Trib. Madhya Pradesh 120.1996.

Small shrubs, up to 3 m high; branches smooth, brown; leaves 2-pinnate; pinnae 2-8 pairs, 12-25 mm long; leaflets 6-12 pairs, linear or oblong, glabrous; stipular spines straight, paired, 7-

18 mm long, connate at base; flowers pale yellow, in axillary, pedunculate, fascicled, globose heads; pods ovate-oblong, 3-7.5 × 0.8-1.5 cm, reticulate, glabrous, dehiscent, 5-6-seeded.

Fl.-Fr.: Dec. – May

Alirajpur, Indore, Khandwa.

Acacia lenticularis Buch.-Ham. ex Benth. in Hook., London J. Bot. 1: 508. 1842; Baker in Hook.f., Fl. Brit. India 2: 296. 1878; Singh in Verma *et al.,* Fl. Madhya Pradesh 1: 433. 1993; Samvatsar, Fl. West. Trib. Madhya Pradesh 120.1996. *Kanta siris*

Tall trees; leaflets 6-8 pairs oblong to oblong-obovate, glabrous to sparsely pubescent gland on rachis absent; flowers white, in dense spikes; pods flat to 25 cm long; seeds 6-9.

Fl.-Fr.: May - Dec.

Rare, **Dhar,** Dharampuri bet.

Acacia leucophloea (Roxb.) Willd., Sp. Pl. 4(2): 1083. 1806; *Mimosa leucophloea* Roxb., Pl. Coromandel 2: 27. t. 150.1800; Baker in Hook.f., Fl. Brit. India 2: 294. 1878; Sanjappa, Legumes of india 41. 1991; Singh in Verma *et al.,* Fl. Madhya Pradesh 1: 434. 1993; Samvatsar, Fl. West. Trib. Madhya Pradesh 120.1996. *Reunja* **(Plate 3, Fig. 6)**

Trees, 6-10 m high, usually with insect galls on the trunk, and branches; pinnae 6-10 pairs, 2.5-4.5 cm long, with concave glands in between, leaflets 15-25 pairs, linear-oblong; stipular spines up to 3.5 cm long; petiolar gland near the middle; heads pale, fragrant, in tomentose panicles; pedicels with a whorl of 4 deciduous bracts towards the base; pods sessile, flat, 6-10 × 0.5-0.8 cm, shortly falcate, pubescent.

Fl.-Fr.: Aug. - Jan.

Very common in wastelands, along roads and on the fringes of forests throughout the area.

Acacia nilotica (L.) Willd. ex Delile

Key to subspecies

1. Canopy rounded, side branches spreading horizontally .... **A. nilotica** subsp. **indica**
1. Canopy vertical, side branches parallel to the main stem .... **A. nilotica** subsp. **cupressiformis**

subsp. **cupressiformis** (J.L. Stewart) Ali & Faruqui, Pakistan J. Bot. 1: 4 1969; *Acacia arabica* var. *cupressiformis* J.L. Stewart in Punjab Pl. 51. 1869; Sanjappa, Legumes of India 41. 1991; Samvatsar, Fl. West. Trib. Madhya Pradesh 120.1996; Khanna *et al.,* Suppl. Fl. Madhya Pradesh 1: 83.2001.

Fl.-Fr.: July-Jan.

Rare, **Dhar** only, at Katnera.

subsp. **indica** (Benth.) Brenan, Kew Bull. 12: 84. 1957; *A. arabica* var. *indica* Benth., London J. Bot. 1: 500, 1842; *A. nilotica* var. *indica* (Benth.) Hill, Bot. Mus. Leafl. 8: 99, 1940; Baker in Hook.f., Fl. Brit. India 2:293, 1878; Sanjappa, Legumes of India 42 . 1991; Singh in Verma *et al.,* Fl. Madhya Pradesh 1: 434. 1993; Samvatsar, Fl. West. Trib. Madhya Pradesh 120.1996. *Babul*

Trees, young branches pubescent; generally leaves with a gland between the first pair of pinnae and one between the terminal pair; pinnae 5-7 pairs; leaflets 18-25 pairs, subsessile, glabrous, obtuse at apex; flowers in heads; pods distinctly stalked; grey pubescent, slightly compressed, moniliform, 10-12 seeded.

Fl.-Fr.: July - April

Common everywhere, especially near streams.

Acacia pennata (L.) Willd. Sp. Pl. 4 (2): 1090-1091. 1806; *Mimosa pennata* L. Sp. Pl. 522. 1753; Baker in Hook.f., Fl. Brit. India 2: 297. 1878; Singh in Verma *et al.,* Fl. Madhya Pradesh 1: 434. 1993; Samvatsar, Fl. West. Trib. Madhya Pradesh 122. 1996. *Paoni*

A prickly scandent shrub; leaves bipinnate with glands above the middle of the petiole and one between each of the 2-4 upper pair of pinnae; leaflets 40-50 pairs, slightly overlapping, sessile, linear, obtuse, ciliate along the margins; flowers in large terminal panicles, heads globose; pods stalked, straight, flat, thin, dry, strap-shaped, rounded or shortly apiculate, glabrous; seeds 8-14, dark brown.

Fl.-Fr.: July - Dec.

Common, all over the area.

Acacia planifrons Wight & Arn. Prodr. Fl. Ind. Orient. 1: 276. 1834; Baker in Hook.f., Fl. Brit. India 2: 293. 1878; Samvatsar, Fl. West. Trib. Madhya Pradesh 122. 1996; Khanna *et al.,* Suppl. Fl. Madhya Pradesh 1: 84.2001.

Armed, 4-6 m tall trees; petiole without gland; leaflets 2-3 mm long, linear, obtuse; flowers white, in globular heads; calyx funnel shaped; pods turgid, circinate, glabrous; 6-9 seeded.

Fl.-Fr.: Oct. -Dec.

Burhanpur, Dhar, Jhabua, Khandwa.

Acacia polyacantha Willd. Sp. Pl. 4(2): 1079–1080.1806; *Mimosa suma* Roxb., Fl. Ind. 2: 563. 1832. *Acacia suma* (Roxb.) Kurz in Brandis, For. Fl. N.W.C. Ind. 187. 1874; Baker in Hook.f., Fl. Brit. India 2: 294. 1878; Singh in Verma *et al.,* Fl. Madhya Pradesh 1: 435. 1993. *Safed babool*

Trees, to 15 m high, bark yellowish-grey; stipular spines paired, straight or hooked; leaves 10-20 cm long; pinnae 10-30 pairs; leaflets 30-50, opposite, sessile; flowers white, 4 mm across, in axillary solitary or paired spikes; pod stipitate, 8 ×1.5 cm, glabrous, grey, woody, oblong, base and apex horned; seeds 5-8.

Fl.-Fr.: April - Dec.

In mixed dry deciduous forests, Indore, Khandwa.

Acacia senegal (L.) Willd., Sp. Pl. 4: 1077. 1806; Baker in Hook.f., Fl. Brit. India 2: 295. 1878; Singh in Verma *et al.,* Fl. Madhya Pradesh 1: 435. 1993. *Mimosa senegal* L., Sp. Pl. 521. 1753. *Kumpta*

Thorny, small trees; leaves 2-pinnate; pinnae 3-5 pairs; petioles armed with numerous prickles with a gland below the lowest pair and one between the last pair; leaflets 8-14 pairs, linear-oblong, 1.5-3 × *ca* 1.2 mm, glabrous; stipular spines in threes, 2 lateral ones straight, third one, hooked, brown, compressed; flowers white, in 5-8 cm long, cylindrical spikes; pods linear-oblong, *ca* 8 × 1.8 cm, indehiscent, 4-6-seeded; seeds sub-orbicular, 8-12 mm across.

Fl.-Fr.: July – Jan.

Khandwa only, planted outside the city, in the forest.

Acacia torta (Roxb.) Craib in Bull. Misc. Inform. Kew 1915(9): 410.1915; *Mimosa torta* Roxb., Fl. Ind. 2: 566. 1832; *Acacia pennata auct. non* (L.) Willd., 1806: Baker in Hook.f., Fl. Brit. India 2: 297. 1878, *p.p;* Singh in Verma *et al.,* Fl. Madhya Pradesh 1: 436. 1993; Samvatsar, Fl. West. Trib. Madhya Pradesh 122. 1996. *Aila*

Woody climbers; leaf rachis with a conical gland and a few smaller glands between upper 2-3 pairs of pinnae; pinnae 8-12 pairs; flowers white; pods to 12 × 2 cm, 5-7 seeded.

Fl.-Fr.: April - March

Alirajpur, Badwani, Burhanpur, Khandwa, Khargone.

AESCHYNOMENE L. W. 170; MP 2; SA 1

Aeschynomene indica L. Sp. Pl. 713-714. 1753; Baker in Hook.f., Fl. Brit. India 2: 151. 1876; Singh in Verma *et al.,* Fl. Madhya Pradesh 1: 318. 1993; Samvatsar, Fl. West. Trib. Madhya Pradesh 85. 1996. *Antwa*

Erect, much branched, marshy annual herbs; stem fistular, warty, hairy on young parts; leaflets 15-35, lanceolate to elliptic, obliquely rounded, with subcordate base; flowers in 1-4 flowered, axillary racemes; calyx membranous, upper lip bifid, lower lip minutely 3 toothed, standard yellow, veined with purple; pods slightly curved or nearly straight, 6-10 jointed.

Fl.-Fr.: Aug. -Dec.

Very common in fields, road-side, wet places, throughout the area.

ALBIZIA Duraz. W. 137; MP 5; SA 4

Key to Species

1. Pinnae more than 6 pairs; leaflets 10-30 pairs, less than 1 cm long, midrib central; seeds up to 8 .... **A. amara**
1. Pinnae less than 6 pairs; leaflets up to 15 pairs, more than 1cm long; midrib eccentric; seeds up to 12
    2. Flowers pedicellate, not in panicles; pinnae 2-3 pairs; leaflets 5-9 pairs; pod 10-30 × 2-4 cm, pale yellow .... **A. lebbeck**
    2. Flowers sessile, in panicles
        3. Midrib towards upper margin; calyx teeth obsolete .... **A. odoratissima**
        3. Midrib towards lower margin; calyx teeth distinct .... **A. procera**

Albizia amara (Roxb.) Boivin, Encycl. du 19 Siecle 2: 34. 1834; *Mimosa amara* Roxb., Pl. Coromandel 2: 13. t. 122. 1799; Baker in Hook.f., Fl. Brit. India 2: 301. 1878; Sanjappa, Legumes of India 54. 1991; Singh in Verma *et al.,* Fl. Madhya Pradesh 1: 437. 1993; Samvatsar, Fl. West. Trib. Madhya Pradesh 122. 1996. *Kastar*

Small deciduous trees; branchlets yellow-pubescent; leaves abruptly bipinnate; petiole gland near middle or above it; rachis with gland between upper pinnae; pinnae 8-10 pairs; leaflets 16-25 pairs, sessile,linear oblong, ciliate along margin; flowers pale yellow, subsessile, in 12-20 flowered heads; pods greyish-brown, faintly veined, pubescent; seeds 6-8.

Fl.-Fr.: April-June

Common at the fringe of forest, road-sides, throughout the area.

Albizia lebbeck (L.) Benth. in Hook., Lond. J. Bot. 3: 87. 1844; *Mimosa lebbeck* L., Sp. Pl. 516. 1753; Baker in Hook.f., Fl. Brit. India 2: 298. 1878; Sanjappa, Legumes of India 56. 1991; Singh in Verma *et al.,* Fl. Madhya Pradesh 1: 437. 1993; Samvatsar, Fl. West. Trib. Madhya Pradesh 122. 1996. *Kala siris* **(Plate 3, Fig. 7)**

Large deciduous tree, 10-25 m high; leaves 2-pinnate; rachis 7-22 cm long, with a large gland near base and 1-3 smaller glands between pinnae; pinnae 2-3 pairs; leaflets 5-9 pairs, oblong, 2-5 × 0.9-3 cm, obtuse, mid rib closer to upper margin; flowers white, pedicellate, axillary, in axillary heads; pods straw-coloured, linear-oblong, 10-30 × 2-5 cm, 8-12-seeded; seeds pale brown, ovoid-oblong, *ca* 1 cm long, compressed.

Fl.-Fr.: March - Oct.

In mixed dry deciduous forests and along roadsides.

Albizia odoratissima (L.f.) Benth. in Hook. London J. Bot. 3: 88. 1844; *Mimosa odoratissima* L.f. Suppl. Pl. 437. 1781; Baker in Hook.f., Fl. Brit. India 2: 299. 1878; Singh in Verma *et al.,* Fl. Madhya Pradesh 1: 437. 1993; Samvatsar, Fl. West. Trib. Madhya Pradesh 123. 1996. *Siris*

Tall trees; young branches pubescent; leaves abruptly 2-pinnate, gland of petiole near the base; rachis with gland between uppermost pair of pinnae; pinnae 4 pairs; leaflets 12-18 pairs, obliquely oblong, obtuse at apex; flowers white, fragrant, sessile, in numerous, small, 10-12 flowered heads; panicles large, terminal; pods shortly stalked, thin, flexible, glabrous, faintly veined, brown; seeds 8-12.

Fl.-Fr.: March - Oct.

Frequent in forest, in all districts.

Albizia procera (Roxb.) Benth. in Hook., London J. Bot. 3: 89. 1844; *Mimosa procera* Roxb., Pl. Coromandel 2: 12. t. 121. 1799; Baker in Hook.f., Fl. Brit. India 2: 299. 1878; Singh in Verma *et al.,* Fl. Madhya Pradesh 1: 438. 1993; Samvatsar, Fl. West. Trib. Madhya Pradesh 123. 1996. *Safed siris*

Tall trees; young branches pale, lenticellate; leaves abruptly 2-pinnate; rachis with gland near base of petiole; pinnae usually 3 pairs; leaflets 5-7 pairs, obliquely ovate, obtuse, pubescent beneath; flowers greenish-white, sessile,in terminal, 10 -20 flowered heads; pods scarcely stalked, thin, flexible, orange-brown, smooth; seeds 6-12.

Fl.-Fr.: Aug. -May

Common forest tree, everywhere in the area.

ALHAGI Gagneb. W. 8; MP 1; SA 1

Alhagi maurorum Medik. in Vorles. Churpfalz. Phys. -Okon. Ges. 2: 397. 1787; *A. camelorum* Fisch., Ind. Hort. Gorenk. ed. 1:72. 1812; *A. pseudoalhagi* (M. Bieb.) Desv., J. Bot. 1:120.1823; *Hedysarum pseudoalhagi* M. Bieb., Fl. Taur. Caue. 2:174. 1808; Baker in Hook.f., Fl. Brit. India 2: 145. 1876; Samvatsar, Fl. West. Trib. Madhya Pradesh 85. 1996; Khanna *et al.,* Suppl. Fl. Madhya Pradesh 43.2001. *Javasa*

Erect, spinous shrub; leaves simple, 10-12 × 4.5-5.5 mm long, oblong, obovate, mucronate, base cuneate, stipules subulate; flowers red, 4-7 on axillary, 2-3 cm long, spinous shoots; calyx teeth triangular, auricled at the base; pods 1.5-2.5 cm, falcate, glabrous, moniliform.

Fl.-Fr.: March-May

Common on the gravelly banks of river, in barren fields, Alirajpur, Dhar, Indore.

ALYSICARPUS Neck. ex Desv. W. 34; MP 15; SA 14

Key to Species

1. Upper leaves 3-foliolate, lower 1 and 3 intermixed .... **A. heterophyllus**
1. All leaves unifoliolate
  2. Calyx shorter or very slightly longer than the first joint of pod
    3. Pod moniliform, not veined .... **A. monilifer**
    3. Pod at the most slightly moniliform, veined
      4. Calyx equalling or slightly exceeding the first joint of pod; stem glabrous .... **A. vaginalis**
      4. Calyx shorter than the first joint of pod; stem densely pubescent, with long, scattered hairs .... **A. hamosus**
  2. Calyx much longer than the first joint of pod
    5. Calyx teeth not imbricating in fruit; rachis villous .... **A. pubescens**
    5. Teeth of calyx imbricate in fruiting stage
      6. Pod 4-angled, exserted .... **A. tetragonolobus**
      6. Pod flattened or subterete
        7. Pod quite smooth, subcylindric, 3-6 jointed .... **A. roxburghianus**
        7. Pod transversely rugose, veined or ribbed
          8. Pod 1-3 jointed .... **A. glumaceous**
          8. Pod 3-5 jointed .... **A. rugosus**
            9. Stem robust, erect; racemes more than 5 cm long; flowers distant .... **A. heyneanus**
            9. Stem diffuse; racemes less than 5 cm long, congested .... **A. scariosus**
              10. Racemes 15-25 cm long, joints reticulately veined .... **A. longifolius**
              10. Racemes3-10 cm long, joints faintly reticulate .... **A. bupleurifolius**

Alysicarpus bupleurifolius (L.) DC. Prodr. 2: 352. 1825; *Hedysarum bupleurifolium* L. Sp. Pl. 745. 1753; Baker in Hook.f., Fl. Brit. India 2: 158, 1876; Singh in Verma *et al.,* Fl. Madhya Pradesh 1: 319. 1993; Samvatsar, Fl. West. Trib. Madhya Pradesh 86. 1996.

Annual, erect herbs; leaves 1.5-3 × 0.5-1.5 cm, lanceolate to oblong, glabrous above, sparsely hairy below; flowers in spicate lax racemes; pods moniliform, 2-4 jointed; joints faintly reticulate to rugose.

Fl.-Fr.: Aug. -Nov.

Common in moist places in fields and grasslands, road-sides, throughout the area.

Alysicarpus glumaceus (Vahl) DC., Prodr. 2: 353. 1825; *Hedysarum glumaceum* Vahl, Symb. Bot. 2: 106. 1791; *Alysicarpus rugosus* (Willd.) DC. Prodr. 2:353. 1825; Baker, in Hook.f., Fl. Brit. India 1:371. 1876; Kothari in N.P. Singh *et al.,* Fl. Maharashtra 1:601.2000.

Annual, erect or diffuse herbs; leaves 1-foliolate, obovate-oblong, 2-6 × 1-2.5 cm, obtuse at both ends, glabrous above, pubescent beneath; flowers pinkish yellow, in 3-10 cm long racemes; pods 6-8 mm long, turgid, 3-6-jointed, transversely ribbed; seeds brown, ellipsoid, *ca* 1.3 mm long, smooth.

Fl.-Fr.: Sept. – Jan.

Common in wastelands and cultivated fields.

Alysicarpus hamosus Edgew. in J. Asiat. Soc. Bengal 21: 171. 1853; Baker in Hook.f., Fl. Brit. India 2: 157. 1876; *Hedysarum procumbens* Roxb., Fl. Ind. 3: 345. 1832, *non* Mill., 1768; *Alysicarpus procumbens* (Roxb.) Schindler, Fedde. Rep. 21:11. T. 43. 1925, *nom. illeg;* Singh in Verma *et al.,* Fl. Madhya Pradesh 1: 319 1993.

Annual, prostrate or procumbent-ascending herbs; leaves 1-foliolate, suborbicular, 1.5-2.5 × 1.5-2.5 cm, obtuse and mucronate at apex, cordate at base, appressed hairy; stipules 6-8 mm long; flowers bluish or purplish, terminal or axillary racemes; pods 10-15 × 2-4 mm, compressed, pubescent, 4-7-jointed, reticulately veined.

Fl.-Fr.: Aug. – Dec.

Common forest undergrowth, everywhere.

Alysicarpus heterophyllus (Baker) Jafri *et al.,* in Biologia 12:33, 1966; *A. vaginalis* (L.) DC. var. *heterophyllus* Baker in Hook.f., Fl. Brit. India 2: 158, 1876; Samvatsar, Fl. West. Trib. Madhya Pradesh 86. 1996; Khanna *et al.,* Suppl. Fl. Madhya Pradesh 45. 2001.

Erect, diffusely branched herbs; upper leaves 3-foliolate, lower 1 and 3 foliolate intermixed; leaflets 1-7 × 0.25-1 cm, lanceolate to oblong-lanceolate, acute; flowers pink, remotely paired in 6-20 flowered racemes; calyx equalling or slightly longer than the first joint of pod; joints 3-5 longer than broad, tetragonal, reticulately veined, minutely hairy.

Fl.-Fr.: Aug. -Oct.

Rare in wastelands, **Alirajpur** only, Kawant road, Phulmal.

Alysicarpus heyneanus Wight & Arn., Prodr. Fl. Ind. Orient. 1: 234. 1834; *Alysicarpus rugosus* (Willd.) DC. var. *heyneanus* (Wight & Arn.) Baker in Hook.f., Fl. Brit. India 2: 159. 1876; Khanna *et al.,* Suppl. Fl. Madhya Pradesh 1: 45. 2001.

Annual, erect or diffuse herbs; leaves 1-foliolate, obovate, oblong or linear-lanceolate, 1.5-14 × 0.4-2 cm, acute or obtuse-retuse at apex, pubescent beneath; flowers red to purple, in terminal and axillary, lax, 10-15 cm long racemes; pods 6-8 mm long, transversely rugose.

Fl.-Fr.: Aug. – Feb.

Common in dry regions of the study area.

Alysicarpus longifolius (Rottler ex Spreng.) Wight & Arn., Prodr. Fl. Ind. Orient. 1: 233. 1834; *Hedysarum longifolium* Rottler ex Spreng., Syst. 3: 319. 1826; Baker in Hook.f., Fl. Brit. India 2: 159, 1876; Singh in Verma *et al.,* Fl. Madhya Pradesh 1: 320.1993; Samvatsar, Fl. West. Trib. Madhya Pradesh 86. 1996. *Jangli gailiya*

Stem erect, ascending glabrous; leaves short petioled, oblong or lanceolate, subcordate, mucronate, hairy on the nerves beneath; pod shortly stalked, apiculate, slightly moniliform, joints 4-6, glabrous, reticulately veined.

Fl.-Fr.: Aug. -Sept.

Common along hedges of fields, throughout the area.

Alysicarpus monilifer (L.) DC. Prodr. 2: 353.1825; *Hedysarum moniliferum* L., Mant. Pl. 1: 102 -103.1767; Baker in Hook.f., Fl. Brit. India 2: 157.1876; Singh in Verma *et al.,* Fl. Madhya Pradesh 1: 320.993; Samvatsar, Fl. West. Trib. Madhya Pradesh 86.1996. **(Plate 3, Fig. 8)**

Perennial, much branched, prostrate herbs; branches with spreading hairs; leaves unifoliolate; leaflets up to 2.5-1.8 cm, base rounded, often cordate, apex rounded or retuse, mucronate; flowers red, in close, axillary racemes; pods 0.5-2.3 cm long, distinctly moniliform, turgid, faintly reticulate, 2-7 seeded; seeds smooth, rather shining, brown.

Fl.-Fr.: Aug. -March

Common in fields as weed, throughout the area.

Alysicarpus pubescens

Key to Varieties

1. Flowers sessile, in dense spikes; pods 3-4 jointed .... var. **pubescens**
1. Flowers pedicellate, in lax spikes; pods 1-jointed .... var. **vasavadae**

var. **pubescens** Law in Wight Icon. t. 450.1840; Baker in Hook.f., Fl. Brit. India 2:150.1876; Naik in Fl. Marathwada 1:147. 1998 *pro parte*

Annual, erect or decumbent herbs, *ca* 30-90 cm tall; leaves 1-foliolate, petioles up to 3 mm long; stipules linear-lanceolate, 10-12 mm long, produced into a sharp point, covered with long white hairs; leaflet oblong-lanceolate, 1-4 × 0.3-1 cm, strongly 3-nerved from the rounded or subcordate base, apex acute, mucronate, clothed with white, long hairs, especially on the nerves beneath, stipels minute, caducous; flowers sessile, in dense, terminal spicate racemes; bracts ovate-acuminate, villous outside; calyx 6-8 mm long, divided nearly to the base, densely plumose; corolla: standard petal yellow, others deep pink; pods 2-4 jointed, moniliform, included in the calyx, joints reticulately veined.

Fl.-Fr.: Sept. -Nov.

Common in grass-lands, throughout the area.

var. **vasavadae** (Hemadri) Sanjappa, in Taxon 32:668. 1983 & Legumes of India 80.1991; *Alysicarpus vasavadae* Hemadri in Indian Forester 97 (1): 65. 1971; Kothari, in N.P. Singh *et al.,* Fl. Maharashtra State 1:604.2000.

Herbs, up to 90 cm tall; stipules minute; leaves ovate-lanceolate, acuminate, 1.5-5.0 × 0.2-0.9 cm; flowers pedicellate, orange-yellow, in terminal spicate racemes; pods 0.5-0.7 cm long, 1-seeded, compressed and conspicuously reticulately veined.

Fl.-Fr.: Sept. -Nov.

Rare, Barwani, Khandwa, Khargone.

Alysicarpus roxburghianus Thoth. & Pramanik in Bull. Bot. Surv. India 21: 189. 1981; *Hedysarum bupleurifolium auct. non* L., 1753: Roxb., Pl. Coromandel 2: t. 194. 1805; Samvatsar, Fl. West. Trib. Madhya Pradesh 87. 1996; Khanna *et al.,* Suppl. Fl. Madhya Pradesh 46.2001.

Perennial, much branched, herbs or undershrubs; leaves alternate, 1-foliolate, linear, 2.8-10 × 0.3-0.6 cm, acute at apex, cuneate at base or sometimes slightly rounded, glabrous above, glabrescent below; stipules triangular with sheathing base, 4-7 mm long, membranous, glabrous; flowers pale-rose or pinkish brown, in terminal, lax, racemes; pods subglobose, 1.5-8 mm long, 3-6-jointed, smooth.

Fl.-Fr.: Sept. – Dec.

Rare, **Alirajpur** only, at Kitty.

Alysicarpus rugosus (Willd.) DC., Prodr. 2: 353. 1825; *Hedysarum rugosum* Willd., Sp. Pl. 3(2): 1172. 1802; Baker in Hook.f., Fl. Brit. India 2: 159. 1876; Singh in Verma *et al.,* Fl. Madhya Pradesh 1: 321. 1993; Samvatsar, Fl. West. Trib. Madhya Pradesh 87. 1996.

Annual or perennial, erect or ascending herbs; leaves alternate, 1-foliolate, ovate-lanceolate, linear-lanceolate or linear, 1.5-11 × 0.2-2.3 cm, acute or mucronate at apex, rounded or subcordate at base, finely puberulous and long hairs beneth; stipules lanceolate, 0.5-3 cm long; flowers pinkish-buff or reddish –purple; pods turgid moniliform, 3-6-joints, deeply transversely plicate, glabrous or pubescent.

Fl.-Fr.: Sept. - Jan.

In wastelands and cultivated fields, throughout the area.

Alysicarpus scariosus (Rottler ex Spreng.) Graham ex Thwaites, Enum. Pl. Zeyl. 88. 1858; *Hedysarum scariosum* Rottler ex Spreng., Syst. Veg. 3: 319. 1926; *Alysicarpus rugosus* var. *styracifolius* Baker in Hook.f., Fl. Brit. India 2: 159. 1876; Singh in Verma *et al.,* Fl. Madhya Pradesh 1:321. 1993.

Diffusely branched herbs; branches 10-30 cm long, pubescent; leaves 1-foliolate, elliptic-oblong, 0.5-2.5 × 0.4-1.5 cm, glabrous above, pilose beneath; flowers pinkish- yellow, in dense, axillary racemes; pods moniliform, 3-jointed, 1.5-2 mm long, transversely ribbed, sparsely pilose.

Fl.-Fr.: July - Dec.

Barwani, Indore, Khargone.

Alysicarpus tetragonolobus Edgew. in J. Asiat. Soc. Beng. 21: 169. 1853; Baker in Hook.f., Fl. Brit. India 2: 159. 1876; Singh in Verma *et al.,* Fl. Madhya Pradesh 1: 321. 1993; Samvatsar, Fl. West. Trib. Madhya Pradesh 87. 1996.

Procumbent, caespitose herbs; stem marked with a line of hairs; leaflets elliptic-oblong or lanceolate, obtuse or subacute, hairy beneath; flowers in lax racemes pods sub-tetragonal due to a long ridge along the middle of each face; distinctly exserted; joints 2-6, transversely ribbed, glabrous.

Fl.-Fr.: Aug. -Oct.

Alirajpur, Burhanpur, Dhar, Indore, Khandwa.

Alysicarpus vaginalis (L.) DC. Prodr. 2: 353. 1825; *Hedysarum vaginale* L. Sp. Pl. 746. 1753; Baker in Hook.f., Fl. Brit. India 2: 158. 1876; Singh in Verma *et al.,* Fl. Madhya Pradesh 1: 322. 1993; Samvatsar, Fl. West. Trib. Madhya Pradesh 87. 1996.

Prostrate or decumbent herbs; leaflets very variable in form on one and the same plant, lanceolate, oblong or ovate-oblong, rounded, acute or truncate at apex; flowers in elongated racemes; pod sub-terete, apiculate, glabrous or so, reticulately veined.

Fl.-Fr.: Oct. -Nov.

Common in fields as a weed, throughout the area.

BAUHINIA L. W. 361; MP 7; SA 6

Key to Species

1. Gigantic tendril climbers; fertile stamens 3; pod broad, velvety brown .... **B. vahlii**

1. Trees

  2. Stamens 10; flowers white, in 5-10 cm long racemes

    3. Calyx limb spathaceous; stigma on a slender style ....**B. racemosa**

    3. Calyx limb split into 5, linear segments ....**B. malabarica**

  2. Stamens 3-5

    4. Flowers small, yellow, in terminal panicles... ....**B. semla**

    4. Flowers large, white, purple or pink, solitary or short, in few flowered racemes

      5. Buds sharply angled; all petals similar; flowering when leafy; stamens 3 ....**B. purpurea**

      5. Buds terete; upper (posterior) petal variegated; deciduous when flowering .... **B. variegata**

Bauhinia malabarica Roxb., Fl. Ind. 2: 321. 1832; Baker in Hook.f., Fl. Brit. India 2: 277. 1878; Singh in Verma *et al.,* Fl. Madhya Pradesh 1: 412. 1993; Samvatsar, Fl. West. Trib. Madhya Pradesh 112. 1996. *Aamta*

Medium-sized shrubs or trees, 4-8 m high; bark greyish brown; leaves slightly broader than long, 5-12 × 5.5-13 cm, divided up to ¼th of its length into rounded lobes, slightly cordate at base, 7-9-nerved; flowers in 2.5-5 cm long, racemose corymbs, often 2-3 in axils; pods brown, 15-30 × 2-2.5 cm, flat, beaked, slightly curved; seeds dark brown, ovoid or globose, *ca* 4.5 mm long.

Fl.-Fr.: Sept. – April

Alirajpur, Barwani, Indore, Khargone.

Bauhinia purpurea L., Sp. Pl. 1: 375. 1753; Baker in Hook.f., Fl. Brit. India 2: 284. 1878; Singh in Verma *et al.,* Fl. Madhya Pradesh 1: 413. 1993; Samvatsar, Fl. West. Trib. Madhya Pradesh 112. 1996. *Kachnar*

Medium-sized trees, 6-10 m high; young parts pubescent; bark ash-coloured to dark brown; leaves longer than broad, 7-18 cm long, cordate at base, cleft half way down; lobes subacute, 9-11-nerved; flowers rose-purple, in few-flowered, terminal, corymbose or paniculate, tomentose racemes; pods 15-25 × 2-2.7 cm, flat; stalk *ca* 2 cm long; seeds 12-15, brown, glabrous, up to 1.3 cm across, compressed, smooth.

Fl.-Fr.: Oct. –March

Alirajpur, Burhanpur, Dhar, Indore, Khandwa.

Bauhinia racemosa Lam., Encycl. 1: 390.1785; Baker in Hook.f., Fl. Brit. India 2: 276. 1878; Singh in Verma *et al.,* Fl. Madhya Pradesh 1: 414. 1993; Samvatsar, Fl. West. Trib. Madhya Pradesh 112. 1996. *Asta* **(Plate 4, Fig. 1)**

Deciduous trees, 3-5 mm tall; bark grey, rough; leaves 2.5-5.5 × 4-7.5 cm, broader than long, usually cordate at base, grey-pubescent beneath; flowers in terminal, leaf-opposed racemes, creamy-white, buds asymetrically fusiform, pointed; stamens 10, all perfect; pods 7-20 × 1.5-2.5 cm, turgid, torulose, falcate, indehiscent, 12-15 seeded; seeds oval, black.

Fl.-Fr.: April-Aug.

Common in forest all over the area.

Bauhinia semla Wunderlin in Taxon 25: 362. 1976, *p.p. excl. syn. cit. B. emarginata* Roxb. ex G. Don; Baker in Hook.f., Fl. Brit. India 2: 279. 1878; Singh in Verma *et al.,* Fl. Madhya Pradesh 1: 414. 1993; Samvatsar, Fl. West. Trib. Madhya Pradesh 113. 1996. *Semla*

Medium-sized trees, 4-8 m high; bark dark brown and pale pink; leaves slightly broader than long, 7.5-15 cm long, retuse at apex, cordate or truncate at base, subcoriaceous, glabrous, 9-11-nerved; petioles up to 8 cm long, thickened at ends; flowers in lax, terminal or axillary, corymbose panicles; pedicels up to 2 cm long, pubescent; pods 10-15 × 3-4 cm, flat; seeds 6-8, dark brown, suborbicular, smooth.

Fl.-Fr.: Sept. - Feb.

Alirajpur, Barwani, Khargone.

Bauhinia vahlii Wight & Arn., Prodr. 297. 1834; Baker in Hook.f., Fl. Brit. India 2: 279. 1878; Singh in Verma *et al.,* Fl. Madhya Pradesh 1: 414. 1993. *Mahul* **(Plate 4, Fig. 2)**

Large woody climbers; branches rusty- brown tomentose; branchlets terminating in a pair of revolute tendrils; leaves alternate, almost orbicular, broader than long, up to 35 cm across, lobed at apex, cordate at base, glabrescent above, densely tomentose beneath, 11-13-nerved; petioles 7-15 cm long; flowers in long peduncled, terminal, corymbose or subcorymbose, densely hairy racemes; pods woody, 22-28 × 5-7 cm, rusty- velvetty, 6-12-seeded; seeds dark brown, flat, compressed, polished.

Fl.-Fr.: April –Jan.

Alirajpur, Dhar, Indore, Khandwa, Khargone.

Bauhinia variegata L., Sp. Pl. 1: 375. 1753; Baker in Hook.f., Fl. Brit. India 2: 284. 1878; Singh in Verma *et al.,* Fl. Madhya Pradesh 1: 416. 1993; Samvatsar, Fl. West. Trib. Madhya Pradesh 113. 1996. *Dhondi*

Small to medium-sized trees, 4-7 m high; branches hairy; bark dark brown, smooth; leaves slightly broader than long, 5-15 cm long, divided 1/3 way down, cordate at base, 9-15-nerved; lobes obtuse; flowers few in lateral corymbose, pubescent racemes; bracts deltoid; pods dark brown, 15-25 × 1-2 cm, flat, 10-15-seeded; seeds brown, ellipsoid, 1-1.2 cm long.

Fl.-Fr.: Sept. -April

Alirajpur, Dhar, Indore.

BUTEA Roxb. ex Willd. W. 4; MP 2; SA 2

Key to Species

1. Trees; lowest calyx tooth shorter than the lateral ones .... **B. monosperma**
1. Lianas; lowest calyx tooth as long as the laterals .... **B. superba**

Butea monosperma (Lam.) Taub. in Engl. & Prantl., Nat. Pflanzenfam. 3(3): 366. f. 131. 1894; *Erythrina monosperma* Lam., Encycl. 2: 391. 1786; Baker in Hook.f., Fl. Brit. India 2: 194. 1876; Singh in Verma *et al.,* Fl. Madhya Pradesh 1: 325. 1993; Samvatsar, Fl. West. Trib. Madhya Pradesh 88. 1996. *Palash*

Medium sized, deciduous trees; leaflets 3, rhomboid or broadly obovate, up to 25 cm in diam., hairy; flowers orange-red, showy, 5-8 cm long, in terminal and lateral racemes; lowest calyx-tooth shortest; pods flat, 12-18 × 2.5-5.5 cm, grey-silky, one-seeded.

Fl.-Fr.: Jan. – May

Very common all over the area, abundance at a place indicates burnt ground there.

Butea superba Roxb. in Asiat. Res. 3: 473-474. 1792; Baker in Hook.f., Fl. Brit. India 2: 195. 1876; Singh in Verma *et al.,* Fl. Madhya Pradesh 1: 325. 1993; Samvatsar, Fl. West. Trib. Madhya Pradesh 88. 1996. *Palash bel*

Large woody twiners, 10-15 m high, with thick trunk, but sometimes branches spreading and partly lying on the ground; leaves pinnately 3-foliolate; leaflets obovate-rhomboid, 20-50 cm long, retuse at apex; petioles 15-20 cm long; flowers 4-6.5 cm long, in 20-30 cm long racemes; pods 10-15 cm long, narrowed at base, 1-seeded.

Fl.-Fr.: March-June

Alirajpur, Dhar, Khandwa.

CAESALPINIA L. W. 162;MP 2; SA 2

Key to Species

1. Pod prickly, obovoid, 1-2 seeded; stipules foliaceous, pinnate .... **C. bonduc**
1. Pod smooth, oblong, 5-6 seeded; stipules simple .... **C. decapetala**

Caesalpinia bonduc (L.) Roxb., Fl. Ind. 2: 362. 1832, *p.p. excl. pl. descr. emend.* Dandy & Excell, 1938; *Guilandina bonduc* L., Sp. Pl. 381. 1753; Baker in Hook.f., Fl. Brit. India 2: 255. 1878; Singh in Verma *et al.,* Fl. Madhya Pradesh 1: 417. 1993; Samvatsar, Fl. West. Trib. Madhya Pradesh 113. 1996. *Gatar* **(Plate 4, Fig. 3)**

Armed shrubs, pinnae 6-8 pairs, with a pair of hooked stipular spines; leaflets 6-9 pairs, 1.2-2 × 1.2-1.5 cm, elliptic-oblong, obtuse, strongly mucronate; racemes often panicled; pods shortly stalked, oblong, densely armed on the faces with wiry prickles, 1-2 seeded.

Fl.-Fr.: July - Nov.

Common, in hedges, village roads, throughout the area.

Caesalpinia decapetala (Roth) Alston in Trimen, Handb. Fl. Ceylon 6: 89. 1931; *Reichardia decapetala* Roth, Nov. Pl. Sp. 212. 1821; *Caesalpinia sepiaria* Roxb., Fl. Ind. 2: 360.1832; Baker in Hook.f., Fl. Brit. India 2: 256. 1878; Singh in Verma *et al.,* Fl. Madhya Pradesh 1: 417. 1993; Samvatsar, Fl. West. Trib. Madhya Pradesh 113. 1996. *Golchuna*

Armed shrubs; young parts brown-pubescent; pinnae 8-10 pairs; leaflets 8-12 pairs, 1-2 × 0.5-0.8 cm, oblong, obtuse at both ends, finely pubescent; racemes simple; pods falcate-oblong, beaked, reticulately veined, 6-8 seeded.

Fl.-Fr.: Oct. – June

Less common than the above species.

CAJANUS DC. W. 32;MP 3; SA 2

Key to Species

1. Pod with deep lines between the seeds; stipules triangular; calyx fulvous pubescent .... **C. scarabaeoides**
1. Pod with sharply defined (not deep) lines between the seeds; stipules ovate; calyx densely hairy .... **C. platycarpus**

Cajanus platycarpus (Benth.) van der Maesen in Agric. Univ. Wageningen Pap. 85 (4): 160.1985; *Atylosia platycarpa* Benth. in Miq., Pl. Jungh. 2:243. 1852; Baker in Hook.f., Fl. Brit. India 2: 216. 1876; Singh in Verma *et al.,* Fl. Madhya Pradesh 1: 324. 1993; Samvatsar, Fl. West. Trib. Madhya Pradesh 88. 1996.

Trailing or twining herbs; branches slender, rusty- brown, tomentose; leaves 3-foliolate, stipulate; leaflets orbicular, 3-6 × 3-6 cm, acute or acuminate at apex, sparingly pubescent on both surfaces, terminal one roundish, acute or cuspidate; petioles 2.5-11.5 cm long; flowers pale yellow, in terminal or axillary racemes; pods linear-oblong, 2.5-4 × 1.2-1.4 cm, flattened, compressed, sparsely pilose, transversely depressed between seeds, 4-7-seeded.

Fl.-Fr.: Aug. -Sept.

Alirajpur, Burhanpur, Indore.

Cajanus scarabaeoides (L.) du Petit-Thou., Dict. Sci. Nat. 6: 617. 1817; *Dolichos scarabaeoides* L., Sp. Pl. 726. 1753; *Atylosia scarabaeoides* (L.) Benth. in Miq., Pl. Jungh. 2: 242. 1852;

Baker in Hook.f., Fl. Brit. India 2: 215. 1876; Singh in Verma *et al.,* Fl. Madhya Pradesh 1: 324. 1993; Samvatsar, Fl. West. Trib. Madhya Pradesh 88. 1996. *Bankulthi*

Pubescent, twining shrubs; leaflets elliptic to elliptic-oblong, acute to obtuse, grey-pubescent beneath, 4-4.5 × 2-2.2 cm, racemes slender 1-2 cm long; flowers yellow; calyx lobes longer than the tube; pod covered with golden spreading hairs, 2.5 × 0.6-0.7 cm, 4-6 seeded.

Fl.-Fr.: Aug. -March

Common along forest paths, all over the area.

CANAVALIA Adans. W. 17; MP 2; SA 1

Canavalia gladiata (Jacq.) DC. Prodr. 2: 404. 1825; *Dolichos gladiatus* Jacq., Coll. Bot. 2: 276 1788; *Canavalia ensiformis* sensu Baker in Hook.f., Fl. Brit. India 2: 195. 1876; Singh in Verma *et al.,* Fl. Madhya Pradesh 1: 328. 1993. *Ban sem*

Perennial shrubby climbers; stems and branches glabrous; petiole 10-12 cm long, leaflets 3, membranous, ovate, 7-14 × 3-9 cm; flowers purple or white tinged with purple in lax many flowered axillary racemes; pods 15-35 × 2-4 cm, linear or oblong; seeds 8-20.

Fl.-Fr.: Aug. -Jan.

Common along village roads on adjacent vegetation, in all districts.

CASSIA L. W. 90; MP 1; SA 1

Cassia fistula L. Sp. Pl. 377. 1753; *C. rhombifolia* Roxb. Fl. Ind. 3: 334. 1832; Baker in Hook.f., Fl. Brit. India 2: 261. 1878; Singh in Verma *et al.,* Fl. Madhya Pradesh 1: 421. 1993; Samvatsar, Fl. West. Trib. Madhya Pradesh 114. 1996. *Amaltas*

Deciduous, small trees; branchlets glabrous; leaves eglandular; leaflets 4-8 pairs, 7-15 × 2-5 cm, ovate, subacute, silvery, pubescent beneath when young; flowers in lax, pendulous racemes; pods long, cylindrical, nearly straight, smooth, shining, brown- black; seeds numerous, discoid.

Fl.-Fr.: March-May

Very common in the forests, and along road-sides.

CHAMAECRISTA Moench W. 330; MP 3; SA 3

Key to Species

1. Leaflets 30-60 pairs; fertile stamens 10, alternately long and short .... **C. mimosoides**
1. Leaflets up to 30 pairs; fertile stamens 5
    2. Leaflets 2 pairs; plant glandular; flowers light pink .... **C. absus**
    2. Leaflets up to 30 pairs; plants not glandular; flowers yellow .... **C. pumila**

Chamaecrista absus (L.) Irwin & Barneby in Mem. New York Bot. Gard. 35(2): 664. 1982; *Cassia absus* L., Sp. Pl. 376. 1753; Baker in Hook.f., Fl. Brit. India 2: 265. 1878; Singh in Verma *et al.,* Fl. Madhya Pradesh 1: 420.1993; Samvatsar, Fl. West. Trib. Madhya Pradesh 114. 1996. *Chaksu*

Annual herbs, clothed with glandular hairs; rachis grooved; leaflets 2-5 × 1-3 cm, obliquely elliptic to obovate; glands linear, between both pairs of leaflets; flowers pink or red, in terminal or leaf opposed, few-flowered racemes; bracts broadly ovate, acute or acuminate; bracteoles linear; stamens 5; pods 3-5.5 × 0.7-0.8 cm, hairy, flat; seeds 4-8, compressed.

Fl.-Fr.: Aug. – Nov.

Common forest undergrowth, in all districts.

Chamaecrista mimosoides (L.) Greene in Pittonia 4: 27. 1899; *Cassia mimosoides* L., Sp. Pl. 379. 1753; Baker in Hook.f., Fl. Brit. India 2: 266. 1878; Singh in Verma *et al.,* Fl. Madhya Pradesh 1: 421. 1993; Samvatsar, Fl. West. Trib. Madhya Pradesh 114. 1996.

Prostrate herbs, leaves 6-10 cm long with a gland at tip of petiole, midrib of leaflets eccentric; flowers solitary, axillary; pods flat, pubescent, 3.5-0.3 cm; seeds 15-20, compressed, polished brown.

Fl.-Fr.: Aug. – Nov.

Common in grasslands all over the area.

Chamaecrista pumila (Lam.) V. Singh in J. Econ. Taxon. Bot. 16(3): 600.1992; *Cassia pumila* Lam., Encycl. 1: 651. 1785; Baker in Hook.f., Fl. Brit. India 2: 266. 1878; Singh in Verma *et al.,* Fl. Madhya Pradesh 1: 422. 1993; Samvatsar, Fl. West. Trib. Madhya Pradesh 115. 1996.

Erect or prostrate herbs, up to 50 cm long; leaves to 10 cm long, leaflets up to 30 pairs; petiolar glands stalked, persistent; often rachis also bears similar, but deciduous glands in between the leaflets; anthers unequal; pods up to 4.5 cm long, straight or slightly curved; seeds subquardate, brown.

Fl.-Fr.: Aug. – Nov.

Common in the same habitat as above, prefers moist ground.

CLITORIA L. W. 66; MP 2; SA 1

Clitoria annua J. Graham, Cat. Pl. Bombay 47. 1839; *Clitoria biflora* Dalzell in Kew Journ. Bot. 2: 35. 1850; Baker in Hook.f., Fl. Brit. India 2: 208. 1876; Singh in Verma *et al.,* Fl. Madhya Pradesh 1: 330.1993; Samvatsar, Fl. West. Trib. Madhya Pradesh 89. 1996.

Annual herbs, up to 60 cm high; leaflets 5, elliptic to ovate-lanceolate, 2.5-15 × 1.2-4 cm, acute; flowers white, axillary, geminate or rarely solitary due to the abortion of one flower, deflexed; bracteoles ovate-lanceolate, aristate; pods flat, 3-5 cm long, pubescent, 5-6 seeded.

Fl.-Fr.: Aug. -Oct.

Alirajpur, Barwani, Burhanpur, Dhar, Indore, Khandwa.

CODARIOCALYX Hassk. W. 2; MP 1; SA 1

Codariocalyx motorius (Houtt.) H. Ohashi in J. Jap. Bot. 40(12): 367–368. 1965; *Hedysarum motorium* Houtt., Nat. Hist. 2(10): 246. 1779; *Desmodium gyrans* (L.f.) DC., Prodr. 2: 326. 1825; Baker in Hook.f., Fl. Brit. India 2: 174. 1876; *Desmodium motorium* (Houtt.) Merr. in

J. Arnold Arbor. 19: 345. 1938; Singh in Verma *et al.*, Fl. Madhya Pradesh 1: 348. 1993. *Telegraph plant*

Herbs or undershrubs, 1-1.5 m high; leaves 1-3-foliolate; lateral leaflets, very small, 1-2.5 cm, lanceolate-oblong, terminal leaflet lanceolate, oblong or elliptic, 8-10 cm long, obtuse or rounded at apex, rounded or acute at base, glabrous above, appressed hairy beneath; flowers in terminal and axillary racemes; pods 2.5-4.5 cm long; joints 6-10, glabrous or hooked hairy, incised along the lower suture.

Fl.-Fr.: Sept. – Jan.

Rare, Burhanpur, Khandwa, as forest undergrowth.

CROTALARIA L. W. 699; MP 33; SA 21

Key to Species

1. Leaves 1-foliolate
  2. Pod included in calyx, glabrous or hairy
    3. Corolla exserted; seeds 4-10; stipules present .... **C. linifolia**
    3. Corolla not exserted; seeds 2-20
      4. Stipules present .... **C. calycina**
      4. Stipules absent
        5. Erect, villous herbs; flowers in 6-12 flowered umbel; seeds 6-8 .... **C. nana**
        5. Prostrate herbs; flowers solitary, axillary; seeds 1 or 2 .... **C. hebecarpa**
  2. Pod exserted
    6. Stipules absent or minute
      7. Pod silky villous; seeds 4-8
        8. Stiff plants; stem angular; leaves 8-12 × 0.5-2.5 cm; racemes 12-50 cm long, many flowered; corolla not exserted; pod 1.5-2.5 cm long, fulvous silky .... **C. leptostachya**
        8. Slender plants; stem terete; leaves 0.75 × 0.2-0.4 cm; racemes 2-4 cm long, few flowered; corolla slightly exserted; pod 0.4-0.7 cm long, silky .... **C. pusilla**
      7. Pod glabrous; seeds 8-20
        9. Corolla exserted; seeds 8-10 .... **C. filipes**
        9. Corolla not exserted
          10. Bracts leafy, lanceolate; racemes 4-5 flowered; seeds 15-20 .... **C. vestita**
          10. Bracts minute, subulate
            11. Plants prostrate; leaf base obliquely rounded; racemes 2-4 flowered .... **C. prostrata**
            11. Plants erect; leaf base acute; racemes 6-20 flowered .... **C. albida**
    6. Stipules present
      12. Pod glabrous; seeds 15-30
        13. Corolla not exserted .... **C. mysorensis**
        13. Corolla exserted
          14. Stipules and bracts subulate; pod cuneate at base .... **C. retusa**

14. Stipules and bracts foliaceous; pod rounded at base .... **C. spectabilis**

12. Pod hairy; seeds 8-20; corolla exserted; racemes 2-3 flowered; pod 2-2.5 cm long

15. Branches triquetrous; seeds 15-20; plants glabrous.. **C. triquetra**

15. Branches terete; seeds 8-10; plants hirsute .... **C. hirsuta**

1. Leaves trifoliolate

16. Pod 2-seeded

17. Low, diffuse herbs; calyx teeth exceeding the tube; pod sub-globose, beaked ....**C. medicaginea**

17. Suffruticose, erect herbs; calyx teeth equalling the tube; pod sub-quadrangular ....**C. notonii**

16. Pod many seeded

18. Bracts large, foliaceous, ovate; pod 8-10 seeded ....**C. orixensis**

18. Bracts minute, setaceous; pod 20-40 seeded .... **C. pallida**

Crotalaria albida Heyne ex Roth, Nov. Pl. Sp. 333. 1821; Baker in Hook.f., Fl. Brit. India 2: 71. 1876; Singh in Verma *et al.,* Fl. Madhya Pradesh 1: 333. 1993; Samvatsar, Fl. West. Trib. Madhya Pradesh 90.1996.

Herbs or much branched, pubescent, undershrubs, *ca* 20-70 cm high; branches cylindrical, silky hairy; leaves simple, linear or oblong, 2.5-5 cm long, obtuse or emarginate at apex, cuneate at base, gland-dotted, glabrescent above and silky hairy beneath; stipules absent; flowers yellow, in terminal, 5-10 cm long racemes; pods oblong-cylindrical, 10-12 mm long, glabrous, sessile, 6-12-seeded.

Fl.-Fr.: Aug. – March

Rare, Alirajpur, Khargone.

Crotalaria calycina Schrank., Pl. Rar. Hort. Monac. t. 12. 1819; Baker in Hook.f., Fl. Brit. India 2: 72. 1876; Singh in Verma *et al.,* Fl. Madhya Pradesh 1: 334. 1993; Samvatsar, Fl. West. Trib. Madhya Pradesh 90.1996.

An erect herb; stem simple or sparsely branched, densely clothed with appressed silky hairs; leaves variable, generally elliptic-lanceolate, appressed villous beneath; flowers large, in terminal 2-12 flowered racemes; pods sessile, glabrous, many seeded.

Fl.-Fr.: Aug. – March

Common along streams and in moist soil, all over.

Crotalaria filipes

Key to varieties

1. Stem filiform, hairy; leaves 0.3-1.6 cm long ....var. **filipes**

1. Stem stout , densely hairy; leaves 1.2-2.5 cm long.. var. **trichophora**

var. **filipes** Benth. in Lond. J. Bot. 2: 475. 1843; Baker in Hook.f., Fl. Brit. India 2: 66. 1876; Samvatsar, Fl. West. Trib. Madhya Pradesh 90.1996; Khanna *et al.,* Suppl. Fl. Madhya Pradesh 49.2001.

Prostrate herbs; stem filiform, hairy; leaves 0.3-1.6 cm long, ovate-acute or obtuse, slightly cordate at base; flowers in 1-3 flowered axillary racemes; pod exserted, glabrous, 8-10 seeded.

Fl.-Fr.: Sep. -Nov.

Rare, in grass-lands and on hill slopes, Alirajpur, Indore.

var. **trichophora** (Benth. ex Baker) Cooke, Fl. Bombay Presi. 1:293. 1958 (BSI Reprint); *Crotalaria trichophora* Benth ex Baker in Baker in Hook.f., Fl. Brit. India 2: 67. 1876; Samvatsar, Fl. West. Trib. Madhya Pradesh 90.1996; Khanna *et al.,* Suppl. Fl. Madhya Pradesh 49.2001.

Densely hairy herbs; leaves ovate-oblong, 1.2-2.5 × 0.5-0.8 cm, slightly cordate at base; flowers yellow, in 1-3 flowered lateral racemes; pods 3-6 mm long, exserted, exserted, glabrous, 8-10 seeded.

Fl.-Fr.: Sep. -Dec.

Alirajpur only, at Bhabra (Kattiwada).

Crotalaria hebecarpa (DC.) Rudd in Phytologia 54(1): 28. 1983; *Goniogyna hebecarpa* DC. in Ann. Sci. Nat. Paris 4: 92. 1825; *Heylandia latebrosa* DC., Mem. Legum. 6: 201. 1826; Baker in Hook.f., Fl. Brit. India 2: 65. 1876; *Goniogyna hirta* (Willd.) Ali in Taxon 16: 463. 1967; Singh in Verma *et al.,* Fl. Madhya Pradesh 1: 363. 1993; Samvatsar, Fl. West. Trib. Madhya Pradesh 96. 1996.

Prostrate, annual herbs, clothed with long, spreading hairs; leaves 0.8-1.5 × 0.5-1 cm, cordate-ovate, entire, ciliate, obtuse; flowers bright yellow with purple veins, solitary, axillary; pods flat, oblong, 1-2 seeded, beaked, hairy.

Fl.-Fr.: July – March

Very common in grass lands, open ground, road sides throughout the area.

Crotalaria hirsuta Willd., Sp. Pl. 3(2): 978. 1803; Baker in Hook.f., Fl. Brit. India 2: 68. 1876; Singh in Verma *et al.,* Fl. Madhya Pradesh 1: 334. 1993; Samvatsar, Fl. West. Trib. Madhya Pradesh 90.1996.

Erect, silky herbs; leaves ovate, acute, membranous; leaves simple, alternate, ovate, elliptic, ovate-elliptic or ovate-lanceolate, 3.0-6.5 × 1.5-4.3 cm, obtuse or rounded at base, acute or acuminate, glabrescent above, sparsely pubescent beneath; stipules linear or subulate lanceolate; flowers yellow, in 2-3 flowered, leaf opposed racemes; pods hirsute, oblong, 8-10 seeded.

Fl.-Fr.: Oct. -Dec.

Alirajpur, Barwani, Burhanpur, Dhar, Indore, Khargone.

Crotalaria leptostachya Benth. in London J. Bot. 2: 562, 1843; Baker in Hook.f., Fl. Brit. India 2: 78. 1876; Samvatsar, Fl. West. Trib. Madhya Pradesh 91. 1996; Khanna *et al.,* Suppl. Fl. Madhya Pradesh 1: 51.2001.

Erect shrubs with tetragonous branches; leaves membranous, mostly linear or linear-oblong, apiculate, fulvous silky hairy on both sides; stipules minute, deciduous; flowers yellow, in 12-30 cm long racemes; pods densely fulvous silky 1.5-2.5 cm long, 4-6 seeded.

Fl.-Fr.: Aug. -Nov.

Alirajpur only, near streams, at Shakarja.

Crotalaria linifolia L.f., Suppl. Pl. 322. 1782; Baker in Hook.f., Fl. Brit. India 2: 72. 1876; Singh in Verma *et al.,* Fl. Madhya Pradesh 1: 336. 1993; Samvatsar, Fl. West. Trib. Madhya Pradesh 91. 1996.

Suffruticose, appressed hairy herbs; leaves oblanceolate or oblong-obtuse, apiculate; flowers yellow, in terminal, 14-20 flowered racemes; pods included, ovoid-oblong, sessile, glabrous, black when ripe, 4-10 seeded.

Fl.-Fr.: Aug. – March

Common in open ground, all over the area.

Crotalaria medicaginea

Key to varieties

1. Flowers 2-6; weak herbs, scarcely 30 cm tall .... var. **medicaginea**
1. Flowers 6-12; stout, stronger plants, 60-80 cm tall .... var. **luxurianus**

var. **luxurianus** (Benth.) Baker in Hook.f., Fl. Brit. India 2: 81. 1876; *Crotalaria luxurianus* Benth. in London J. Bot. 2: 578. 1843; Samvatsar, Fl. West. Trib. Madhya Pradesh 91. 1996; Khanna *et al.,* Suppl. Fl. Madhya Pradesh 51.2001.

Perennial, robust herbs, up to 1 m high; leaves 3-foliolate, leaflets oblanceolate, 2.5 × 0.7 cm, rounded, apiculate, silky beneath; flowers in terminal and leaf opposed, 6-12 flowered racemes; pods obliquely subglobose, pubescent, beaked, 2-seeded.

Fl.-Fr.: Nov. – Dec.

Common along hill slopes, forest areas, all over.

var. **medicaginea** Lam., Encycl. 2(1): 201. 1786; Baker in Hook.f., Fl. Brit. India 2: 81. 1876; Singh in Verma *et al.,* Fl. Madhya Pradesh 1: 336. 1993; Samvatsar, Fl. West. Trib. Madhya Pradesh 91. 1996. *Gulabi*

Prostrate, appressed hairy herbs; leaves 3-foliolate, leaflets oblanceolate, rounded, apiculate, silky beneath; flowers yellow, in terminal and leaf opposed, 2-6 flowered racemes; pods obliquely subglobose, pubescent, beaked, 2-seeded.

Fl.-Fr.: Sept. – April

Common, same habitat and area, as above.

Crotalaria mysorensis Roth, Nov. Pl. Sp. 338. 1821; Baker in Hook.f., Fl. Brit. India 2: 70.1876; Singh in Verma *et al.,* Fl. Madhya Pradesh 1: 336. 1993; Samvatsar, Fl. West. Trib. Madhya Pradesh 91. 1996.

Erect, sparesely branched annual herbs; stem densely clothed with spreading hairs; leaves subsessile, linear oblong, rounded at both ends, apiculate, silky on both surfaces; pods scarcely stalked, truncate at the apex, glabrous when ripe, 20-50 seeded.

Fl.-Fr.: Sept. -Oct.

Occasional along streams and fields, Dhar, Indore.

Crotalaria nana Burm.f., Fl. Indica 156. t. 48. f. 2. 1768; Baker in Hook.f., Fl. Brit. India 2: 71, 1876; Singh in Verma *et al.,* Fl. Madhya Pradesh 1: 337. 1993; Samvatsar, Fl. West. Trib. Madhya Pradesh 91 1996.

Erect, villous herbs; leaves subsessile, closely and minutely pellucid-punctate, oblong or oblanceolate, subacute, apiculate, silky hairy on both surfaces; flowers in 4-12 flowered, crowded umbels; pods globose-ovoid, glabrous, black when ripe, equalling or included in calyx, 6-8 seeded.

Fl.-Fr.: Aug. -Nov.

Rare, in wastelands, along the road-sides and in shade, Alirajpur, Dhar.

Crotalaria notonii Wight & Arn. Prodr. 1: 192. 1834; Baker in Hook.f., Fl. Brit. India 2: 82, 1876; Singh in Verma *et al.,* Fl. Madhya Pradesh 1: 337. 1993; Samvatsar, Fl. West. Trib. Madhya Pradesh 91. 1996.

Suffruticose herbs; branches and leaves yellowish tomentose; leaves 3-foliolate; leaflets obovate, 2.5 × 0.8 cm, glabrous above, adpressed-silky hairy below; flowers yellow, in terminal or leaf opposed, 12-20 long racemes; pods subquadrangular, thinly silky, 2-seeded.

Fl.-Fr.: Oct. -Nov.

Not common, in wastelands and forest undergrowth, Alirajpur, Dhar, Indore.

Crotalaria orixensis Willd. in Ges. Naturf. Freunde Berlin Neue Schriften 4: 217. 1803; Baker in Hook.f., Fl. Brit. India 2: 1876; Singh in Verma *et al.,* Fl. Madhya Pradesh 1: 337. 1993; Samvatsar, Fl. West. Trib. Madhya Pradesh 91. 1996. *Jansru*

Diffuse branched herbs with spreading hairs; leaves 3-foliolate; leaflets oblong-obovate, 2.5-4 cm long, obtuse at apex, gland- dotted, glabrous above, silky beneath; stipules linear-lanceolate; flowers yellow, in 3-12 flowered racemes; pods long-stipitate, glabrous, seeds 8-10.

Fl.-Fr.: Sept. -Oct.

Common along edges of ponds, marshy ground, along water courses, margins of ponds, bunds of fields, all over the area.

Crotalaria pallida Aiton, Hort. Kew. 3: 20-21. 1789; *C. striata* DC., Prodr. 2: 131. 1825; Baker in Hook.f., Fl. Brit. India 2: 84. 1876.

Annual or perennial herbs; young branches slightly silky hairy; leaves 3-foliolate; leaflets oblong or obovate, 4-7 × 1-3 cm, rounded to emarginate at apex, glabrous above, pubescent beneath; stipules setaceous; flowers yellow, in terminal, 15-36 cm long, 20-50 or more-flowered racemes; pods 2.5-6 cm long, glabrescent, deflexed, stalked, 20-30-seeded; seeds light- brown or olive- green, *ca* 3 mm long.

Fl.-Fr.: Oct. – March

Rare, **Khandwa** only, in open situations, at Mortakka.

Crotalaria prostrata Roth ex Willd. Enum. Hort. Berol. 2: 747. 1809; Baker in Hook.f., Fl. Brit. India 2: 67. 1876; Singh in Verma *et al.,* Fl. Madhya Pradesh 1: 338. 1993; Samvatsar, Fl. West. Trib. Madhya Pradesh 92. 1996. *Kartik jhumka*

Suberect herbs; leaves subsessile, oblong, rounded at apex, yellow or yellowish-brown, hairy on both surfaces; flowers yellow, in leaf opposed, 2-5 flowered racemes; pods glabrous, exserted, 16-20 seeded.

Fl.-Fr.: Oct. -Nov.

Common in hilly regions all over the area.

Crotalaria pusilla Heyne ex Roth, Nov. Pl. Sp. 335. 1821; Baker in Hook.f., Fl. Brit. India 2: 70.1876; Singh in Verma *et al.,* Fl. Madhya Pradesh 1: 338. 1993; Samvatsar, Fl. West. Trib. Madhya Pradesh 92. 1996.

Branched leafy herbs; branches clothed with brownish-silky hairs; leaves subsessile, linear-oblong, densely silky, obtuse or subacute; flowers yellow, in terminal, 6-8 flowered racemes; pods oblong, densely silky-villous; seeds 4-8.

Fl.-Fr.: Sept. -Oct.

Alirajpur, Burhanpur, Indore, Khandwa.

Crotalaria retusa L., Sp. Pl. 175. 1753; Baker in Hook.f., Fl. Brit. India 2: 75. 1876; Singh in Verma *et al.,* Fl. Madhya Pradesh 1: 339. 1993; Samvatsar, Fl. West. Trib. Madhya Pradesh 92. 1996. *Jhun jhuna*

Erect undershrubs; branches striate, appressedly white pubescent; leaves obovate, oblanceolate, silky pubescent beneath; flowers yellow, in terminal, 15-30 flowered racemes; pod exserted glabrous, stalked, linear-oblong; seeds 15-30.

Fl.-Fr.: Sept. -Nov.

Alirajpur, Dhar, Indore.

Crotalaria spectabilis Roth., Nov. Pl. Sp. 311-342. 1821; *C. sericea* Retz. Observ. Bot. 5: 26. 1789 non Burm.f. 1768; Baker in Hook.f., Fl. Brit. India 2: 75. 1876; Singh in Verma *et al.,* Fl. Madhya Pradesh 1: 339. 1993; Samvatsar, Fl. West. Trib. Madhya Pradesh 92. 1996. *Dhundani***(Plate 4, Fig. 4)**

Erect glabrous herbs; leaves obovate, cuneate or lanceolate-oblong, acute or subacute, mucronate, finely pubescent; flowers yellow, in terminal, 20-50 flowered racemes; pods glabrous, linear-oblong, stalked; seeds 20-30.

Fl.-Fr.: Oct. -Dec.

Alirajpur, Barwani, Dhar, Indore, Khandwa, Khargone.

Crotalaria triquetra Dalzell in Hooker's J. Bot. Kew Gard. Misc. 2: 34. 1850; Baker in Hook.f., Fl. Brit. India 2: 71. 1876; Singh in Verma *et al.,* Fl. Madhya Pradesh 1: 340.1993; Samvatsar, Fl. West. Trib. Madhya Pradesh 92. 1996.

Suffruticose herbs; branches 3-4 angled, sparsely pubescent; leaves oblong, 3 × 0.7 cm, obtuse, sparsely hairy beneath; stipules lanceolate; flowers yellow, in 1-flowered racemes; pods oblong-cylindrical with appressed silky hairs; seeds 15-20.

Fl.-Fr.: Oct. -Dec.

Alirajpur, Dhar in waste lands and forest.

Crotalaria vestita Baker in Hook.f., Fl. Brit. India 2(4): 67. 1876; Singh in Verma *et al.,* Fl. Madhya Pradesh 1: 340.1993; Samvatsar, Fl. West. Trib. Madhya Pradesh 92. 1996.

Branched herbs; stems and branches terete, slender, clothed with silky hairs; leaves simple, subsessile, linear-oblong or oblong-obovate, 12-37 × 6-15 mm; flowers yellow, in lateral, leaf-opposed, 4-5-flowered racemes; pods oblong, glabrous, 15-20 seeded.

Fl.-Fr.: Sept. -Jan.

Alirajpur, Indore, as a weed of fields.

CULLEN Medik. W. 13; MP 1; SA 2

Key to Species

1. Leaves unifoliolate .... **C. corylifolium**
1. Leaves trifoliolate .... **C. plicatum**

Cullen corylifolium (L.) Medik. in Vorles. Churpfälz. Phys. -Okon. Ges. 2: 381. 1787; *Psoralea corylifolia* L., Sp. Pl. 2: 764. 1753; Baker in Hook.f., Fl. Brit. India 2: 103. 1876; Singh in Verma *et al.,* Fl. Madhya Pradesh 1: 386. 1993; Samvatsar, Fl. West. Trib. Madhya Pradesh 104. 1996. *Baochi*

Erect annual herbs; stem and branches grooved, hairy; leaves 2-11 × 1-9 cm, simple, broadly elliptic, rounded and mucronate at apex, nigro punctate; flowers blue, axillary, solitary or few in dense racemes; pods ovoid oblong, closely pitted, mucronate, black, glabrous, seed 1.

Fl.-Fr.: Aug. -Nov.

Common weed of fields, road-sides, all over the area.

Cullen plicatum (Delile) C. H. Stirt. in Bothalia 13: 317. 1981; *Psoralea plicata* Delile, Descr. Egypte, Hist. Nat. 252. t. 37. f. 3. 1813; Baker in Hook.f., Fl. Brit. India 2: 103. 1876; Sanjappa, Legumes of India 132. 1991.

Perennial, much branched, undershrubs; stems striate, glabrous, woody; branches rigid, often spiny; leaves 3-foliolate; terminal leaflet oblong, 6-15 mm long, crenate; lateral ones smaller, silvery pubescent on both the surfaces and gland-dotted; flowers yellow, shortly pedunculate, 6-12-flowered racemes; pods densely hairy, indehiscent, as long as calyx, 1-seeded.

Fl.-Fr.: April – May

Khandwa city only, along open sewage lines, not common.

DALBERGIA L.f. W. 287; MP 5; SA 5

Key to Species

1. Stamens 10, diadelphous, (5) + (5)
   2. Large, woody twiners .... **D. volubilis**

2. Trees, tall or medium .... **D. lanceolaria**

1. Stamens 9, monoadelphous

3. Leaflets 3-5, acuminate; rachis zigzag .... **D. sissoo**

3. Leaflets 5-7, obtuse; rachis straight .... **D. latifolia**

Dalbergia lanceolaria

Key to Subsp.

1. Standard broader than long; leaflets 7-11, nerves prominent; pod long-stipitate .... **D. lanceolaria** subsp. **lanceolaria**

1. Standard longer than broad; leaflets 9-11, nerves obscure; pod short stipitate .... **D. lanceolaria** subsp. **paniculata**

subsp. **lanceolaria** L.f., Suppl. Pl. 316. 1782; Baker in Hook.f., Fl. Brit. India 2: 235. 1876; Singh in Verma *et al.,* Fl. Madhya Pradesh 1: 342. 1993; Samvatsar, Fl. West. Trib. Madhya Pradesh 93. 1996. *Takoli*

Large, deciduous trees; branches glabrous; young parts rusty-brown tomentose; leaves imparipinnate; leaflets 7-11, elliptic, broadly oblong or somewhat obovate, 2.5-5 × 1-2.5 cm, emarginate at apex, rounded at base, glabrous above, pilose beneath; flowers in large, axillary or terminal panicles with corymbose branches and unilaterally arranged flowers; panicle-branches silky or subglabrous; pods shining brown, 5-10 × 1-2 cm, narrowed at ends, glabrous, stipitate, usually 1-seeded; seeds orbicular.

Fl.-Fr.: April – Jan.

Common forest tree, all over the area.

subsp. **paniculata** (Roxb.) Thoth. in Bull. Bot. Surv. India 25 (1-4): 171. 1985; *Dalbergia paniculata* Roxb., Pl. Coromandel 2: 8. t. 114. 1799; Baker in Hook.f., Fl. Brit. India 2: 236. 1876; Singh in Verma *et al.,* Fl. Madhya Pradesh 1: 342. 1993; Samvatsar, Fl. West. Trib. Madhya Pradesh 93. 1996. *Phansi*

Deciduous trees, up to 15 m tall; leaflets 9-15, oblong or obovate, rounded or emarginated; flowers white, in pubescent, terminal panicles; pods narrowed at both ends, reticulately veined, 1 or 2 seeded.

Fl.-Fr.: April – Jan.

Common in mixed forests, hill slopes, in the entire area.

Dalbergia latifolia Roxb., Pl. Coromandel 2: 7. t. 113. 1798 & Fl. Ind. 3: 221. 1832; Baker in Hook.f., Fl. Brit. India 2: 131. 1876; Singh in Verma *et al.,* Fl. Madhya Pradesh 1: 342. 1993; Samvatsar, Fl. West. Trib. Madhya Pradesh 93. 1996. *Sheesham*

Deciduous trees, up to 10 m high; leaflets broadly elliptic or elliptic-obovate, 2-5-1.5-4 cm, glabrous; flowers pale- white, in lax, axillary panicles; pods oblong-lanceolate, 4-8 × 1.5-2 cm, flat, narrowed at both ends, 1-3 seeded, glabrous, black when ripe.

Fl.-Fr.: Sept. – April

Common forest tree, all over, but the number of individuals far less.

Dalbergia sissoo Roxb. ex DC., Prodr. 2: 416. 1825; Baker in Hook.f., Fl. Brit. India 2: 231. 1876; Singh in Verma *et al.,* Fl. Madhya Pradesh 1: 343. 1993; Samvatsar, Fl. West. Trib. Madhya Pradesh 93. 1996. *Sissoo*

Large trees, deciduous, up to 25 m high; young branches pubescent; bark grey or pale brown; wood brown, very hard; leaves imparipinnate, 10-20 cm long; leaflets 3-5, alternate, broadly ovate or suborbicular, 2.5-6 cm long, cuspidate or acuminate at apex, glabrous; flowers pale white, in dense, 5-8 cm long, axillary, paniculate racemes; pods flat, strap-shaped, 3.5-10 × 0.5-1.5 cm, cuneate at base, 1-3-seeded.

Fl.-Fr.: Feb. – July

Along road sides in villages, planted, not wild in this area.

Dalbergia volubilis Roxb., Pl. Coromandel 2: 48. t. 191. 1805 & Fl. Ind. 3: 231. 1832; Baker in Hook.f., Fl. Brit. India 2: 235. 1876; Singh in Verma *et al.,* Fl. Madhya Pradesh 1: 343. 1993; Samvatsar, Fl. West. Trib. Madhya Pradesh 93. 1996. *Dhoban bel*

Large woody, scandent shrubs or climbers; leaves imparipinnate, 10-25 cm long; leaflets 7-13, alternate, obovate, ovate-oblong or elliptic-oblong, 2.5-5 × 1.2-2.8 cm, obtuse or emarginate at apex, cuneate at base, glabrous above, appressed hairy beneath; flowers in terminal and axillary, dense panicles; branches of panicles rusty-brown pubescent; pods oblong, 4-7 × 2-2.5 cm, obtuse, glabrous, 1-seeded.

Fl.-Fr.: Dec. – July

Alirajpur only, in forest.

DELONIX Rafin. W. 11; MP 1; SA 1

Delonix elata (L.) Gamble, Fl. Pres. Madras 1: 280.1957 (BSI Reprint); *Poinciana elata* L. Cent. Pl. 2: 16. 1756; Baker in Hook.f., Fl. Brit. India 2: 260.1878; Singh in Verma *et al.,* Fl. Madhya Pradesh 1: 425. 1993; Samvatsar, Fl. West. Trib. Madhya Pradesh 117. 1996. *Sankesar*

Trees with greyish bark, 5-10 m tall; leaves 5-20 cm long; pinnae 4-5 cm long; leaflets linear-oblong 7-15 × 2-3 mm; racemes terminal; flowers pale to yellow in terminal, few flowered, corymbiform racemes; pods 12-20 cm long; 4-8 seeded.

Fl.-Fr.: Nov. -June

Barwani, Burhanpur, Dhar, Jhabua, Khandwa, Khargone, hedges of fields and along village roads, not in forest.

DESMODIASTRUM A. Pramanik & Thoth. W. 5; MP 1; SA 1

Desmodiastrum racemosum (Benth.) Pramanik & Thoth. in J. Indian Bot. Soc. 65: 376. 1986; *Alysicarpus racemosus* Benth. in Linnaea 24: 642. 1852; *Desmodium parviflorum* (Dalzell) Baker in Hook.f. Fl. Brit. India 2:172. 1875; *A. belgaumensis* var. *racemosus* Baker in Hook.f. Fl. Brit. India 2:168. 1876; *Desmodium alysicarpoides* van Meeuwen, Reinwardtia 6:246. 1962; Sanjappa, Legumes of India. 1991.

Herbs up to 60 cm tall; leaves 1 and 3 foliolate, intermixed; leaflets appressed pubescent beneath; petiole of 1-foliolate leaf longer than that of 3-foliolate leaf; flowers in lax, terminal racemes; pods falcate, reticulately veined, joints 4-6, rounded on both margins.

Fl.-Fr.: Aug. -Nov.

Not common, Alirajpur and Dhar, in grasslands.

DESMODIUM Desv. W. 326; MP 14; SA 9

Key to Species

1. Leaves 3-foliolate
   2. Trees; leaflets 6-15 × 5-15; flowers rosy or white, in fascicled racemes; pod 2-5 jointed; joints 2-3 times longer than broad .... **D. oojeinense**
   2. Herbs or undersrubs
      3. Pod distinctly divided into 1-seeded segments
         4. Joints of pod much longer than broad
            5. Straggling herbs; flowers lemon yellow .... **D. scorpiurus**
            5. Erect herbs
               6. Joints of pod 2-4; upper suture more indented than lower; leaflets 5-6 × 2-3 cm; corolla uniformly pink .... **D. neomexicanum**
               6. Joints of pod 6-10; lower suture nearly straight; leaflets 12-14x7-9 cm; standard white, other petals blue .... **D. laxiflorum**
         4. Joints of pod nearly as long as broad; stipules foliaceous, auricled; pods indented on both the sutures .... **D. dichotomum**
      3. Pod not distinctly divided into 1-seeded segments
         7. Flowers axillary; pod indehiscent, indented along ventral suture and longer than broad .... **D. triflorum**
         7. Flowers in short racemes; pod dehiscent, indented along both sutures, broader than long .... **D. microphyllum**
1. Leaves 1-foliolate; lower suture of pod deeply indented, upper nearly straight joints 4-7, as long as broad; flowers 2 per bract
   8. Leaves fulvous tomentose beneath; petiole 1.5 cm long; racemes 10-20 cm long, 20-30 flowered; branches terete .... **D. velutinum**
   8. Leaves sparsely grey- pubescent beneath; petiole 3.5 cm long; racemes 15-30 cm long, 40-50 flowered; branches angled .... **D. gangeticum**

Desmodium dichotomum (Willd.) DC., Prodr. 2: 336. 1825; *Hedysarum dichotomum* Willd., Sp. Pl. 3 (2): 1180-1181. 1802; *Desmodium diffusum* (Willd.) DC., Prodr. 2: 336. *non* 335. 1825; Baker in Hook.f., Fl. Brit. India 2: 169. 1876; Singh in Verma *et al.,* Fl. Madhya Pradesh 1: 364. 1993; Samvatsar, Fl. West. Trib. Madhya Pradesh 94. 1996.

Undershrubs, up to 1 m high, with deeply grooved stem; stem and branches clothed with white, spreading hairs; leaflets ovate-oblong or elliptic-oblong, 3-8 × 1.5-4.5; stipules foliaceous, sub-amplexicaul; flowers pink, fascicled in racemes or panicles; pods 1.3-2 cm long, flat, indented into 3-8, 1-seeded rectangular joints, hooked-hairy.

Fl.-Fr.: Aug. -Oct.

Common along streams and as a weed in fields, all over the area.

Desmodium gangeticum (L.) DC., Prodr. 2: 327. 1825; *Hedysarum gangeticum* L., Sp. Pl. 746. 1753; Baker in Hook.f., Fl. Brit. India 2: 168. 1876; Singh in Verma *et al.,* Fl. Madhya Pradesh 1: 346. 1993; Samvatsar, Fl. West. Trib. Madhya Pradesh 94. 1996. *Salparni*

Undershrubs, up to 1 m high; leaves 1-foliolate; leaflets ovate-oblong, 3-15 × 2-8 cm, hairy beneath, entire or repand; stipules lanceolate; flowers purple-pink, clustered in racemes; pods linear, flat, falcate, distinctly indented on the lower suture into 4-8, 1-seeded joints, hooked-hairy; seeds sub-orbicular, brown.

Fl.-Fr.: July – April

Common forest undergrowth in all districts.

Desmodium laxiflorum DC. in Ann. Sci. Nat. Paris 4: 100.1825 & Prodr. 2: 335. 1825; Baker in Hook.f., Fl. Brit. India 2: 164. 1876; Singh in Verma *et al.,* Fl. Madhya Pradesh 1: 348. 1993; Samvatsar, Fl. West. Trib. Madhya Pradesh 94. 1996. *Latkni*

Undershrubs; leaves 3-foliolate; leaflets ovate-elliptic, acute to acuminate, silky pubescent beneath; flowers in terminal and axillary racemes, 2-5 per bract; pods linear, sutures straight; joints 6-10, more than twice as long as broad, with hooked hairs.

Fl.-Fr.: Sept. - Nov.

Common in moist and shady places, throughout the area.

Desmodium microphyllum (Thunb.) DC., Prodr. 2: 337. 1825; *Hedysarum microphyllum* Thunb., Fl. Jap. 284. 1784; *Desmodium parvifolium* DC. in Ann. Sci. Nat. Paris 4: 100.1825 & Prodr. 2: 334. 1825; Baker in Hook.f., Fl. Brit. India 2: 174. 1876; Singh in Verma *et al.,* Fl. Madhya Pradesh 1: 348. 1993.

Perennial, ascending or trailing, slender, much branched herbs; leaves 3-foliolate; leaflets oblong, 5-7 mm long, rounded, retuse or mucronate at apex, subcordate at base, glabrous above, pubescent beneath; lateral leaflets often smaller; stipules linear, setaceous; flowers pale blue, in terminal, 4-8-flowered racemes; rachis hooked hairy; pods 5-12 mm long, hooked hairy; 1-4-jointed.

Fl.-Fr.: Nov.

In waste lands, occasional, Burhanpur, Khandwa.

Desmodium neomexicanum A. Gray, Pl. Wright. 1: 53. 1852; *D. spirale auct. non* (Sw.) DC., 1825: Baker in Hook.f., Fl. Brit. India 2: 164. 1876; Samvatsar, Fl. West. Trib. Madhya Pradesh 95. 1996; Khanna *et al.,* Suppl. Fl. Madhya Pradesh 55. 2001.

Annual, erect or straggling herbs; stems and branches slightly hairy; leaves 3-foliolate; leaflets membranous, the terminal largest, ovate-lanceolate; flowers white with blue tinge, in axillary and terminal racemes or panicles; pods 2-3 cm long, reticulately veined, indented deeply on both the sutures, 4-6-jointed.

Fl.-Fr.: Aug. – Nov.

Common in hilly regions, along streams, all over the area.

Desmodium oojeinense (Roxb.) Ohashi in Ginkgoana 1: 117. 1973; *Dalbergia oojeinensis* Roxb., Fl. Ind. 3: 220.1832. *Ougeinia oojeinensis* (Roxb.) Hochr. in Bull. Soc. Bot. Geneve 13-14: 51. 1909; Baker in Hook.f., Fl. Brit. India 2: 161. 1876; Singh in Verma *et al.,* Fl. Madhya Pradesh 1: 382. 1993; Samvatsar, Fl. West. Trib. Madhya Pradesh 103. 1996. *Tinsa*

Trees; leaves pinnately 3 foliolate; leaflets 6.5-18 × 4.5-15 cm, broadly ovate, rigidly coriaceous, sparsely pubescent beneath; flowers white or rosy, numerous in fascicled racemes; pods 2-5 jointed, joints reticulately veined, 2-3 times as long as broad.

Fl.-Fr.: Feb.-May

Occasional on road sides, hill slopes, hedges of fields, in all districts.

Desmodium scorpiurus (Sw.) Poir. in Dict. Sci. Nat. ed. 2. 13: 110.1819; *Hedysarum scorpiurus* Sw., Prodr. 107. 1788; Vartak & Kumbhojkar, J. Bombay Nat. Hist. Soc. 81:224. 1984; Kothari, in N.P. Singh *et al.,* Fl. Maharashtra 1:675. 2000.

Spreading ascending or diffuse herbs with hooked hairy stems; leaves 3-foliolate; leaflets elliptic or less often round or oblong, 0.5-4.5 × 0.2-1.8 cm, base obtuse, apex obtuse rounded or emarginate, both surfaces apressed hairy; stipules amplexicaul at the base; flowers pink or pale purple, in laxly flowered, terminal and axillary racemes; pods linear, slender, 3–8-jointed.

Fl.-Fr.: Oct. - May

Rare, **Burhanpur** only, along raiway tracks, Asirgad, Chandni and banks of streams.

Desmodium triflorum (L.) DC., Prodr. 2: 334. 1825; *Hedysarum triflorum* L., Sp. Pl. 2: 749-750.1753; Baker in Hook.f., Fl. Brit. India 2: 173. 1876; Singh in Verma *et al.,* Fl. Madhya Pradesh 1: 350.1993, Samvatsar, Fl. West. Trib. Madhya Pradesh 95. 1996. *Jangli methi*

Creeping herbs, more or less glabrous; leaves trifoliolate; leaflets obovate, retuse or emarginate at apex, 5-6 mm long; flowers small, purple, solitary or racemed in the axils; pod 3-5 seeded, 1-1.5 cm long, straight on the dorsal suture, reticulate, glabrous.

Fl.-Fr.: Sept. – April

Very common throughout the area.

Desmodium velutinum (Willd.) DC., Prodr. 2: 228. 1825; *Hedysarum velutinum* Willd. Sp. Pl. 3(2): 117. 1802; *D. latifolium* (Roxb.) DC. Prodr. 2: 328, 1825; Baker in Hook.f., Fl. Brit. India 2: 168. 1876; Singh in Verma *et al.,* Fl. Madhya Pradesh 1: 351. 1993; Samvatsar, Fl. West. Trib. Madhya Pradesh 95. 1996.

Erect, hairy undershrubs, 1-2 m high; leaflet one, ovate, 3-12 × 3-11 cm, entire or sinute-repand, pubescent above, tomentose beneath; stipules ovate, acuminate; flowers pink to purple, fascicled in axillary and terminal, spicate racemes; bracts ovate, acuminate; pod 3-6 joints, as broad as long, clothed with hooked hairs.

Fl.-Fr.: Aug. -Nov.

Common forest undergrowth, all over.

DICHROSTACHYS (DC.) Wight & Arn. W. 17; MP 1; SA 1

Dichrostachys cinerea (L.) Wight & Arn., Prodr. 1: 271. 1834; *Mimosa cinera* L., Sp. Pl. 1: 520.1753; Baker in Hook.f., Fl. Brit. India 2: 288. 1878; Singh in Verma *et al.,* Fl. Madhya Pradesh 1: 438. 1993; Samvatsar, Fl. West. Trib. Madhya Pradesh 123. 1996. *Virtuli*

Polygamous shrubs or small trees; branches ending into thorns; leaves 2-pinnate; pinnae 8-14 pairs, with stalked glands in between; leaflets 12-20 pairs, linear-oblong, oblique-based; flowers in dense, axillary, pedunculate spikes; upper flowers fertile, yellowish white; lower ones sterile, white, purple or rose-coloured; pods flat, 4-7 cm long, glabrous 6-10 seeded.

Fl.-Fr.: Aug. -Oct.

Common in open forest, village roads and on the hill slopes, in all districts.

ELEIOTIS DC. W. 2; MP 2; SA 1

Eleiotis monophyllos (Burm.f.) DC. *Hedysarum sororium* L., Mant. Pl. 270.1771; Baker in Hook.f., Fl. Brit. India 2: 153. 1876; *Eleiotis monophylla* (Burm.f.) DC., Mém. Légum. 7: 350.1825; Singh in Verma *et al.,* Fl. Madhya Pradesh 1: 352. 1993.

Slender, trailing annual herbs with triangular branches; leaves unifoliolate (occasionally 3-foliolate on the same plant), orbicular, 2-4 cm, obtuse or emarginate; flowers small, reddish, in axillary racemes; pod boat-shaped with one joint, up to 7 mm long, reticulately veined, 1-seeded.

Fl.-Fr.: Aug.-Dec.

Rare, as forest undergrowth, Burhanpur, Indore, Khandwa.

ENDOSAMARA R. Geesink Monotypic

Endosamara racemosa (Roxb.) R. Geesink, Leiden Bot. Ser 8: 93. 1984; *Robinia racemosa* Roxb., Hort. Bengal. 56. 1814; *Millettia racemosa* Benth. in Miq. Pl. Junag. 249. 1852; Singh in Verma *et al.,* Fl. Madhya Pradesh 1: 378. 1993; Samvatsar, Fl. West. Trib. Madhya Pradesh 102. 1996. *Ardhagi*

Large woody climbing shrubs, up to 3-4 m high; leaves imparipinnate, up to 30 cm long; leaflets 5-17, obovate-oblong, 5-8 cm long, acute at apex, obtusely cuspidate, often wavy, glabrous, young ones slightly silky beneath, terminal leaflet 7.5-10 × 3.5-5 cm; stipules large, filiform; flowers pale yellow or reddish, in paniculate racemes; pods flat, linear, 10-20 × 1-2 cm, torulose, dehiscent, 4-5-seeded.

Fl.-Fr.: April – Dec.

Rare, Dhar, Indore, as forest undergrowth.

ERYTHRINA L. W. 129; MP 2; SA 1

Erythrina suberosa Roxb. Fl. Indica 3: 253. 1832; Baker in Hook.f., Fl. Brit. India 2: 189. 1876; Singh in Verma *et al.,* Fl. Madhya Pradesh 1: 353. 1993; Samvatsar, Fl. West. Trib. Madhya Pradesh 96. 1996. *Gadha palas* **(Plate 4, Fig. 5)**

Small trees, with corky, deeply fissured bark; leaflets 3, rhomboid-ovate, 10-15 cm long and broad, pubescent below, stipellate; flowers scarlet, 1.5-3 cm long, in racemes gathered at the ends of branches; calyx 2-lipped; pods linear, 6-15 × 0.3-0.8 cm, straight, torulose, 2-5 seeded.

Fl.-Fr.: March-June

Occasional, in wasteland and on the fringes of forests, everywhere, but number of individuals is less.

FLEMINGIA Roxb. ex W. Aiton & W.T. Aiton W. 37; MP 9; SA 2

Key to Species

1. Leaves 3-foliolate; flowers reddish, in axillary racemes or clustered; pods with powdery coating .... **F. macrophylla**
1. Leaves unifoliolate; flowers axillary or terminal in branched racemes; plants glabrous ...... **F. strobilifera**

Flemingia macrophylla (Willd.) Kuntze ex Merr. in Philipp. J. Sci. 5: 130.1910; *Crotalaria macrophylla* Willd., Sp. Pl. 3: 982. 1802; *Flemingia congesta* Roxb. ex W. T. Aiton, Hortus Kew. ed. 2.4: 349. 1812; Baker in Hook.f., Fl. Brit. India 2: 228. 1876; Singh in Verma *et al.,* Fl. Madhya Pradesh 1: 357. 1993. *Bara salpan*

Erect shrubs; leaves digitately 3-foliolate; leaflets ovate-oblong or lanceolate, 10-20 × 3-15 cm, acute or obtuse at apex, cuneate at base, glabrous above, brown pubescent on nerves and black gland dotted beneath; stipules linear-lanceolate, 10-13 mm long; flowers purplish, in dense, axillary racemes or panicles; bracts 3-8 mm long; pods 10-15 mm long, brown pubescent, dotted with black glands, 2-seeded.

Fl.-Fr.: Sept. – May

Rare, **Khandwa** only, as forest undergrowth, in Kalibhit and Sunderdev.

Flemingia strobilifera (L.) Aiton & W.T. Aiton, Hortus Kew. ed. 2.4: 350.1812; *Hedysarum strobiliferum* L., Sp. Pl. 746. 1753; Baker in Hook.f., Fl. Brit. India 2: 227. 1876; Singh in Verma *et al.,* Fl. Madhya Pradesh 1: 359. 1993; Samvatsar, Fl. West. Trib. Madhya Pradesh 96. 1996. **(Plate 4, Fig. 6)**

Erect shrubs; leaves 1-foliolate; leaflets oblong to ovate-lanceolate, 8-15 × 3-7.5 cm, acute or acuminate at apex, rounded at base, thinly pubescent above, silky pubescent and gland-dotted below; stipules 5-10 mm long; flowers pale-yellow with red streaks, in axillary and terminal, 7-12 cm long racemes or panicles; pods oblong, 6-10 mm long, densely pubescent, 2-seeded.

Fl.-Fr.: March – June

Alirajpur, Barwani, Burhanpur, Dhar, Khargone.

GALACTIA R. Br. W. 113; MP 1; SA 1

Galactia tenuiflora (Klein ex Willd.) Wight & Arn., Prodr. 1: 206. 1834; *Glycine tenuiflora* Klein ex Willd., Sp. Pl. 3: 1059. 1802; Baker in Hook.f., Fl. Brit. India 2:192. 1876; Singh in Verma *et al.,* Fl. Madhya Pradesh 1: 360.1993; Samvatsar, Fl. West. Trib. Madhya Pradesh 96. 1996.

A twining herb; leaves pinnately trifoliolate; petiole 1.5-3.5 cm long; stipules small; stipels minute; leaflets 4-5.5 × 1.5-2.5 cm, silky beneath, ovate-lanceolate; flowers bluish-purple in 6-12 cm long racemes; pods 3-5 × 0.75-1 cm, glabrescent, seeds 6-9.

Fl.-Fr.: Aug. – Jan.

Alirajpur, Burhanpur, Dhar, Khandwa.

HARDWICKIA Roxb. W. 2; MP 1; SA 1

Hardwickia binata Roxb., Pl. Coromandel 3: 6. t. 209. 1819; Baker in Hook.f., Fl. Brit. India 2: 270.1878; Singh in Verma *et al.,* Fl. Madhya Pradesh 1: 426. 1993; Samvatsar, Fl. West. Trib. Madhya Pradesh 117. 1996. *Anjan*

Trees; leaflets 1-pair, 2.5-4.5 × 1-1.5 cm, subsessile, ovate, trapezoid, oblique at base, obtuse; flowers in axillary and terminal panicles; pods strap-shaped, veined, glabrous; seed 1 near the tip.

Fl.-Fr.: Oct. -Jan.

Common in forests, hill slopes all over, but endangered due to lopping of branches for fodder, on a daily basis.

INDIGASTRUM Jaub. ex Spach. W. 9; MP 1; SA 1

Indigastrum parviflorum (F. Heyne ex Hook. & Arn.) Schrire in Bothalia 22(2): 168. 1992; *Indigofera parviflora* F. Heyne ex Hook. & Arn., 1: 201. 1834; Baker, in Hook.f., Fl. Brit. India 2:97. 1876; Sanjappa, Legumes of India 193. 1991; Samvatsar, Fl. West. Trib. Madhya Pradesh 100j. 1996; Khanna *et al.,* Suppl. Fl. Madhya Pradesh 67.2001.

Erect, branched herbs; stems and branches silvery canescent; leaves 7-9-foliolate; leaflets elliptic-oblong, 1.2 × 0.5 cm, membranous, grey pubescent; flowers reddish- purple, in 6-8-flowered racemes; pods linear, deflexed, hairy, 12-16-seeded.

Fl.-Fr.: Sept. –Jan.

Rare, **Alirajpur** only, as forest undergrowth, at Aamkhut.

INDIGOFERA L. W. 674; MP 23; SA 17

Key to Species

1. Leaves simple
    2. Leaves linear; pod 1-seeded; corolla much exserted from calyx .... **I. linifolia**
    2. Leaves ovate-cordate; pod 2-seeded; corolla not exserted from calyx .... **I. cordifolia**
1. Leaves compound
    3. Leaves 3-5 foliolate
        4. Leaflets gland-dotted beneath
            5. Leaflets nigro-punctate beneath; pods globose, margined with blunt teeth, 1-2 seeded .... **I. glandulosa**
            5. Leaflets brown-orange punctate beneath; pods elongated, tetragonous, sharply angled with narrow wings, up to 1.5 cm long, 3-5 seeded

6. Plants slender; pods 1 mm thick .... **I. prostrata**
6. Plants stout; pods 2-3 mm thick .... **I. trifoliata**

4. Leaflets not gland-dotted
7. Pods tetragonous, 6-10 seeded, spine tipped; terminal leaflet stalked .... **I. trita**
7. Pods torulose
8. Leaflets opposite
9. Seeds blackish-brown; leaflets normally 5 .... **I. angulosa**
9. Seeds yellowish-brown; leaflets 3 .... **I. deccanensis**
8. Leaflets alternate
10. Flowers 10-20 in dense, sessile, villous heads; seeds 3-4; leaflets 5-7 ...... **I. trigonelloides**
10. Flowers more than 20 in elongated racemes; seeds 5-6; leaflets 1-3-5, on the same plant .... **I. oblongifolia**

3. Leaves more than 5-foliolate
11. Plants viscous .... **I. colutea**
11. Plants not viscous
12. Pods 2-seeded, subovoid; leaflets alternate; plants prostrate .... **I. linnaei**
12. Pods more than 2-seeded, elongated; leaflets opposite; plants erect
13. Racemes shorter than the leaves; leaflets 6-7; pods torulose, segmented .... **I. coerulea**
13. Racemes longer than the leaves
14. Pods tetragonous, deflexed, densely hirsute, 6-seeded; leaflets opposite .... **I. astragalina**
14. Pods not tetragonous, more than 2 cm long
15. Leaflets 7-13, opposite; bracts setaceous; calyx not silky brown .... **I. tinctoria**
15. Leaflets 13-31, subopposite; bracts boat-shaped; calyx silky brown .... **I. cassioides**

Indigofera angulosa Edgew. in J. Asiat. Soc. Bengal 21:168. 1852; Baker in Hook.f. Fl. Brit. India 2:97. 1876; Singh in Fl. Madhya Pradesh 1:364. 1993; Samvatsar, Fl. West. Trib. Madhya Pradesh 98. 1996.

Suffruticose herbs up to 1.5 m tall; leaflets 5, 1.5-3 × 0.5-1.5 cm, elliptic-oblong, lanceolate or obovate; flowers red, in axillary, 10-15 cm long racemes; pods torulose, rectangular, 10-15 mm long; seeds 1-4, ellipsoid, blackish-brown, concave with a raised ridge in the centre.

Fl.-Fr.: Aug. -Oct.

Rare, **Indore** only, Bai gram.

Indigofera astragalina DC., Prodr. 2: 228. 1825; *I. hirsuta auct. non* L. 1753: Baker in Hook.f., Fl. Brit. India 2: 98. 1876, *p.p.* Singh in Verma *et al.,* Fl. Madhya Pradesh 1: 365. 1993; Samvatsar, Fl. West. Trib. Madhya Pradesh 98. 1996.

Annual, erect herbs; stems and branches covered with brown stiff hairs; leaves 7-9-foliolate, 8-15 cm long; leaflets opposite, oblong to obovate, elliptic or elliptic-oblong, 1-4 × 1-2 cm, mucronate at apex, pubescent on both the surfaces; flowers light pink to brick red, in axillary, 8-20 cm long racemes; pods subquadrangular, 10-18 × 3-3.5 mm, reflexed, hirsute, 3-6-seeded; seeds brown, tetragonus, pitted.

Fl.-Fr.: Aug. – Sept.

Alirajpur, Burhanpur, Dhar, Indore, Khandwa, Khargone.

Indigofera cassioides Rottler ex DC., Prodr. 2: 225. 1825. *I. pulchella* Roxb., Fl. Ind. 3: 382. 1832; Baker in Hook.f., Fl. Brit. India 2: 101. 1876; Singh in Verma *et al.,* Fl. Madhya Pradesh 1: 365. 1993; Samvatsar, Fl. West. Trib. Madhya Pradesh 98. 1996. *Baroli*

Undershrubs, branches striate; leaves 8-25 cm long, imparipinnate; leaflets 13-20, elliptic-oblong to obovate, 1.5-2.5 × 0.5-1.5 cm; flowers pink to lilac red, in axillary lax racemes; pods 3-6 cm long, turgid, reflexed, 8-10 seeded.

Fl.-Fr.: Oct. – May

Alirajpur, Barwani, Dhar, Indore, Khargone.

Indigofera coerulea Roxb. Fl. Ind. 3:377. 1852; *I. articulata* sensu Cooke Fl. Bombay Presi. 1:336. 1958 (B.S.I. Reprint); Baker in Hook.f. Fl. Brit. India 2:97 1876; Singh in Verma *et al.,* Fl. Madhya Pradesh 1: 364. 1993.

Shrubs, about 1m tall, argento-canescent; leaves 5-8 cm long; leaflets 5-9, obovate-oblong, 1.2-3 × 0.6-2 cm; flowers red, in axillary racemes; pedicels very short; pods 2-4 mm long, curved, torulose, 3-4 seeded.

Fl.-Fr.: Sept. -Jan.

Rare, Burhanpur, Khandwa.

Indigofera colutea (Burm.f.) Merr. in Philipp. J. Sci. 19: 355. 1921; *Galega colutea* Burm.f., Fl. Ind. 172. 1768; *Indigofera viscosa* Lam., Encycl. 3: 249. 1789; Baker in Hook.f., Fl. Brit. India 2: 95. 1876; Samvatsar, Fl. West. Trib. Madhya Pradesh 98. 1996.

Annual herbs, clothed with viscous, gland-tipped hairs; leaflets 7-11, 0.5-1 × 0.2-0.4 cm, obovate-elliptic, obtuse, apiculate, appressed hairy; stipules filiform; flowers pink, in 6 to 12 flowered, glandular racemes not exceeding the leaves; calyx teeth setaceous; pods subcylindric, faintly torulose, glandular hairy.

Fl.-Fr.: April – Dec.

Alirajpur, Burhanpur, Dhar, Indore.

Indigofera cordifolia Heyne ex Roth, Nov. Pl. Sp. 357. 1821; *Heylandia cordifolia* Graham in Wall., Numer. List 5343. 1831; Baker in Hook.f., Fl. Brit. India 2: 93. 1876; Singh in Verma *et al.,* Fl. Madhya Pradesh 1: 366. 1993; Samvatsar, Fl. West. Trib. Madhya Pradesh 98. 1996. **(Plate 4, Fig. 7)**.

A diffuse prostrate herb; leaves alternate, simple, broadly ovate, 0.7-2 × 0.5-1.5 cm, cordate, mucronate, hairy on both surfaces; flowers pink, in dense, axillary clusters; pods cylindric, oblong, straight, densely pubescent, 2-seeded.

Fl.-Fr.: Aug. – Oct.

Common everywhwre.

Indigofera deccanensis Sanjappa in J. Econ. Taxon. Bot. 4:282. 1983, in Legumes of India 188. 1991, in Hajra *et al.,* Fasc. Fl. India 21:51, f. 19. 1995; *I. torulosa* Wadood Khan, in Indian

Forester 108:516. 1982, *non* E. Meyer, 1836; Samvatsar, Fl. West. Trib. Madhya Pradesh 98. 1996.

Perennial, much branched, suffruticose, grey pubescent shrubs; leaflets 5, in opposite pairs and an odd one, 1 to 1.5 cm long, terminal larger, obovate; racemes equalling or longer than the leaves; flowers pink; pods deflexed, curved upwards, pointed, torulose; seeds 2-4.

Fl.-Fr.: Sept. -Dec.

Common along forest edges, ravines and more exposed situations within forests.

Indigofera glandulosa

Key to Varieties

1. Pods oblong, 2-seeded .... **I. glandulosa** var. **glandulosa**
1. Pods orbicular, 1-seeded .... **I. glandulosa** var. **sykesii**

Indigofera glandulosa var. **glandulosa** Roxb. ex Willd., Sp. Pl. 3: 1227. 1803; Baker in Hook.f., Fl. Brit. India 2: 94. 1876; Singh in Verma *et al.,* Fl. Madhya Pradesh, 1: 366. 1993; Samvatsar, Fl. West. Trib. Madhya Pradesh 99. 1996. *Barbata*

An erect, annual herb; stem 20-45 cm tall, branches clothed with long, spreading hairs when young; leaves 3-foliolate; leaflets 0.8-3.5 × 0.04-1.2 cm, terminal largest, obovate, retuse, obliquely cuneate at the base; stipules linear, hairy; flowers reddish-purple, 8-10 in axillary, racemose heads; pods angled, deflexed, 3-5 mm long, angles slightly winged and toothed; seeds 2, smooth, sub-reniform.

Fl.-Fr.: Aug.-Dec.

Very common in fields, road-sides, grass-lands.

Indigofera glandulosa var. **sykesii** Baker ex B. K.V. Kumar & N. Ramayya, Curr. Sci. 52 (9):429. 1983; Naik, Fl. Marathwada 1:280.1998. (We have followed the citation as given by the International Plant Names Index, as some floras have given the authorship of this variety to Griff.).

Differs from the typical variety as shown in the key.

Fl.-Fr.: Sept. -Nov.

Common in the same habitat as above.

Indigofera linifolia (L.f.) Retz., Observ. Bot. 4: 29. 1786; *Hedysarum linifolium* L.f., Suppl. Pl. 331. 1781; Baker in Hook.f., Fl. Brit. India 2: 92. 1876; Singh in Verma *et al.,* Fl. Madhya Pradesh 1: 367. 1993; Samvatsar, Fl. West. Trib. Madhya Pradesh 99. 1996. *Torki*

Annual or perennial, much-branched, diffuse herbs; leaves linear to oblong-lanceolate, 20-45 × 2-4 mm, sharp-tipped, appressed-pubescent on both surfaces; flowers pink, 6-8 flowered racemes; pods silvery, globose, mucronate.

Fl.-Fr.: Sept. -Dec.

Common throughout the area.

Indigofera linnaei Ali, Bot. Notiser 111: 549. 1958; *I. enneaphylla* L. Mant. 2: 272. 1771, nom. illegit. Baker in Hook.f., Fl. Brit. India 2: 94. 1876; Singh in Verma *et al.*, Fl. Madhya Pradesh 1: 368. 1993; Samvatsar, Fl. West. Trib. Madhya Pradesh 99. 1996. *Atahai* **(Plate 4, Fig. 8)**

Trailing herbs; leaves 5-7 foliolate; leaflets alternate, 0.2-0.5 × 0.1-3 cm, obovate-obtuse, appressed hairy; flowers bright red in short, dense, axillary racemes; pods minutely hairy.

Fl.-Fr.: Sept. -Nov.

Common along road-sides and in grassland, throughout.

Indigofera oblongifolia Forssk., Fl. Aegypt. -Arab. 137. 1775; *I. paucifolia* Delile, Fl. Aegypt. -Arab. 107. t. 37 fig. 22. 1813; Baker in Hook.f., Fl. Brit. India 2: 97. 1876; Singh in Verma *et al.,* Fl. Madhya Pradesh 1: 368. 1993; Samvatsar, Fl. West. Trib. Madhya Pradesh 99. 1996. *Jhilla*

A low shrub with numerous branches from the base; stem woody, stout, silvery white; leaflets 1-4, normally 5, on one and the same branch, alternate, 1-2 × 0.5-1 cm normally oblanceolate, densely white hairy below; flowers red, many, very small, in elongated racemes, exceeding the leaves; pods numerous, curved outwards, torulose; seeds 6-8.

Fl.-Fr.: Sep. -Dec.

Occasional along streams, Burhanpur, Dhar, Indore.

Indigofera prostrata Willd., Sp. Pl. 3: 1226. 1802; *Indigofera trifoliata* auct. non L. 1753; Baker in Hook.f., Fl. Brit. India 2: 96. 1876; Singh in Verma *et al.,* Fl. Madhya Pradesh 1: 369. 1993; Samvatsar, Fl. West. Trib. Madhya Pradesh 100.1996.

Erect or trailing, herbs or undershrubs, 30-60 cm long; branches covered with apressed white hairs; leaves digitately 3-foliolate; leaflets obovate to obovate-oblong, 10-30 × 8-12 mm, rounded to obtuse at apex, cuneate to obtuse at base, globrous above, densely appressed hairy and brownish-orange or black gland-dotted beneath; flowers pink or red, in 5-8 mm long, globose racemes; pods deflexed, linear, 10-20 mm long, tetragonous, apiculate, broadly winged along sutures, 5-7-seeded; seeds reddish- brown, ellipsoid, *ca* 1 mm long.

Fl.-Fr.: July – Nov.

Alirajpur only, along hill slopes and open spaces in the forest, Kattiwada.

Indigofera tinctoria L. Sp. Pl. 2: 751. 1753; Baker in Hook.f., Fl. Brit. India 2: 99, 1876; Singh in Verma *et al.,* Fl. Madhya Pradesh 1: 369. 1993; Samvatsar, Fl. West. Trib. Madhya Pradesh 100.1996. *Neel*

Grey appressed-pubescent undershrubs; leaves compound, 2.5-11 cm long; leaflets 9-13, 0.5-2 × 0.2-1 cm, opposite, obovate, rounded, apiculate, greenish-black on drying; flowers red, in 15-20 flowered racemes; pods linear, slightly curved, sparsely hairy, deflexed, 6-12 seeded.

Fl.-Fr.: Nov.-Dec.

Common along road-sides and on forest paths, all over the area.

Indigofera trifoliata

Key to Varieties

1. Pod distinctly winged, 2-2.5 cm long .... **I. trifoliata** var. **duthiei**
1. Pod not winged, 1.5-2 cm long .... **I. trifoliata** var. **trifoliata**

var. **duthiei** (Drumm. ex Nailk) Sanjappa in Legumes of India 196. 1991 & in Hajra *et al.,* Fasci. Fl. India 141. 1995; *I. duthiei* Drumm. ex Naik, Fl. Usmanabad 91. 1979; Sanj. in J. Econ. Tax. Bot. 5:1031. 1984.

Suffruticose herbs or undershrubs, 30-60 cm tall; leaves 3-foliolate; leaflets 1.0-1.5 × 0.4-0.7 cm, oblanceolate, pellucid-punctate beneath; flowers red, sessile, 4-8 in short racemes; pods deflexed, tetragonous, narrowly winged.

Fl.-Fr.: Aug. -Dec.

Alirajpur, Burhanpur, Dhar, Indore, Khandwa.

var. **trifoliata** L. in Torner, Cent. Pl. 2: 29. 1756; Baker in Hook.f., Fl. Brit. India 2: 96. 1876; Singh in Verma *et al.,* Fl. Madhya Pradesh 1: 369. 1993; Samvatsar, Fl. West. Trib. Madhya Pradesh 100.1996.

Perennial herbs, glabrous or clothed with few appressed hairs; leaves compound; leaflets 3, 1.2-2.5 × 0.3-1.2 cm, obovate-oblanceolate, retuse or rounded; flowers red, in axillary, 6-12 flowered racemes, which are shorter than leaves; pods deflexed, linear, 1.5-2 cm long, not torulose, with two narrow wings on either side of the suture, 6-10 seeded.

Fl.-Fr.: July-Nov.

Common in wastelands and open forests in wet and shady habitats, everywhere.

Indigofera trigonelloides Jaub. & Spach, Ill. Pl. Orient. 5: 92, pl. 482. 1856; Baker in Hook.f., Fl. Brit. India 2: 94. 1876; Singh in Verma *et al.,* Fl. Madhya Pradesh 1: 370.1993; Samvatsar, Fl. West. Trib. Madhya Pradesh 100.1996.

Annual, trailing, densely silvery covered herbs; leaves 5-7-foliolate; leaflets 0.5-1.25 × 0.3-0.5 cm, oblanceolate, silvery white on both sides; flowers red, in dense sessile heads; pods linear, 0.6-0.8 cm, patent, silvery hairy, 3-4 seeded.

Fl.-Fr.: Aug. -Dec.

Rare, **Dhar** only, Jama masjid, Mandu plateau.

Indigofera trita L.f. Suppl. 335. 1781; Baker in Hook.f., Fl. Brit. India 2:96. 1876; Singh in Verma *et al.,* Fl. Madhya Pradesh 1: 370.1993; Samvatsar, Fl. West. Trib. Madhya Pradesh 101. 1996.

Undershrubs with fine appressed hairs; leaflets 0.5-4.5 × 0.4-2 cm, obovate-oblong, obtuse or retuse, mucronate; flowers red, in spicate racemes; pods up to 2.5 cm long, patent, sometimes deflexd, appressed pubescent, shortly beaked with a sharp point, 6-10 seeded; seeds oblong, brown.

Fl.-Fr.: Sept.-Feb.

Common along water streams and in shady and moist places, in all districts.

LATHYRUS L. W. 159; MP 2; SA 1

Lathyrus aphaca L., Sp. Pl. 729. 1753; Baker in Hook.f., Fl. Brit. India 2: 179. 1876; Singh in Verma *et al.,* Fl. Madhya Pradesh 1: 371. 1993; Samvatsar, Fl. West. Trib. Madhya Pradesh 101. 1996. *Jangli matar*

Annual, slender or trailing, glabrous, 20-45 cm, high herbs; leaves reduced to coiled tendrils; stipules foliaceous at the base of tendrils, ovate-hastate, 1.3-2.5 × 0.8-1.7 cm, truncate, entire; flowers axillary,yellow, 1 or 2; pods black, linear, oblong or subfalcate, 1.5-3 × 0.3-0.4 cm, wingless, glabrous, 4-7-seeded.

Fl.-Fr.: Jan. – March

Common weed of gardens, not seen outside.

MACROTYLOMA (Wight & Arn.) Verdc. W. 25; MP 1; SA 1

Macrotyloma uniflorum (Lam.) Verdc. in Kew Bull. 24(2): 322.401. 1970; *Dolichos uniflorus* Lam., Encycl. 2: 299. 1786; *Dolichos biflorus* Murray, Syst. Nat. ed. 13.548. 1774; Baker in Hook.f., Fl. Brit. India 2: 210.1876; Singh, in Verma et al. Fl. Madhya Pradesh 1:375. 1993.

Slender climbing or twining herbs; leaves 3-foliolate; leaflets ovate-elliptic to rhomboid, 2-5 × 1.5-3 cm, acute at apex, cuneate at base, glabrescent; flowers yellow or greenish-yellow with violet blot on the standard, usually solitary, occasionally 2-3 in the axil of leaves; pods linear, 4-6 × 0.6-0.8 cm, compressed, straight or falcate, densely pubescent, 4-7-seeded; seeds dark or variously coloured, oblong or subreniform, 3-4 mm long.

Fl.-Fr.: Nov. – April

Burhanpur, Khandwa.

MEDICAGO L. W. 103; MP 2; SA 2

Key to Species

1. Edges of fruit with small tubercles; fruit 1-seeded, distant end coiled .... **M. lupulina**
1. Edges of fruit spiny, partitioned between seeds, coiled .... **M. polymorpha**

Medicago lupulina L., Sp. Pl. 2: 779. 1753; Baker in Hook.f., Fl. Brit. India 2: 90.1876; Singh in Verma *et al.,* Fl. Madhya Pradesh 1: 376. 1993.

Annual or biennial, prostrate or decumbent herbs; stems with simple or glandular pubescent hairs; leaflets obovate, 5-20 × 4-8 mm, shallowly notched, mucronate at apex, cuneate-deltoid at base, dentate towards apex, glabrous or hairy beneath; stipules lanceolate, 2-3 mm long, acuminate, entire or dentate; flowers yellow, in axillary, pedunculate racemes; pods 2-3 mm long, curved, 1-seeded.

Fl.-Fr.: Dec. – March

Indore, Khargone.

Medicago polymorpha L., Sp. Pl. 2: 779. 1753; *M. denticulata* Willd. Sp. Pl. 3: 1414, 11803; Baker in Hook.f., Fl. Brit. India 2: 90, 1876; Singh in Verma *et al.,* Fl. Madhya Pradesh 1: 376. 1993; Samvatsar, Fl. West. Trib. Madhya Pradesh 101. 1996.

A glabrous annual herb with erect and prostrate branches; leaves 3-foliolate; leaflets, 1-3 × 5-2 cm, obovate, cuneate, faintly toothed, sometimes retuse or emarginate at the apex; peduncles much shorter than the petioles; bracts scarious; pods subglobose, of 2-4 spirals, deeply reticulate, margined with double row of blunt spines, many seeded.

Fl.-Fr.: Oct. -Dec.

Very common along irrigation canals and moist places, throughout the area.

MELILOTUS Miller W. 22; MP 2; SA 2

Key to Species

1. Flowers white, 4 mm long; pod 2-seeded; robust hersbs .... **M. albus**
1. Flowers yellow, 2 mm long; pod 1-seeded; slender herbs .... **M. indicus**

Melilotus albus Medik. Vorles. Churpfälz. Phys. -Ökon. Ges. 2: 382. 1787; Baker in Hook.f., Fl. Brit. India 2: 89. 1876; Singh in Verma *et al.,* Fl. Madhya Pradesh 1: 377. 1993; Samvatsar, Fl. West. Trib. Madhya Pradesh 102. 1996.

Annual herbs up to 60 cm high; leaflets 1.7-3 × 1-1.8 cm, oblanceolate, elliptic or obovate, inciso-serrate in upper half, cuneate at the base; flowers white, in axillary racemes; pods oblong, 2 seeded, reticulately veined.

Fl.-Fr.: Jan. -April

Common weed in fields and other moist habitats, throughout.

Melilotus indicus (L.) All., Fl. Pedem. 1: 308. 1785; *Trifolium indicum* L. Sp. Pl. 2: 765. 1753; *Melilotus parviflorus* Desf. Fl. Atlant. 2: 192. 1799; Baker in Hook.f., Fl. Brit. India 2: 89. 1876; Singh in Verma *et al.,* Fl. Madhya Pradesh 1: 377. 1993; Samvatsar, Fl. West. Trib. Madhya Pradesh 102. 1996. *Ban methi*

Annual herbs; leaflects 1.5-2.5 × 0.5-1.5 cm, obovate, oblong, glabrous or with few hairs beneath; flowers yellow, in 2-6 cm long racemes; pod ellipsoid, compressed, tapering at both ends glabrous.

Fl.-Fr.: Nov. -Feb.

Common in fields as winter weed.

MILLETIA Wight & Arn. W. 202; MP 2; SA 1

Millettia extensa (Benth.) Benth. ex Baker in Hook.f., Fl. Brit. India 2(4): 109. 1876; *Otosema extensa* Benth. in Miq. Pl. Junag. 249. 1852; *M. auriculata* Baker ex Brandis, For. Fl. 138. 1874; Baker in Hook.f., Fl. Brit. India 2: 108. 1876; Singh in Verma *et al.,* Fl. Madhya Pradesh 1: 378. 1993; Samvatsar, Fl. West. Trib. Madhya Pradesh 102. 1996. *Agar bel*

A large, subrect or climbing shrub; leaflets 7-9, obovate-oblong, cuspidate; racemes densely silky; pods linear, flattened, velvety, sutures thickened.

Fl.-Fr.: March-May

Common forest element in all districts.

MIMOSA L. W. 708; MP 5; SA 2

Key to Species

1. Stamens 10; rachis up to 5 cm long; leaflets 6-8 pairs; pinnae less than 3 cm long; pod prickly along both sutures .... **M. hamata**
1. Stamens 8; rachis 10-12 cm long; leaflets 12-15 pairs; pinnae more than 3 cm long; pod smooth .... **M. rubicaulis**

Mimosa hamata Willd., Sp. Pl. 4: 1033. 1806; *Mimosa armata* Rottler ex Spreng., Syst. Veg. 2: 206. 1825; Baker in Hook.f., Fl. Brit. India 2: 291. 1878; Singh in Verma *et al.,* Fl. Madhya Pradesh 1: 440.1993; Samvatsar, Fl. West. Trib. Madhya Pradesh 123. 1996. *Arail*

Straggling shrubs, much branched, prickly, 2-3 m high; prickles 4-5 mm long, hooked; leaves 2-pinnate; rachis 1.2-5 cm long; pinnae 3-6 pairs, 7-25 mm long, often prickly; leaflets 6-10 pairs, ovate-oblong, acute, hairy on both the surfaces; flowers pink, in solitary or paired, globose heads, tetramerous; pods linear, stalked, flat, curved, velvetty, prickly along sutures, 3-8-jointed, 3-8-seeded.

Fl.-Fr.: Feb. –Dec.

Common in hedges of fields, everywhere.

Mimosa rubicaulis Lam., Encycl. 1: 20.1783; Baker in Hook.f., Fl. Brit. India 2: 291. 1878; Samvatsar, Fl. West. Trib. Madhya Pradesh 123. 1996.

Prickly shrubs, branches glabrous; leaves 2-pinnate; pinnae 5-11 pairs; leaflets 8-15 pairs, linear-oblong, obtuse, mucronate, pubescent beneath; flowers pink, in globose heads; pods flat, falcate, glabrous, with 3-10 one-seeded joints.

Fl.-Fr.: Aug-Oct.

Alirajpur, Burhanpur, Dhar, Indore, Khandwa.

MUCUNA Adans. W. 82; MP 2; SA 1

Mucuna pruriens (L.) DC., Prodr. 2: 405. 1825; *Dolichos pruriens* L. in Stickman, Diss. Herb. Amb. 23. 1754; *Mucuna prurita* Hooker in Bot. Misc. 2: 348. 1830, *nom. superfl*; Baker in Hook.f., Fl. Brit. India 2: 187. 1876; Singh in Verma *et al.,* Fl. Madhya Pradesh 1: 380.1993; Samvatsar, Fl. West. Trib. Madhya Pradesh 103. 1996. *Kounch* **(Plate 5, Fig. 1)**

Twining, hirsute, annual herbs; leaves 3-foliolate; leaflets rhomboid-ovate, 6-9 × 3.5-8 cm, lateral leaflets very oblique; flowers purple, 2-3 cm long, in drooping, 6-30 flowered racemes; standard shorter than keel; pods slightly curved at both ends in opposite direction, clothed with irritant bristles, 6-10 × 1.2-2 cm.

Fl.-Fr.: Aug. -Oct.

Common in wastelands and open forests among the clumps of trees and shrubs, hedges of fields.

NEONOTONIA (Wight & Arn.) Lackey W. 2; MP 1; SA 1

Neonotonia wightii (Wight & Arn.) J. A. Lackey in Phytologia 37: 210.1977; *Johnia wightii* Wight & Arn., Prodr. Fl. Ind. Orient. 449. 1834; *Glycine javanica* L., Sp. Pl. 2: 754. 1753; Baker in Hook.f., Fl. Brit. India 2: 183. 1876; Samvatsar, Fl. West. Trib. Madhya Pradesh 103. 1996; Khanna *et al.,* Suppl. Fl. Madhya Pradesh 1: 65. 2001.

Climbing vines, up to *ca* 4 m long; stems angular, pubescent; leaves 3-foliolate; leaflets 9-10.5 × 7-8 cm, acute or obtuse at apex, cuneate, rounded or truncate to cordate at base, pubescent on both the surfaces; flowers dull purplish white, in axillary, up to 30 cm long racemes; pods oblong, 2-3 × 0.4-0.6 cm, compressed, septate, 3-5-seeded; seeds black, oblong-truncate, *ca* 3.5 mm across, shining.

Fl.-Fr.: Dec- Feb.

Alirajpur, Dhar.

NEPTUNIA Lour. W. 12; MP 2; SA 2

Key to Species

1. Stipules membranous, 5-15 mm long; pods2-4 cm long; seeds 4-5 mm long; plants rooting at nodes .... **N. oleracea**
1. Stipules not membranous, 2-4 mm long; pods 2-2.5 cm long; plants not rooting at nodes .... **N. triquetra**

Neptunia oleracea Lour., Fl. Cochinch. 2: 654. 1790; Baker in Hook.f., Fl. Brit. India 2: 285. 1878; Singh in Verma *et al.,* Fl. Madhya Pradesh 1: 441. 1993. *Lajalu*

Free floating herbs; stems creeping or floating, rooting at nodes; leaves alternate, 2-pinnate; leaflets 8-13 pairs, linear, 3-8 × 1.5-2 mm, obtuse at apex, truncate at base; flowers yellow, in axillary heads; pods oblong, 1.5-2.5 cm long, oblique, beaked, 6-10-seeded; seeds ovoid, *ca* 5 mm, compressed.

Fl.-Fr.: Sept. – April

Rare, **Indore** only, in ditches and ponds, at Gautampura.

Neptunia triquetra (Vahl) Benth. in Hook. Lond. J. Bot. 4: 355. 1842; *Mimosa triquetra* Vahl, Symb. Bot. 3: 102. 1797; *Desmanthus triquetrus* Willd. Sp. Pl. 4: 1045, 1806; Baker in Hook.f., Fl. Brit. India 2: 286. 1878; Singh in Verma *et al.,* Fl. Madhya Pradesh 1: 442. 1993; Samvatsar, Fl. West. Trib. Madhya Pradesh 124. 1996.

Diffuse, perennial herbs; leaves 2.5-5 cm long; pinnae 1-3 pairs; leaflets 12-18 pairs, linear-oblong, obtuse, ciliate on margins, flowers yellowish, in globose heads; peduncle with 2 large bracts; pods 1-2.5 × 0.6-0.8 cm, oblong; seeds sub-orbicular, compressed, blackish-brown.

Fl.-Fr.: Aug. -Oct.

Alirajpur, Dhar.

PARACALYX Ali W. 6; MP 1; SA 1

Paracalyx scariosus (Roxb.) Ali in Univ. Stud. Karachi 5(3): 95. 1968; *Cylista scariosa* Roxb., Pl. Coromandel 1: 64. t. 92. 1798. Baker in Hook.f., Fl. Brit. India 2: 219. 1876; Singh in Verma *et al.,* Fl. Madhya Pradesh 1: 383. 1993; Samvatsar, Fl. West. Trib. Madhya Pradesh 103. 1996.

Woody, tomentose twiners; leaves 3-foliolate; leaflets 5-10 × 4-5 cm, rhomboid-ovate, acuminate; flowers yellow, in axillary, peduncled racemes; pods small, oblique, pubesent, glandular, 1-seeded.

Fl.-Fr.: Nov. – Feb.

Alirajpur, Dhar, Indore.

PARKINSONIA L. W. 9; MP 1; SA 1

Parkinsonia aculeata L. Sp. Pl. 375. 1753; Baker in Hook.f., Fl. Brit. India 2: 260.1878; Singh in Verma *et al.,* Fl. Madhya Pradesh 1: 426. 1993; Samvatsar, Fl. West. Trib. Madhya Pradesh 117. 1996. *Vilayati babool*

Armed bush or small trees; pinnae 1-3 pairs; pinnules numerous, minute, oblanceolate, obtuse; flowers yellow, in lax, axillary racemes; pods moniliform, attenuated at both ends, finely veined, glabrous; seeds 1-6.

Fl.-Fr.: Jan. -May

Common along road-sides, in all districts, but the number of plants few and far between.

PHYLLODIUM Desv. W. 6; MP 1; SA 1

Phyllodium pulchellum (L.) Desv. in J. Bot. 2. 1: 124. t. 5. f. 24. 1813; *Hedysarum pulchellum* L., Sp. Pl. 747. 1753; *Desmodium pulchellum* (L.) Benth., Fl. Hongk. 83. 1861; Baker in Hook.f., Fl. Brit. India 2: 162. 1876; Singh in Verma *et al.,* Fl. Madhya Pradesh 1: 349. 1993; Samvatsar, Fl. West. Trib. Madhya Pradesh 95. 1996. *Chipti*

Undershrubs or shrubs, erect, much branched, 1-1.5 m high; leaves 3-foliolate; leaflets ovate or elliptic-lanceolate, obtuse or mucronate at apex, rounded at base, glabrous above, appressed hairy beneath; flowers white or yellow with pink tinge, in terminal or axillary, 10-50 cm long racemes; pods 7-8 mm long, 1-2-jointed, constricted on both sutures.

Fl.-Fr.: Sept. -Feb.

Alirajpur, Barwani, Khargone.

PITHECELLOBIUM Mart. W. 79; MP 1; SA 1

Pithecellobium dulce (Roxb.) Benth. in Hook., Lond. J. Bot. 3: 199. 1844; *Mimosa dulcis* Roxb., Pl. Coromandel 1: 67. t. 99. 1798; Baker in Hook.f., Fl. Brit. India 2: 302. 1878; Singh in Verma *et al.,* Fl. Madhya Pradesh 1: 443. 1993; Samvatsar, Fl. West. Trib. Madhya Pradesh 124. 1996. *Jungle jalebi*

Evergreen, trees, 5-12 m high, armed with pairs of straight, stipular spines on the trunk; leaves 2-pinnate; leaflets 2 on each pinnae, obovate or obliquely elliptic-oblong, 2-4.5 × 1-2 cm; flowers white, in small globose, axillary, paniculate heads; pods reddish brown, 10-12 × 0.5-0.8 cm, turgid, spirally twisted, 6-8-seeded; seeds shining brown or black, ovoid, embedded in whitish pulp, with edible funiculus (seed stalk).

Fl.-Fr.: Jan. – May

Common along the road sides, planted as living hedge also.

PONGAMIA Venten. Monotypic

Pongamia pinnata (L.) Pierre, Fl. For. Coch. Sub. t. 385. 1899; *Cytisus pinnatus* L., Sp. Pl. 741. 1753; *Galedupa indica* Lam. Encycl. 2: 594. 1788; Baker in Hook.f., Fl. Brit. India 2: 240.1876; Singh in Verma *et al.,* Fl. Madhya Pradesh 1: 385. 1993; Samvatsar, Fl. West. Trib. Madhya Pradesh 104. 1996. *Karanj*

Trees; leaves 5-9 foliolate; leaflets imparipinnate, 7-25 cm long, 4-13 × 2-9 cm, ovate or elliptic, glabrous; flowers purplish-white, in axillay racemes; pods 3-7 × 1.5-3.5 cm, compressed, obliquely oblong; seed solitary, brown, rugose.

Fl.-Fr.: March - Feb.

Common along banks of rivers, also along roadsides in villages.

PROSOPIS L. W. 47; MP 2; SA 2

Key to Species

1. Plants armed all over with conical prickles and spiny stipules; pod cylindric, moniliform .... **P. cineraria**
1. Plants armed with spiny stipules only; pod compressed, sutures straight .... **P. juliflora**

Prosopis cineraria (L.) Druce in Rep. Bot. Exch. Club Soc. Brit. Isles 3:422. 1914; *Mimosa cineraria* L., Sp. Pl. 517. 1753; *Prosopis spicigera* L. Mant. 68. 1767; Baker in Hook.f., Fl. Brit. India 2: 288, 1878; Singh in Verma *et al.,* Fl. Madhya Pradesh 1: 444. 1993; Samvatsar, Fl. West. Trib. Madhya Pradesh 124. 1996. *Shami*

A tree; branches with scattered prickles; leaves 2-pinnate; pinnae usually 2 pairs; leaflets 7-12 pairs, subsessile, obliquely rounded and mucronate; flowers yellow, in axillary spikes and terminal panicles; pods turgid, straight, slender, cylindric, torulose, glabrous; seeds 10-15, dull brown.

Fl.-Fr.: Dec. -March

Alirajpur, Burhanpur, Dhar, Indore.

Prosposis juliflora (Swartz.) DC., Prodr. 2: 447. 1825; *Mimosa juliflora* Swartz., Prodr. 85. 1788; Singh in Verma *et al.,* Fl. Madhya Pradesh 1: 444. 1993; Samvatsar, Fl. West. Trib. Madhya Pradesh 124. 1996.

Shrubs to small trees; branches with stipular spines; pinnae 1-2 pairs; leaflets 15-20 pairs, oblong, obtuse; spikes 1-few, fascicled; pods pale yellow, straight or curved, thick, glabrous, sutures parallel or slightly constricted.

Fl.-Fr.: Sept. -April

Introduced and now naturalized and extensively planted along road-side, throughout the area.

PTEROCARPUS Jacq. W. 66; MP 1; SA 1

Pterocarpus marsupium Roxb., Pl. Coromandel 2: 9. t. 116. 1799; subsp. **acuminatus** (Prain) Thoth., Tax. Rev. Trib. Dalberg. Indian Subcont. 227. 1987; *P. marsupium* Roxb., var. *acuminata* Prain, in J. Asiat. Soc. Bengal 66:455. 1897; Baker in Hook.f., Fl. Brit. India 2: 239. 1876; Singh in Verma *et al.,* Fl. Madhya Pradesh 1: 387. 1993; Samvatsar, Fl. West. Trib. Madhya Pradesh 104. 1996. *Bija*

Deciduous tree with yellow-grey bark; leaflets 5-7, alternate, 5-12 × 4-7 cm, coriaceous, oblong, obtuse or emarginate; flowers pale-yellow, in short lateral and terminal panicles; pods samaroid, globular, pubescent.

Fl.-Fr.: April – Feb.

Alirajpur, Barwani, Burhanpur, Dhar, Indore, Khargone.

PUERARIA DC. W. 20; MP 2; SA 2

Key to Species

1. Stout perennial, stem 12 cm in diameter; terminal leaflet 18x16 cm; raceme 15-30 cm long; petals 1.5 cm long; pods 5-7 cm long, densely brown-hirsute; seeds 3-6; root very large, tuberous .... **P. tuberosa**
1. Herbaceous annual with slender stem; terminal leaflet 6-10 cm long; petals 0.8-1.2 cm long; pods 5-8 cm long; seeds 15-20 .... **P. phaseoloides**

Pueraria phaseoloides (Roxb.) Benth. in J. Linn. Soc., Bot. 9: 125. 1867; *Dolichos phaseoloides* Roxb., Fl. Ind. 3: 316. 1832; Baker in Hook.f., Fl. Brit. India 2: 199. 1876; Khanna *et al.,* Suppl. Fl. Madhya Pradesh 1: 67.2001.

Slender twiners, often creeping in the lower part; leaves 3-foliolate, 15-20 cm long; lateral leaflets ovate-oblong, up to 15 × 7 cm, terminal leaflet rhomboid, mucronate at apex, appressed pubescent above and appressed hairy on the lower surface; flowers pale blue or white, in a long pedunculate, axillary, puberulous racemes; pods 5-8 cm long, compressed, slightly curved, appressed pubescent, 17-20-seeded; seeds dark brown, minutely rugose.

Fl.-Fr.: Feb. – June

Rare, **Indore** only, Chhapariya forest.

Pueraria tuberosa (Roxb. ex Willd.) DC. in Ann. Sci. Nat. 4: 97. 1825 & Prodr. 2: 240.1825; *Hedysarum tuberosum* Roxb. ex Willd. Sp. Pl 3: 1197. 1803; Baker in Hook.f., Fl. Brit. India 2 : 197. 1876; Singh in Verma *et al.,* Fl. Madhya Pradesh 1: 388. 1993; Samvatsar, Fl. West. Trib. Madhya Pradesh 104. 1996. *Vidari kand*

Large twiners; leaves 3-foliolate; leaflets 9-24 × 6-19 cm, ovate, acuminate, glabrescent above, pubescent beneath; flowers bluish-purple, in lax, leafless racemes; pods membranous, flat, constricted between the seeds, covered with bristly, brown hairs.

Fl.-Fr.: Feb. -June

Occassional , Alirajpur, Dhar, Indore.

RHYNCHOSIA Lour. W. 251; MP 5; SA 7

Key to Species

1. Seeds strophiolate
   2. Racemes 20-30 cm long; stem, petiole and calyx gland-dotted; standard petal yellow, blotched with purple, margins only yellow .... **R. rothii**
   2. Racemes much shorter, few flowered and lax or many flowered, capitate; stem, petiole and calyx not gland-dotted
      3. Racemes few flowered; peduncles shorter than leaves; standard petal yellow, with purple stripes; pubescence not fulvous hairy .... **R. aurea**
      3. Racemes many flowered, capitates; peduncles longer than leaves; standard petal uniformly yellow; pubescence fulvous-hairy .... **R. capitata**
1. Seeds not strophiolate
   4. Leaflets small, obtuse, under 4 cm across; pods not tipped with persistent base of style ......**R. minima**
   4. Leaflets moderately large; pods tipped with persistent base of style
      5. Pods with bubous-based, setose hairs, as well as glandular pubescence; petals uniformly yellow; bracts minute, deciduous .... **R. viscosa**
      5. Pods neither with such hairs, nor glandular
         6. Leaflets rhomboid-ovate, long acuminate; petals with red veins .... **R. acutissima**
         6. Leaflets deltoid, shortly acuminate; petals uniformly yellow .... **R. bracteata**

Rhynchosia acutissima Thwaites, Enum. Pl. Zeyl. 5: 413. 1864; Baker, in Hook.f., Fl. Brit. India 2:226. 1876; Gamble, Fl. Pres. Madras 1:265. 1957 (BSI reprint)

Branchlets pubescent; leaves 3-foliolate; leaflets 11-15 × 8-11 cm, ovate-rhomboid, apex caudate-acuminate, base truncate, 3-ribbed, puberulent above and sericeous below, intercostae reticulate; flowers pale yellow with red veins, in axillary, many flowered racemes; pods villous, greenish-yellow, to 2 cm long; seeds bluish-black.

Fl.-Fr.: Nov. -April

Rare, **Indore** only, ravines of Jagjivan Ram nagar .

Rhynchosia aurea (Willd.) DC., Prodr. 2: 386. 1825; *Glycine aurea* Willd., Nov. Act. Ber. 4: 18. 1803; *Nomismia aurea* (Willd.) Wight & Arn., Prodr. Fl. Ind. Orient. 1: 237. 1834; Baker in Hook.f., Fl. Brit. India 2: 225. 1876; Samvatsar, Fl. West. Trib. Madhya Pradesh 105. 1996.

Annual trailing herbs; branches slender with spreading pubescence; leaflets 4-5 × 3.5-4 cm, rhomboid ovate, acute or subobtuse; racemes 6-10 flowered; pods suborbicular, sparsely pubescent, transversely striate with close parallel veins.

Fl.-Fr.: Aug. -Jan.

Common in moist areas, gullies, all over the study area.

Rhynchosia bracteata Benth. ex Baker in Hook.f., Fl. Brit. India 2: 225. 1876; Singh in Verma *et al.,* Fl. Madhya Pradesh 1: 389. 1993

Climbing or trailing, annual or perennial, suffruticose herbs; leaves 3-foliolate; leaflets rhomboid, 5-10 × 5-10 cm, cuspidate, pubescent above, densely pubescent and gland-dotted below; flowers yellow, 8-13 mm long, in 12-13 cm long racemes; pods obovoid or oblanceolate-oblong, 10-35 × 10-12 mm, compressed, glandular pubescent, velvety, 2-seeded.

Fl.-Fr.: March – June

Burhanpur, Khandwa.

Rhynchosia capitata (Heyne ex Roth) DC., Prodr. 2: 386. 1825; *Glycine capitata* Heyne ex Roth, Nov. Pl. Sp. 346. 1821; *Rhynchosia aurea auct. non* (Willd.) DC., 1825: Baker in Hook.f., Fl. Brit. India 2: 221. 1876; Singh in Verma *et al.,* Fl. Madhya Pradesh 1: 389. 1993; Samvatsar, Fl. West. Trib. Madhya Pradesh 105. 1996.

Annual, climbing or trailing herbs; leaves 3-foliolate; leaflets ovate-rhomboid or orbicular, 2-5 × 2-5 cm, acute or obtuse at apex, cuneate at base, lateral ones obliquely rhomboid-ovate, finely pubescent on both the surfaces; flowers yellow, in few to many-flowered, 7-10 cm long, axillary, capitate racemes; pods globose, oblong or orbicular, flat, sparsely pubescent with long spreading hairs, 2-seeded; seeds globose or oblong-rounded.

Fl.-Fr.: Aug. – Jan.

Burhanpur, Dhar, Indore, Khandwa.

Rhynchosia minima (L.) DC., Prodr. 2: 385. 1825; *Dolichos minimus* L., Sp. Pl. 2: 726. 1753; Baker in Hook.f., Fl. Brit. India 2: 223. 1876; *Rhynchosia minima* (L.) DC. var. *laxiflora* (Camb.) Baker in Hook.f., Fl. Brit. India 2: 223. 1876; Singh in Verma *et al.,* Fl. Madhya Pradesh 1: 390.1993; Samvatsar, Fl. West. Trib. Madhya Pradesh 105. 1996.

Twining herbs, branches slender, glabrescent, leaves 3-foliolate, leaflets 3.5-4.5 × 2.7-3.5 cm, rhombodid, acute, pubescent on veins, and dotted beneath; flowers yellow, in 6 12-flowered racemes; pods oblong, slightly bent upwards, sparsely pubescent; seeds 2.

Fl.-Fr.: July – April

Very common throughout the area, twining on support in hedges of fields and other vegetation.

Rhynchosia rothii Benth. ex Aitchison, Cat. Pl. Punjab & Sindh 50.1869; *R. sericea* Span., Linnaea 15: 195. 1841, *non* Guill. ex Hook. & Arn., 1833; Baker in Hook.f., Fl. Brit. India 2: 225. 1876; Singh in Verma *et al.,* Fl. Madhya Pradesh 1: 390.1993.

Twining, perennial herbs, with woody root stock; leaves 3-foliolate; petioles 5-10 cm long; stipules lanceolate, 6-9 mm long; leaflets rhomboid or suborbicular, 5-7.5 × 5-7.5 cm, obtuse or acute at apex, softly pubescent on both the surfaces; flowers purple, in axillary, many-flowered racemes; pods 25-35 × 6-8 mm, recurved at tip, pubescent, 2-seeded.

Fl.-Fr.: July – April

Barwani, Indore, Khargone.

Rhynchosia viscosa (Roth) DC., Prodr. 2: 387. 1825; *Glycine viscosa* Roth, Nov. Pl. Sp. 349. 1821, *non* Moench, 1802; Baker in Hook.f., Fl. Brit. India 2: 225. 1876; Khanna *et al.*, Suppl. Fl. Madhya Pradesh 1: 68. 2001.

Twining herbs, up to 2.5 m long; leaves 3-foliolate; leaflets broadly ovate, 5-9 × 4-8.5 cm, often slightly lobed, acute to acuminate at apex, cuneate or truncate at base, hairy on both the surfaces, terminal one rhomboid; flowers brownish- purple, in axillary or terminal, 5-35 cm long, many-flowered racemes; pods oblong, falcate, 2.5-3 cm long, 2-seeded; seeds black, globose or orbicular.

Fl.-Fr.: Jan. – March

Burhanpur, Indore, Khandwa.

ROTHIA Pers. W. 2; MP 1; SA 1

Rothia indica (L.) Druce in Rep. Bot. Exch. Club Soc. Brit. Isles 3: 423. 1914; *Trigonella indica* L., Sp. Pl. 2: 778. 1753; *Dillwynia trifoliata* Pers., Syn. Pl. 2:302. 1807; *Rothia trifoliata* (Pers.) DC., Prodr. 2:382. 1825; Syn. Pl. 2: 302, 659. 1807; Baker in Hook.f., Fl. Brit. India 2: 63. 1876; Singh in Verma *et al.,* Fl. Madhya Pradesh 1: 391. 1993.

Prostrate, much branched, hairy, annual herbs; leaves 3-foliolate, petiolate; leaflets oblanceolate to oblong, 10-25 mm long, obtuse or slightly mucronate at apex, fleshy and silky hairy, sessile; flowers axillary, yellow, 1-4, shortly pedicellate; pods linear, 3-6 cm long, compresed, mucronate, silky, many-seeded; seeds purple, reniform, *ca* 1 mm across.

Fl.-Fr.: Aug. – Dec.

Rare, **Barwani** only, Bhatki village.

SENNA Miller W. 272; MP 6; SA 6

Key to Species

1. Leaflets 3 pairs
    2. Glands 2, one each between lowest 2 pairs of leaflets; sepals glabrous .... **S. tora**
    2. Gland one, between the lowest pair of leaflets; sepals ciliate, finely pubescent ....**S. obtusifolia**
1. Leaflets 4 or more pairs
    3. Glands equal to the pairs of leaflets
        4. Stipules auriculate; pods 10-20 seeded .... **S. auriculata**
        4. Stipules not auriculate; pods 4-10 seeded .... **S. uniflora**
    3. Gland one only, present near the base of rachis
        5. Leaflets more than 8 pairs; bracts obtuse; racemes 4-10 flowered .... **S. sophera**
        5. Leaflets less than 6 pairs; bracts acute; racemes 2-4 flowered.................................... **S. occidentalis**

Senna auriculata (L.) Roxb., Fl. Ind. 2: 349. 1832; *Cassia auriculata* L., Sp. Pl. 1: 379. 1753; Baker in Hook.f., Fl. Brit. India 2: 263. 1878; Singh in Verma *et al.,* Fl. Madhya Pradesh 1: 420.1993; Samvatsar, Fl. West. Trib. Madhya Pradesh 114. 1996. *Amoli*

Finely pubescent shrubs, 1-2 m tall; leaves 2.5-10 cm long; leaflets 5-12 pairs, 1.5-3 × 0.5-1.3 cm, oval or oblong-obovate, obtuse or emarginate, apiculate; flowers in corymbose racemes, large, showy, yellow; petals long clawed; pods 7-12.5 × 1.4-1.8 cm, oblong, mucronate, deeply impressed and septate between the seeds; seeds 10-20, oblong-ovate, blackish-brown, areolate.

Fl.-Fr.: Aug. -Nov.

Common along banks of streams or in the vicinity of streams, everywhere.

Senna obtusifolia (L.) Irwin & Barneby in Mem. New York Bot. Gard. 35(1): 252. 1982; *Cassia obtusifolia* L., Sp. Pl. 1: 377. 1753; *C. tora* sensu Benth. in Trans. Linn. Soc. London 27: 532. 1871, *p.p. non* L., 1753; Baker in Hook.f., Fl. Brit. India 2: 263. 1878; Singh in Verma *et al.,* Fl. Madhya Pradesh 1: 422. 1993; Samvatsar, Fl. West. Trib. Madhya Pradesh 114. 1996. *Puvadia*

Annual herbs; leaves 7-10 cm long, with a single subulate gland between the lowest pair of leaflet; leaflets 3 pairs, obovate or oblong, 1.5-5 × 1-2.5 cm, rounded or mucronate at apex, cuneate at base, glabrous above, appressed hairy beneath; flowers yellow, in axillary, few flowered clusters; pods margined, 5-16 × 0.4-0.5 cm, erect on their pedicels; seeds somewhat rhomboid, areolate.

Fl.-Fr.: Sept. -Nov.

Common, throughout the area.

Senna occidentalis (L.) Link, Handbuch 2: 140.1831; *Cassia occidentalis* L., Sp. Pl. 1: 377. 1753; Baker in Hook.f., Fl. Brit. India 2: 262. 1878; Singh in Verma *et al.,* Fl. Madhya Pradesh 1: 422. 1993; Samvatsar, Fl. West. Trib. Madhya Pradesh 115. 1996. *Bari kasondi*

Branched, glabrous shrubs; leaves 10-20 cm long, with a single gland at the base of petiole; leaflets 3-5 pairs, ovate to ovate-oblong, 4-12 × 2.5-4.5 cm, acuminate at apex, with a single gland between the uppermost pair; flowers yellow, in terminal racemes; pods narrowly oblong, flat; seeds ovoid, compressed.

Fl.-Fr.: Oct. -March

Common on road-sides and waste places, in all districts.

Senna sophera (L.) Roxb., Fl. Ind. 2: 347. 1832; *Cassia sophera* L., Sp. Pl. 1: 379. 1753; Baker in Hook.f., Fl. Brit. India 2: 262. 1878; Singh in Verma *et al.,* Fl. Madhya Pradesh 1: 423. 1993; Samvatsar, Fl. West. Trib. Madhya Pradesh 115. 1996. *Kasondi*

Erect, undershrubs; leaves 15-25 cm long; rachis with conical glands at base; leaflets 5-9 pairs, ovate-lanceolate or elliptic-lanceolate, 3-6 × 1.5-2 cm, acute or acuminate at apex, oblique at base, ciliolate or glabrous; flowers yellow, in axillary and terminal corymbose racemes; pods subterete, 6-10 cm long, straight or curved, inflated along sutures; seeds dark brown, broadly ovoid, 2-2.5 mm long, compressed.

Fl.-Fr.: Sept. – Jan.

Alirajpur, Burhanpur, Dhar, Indore, Khandwa.

Senna tora (L.) Roxb., Fl. Ind. 2: 340.1832. *Cassia tora* L., Sp. Pl. 376. 1753; Baker in Hook.f., Fl. Brit. India 2: 263. 1878, *p.p;* Singh in Verma *et al,* Fl. Madhya Pradesh 1: 424. 1993; Samvatsar, Fl. West. Trib. Madhya Pradesh 115. 1996. *Chakunda*

Woody annuals; leaves 5-10 cm long; leaflets 3 pairs, obovate or ovate to oblong, 2-5 × 1.5-2.5 cm, obtuse, rounded or mucronate at apex, glabrous above, appressed hairy beneath; glands between the two lowest pair of leaflets; flowers bright yellow, in axillary, few flowered clusters; pods linear; seeds oblong.

Fl.-Fr.: Aug. – Dec.

Common throughout the area.

Senna uniflora (Mill.) H. S. Irwin & Barneby in Mem. New York Bot. Gard. 35(1): 258. 1982; *Cassia uniflora* Mill. in Gard. Dict. (ed. 8.) 5. 1768; *C. sericea* Sw., Prodr. 66. 1788; Khanna *et al.,* Suppl. Fl. Madhya Pradesh 81.2001. Samvatsar, Fl. West. Trib. Madhya Pradesh 115. 1996.

Annual herbs, 15-25 m high; leaves 6-12 cm long; leaflets 3-5 pairs, 2.5-4.5 × 1.2-2.5 cm, obovate or obovate-oblong to elliptic; flowers yellow, in axillary, 3 to 4-flowered racemes; pods 2.8-4.5 × 0.3-0.4 cm, linear, clothed with reddish- brown or yellow, pubescent, septate, dehiscent, 6-10 seeded; seeds 2.5-3.5 × 1-2.5 mm, rhomboid or subquadrangular, glabrous.

Fl.-Fr.: Sept. –Dec.

Burhanpur, Dhar, Jhabua, Khandwa.

SESBANIA Scop. W. 56; MP 2; SA 2

Key to Species

1. Armed plants; corolla yellow with purple spots .... **S. bispinosa**
1. Unarmed plants; corolla, uniformly yellow ....**S. cannabina**

Sesbania bispinosa (Jacq.) W. F. Wight in U.S. Dept. Agric. Bur. Pl. Ind. Bull. 137: 15. 1909; *Aeschynomene bispinosa* Jacq., Icon. Pl. Rar. 3: 13. t. 564. 1792; *Sesbania aculeata* (Willd.) Poir. Encycl. 7: 128. 1806; Baker in Hook.f., Fl. Brit. India 2: 114. 1876; Singh in Verma *et al.,* Fl. Madhya Pradesh 1: 391. 1993; Samvatsar, Fl. West. Trib. Madhya Pradesh 107. 1996.

Erect shrubs; branches minutely prickly; leaves paripinnate, 20-35 cm long; rachis muricate or prickly; leaflets 20-100, linear, 5-20 mm long, obtuse-apiculate, glabrous above, sparsely hairy along margins and nerves beneath; flowers yellow with black or red margins outside, in 8-13 cm long, 3-12-flowered racemes; pods slightly torulose, 35-40-seeded.

Fl.-Fr.: Sept. -Nov.

Alirajpur, Barwani, Burhanpur, Dhar, Indore, Khandwa.

Sesbania cannabina (Retz.) Pers., Encycl. 7:130.1806; *Aeschynomene cannabina* Retz., Obs. Fasc. 5:26. 1789; *S. aculeata* Pers. var. *cannabina* (Retz.) Baker in Hook.f., Fl. Brit. India 2: 115. 1876; Samvatsar, Fl. West. Trib. Madhya Pradesh 107. 1996; Khanna *et al.,* Suppl. Fl. Madhya Pradesh 1: 81.2001.

Erect, annual shrubs; stems with spreading hairs when young, more or less glabrous later; leaves compound; leaflets narrowly-oblong, tip blunt or flat-apiculate, 7-8 × 1.5-3 mm, hairless or with a few appressed hairs; flowers yellow, in 1-3-flowered racemes; pods slender, 12-20 cm long, 2.5-3 mm, with a slender tip; seeds dark brown, about 4 mm long.

Fl.-Fr.: Dec. -Jan.

Alirajpur, Dhar, Indore.

SMITHIA Aiton W. 22; MP 6; SA 1

Smithia conferta J. E. Smith in Rees. Cyclop. 33: n. 2, 1819; Singh in Verma *et al.,* Fl. Madhya Pradesh 1: 394. 1993; *Smithia geminiflora* Roth var. *conferta* Baker in Hook.f., Fl. Brit. India 2: 149. 1876; Singh in Verma *et al.,* Fl. Madhya Pradesh 1: 394. 1993; Samvatsar, Fl. West. Trib. Madhya Pradesh 107. 1996. *Chhui mui*

Annual diffuse herbs; leaves small, nearly sessile, abruptly pinnate; leaflets 2-8 pairs, 0.8-1.2 × .2-.4 cm, linear-oblong, obtuse, densely bristly on the margins and on the midrib beneath; flowers yellow, in pairs, clustered in the axils of terminal leaves of lateral branches; pods 3-6 jointed; joints small, turgid, papillose.

Fl.-Fr.: Nov. -Dec.

Common along road side ditches, wastelands, throughout the area.

SPATHOLOBUS Hassk. W. 34; MP 1; SA 1

Spatholobus parviflorus (Roxb. ex DC.) Kuntze, Revis. Gen. Pl. 1: 205. 1891; *Butea parviflora* Roxb. ex DC., Prodr. 2: 451. 1825; *Spatholobus roxburghii* Benth. in Miq., Pl. Jungh. 238. 1852, *nom. illeg;* Baker in Hook.f., Fl. Brit. India 2: 193. 1876; Singh in Verma *et al.,* Fl. Madhya Pradesh 1: 326. 1993; Samvatsar, Fl. West. Trib. Madhya Pradesh 88. 1996.

Large lianas; branches densely grey downy; leaflets broadly obovate-ovate, *ca* 20 × 13 cm, obtuse or abruptly acute at apex, broadly cuneate or rounded at base, silky pubescent; flowers very small, white, in axillary racemes; pods 10-12.5 × 3.5-4 cm, apical portion narrower, covered with ferruginous tomentum; seeds sub reniform to obovate, *ca* 12 mm long.

Fl.-Fr.: April – Sept.

Alirajpur, Dhar.

TADEHAGI Ohashi W. 4; MP 1; SA 1

Tadehagi triquetrum (L.) H. Ohashi in Ginkgoana 1: 290.1973; *Hedysarum triquetrum* L. Sp. Pl. 2: 746. 1753; *Desmodium triquetrum* (L.) DC., Prodr. 2: 326. 1824; Baker in Hook.f., Fl. Brit. India 2: 163. 1876, *p.p;* Singh in Verma *et al.,* Fl. Madhya Pradesh 1: 428. 1993.

Erect, herbs or undershrubs; branches grooved, 3-quetrous; leaves 1-foliolate; leaflets ovate-lanceolate to oblong, 5-15 × 3-4.5 cm, acute to subacuminate at apex, subcordate at base, glabrous above, appressed hairy on nerves beneath; flowers purple, in axillary and terminal, 15-25 cm long racemes; pods 25-50 × 4-8 mm; joints 6-8, slightly broader than long.

Fl.-Fr.: Aug. – March

Burhanpur only, Pandhana, Tidgaon.

TAMARINDUS L. Monotypic

Tamarindus indica L. Sp. Pl. 1: 34. 1753; Baker in Hook.f., Fl. Brit. India 2: 273, 1878; Singh in Verma *et al.,* Fl. Madhya Pradesh 1: 428. 1993; Samvatsar, Fl. West. Trib. Madhya Pradesh 118. 1996. *Imli*

Large trees; leaves paripinnate, alternate; leaflets 11-20 pairs, 0.8-1.5 × 0.5-0.6 cm, oblong, obtuse, oblique at base; flowers yellow with reddish-pink dots, in lax terminal racemes; pods brown, slightly curved, subcompressed; seeds 3-12, smooth, brown, shining.

Fl.-Fr.: April-June

Around villages, naturalized,throughout the area.

TAVERNIERA DC. W. 16; MP 1; SA 1

Taverniera cuneifolia (Roth) Arn., Nova Acta Phys. -Med. Acad. Caes. Leop. -Carol. Nat. Cur. 18(1): 332. 1836; *Hedysarum cuneifolium* Roth, Nov. Pl. Sp. 348. 1825; Baker in Hook.f., Fl. Brit. India 2: 140.1876; Khanna *et al.,* Suppl. Fl. Madhya Pradesh 73.2001. *Mulethi*

Shrub, 2-3 ft tall, branches velvet-hairy; leaves unifoliolate; leaflets 0.3-1.5 cm long, obovate to orbicular, entire, mucronate, velvety, becoming hairless; flowers purple, in up to 10 cm long racemes; fruit 1-3 jointed, 1-seeded, joints prickly and ovoid, velvety-hairy.

Fl.-Fr.: Nov. -Dec.

Burhanpur, Khandwa, Khargone.

TEPHROSIA Pers. W. 362; MP 9; SA 9

Key to Species

1. Leaves simple, linear; flowers 1-2, axillary .... **T. strigosa**
1. Leaves compound, not linear
    2. Pod woolly-tomentose .... **T. villosa**
    2. Pod glabrescent
        3. Style bearded
            4. Prostrate herbs .... **T. maxima**
            4. Erect herbs or shrubs
                5. Leaflets equal, 5-30; flowers 1-6, in pseudoracemes
                    6. Flowers 1-2, in leaf axils; leaflets usually 5 (3-11)
                        7. Calyx teeth longer than the tube; pods 8-10 × 5-6 mm; 6-8 seeded .... **T. senticosa**
                        7. Calyx teeth equalling the tube; pods up to 45 mm long, 7-9 seeded .... **T. uniflora** subsp. **petrosa**
                    6. Flowers 2-6, in pseudoracemes, 1.7-2.5 cm long, white; calyx teeth shorter than the tube; pod densely pubescent, 6-10 × 8-9 mm, 5-13 seeded; leaflets 11-12 ...**T. candida**
                5. Leaflets unequal, 1-13; flowers 3-12, in axillary racemes .... **T. tinctoria**

3. Style not bearded
   8. Flowers 1-3, in short, axillary racemes, white to pink .... **T. pumila**
   8. Flowers many in raceme, bluish-purple .... **T. purpurea**

Tephrosia candida (Roxb.) DC., Prodr. 2: 249. 1825; *Robinia candida* Roxb., Hort. Bengal. 56. 1814; Baker in Hook.f., Fl. Brit. India 2: 111. 1876; Samvatsar, Fl. West. Trib. Madhya Pradesh 108. 1996. Khanna *et al.,* Suppl. Fl. Madhya Pradesh 73.2001.

Shrubs, erect, suffruticose, up to 3.5 m high; leaves imparipinnate, 11-27-foliolate, 15-22 cm long; leaflets oblong to lanceolate, 2.5-7 × 0.5-1.5 cm, obtuse, acute or mucronate at apex, glabrous above, densely silky beneath; flowers white, solitary or in fascicles of 2-6, on 5-20 cm long axillary, terminal or leaf-opposed racemes; pods linear-oblong, 6-10 × 0.7-1 cm, densely pubescent, slightly falcate, 5-13-seeded; seeds brown, reniform.

Fl.-Fr.: Aug. -Dec.

Burhanpur, Dhar.

Tephrosia maxima (L.) Pers. in Syn. Pl. 2:329. 1807; *Galega maxima* L., Syst. Nat. ed. 10:1172. 1759; *T. purpurea* (L.) Pers. var. *maxima* (L.) Baker in in Hook.f., Fl. Brit. India 2: 113. 1876; Khanna *et al.,* Suppl. Fl. Madhya Pradesh 73.2001.

Prostrate undershrubs; branches sparsely villous hairy; leaves imparipinnate, 4-13 cm long; leaflet 9-17, 1-2.4 × 0.5-1 cm, obovate-oblong, base pointed to wedge-shaped, apex blunt or flat, apiculate, hairless above, appressed, velvet-hairy beneath, thin leathery; flowers bright purple, in short, 2-3 flowered racemes; pods linear, puberulous; seeds 8-12, smooth.

Fl.-Fr.: Aug. -Jan.

Rare, in open and dry places, **Burhanpur** only, Jambupani and Shahpur.

Tephrosia pumila (Lam.) Pers., Syn. Pl. 2: 230.1807; *Galega pumila* Lam., Encycl. 2: 599. 1786; *Tephrosia purpurea* (L.) Pers. var. *pumila* (Lam.) Baker in Hook.f., Fl. Brit. India 2: 113. 1876; Singh in Verma *et al.,* Fl. Madhya Pradesh 1: 397. 1993; Samvatsar, Fl. West. Trib. Madhya Pradesh 108. 1996.

Procumbent perennial herbs, up to 30 cm high; leaves imparipinnate; leaflets 7-13, oblong or oblanceolate, 10-20 × 6-8 mm, truncate or retuse; flowers reddish-purple, in terminal or leaf-opposed, 1 to 3 flowered racemes; pods 3-4 × 0.4 cm, pubescent, 8-14 seeded.

Fl.-Fr.: Aug. -Nov.

Common in forests, in shady habitats, in all districts.

Tephrosia purpurea (L.) Pers., Syn. Pl. 2: 329. 1807; *Cracca purpurea* L., Sp. Pl. 2: 752. 1753; Baker in Hook.f., Fl. Brit. India 2: 112. 1876; Singh in Verma *et al.,* Fl. Madhya Pradesh 1: 397.1993; Samvatsar, Fl. West. Trib. Madhya Pradesh 108.1996. *Sarpunkha* **(Plate 5, Fig. 2)**

Appressed hairy undershrubs, 20-90 cm high; leaves imparipinnate; leaflets 5-15, 0.5-2 × 0.2-1 cm, elliptic-oblong, oblanceolate or obovate; flowers bright pink or purplish-red, in leaf opposed, lax racemes; pods 3-6 cm long, linear, slightly upcurved, mucronate, appressed pubescent, 6-9 seeded; seeds subcylindric, smooth, dark brown.

Fl.-Fr.: Throughout the year.

Very common along road-sides, throughout the area.

Tephrosia senticosa (L.) Pers. Syn. Pl. 2: 330.1807; *Cracca senticosa* L., Sp. Pl. 2: 752. 1753; Baker in Hook.f., Fl. Brit. India 2:212. 1876; Samvatsar, Fl. West. Trib. Madhya Pradesh 108. 1996.

A much branched, small shrub, very much woody below; branches silvery-white; leaflets 3-11, more or less equal in size, obovate-oblong, emarginate, cuneate, 1-3 × 0.5-1.25 cm; flowers reddish; pods linear, oblong, straight; seeds ellipsoid.

Fl.-Fr.: Sept.-Dec.

Rare, **Dhar** only, among rocks at Dharampuri bet.

Tephrosia strigosa (Dalzell) Santapau & Maheshw. in J. Bombay Nat. Hist. Soc. 54: 805. 1956; *Macronyx strigosus* Dalzell in Hook., J. Bot. Kew Gard. Misc. 2: 35. 1850; *Tephrosia tenuis* Wall. ex Dalzell & Gibson, Bombay Fl. 61. 1861; Baker in Hook.f., Fl. Brit. India 2: 111. 1876; Singh in Verma *et al.,* Fl. Madhya Pradesh 1: 398. 1993; Samvatsar, Fl. West. Trib. Madhya Pradesh 109. 1996.

Annual, appressed hairy herbs, 20-30 cm high; leaves simple, linear or elliptic, 2-5 × 0.3-0.6; acute or apiculate at apex, glabrescent above, silky beneath; flowers bluish-yellow, 1-2 in the axils of leaves; standard pubescent on the back; pods linear, flat, 1-2.5 × 0.3-0.4 cm, 6-10 seeded.

Fl.-Fr.: Aug.-Oct.

Common on hill slopes, dry ground, all over the area.

Tephrosia tinctoria (L.) Pers., Syn. Pl. 2(2): 329. 1807; *Cracca tinctoria* L., Sp. Pl. 2: 752. 1753; Baker in Hook.f., Fl. Brit. India 2: 111. 1876; Singh in Verma *et al.,* Fl. Madhya Pradesh 1: 398. 1993; Samvatsar, Fl. West. Trib. Madhya Pradesh 109. 1996.

Velvety brown, undershrubs; leaves 5-10 cm long; leaflets 3-5 or sometimes, solitary, each 3-9 × 11.2-2.5 cm, terminal leaflet larger, oblong or obovate, apex rounded, mucronate, base rounded; flowers bright pink to reddish, in 3-12 flowered, axillary racemes; pods apiculate 5-5.5 cm long, 8-112 seeded; seeds brown.

Fl.-Fr.: Sept. -Nov.

Rare**, Dhar** only, Shri Ram Ghat.

Tephrosia uniflora Pers. subsp. **petrosa** (Blatt. & Hallb.) J.B. Gillet & Ali in Kew Bull. 13(1): 114. 1958; *T. petrosa* Blatt. & Hallb. in J. Bombay Nat. Hist. Soc. 26 (1): 239. 1918; *T. pauciflora* Graham ex Baker in Hook.f. Fl. Brit. India 2:114. 1876; Bhandari in Fl. Indian Desert 121. 1995 (reprint ed.) Kothari, in N.P. Singh *et al.,* Fl. Maharashtra 1:754.2000.

Perennials, branching from the base; leaves 3-8 cm long; leaflets 3-7, usually 5; petioles 1-3 cm long; flowers rose or purple-red, axillary, usually in pairs; style flattened; stigma knobbed; pods up to 4.5 cm long, 7-9 seeded; seeds olive-green with darker blotches radiating from the hilum.

Fl.-Fr.: Aug. -Jan.

Khandwa only, at Jai Bai.

Tephrosia villosa (L.) Pers., Syn. Pl. 2(2): 329. 1807; *Cracca villosa* L., Sp. Pl. 2: 752. 1753; Baker in Hook.f., Fl. Brit. India 2: 113. 1876; *Tephrosia villosa* (L.) Pers. var. *incana* (Roxb.) Baker in Hook.f., Fl. Brit. India 2: 113. 1876; Singh in Verma *et al.*, Fl. Madhya Pradesh 1: 399. 1993; Samvatsar, Fl. West. Trib. Madhya Pradesh 109. 1996.

Erect, tomentose herbs; leaves imparipinnate, 5-8 cm long; leaflets 11-19, 1-2 × 0.5-1.3 cm, obovate, obtuse, silky beneath; flowers purple, in lax elongated racemes; pods 2.5-5 × 0.5-0.8 cm, flat, woolly tomentose, curved, 6-8-seeded.

Fl.-Fr.: Oct. - Nov.

Alirajpur, Barwani, Burhanpur, Dhar, Indore, Khandwa.

TERAMNUS R. BR.W. 10; MP 2; SA 2

Key to Species

1. Stem and pods clothed with a few short hairs; leaflets membranous; calyx teeth as long as the tube .... **T. labialis**
1. Stem and pods clothed with long, spreading hairs; leaflets sub-coriaceous; calyx teeth shorter than the tube .... **T. mollis**

Teramnus labialis (L.f.) Spreng., Syst. Veg. 3: 235. 1826; *Glycine labialis* L.f., Suppl. Pl. 325. 1782; Baker in Hook.f., Fl. Brit. India 2: 184. 1876, *p.p;* Singh in Verma *et al.*, Fl. Madhya Pradesh 1: 399. 1993; Samvatsar, Fl. West. Trib. Madhya Pradesh 109. 1996. *Mash parni*

Twining herbs; leaves 3-foliolate; leaflets 4. 7-7.5 × 3-5 cm, ovate-oblong or oblong-lanceolate, pubescent beneath; flowers pink, in few flowered racemes; pods straight or slightly curved, apex hooked; sparsely appressed hairy, 8-12 seeded; seeds orange-brown.

Fl.-Fr.: Sept. -Dec.

Common forest undergrowth, all over.

Teramnus mollis Benth. in J. Linn. Soc., Bot. 8: 265. 1865 *non Glycine mollis* Wight & Arn., 1834; *Teramnus labialis* (L.f.) Spreng. var. *mollis* Baker in Hook.f., Fl. Brit. India 2: 184. 1876; Singh in Verma *et al.,* Fl. Madhya Pradesh 1: 400.1993; Samvatsar, Fl. West. Trib. Madhya Pradesh 109. 1996.

Twiners; leaves trifoliate with leaflets rounded, elliptic, ovate, obovate or even narrowly oblong or lanceolate, 1-8 cm long, 0.5 to 4 cm wide, smooth to densely covered with hairs; flowers white, pink or purplish, axillary racemes; pods densely hairy, 8-10 seeded; seeds black.

Fl.-Fr.: July - Dec.

Common in hedges, wet and shady places, throughout the area.

TRIFOLIUM L. W. 244; MP 2; SA 1

Trifolium fragiferum L., Sp. Pl. 2: 772. 1753; Baker in Hook.f. Fl. Brit. India 2:86. 1876.

Annual or perennial, procumbent herbs; leaves 3-foliolate; leaflets toothed, obovate-elliptic, *ca* 27 x16 mm; flowers pinkish, in dense heads; pods small, indehiscent, 1-6 seeded, enclosed in the inflated calyx.

Fl.-Fr.: Aug. -Oct.

Rare, **Khandwa** only, Mortakka, a weed of cultivated fields, river beds.

TRIGONELLA L. W. 95; MP 3; SA 2

Key to Species

1. Flowers 3-4 in axillary clusters; pods 1-2 seeded .... **T. occulta**
1. Flowers 6-15, in dense, axillary racemes; pods 4-8 seeded .... **T. corniculata**

Trigonella corniculata (L.) L., Syst. Nat. 2: 1180.1759; *Trifolium corniculatum* L. Sp. Pl. 2: 766. 1753; Baker in Hook.f. Fl. Brit. India 2:88. 1876; Singh in Verma *et al.,* Fl. Madhya Pradesh 1: 401. 1993.

Erect or suberect, much branched herbs; leaves 3-foliolate; leaflets obovate, 1-4 × 0.5-3 cm, obtuse to emarginate at apex, cuneate at base, dentate, glabrous above, sparsely pilose along the nerves beneath; flowers bright yellow, in dense racemes; pods falcately oblong, 1-2 × *ca* 0.25 cm, recurved, transversely veined, 5-8-seeded; seeds ovoid, compressed, dotted.

Fl.-Fr.: Feb. – March

Burhanpur only, Nepanagar, weed in cultivated fields.

Trigonella occulta Delile, Descr. Egypte, Hist. Nat. 71. 1812; Baker in Hook.f., Fl. Brit. India 2: 87, 1876; Singh in Verma *et al.,* Fl. Madhya Pradesh 1: 402. 1993; Samvatsar, Fl. West. Trib. Madhya Pradesh 110.1996.

Diffuse herbs; leaves pinnately 3-foliolate; leaflets oblanceolate, cuneate, sharply toothed, sparesely silky beneath; flowers yellow, 3-4 in axillary clusters; pods membranous, oblong ellipsoid, sparsely hairy; seeds 1-2.

Fl.-Fr.: Feb. –March

Common in drying ponds, ditches, where moisture is retained.

URARIA Desv. W. 25; MP 4; SA 2

Key to Species

1. Plants erect, up to 1 m tall; leaflets 5-9, lower often 1-3 foliolate, variegated; racemes compact, 10-12 cm long, not glandular; segments of pod 3-6 .... **U. picta**
1. Plants straggling; leaves 1-3 foliolate, not variegated; racemes up to 10-20 cm long, glandular; segments of pod 4-8 .... **U. rufescens**

Uraria picta (Jacq.) Desv. ex DC., Prodr. 2: 324. 1825; *Hedysarum pictum* Jacq., Collectanea 2: 262. 1788; Baker in Hook.f., Fl. Brit. India 2: 155. 1876; Singh in Verma *et al.,* Fl. Madhya Pradesh 1: 403. 1993; Samvatsar, Fl. West. Trib. Madhya Pradesh 110.1996. *Prishta parni* **(Plate 5, Fig. 3)**

Perennial, erect, suffruticose, herbs or undershrubs; lower leaves 1-3-foliolate, upper ones 5-9-foliolate; leaflets linear-lanceolate, rigidly coriaceous, 3-20 × 0.4-3 cm, curved on margins, glabrous above, minutely pubescent beneath, with a median glaucous band on the upper surface; flowers violet or purple, in cylindrical, 20-35 cm long racemes; pods 8-10 mm long, glabrous, 3-6-jointed, folded on one another, seeds shining white.

Fl.-Fr.: Aug. - Jan.

Alirajpur, Dhar, Indore, as forest undergrowth.

Uraria rufescens (DC.) Schindl. in Repert. Spec. Nov. Regni Veg. 21 (581-587): 14. 1925; *Desmodium rufescens* DC. in Ann. Sci. Nat. (Paris) 4: 101. 1825. *Uraria hamosa* (Roxb.) Wight & Arn., Prodr. Fl. Ind. Orient. 222. 1834; Baker in Hook.f., Fl. Brit. India 2: 156. 1876; Singh in Verma *et al.,* Fl. Madhya Pradesh 1: 404. 1993.

Perennial,straggling or diffuse, undershrubs; leaves 1-3-foliolate; leaflets elliptic, ovate or oblong, 5-15 × 2-8 cm, obtuse or emarginate at apex, rounded or semi-cordate at base, glabrous above, downy beneath; flowers purplish, in terminal or axillary racemes or lax panicles; pods rounded, *ca* 2 mm across, with 5-6 segments, tightly folded on each other.

Fl.-Fr.: Aug. -Oct.

Khandwa only, Bhagpura, Mahlu area.

VICIA L. W. 232; MP 5; SA 3

Key to Species

1. Flowers in axillary, pedunculate,1-5 flowered racemes; pod 1 cm long, 2-seeded .... **V. hirsuta**
1. Flowers axillary, solitary or 2; pod longer than 1 cm, more than 2-seeded
2. Leaflets 6-10 pairs, opposite; pod straight, 3.5 -5 cm long, 8-10 seeded .... **V. sativa**
2. Leaflets 5-6 pairs, alternate and opposite; pod deflexed, 2.5-3 cm long, 5-6 seeded .... **V. peregrina**

Vicia hirsuta (L.) Gray, Nat. Arr. Brit. Pl. 2: 614. 1821; *Ervum hirsutum* L., Sp. Pl. 2: 738. 1753; Baker in Hook.f., Fl. Brit. India 2: 177. 1876; Singh in Verma *et al.,* Fl. Madhya Pradesh 1: 405. 1993; Samvatsar, Fl. West. Trib. Madhya Pradesh 110.1996.

Much branched, climbing herbs; leaves paripinnate, ending in 3-fid tendrils; leaflets 5-10 pairs, linear-oblong to lanceolate, 8-15 × *ca* 2.5 mm, mucronate at apex, truncate at base, sparsely hairy; flowers bluish pink, 2-4, on 1-2 cm long, pedunculate, axillary racemes; pods elliptic-oblong to rhomboid, *ca* 1 × 0.3-0.4 cm, pubescent, 2-seeded; seeds black, *ca* 3 mm long, with distinct aril.

Fl.-Fr.: Nov. –March

Common winter weed, in all districts.

Vicia peregrina L., Sp. Pl. 2: 737. 1753; Baker in Hook.f., Fl. Brit. Indian 2: 178. 1876; Samvatsar, Fl. West. Trib. Madhya Pradesh 110.1996; Khanna *et al.,* Suppl. Fl. Madhya Pradesh 1: 77.2001.

Erect, annuals slender herbs; leaflets 5-6 pairs, 1.25-2.5 cm; stipules 2-fid, divisions linear; flowers purplish-red, in pairs; pods geminate, apiculate, with a sharp, curved point, pale brown; seeds 5-6, oblong, truncate, dark brown.

Fl.-Fr.: Jan. -April

Rare, **Alirajpur**, occurs only at Shakarja, Narmada river bed.

Vicia sativa L., Sp. Pl. 2: 736. 1753; Baker in Hook.f., Fl. Brit. India 2: 178. 1876; Singh in Verma *et al.,* Fl. Madhya Pradesh 1: 405. 1993; Samvatsar in Fl. West. Trib. Madhya Pradesh 110.1996.

Annual, erect or suberect, climbing herbs; leaves pinnately compound; rachis terminating in 2-3-fid tendrils; leaflets 4-6 pairs, alternate, obovate or lanceolate-oblong, 10-30 × 2-8 mm, acute, obtuse or emarginate at apex, entire; flowers purplish violet, usually solitary, axillary, rarely 2-3; pods linear, 3.5-7 cm long, glabrous or pubescent, 6-10-seeded; seeds blackish, 3-4 mm across.

Fl.-Fr.: Jan. –March

Common in fields as a weed, everywhere.

VIGNA Savi W. 118; MP 6; SA 5

Key to Species

1. Stipules 2-lobed or cordate at base; keel prolonged into a continuous incurved beak .... **V. vexillata**
1. Stipules peltate at base
   2. Flowers in dense, terminal clusters .... **V. indica**
   2. Flowers in axillary racemes or subumbellate clusters
      3. Corolla less than 10 mm long; leaflets 3-5 lobed; lobes spathulate; stem 30-60 cm long **V. trilobata**
      3. Corolla more than 10 mm long; leaflets entire
         4. Pods hispid with brownish hairs; corolla usually tinged with pink .... **V. radiata**
         4. Pods glabrous; corolla uniformly yellow .... **V. umbellata**

Vigna indica Dixit, T.M., S.P. Sutar, S.R. Yadav, K.V. Bhat & S.R. Rao, in Rheedea 21(1) 1:7.2011.

Trailing or twining herbs, 15-60 cm tall; leaves pinnately 3-foliolate or deeply 3-lobed; lobes ovate, retuse, 1.5-3.0 × 1.2-2.5 cm; flowers yellow, 5-10 in dense terminal clusters; pods cylindrical, hairy; seeds 5-10, brown, granulate, truncate at both ends.

Fl.-Fr.: Aug. -Oct.

Burhanpur only.

Vigna radiata (L.) R. Wilczek, var. **sublobata** (Roxb.) Verdc., in Kew Bull. 24:559. 1970; *Phaseolus radiatus* L., Sp. Pl. 725. 1753; *P. sublobatus* Roxb., Fl. Ind. 3:288. 1832; *P. mungo* L. var. *radiatus* (L.) Baker in Hook.f., Fl. Brit. India 2: 203. 1876, *p.p;* Singh in Verma *et al.,* Fl. Madhya Pradesh 1: 408. 1993; Samvatsar, Fl. West. Trib. Madhya Pradesh 111. 1996. *Jangli moong*

Annual, erect or climbing herbs; leaves 3-foliolate; leaflets elliptic, ovate or rhomboid, 5-15 × 3-10 cm, acute or acuminate at apex, cuneate or truncate at base, glabrous or pilose, entire or slightly lobed; laterals oblique; flowers yellow, in 4-8-flowered, subumbellate clusters; pods linear, cylindrical, 4-10 cm long, bristly hairy, 8-15-seeded; seeds greenish brown or blackish, oblong.

Fl.-Fr.: Oct. – Dec.

Alirajpur, Dhar, Indore.

Vigna trilobata (L.) Verdc. in Taxon 17(2): 172. 1968; *Dolichos trilobatus* L., Mant. Pl. 1: 101. 1767; Baker in Hook.f., Fl. Brit. India 2: 201. 1876; Singh in Verma *et al.,* Fl. Madhya Pradesh 1: 409. 1993; Samvatsar, Fl. West. Trib. Madhya Pradesh 111. 1996. *Jangli moong*

Annual or perennial, trailing herbs; leaves 3-foliolate; leaflets palmately 3-lobed, middle lobe larger, broadly oblong or spathulate, 1.5 × 1.5 cm, obtuse, lateral ones usually 1-lobed on the outer margin, glabrous or sparsely pubescent; flowers yellow, axillary, 5.5-13 cm long, 2-3-flowered racemes; pods cylindrical, 2.5-5 × 0.2-0.3 cm, glabrous, septate, 6-13-seeded; seeds oblong-rounded.

Fl.-Fr.: Sept. –Oct.

Alirajpur, Barwani, Burhanpur, Dhar, Indore, Khandwa.

Vigna umbellata (Thunb.) Ohwi & H. Ohashi in J. Jap. Bot. 44: 31. 1969; *Dolichos umbellatus* Thunb. in Trans. Linn. Soc. 2: 339. 1794; *Phaseolus calcaratus* Roxb., Fl. Ind. 3: 289. 1832; Baker in Hook.f., Fl. Brit. India 2: 203. 1876; Singh in Verma *et al.,* Fl. Madhya Pradesh 1: 409. 1993. *Sutri*

Suberect or twining herbs; leaves 3-foliolate; leaflets ovate-deltoid, 4-12 × 2-8 cm, acuminate at apex, subcuneate or subtruncate at base, entire or obscurely lobed, puberulous; petioles 2-15 cm long, pilose; flowers yellow, in axillary, tubercled, 5-8 cm long, 5-20-flowered racemes; pods linear, 4-9 cm long, up to 5 mm broad, glabrous, septate, 4-12-seeded; seeds dark brown, ovoid.

Fl.-Fr.: Oct. – Nov.

Barwani, Burhanpur, Indore, Khandwa.

Vigna vexillata (L.) A. Rich., Hist. Fis. Cuba, Bot. 10: 191. 1845; *Phaseolus vexillatus* L., Sp. Pl. 2: 724. 1753; Baker in Hook.f., Fl. Brit. India 2: 206. 1876; Singh in Verma *et al.,* Fl. Madhya Pradesh 1: 410.1993.

Perennial, trailing or twining herbs, with a tuberous rootstock; leaves 3-foliolate; leaflets ovate-lanceolate or oblong, 2.5-16.5 × 0.4-8.3 cm, acute or acuminate at apex, subcuneate at base, pubescent on both the surfaces; flowers yellow or purplish, in axillary 2-6-flowered racemes; pods linear, up to 15 cm long, brown-hairy, 8-18-seeded; seeds quadrate or oblong, reniform.

Fl.-Fr.: Oct. -Nov.

Khandwa only, Kalibhit and Battu Bagra forest.

ZORNIA J.F. Gmelin W. 86; MP 1; SA 1

Zornia gibbosa Span. in Linnaea 15: 192. 1841; *Z. diphylla* auct. plur. non Pers. 1807 Baker in Hook.f., Fl. Brit. India 2: 147. 1876; Singh in Verma *et al.,* Fl. Madhya Pradesh 1: 411. 1993; Samvatsar, Fl. West. Trib. Madhya Pradesh 111. 1996.

Diffuse wiry herbs; leaves 2-foliolate; leaflets 0.7-3 × 0.3-1 cm, lanceolate, acute, puncticulate; stipules and bracts peltate; flowers pale yellow, sessile, exserted from the bracts; pods not reticulate, with retrorsely scabrous, glochidiate bristles.

Fl.-Fr.: Aug. -Nov.

Common on the grassy slopes during rainy season, throughout the area.

ROSACEAE 133/1966; 2/2;1/1

POTENTILLA L. W. 179; MP 2; SA 1

Potentilla supina L., Sp. Pl. 1: 497. 1753; Hook.f., Fl. Brit. India 2: 359. 1878; Anand Kumar in Verma *et al.,* Fl. Madhya Pradesh 1: 446. 1993; Samvatsar, Fl. West. Trib. Madhya Pradesh 125. 1996. *Nak chhikni*

A prostrate or procumbent herbs; basal and lower cauline leaves larger, with more leaflets, upper cauline leaves nearly sessile, smaller with fewer leaflets; leaflets 3-9, subsessile, 0.5-1 × 0.3-0.5 cm, obovate, cuneate, coarsely serrate-dentate, appressedly hairy, petiolate; flowers yellow, axillary, solitary; bracteole ovate-oblong, pilose; achenes minute, brown, ovoid.

Fl.-Fr.: Nov. -March

Common, in moist soil everywhere.

VAHLIACEAE 1/3; 1/1; 1/1

VAHLIA Thunb. W. 3; MP 1; SA 1

Vahlia digyna (Retz.) Kuntze, Revis. Gen. Pl. 1: 227. 1891; *Oldenlandia digyna* Retz., Observ. Bot. 4: 23. 1786; *Vahlia viscosa* Roxb., Fl. Ind. 2: 89. 1832; C.B. Clarke in Hook.f., Fl. Brit. India 2: 399. 1878; Tripathi in Verma *et al.,* Fl. Madhya Pradesh 1: 449. 1993; Samvatsar, Fl. West. Trib. Madhya Pradesh 125. 1996.

A small glandular pubescent herb; leaves subsessile, oblong-lanceolate, 1.25-3 cm long; flowers small, white, solitary or in pairs in the leaf axil; sepals 5, glandular pubescent; petals and stamens 5, epigynous; filaments with a minute hairy scale at the base; capsule subglobose; seeds many.

*Fl.-Fr.:*Aug. -Nov.

Wet soil, **Dhar** only, Katnera.

CRASSULACEAE 48/ 1312; 1/1; SA 1/1

KALANCHOE Adans. W. 73; MP 1; SA 1

Kalanchoe olivacea Dalzell in Dalzell & Gibs., Bombay Fl. 313. 1861; Samvatsar, Fl. West. Trib. Madhya Pradesh 125. 1996; Khanna *et al.,* Suppl. Fl. Madhya Pradesh 93.2001.

Succulent herbs, up to 30 cm high; leaves simple, ovate-lanceolate up to 12.5 × 5.5 cm, glabrous, dotted with red spots; flowers white, in paniculate cymes; follicles glabrous.

Fl.-Fr.: Oct. -Dec.

Rare, **Alirajpur** only, among boulders, at Chatliya pani (Kattiwada).

COMBRETACEAE 17/480; 4/15; 3/8

Key to Genera

1. Petals present; fruits 4-5 winged; shrubby twiners .... COMBRETUM
1. Petals absent; trees
   2. Flowers in globose heads; fruits 2-winged .... ANOGEISSUS
   2. Flowers in spikes or racemes; fruits smooth, angled or 5-winged .... TERMINALIA

ANOGEISSUS (DC.) Guillem. & Perr.W. 8; MP 4; SA 3

Key to Species

1. Beak of fruit shorter than its nucleus; branches drooping, young ones not golden-villous; leaves 1-4 cm long .... **A. pendula**
1. Beak of fruit equal to or longer than its nucleus
   2. Fruit (including wings) longer than broad; wing margins entire; peduncle branched .... **A. latifolia**
   2. Fruit (including wings) broader than long; wing margins jagged; peduncle not branched ..... **A. sericea**

Anogeissus latifolia (Roxb. ex DC.) Wall. ex Bedd., Fl. Sylv. S. India 1 : 15. 1869; *Conocarpus latifolia* Roxb. ex DC., Prodr. 3: 16. 1828; *Anogeissus latifolia* var. *glabra* C.B. Clarke in Hook.f., Fl. Brit. India 2: 450.1878; *A. latifolia* var. *villosa* C.B. Clarke in Hook.f., Fl. Brit. India 2: 450.1878; Anand Kumar in Verma *et al.,* Fl. Madhya Pradesh 1: 455. 1993. Samvatsar, F. West. Trib. Madhya Pradesh 126. 1996. *Dhawada* **(Plate 5, Fig. 4)**

Trees, up to 20 m high; bark ash-coloured, smooth; leaves alternate or subopposite, elliptic, elliptic-oblong, elliptic-obovate or suborbicular, 4-8.5 × 2-5 cm, obtuse at apex and base, glabrous; flowers yellow, sessile, in dense heads. fruits 3.5-5 × 4-6 mm, 2-winged; persistent calyx-tube 4-6 mm long.

Fl.-Fr.: May–Dec.

Common in all districts.

Anogeissus pendula Edgew. in J. Asiat. Soc. Bengal 21: 171. 1853; C.B. Clarke in Hook.f., Fl. Brit. India 2: 451. 1878; Anand Kumar in Verma *et al.,* Fl. Madhya Pradesh 1: 456. 1996; Samvatsar, Fl. West. Trib. Madhya Pradesh 126. 1996. *Kardhai*

Trees, 4-10 m, with pendulous branches; bark grey; leaves 1-4 × 0.5-2 cm, elliptic or obovate, base narrowed, apex obtuse or acute, usually mucronate; flowers greenish-yellow; fruits packed in dense heads, upper portion of nucleus and beak hairy; seed ovoid or ellipsoid.

Fl.-Fr.: April – Aug.

Alirajpur, Barwani, Dhar, Indore, Khandwa, Khargone.

Anogeissus sericea Brandis in Indian Forester 25: 287. 1899; *A. latifolia* (Roxb. ex DC.) Wall. ex Guill. & Perr. var. *parvifolia* C.B. Clarke in Hook.f., Fl. Brit. India 2: 450.1878; Anand Kumar in Verma *et al.,* Fl. Madhya Pradesh 1: 456. 1996. *Indu koda*

Trees, up to 10 m high; branches pendulous, silky pubescent when young; leaves subopposite or opposite, elliptic, 2-3.5 × 1-2.5 cm, acute or obtuse at apex, cuneate at base, glabrous or nearly so above, silky pubescent beneath; flowers yellow, in dense, globose heads; fruits 2-3 × 4-5 mm, 2-winged; wings jagged on margin; persistent calyx-tube 4-5 mm long.

Fl.-Fr.: Jan. – May

Indore only, Nahar jhabua forest.

COMBRETUM Loefl. W. 289; MP 3; SA 2

Key to Species

1. Sepals and petals 5; bracteoles and bracts large, conspicuous; leaves abruptly acuminate, veins many, transversely horizontal; fruit 5-winged, 2.5 cm long .... **C. decandrum**
1. Sepals and petals 4; bracteoles and bracts minute; leaves ovate to lanceolate, veins few, irregularly branching; fruit 4-winged, 1.8 cm long .... **C. ovalifolium**

Combretum decandrum Jacq., Enum. Syst. Pl. 19. 1760; *C. roxburghii* Spreng., Syst. Veg. 2: 331. 1825; C.B. Clarke in Hook.f., Fl. Brit. India 2: 452. 1878; Anand Kumar in Verma *et al.,* Fl. Madhya Pradesh 1: 458. 1993; Samvatsar, Fl. West. Trib. Madhya Pradesh 126. 1996. *Bellora*

Scandent shrubs or climbers; leaves opposite, elliptic-oblong or ovate-elliptic, 6-13.5 × 3-8 cm, subcaudate at apex, rounded at base, glabrous, hairy on nerves beneath; flowers greenish white, 5-merous, in terminal panicles; fruits 5-winged.

Fl.-Fr.: Nov. -May

Barwani, Burhanpur, Khandwa, Khargone.

Combretum ovalifolium Roxb., Fl. Ind. 2: 226. 1832; C.B. Clarke in Hook.f., Fl. Brit. India 2: 458. 1878; *C. albidum* G. Don in Trans. Linn. Soc. London 15: 429. 1827; Anand Kumar in Verma *et al.,* Fl. Madhya Pradesh 1: 457. 1993; Samvatsar, Fl. West. Trib. Madhya Pradesh 126. 1996. *Hathisadan*

Large, climbing shrubs; leaves opposite 3-11 × 2.5-7 cm elliptic or ovate, often nearly orbicular; flowers 4-merous, in axillary, panicled spikes and often in terminal panicles; calyx-limb deciduous, cup-shaped; fruits golden yellow, 2-2.5 × 2-3 cm, winged, 4-seeded.

Fl.-Fr.: Nov. –May

Common twiner in all districts.

TERMINALIA L. W. 109; MP 6; SA 5

Key to Species

1. Fruits with 5 wings; flowers bisexual, in panicles of spikes; leaves with glands
    2. Leaves with a cordate base; glands stalked; calyx glabrous outside

3. Fruits with sharp, hard angles or wings, usually notched at apex, nerves on wings oblique and curving upwards .... **T. arjuna**

3. Fruits with long, papery wings, apex rounded, nerves straight and horizontal .... **T. crenulata**

2. Leaves not cordate at base; glands subsessile; calyx hairy outside .... **T. tomentosa**

1. Fruits not winged

4. Flowers in simple spikes; leaves alternate, clustered at the apex of branches .... **T. bellirica**

4. Flowers in panicles; leaves not clustered at the ends of branches .... **T. chebula**

Terminalia arjuna (Roxb. ex DC.) Wight & Arn., Prodr. Fl. Ind. Orient. 314. 1834; *Pentaptera arjuna* Roxb. ex DC., Prodr. 3: 14. 1828; C.B. Clarke in Hook.f., Fl. Brit. India 2: 447. 1878; Anand Kumar in Verma *et al.,* Fl. Madhya Pradesh 1: 460.1993; Samvatsar, Fl. West. Trib. Madhya Pradesh 127. 1996. *Arjun*

Large trees, with greenish-white to black bark, flaking off in flat pieces; leaves elliptic-oblong, subopposite; petioles with 1 or 2 glands at the top; flowers greenish-yellow, in axillary spikes and terminal panicles; calyx-tube campanulate; teeth 3-angular; drupes woody, ovoid or obovoid.

Fl.-Fr.: April–Dec.

Common along banks of srteams and rivers.

Terminalia bellirica (Gaertn.) Roxb., Pl. Coromandel 2: 54. t. 198. 1805; *Myrobalanus bellirica* Gaertn., Fruct. Sem. Pl. 2: 90. t. 97. 1790; C.B. Clarke in Hook.f., Fl. Brit. India 2: 445. 1878; Anand Kumar in Verma *et al.,* Fl. Madhya Pradesh 1: 461. 1993; Samvatsar, Fl. West. Trib. Madhya Pradesh 127. 1996. *Baheda*

Deciduous trees, up to 20 m high; leaves elliptic-obovate, alternate, clustered at the ends of branches; flowers greenish-yellow, in axillary spikes with pedicellate male flowers in upper part and sub-sessile, bisexual and female flowers in the lower; drupes ovoid or pyriform, tomentose, to 3.0-2.5 cm.

Fl.-Fr.: April– Feb.

Common forest tree in the study area.

Terminalia chebula Retz., Obs. Bot. 5:31. 1788; Clarke in Hook.f. Fl. Brit. India 2:446. 1878; Anand Kumar in Verma *et al.,* Fl. Madhya Pradesh 1:461. 1993. *Harad*

Trees up to 18 m tall; leaves subopposite, 7-12 × 4-6.5 cm, coriaceous, glands 2; spikes 5-7, occasionally branched; flowers creamy-white; calyx hairy within, lobes triangular; drupes oblong-ellipsoid, 4 × 2.5 cm, glossy, faintly angled.

Fl.-Fr.: April-June

Khargone only, Peepal zopa.

Terminalia crenulata Roth, Nov. Pl. Sp. 380.1821; *Terminalia tomentosa* Wight & Arn. var. *crenulata* (Roth) Clarke in Hook.f., Fl. Brit. India 2: 448. 1878; Anand Kumar in Verma *et al.,* Fl. Madhya Pradesh 1: 462. 1993; Samvatsar, Fl. West. Trib. Madhya Pradesh 127. 1996. *Sadad*

Medium sized trees; leaves sub-opposite or alternate, elliptic-obovate, 6-18 × 4-7 cm, entire or crenulate, with two stalked glands from lower midrib; flowers pale yellow, in branched spikes; calyx glabrous; fruits 5-winged, about 3 × 3.5 cm, reddish.

Fl.-Fr.: Apr. -Aug.

Rare, Alirajpur, Dhar, Indore, in hilly areas.

Terminalia tomentosa (Roxb. ex DC.) Wight & Arn., Prodr. Fl. Ind. Orient. 314. 1834; *Pentaptera tomentosa* Roxb. ex DC., Prodr. 3: 14. 1828; *Terminalia alata* Heyne ex Roth, Nov. Pl. Sp. 379. 1821; C.B. Clarke in Hook.f., Fl. Brit. India 2: 447. 1878; Anand Kumar in Verma *et al.,* Fl. Madhya Pradesh 1: 460.1993; Samvatsar, Fl. West. Trib. Madhya Pradesh 127. 1996. *Saja*

Deciduous trees, 20-40 m high; leaves 8-15 × 4-7 cm, elliptic- lanceolate or obovate acute, emarginate, entire; flowers yellow in racemes; fruit 4-5 cm long, glabrous, wings transversely striated, crenulate.

Fl.-Fr.: April-March

Very common in the entire area.

MYRTACEAE 145/5970; 1/4; 1/2

SYZYGIUM Gaertn. W. 1123; MP 4; SA 2

Key to Species

1. Leaves 3-6 cm broad; petiole 1.5-2 cm long; large trees .... **S. cumini**
1. Leaves less than 3 cm broad; petiole 0.5 cm long; trees of medium size, along river and stream banks .... **S. salicifolium**

Syzygium cumini (L.) Skeels in Bull. Bur. Pl. Industr. U.S.D.A. 248: 25. 1912; *Myrtus cumini* L., Sp. Pl. 1: 471. 1753; *Eugenia jambolana* Lam., Encycl. 3: 198. 1789; Duthie in Hook.f., Fl. Brit. India 2: 499. 1879; Mukherjee in Verma *et al.,* Fl. Madhya Pradesh 1: 466. 1993; Samvatsar, Fl. West. Trib. Madhya Pradesh 128. 1996. *Jamun*

Large trees; leaves sub-coriaceous 10-14 × 3-6 cm, lanceolate, acute, subobtuse or shortly acuminate, lateral veins prominent, close, uniting in an intramarginal nerve; flowers greenish-white, crowded in heads on the ends of panicled cymes; berries subglobose, purple, juicy, 1-seeded.

Fl.-Fr.: March – Aug.

Common tree in the study area.

Syzygium salicifolium (Wight) J. Graham, Cat. Pl. Bombay 73. 1839; *Eugenia salicifolia* Wight, Ill. Ind. Bot. 2: 16. 1841; *Syzygium heyneanum* Wallich ex Wight & Arnott, Prodr. 1: 329. 1834; *Eugenia heyneana* Duthie, in Hook.f., Fl. Brit. India 2:500.1879; Mukherjee in Verma *et al.,* Fl. Madhya Pradesh 1: 466. 1993; Samvatsar, Fl. West. Trib. Madhya Pradesh 128. 1996. *Kat jamun*

A shrub, or a medium tall tree; leaves 7-14 × 2-4 cm, oblong lanceolate, subobtuse, glabrous, nerves uniting in an intramarginal nerve; flowers white, in small heads at the ends of branches of cymes; berries oblong-ellipsoid.

Fl.-Fr.: March – Aug.

Common along banks of rivers and streams.

LECYTHIDACEAE 24/319; 2/3; 1/1

CAREYA Roxb. W. 3; MP 2; SA 1

Careya arborea Roxb., Hort. Beng. 54. 1814 & Pl. Coromandel 3: 14. t. 218. 1811; Clarke in Hook.f., Fl. Brit. India 2: 511. 1879; Mukherjee in Verma *et al.,* Fl. Madhya Pradesh 1: 468. 1993; Samvatsar, Fl. West. Trib. Madhya Pradesh 128. 1996. *Kumbhi* **(Plate 5, Fig. 5)**

Deciduous trees with thick, dark brown bark; leaves 15-30 cm × 10-15 cm, clustered at the ends of branchlets, obovate, crenulate; flowers pale white in short spikes; each flower with a central elliptic bract and 2 linear lateral ones; stamens indefinite; berries globose 4-8 cm across, crowned with lobes, many seeded.

Fl.-Fr.: March - July

Occasional along water streams, on the hill slopes, fringe of forest, Alirajpur, Barwani, Burhanpur, Dhar, Khandwa, Khargone.

LYTHRACEAE 31/604; 6/17; 6/14

Key to Genera

1. Small, marshy herbs; flowers very small; calyx membranous
   2. Placenta not continuous with the style
      3. Fruit dehiscent, opening by valves .... ROTALA
      3. Fruit indehiscent, opening irregularly .... AMMANNIA
   2. Placenta continuous with the style; fruit dehiscing operculately or irregularly .... NESAEA
1. Trees or shrubs; flowers larger; calyx moderately thick
   4. Flowers tetramerous; stamens 8; seeds angular, pyramidal .... LAWSONIA
   4. Flowers trimerous; bracteoles 2; stamens more than 8; seeds not angular
      5. Stamens 12; leaves black-dotted below; panicle axillary; ovary 2-celled; capsule included; seeds not winged .... WOODFORDIA
      5. Stamens indefinite; leaves not punctate; panicles terminal; ovary 3-6 celled; capsule exserted; seeds winged .... LAGERSTROEMIA

AMMANNIA L. W. 26; MP 4; SA 4

Key to Species

1. Petals absent; cymes sessile or sub-sessile; styles absent or minute
   2. Leaves tapering at base; flowers in dense axillary clusters .... **A. baccifera**

2. Leaves rounded or cordate at base; flowers fewer .... **A. aegyptiaca**

1. Petals present; cymes pedunculate; styles distinct; flowers in 2 rows; capsule subglobose or lobed

3. Peduncles and pedicels filiform; stamens as long as sepals; style as long as ovary; capsule lobed, 1.5 mm in diameter .... **A. multiflora**

3. Peduncles and pedicels stout; stamens exceeding sepals; style twice as long as ovary; capsule globose, 2-3.5 mm in diameter .... **A. auriculata**

Ammannia aegyptiaca Willd., Hort. Berol. t. 6. 1809; *A. salicifolia* Hiern in Oliver, Fl. Trop. Afr. 2: 478. 1871, *excl. syn., non* Monti *ex* Blume, 1767; C.B. Clarke in Hook.f., Fl. Brit. India 2: 569. 1879; *Ammannia baccifera* L. subsp. *aegyptiaca* (Willd.) Koehne in Engl., Bot. Jahrb. Syst. 1: 260.1880; Srivastava in Verma *et al.,* Fl. Madhya Pradesh 1: 473. 1993; Samvatsar, Fl. West. Trib. Madhya Pradesh 129. 1996. *Dad mari*

Profusely branched herbs, 90-100 cm tall; leaves 1.5-6 × 6-1 cm, elliptic oblong, obtuse at base; flowers green; calyx campanulate, teeth triangular; capsule dehiscing irregularly.

Fl.-Fr.: Aug.-Dec.

Common, along water streams and in marshy land, throughout the area.

Ammannia auriculata Willd. Hort. Berol. 1: t. 7. 1806; *A. senegalensis* DC. Prodr. 3: 77. 1828 non Lam. 1791; Clarke in Hook.f., Fl. Brit. India 2: 570 1879; Samvatsar, Fl. West. Trib. Madhya Pradesh 129. 1996; Khanna *et al.,* Suppl. Fl. Madhya Pradesh 1: 97.2001.

Erect, annual herbs, up to 40 cm high, with 4-angled stem; leaves elongated-oblong, 2-5 × 0.3-0.6 cm, with auricled base; flowers pink, in axillary, pedunculate cymes shorter than leaves; capsule subglobose, about 2 mm in diameter; seeds excavated on ventral side.

Fl.-Fr.: Aug. -Oct.

Rare, in water logged soils, very close to *A. multiflora* Roxb., which, however, bears smaller leaves and longer inflorescence, Alirajpur, Barwani, Dhar, Khandwa, Khargone.

Ammannia baccifera L., Sp. Pl. 120.1753; C.B. Clarke in Hook.f., Fl. Brit. India 2: 569. 1879; Srivastava in Verma *et al.,* Fl. Madhya Pradesh 1:473. 1993; Samvatsar, Fl. West. Trib. Madhya Pradesh 311. 1996. *A. salicifolia auct. non* Blume, 1856: C.B. Clarke in Hook.f., Fl. Brit. India 2: 569. 1879. *Dad mari*

Much branched herbs; leaves 1-5 × 0.5-0.8 cm, linear- lanceolate, acute, glabrous; flowers minute, in dense, axillary clusters; capsule irregularly dehiscing; seeds many.

Fl.-Fr.: Sept. -Dec.

Very common in marshy fields or on the banks of shallow ponds, throughout the area.

Ammannia multiflora Roxb., Fl. Ind. 1: 447. 1820; Clarke in Hook.f., Fl. Brit. India 2: 570.1879; Srivastava in Verma *et al.,* Fl. Madhya Pradesh 1:473. 1993; Samvatsar, Fl. West. Trib. Madhya Pradesh 131. 1996.

Erect, annual herbs, up to 60 cm high; stem scabrid on angles; leaves 2-4 × 0.3-0.5 cm, elliptic oblong, with auricled base; flowers red, in 2 to many flowered, axillary, pedunculate cymes; capsule lobed, enclosed in the calyx-tube or slightly exserted; seeds many, minute, red, smooth.

Fl.-Fr.: Oct. -Feb.

Common in wet and shady localities in forest.

(The juice of all species of **Ammannia** cause blisters on skin).

LAGERSTROEMIA L. W. 20; MP 2; SA 1

Lagerstroemia parviflora Roxb., Pl. Coromandel 1: 47. t. 66. 1795; C.B. Clarke in Hook.f., Fl. Brit. India 2: 575. 1879; Srivastava in Verma *et al.,* Fl. Madhya Pradesh 1: 474. 1993; Samvatsar, Fl. West. Trib. Madhya Pradesh 131. 1996. *Lendia* **(Plate 5, Fig. 6)**

A large tree; leaves 3-12 × 2-6.5 cm, elliptic-oblong, acute or acuminate; flowers white, fragrant in axillary and terminal, many or few flowered panicles; capsule ellipsoid; seeds winged, the wings much longer than the nucleus, thin on the inside edge, thick on the outer.

Fl.-Fr.: April - Oct.

Very common in the forests.

LAWSONIA L. W. 2; MP 1; SA 1

Lawsonia inermis L., Sp. Pl. 340.1753; Srivastava in Verma *et al.,* Fl. Madhya Pradesh 1: 475. 1993; Samvatsar, Fl. West. Trib. Madhya Pradesh 131. 1996. *L. alba* Lam., Illustr. t. 296. f. 2. 1792; C.B. Clarke in Hook.f., Fl. Brit. India 2: 573. 1879. *Mehndi*

Glabrous, much branched shrubs; leaves 2-4 × 0.8-1.6 cm, usually elliptic, narrowed into a short petiole; flowers creamish, fragrant; petals 4, very shortly clawed, suborbicular or subreniform; stamens 8; capsule globose or depressed-globose, veined, dark brown; seeds many pyramidal, brown.

Fl.-Fr.: April - Nov.

Common along village roads, not wild, naturalized.

NESAEA Kunth W. 25; MP 1; SA1

Nesaea lanceolata Koehne in Bot. Jahrb. Syst. 3: 325. 1882; *Ammannia lanceolata* Heyne ex Clarke in Hook.f., Fl. Brit. India 2: 570, 1879; Srivastava, in Verma *et al.,* Fl. Madhya Pradesh 1:475. 1993; Samvatsar, Fl. West. Trib. Madhya Pradesh 131. 1996.

Annual, 10-20 cm tall herb; stem winged in apical portion, shortly angled in the lower, scabrid on angles; leaves 1.5-3.5 × 0.5-1.2 cm oblong or narrowly lanceolate; flowers rosy, sessile or subsessile, 2-5 in each axil; capsule subglobose.

Fl.-Fr.: Sept. -Dec.

Rare, Alirajpur, Dhar.

ROTALA L. W. 24; MP 9; SA 6

Key to Species

1. Leaves in whorls of 4-10; flowers trimerous .... **R. verticillaris**
1. Leaves decussate
    2. Flowers pentamerous; bracteoles exceeding the calyx; flowering branches compact; calyx lobes sharply triangular
        3. Calyx with linear-subulate accessory lobes; fruits *ca* 1.5 mm long; leaves up to 3 cm long, never whorled .... **R. densiflora**
        3. Calyx without such lobes; fruits *ca* 1 mm long; leaves less than 1.5 cm long, sometimes 3-4 whorled .... **R. mexicana**
    2. Flowers tetramerous
        4. Calyx appendages linear; leaves cordate.................... **R. illecebroides**
        4. Calyx appendages absent; leaves not cordate
            5. Flowers solitary and in axillary spikes; sepals linear, persistent; stamens inserted above the middle of calyx tube .... **R. indica**
            5. Flowers in dense, spiciform racemes; petals obovate, deciduous; stamens inserted at the base of calyx tube; capsule 2-valved .... **R. serpyllifolia**

Rotala densiflora (Roth) Koehne in Bot. Jahrb. Syst. 1(2):164. 1880; *Ammannia densiflora* Roth ex Roem. & Schult., Syst. Veg. 3: 304. 1818; *A. pentandra auct. non* Roxb., 1820; C.B. Clarke in Hook.f., Fl. Brit. India 2: 568. 1879; Srivastava in Verma *et al.,* Fl. Madhya Pradesh 1: 477. 1993; Samvatsar, Fl. West. Trib. Madhya Pradesh 132. 1996.

A delicate herb; stem tetragonous with numerous slender branches; leaves sessile, oblong, smaller and bract like; flowers sessile, one flower per bract; bracteoles subulate; capsule subglobose, 3- valved.

Fl.-Fr.: Sept. – May

Rare, in moist places, **Alirajpur**, Aamkhut forest.

Rotala illecebroides (Arn. ex C.B. Clarke) Koehne in Bot. Jahrb. Syst. 1: 161. 1881*; Ammannia pentandra* var. *illecebroides* Clarke in Hook.f. Fl. Brit. India 2:569. 1879; Srivastava in Verma *et al.,* Fl. Madhya Pradesh 1:477. 1993.

About 10 cm tall herbs; leaves sessile, opposite-decussate, 2.5-3.5 × 1.2-3 mm; flowers tetramerous, solitary, axillary; calyx appendages 4, linear, lobes 4, 1 mm long; petals 2, linear; stamens 4, inserted above the calyx tube; ovary 3-4 lobed; capsule subglobose, 0.5 mm, included within the calyx tube.

Fl.-Fr.: Oct. -Dec.

In moist ground at **Khandwa** only, Chandgarh range.

Rotala indica (Willd.) Koehne in Bot. Jahrb. Syst. 1: 172. 1880; Srivastava in Verma *et al.,* Fl. Madhya Pradesh 1: 478. 1993; *Peplis indica* Willd., Sp. Pl. 2: 244. 1799. *Ammannia peploides* Spreng., Syst. 1: 444. 1824; C.B. Clarke in Hook.f., Fl. Brit. India 2: 566. 1879.

Herbs, annual, erect or decumbent, 12-25 cm high; branches 4-gonous; leaves opposite; those on main stem oblanceolate-obovate, 1-1.8 × 0.5-0.8 cm, rounded at apex, narrowed at base; those on spikes linear-oblong, 4-6 × 1-2 mm; flowers pink, minute, solitary or in 2-3 cm long, leafy spikes; capsules ellipsoid, *ca* 1.5 mm long, 2-valved; seeds many, ellipsoid.

Fl.-Fr.: Sept. – Nov.

Common, in moist ground, all over the area.

Rotala mexicana Schltdl. & Cham. in Linnaea 5: 567. 1830; *Ammannia pygmaea* Kurz in Seem., J. Bot. 5: 376. 1867; C.B. Clarke in Hook.f., Fl. Brit. India 2: 568. 1879; Srivastava in Verma *et al.,* Fl. Madhya Pradesh 1: 478. 1993; Samvatsar, Fl. West. Trib. Madhya Pradesh 132. 1996.

Tufted annual herbs; stem erect or creeping; branches erect, quadrangular; leaves linear, 0.3-0.6 × .05-.06 cm., 1-nerved; flowers solitary, axillary, very minute, bract subulate; stamens 2; capsule 0.1 cm long, 3-valved, globose.

Fl.-Fr.: Aug. – Nov.

Rare, **Alirajpur** only, Kadwal forest.

Rotala sepyllifolia (Roth) Bremek. in Acta Bot. Neerl. 3(1):149. 1954; *Micranthus serpyllifolius* Roth, Nov. Pl. Sp. 282. 1821; *Ameletia tenuis* Wight, Pl. Ind. Ori. t. 257. 1840; *Ammannia tenuis* (Wight) Clarke in Hook.f. Fl. Brit. India 2:567. 1879; Srivastava in Verma *et al.,* Fl. Madhya Pradesh 1:479. 1993; Samvatsar, Fl. West. Trib. Madhya Pradesh 132. 1996.

Small annual herbs commonly found on rocks in stream beds, rivers; stem wiry, red; leaves broadly ovate to elliptic, obtuse or acute; flowers red in dense, terminal spicate racemes; calyx tube campanulate with acute teeth; petals 4,obovate; capsules ellipsoid.

Fl.-Fr.: Nov. -Feb.

Common throughout, in all districts in rocky river beds.

Rotala verticillaris L. Mantis Pl. 2: 175. 1771*, p.p., Ammannia rotala* F. Muell. Fragm. 3. 108. 1803; Clarke in Hook.f. Fl. Brit. India 2:567. 1879; Srivastava in Verma *et al.,* Fl. Madhya Pradesh 1:479. 1993; Samvatsar, Fl. West. Trib. Madhya Pradesh 132. 1996.

Herbs up to 10 cm tall; leaves in whorls of 4-10, linear-lanceolate, 3-10 × 1 mm, base cuneate, apex acute; flowers solitary, *ca* 1.5 mm across; calyx tube 2 mm long; stamens 3, inserted above the calyx tube; capsules exceeding the calyx tube; seeds numerous, ovoid, compressed, about 0.2 mm long.

Fl.-Fr.: Dec. -Feb.

In water-logged areas, **Dhar** only, Mordundia, near Katnera.

WOODFORDIA Salisb. W. 2; MP 1; SA 1

Woodfordia fruticosa (L.) Kurz in J. Asiat. Soc. Bengal, Pt. 2, Nat. Hist. 40(2): 56. 1871; *Lythrum fruticosum* L., Syst. Nat. ed. 10.1045. 1759; Srivastava in Verma *et al.,* Fl. Madhya Pradesh 1: 479. 1993; Samvatsar, Fl. West. Trib. Madhya Pradesh 132. 1996; *Woodfordia* *floribunda* Salisb., Parad. Lond. t. 42. 1806; C.B. Clarke in Hook.f., Fl. Brit. India 2: 572. 1879. *Dhawai*

Semi-deciduous shrubs or undershrubs; leaves 5-10 cm long opposite, very short petioled, ovate-lanceolate, nigro-punctate below; flowers red, in racemes,arising from the axils of fallen leaves; capsule ellipsoid, papery; seeds obovoid, brown.

Fl.-Fr.: Jan. – May

Common in forest clearings, freshly cut paths, all over the area.

ONAGRACEAE 43/720; 1/5; 1/5

LUDWIGIA L. W. 74; MP 5; SA 5

Key to Species

1. Plants with pneumatophores developing from nodes of floating branches .... **L. adscendens**

1. Plants without pneumatophores, inhabiting marshy areas

  2. Stamens twice as many as sepals

    3. Flowers more than 2 cm across; capsules 2-4 cm long; seeds multiseriate in the upper part of capsule, uniseriate in the lower half and embedded in endocarp .... **L. octovalvis**

    3. Flowers less than 2 cm across; capsules 1.5-2 cm long; seeds uniseriate, embedded in endocarp at the base .... **L. hyssopifolia**

  2. Stamens as many as sepals; capsules 0.8-1.2 cm long

    4. Capsules inflated; seeds in many rows in each locule .... **L. perennis**

    4. Capsules slender; seeds in one row in each locule .... **L. prostrata**

Ludwigia adscendens (L.) Hara in J. Jap. Bot. 28: 290.1953; *Jussiaea adscendens* L., Mant. Pl. 69. 1767; *J. repens* L., Sp. Pl. 388. 1753; C.B. Clarke in Hook.f., Fl. Brit. India 2: 587. 1879; Tripathi, in Verma *et al.,* Fl. Madhya Pradesh 1:481. 1993.

Herbs, creeping or floating, rooting at the nodes; leaves obovate or lanceolate, 2.5-5.5 × 1-2 cm, sparsely pubescent, rounded at apex, narrowed at base; flowers whitish yellow, solitary, axillary, 1-1.5 cm across; capsules cylindrical, 1.5-3 cm long, 10-ribbed, glabrous or sparsely pubescent, irregularly dehiscent; seeds uniseriate in each locule, quadrate with a white corky reticulate testa.

Fl.-Fr.: Nov. – June

Rare, along ponds, **Indore** only, Depalpur tank.

Ludwigia hyssopifolia (G. Don) Exell, Garcia de Orta 5:471. 1957; *Jussiaea hyssopifolia* G. Don, Gen. Hist. 2: 293. 1832. *J. fissendocarpa* Haines in J. Asiat. Soc. Bengal, Pt. 2, Nat. Hist. 15: 313. 1920; Tripathi in Verma *et al.,* Fl. Madhya Pradesh 1: 482. 1993; Samvatsar, Fl. West. Trib. Madhya Pradesh 134. 1996.

Glabrous up to 90 cm tall herb, leaves oblong-lanceolate 2-6 cm long; flowers yellow, 1 cm across;, calyx lobes 4, lanceolate; petals 4, obovate; stamens 8; capsule 8-ribbed, puberulous; seeds dimorphic, many seriate, at apex, uniseriate below.

Fl.-Fr.: Oct. – May

Rare, **Alirajpur** only, Akadia village, near water body.

Ludwigia octovalvis (Jacq.) Raven in Kew Bull. 15: 476. 1962; *Oenothera octovalvis* Jacq., Enum. Syst. Pl. 19. 1760; *Jussiaea suffruticosa* L., Sp. Pl. 388. 1753; C.B. Clarke in Hook.f., Fl. Brit. India 2: 587. 1879; Tripathi in Verma *et al.,* Fl. Madhya Pradesh 1: 482. 1993; Samvatsar, Fl. West. Trib. Madhya Pradesh 134. 1996.

Densely tomentose, woody plants; leaves 2-8 × 1-5 cm, elliptic to lanceolate, depressed pubescent; flowers large, yellow; pedicels very short; capsules 8-ridged, adpressed tomentose, about 4 cm long; seeds globose.

Fl.-Fr.: Oct. – June

Common along ponds and ditches, all over the area.

Ludwigia perennis L., Sp. Pl. 119. 1753; *L. parviflora* Roxb., Fl. Ind. 1: 440.1820; C.B. Clarke in Hook.f., Fl. Brit. India 2: 588. 1879; Tripathi in Verma *et al.,* Fl. Madhya Pradesh 1: 483. 1993; Samvatsar, Fl. West. Trib. Madhya Pradesh 134. 1996.

Herbs, annual, up to 50 cm high; stem subangular; leaves lanceolate or narrowly. elliptic, 3-12 × 0.5-2 cm, narrowed at both ends; petioles 0.2-1cm long; flowers axillary, solitary or paired; calyx-tube short, 4-lobed, ovate, acute; petals 4, elliptic-oblong; stamens 4; ovary oblong; style short, stout; stigma capitate; capsules oblong 1-2 cm long, obscurely 4-angled; seeds pluriseriate in each locule, ellipsoid, shining.

Fl.-Fr.: Sept. – April

Common in all districts on the banks of ditches or within them.

Ludwigia prostrata Roxb., Fl. Ind. (Carey & Wall. ed.) 1: 441. 1820; C.B. Clarke in Hook.f., Fl. Brit. India 2: 588. 1879; Tripathi in Verma *et al.,* Fl. Madhya Pradesh 1: 483. 1993; Samvatsar, Fl. West. Trib. Madhya Pradesh 134. 1996.

Annual, up to 50 cm high, herbs; stem prostrate or decumbent at base; leaves linear-lanceolate, lanceolate or narrowly elliptic, 4-7 × 1-2 cm, narrowed at both ends; flowers sessile, yellow; capsules linear, 1-3 cm long, 4-gonous; seeds uniseriate in each locule, ovoid.

Fl.-Fr.: Nov. – April

Rare, along banks of ponds, ditches, **Alirajpur** only, at Chatliya pani, in shady, wet areas.

PASSIFLORACEAE 27/694; 1/2; SA 1

PASSIFLORA L. W. 361; MP 2; SA 1

Passiflora foetida L. Sp. Pl. 959, 1753; Srivastava in Verma *et al.,* Fl. Madhya Pradesh 1: 486. 1993; Samvatsar, Fl. West. Trib. Madhya Pradesh 135. 1996. *Rakhi*

A slender, herbaceous, foetid, tendrillar climber; leaves 2-12 × 1.5-6 cm, 3-lobed, ovate-triangular, ciliate on margins; flowers white, axillary, solitary, bracts and bracteoles deeply pinnatisect; berries 1.5-2.5 cm across, globose, glabrous, orange red when ripe; seeds oblong, deeply pitted on faces, glabrous, pale to dark brown.

Fl.-Fr.: Aug. -Dec.

Alirajpur, Barwani, Burhanpur, Dhar, Jhabua, Khargone.

CUCURBITACEAE 134/965; 13/25; 11/20

Key to Genera

1. Flowers white
   2. Corolla lobes divided almost to the base; tendrils 2-5 fid; petals fimbriate; petiole without glands .... TRICHOSANTHES
   2. Corolla campanulate, lobes not deeply divided; tendril simple .... COCCINIA
1. Flowers yellow
   3. Tendrils 2-5 fid; corolla deeply divided
      4. Male flowers in racemes; fruits dry at maturity, dehiscence operculate .... LUFFA
      4. Male flowers solitary or in fascicles; fruits fleshy
         5. Pollen muricate; pistillode absent .... DIPLOCYCLOS
         5. Pollen smooth; pistillode present, glandular .... CITRULLUS
   3. Tendrils simple
      6. Male peduncle with spathaceous bract; calyx with scales .... MOMORDICA
      6. Male peduncle without such bract; calyx without scales
         7. Petiole with conspicuous, stipuliform, fimbriate bract .... CTENOLEPIS
         7. Petiole without such bract
            8. Corolla campanulate, divided below the middle .... CORALLOCARPUS
            8. Corolla 5-partite
               9. All anthers 1-celled; ripe fruits yellow with 9-11 green stripes .... CUCUMIS
               9. Two anthers 2-celled, one anther 1-celled; ripe fruit red
                  10. Fruits 0.4-0.5 cm across, smooth; plants monoecious; leaves scabrid on both surfaces; stem hispid .... MUKIA
                  10. Fruits 2-2.5 cm across; plants dioecious; leaves scabrid on upper surface; stem glabrous .... SOLENA

CITRULLUS Schrad. W. 4; MP 1; SA 1

Citrullus colocynthis (L.) Schrad. ex Eckl. & Zeyh., Enum. Pl. Afr. Austr. 2: 279. 1836; *Cucumis colocynthis* L., Sp. Pl. 1011. 1753; C.B. Clarke in Hook.f., Fl. Brit. India 2: 620.1879; Samvatsar, Fl. West. Trib. Madhya Pradesh 136. 1996. *Indrayan* **(Plate 5, Fig. 7)**

Stems diffuse, or creeping, angled, hirsute or scabrid, tendril 2-fid; leaves variable, deltoid, pale green above, ashy beneath, scabrid on both surfaces, 3-7 lobed; fruit globular, slightly depressed, variegated green and white, epicarp thin; seeds pale brown.

Fl.-Fr.: Nov. -Jan.

Common in fields and on hedges, as a weed, throughout the area.

COCCINIA Wight & Arn. W. 31; MP 1; SA 1

Coccinia grandis (L.) Voigt, Hort. Suburb. Calc. 59. 1845; *Bryonia grandis* L., Mant. Pl. 126. 1767; *Coccinia indica* Wight & Arn., Prodr. Fl. Ind. Orient. 1: 347. 1834; *Cephalandra indica* Naudin in Ann. Sci. Nat., Bot. ser. 5, 5: 16. 1859; C.B. Clarke in Hook.f., Fl. Brit. India 2: 621. 1879; Srivastava in Verma *et al.,* Fl. Madhya Pradesh 1: 491. 1993; Samvatsar, Fl. West. Trib. Madhya Pradesh 136. 1996. *Tindori*

Slender, dioecious climbers; nodes on the mature stem annulate; leaves entire or deeply 3-5 lobed, cordate; flowers white, usually solitary, axillary; fruits oblong-obtuse, 5-6 cm long, green, striped with white when young, red at maturity.

Fl.-Fr.: Aug. – Dec.

Very common throughout the area.

CORALLOCARPUS Welw. W. 16; MP 2; SA 2

Key to Species

1. Berry stalked, abruptly contracted into a beak, scarlet in the middle, apex and base green; seeds margined .... **C. epigaeus**
1. Berry sessile, gradually contracting in a beak, orange-red, except the base; seeds not margined .... **C. conocarpus**

Corallocarpus conocarpus Clarke in Hook.f., Fl. Brit. India 2: 628, 1879; *Aescmandra conocarpa* Dalzell & Gibs. Bombay Fl. 100.1861; Bhandari in Fl. Ind. Des. 148. 1995; Samvatsar, Fl. West. Trib. Madhya Pradesh 137. 1996; Khanna *et al.,* Suppl. Fl. Madhya Pradesh 1: 98. 2001.

Climbers; stem striate, glabrous; tendrils simple; leaves palmately 3-5 lobed, cordate at base, lobes oblong-lanceolate, more or less sparsely hairy on both surfaces; fruit glabrous to glabrescent; seeds dark brown.

Fl.-Fr.: June-Aug.

Very rare, in hedges, Barwani, Dhar

Corallocarpus epigaeus (Rottl. & Willd.) Hook.f. in Gen. Pl. 1: 831. 1862; *Bryonia epigaea* Rottl. & Willd. Neue Schrift. Ges. Naturf. Freunde Berlin 4: 223. 1803; Clarke in Hook.f., Fl. Brit. India 2:628. 1879; Samvatsar, Fl. West. Trib. Madhya Pradesh 137. 1996; Khanna *et al.,* Suppl. Fl. Madhya Pradesh 1: 99.2001.

Prostrate or climbing; stem slender, grooved, zigzag; tendrils simple; leaves suborbicular, pubescent on both surfaces, deeply 3-5 lobed; fruit stalked, ellipsoid or ovate, beaked, scarlet in the middle, green at base and beak; seeds 6-9, pyriform, turgid, brown.

Fl.-Fr.: Aug. -Nov.

In hedges, not common, Alirajpur, Dhar.

CTENOLEPIS Hook.f. W. 3; MP 2; SA 2

Key to Species

1. Leaves 3-5 lobed; berry ob-reniform, white-dotted; seeds plano-convex .... **C. garcinii**
1. Leaves 5-7 lobed; berry globose or obscurely subquadrate; seeds much compressed......... **C. cerasiformis**

Ctenolepis cerasiformis (Stocks) C.B. Clarke in Hook.f., Fl. Brit. India 2: 630.1879; *Zehneria cerasiformis* Stocks in Hooker's J. Bot. Kew Gard. Misc. 4: 49 1852; *Blastania fimbristipula* Kotschy & Peyr. Pl. Tinn. 15: t. 7. 1867; *Bryonia fimbristipula* Fenzl & Hook.f., Fl. Trop. Afr. 2:558. 1871; Madhusudan Rao, in N.P. Singh *et al.,* Fl. Maharashtra 2:54.2001.

Monoecious climbers; tendrils simple; leaves 3-partite, 3-10 × 2.5-9 cm, membranous, scabrid, white, punctate on both surfaces, segments ovate-oblong or lanceolate, margin crenulate or denticulate, acute at apex, narrowed at base; bracts orbicular or reniform, 0.5-1.2 cm long, ciliate; flowers minute; male flowers 5-10 at the apex of 1.5-3 cm long filiform peduncle, female flowers solitary, on very short peduncles; fruits globose or ellipsoid, *ca* 1 cm across; seeds 2, concave on one side and convex on the other.

Fl.-Fr.: July – Oct.

Rare, in hedges, **Burhanpur** only, at Lalbaug.

Ctenolepis garcinii (L.) C.B. Clarke in Hook.f., Fl. Brit. India 2: 629. 1879; *Sicycos garcinii* Burm.f., Fl. Ind. 211. 1768; *Blastania garcinii* (Burm.f.) Cogn. in A. DC. & C. DC., Monogr. Phan. 3: 629. 1881; Srivastava in Verma *et al.,* Fl. Madhya Pradesh 1: 489. 1993; Samvatsar, Fl. West. Trib. Madhya Pradesh 136. 1996. *Ankhphorni*

Annual, monoecious climbers; leaves orbicular in outline, 3 to 5 lobed, 2-3.5 cm in diameter, scabrous; male flowers 3-4, on each axillary peduncle; female ones solitary on short axillary pedicels in the same axils with males; fruits fleshy, obreniform, about 1 cm in diameter, white dotted, red when ripe; seeds concavo-convex or plano-convex, smooth, with thick and obtuse edges.

Fl.-Fr.: Aug. – Jan.

Common in hedges all over the area.

CUCUMIS L. W. 52; MP 3; SA 3

Key to Species

1. Berry softly echinate or tuberculate ....**C. prophetarum**
1. Berry glabrous
   2. Ovary hispidulous; berry elongated or cylindric ....**C. sativus**
   2. Ovary hairy or densely white-woolly; berry polymorphic ....**C. melo**

Cucumis melo L., Sp. Pl. 1011. 1753; C.B. Clarke in Hook.f., Fl. Brit. India 2: 620.1879; *Cucumis callosus* (Rottler) Cogn. Pflanzenr. IV, 275 II: 129 1924; Srivastava in Verma *et al.,* Fl. Madhya Pradesh 1: 493. 1993; Samvatsar, Fl. West. Trib. Madhya Pradesh 137. 1996. *Ban kachriya*

Annual, prostrate, monoecious herbs; tendrils simple; leaves broadly ovate-rhomboid, 3-9.5 × 3-9.2 cm, villous or hirsute, 5-angled or 3-7- lobed, lobes sub-oribicular, margin denticulate, cordate at base; flowers yellow, solitary; fruits polymorphous, glabrous or more or less hairy.

Fl.-Fr.: May – Sept.

Alirajpur, Barwani, Burhanpur, Indore, Khandwa.

Cucumis prophetarum L., Cent. Pl. I 32 1755; C.B. Clarke in Hook.f., Fl. Brit. India 2: 619. 1879; Srivastava in Verma *et al.,* Fl. Madhya Pradesh 1: 493. 1993.

Perennials; stems brittle, angular, sulcate; leaves sub-orbicular, 2-4 cm long, 3-5-lobed, scabrid; tendrils filiform; flowers yellow, solitary; fruits ellipsoid-globose, echinate, stripes green and white.

Fl.-Fr.: April – Oct.

Rare, **Khandwa** only, Kalibhit and Mahlu.

Cucumis sativus L., Sp. Pl. 1012. 1753; C.B. Clarke in Hook.f., Fl. Brit. India 2: 620.1879; *Cucumis setosus* Cogn. in Monogr. Phan. 3: 491. 1881; Srivastava in Verma *et al.,* Fl. Madhya Pradesh 1: 494. 1993; Samvatsar, Fl. West. Trib. Madhya Pradesh 137. 1996. *Khira.*

Annual, prostrate, monoecious herbs; leaves broadly ovate, cordate, 7-14.5 × 6.8-15.5 cm, villous-hispid, palmately 3-5 lobed, lobes triangular, margin dentate, acute or acuminate at apex; flowers yellow, solitary or fasciculate; fruits cylindrical or obscurely 3-angled, 10-24.5 × 2.5 - 5 cm, tuberculate when young.

Fl.-Fr.: Feb. – Oct.

Alirajpur, Barwani, Dhar, Indore, Khargone.

DIPLOCYCLOS (Endl.) T. Post & Kuntze W. 5; MP 1; SA 1

Diplocyclos palmatus (L.) Jeffrey in Kew Bull. 15: 352. 1962; *Bryonia palmata* L. Sp. Pl. 1012. 1753; Srivastava in Verma *et al.,* Fl. Madhya Pradesh 1: 496. 1993; Samvatsar, Fl. West. Trib. Madhya Pradesh 137. 1996. *Shivlingi*

Slender, delicate climbers; leaves 8-15 cm across, deeply 3-7 lobed, cordate, denticulate; flowers yellow; fruits globose, striped, 1.5 cm across; seeds with a crenulate, transverse ridge.

Fl.-Fr.: Sept. -Nov.

Common during rainy season on hedges of fields, throughout the area.

LUFFA Miller W. 5; MP 3; SA 3

Key to Species

1. Berry up to 1.8 cm long; flowers yellow .... **L. tuberosa**
1. Berry more than 2 cm long
    2. Flowers white; berry 2.5-3 cm long, densely covered with bristles .... **L. echinata**
    2. Flowers yellow; berry clavate, acutely 10-angled .... **L. acutangula**

Luffa acutangula (L.) Roxb., Fl. Ind. 3: 713. 1832; C.B. Clarke in Hook.f., Fl. Brit. India 2: 615. 1879; Srivastava in Verma *et al.,* Fl. Madhya Pradesh 1: 499. 1993. *Cucumis acutangula* L., Sp. Pl. 1011. 1753. *Kadwi torai*

Annual, monoecious climbers; tendrils often 3-fid; leaves orbicular, 12-16.5 × 11.5 - 15.5 cm, scabrid, palmately 5-7-angled or sublobate, margin serrate-dentate, cordate at base; flowers yellow, solitary; fruits clavate-oblong, 12.5-25 × 5-8.5 cm, with 10 sharp angles; seeds ovate, compressed.

Fl.-Fr.: May – Sept.

Very common in hedges.

Luffa echinata Roxb., Fl. Ind. 3: 716. 1832; C.B. Clarke in Hook.f., Fl. Brit. India 2: 615. 1879; Srivastava in Verma *et al.,* Fl. Madhya Pradesh 1: 500.1993.

Annual, dioecious climbers; tendrils 2-fid; leaves suborbicular-reniform, 3-7.5 × 4-9 cm scabrous, margin denticulate, 5-angled or deeply 5-lobed, lobes obtuse or subacute at apex; flowers white, solitary an 1-2.5 cm long peduncle; fruits ovoid, 5-8 × 4 - 6 mm, densely covered with bristles; seeds many, ovate verrucose.

Fl.-Fr.: Sept. – Dec.

Rare, **Indore** only, on open ground in Manpur, climbing when support available.

Luffa tuberosa Roxb., Fl. Ind. ed. 1832 3: 717 1832; Khanna *et al.,* Suppl. Fl. Madhya Pradesh 1: 99.2001.

Perennial, monoecious climbers; tendrils simple, filiform; leaves orbicular-reniform, 5-7-lobed, entire; male flowers in 2-5-flowered racemes; female flowers with 2-3 stigmas; fruits angular, pyriform, with 8 sharp ridges; seeds blackish grey, broadly ovoid.

Fl.-Fr.: July – Oct.

Occasional in hedges, Burhanpur, Indore, Khandwa.

MOMORDICA L. W. 36; MP 3; SA 3

Key to Species

1. Plants dioecious; berry softly, densely prickly .... **M. dioica**
1. Plants monoecious; berry with length-wise, irregular ridges
   2. Bract of male flowers nearly at apex of the pedicel .... **M. balsamina**
   2. Bract of male flowers about the middle of the pedicel .... **M. charantia**

Momordica balsamina L., Sp. Pl. 1009. 1753; C.B. Clarke in Hook.f., Fl. Brit. India 2: 617. 1879; Madhusudan Rao, in N.P. Singh *et al.,* Fl. Maharashtra 2:63.2001.

Annual, monoecious climbers; tendrils simple, filiform; leaves orbicular, 3-8 × 3-7.5 cm, cordate at base, 3-5-lobed, lobes rhomboid, lobulate, acute at apex; male flowers solitary, peduncles bracteate at apex; bract nearly at the base of pedicel; female flowers solitary; fruits ovoid, 2.5-6.5 × 1.5-2.5 cm, rostrate, smooth or tuberculate; seeds ovate, slightly verrucose, margin tuberculate.

Fl.-Fr.: July – Nov.

Rare, **Burhanpur** only, in vicinity of Tapti river bridge.

Momordica charantia L., Sp. Pl. 1009. 1753; C.B. Clarke in Hook.f., Fl. Brit. India 2: 616. 1879; Srivastava in Verma *et al.,* Fl. Madhya Pradesh 1: 502. 1993; Samvatsar, Fl. West. Trib. Madhya Pradesh 138. 1996. *Karela*

Annual, monoecious climbers; tendrils simple, pubescent; leaves suborbicular, 4-10.2 × 4-9. 8 cm, cordate at base, 5-7-lobed, lobes ovate-oblong, margin coarsely spinuous dentate, acute or subacute at apex; male flowers yellow, solitary, on 5-10 cm long peduncles with an orbicular or reniform bract at or below the middle; fruits fusiform, 4-17 cm long, ribbed, muricate-tuberculate; seeds compressed, sculptured on both surfaces, sub-tridentate at apex and base.

Fl.-Fr.: In most parts of the year

Very common, the wild form has small fruits, up to 2.5 cm long.

Momordica dioica Roxb. ex Willd., Sp. Pl. 4: 605. 1805; C.B. Clarke in Hook.f., Fl. Brit. India 2: 617. 1879; Srivastava in Verma *et al.,* Fl. Madhya Pradesh 1: 503. 1993; Samvatsar, Fl. West. Trib. Madhya Pradesh 138. 1996. *Kakoda*

Perennial, dioecious climbers; tendrils simple, filiform; leaves broadly ovate-cordate, 3-5 lobed; male peduncles 1-flowered with orbicular or boat shaped, spathaceous bract at the top; bracts on the female peduncles in the middle; fruits ellipsoid, shortly beaked, densely echinate with soft spines; seeds many.

Fl.-Fr.: July – Oct.

Common throughout the area.

MUKIA Arnott Monotypic

Mukia maderaspatana (L.) M. Roem. in Fam. Nat. Syn. Monogr. 2: 47. 1846; *Cucumis maderaspatanus* L., Sp. Pl. 1012. 1753; *Bryonia scabrella* L.f., Suppl. Pl. 424. 1781; *Mukia scabrella* (L.f.) Arn. in Hook., J. Bot. 3: 276. 1841; C.B. Clarke in Hook.f., Fl. Brit. India 2: 623. 1879; *Melothria maderaspatana* (L.) Cogn. in A. DC. & C. DC., Monogr. Phan. 3: 623. 1881; Srivastava in Verma *et al.,* Fl. Madhya Pradesh 1: 501. 1993; Samvatsar, Fl. West. Trib. Madhya Pradesh 140.1996. *Charak chhebdi*

Annual climbers with wiry branches clothed with white hairs; leaves orbicular, 5-angled or 3-5 lobed, very scabrid on surfaces, hispid beneath; fruit globose, the size of pea, red when ripe; seeds 1-4 ellipsoid, turgid, white, scabrous.

Fl.-Fr.: July-Feb.

Common, in hedges everywhere.

SOLENA Lour. W. 2; MP 1; SA 1

Solena amplexicaulis (Lam.) Gandhi, Fl. Hassan Dist. 179. 1976; *Bryonia amplexicaulis* Lam., Encycl. 1: 496 178; *Zehneria umbellata* (Klein ex Willd.) Thwaites, Enum. Pl. Zeyl. 125 1859; C.B. Clarke in Hook.f., Fl. Brit. India 2: 625. 1879; *Melothria heterophylla* (Lour.)

Cogn., Monogr. Phan. 3: 618. 1881; Srivastava in Verma *et al.,* Fl. Madhya Pradesh 1: 501. 1993; Samvatsar, Fl. West. Trib. Madhya Pradesh 140.1996.

Annual, monoecious climbers; tendrils simple, striate; leaves ovate-deltoid, 3-7.5 × 5 - 10 cm, scabrid, margin dentate, entire, 5-angled or 3-5 - lobed, acute at apex, cordate at base; male flowers in fascicles on short peduncles; female flowers solitary or subfasciculate; fruits globose, 4-9 mm across; seeds ovoid-oblong or ellipoid, scrobiculate, indistinctly marginate.

Fl.-Fr.: Aug. – Jan.

Common in hedges, everywhere.

TRICHOSANTHES L. W. 37; MP 3; SA 2

Key to Species

1. Stem glabrous; leaves acuminate; bracts of male flowers acutely laciniate ... .**T. tricuspidata**
1. Stem glandular; leaves obtuse; bracts absent.. …………**T. cucumerina**

Trichosanthes cucumerina L., Sp. Pl. 1008. 1753; C.B. Clarke in Hook.f., Fl. Brit. India 2: 609. 1879; Srivastava in Verma *et al.,* Fl. Madhya Pradesh 1: 504. 1993; Samvatsar, Fl. West. Trib. Madhya Pradesh 141. 1996. *Jangli chichinda* **(Plate 5, Fig. 8)**

Annual climbing herbs; leaves 5-10 × 50-10 cm ovate to suborbicular, 5-lobed, pubescent beneath; male flowers in 8-15 flowered racemes; female solitary; fruits ovoid, fusiform, beaked, yellow when ripe; seeds elliptic, compressed.

Fl.-Fr.: July – Nov.

Common on hedges and in forest, throughout the area.

Trichosanthes tricuspidata Lour., Fl. Cochinch 2: 589. 1790; *Modecca bracteata* Lam. Encycl. 4: 210.1797; *T. palmata* Roxb. Fl. Ind. 3: 704. 1832; *Trichosanthes bracteata* (Lam.)Voigt, Hort. Sub. Calc. 58. 1845; Clarke in Hook.f., Fl. Brit. India 2: 606. 1879; Srivastava in Verma *et al.,* Fl. Madhya Pradesh 1: 504. 1993; Samvatsar, Fl. West. Trib. Madhya Pradesh 140.1996. **(Plate 6, Fig. 1)**

Annual climbers; leaves 5-25 cm in diameter, orbiccular, 3-7 lobed glandular beneath, scabrid with white tubercles above; male racemes 4-6 flowered; bracts obovate; female flowers solitary, ebracteate; fruits globose, red when ripe, with 10 orange streaks; seeds numerous, ellipsoid smooth.

Fl.-Fr.: Apr. -July

Rare, in hedges of fields, Alirajpur, Dhar, Indore, Khargone.

CACTACEAE 176/2233; 3/6; 2/2

Key to Genera

1. Phylloclades densely spiny .... OPUNTIA
1. Phylloclades not spiny, soft .... …………NOPALEA

NOPALAEA Salm-Dyck W. 7; MP 1; SA 1

Nopalea cochenillifera (L.) Salm-Dyck, Cact. Hort. Dyck ed. 2:64. 1850; *Cactus cochenillifera* L., Sp. Pl. 468. 1753; *Opuntia cochenillifera* (L.)Mill. Gard. Dict. ed. 8, 6:1768; Bailey, Man. Cult. Pl. 703. 1949; Naik, Fl. Marathwada 1:413. 1998.

Erect shrubs, 1–2 m high, with a definite trunk, joints narrow, 10 to 15 cm. long, only moderately thick, often drooping, bright green even in age, bearing usually two somewhat spreading spines at an areole; spines at first pale yellow or pinkish, in age gray, the longest 4 cm long; flowers dark red, 4–5 cm long; sepals obtuse; petals erect, dark red; stamens and style long-exerted, dark red.

Fl.-Fr.: Oct. –March

Common as live hedge in all districts, as well as escape.

OPUNTIA Miller W. 194; MP 2; SA 1

Opuntia elatior Mill., Gard. Dict. ed. 8. n. 4. 1768; *O. dillenii auct. non* (Ker Gawl.) Haw., 1819: J. Graham, Cat. Pl. Bombay 546. 1839; C.B. Clarke in Hook.f., Fl. Brit India 2: 657. 1879; Mukherjee in Verma *et al.,* Fl. Madhya Pradesh 1: 507. 1993; Samvatsar, Fl. West. Trib. Madhya Pradesh 141. 1996. *Nag phani*

Succulent shrubs, 1-2 m high; stem-joints 10-30 cm long, broadly obovate, undulate; areoles bearing 3-8, 2-5 cm long, stout sharp, yellowish-brown spines; glochidia yellowish, up to 1 cm long; flowers yellow, tinged with orange, 4-6 cm in diameter; filaments pink or red; anthers apiculate; berries pear-shaped, purple, marked with areoles, but bristles and spines deciduous.

Fl.-Fr.: June – Jan.

Common throughout the area, escape or hedge plant.

AIZOACEAE 146/2271; 2/4; 2/3

Key to Genera

1. Styles 2, free; capsule dehiscing longitudinally by valves .... ZALEYA
1. Style usually 1; capsule dehiscing circumscissusly .... TRIANTHEMA

TRIANTHEMA L. W. 19; MP 2; SA 1

Trianthema portulacastrum L., Sp. Pl. 223. 1753; *T. monogyna* L., Mant. Pl. 69. 1767; C.B. Clarke in Hook.f., Fl. Brit. India 2: 660.1879; Tripathi in Verma *et al.,* Fl. Madhya Pradesh 1: 508. 1993; Samvatsar, Fl. West. Trib. Madhya Pradesh 142. 1996. *Safed punarnava*

Succulent, prostrate herbs; leaves obovate or suborbicular, 1-3 cm in diameter, pairs of leaves unequal-sized; flowers solitary, sessile, almost concealed by the pouch of the petiole; perianth-tube adnate to the petiole; stamens 10 or more; style one; capsule membranous, circumscissile; lid truncate, with 2 spreading teeth; seeds reniform, black, muriculate.

Fl.-Fr.: July – Oct.

Common in open, moist ground, all over the area.

ZALEYA Burm.f. W. 4; MP 2; SA2

Key to Species

1. Stamens 5; capsule 2-seeded .... **Z. pentandra**
1. Stamens 10; capsue 4-seeded .... **Z. decandra**

Zaleya decandra (L.) Burm.f., Fl. Ind. 110.1768; *Trianthema decandra* L., Mant. Pl. 70.1767; C.B. Clarke in Hook.f., Fl. Brit. India 2: 661. 1879; Khanna *et al.,* Suppl. Fl. Madhya Pradesh 1: 103.2001.

Prostrate herbs with few, striate branches; leaves opposite, elliptic-ovate or elliptic-oblong, 1.5-3.5 × 0.6-1.3 cm, rounded and apiculate at apex, slightly succulent; flowers in axillary, sub-umbellate clusters; petals absent; capsules 2-seeded, larger than calyx-tube; seeds black, orbicular-reniform, striate.

Fl.-Fr.: Oct. – April

Rare, **Burhanpur** only, in wet soil, Shahi kila and Jainabad.

Zaleya pentandra (L.) C. Jeffrey, Kew Bull. 14: 238 1960*; Trianthema pentandra* L., Mant. Pl. 70 1767*;* C.B. Clarke, in Hook.f., Fl. Brit. India 2: 660.1879; Samvatsar, Fl. West. Trib. Madhya Pradesh 142. 1996; Khanna *et al.,* Suppl. Fl. Madhya Pradesh 1: 103.2001.

Diffuse, up to 30 cm high herbs; branches angled and grooved; leaves elliptic-oblong or obovate, 1-3 × 0.5-1.3 cm, obtuse at apex, narrowed at base; petioles up to 1.5 cm long, swollen at base; flowers sessile, in axillary fascicles; capsules 4-seeded, 4-6 mm long, lid with deflexed horns; seeds black, orbicular-reniform, 1.5-2 mm across, compressed, rugulose.

Fl.-Fr.: Aug. – March

Rare, **Dhar** only, Dharam puri and Katnera, in same habitat as above species.

MOLLUGINACEAE 11/103; 2/6; 2/4

Key to Genera

1. Seeds strophiolate, strophiole appendaged; flowers in axillary fascicles; leaves stellately hairy or glabrous .... GLINUS
1. Strophiole absent; flowers in terminal cymes; leaves glabrous ....MOLLUGO

GLINUS L. W. 10; MP 2; SA 2

Key to Species

1. Styles 5-6; pedicels short; plant stellately hairy .... **G. lotoides**
1. Styles 3; pedicels long; plant not stellately hairy .... **G. oppositifolius**

Glinus lotoides L., Sp. Pl. 463. 1753; *Mollugo hirta* Thunb., Prodr. Fl. Cap. 1: 24. 1794; C.B. Clarke in Hook.f., Fl. Brit. India 2: 662. 1879; Tripathi in Verma *et al.,* Fl. Madhya Pradesh 1: 509. 1993; Samvatsar, Fl. West. Trib. Madhya Pradesh 141. 1996.

Prostrate or decumbent ascending, much branched, hirsute, annual herbs; stems stellately pubescent; leaves 0.5-2 × 0.3-0.5 cm, opposite or ternate, unequal, obovate-rounded, obtuse, glabrous or stellately hairy; flowers 3-8 together; tepals oblong-lanceolate, acutely mucronate, stellate-pubescent on the outside; capsule 5-valved.

Fl.-Fr.: March – Oct.

Common in drying ponds and ditches, all over the area.

Glinus oppositifolius (L.) A. DC. in Bull. Herb. Boissier ser. 2. 1: 552. 1901; *Mollugo oppositifolia* L., Sp. Pl. 89. 1753; *M. spergula* L., Syst. Nat. ed. 10.881. 1759; C.B. Clarke in Hook.f., Fl. Brit. India 2: 662. 1879; Tripathi in Verma *et al.,* Fl. Madhya Pradesh 1: 509. 1993; Samvatsar, Fl. West. Trib. Madhya Pradesh 142. 1996.

Erect or ascending annual herbs, usually profusely branched from the base, sometimes prostrate; leaves 1-2 × 0.1-0.5 cm oblong-spathulate; flowers in axillary fascicles of 2-6, tepals obtuse; capsules 3-4 valved; seeds numerous, sub-reniform.

Fl.-Fr.: July – Dec.

Common in the same habitat and area, as above.

MOLLUGO L. W. 11; MP 4; SA 2

Key to Species

1. Leaves radical, spathulate; flowers in trichotomously branched cymes .... **M. nudicaulis**
1. Leaves cauline, in whorls at nodes; flowers in dichotomously branched cymes ....**M. pentaphylla**

Mollugo nudicaulis Lam., Encycl. 4: 234. 1797; C.B. Clarke in Hook.f., Fl. Brit. India 2: 664. 1879; Tripathi in Verma *et al.,* Fl. Madhya Pradesh 1: 510.1993; Samvatsar, Fl. West. Trib. Madhya Pradesh 142. 1996.

Slender annual herbs; leaves in a radical cluster, 0.5-5 × 0.3-1 cm, oblong-spathulate; flowers white in profusely branched cymes, tepals scarious margined; capsule ellipsoid; seeds shining reddish-black, tubercled.

Fl.-Fr.: Sept. – Dec.

Common as a weed in fields all over the area.

Mollugo pentaphylla L., Sp. Pl. 89. 1753; *M. stricta* L., Sp. Pl. ed. 2. 131. 1762; C.B. Clarke in Hook.f., Fl. Brit. India 2: 663. 1879; Tripathi in Verma *et al.,* Fl. Madhya Pradesh 1: 511. 1993; Samvatsar, Fl. West. Trib. Madhya Pradesh 142. 1996.

Slender annual herbs, usually much branched; leaves in pseudo-whorls of 3-5, 1-2.5 × 0.5-0.8 cm lower ones rosulate, spathulate, fugacious, upper linear-lanceolate; flowers white, in terminal and leaf opposed cymes; capsule ellipsoid; seeds brownish-black, tubercled.

Fl.-Fr.: July – March

Weed of fields and forest clearings, common in all districts.

APIACEAE 418/3257; 10/14; 6/6

Key to Genera

1. Umbels simple; leaves simple; vittae absent; prostrate herbs
   2. Bracts absent; leafbase not sheathing ....HYDROCOTYLE
   2. Bracts present; leafbase sheathing.. ....CENTELLA
1. Umbels compound; leaves simple to compound or decompound; vittae present
   3. Fruit strongly dorsally compressed
      4. Vittae extending about 2/3 length of the fruit .... HERACLEUM
      4. Vittae 2, extending to the base of the fruit; fruit muricate ..............................TRACHYSPERMUM
   3. Fruit compressed laterally, slightly or so
      5. Vittae 2 or 3; fruit hispid or pubescent .... PIMPINELLA
      5. Vittae 1; rays numerous; pedicels almost equal .... SESELI

CENTELLA L. W. 53; MP 1; SA 1

Centella asiatica (L.) Urban in Mart., Fl. Bras. 11: 287. t. 78. f. 1. 1879; *Hydrocotyle asiatica* L., Sp. Pl. 234. 1753; C.B. Clarke in Hook.f., Fl. Brit. India 2: 669. 1879; Khanna in Verma *et al.,* Fl. Madhya Pradesh 1: 514. 1993; Samvatsar, Fl. West. Trib. Madhya Pradesh 143. 1996. *Manduk parni*

A creeping , prostrate, herb; rooting at nodes; leaves 2-4 cm in diameter, orbicular-reniform, glabrous on both sides; flowers in fascicled umbels; each umbel consisting of 3-4 pink, sessile flowers; bracts ovate, acute; fruits ovoid, hard, rugose; primary and secondary ridges distinct.

Fl.-Fr.: Oct. – April

Common along streams and ditches, in all districts but not abundant.

HERACLEUM L. W. 52; MP 1; SA 1

Heracleum nepalense D. Don, Prodr. Fl. Nepal. 185 1825*;* C.B. Clarke in Hook.f., Fl. Brit. India 2: 714. 1879; Khanna in Verma *et al.,* Fl. Madhya Pradesh 1: 517. 1993.

Annual, up to 75 cm high, sparingly pubescent herbs; upper leaves acutely 3-lobed; lower leaves pinnate; segments deeply lobed, serrate; bracts 3-5, linear; petals white; fruits obovate, dorsal and intermediate ridges hardly excurrent; wing broad.

Fl.-Fr.: July–Oct.

Rare, **Khandwa** only, at Mortakka.

HYDROCOTYLE L. W. 85; MP 1; SA 1

Hydrocotyle sibthorpioides Lam., Encycl. 3: 153. 1789; *H. rotundifolia* Roxb., Fl. Ind. 2: 88. 1824; C.B. Clarke in Hook.f., Fl. Brit. India 2: 668. 1879; Khanna in Verma *et al.,* Fl. Madhya Pradesh 1: 518. 1993.

Perennial, prostrate herbs; leaves orbicular, 0.6-2.5 × 0.5-2.5 cm, 5-7-lobed; lobes cordate at base, crenate, hirsute; flowers arranged in simple, 10 or few flowered umbels; petals ovate, acute; fruits suborbicular, 1-1.5 mm across; mericarps with prominent lateral ribs and slender dorsal ribs, vittae obscure, commissural face slightly concave; carpophore absent.

Fl.-Fr.: March – Sept.

Burhanpur, Pama khedi (Sitala Mata temple).

PIMPINELLA L. W. 72; MP 4; SA 1

Pimpinella heyneana (DC.) Kurz. in J. Asiat. Soc. Beng. 46. 115. 1877; *Helosciadium heyneanum* DC., Prodr. 4: 106. 1830; Clarke in Hook.f., Fl. Brit. India 2: 684. 1879. Khanna in Verma *et al.,* Fl. Madhya Pradesh 1:520.1993; Samvatsar, Fl. West. Trib. Madhya Pradesh 143. 1996.

Annual erect herbs; leaves 3-foliolate; leaflets to 4 × 1.5 cm; primary umbels 6-20 rayed, rays unequal, very slender, glabrous; bracts absent; partial umbels 4-12 flowered; pedicels filiform, very unequal; bracteoles 1-2, linear or absent; fruits subglobose, glabrous; primary ridges slender.

Fl.-Fr.: Aug. -Oct.

Occassional on hill slopes and wet places, in all districts.

SESELI L. W. 77; MP 1; SA 1

Seseli diffusum (Roxb. ex Sm.) Santapau & Wagh in Bull. Bot. Surv. India 5: 108. 1963; *Ligusticum diffusum* Roxb. ex Sm. in Rees, Cycl. 21: n. 11. 1812; *Seseli indicum* Wight & Arn., Prodr. Fl. Ind. Orient. 371. 1834; C.B. Clarke in Hook.f., Fl. Brit. India 2: 693. 1879. Khanna in Verma *et al.,* Fl. Madhya Pradesh 1:521. 1993.

Annual, diffusely branched, more or less villous, herbs, up to 25 cm high; leaves oblong, 3.5-7.5 × 1.2-2.2 cm, bipinnate; leaflets ovate-cuneate, 3-6 × 2-4.5 mm, pinnately lobed to pinnatifid; lobes obtuse, apiculate; flowers arranged in compound umbels; petals white, broadly ovate; fruits broadly ovoid, 1-2.5 × 2-3 mm, slightly laterally compressed; mericarps glabrous or stellate-hispid; ribs prominent, acute.

Fl.-Fr.: March – June

Indore only, normally found below road bridges, where soil is moist, Kshipra and Chambal rivers.

TRACHYSPERMUM Link W. 5; MP 1; SA 1

Trachyspermum roxburghianum (DC.) H. Wolff Pflanzenr. IV 228(Heft 90): 129. 1927; *Ptychotis roxburghiana* DC. Prodr. 4: 109. 1830; *Carum roxburghianum* Kurz in J. Asiat. Soc. Bengal, Pt. 2, Nat. Hist. 46(2): 114. 1877; C.B. Clarke in Hook.f., Fl. Brit. India 2: 682. 1879; Khanna *et al.,* Suppl. Fl. Madhya Pradesh 1: 105. 2001. *Jangli ajwain*

Erect, slender herbs; leaves triangular-ovate, 3-8 cm long, bipinnate or ternately pinnate; segments narrowly-oblong; peduncles terminal and lateral, 5-8 cm long; flowers white or pink; fruits ovoid, laterally compressed.

Fl.-Fr.: Jan. – May

Common in hilly areas of all districts.

CORNACEAE 10/115; 1/1; 1/1

ALANGIUM Lam. W. 27; MP 1; SA 1

Alangium salvifolium (L.f.) Wangerin in Engl. Pfreich. 41: 9. f. 2A-E. 1910; *Grewia salvifolia* L.f., Suppl. 409. 1781; *Alangium lamarckii* Thw. Enum. Pl. Zey. 133. 1859; Clarke in Hook.f., Fl. Brit. India 2: 741. 1879; Srivastava in Verma *et al.,* Fl. Madhya Pradesh 1: 524. 1993; Samvatsar in Fl. West. Trib. Madhya Pradesh 144. 1996. *Ankol*

Woody shrubs or small trees; branches with sharp spines; leaves 10-14 × 3-6 cm elliptic to oblanceolate, acuminate, 3-5 veined from base; flowers white, pubescent outside, in axillary clusters; fruits ovoid, 2-2.5 cm long, rusty tomentose; crowned with calyx, endocarp bony.

Fl.-Fr.: Feb. -May

Common throughout the area.

RUBIACEAE 609/13673; 20/42; 15/27

Key to Genera

1. Leaves whorled; stipules absent; fruit fleshy, globose .... RUBIA
1. Leaves opposite; stipules present
    2. Herbs; stipules pectinate or bristly
        3. Ovule solitary per cell .... SPERMACOCE
        3. Ovules more than one per cell
            4. Petals toothed; calyx lobes 5 .... DENTELLA
            4. Petals not toothed; calyx lobes 4
                5. Seeds angular, numerous, testa pitted
                    6. Calyx teeth erect, contiguous; corolla tube short .... OLDENLANDIA
                    6. Calyx teeth spreading, not contiguous; corolla tube long, slender .... KOHAUTIA
                5. Seeds plano-convex, few, with a central cavity, testa smooth .... NEANOTIS
    2. Shrubs or trees
        7. Stipules intrapetiolar; placentas 5; ovary 1-celled; seeds smooth; fruit berry
            8. Leaves dimorphic; plants armed .... CERISCOIDES
            8. Leaves monomorphic; plants not armed .... GARDENIA
        7. Stipules interpetiolar; ovary 2-celled; plants armed; fruit baccate
            9. Leaves up to 20 cm long; spines at the ends of arrested branchlets; flowers dimorphic; fruit up to 7 cm long .... TAMILNADIA

9. Leaves up to 4 cm long; spines axillary; flowers monomorphic; fruit up to 2 cm long .... CATUNAREGAM
   10. Ovule 1 per cell
      11. Style branches 3; ovary 5-celled; fruit not syncarpous; flowers white or blue .... SPERMADICTYON
      11. Style branches 2; ovary 2-celled; fruit syncarpous or not
         12. Fruit syncarpous, globose; corolla lobes valvate .... MORINDA
         12. Fruit not syncarpous, didymous; inflorescence always terminal; corolla lobes twisted .... IXORA
   10. Ovules more than 1 per cell; fruit capsular; seeds winged; corolla valvate
      13. Flowers in panicles; capsule up to 2 cm long; flowers greenish-white; capsules drooping .... HYMENODICTYON
      13. Flowers in globose heads; capsule much smaller; flowers yellow- orange
         14. Stigma clavate; leaves 10-25 cm long, cordate; petiole 5-10 cm long; calyx divided half way down .... HALDINA
         14. Stigma mitriform (turban-shaped); leaves 5-12 cm long, not cordate; petiole 1-2 cm long; calyx truncate . ... MITRAGYNA

CATUNAREGAM Adans. ex V. Wolff W. 12; MP 1; SA 1

Catunaregum spinosa (Thunb.) Tirveng. in Taxon 27: 515. 1978 & Bull. Mus. Hist. Nat. (Paris) ser. 3, 35: 13. 1978; *Gardenia spinosa* Thunb., Gardenia 7:16. 1780; *Randia dumetorum* (Retz.) Lam. Tabl. Encycl. 2:227. 1819; Hook.f., Fl. Brit. India 3:110.1880; *Xeromphis spinosa* (Thunb.) Keay, Bull. Jard. Bot. Etat. 28:37. 1958; Khanna in Verma *et al.,* Fl. Madhya Pradesh 1: 530.1993; Samvatsar, Fl. West. Trib. Madhya Pradesh 146. 1996. *Mendal*

Much branched, spinous shrubs; thorns axillary, stout; leaves 5-5.5 × 2-2.8 cm, elliptic to obovate, hairy on the veins beneath; flowers white, turning yellowish, solitary or fascicled in the axils; corolla lobes obtuse, reflexed, hairy outside; seeds embedded in pulp.

Fl.-Fr.: Apr. -May

Occasional in forest, in all districts. The fruits are an absolute essential in Adibasi marriages.

CERISCOIDES (Hook.f.) Tirveng. W. 11; MP 1; SA1

Ceriscoides turgida (Roxb.) Tirveng. in Bull. Mus. Natl. Hist. Nat., Ser. 3, Bot. 35: 15. 1978; *Gardenia turgida* Roxb., Fl. Ind. 2: 557. 1824; Hook.f., Fl. Brit. India 3: 118. 1880; Khanna in Verma *et al.,* Fl. Madhya Pradesh 1: 534. 1993; Samvatsar, Fl. West. Trib. Madhya Pradesh 147. 1996. *Phendra* **(Plate 6, Fig. 2)**

Small, deciduous trees; branches armed with straight, opposite, thorns; leaves 2.5-10 × 3-5 cm, elliptic to obovate, entire, glabrous; flowers white, unisexual, males fascicled, females solitary; berry globose, 3-8 cm in diameter; seeds angular.

Fl.-Fr.: March – Dec.

Commonly found in all districts, but the number of individuals is just 2-3, everywhere.

DENTELLA J. R. & G. Forst. W. 8; MP 2; SA2

Dentella repens (L.) J. R. Forst. & G. Forst., Char. Gen. Pl. 26. t. 13. 1776; *Oldenlandia repens* L., Mant. Pl. 40.1767; Hook.f., Fl. Brit. India 3: 42. 1880; Khanna in Verma *et al.,* Fl. Madhya Pradesh 1: 532. 1993; Samvatsar, Fl. West. Trib. Madhya Pradesh 147. 1996.

Key to Varieties

1. Ovaries and fruits densely pilose .... **D. repens** var. **repens**
1. Ovaries and fruits glabrous .... **D. repens** var. **serpyllifolia**

D. repens var. **repens**

Annual, prostrate, dichotomously branched, herbs, rooting at nodes; leaves opposite, elliptic-oblong or subspathulate, 3-9 × 1-3 mm, acute at apex, cuneate at base, entire, glabrous, sessile or subsessile; stipules minute, connate; flowers solitary, axillary or in the fork of branches; calyx campanulate, persistent; limb tubular; lobes equal, linear, *ca* 2 mm long; corolla white, infundibuliform; lobes ovate, 1.5-2 mm long; stamens 5, inserted at the middle of corolla-tube; ovary 2-locular, with many ovules; capsules globose or obliquely ellipsoid, dry, glabrous or densely pilose; seeds angular, rugose, finely echinulate.

Fl.-Fr.: Jan. – July

In moist places of wastelands, along ponds and fields; usually in sandy alluvial soil.

D. repens var. **serpyllifolia** (Wall. *ex* Craib) Verdc. in Kew Bull. 37: 545. 1983; *D. serpyllifolia* Wall. *ex* Craib, Fl. Siam. 2: 27. 1932; *D. repens auct. non* (L.) J. R. Forst. & G. Forst., 1776: Hook.f., Fl. Brit. India 3: 42. 1880; Khanna in Verma *et al.,* Fl. Madhya Pradesh 1: 532. 1993.

Annual, small, prostrate, dichotomously branched herbs; leaves 4-9 × 1.5-3 mm, sessile or subsessile, oblong-elliptic or obovate-lanceolate; stipules connate, small; flowers solitary, axillary and in forks of branches; calyx persistent, glabrous; corolla white to purple, infundibuliform; capsules dry, indehiscent; seeds many, minute, reticulate.

Fl.-Fr.: March – Oct.

In moist places, **Khandwa**, in moist places along railway tracks.

GARDENIA Ellis W. 134; MP 3; SA 2

Key to Species

1. Placentas 2; leaves petiolate, main nerves 20-25 pairs; calyx teeth erect .... **G. resinifera**
1. Placentas 4-5; leaves sessile, main nerves 10-20 pairs; calyx teeth curved .... **G. latifolia**

Gardenia latifolia Aiton, Hort. Kew. 1: 294. 1789; Hook.f., Fl. Brit. India 3: 116. 1880; Khanna in Verma *et al.,* Fl. Madhya Pradesh 1: 533. 1993; Samvatsar, Fl. West. Trib. Madhya Pradesh 147. 1996. *Safed papra* **(Plate 6, Fig. 3)**

Deciduous, 5-8 m high, trees, young parts with yellow resinous exudation; branches divergent; leaves opposite or ternate, broadly elliptic or obovate-orbicular, 15-25 × 8-18 cm, acute or shortly acuminate at apex, narrowed at base, greenish above, paler beneath, with 16-22 pairs of nerves; petioles 1-2 cm long; flowers solitary, fragrant, subsessile; calyx-lobes lanceolate, subulate, recurved; corolla white, turning yellow; lobes 5-9, oblong-obovate; tube 3-5 cm long, hirsute; stamens 9, epipetalous; ovary 1-locular, with many ovules; stigma 2-lamellate; fruits greenish- yellow when ripe, subglobose, 3-5 cm long; seeds pale brown.

Fl.-Fr.: March – June

Barwani, Dhar, Indore, Khandwa, Khargone.

Gardenia resinifera Roth, Nov. Pl. Sp. 150.1821; *G. lucida* Roxb., Fl. Ind. 2: 553. 1824; Hook.f., Fl. Brit. India 3: 115. 1880; Khanna in Verma *et al.*, Fl. Madhya Pradesh 1: 533. 1993; Samvatsar, Fl. West. Trib. Madhya Pradesh 147. 1996. *Dikendi*

Shrubs, 2-2.5 m high, exuding gum from apical buds; bark greenish-grey; leaves elliptic or elliptic-ovate, 6-15 × 3-7 cm, obtuse at apex, rounded at base; lateral nerves 20-30 pairs; stipules ovate, connate; flowers solitary, axillary, fragrant; calyx tubular; lobes 5, equal or unequal, lanceolate, 7-10 mm long, puberulous outside; tube *ca* 0.6 cm long; corolla white; lobes 5, oblong; tube 4-5 cm long; stamens 5, epipetalous; fruits brownish black, ellipsoid, *ca* 2 cm long, not ribbed; seeds reddish.

Fl.-Fr.: March – Nov.

Occasional, Dhar, Indore, Khandwa.

HALDINA Ridsdale Monotypic

Haldina cordifolia (Roxb.) Ridsd. in Blumea 24: 361. 1978; *Nauclea cordifolia* Roxb., Pl. Coromandel 1: 40, t. 53. 1796; *Adina cordifolia* (Roxb.) Hook.f. *ex* Brandis, Forest Fl. N. W. India 263. t. 33. 1874; Hook.f., Fl. Brit. India 3: 24. 1880; Khanna in Verma *et al.,* Fl. Madhya Pradesh 1: 533. 1993; Samvatsar, Fl. West. Trib. Madhya Pradesh 145. 1996. *Haldu*

Large trees; leaves 2-25 × 1.5-25 cm, broadly ovate to suborbicular, pubescent beneath; stipules suborbicular, depressed, entire, pubescent, keeled, caducous; flowers yellow in globose, axillary heads; seeds many, oblong, brown, winged.

Fl.-Fr.: July – April

Common in all districts.

HYMENODICTYON Wall. W. 23; MP 2; SA 1

Hymenodictyon orixense (Roxb.) Mabb. in Taxon 31: 66. 1982; *Cinchona orixensis* Roxb., Bot. Descr. Swietenia 21. 1793 & in Med. Facts Obs. 6: 152. 1793; *Hymenodictyon excelsum* (Roxb.) Wall. in Roxb., Fl. Ind. 2: 149. 1824; Hook.f., Fl. Brit. India 3: 35. 1880; Khanna in Verma *et al.,* Fl. Madhya Pradesh 1: 537. 1993; Samvatsar, Fl. West. Trib. Madhya Pradesh 149. 1996. *Bhor sal*

Trees; bark smooth; leaves elliptic-oblong 10-22 × 5-12 cm, abruptly acuminate, pubescent on both surfaces; flowers greenish-white, fragrant in cymes, along rhachis of dense cylindrical, pubescent, spiciform racemes; calyx narrowly campanulate; corolla campanulate; capsule ellipsoid, 2-2.5 cm long, pendulous on decurved peduncles, 2-valved, brown when ripe; seeds winged all round.

Fl.-Fr.: June – Feb.

Alirajpur, Barwani, Dhar, Indore, Khargone.

IXORA L. W. 544; MP 2; SA 1

Ixora pavetta Andr., Bot. Repos. t. 78. 1799; *I. parviflora* Vahl, Symb. Bot. 3: 11. t. 52. 1794, *non* Lam., 1791 *I. arborea* Roxb. ex Sm. in Rees, Cycl. 19: 5. 1811; Hook.f., Fl. Brit. India 3:142. 1880; Khanna in Verma *et al.,* Fl. Madhya Pradesh 1: 538. 1993; Samvatsar, Fl. West. Trib. Madhya Pradesh 149. 1996. *Lokhandi* **(Plate 6, Fig. 4)**

Evergreen shrubs or small trees; leaves oblong or elliptic-lanceolate, 7-14 × 4-6 cm, subacute at apex, rounded or cordate at base, entire, glabrous, shining, coriaceous; flowers fragrant, in terminal, sessile, subglobose, clustered cymes; corolla white or pink; lobes linear-oblong; berries globose, didymous, 6-7 mm across.

Fl.-Fr.: March – June

Barwani, Dhar, Indore, Khandwa, Khargone.

KOHAUTIA Cham. ex Schltdl. W. 27; MP 2; SA 2

Key to Species

1. Leaves whorled; stem, branches and capsule scabrid .... **K. aspera**
1. Leaves opposite; stem, branches and capsule smooth .... **K. gracilis**

Kohautia aspera (Heyne *ex* Roth) Bremek. in Verh. Kon. Nederl. Akad. Wet. Afd. Ser. 2, 48(2): 113. 1952; *Hedyotis aspera* Heyne *ex* Roth, Nov. Pl. Sp. 94. 1821; *Oldenlandia aspera* (Heyne *ex* Roth) DC., Prodr. 4: 428. 1830; Hook.f., Fl. Brit. India 3: 68. 1880; Khanna in Verma *et al.,* Fl. Madhya Pradesh 1: 539. 1993; Samvatsar, Fl. West. Trib. Madhya Pradesh 148. 1996.

Annual herbs; stem 15-30 cm tall, stem and branches scabrid; leaves 1-4 × 0.3-0.5 cm linear, acute, margins slightly recurved; flowers bluish, small, in terminal, branched cymes; calyx teeth triangular, subulate; capsule 0.5-0.6 cm across, globose, somewhat didymous, crowned by the distant calyx teeth.

Fl.-Fr.: Dec. – April

Alirajpur, Dhar, Indore, Khandwa.

Kohautia gracilis (Wall.) DC., Prodr. 4: 430.1830; *Hedyotis gracilis* Wall. in Roxb., Fl. Ind. 1: 371. 1820; *Oldenlandia gracilis* (Wall.) Hook.f., Fl. Brit. India 3: 68. 1880; Khanna in Verma *et al.,* Fl. Madhya Pradesh 1: 540.1993; Samvatsar, Fl. West. Trib. Madhya Pradesh 148. 1996.

Perennial herbs, branching below, erect and unbranched above; leaves linear, 5-8 × 0.2-0.5 cm, acuminate at apex, narrowed at base, membranous, glabrous; flowers in terminal, monochasial cymes; corolla pale brown or brownish red; capsules green or dull brown, oblong, 4-6 mm long, didymous; seeds many, purple or black, triangular, narrowly winged.

Fl.-Fr.: May – Sept.

Alirajpur, Dhar, Khandwa.

MITRAGYNA Korth. W. 7; MP 1; SA 1

Mitragyna parvifolia (Roxb.) Korth., Observ. Naucl. Indic. 19. 1839; *Nauclea parvifolia* Roxb., Pl. Coromandel 1: 40. t. 52. 1796; *Stephegyne parvifolia* Korth., Verh. Nat. Gesch. Ned. Bezitt. Bot. 161. 1842; Hook.f., Fl. Brit. India 3: 25. 1880; Khanna in Verma *et al.,* Fl. Madhya Pradesh 1: 541. 1993; Samvatsar, Fl. West. Trib. Madhya Pradesh 150.1996. *Kalam*

Trees, 10-15 m high; leaves elliptic-oblong, panduraeform or ovate to orbicular, 3.6-18 × 2.5-11 cm, stipules caducous, broadly oblong, enclosing the terminal bud; flowers yellow; heads solitary, pedunculate, about 2 cm, in diameter; calyx-limb truncate; stigma mitriform; capsule oblong-ellipsoid, 10-ribbed, separating into 2, many-seeded cocci; seeds winged.

Fl.-Fr.: June – March

Very common, all over the study area.

MORINDA L. W. 112; MP 1; SA 1

Morinda coreia Buch. -Ham., Trans. Linn. Soc. London 13: 537 1822; *Morinda pubescens* Sm. in Rees, Cycl. 24. n. 3: 1813; *M. tinctoria* Roxb., Fl. Ind. 1: 543. 1832; Hook.f., Fl. Brit. India 3: 156. 1880; Khanna, in Verma *et al.,* Fl. Madhya Pradesh 1:543. 1993; Samvatsar, Fl. West. Trib. Madhya Pradesh 150.1996. *Aal*

Small trees; leaves 10-16 × 5-8 cm, elliptic, acuminate, tomentose on both surfaces; petiole 0.6-1.2 cm long densely hairy; flowers white in globose heads; peduncles solitary, leaf opposed; calyx truncate; corolla tube hairy outside, lobes oblong, obtuse or subacute, hairy on back; fruit syncarpic, globose or ellipsoid, fleshy.

Fl.-Fr.: April – Feb.

Common along roadsides of villages, all over the area.

NEANOTIS W. H. Lewis W. 30; MP 6; SA 4

Key to Species

1. Annuals
   2. Calyx teeth bristle-tipped; capsules 1-3 seeded; stem hispid; leaves ovate-lanceolate .... **N. montholonii**
   2. Calyx teeth not bristle-tipped; capsules 1-6 seeded; stem glabrous; leaves linear .... **N. subtilis**
1. Perennials
   3. Leaves sessile, 1-6 cm long; flowers solitary or more, on reduced axillary shoots; corolla tube purple .... **N. lancifolia**
   3. Leaves 4-11 cm long; flowers more on 1-4.5 cm long peduncle; corolla uniformly white..... **N. ingrata**

Neanotis ingrata (Wall. ex Hook.f.) W. H. Lewis in Ann. Missouri Bot. Gard. 53: 39. 1966; *Anotis ingrata* Wall. ex Hook.f., Fl. Brit. India 3: 71 1880; Khanna in Verma *et al.,* Fl. Madhya Pradesh 1: 544. 1993.

Perennial, branched herbs; leaves petiolate, 2-4.5 cm × 0.6-1.7 cm, ovate-lanceolate, puberulous above and beneath, acuminate; flowers white, in terminal and subterminal cymes; capsule hemispherical, slightly compressed, glabrous or hairy; cells many seeded.

Fl.-Fr.: July-Sept.

Burhanpur, Indore, Khandwa.

Neanotis lancifolia (Hook.f.) W. H. Lewis in Ann. Missouri Bot. Gard. 53: 39. 1966; *Anotis lancifolia* Hook.f., Fl. Brit. India 3: 73 188; Khanna *et al.,* Suppl. Fl. Madhya Pradesh 1: 116.2001

Perennial, erect herbs; leaves 3-6 × 1.5-2 cm, ovate-lanceolate, acute at apex, more or less hairy; flowers purplish, in terminal and subterminal many-flowered, corymbose cymes; capsules hispid, with recurved calyx-teeth.

Fl.-Fr.: Aug. -Nov.

Rare, a monsoon weed, in **Indore** only, Berchha pond (Mhow).

Neanotis montholonii (Hook.f.) W. H. Lewis in Ann. Missouri Bot. Gard. 53: 39. 1966; *Anotis montholonii* Hook.f. in . Fl. Brit. India 3: 73. 1880; Samvatsar, Fl. West. Trib. Madhya Pradesh 151. 1996; Khanna *et al.,* Suppl. Fl. Madhya Pradesh 1: 116.2001.

Annuals; stem and branches more or less hispid; leaves 3.5-5 × 1.5-2 cm, ovate or ovate-lanceolate, hairy on both surfaces; petiole 0.2-0.6 cm long; flowers in terminal and axillary peduncled, many flowered cymes; peduncles hairy, clustered, didymous with a large double convex crown giving the appearence of two hemispheric bosses, 1-3 seeded; seeds globose, pitted black.

Fl.-Fr.: Aug. -Dec.

Common as a weed in the fields during rains and in the plains.

Neanotis subtilis (Miq.) Govaerts ex Punekar & Lakshmi., Fl. Ainshi Natl. Park 255. 2011; *Hedyotis subtilis* Miq., Anal. Bot. Ind. 3:12. 1852; *Neanotis foetida* (Dalzell) Lewis, Ann. Missouri Bot. Gard. 53: 38 1966; *Anotis foetida* (Dalz.) Hook.f. in Fl. Brit. India 3: 74. 1880; *Neanotis hohenackeri* Daniel & Vajr. in J. Econ. Taxon. Bot. 3:675. 1982; Samvatsar, Fl. West. Trib. Madhya Pradesh 150.1996; Khanna *et al.,* Suppl. Fl. Madhya Pradesh 1:116.2001.

Slender annuals; leaves sessile, 1.5-2.5 cm × 0.1-0.4 cm, linear to linear-lanceolate, scaberulous above, margins recurved; flowers sub-sessile, in capitates cymes; calyx teeth narrowly triangular; corolla funnel shaped, purple, externally papillose; capsule 2-4 mm long, sessile didymous, crown very prominent, 1-6 seeded; seeds elliptic or orbicular, pitted, black.

Fl.-Fr.: July-Oct.

Common as a weed in fields during rains; everywhere in the plains.

OLDENLANDIA L. W. 246; MP 7; SA 7

Key to Species

1. Peduncles axillary and terminal, few to many flowered
   2. Stem densely pilosulous-glandular in the upper nodes; leaves appearing whorled due to reduced internodes .... **O. ovatifolia**

2. Stem not as above
   3. Stem sharply angled; leaves linear; flowers very small, white; peduncles not filiform .. **O. attenuata**
   3. Stem terete or nearly so; leaves linear to lanceolate; flowers larger, bluish or white; peduncles filiform .... **O. affinis**

1. Peduncles axillary
   4. Calyx teeth touching the capsule; peduncles filiform, equalling the leaves; capsules pyriform, ribbed .... **O. pumila**
   4. Calyx teeth not touching the capsule
      5. Anthers and stigmas included; stem glabrous or so; apex of capsule truncate .... **O. corymbosa**
      5. Anthers and stigmas exserted; flowers 1-3, axillary; stem puberulent or scabrid, terete or flattened
         6. Peduncles up to 2 cm long .... **O. diffusa**
         6. Peduncles up to 3 cm long .... **O. herbacea**

Oldenladia affinis (Roem. & Schult.) DC. Prodr. 4: 428. 1830; *Hedyotis affinis* Roem. & Schult. in Syst. 3: 194. 1818; *Oldenlandia dichotoma* (K. D. Koenig ex Roth) Hook.f., Fl. Brit. India 3:68. 1880; Khanna in Verma *et al.,* Fl. Madhya Pradesh 1:546. 1993; Samvatsar, Fl. West. Trib. Madhya Pradesh 148. 1996.

A very slender annual, 20-40 cm tall; stem terete, usually glabrous, branches filiform, branchlets like fine hairs; leaves linear-lanceolate, acute, flat, lower leaves larger than upper, 1.5-3 cm, while uppermost 1-3 cm cm; flowers white, on long capillary pedicels, sometimes arranged in terminal and axillary, slender, lax panicles; calyx teeth subulate, capsule 1-2 mm across, globose.

Fl.-Fr.: Aug. -Oct.

Alirajpur, Dhar, Indore.

Oldenlandia attenuata (Willd.) M.R. Almeida, Fl. Maharashtra 3A: 43.2001; *Hedyotis attenuata* Willd., Neue Schriften Ges. Naturf. Freunde Berlin 4: 216 1803; *Oldenlandia pusilla* Rottl. 1803, nom. invalid; *Oldenlandia brachiata* (Wight) Hook.f., Fl. Brit. India 3:66. 1880; Khanna in Verma *et al.,* Fl. Madhya Pradesh 1: 547. 1993; Samvatsar, Fl. West. Trib. Madhya Pradesh 149. 1996.

Annual slender herbs with acutely 4-angled stem; leaves 2-4 × 0.2-0.3 cm, linear stiff; peduncles in lower axil, solitary, in upper and terminal many flowered; flowers white, minute; calyx lobes minute; capsule small, broadly didymous.

Fl.-Fr.: Aug. -Nov.

Alirajpur only, Shakarja.

Oldenlandia corymbosa L. Sp. Pl. 119. 1753; *Hedyotis corymbosa* (L.) Lam. Tabl. Encl. 1: 272. 1791; *H. pusilla* Hochst. ex A. Rich.,Tent. Fl. Abyss. 1:362. 1848; Hook.f., Fl. Brit. India 3:64. 1880; Khanna, in Verma *et al.,* Fl. Madhya Pradesh 1:546. 1993; Samvatsar, Fl. West. Trib. Madhya Pradesh 148. 1996. *Dhaman papar*

Annuals, erect or decumbent, or prostrate herbs; stem acutely angular, glabrous or scaberulous; leaves 1-3 cm, sessile, linear-lanceolate, to oblong, acute, glabrous, scabrid on margins; flowers white, in axillary cymes; calyx lobes ovate-triangular, 2-3 mm across; capsule glabrous with a truncate mouth.

Fl.-Fr.: Sept. -Dec.

Very common throughout the area.

Oldenlandia diffusa (Willd.) Roxb., Hort. Bengal. 11. 1814; *Hedyotis diffusa* Willd., Sp. Pl. 1:566. 1798; Hook.f., Fl. Brit. India 3: 65. 1880; Khanna in Verma *et al.,* Fl. Madhya Pradesh 1: 546. 1993.

Annuals herbs; stem usually prostrate; leaves 1-2.5 cm,margins revolute, narrow at base; flowers white, axillary, usually solitary or 2-3 together; calyx lobes ovate-acute, subequal; capsule glabrous, *ca* 3 mm long.

Fl.-Fr.: July. -Oct.

Common in open, moist places, all over the area.

Oldenlandia herbacea (L.) Roxb., Fl. Ind. 1: 445. 1820; *Hedyotis herbacea* L., Sp. Pl. 102. 1753; *Oldenlandia heynei* G. Don, Gen. Hist. 3: 531. 1834; Hook.f., Fl. Brit. India 3: 65. 1880; Khanna, in Verma *et al.,* Fl. Madhya Pradesh 1:547. 1993; Samvatsar, Fl. West. Trib. Madhya Pradesh 148. 1996.

Slender, branched herbs; leaves linear to linear-lanceolate, 1.5-5 cm; flowers pale purple, 1 or 2, in axillary peduncles; capsule ovoid to sub-globose; seeds many, ellipsoid.

Fl.-Fr.: Oct. - Dec.

Alirajpur, Burhanpur, Dhar, Khandwa.

Oldenlandia ovatifolia (Cav.) DC., Prodr. 4: 427. 1830; *Hedyotis ovatifolia* Cav., Descr. Pl. Ic. 6: 52. 1802; *Oldenlandia nudicaulis* Roth, Nov. Pl. Sp. 95. 1821; Hook.f., Fl. Brit. India 3: 70.1880; *Thecagonum ovatifolium* (Cav.) Babu in Bull. Bot. Surv. India 11: 214. 1969; Khanna in Verma *et al.,* Fl. Madhya Pradesh 1: 551. 1993.

Annual, erect, 5-15 cm high herbs; leaves ovate or ovate-oblong, 1.5-7 × 1-4 cm, obtuse at apex, attenuate at base, subsessile or shortly petiolate; flowers white, in terminal, dichotomously branched, loose corymbose cymes; capsules ovoid, 1.5-2 mm long, 4-angled, crowned with subulate calyx-lobes; seeds dark brown or purple, subglobose, reticulate.

Fl.-Fr.: July - Dec.

Rare, **Khandwa** only, Kalibhit forest.

Oldenlandia pumila (L.f.) DC., Prodr. 4: 425. 1830; *Hedyotis pumila* L.f., Suppl. Pl. 119. 1782; *Oldenlandia crystallina* Roxb., Fl. Ind. 1: 443. 1820; Hook.f., Fl. Brit. India 3: 65. 1880; Khanna, in Verma *et al.,* Fl. Madhya Pradesh 1:547. 1993; Samvatsar, Fl. West. Trib. Madhya Pradesh 149. 1996.

Annual, diffusely branched, *ca* 20 cm high herbs; rootstock woody; leaves elliptic or elliptic-lanceolate, 0.7-1.5 cm, subacute at apex, narrowed at base, membranous, subsessile; flowers white, axillary, usually solitary, or rarely in 2-flowered cymes; capsules obovoid, laterally compressed, 2-3 mm long, didymous, glabrous; seeds brownish, angular.

Fl.-Fr.: Sept. – Feb.

Alirajpur, Burhanpur, Dhar, Indore, Khandwa.

RUBIA L. W. 82; MP 1; SA 1

Rubia cordifolia L. Mant. Pl. 197, 1767; Hook.f., Fl. Brit. India 3: 202, 1881; Khanna in Verma *et al.,* Fl. Madhya Pradesh 1: 549. 1993; Samvatsar, Fl. West. Trib. Madhya Pradesh 151. 1996. *Manjith*

Perennial climbers; stem acutely 4-angled with minute, recurved prickles on the angles; leaves in whorls of 3-4, ovate or ovate-oblong to lanceolate, with a cordate or rounded base, glabrous, except prickly on nerves beneath; flowers greenish-yellow in cymes; fruit glabrous, 1-seeded, bluish black.

Fl.-Fr.: July-Dec.

Alirajpur, Dhar.

SPERMACOCE L. W. 284; MP 2; SA 2

Key to species

1. Prostrate or diffuse herbs; flowers 4-6 in each cluster; stigmas 2, distinct; capsules with a persistent septum after dehiscence .... **S. articularis**
1. Erect herbs; flowers many in each cluster; stigmas obscurely 2-lobed; capsules without any septum after dehiscence .... **S. pusilla**

Spermacoce articularis L.f., Suppl. Pl. 119. 1782; *S. hispida auct. non* L., 1753: Hook.f., Fl. Brit. India 3: 200.1881; *Borreria articularis* (L.f.) F.N. Will. Bull. Herb. Boiss. Ser. 5, 2: 956. 1905; Khanna, in Verma *et al.,* Fl. Madhya Pradesh 1:549. 1993; Samvatsar, Fl. West. Trib. Madhya Pradesh 146. 1996.

Prostrate or procumbent herbs; leaves 1-4 × 0.5-1 cm elliptic, lanceolate to obovate-obtuse, margins flexuous or not; flowers pink, in axillary clusters; corolla long, funnel-shaped to short, campaunlate.

Fl.-Fr.: Aug. – April

Common in open areas all over.

Spermacoce pusilla Wall. in Roxb., Fl. Ind. 1: 379. 1820; *S. stricta auct. mult. non* L.f., 1782: Hook.f., Fl. Brit. India 3: 200.1881; *Borreria pusilla* (Wall.) DC., Prodr. 4: 543. 1830; Khanna, in Verma *et al.,* Fl. Madhya Pradesh 1:550.1993; Samvatsar, Fl. West. Trib. Madhya Pradesh 146. 1996.

Erect, branched, plants; stem sharply 4-angular, angles hispid; leaves 2-5 × 0.5-1 cm. narrowly elliptic or lanceolate, sessile, margins not prominently incurved, lateral veins obvious, 2-3 pairs; flowers white; capsules obovoid or ellipsoid, densely hairy outside.

Fl.-Fr.: Sept –Dec.

Common in the same habitat as the above species, but the plants more in numbers.

SPERMADICTYON Roxb. Monotypic

Spermadictyon suaveolens Roxb., Pl. Coromandel 3: 32. t. 236. 1815; *Hamiltonia suaveolens* Roxb., Fl. Ind. 2: 223. 1824; Hook.f., Fl. Brit. India 3: 197. 1881; Khanna in Verma *et al.*, Fl. Madhya Pradesh 1: 550.1993; Samvatsar, Fl. West. Trib. Madhya Pradesh 151. 1996. *Bercha*

A much branched shrub with divaricate, glabrous branches; leaves 15-18 × 4.5-8 cm, elliptic, lanceolate, oblong, acute, slightly cuneate at base; flowers blue or purple or white, fragrant, in terminal, subglobose, corymbiform heads, on short, trichotomous branches; capsule ellipsoid, 3-4 mm black; seeds glabrous, trianglular.

Fl.-Fr.: Aug. – March

Common as forest undergrowth, in all districts, but the number of plants few at a place.

TAMILNADIA Tirveng. & Sastre Monotypic

Tamilnadia uliginosa (Retz.) Tirveng. & Sastre in Mauritius Inst. Bull. 8: 85. 1979; *Gardenia uliginosa* Retz., Observ. Bot. 2: 14. 1781; *Randia uliginosa* (Retz.) Poir. in Lam., Encycl. Suppl. 2: 829. 1811; Hook.f., Fl. Brit. India 3: 110.1880; *Catunaregam uliginosa* (Retz.) Sivarajan in Manilal & Sivarajan, Fl. Calicut 132. 1982; Khanna in Verma *et al.,* Fl. Madhya Pradesh 1: 530.1993; Samvatsar, Fl. West. Trib. Madhya Pradesh 151. 1996. *Kala phetra* **(Plate 6, Fig. 5)**

Small, armed trees; spines terminal, at the end branchlets; leaves crowded towards the ends of short branchlets, oblong or oblanceolate, 11-22 × 4-9 cm, obtuse or acute at apex, narrowed at base into a slender petiole; flowers white, solitary, sessile or pedicellate; fruits yellow, ovoid, *ca* 5 cm long, smooth; seeds many, compressed, embedded in a pulp.

Fl.-Fr.: May – Aug.

Alirajpur, Barwani, Dhar, Khandwa, Khargone.

COMPOSITAE 1911/32913; 64/118; 47/73

Key to Genera

1. Male and female heads separate; involucres of female florets closed, spiny and with 2 achenes; petals absent; anthers free .... XANTHIUM
1. Male and female florets, either in the same or separate heads, present in open involucres; petals present; anthers united
  2. Heads compound, each floret individually subtended by its involucre and all florets in the head subtended by a common involucre
    3. Heads 1-flowered, of disc florets only
      4. Plants spiny all over, including the outer common involucres .... ECHINOPS
      4. Plants not spiny, neither the involucres
        5. Leaves alternate, serrate, base amplexicaul; heads sessile, 2-bracteate, axillary; epaleate; florets lilac ....CAESULIA

5. Leaves opposite, margin entire; heads peduncled; involucral bracts 4-5; florets white .... LAGASCEA

3. Heads many flowered, consisting of disc and ray florets

6. Stem winged; leaves alternate, decurrent; heads heterogamous; outer female florets tubular, multi- seriate, inconspicuous; all purple; achenes angled; pappus absent .... SPHAERANTHUS

6. Stem not winged

7. Heads axillary, sessile; leaves 3-nerved, not in rosettes .... FLAVERIA

7. Heads terminal, long peduncled, surrounded by 3, large, crenate, foliaceous bracts; leaves in rosettes .... ELEPHANTOPUS

2. Heads not compound

8. Heads homogamous, composed either disc or ray florets

9. Heads composed of disc florets

10. Leaves opposite; florets white

11. Anthers without apical appendages; achenes with wart-shaped glands; pappus of 3-4, thick, clavate, sticky hairs .... ADENOSTEMMA

11. Anthers with apical appendages

12. Pappus of 3-5 scales .... AGERATUM

12. Pappus of hairs .... ACILEPIS

10. Leaves alternate; florets purple or yellowish-green

13. Heads yellowish-green on long peduncles; plants fleshy .... KLEINIA

13. Heads purple; plants not fleshy

14. Radical leaves present; cauline leaves amplexicaul, auricled; heads long peduncled; epaleate; involucres 1-seriate; achenes 5-ribbed, scabrid on ribs; pappus white .... EMILIA

14. Radical leaves absent; cauline leaves amplexicaul or not; heads not peduncled; achenes not scabrid

15. Receptacle densely bristly; leaves dotted with black glands, margin spiny; pappus many seriate

16. Heads solitary; achenes faintly ribbed, areole at the base lateral or oblique .... TRICHOLEPIS

16. Heads many; achenes strongly ribbed (ribs more than 20), areole at the base horizontal .... GONIOCAULON

15. Receptacle naked or pitted; leaves not dotted with black glands

17. Pappus 2-seriate

18. Outer series with long and white, hairs; achenes terete .... CYANTHILLIUM

18. Outer series brown-plumose, inner white plumose; achenes 10-ribbed .... BACCHAROIDES

17. Pappus 1-seriate; leaves white-woolly beneath or not

19. Leaves white-woolly beneath; heads with foliaceous bracts; pappus of copius hairs .... PHYLLOCEPHALUM

19. Leaves not white-woolly beneath; heads without foliaceous bracts; pappus short, reddish; achenes 8-10 ribbed. .... CENTRATHERUM

9. Heads with ray florets only; pappus dimorphic-fine hairy and downy; sap milky or yellow

20. Florets blue; leaves radical and cauline; heads in axillary and terminal clusters; involucres 2-seriate; achenes 5-angled; pappus of 2-3 scales .... CICHORIUM

20. Florets yellow

21. Sap milky; achenes compressed or 5-ribbed, base narrow, apex truncate leaves radical and cauline .... SONCHUS

21. Sap yellow; achenes ribbed, truncate at both ends; leaves in rosettes only ....LAUNAEA

8. Heads heterogamous

22. Involucral bracts all, or outer only spiny or prickly at tip

23. Florets yellow; achenes with hooked spinules and 2 apical horns; pappus absent .... ACANTHOSPERMUM

23. Florets purple; achenes acutely angled; pappus hairs unequal in length ...... OLIGOCHAETA

22. Involucral bracts not spiny

24. Leaves all opposite or upper alternate

25. Leaves opposite

26. Heads solitary, or panicled, terminal or axillary on long peduncles; receptacle paleate

27. Receptacle conical; heads pedunculate, panicled; florets yellow or white involucral bracts many; pappus absent .... ACMELLA

27. Receptacle flat; florets yellow, white or many coloured

28. Leaves peninerved; heads white; involucral bracts 8; pappus absent ...... ECLIPTA

28. Leaves 3-nerved at base

29. Leaves deeply cordate, margin entire; ray florets numerous, colours various; pappus absent .... ZINNIA

29. Leaves cuneate at base, margin dentate; ray florets 5, white; pappus bristly .... TRIDAX

26. Heads clustered

30. Leaves pinnately 3-foliolate; achenes glabrous; pappus of 3-4, rigid, retrorsely hispid awns .... BIDENS

30. Leaves simple, base 3-nerved; achenes viscous hairy; pappus of few, linear, scarious, ciliate scales ....GALINSOGA

25. All leaves not opposite, atleast lower opposite, base 3-5 nerved; florets yellow; receptacle paleaceous

31. Ray florets 3, sterile; disc florets fertile, achenes enclosed in the concave palea; pappus absent .... SCLEROCARPUS

31. Ray florets 5, fertile, disc florets also fertile; palea not as above; pappus of 3-5 bristles .... BLAINVILLEA

24. All leaves alternate

32. Leaf margin pinnati -partite or sinuately lobed

33. Heads 1-2; florets yellow

34. Ray floret one, ligulate, ligule 2-fid; disc florets 3-4 with a 4-fid limb; receptacle paleate; achenes with 2 horns .... GLOSSOCARDIA

34. Ray florets 1-few seriate, eligulate; disc florets many with 2-4 fid limb

35. Pappus absent; corolla of disc floret 4-lobed; leaves sinuate ....CENTIPEDA

35. Pappus present

36. Pappus tubular with fimbriate margin; margins of involucral bracts not scarious; achenes without wings .... GRANGEA

36. Pappus a small auricle or absent; margins of involucral bracts scarious; achenes winged ....COTULA

33. Heads clustered; florets white or purple

37. Florets white; ray florets 5; pappus of 2 awns; receptacle flat ....PARTHENIUM

37. Florets purple; ray florets indefinite; pappus absent; receptacle conical .... CYATHOCLINE

32. Leaf margin entire or nearly so

38. Heads solitary on short or long peduncles; florets yellow

39. Pappus of hairs only; plants glandular .... PENTANEMA

39. Pappus of hairs and scales; plants pubescent .... PULICARIA

38. Heads in clusters; florets yellow or not

40. Plants white- woolly, confined to marshy habitat; female florets numerous, bisexual few, achenes glandular; annuals

41. Heads in leafless corymbose clusters .... LAPHANGIUM

41. Heads in leafy spikes or clusters

42. Heads in leafy spikes .... GNAPHALIUM

42. Heads in axillary or terminal clusters .... GNOMOPHALIUM

40. Plants not white-woolly, comparatively of dryer habitat

43. Stem winged; leaf bases decurrent .... LAGGERA

43. Stem not winged

44. Ray florets ligulate, purple, not tubular; disc florets yellow .... ERIGERON

44. Ray florets not ligulate, tubular, both florets of same colour

45. Achenes flat, compressed .... CONYZA

45. Achenes terete, ribbed or angled .... BLUMEA

46. Achenes terete, pubescent .... CHRYSANTHELLUM

46. Achenes dorsally compressed, margin lacerate .... SYNEDRELLA

ACANTHOSPERMUM Schrank W. 6; MP 1; SA1

Acanthospermum hispidum DC., Prodr. 5: 522. 1836; Chowdhery in Hajra *et al.,* Fl India 12: 361. 1995; Shukla in Verma *et al.,* Fl. Madhya Pradesh 1: 559. 1993; Samvatsar, Fl. West. Trib. Madhya Pradesh 154. 1996.

Densely hairy herbs; branches forking; leaves 0.6-3.5 × 0.8-0.6 cm, ovate or elliptic, hairy on both surfaces; heads yellow, solitary, sessile, at the forks of branches; outer ligulate florets female; achenes obconical, with 2 sharp horns at the tip, compressed and hairy.

Fl.-Fr.: July – Nov.

Very common in open land, all over the area.

ACILEPIS D. Don W. 32; MP 2; SA 1

Acilepis divergens (Roxb.) H. Rob. & Skvarla, Proc. Biol. Soc. Wash. 122(2): 140.2009; *Eupatorium divergens* Roxb., Fl. Ind. (Roxburgh) 3: 414. 1832, *nom illeg; Conyza divergens* Wall., Numer. List [Wallich] n, 3027 3027. 1831; nom. nud; *Vernonia divergens* (Roxb.) Edgew., J. Asiat. Soc. Bengal 2: 172. 1853; Hook.f., Fl. Brit. India 3: 234. 1881; Shukla in Verma *et al.,* Fl. Madhya Pradesh 1: 616. 1993.

Perennial, erect under shrubs 1–1.5 m high stems stout, pubescent, branches divergent; leaves 6–15 × 2.5–6 cm elliptic–ovate or lanceolate, acute serrate, prominently nerved and reticulate beneath; heads *ca* 2.5 mm across in corymbose panicles; achenes *ca* 2.5 mm long, 10-ribbed, truncate at both ends, minutely hairy.

Fl.-Fr.: Jan – Feb.

Rare, Barwani, Khargone.

ACMELLA A. Rich. ex Pers. W. 31; MP 3; SA 2

Key to Species

1. Heads yellow, narrow .... **A. paniculata**
1. Heads white, broader .... **A. radicans**

Acmella paniculata (Wall. ex DC.) R. K. Jansen in Syst. Bot. Monogr. 8 67 1985*; Spilanthes paniculata* Wall. ex DC., Prodr. 5: 625. 1836; *Spilanthes acmella* var. *paniculata* (DC.) C.B. Clarke, Comp. Ind. 139. 1876; Hook.f., Fl. Brit. India 3: 307. 1881; Shukla in Verma *et al.,* Fl. Madhya Pradesh 1: 612. 1993; Chowdhery in Hajra *et al.,* Fl. India 12: 410.1995. *Akal kara*

Annual erect or ascending stout herbs 20–50 cm high; leaves 1.5-7 × 1-5 cm, ovate – elliptic to elliptic acute or obtuse; heads 6–8 mm across in axillary and terminal panicles; achenes black, *ca* 2 mm long dorsally compressed, ciliate along the margins; pappus of 2–3 weak strigose awns.

Fl.-Fr.: March – April

Common in wet or moist soil, throughout.

Acmella radicans (Jacquin) R. K. Jansen, Syst. Bot. Monogr. 8: 69. 1985; *Spilanthes radicans* Jacquin, Collect. Bot. Chem. Hist. Nat. 11(3): 1714. 1804; Schrader, in DC., Prodr. 5:624. 1836; Sivarajan & Matthew in Anc. Sci. Life 3:169. 1984; Chowdhery in Hajra *et al.,* Fl. India 12: 413. 1995; Raja *et al.,* Int. J. Int. sci. Inn. Tech. 2(6):34-35. 2013.

Erect herbs; stems terete, minutely pubscent; leaves opposite, ovate, *ca* 7 × 4 cm, obtuse at base, acute at apex, serrulate at margin, glabrous; heads white, discoid, conical, axillary and terminal, 8 – 10 mm across; achenes dimorphic, ellipsoid, obcordate, with corky and densely ciliate margins, marginal ones trigonous, brownish-black, laterally compressed, pappus 2, with subequal bristles, persistent.

*Fl.-Fr.:*Feb. -Sept.

Rare, **Alirajpur** only, weed in forest nursery, in the same habitat as above.

ADENOSTEMMA J. R. & G. Forst. W. 27; MP 2; SA 1

Adenostemma lavenia (L.) Kuntze, Rev. Gen. Pl. 1: 304: 1891; *Verbesina lavenia* L., Sp. Pl. 902. 1753; *Adenostemma viscosum* J. R. Forst. Char. Gen. Pl. 90.1776; C.B. Clarke in Hook.f., Fl. Brit. India 3: 242. 1881; Shukla in Verma *et al.,* Fl. Madhya Pradesh 1: 560.1993; Uniyal in Hajra *et al.,* Fl. India 12: 346. 1995; Samvatsar, Fl. West. Trib. Madhya Pradesh 154. 1996.

Annual or perennial, erect, 30-80 cm high herbs; leaves petiolate, variously shaped 5–20 × 1.3–10 cm., ovate, cuneate at base, acuminate at apex, entire or serrate along margins, glabrous; heads *ca* 5 mm across in corymbose panicles, peduncles glandular pubescent; involucral bracts biseriate; outer bracts *ca* 4mm long, linear – spathulate, herbaceous; inner bracts *ca* 5 mm long, acute with ciliate margins; corolla 3–4 mm long, tubular, glandular at base, pubscent in upper part; achenes 4–5 mm long obovate muricate or tuberculate; pappus hairs ending in a knob.

Fl.-Fr.: Nov – Dec.

Along banks of streams, Alirajpur, Indore.

AGERATUM L. W. 51; MP 2; SA 1

Ageratum conyzoides L., SP. Pl. 839. 1753; Hook.f., Fl. Brit. India 3:243. 1881; Shukla in Verma *et al.,* Fl. Madhya Pradesh 1: 561. 1993; Uniyal in Hajara *et al.,* Fl. India 12: 348. 1995; Samvatsar, Fl. West. Trib. Madhya Pradesh 154. 1996. *Ajgandha*

Annual, erect, 40-60 cm high herbs; leaves 5–7 × 2.5-5 cm, sub rhomboid, ovate, 3– nerved from base, usually crenate, sparsely pilose above, glandular punctate beneath; heads 4–4.5 mm across in corymbs; achenes *ca* 1 mm long. black, short bristly on the ribs; pappus scales 5, *ca* 1.5 mm long, lanceolate.

Fl.-Fr.: Through out the year.

Very common in all districts, along village roads, near streams etc.

BACCHAROIDES L. ex Moench W. 23; MP 1; SA 1

Baccharoides anthelmintica (L.) Moench, Method. 578. 1794; *Conyza anthelmintica* L., Sp. Pl. ed 2, 1207. 1763; *Vernonia anthelmintica* (L.) Willd., Sp. Pl. 3: 1634. 1803; Hook.f., Fl. Brit. India 3: 236. 1881; Shukla in Verma *et al.,* Fl. Madhya Pradesh 1: 615. 1993; Uniyal in Hajra *et al.,* Fl. India 13: 357. 1995; Samvatsar, Fl. West. Trib. Madhya Pradesh 157. 1996. *Ghoda jeera*

Annual, 50-150 cm, erect herbs; stems striate, pubescent; leaves 5–10 × 2–4.5 cm lanceolate, oblanceolate or elliptic–lanceolate, acute, coarsely serrate, pubescent on both surfaces; heads *ca* 10 mm across, sub corymbose; achenes 4–5 mm long, oblong–cylindric, black, 10– ribbed, pubescent on ribs, glandular between grooves; pappus 6–7 mm long, tawny white, inner short, paleaceous.

Fl.-Fr.: Sept. – March

Occasional, Alirajpur, Indore, a weed in fields.

BIDENS L. W. 249; MP 2; SA 1

Bidens biternata (Lour.) Merr. & Sherff in Bot. Gaz. 88: 293. 1929; *Coreopsis biternata* Lour., Fl. Cochinch. 508. 1790; *Bidens pilosa auct. non* L., Hook.f., Fl. Brit. India 3: 309. 1881; Shukla in Verma *et al.,* Fl. Madhya Pradesh 1: 563. 1993; Chowdhery in Hajra *et al.,* Fl. India 12: 367. 1995; Samvatsar, Fl. West. Trib. Madhya Pradesh 154. 1996.

Annual, erect herbs 20–70 cm high, stem angled, sparsely pubescent when young; leaves opposite, bipinnatifid, compound 1–2 cm long; lateral: leaflets 1.5 – 3 × 0.4 – 2 cm ovate, deeply dentate or pinnatifid, acute subsessile: terminal leaflets 1.5–6 × 1–2.5 cm elliptic, with 3–5 pinnatifid lobes on each side; heads *ca* 6 mm across, on dichotomously branched peduncles; achenes 7–15 mm long, tetragonous, black; pappus bristles 3.5 – 4 mm long, retrorsely barbed, yellow.

Fl.-Fr.: Jan. –Feb.

Common forest undergrowth, very troublesome due to the retrorsely barbed bristles of the achenes.

BLAINVILLEA Cass. W. 10; MP 1; SA1

Blainvillea acmella (L.) Philipson in Blumea 6: 350.1950; *Verbesina acmella* L., Sp. Pl. 901. 1753; *Blainvillea latifolia* (L.f.) DC. ex Wight, Contr. Bot. Ind. 71. 1834; Hook.f., Fl. Brit. India 3: 305. 1881; Shukla in Verma *et al.,* Fl. Madhya Pradesh 1: 564. 1993; Chowdhery in Hajra *et al.,* Fl. India 12: 377. 1995; Samvatsar, Fl. West. Trib. Madhya Pradesh 154. 1996.

A 30-60 cm tall hispid herb; leaves 4-10 × 2-6 cm, ovate or ovate-lanceolate, serrate; heads white or yellow, solitary, sessile or peduncled; involucral bracts foliaceous, stronglyveined; pappus setae 2-5, plumose; achenes triquetrous, obovoid-cuneate, tubercled, pubescent, those of disk florets slightly and transversely rugose.

Fl.-Fr.: Aug. – Nov.

Common along hedges of fields, open ground, everywhere.

BLUMEA DC. W. 107; MP 14; SA 12

Key to Species

1. Plants densely white woolly (rarely glabrate in **B. lacera**)

    2. Heads in lax panicles; achenes subangular, terete

        3. Heads in corymbose panicles; corolla yellow . ... **B. lacera**

        3. Heads in narrow, spiciform panicles; corolla purple .... **B. axillaris**

    2. Heads in corymbs; achenes 8-10 ribbed.......... **B. laciniata**

1. Plants glabrous, never white-woolly

    4. Achenes ribbed or angled

5. Receptacle glabrous; heads pedunculate or subsessile; corolla yellow
   6. Female florets pubescent .... **B. eriantha**
   6. Female florets glabrous
      7. Leaves membranous .... **B. membranacea**
      7. Leaves not membranous .... **B. virens**
5. Receptacle pilose
   8. Pappus purple or yellowish .... **B. lanceolaria**
   8. Pappus white; heads sessile .... **B. fistulosa**

4. Achenes neither ribbed, nor angled
   9. Stem unbranched .... **B. belangeriana**
   9. Stem branched
      10. Upper leaves distinctly sub-opposite .... **B. bifoliata**
      10. Upper leaves alternate
         11. Erect herbs; female florets glabrous .... **B. obliqua**
         11. Prostrate herbs; female florets pubescent .... **B. oxydonta**

Blumea axillaris (Lam.) DC., Prodr. 5:434. 1836; *Conyza axillaris* Lam., Encycl. 284. 1786; *Erigeron mollis* D. Don, Prodr. Fl. Nepal. 172. 1825; *Blumea mollis* (D. Don) Merr., Phillip. J. Sci., (Bot.) 5:395. 1910; *Blumea solidaginoides* (Poir.) DC., Prodr. 5:443. 1836; *Gnaphalium solidaginoides* Poir., Encycl. Suppl. 2:805. 1811; *Blumea neilgherriensis* Hook.f., F. Brit. India 3:261. 1881; Shukla, in Verma *et al.,* Fl. Madhya Pradesh 1:571. 1993; Samvatsar, Fl. West. Trib. Madhya Pradesh 157. 1996.

Erect, up to 60 cm high annual herbs; stem softly glandular hairy; leaves 3.5-6 × 1.5-2.5 cm, obovate or elliptic-oblong, margin dentate or irregularly toothed, white-pilose on both surfaces, lower pedicellate, upper sessile; capituli purple, many in axillary or terminal, spiciform panicles; involucral bracts 3-seriate, outer *ca* 2 mm long, linear, glandular hairy outside, inner bracts *ca* 5 mm long, linear with scarious margin; receptacle glabrous; ray florets 2-2.5 mm long, tubular; disc florets 5-lobed, 3-4 mm long, papillate at base; achenes *ca* 5 mm long, angular; pappus *ca* 3 mm long, white.

*Fl.-Fr.:*Dec. -Feb.

Along streams, Burhanpur, Indore, Khandwa.

Blumea belangeriana DC. Prodr. 5:414. 1836; Hook.f., Fl. Brit. India 3: 266. 1881; Randeria in Blumea 10: 277. 1960; S. Kumar in Hajra *et al.,* Fl. India 13: 135. 1995; Samvatsar, Fl. West. Trib. Madhya Pradesh 156. 1996; Khanna *et al.,* Suppl. Fl. Madhya Pradesh 1: 121.2001.

Herbs; stem terete, hispid at base, softly villous at the apex; leaves 5-8 × 2-4 cm, obovate to elliptic oblong, closely dentate, sparsely hairy, upper subsessile; heads axillary, sessile, solitary or clustered; receptacle glabrous; involucral bracts linear, woolly; achenes terete, nearly glabrous.

Fl.-Fr.: Dec. –March

Occasional on banks of streams, Alirajpur, Dhar.

Blumea bifoliata (L.) DC. in Wight, Contrib. Bot. Ind. 14.1834; *Conyza bifoliata* L., Sp. Pl. 2: 1207.1753; Hook.f., Fl. Brit. India 3: 261.1881; Shukla in Verma *et al.,* Fl. Madhya Pradesh 1: 567.1993; S. Kumar in Hajra *et al.,* 1: 1.18. 1905: Fl. India 13: 120.1995.

Annual, herbs, 5–60 cm high, stem branched at the base; leaves 1–5 × 0.4–2.2 cm oblong–spathulate membranous, margins entire or distantly serrate, villous and stipitate glandular on both surfaces; heads *ca* 6 mm across, terminal and axiallary; achenes oblong elliptic, puberulous; pappus *ca* 3 mm long, yellowish–white.

Fl.-Fr.: Jan. – March

On damp floor, confined to **Indore**, Gautampura.

Blumea eriantha DC. in Wight, Contrib. Bot. Ind. 15. 1834; Hook.f., Fl. Brit. India 3: 266. 1881; Shukla in Verma *et al.,* Fl. Madhya Pradesh 1: 567. 1993; S. Kumar in Hajra *et al.,* Fl. India 13: 122. 1995; Samvatsar, Fl. West. Trib. Madhya Pradesh 156. 1996. *Jangli kapoor*

Stem 30-40 cm tall, dichotomously branched, more or less pubescent; leaves 3-8 × 1.5-3 cm irregularly toothed, lower petiolate, obovate, upper sessile or nearly so, elliptic-oblong, pubescent; heads yellow, on long, slender, hairy peduncles of dichotomous cymes; involucral bracts linear, silky pubescent; receptacles glabrous; achenes minute, angled, sparsely hairy on angles.

Fl.-Fr.: Jan. – March

Common in open ground, dry areas, throughout the area.

Blumea fistulosa (Roxb.) Kurz in J. Asiat. Soc. Bengal 47 (2): 187. 1877; *Conyza fistulosa* Roxb., Fl. Ind. 3: 429. 1832; *B. glomerata* DC. in Wight, Contrib. Bot. Ind. 15. 1834; Hook.f., Fl. Brit. India 3: 262. 1881; Shukla in Verma *et al.,* Fl. Madhya Pradesh 1:569. 1993; S. Kumar in Hajra *et al.,* Fl. India 13: 124. 1995.

Annual, erect herbs, 30–80 cm high, stem puberulous to densely velutinous at the apex; lower leaves 3–15 × 0.5 6 cm obovate with long tapering base, serrate or lyrate–pinnatifid, petioled; upper leaves smaller, oblong–lanceolate to broadly oblanceolate, acute to apiculate, coarsely serrate–dentate with mucronate teeth, tomentose to puberulous on both surfaces; heads 4–5 mm across, in compact globose clusters arranged in interrupted spicate panicles; pappus *ca* 3.5 mm long, white.

Fl.-Fr.: Dec. –Feb.

Along forest paths, Barwani, Burhanpur, Khandwa, Khargone.

Blumea lacera (Burm.f.) DC. in Wight, Contrib. Bot. Ind. 14. 1834; *Conyza lacera* Burm.f., Fl. Ind. 180. t. 59. f. 1. 1768; *Blumea subcapitata* DC., Prodr. 5: 439. 1836; *B. lacera* DC. var. *cinerascens* Hook.f., Fl. Brit. India 3: 263. 1881 & var. *glandulosa* Hook f., Fl. Brit. India 3: 263. 1881; Shukla in Verma *et al.,* Fl. Madhya Pradesh 1: 569. 1993; S. Kumar in Hajra *et al.,* Fl. India 13: 128. 1995; Samvatsar, Fl. West. Trib. Madhya Pradesh 156. 1996. *Kalhar*

Annuals with an aromatic smell; stem 30-100 cm tall, usually glandular pubescent; leaves 5-20 × 2-7 cm, lower petiolate; upper subsessile, elliptic-oblong or obovate, silky pubescent, on both surfaces, serrate-dentate; heads yellow, numerous in short, axillary cymes and terminal spiciform panicles; involucral bracts densely silky villous; receptacles glabrous; pappus white; achenes sub-tetragonous, not ribbed.

Fl.-Fr.: Nov. – Jan.

Common along roadsides, open places, shady areas, all over.

Blumea laciniata (Roxb.) DC., Prodr. 5: 436. 1836; *Conyza laciniata* Roxb., Fl. Ind. 3:428. 1832; Hook.f., Fl. Brit. India 3: 264. 1881; S. Kumar in Hajra *et al.,* Fl. India 13: 128. 1995; Samvatsar, Fl. West. Trib. Madhya Pradesh 156. 1996.

Erect, 30-80 cm high, annual herbs, clothed with stalked glands; lower leaves petiolate, lyrate-pinnatifid; upper ones obovate, entire to coarsely toothed; heads yellow, combined into a terminal panicle; involucral bracts acute, hairy; achenes 8-10 ribbed, thinly hairy; pappus white.

*Fl.-Fr.:*Jan. -May

Common along roadsides, Alirajpur, Burhanpur, Dhar, Indore, Khandwa.

Blumea lanceolaria (Roxb.) Druce, Rep. Bot. Exch. Club. Brit. Isles 4: 609. 1917; *Conyza lanceolaria* Roxb., Fl. Ind. 3:432. 1832; Hook.f., Fl. Brit. India 3:369. 1881; Shukla in Verma *et al.,* Fl. Madhya Pradesh 1: 570.1993.

Annual or perennial, erect herbs 60–100 cm high; stem woody at base; leaves oblanceolate, semi-amplexicaul 6–30 × 0.8-8 cm, acute to rounded, minutely serrate-dentate, pubescent beneath; heads *ca* 7 mm across, in terminal pyramidal panicles; pappus *ca* 4 mm long, yellowish-red.

Fl.-Fr. : Feb. –April

In moist places, all over the area.

Blumea membranacea DC., Prodr. 5: 440.1836; Hook.f., Fl. Brit. India 3: 265. 1881; Shukla in Verma *et al.,* Fl. Madhya Pradesh 1: 571. 1993; S. Kumar in Hajra *et al.,* Fl. India 13: 134. 1995; Samvatsar, Fl. West. Trib. Madhya Pradesh 156. 1996.

Annual, erect herbs 50–125 cm high, stem pilose with stipitate glands on younger parts; leaves membranous 2–20 × 1–6. 3 cm, oblanceolate to obovate lanceolate, tapering at base, acute at apex margin serrate–dentate, sparsely pilose on both surfaces; lower leaves lyrate pinnatifid. with terminal elliptic lobes; upper entire; heads *ca* 5 mm across in terminal and axillary panicles; achenes *ca* 1 mm long linear-oblong 5- ribbed, pubescent, *ca* 5 mm long, white.

Fl.-Fr.: Jan. – Feb.

Common as forest undergrowth, all over the area.

Blumea obliqua (L.) Druce, Rep. Bot. Exch. Club. Brit. Isles 4: 609. 1916; *Erigeron obliquus* L., Mant. Pl. 2: 573. 1771; *Blumea amplectens* DC. in Wight, Contrib. Bot. Ind. 13. 1834; Hook.f., Fl. Brit. India 3: 260.1881; Shukla in Verma *et al.,* Fl. Madhya Pradesh 1: 573. 1993; S. Kumar in Hajra *et al.,* Fl. India 13: 137. 1995; Samvatsar, Fl. West. Trib. Madhya Pradesh 156. 1996.

Annual, erect dichotomously branched herbs, 30–50 cm high; lower leaves oblong–lanceolate or oblanceolate–spathulate obtuse or subacute; upper leaves ovate–oblong, auriculate at base, entire or coarsely dentate at margins, densely pubescent on both surfaces; heads *ca* 6 mm across in solitary, terminal and axillary, pedunculate; achenes dark brown; pappus *ca* 4 mm long, yellowish–white.

Fl.-Fr.: Feb. – March

Common in dry areas, open ground, all over.

Blumea oxyodonta DC. in Wight, Contrib. Bot. Ind. 15. 1834; Hook.f., Fl. Brit. India 3: 266. 1881; Shukla in Verma *et al.,* Fl. Madhya Pradesh 1: 573. 1993; S. Kumar in Hajra *et al.,* Fl. India 13: 139. 1995; Samvatsar, Fl. West. Trib. Madhya Pradesh 157. 1996.

Prostrate or procumbent herbs, 6–40 cm high, with a deep tap root, branches arising from base, villous with long, whitish silky hairs; leaves often crowded at the base, 0.7-7 × 0.3-1.5 cm, obovate to obovate–oblong, serrate–dentate, white silky hairy on both surfaces; heads *ca* 4 mm across, in terminal or axillary corymbs; achenes, oblong, brown, pubescent; pappus *ca* 3.5 mm long, white, deciduous.

Fl.-Fr.: Jan – March

Common, in dry, sandy localities, in all districts.

Blumea virens Wall. ex DC. in Wight, Contrib. Bot. Ind. 14. 1834; Hook.f., Fl. Brit. India 3: 264. 1881; Shukla in Verma *et al.,* Fl. Madhya Pradesh 1: 573. 1993; Samvatsar, Fl. West. Trib. Madhya Pradesh 157. 1996.

Erect, 30-100 cm tall herbs; stem grooved, glabrous; leaves 3-15 × 0.5-10 cm, lower leaves larger, petiolate, obovate, or sometimes lyrate, irregularly or sinuately toothed; upper leaves smaller, elliptic-oblong; heads yellow, on capillary, glabrous peduncles in panicles; involucral bracts slender; inner pubescent at tips; receptacles glabrous; achenes oblong, hairy, ribbed; pappus shorter than inner bracts.

Fl.-Fr.: Dec. –Feb.

Common throughout the area on bunds of fields, forest edges, waste places.

CAESULIA Roxb. Monotypic

Caesulia axillaris Roxb., Pl. Corom. 1: 64. 1768; Hook.f., Fl. Brit. India 3: 291. 1881; Shukla in Verma *et al.,* Fl. Madhya Pradesh 1: 573. 1993; Pant in Hajra *et al.,* Fl. India 13: 2. 1995; Samvatsar, Fl. West. Trib. Madhya Pradesh 157. 1996.

Annual, erect or decumbent glabrous herbs; leaves semi–amplexicaul, dilated and sheathing at base, linear–lanceolate, remotely denticulate 4–10 × 0.4 –1 cm glabrous; heads 2–3 united to form a compound head; achenes *ca* 3 mm long, dark brown, ribbed, winged, shortly hairy at top with 2 pappus scales.

Fl.-Fr.: Sept. – Nov.

Common all over near streams and in moist soil.

CENTIPEDA Lour. W. 11; MP 1; SA 1

Centipeda minima (L.) A. Br. & Aschers., Ind. Sem. Fl. Berol. App. 6, 1867; *Artemisia minima* L., Sp. Pl. 848. 1753; *Centipeda orbiculoris* Lour. Fl. Cochinch. 493. 1790; Hook.f., Fl. Brit. India 3: 317. 1881; Naithani in Hajra *et al.,* Fl. India 12: 48. 1995.

Annual, prostrate hairy herbs 30–40 cm high with radiating branches from rootstock; leaves subsessile 1–1.2 cm long, obovate– spathulate, toothed; heads 2.5–4 mm across, solitary discoid, globose, yellow; involucral bracts 1–2 seriate, small, oblong–obtuse with scarious margins; ray florets *ca* 0.3 mm long, many seriate, with minute tubular corolla and 2–3 dentate ligule; disc florets *ca* 0.5 mm long, bisexual, companulate 4– fid; achenes *ca* 1 mm long, 4–angled, hairy on angles; pappus absent.

Fl.-Fr.: – Nov. –Feb.

Rare, in open grassland, **Khandwa** only, Chandgarh range.

CHRYSANTHELLUM Rich. W. 11; MP 1; SA 1

Chrysanthellum indicum DC., Prodr. 5: 631. 1836., Hook.f., Fl. Brit. India 3: 310.1881; *Chrysanthellum americanum* (L.) Vatke, Abh. Nat. Ver. Brem. 9: 122. 1887; Shukla in Verma *et al.,* Fl. Madhya Pradesh 1: 576. 1993; Chowdhery in Hajra *et al.,* Fl. India 12: 379. 1995.

Annual, diffuse to procumbent, 10-20 cm tall herbs, stems quadrangular, grooved; leaves alternate, pinnatifid, segments oblong or cuneiform, obtusely lobed, membranous at margins, petioles 5–15 mm long; heads terminal, solitary, peduncled, *ca* 4 mm across; achenes 2–3 mm long oblong, cuneate, outer margin thick, obtuse, inner flat with winged margins.

Fl.-Fr.: Aug. – Nov.

Rare, **Indore**, at Bijasan hill.

CICHORIUM L. W. 10; MP 2; SA 1

Cichorium intybus L., SP. Pl. 813. 1753; Hook.f., Fl. Brit. India 3: 391. 1881; Mamgain Rao, in Hajra *et al.,* Fl. India 12: 248. 1995; Samvatsar, Fl. West. Trib. Madhya Pradesh 158. 1996; Khanna *et al.,* Suppl. Fl. Madhya Pradesh 1: 124.2001.

Perennial, erect, 30-80 cm high, glabrous or hairy herbs; lower leaves variable 5–20 × 1–4.3 cm oblong, oblong –lanceolate or oblanceolate, pinnatifid, runcinate–pinnatifid, dentate toward margins, glabrous or sparsely hairy; upper leaves cordate–amplexicaul at base; heads usually solitary or clustered, terminal and axillary, *ca* 7 mm across; achenes *ca* 2 mm long sub-compressed, angled, smooth, many ribbed, base contracted; pappus pale, very short.

Fl.-Fr.: March – June

Alirajpur, Dhar, Indore, Khandwa.

CONYZA Less. W. 154; MP 4; SA 1

Conyza japonica (Thunb.) Less. ex DC., Prodr. 5: 382. 1836; *Erigeron japonicum* Thunb. in Fl. Jap. 312. 1784; Hook.f., Fl. Brit. India 3: 258. 1881; Shukla in Verma *et al.,* Fl. Madhya Pradesh 1: 578. 1993; Hajra in Hajra *et al.,* Fl. India 12: 105. 1995; Samvatsar, Fl. West. Trib. Madhya Pradesh 158. 1996.

Annual or perennial erect villous herbs, 50–120 cm high; lower leaves sessile, in rosettes, obovate; cauline ones cuneate at base, serrate or crenate, obtuse or acute 1–4 × 0.5 –2 cm; heads *ca* 5 mm across, densely glomerate, in branched, terminal panicles; achenes *ca* 1 mm long, thinly hairy; pappus *ca* 3 mm long, reddish- brown.

Fl.-Fr.: Feb. – March

Common, in nurseries and escape, everywhere.

COTULA L. W. 62; MP 1; SA 1

Cotula anthemoides L. Sp. Pl. 891. 1753; Hook.f., Fl. Brit. India 3: 316. 1881; Naithani in Hajra *et al.,* Fl. India 12: 52. 1995; Samvatsar, Fl. West. Trib. Madhya Pradesh 159. 1996; Khanna *et al.,* Suppl. Fl. Madhya Pradesh 1: 126.2001.

Annual, diffuse, procumbent or decumbent, branched, 5-20 cm high herbs; leaves slightly thick, 1–5 cm long, 2–pinnatisect or pinnatifid; segments narrowly oblong, slightly curved, ultimate segments obtuse, shortly mucronate; heads 3–5 mm across, yellow, solitary; achenes 0.5 – 0.8 mm, ovate with thick narrow wings, that of ray florets broad, flat.

Fl.-Fr.: Jan. – March

Rare, in fields and waste places, **Alirajpur**, Ranapur village.

CYANTHILLIUM Blume W. 8; MP 1; SA 1

Cyanthillium cinereum (L.) H. Rob., Proc. Biol. Soc. Wash. 103: 252. 1990; *Conyza cinerea* L., Sp. Pl. 862. 1753; *Vernonia cinerea* (L.) Less, Linnaea 4: 291. 1829; Hook.f., Fl. Brit. India 3:233. 1881; Shukla in Verma *et al.,* Fl. Madhya Pradesh 1:616. 1993; Samvatsar, Fl. West. Trib. Madhya Pradesh 167. 1996. *Sahadevi*

Annual or perennial, erect herbs, 15–80 cm high, stems terete, ribbed, grayish pubescent; leaves sub-spathulate, lanceolate or elliptic, 2–8 × 1–4 cm, ovate, obtuse or acute, attenuate at base, pubescent on both surfaces; heads 2–3 mm across, in terminal corymbs; achenes *ca* 1.5 mm long, obscurely 4–5 angled, white, puberulous; pappus 4–5 mm long, white, biseriate.

Fl.-Fr.: Throughout the year.

Very common in all districts.

CYATHOCLINE Cass. W. 4; MP 1; SA 1

Cyathocline purpurea (Ham. ex D. Don) Kuntze, Rev. Gen. Pl. 338. 1891; *Tanacetum purpureum* Ham. ex D. Don, Prodr. Fl. Nepal 181, 1825; *Cyathocline lyrata* Cass. in Ann. Sci. Nat. ser. 1, 17: 420, 1829; Hook.f., Fl. Brit. India 3: 246, 1881; Shukla in Verma *et al.,* Fl. Madhya Pradesh 1: 581, 1993; Hajra in Hajra *et al.,* Fl. India 12: 111. 1995; Samvatsar, Fl. West. Trib. Madhya Pradesh 159. 1996. *Okhrad gangotri*

Erect, annual, delicate, scabrous or pubescent herbs, often tinged with reddish-purple; leaves 4-12 cm long, lyrately pinnatifid; heads purple, hemispheric, 3 × 6 mm, in corymbose panicles; receptacles naked, convex at the top; corolla of outer florets filiform, longer than inner; achenes oblong, up to 3 mm long, smooth with thickened margins; pappus hairs white.

Fl.-Fr.: Oct. –April

Common, throughout the area.

ECHINOPS L. W. 193; MP 1; SA 1

Echinops echinatus Roxb. Fl. Ind. 3: 547, 1832; Hook.f., Fl. Brit. India 3: 358. 1881; Shukla in Verma *et al.*, Fl. Madhya Pradesh 1: 583. 1993; Samvatsar, Fl. West. Trib. Madhya Pradesh 159. 1996. *Bajar battu*

Erect, much-branched 50-100 cm, tall herbs; leaves 7-15 × 2-6 cm deeply pinnatifid, lobes sinuate and spinescent; heads homogamous, white or pale blue, globose; involucre with white bristles; middle bract with a sharp spine exceeding the compound head, inner connate for more than half their length into a tube which surrounds the achene; achenes obconic, densely villous; pappus short, antrorsely hispid, yellowish, forming a cylindric bunch on the achenes.

Fl.-Fr.: Oct. -March

Very common in open ground, all over the area.

ECLIPTA L. W. 11; MP 1; SA 1

Eclipta prostrata (L.) L., Mant. Pl. 2: 286. 1771; *Verbesina prostrata* L., Sp. Pl. 902. 1753; *Eclipta alba* (L.) Hassk., Pl. Jav. Rav. 528. 1848; Hook.f., Fl. Brit. India 3: 304. 1881; Shukla in Verma *et al.,* Fl. Madhya Pradesh 1: 584. 1993; Chowdhery in Hajra *et al.,* Fl. India 12: 381. 1995; Samvatsar, Fl. West. Trib. Madhya Pradesh 159. 1996. *Bhangra*

Annual or perennial, prostrate, decumbent or erect hairy herbs, 20–40 cm high; leaves opposite, subsessile 2–5 × 1–1.5 cm, elliptic – lanceolate, mucronulate, cuneate at base, appressed pilose on both surfaces; heads 3–5 mm across, white, solitary or fascicled, axillary, on 0.3–1 cm long peduncles or sessile; achenes 3–3.5 mm long, 3–angled in female and 4–angled in bisexual florets; pappus absent,only scales present.

Fl.-Fr.: Throughout the year.

Common around wells in moist soil, near streams, throughout the area.

ELEPHANTOPUS L. W. 29; MP 1; SA 1

Elephantopus scaber L., Sp. Pl. 814. 1753; Hook.f., Fl. Brit. India 3: 242. 1881; Shukla in Verma *et al.,* Fl. Madhya Pradesh 1: 584. 1993; Uniyal in Hajra *et al.,* Fl. India 13: 333. 1995. *Gojiva*

Perennial, subscapose, erect, 30-40 cm high herbs; leaves usually in basal rosettes, 3–15 × 0.5 –7 cm, obovate oblong, spathulate, tapering at base, obtuse at apex, sparsely hairy beneath; cauline leaves smaller; heads axillary and terminal, on 4–15 cm long peduncles; bracts foliaceous, 3, at the top, 1–2 cm long, 1– nerved; achenes *ca* 5 mm long, 10– ribbed, slightly pubescent; pappus bristles 5, hairy, *ca* 5 mm long with dilated base.

Fl.-Fr.: Oct. – Dec.

Barwani, Khandwa, Khargone.

EMILIA Cass. W. 115; MP 1; SA 1

Emilia sonchifolia (L.) DC. in Wight, Contrib. Bot. Ind. 24. 1834; *Cacalia sonchifolia* L., Sp. Pl. 835. 1753; Hook.f., Fl. Brit. India 3: 336. 1881; Shukla in Verma *et al.,* Fl. Madhya Pradesh

1: 585. 1993; Mathur in Hajra *et al.,* Fl. India 13: 212. 1995; Samvatsar, Fl. West. Trib. Madhya Pradesh 160.1996.

Annual, erect, branched, glaucescent, 30-40 cm high herbs; leaves sessile; lower leaves 6–11 cm long, lyrately pinnatifid with triangular–ovate, terminal lobe and irregularly toothed lateral lobes; upper leaves smaller, lanceolate, amplexicaul with lyrate auricles; heads 7–9 mm long, on 3–5 cm long peduncles, combined into corymbs; achenes 3–4 mm long, with 5 hairy ribs, alternating with faint glabrous ribs; pappus white,7–8 mm long.

Fl.-Fr.: Feb. – March

Common, grows in waste lands and sandy soil.

ERIGERON L. W. 476; MP 3; SA 2

1. Heads *ca* 3 mm across; pappus *ca* 2 mm long reddish .... **E. trilobus**
1. Heads *ca* 10 mm; pappus *ca* 4 mm long dirty white .... **E. sublyratus**

Erigeron sublyratus DC. in Wight. Contr. Bot. Ind. 9. 1834; *E. asteroides* Roxb. Fl. Ind. ed. 2.3:432. 1832; Hook.f., Fl. Brit. India 3: 254. 1881; Shukla in Verma *et al.,* Fl. Madhya Pradesh 1: 580.1993 and 1:586. 1993; Hajra in Hajra *et al.,* Fl. India 12: 125. 1995; Samvatsar, Fl. West. Trib. Madhya Pradesh 160.1996.

Annual, erect or decumbent, pubescent, 20-35 cm high herbs; basal leaves petioled; cauline ones 1.2 – 5 × 0.5–2.4 cm, obovate or oblong or spathulate, acute or obtuse at apex, semi-amplexicaul, dentate, lobulate along margins, puberulous on both surfaces; heads *ca* 10 mm across, solitary or in subcorymbs; achenes *ca* 0.5 mm long, obovoid, flattened or compressed; pappus *ca* 4 mm long, dirty white.

Fl.-Fr.: Aug. – Feb.

Alirajpur, Burhanpur, Dhar, Indore, Khandwa.

Erigeron trilobus (Decne.) Boiss., Fl. Orient. 3 168 187; *Conyza triloba* Decne., Ann. Sci. Nat., Bot. sér. 2, 2: 261. 1834; *Conyza stricta* Willd., Sp. Pl. 3: 1922. 1803; Hook.f., Fl. Brit. India 3: 258. 1881; Shukla in Verma *et al.,* Fl. Madhya Pradesh 1: 580.1993; Hajra in Hajra *et al.,* Fl. India 12: 108. 1995; Samvatsar, Fl. West. Trib. Madhya Pradesh 158. 1996.

Annual, erect, pubescent, 20-60 cm high herbs; stem ribbed, glandular–villous; leaves in lower part oblanceolate–spathulate, serrate dentate, apiculate, glandular hairy on both surfaces; in upper part linear; heads *ca* 3 mm across, in corymbose panicles; ray florets *ca* 1.5 mm long, ligulate, bidentate; disc florets *ca* 2 mm long, tubular, 5–fid, hairy; achenes *ca* 0.5 mm long; pappus *ca* 2 mm long, reddish.

Fl.-Fr.: Dec. –Jan.

Rare, in moist places, **Alirajpur** only, at Kattiwada.

FLAVERIA A. Juss. W. 24; MP 1; SA 1

Flaveria trinervia (Spreng.) C. Mohr, Contr. U.S. Nat. Herb. 6: 810.1901; *Oedera trinervia* Spreng., Bot. Gard. Halle 63. 1800; *Flaveria australasica* sensu Gamble, Fl. Presi. Madras 2:501. 1957 (BSI Reprint); Shukla in Verma *et al.,* Fl. Madhya Pradesh 1: 587. 1993;

Chowdhery in Hajra Fl. India 12: 386. 1995; Samvatsar, Fl. West. Trib. Madhya Pradesh 160.1996.

Erect, dichotomously branched herbs or undershrubs; leaves 2-6 × 1.5-3 cm, elliptic,broadly elliptic or oblanceolate, 3-nerved, acute, coarsely serrate; petiole dilated at base; heads yellow, fasciculate; involucral bracts 2 per head, connivent, oblong; achenes 2-2.5 mm long, obovoid, 8-10 ribbed, black; pappus absent.

Fl.-Fr.: Nov. -Feb.

Common weed of fields and gardens, in the whole area.

GALINSOGA Ruiz & Pavon W. 15; MP 2; SA 1

Galinsoga parviflora Cav., Ic. Descr. Pl. 3: 41. t. 281. 1795; Hook.f., Fl. Brit. India 3: 311. 1881; Shukla in Verma *et al.,* Fl. Madhya Pradesh 1:588. 1993; Chowdhery in Hajra *et al.,* Fl. India. 12: 388. 1995; Samvatsar, Fl. West. Trib. Madhya Pradesh 160.1996.

Annual, erect, flaccid, simple or branched herbs, 15–40 cm high; leaves opposite, 2–4 × 1–3.5 cm ovate or oblong–ovate, base rounded or cuneate, obtuse or acute at apex, margins undulate–serrate to entire, sparsely pilose on both surfaces; heads *ca* 6 mm across, in terminal and axillary panicles; achenes *ca* 2 mm long, ovate, angular; pappus scales *ca* 2 mm long ciliate; palea oblong – spathulate, mostly with 2, lateral, acute lobes.

Fl.-Fr.: Feb. – March

Alirajpur, Burhanpur, Dhar, Indore, Khandwa, in hedges of fields, moist ground.

GLOSSOCARDIA Cass. W. 12; M.P. 1; SA 1

Glossocardia bosvallea (L.f.) DC. in Wight, Contrib. Bot. Ind. 19. 1834; *Verbesina bosvallea* L.f. Suppl. Pl. 379. 1781; *Glossocardia linearifolia* Cass., Dict. Sci. Nat. 19: 62. 1821; Hook.f., Fl. Brit. India 3: 308. 1881; Shukla in Verma *et al.,* Fl. Madhya Pradesh 1: 589. 1993; Chowdhery in Hajra *et al.,* Fl. India 12: 391. 1995; Samvatsar, Fl. West. Trib. Madhya Pradesh 161. 1996. *Patthar suwa*

Annual, diffuse, erect or prostrate herbs, 10–30 cm high, sometimes woody at base, stem and branches grooved; leaves alternate pinnatisect; segments 1–5 cm long, linear, apiculate; heads *ca* 5 mm across, terminal or axillary; achenes 8–9 mm long, narrowly oblong, densely ciliate on angles; pappus consisting of small stiff awns.

Fl.-Fr.: Sept. – Nov.

Common in open, dry, grassy places all over the area.

GNAPHALIUM L. W. 122; MP 1; SA 1

Gnaphalium polycaulon Pers., Syn. Pl. 2: 421. 1807; *G indicum* auct. non L. (1753); Hook.f., Fl. Brit. India 3: 289. 1881; Shukla in Verma *et al.,* Fl Madhya Pradesh 1: 590.1993; Pant in Hajra *et al.,* Fl. India 13: 91. 1995; Samvatsar, Fl. West. Trib. Madhya Pradesh 161. 1996. *Munderi*

Annual, erect, decumbent 10-20 cm high, woolly herbs; leaves 1.5–3 × 0.4–0.7 cm, semiamplexicaul, linear, obovate or spathulate, woolly on both surfaces; heads *ca* 3 mm across, golden- yellow, in leafy spikes; achenes *ca* 0.5 mm long, oblong, minutely papillose; pappus hairs white, *ca* 1.8 mm long .

Fl.-Fr.: March –April

Common in drting ditches, along streams, all over the area.

GNOMOPHALIUM Greuter Monotypic

Gnomophalium pulvinatum (Delile) Greuter, in Willdenowia 33: 242.2003; *Gnaphalium pulvinatum* Delile, Descr. Égypte, Hist. Nat. 266. 1812; Hook.f., Fl. Brit India 3:289. 1881; Shukla in Verma *et al.,* Fl. Madhya Pradesh 1: 591. 1993; Pant in Hajra *et al.,* Fl. India 13: 92. 1995; Samvatsar, Fl. West. Trib. Madhya Pradesh 161. 1996.

Annual, prostrate, diffusely branched, 10-30 cm high, woolly herbs; leaves 0.5 –1.3 × 0.2–0.6 cm, spathulate mucronate, attenuate at base, hairy above and densely woolly beneath; heads *ca* 2 mm across, in axillary or terminal clusters among terminal leaves; achenes long, oblong, minutely papillose; pappus hairs white.

Fl.-Fr.: Dec. – April

Alirajpur, Burhanpur, Dhar, Indore, Jhabua, Khandwa.

GONIOCAULON Cass. Monotypic

Goniocaulon indicum (Klein ex Willd.) Benth ex C.B. Clarke, Compos. Ind. 236. 1876; *Serratula indica* Klein ex Willd; Sp. Pl. 3: 1642. 1804; *Goniocaulon glabrum* Cass. in Bull. Soc. Philom. 34. 1817; Hook.f., Fl. Brit. India 3: 377. 1881; Shukla in Verma *et al.,* Fl. Madhya Pradesh 1: 591. 1993; Hajra in Hajra *et al.,* Fl. India 12: 181. 1995; Samvatsar, Fl. West. Trib. Madhya Pradesh 161. 1996.

Annual, erect herbs, 30–90 cm high, stem 4–8 angled; leaves 4–12 × 2–3 cm, linear–obovate or sub-spathulate, amplexicaul, toothed or serrate at margins with black dots on both surfaces; heads *ca* 8 mm across, in fascicles of compound corymbs; achenes *ca* 5 mm long, oblong, many-ribbed, glabrous; pappus scales many – seriate, unequal, light brown.

Fl.-Fr.: Nov. –Dec.

Common weed of fields, throughout the area.

GRANGEA Adans. W. 8; MP 1; SA 1

Grangea maderaspatana (L.) Poir., in Lamk, Encycl. Suppl. 2: 825. 1812; *Artemisia maderaspatana* L., Sp. Pl. 849. 1753; Hook.f., Fl. Brit. India 3: 247. 1881; Shukla in Verma *et al.,* Fl. Madhya Pradesh 1:592. 1993; Hajra in Hajra *et al.,* Fl. India 12:127. 1995; Samvatsar, Fl. West. Trib. Madhya Pradesh 162. 1996.

Annual, prostrate, tufted leafy herbs, branches 5–20 cm long, forming circular patches; leaves 2–6 cm long, pinnatifid, opposite in subopposite lobes, largest at apex, coarsely serrate-dentate at margins, pubescent on both surfaces; heads solitary, 5–6 mm aross; achenes 1–1.5 mm long, compressed, glandular, turbinate, truncate at tip with 2 hairy tubular pappus scales.

Fl.-Fr.: Dec. –May

Common in moist, open soil, all over the area.

KLEINIA Jacq. W. 59; MP 1; SA 1

Kleinia grandiflora (Wall. ex DC.) N. Rani in Fl. Tamilnadu Carnatic 3(1): 801. 1983; *Notonia grandiflora* Wall. ex DC., Arch. Bot. (Paris) 2: 518. 1833; Hook.f., Fl. Brit. India 3: 337. 1881; Hajra in Hajra *et al.,* Fl. India 12: 127. 1995; Samvatsar, Fl. West. Trib. Madhya Pradesh 163. 1996; Khanna *et al.,* Suppl. Fl. Madhya Pradesh 1: 135. 2001.

Sub-shrubs; branchlets angled above; leaves alternate, to 14 × 8 cm, obovate to oblanceolate, base cuneate, apex obtuse, entire; corymbs terminal. Heads homogamous, cylindric, 1.5 cm; florets to 2.5 cm long; corolla cream or yellowish, tube to 18 mm; achenes linear or oblong, to 5.5 mm, ribbed; pappus 1.5 cm.

Fl.-Fr.: Aug – May

Alirajpur, Dhar, Indore.

LAGASCEA Cav. W. 8; MP 1; SA 1

Lagascea mollis Cav., in Anal. Sci. Nat. 6: 332. t. 44. 1803; Hook.f., Fl. Brit. India 3: 302. 1881; Shukla in Verma *et al.,* Fl. Madhya Pradesh 1: 596. 1993; Chowdhery in Hajra *et al.,* Fl. India 12: 397. 1995; Samvatsar, Fl. West. Trib. Madhya Pradesh 162. 1996.

A remotely branched, villous, annual, herb; leaves ovate, 3-8 × 1.5-5 cm, acute entire or serrulate; heads white, clustered at the ends of branches; clusters subtended by 5, ovate, leafy, free or connate bracts; involucre tubular, pilose; 4-5 lobed; achenes cuneate, compressed, enclosed within involucre; pappus forming a lanciniate cup.

Fl.-Fr.: Aug. – Dec.

Very common in hedges of fields, roadsides of villages, throughout the area.

LAGGERA Sch. -Bip ex Koch W. 9; MP 3; SA 3

Key to Species

1. Stem not winged; heads on long peduncles .... **L. aurita**
1. Stem winged; heads on short peduncles
    2. Wings of stem entire and continuous; anther bases sagittate .... **L. alata**
    2. Wings of stem deeply lobed or toothed, often discontinuous; anther bases obtuse .... **L. crispata**

Laggera alata (D. Don) Sch.-Bip. ex Oliver, in Trans. Linn. Soc. London 29: 94. 1873; *Erigeron alatum* D. Don, Prodr. Fl. Nepal. 171. 1825; Hook.f., Fl. Brit. India 3: 271. 1881; Shukla in Verma *et al.,* Fl. Madhya Pradesh 1: 597. 1993; S. Kumar in Hajra *et al.,* Fl. India 13: 150.1995.

Annual, tall herbs; stem broadly winged; leaves sessile, oblong-lanceolate, decurrent on to wings of stems, dentate-glandular; heads purple; achenes *ca* 1 mm long, cylindric, ribbed, hairy; pappus *ca* 8 mm long, white.

Fl.-Fr.: Nov. - Feb.

Rare, **Burhanpur** only, Asirgad

Laggera aurita L.f., Suppl. 367. 1781; Hook.f., Fl. Brit. India 3: 271. 1881; Shukla in Verma *et al.,* Fl. Madhya Pradesh 1: 597. 1993; S. Kumar in Hajra *et al.,* Fl. India 13: 150.1995.

Annual, erect herbs, *ca* 30–90 cm high, stem viscous hairy, not winged; leaves 3–8 × 0.5–1.2 cm, obovate or oblanceolate, sometimes auricled at base, toothed, lobed or pinnatifid, decurrent, viscidly pubescent; upper leaves much smaller after bract line, sessile; heads *ca* 1 mm across, in corymbose or diffused panicles; achenes *ca* 1 mm long, cylindric, obscurely ribbed, thinly hairy; pappus *ca* 4 mm long, white.

Fl.-Fr.: Feb. – March

Alirajpur, Burhanpur, Dhar, Indore.

Laggera crispata (Vahl) Hepper & Wood in Kew Bull. 38: 83. 1983; *Conyza crispata* Vahl, Symb. Bot. 1: 71. 1790; *Laggera pterodonta* (DC.) Sch. -Bip. ex Oliver in Trans. Linn Soc. 29: 94. 1873; Hook.f., Fl. Brit. India 3: 271. 1881; Shukla in Verma *et al.,* Fl. Madhya Pradesh 1: 598. 1993; Kumar in Hajra *et al.,* Fl. India 13: 151. 1995.

Aromatic, annual or perennial, viscid pubescent, much branched herbs; leaves oblanceolate, obovate or oblong, 3.0-15 × 0.4-0.5 cm, shortly auriculate, subacute and apiculate at apex, serrate-dentate on margins; heads purplish, *ca* 6 mm in diameter, in terminal, leafy, corymbiform panicles, on 2.0-5.0 cm long peduncles; achenes c 1.2 mm long, faintly ribbed, hairy; pappus white, 4-5 mm long.

*Fl.-Fr.:*Jan. -March

Khandwa only, weed in wheat fields.

LAPHANGIUM (Hillard & B.L. Burtt.) Tzvelev W. 3; MP 2; SA 2

Key to Species

1. Leaves 2-8 cm long; involucral bracts 2-3 seriate; pappus white .... **L. luteoalbum**
1. Leaves up to 6 cm long; involucral bracts 3-4 seriate; pappus white or yellow .... **L. affine**

Laphangium affine (D. Don) Tzvelev in Byull. Moskovsk. Obshch. Isp. Prir. Otd. Biol. 98(6): 105. 1993; *Gnaphalium affine* D. Don, Prodr. Fl. Nepal. 173. 1825; *G. luteo-album* L. var. *multiceps* DC., Prodr. 6: 222. 1838; Hook.f., Fl. Brit. India 3: 288. 1881; *Gnaphalium luteo–album* L. subsp. *affine* (D. Don) Koster in Blumea 4(3): 484. 1941; Shukla, in Verma *et al.,* Fl. Madhya Pradesh 1:590.1993; Pant in Hajra *et al.,* Fl. India 13: 89. 1995.

Annual, erect or ascending herbs; leaves sessile, oblong-spathulate, obtuse; heads yellowish-brown, in corymbose leafless clusters; involucral bracts golden yellow; achenes oblong, brown; pappus *ca* 2 mm long, yellowish.

Fl.-Fr.: Throughout the year.

Barwani, Khargone.

Laphangium luteoalbum (L.) Tzvelev in Byull. Moskovsk. Obshch. Isp. Prir. Otd. Biol. 98(6): 105. 1993; *Gnaphalium luteo–album* L., Sp. Pl. 851. 1753; Hook.f., Fl. Brit. India 3: 288.

1881; Pant in Hajra *et al.,* Fl. India 13: 87. 1995; Samvatsar, Fl. West. Trib. Madhya Pradesh 161. 1996; Khanna *et al.,* Suppl. Fl. Madhya Pradesh 1: 134.2001. *Zinki mundi*

Annual, erect herbs; leaves sessile, oblong, woolly on both surfaces; heads white to yellow, in dense leafless corymbose panicles; involucral bracts reddish-brown; achenes oblong, pappilose; pappus hairs white, shorter than involucres.

Fl.-Fr.: Almost throughout the year.

Common in moist, shady areas, all over.

LAUNAEA Cass. W. 59; MP 3; SA 3

Key to Species

1. Heads terminal; achenes ellipsoid, strongly ribbed; pappus hairs white, united at base........**L. asplenifolia**
1. Heads axillary, solitary or in fascicles or racemose; pappus white
   2. Stem rooting at nodes; heads in the axils of cauline leaves; achenes obtusely tetragonous, truncate at apex .... **L. sarmentosa**
   2. Stem not rooting at nodes; heads in racemes; achenes ribbed, rugulose .... **L. procumbens**

Launaea aspleniifolia (Willd.) Hook.f., Fl. Brit. India 3: 415. 1881; *Prenanthes asplenifolia* Willd., Sp. Pl. 3:1540.1803; Mamgain & Rao in Hajra *et al.,* Fl. India 12: 306. 1995; Samvatsar, Fl. West. Trib. Madhya Pradesh 162. 1996; Khanna *et al.,* Suppl. Fl. Madhya Pradesh 1: 137.2001. *Gilobi*

Perennial herbs, with thick rootstock, stem 5–30 cm high, branched from the base, glabrous; leaves radical, 4–9 × 1–3.5 cm, oblanceolate, narrowly obovate or pinnatifid or sinuate lobed, sometimes runcinate pinnatifid, shortly petioled or sessile; heads 9–15 mm long in terminal, paniculate on dichotomous branches; achenes 2–2.2 mm long, columnar, angled and ribbed; pappus 8–9 mm long, white, deciduous.

Fl.-Fr.: March–Oct.

Alirajpur, Dhar, Indore.

Launaea procumbens (Roxb.) Ramayya & Rajagopal in Kew Bull. 23 (3): 465. t. 1969; *Prenanthes procumbens* Roxb., Fl. Ind. 3: 404. 1832; *Launaea nudicaulis sensu auct. plur. non* L. *nudicaulis* (L.) Hook.f., *sensu stricto*. Fl. Brit. India 3: 416. 1881; Shukla in Verma *et al.,* Fl. Madhya Pradesh 1: 599. 1993; Mamgain & Rao in Hajra *et al.,* Fl. India 12: 309. 1995; Samvatsar, Fl. West. Trib. Madhya Pradesh 163. 1996. *Bhun patri*

Annual, much branched, glabrous, 10-40 cm high herbs; stem naked or leaf–bearing; leaves in basal rosette, 4–20 cm long, pinnatifid, obovate–oblong or spathulate, dentate, teeth cartilaginous; heads 12–18 mm long, cylindric, remotely sub-racemose along the branches; achenes 2.5–3 mm long, compressed, thick, 4– ribbed, rugulose; pappus 6–7 mm long, soft white.

Fl.-Fr.: March – Sept.

Common in fields and waste land, throughout the area.

Launaea sarmentosa (Willd.) Sch.-Bip. ex Kuntze, Rev. Gen. Pl. 1: 350.189; *Prenanthes sarmentosa* Willd. Sp. Pl. 3: 1540.1804; *Launaea pinnatifida* Cass. in Ann. Sc. Nat. Ser. 1:

23: 85, 1831; Hook.f., Fl. Brit. India 3: 416. 1881; Samvatsar, Fl. West. Trib. Madhya Pradesh 163. 1996. Khanna *et al.,* Suppl. Fl. Madhya Pradesh 1: 138. 2001.

Creeping, slender, radically spreading herbs; leaves 2.5-8 cm long, oblong, runcinate-pinnatifid or sinuate toothed, with cartilaginous teeth; heads bright yellow, solitary or 2-3, in fascicles; involucral bracts white, membranous margined; achenes obtusely tetragonous, truncate at apex; pappus hairs soft, white.

Fl.-Fr.: Nov. -Feb.

Alirajpur, Dhar.

OLIGOCHAETA (DC.) C. Koch W. 4; MP 1; SA 1

Oligochaeta ramosa (Roxb.) Wagenitz in Verroffent. Geobot. Inst. 37: 323. 1962; *Carduus ramosus* Roxb. Fl. Ind. 3: 407. 1832; *Volutarella divaricata* Benth. in Benth. & Hook.f. Gen. Pl. 2: 476, 1873; Hook.f., Fl. Brit. India 3: 383. 1881; *Amberboa ramosa* (Roxb.) Jafri in Scientist 3:29. 1959; Shukla in Verma *et al.,* Fl. Madhya Pradesh 1: 600.1993; Hajra in Hajra *et al.,* Fl. India 12: 153. 1995. Samvatsar, Fl. West. Trib. Madhya Pradesh 163. 1996. *Bramha dandi*

Annual, procumbent or straggling, dichotomously branched herbs, 30–60 cm high. Leaves 2–4 cm long oblong or obocvate, toothed, usually pinnatifid. Heads 1.2–1.5 cm across avoid. Achenes 4–5 mm long, 4–5 angled, grooved and punctate between angles; pappus *ca* 1.2 cm long, silvery brown.

Fl.-Fr.: Oct. –April

Common in waste lands, in dry habitat, in all districts.

PARTHENIUM L. W. 14; MP 1; SA 1

Parthenium hysterophorus L., SP. Pl. 988. 1753; Rao in J. Bombay Nat. Hist. Soc. 54: 218. 1956; Shukla in Verma *et al.,* Fl. Madhya Pradesh 1: 600.1993; Chowdhery in Hajra *et al.,* Fl. India 12: 403. 1995; Samvatsar, Fl. West. Trib. Madhya Pradesh 164. 1996. *Gajar ghas*

Annual, erect diffusely branched, 50-80 cm high herbs; stem longitudinally grooved, covered with white hairs; leaves deeply 1–2 pinnately dissected 2–12 × 3 – 5.5 cm; pinules oblong, lanceolate, entire to deeply lobed, puberulous, acute with decurrent base; inflorescence a lax panicle of numerous small heads; heads 4–6 mm across; achenes *ca* 2 mm long obovate, black, crowned with remnant of corolla appendage and style.

Fl.-Fr.: Oct. –March

Throughout the area.

PENTANEMA Cass. W. 21; MP 2; SA 2

Key to Species

1. Leaves 1.5-7 × 0.8-2 cm, lower subsessile, upper sessile; heads 1.2-1.8 cm across; ray florets 20-40, 3-toothed at apex; achenes 0.6 mm long .... **P. cernuum**

1. Leaves 7-12 × 2-3.5 cm, all sessile and auriculate basally; heads 1.2-2 cm across; ray florets absent; achenes 0.1-0.2 mm long .... **P. indicum**

Pentanema cernuum (Dalzell) Ling, in Acta Phyt. Sin. 10: 180.1965; *Vicoa cernua* Dalzell in Dalz. & Gibs. Bomb. Fl. 126. 126 1861; Hook.f., Fl. Brit. India 3: 279. 1881; Shukla in Verma *et al.,* Fl. Madhya Pradesh 1: 601. 1993; S. Kumar in Hajra *et al.,* Fl. India 13: 28. 1995.

Annual, erect herbs, 15–40 cm high; leaves elliptic–lanceolate, 1.5–5 × 0.7–2 cm, acuminate, serrate, sparsely hairy on both surfaces; upper leaves sessile with small auricles at base; heads 0.8–1.5 cm across, drooping, often with a small foliaceous bract above the middle; pappus 3–4 mm long, white.

Fl.-Fr.: Nov. – Feb.

Common in open grassland everywhere.

Pentanema indicum (L.) Ling in Acta Phyt. Sin. 10: 179. 1965; *Inula indica* L., Sp. Pl. ed. 2. 1236. 1763; *Vicoa auriculata* Cass., Ann. Sci Nat. Ser. 1, 17: 418. 1929; Hook f., Fl. Brit. India 3: 297. 1881; Shukla in Verma *et al.,* Fl. Madhya Pradesh 1: 601. 1993; S. Kumar in Hajra *et al.,* Fl. India 13: 28. 1995; Samvatsar, Fl. West. Trib. Madhya Pradesh 164. 1996. *Sonsali*

Annual, suffruticose herbs, 30–60 cm high, stems reddish- brown, hairy, glabrous with age; leaves sessile, 2–5 × 0.5–0.6 cm, ovate-lanceolate, acute, cordate-hastate at base, scabrid on margins, pubescent on both surfaces; heads 1.3 –1.5 cm across, usually solitary; achenes *ca* 1 mm long, thinly hairy in disc florets, glabrous in ray florets; pappus hairs white, *ca* 2.5 mm long.

Fl.-Fr.: Sept. –March

Common weed of waste lands, fields, outskirts of forest, all over the area.

PHYLLOCEPHALUM Blume W. 13; MP 3; SA 2

Key to Species

1. Plants not fleshy; leaves 7-12 × 2-3.5 cm, white –cottony between nerves and reticulations beneath; heads 1.5-2 cm across; involucral bracts many seriate; achenes 0.1-0.2 mm long, 8-10 ribbed; pappus 2-seriate . ... **P. indicum**
1. Plants fleshy; leaves 9-12 × 3-4 cm, white-woolly tomentose with brown nerves beneath; achenes 2 mm long, 10-ribbed; pappus 1-seriate .... **P. phyllolaenum**

Phyllocephalum indicum (Less.) K. Kirkman in Rhodora 83: 12. 1981; *Amphiraphis indica* Less., in Linnaea 6: 686. 1831; *Centratherum reticulatum* (Wight) Benth. & Hook.f. in Benth. & Hook.f., Gen. Pl. 2: 225. 1873; *Decaneurum reticulatum* Wight in Wight, Contr. Bot. India 7: 1834; Hook f., Fl. Brit. India 3: 227. 1881; Shukla in Verma *et al.,* Fl. Madhya Pradesh 1: 602. 1993; Uniyal in Hajra *et al.,* Fl. India 12: 341. 1995.

Annual, erect, branched herbs, up to 1 m high; leaves elliptic-ovate, 7-12 × 2-3.5 cm, acuminate at apex, attenuate at base, margins serrate, hairy above and on nerves below, white cottony between the prominently reticulated veins below; upper leaves sessile to slightly petioled, lower ones long petioled; heads in terminal and axillary corymbs, 1.5-2 cm across; involucre bracts many-seriate, outer ones leafy, shorter; achenes 0.1-0.2 cm long, 8-10-ribbed; pappus yellow, of barbellate hairs, 0.25-0.3 cm long.

Fl.-Fr.: Sept. – March

Rare, **Khandwa** only, Kalibhit forest.

Phyllocephalum phyllolaenum (DC.) Narayana, in Curr. Sci. 51: 438. 1982; *Decaneurum phyllolaenum* DC., Prodr. 7: 264. 1839; *Centratherum phyllolaenum* (DC.) Benth. ex C.B. Clarke, Comp. Ind. 4. 1876; Hook.f., Fl. Brit. India 3: 228. 1881; Uniyal in Hajra *et al.,* Fl. India 12: 341. 1995; Samvatsar, Fl. West. Trib. Madhya Pradesh 164. 1996; Khanna *et al.,* Suppl. Fl. Madhya Pradesh 138. 2001.

Annual, branched herbs, up to 1 m high; leaves elliptic, up to 15 × 6 cm, acute or acuminate at apex, sparsely hairy above, white tomentose beneth; heads solitary, axillary and terminal; outer involucral bracts up to 4 cm long; achenes obconic, 10-ribbed; pappus hairs reddish.

Fl.-Fr.: Sept. –Nov.

Alirajpur, Dhar.

PULICARIA Gaertn. W. 68; MP 4; SA 3

Key to Species

1. Leaves semiamplexicaul; heads many; ray florets tubular; bracts of involucres slender; outer row of pappus a minute, toothed cup, inner a row of white hairs . ... **P. foliolosa**

1. Leaves not amplexicaul; heads solitary; ray florets ligulate; involucral bracts sub-foliaceous; outer row of pappus a setulose-laciniate cup, inner row of barbellate hairs, turning reddish

 2. Plants up to 30 cm tall; peduncles not thickened upwards; leaves distant; heads 0.8-1.2 cm across .... **P. angustifolia**

 2. Plants 30-60 cm tall; peduncles thickened upwards; leaves crowded; heads 1.2-1.8 cm across .... **P. wightiana**

Pulicaria angustifolia DC., Prodr. 5: 479. 1836; Hook.f., Fl. Brit. India 3: 299. 1881; Shukla in Verma *et al.,* Fl. Madhya Pradesh 1: 603. 1993; S. Kumar in Hajra *et al.,* Fl. India 13: 33. 1995; Samvatsar, Fl. West. Trib. Madhya Pradesh 164. 1996.

Annual, erect herbs, 7–30 cm high, with woody rootstock; leaves sessile 1–3 × 0.3–1 cm, linear–oblong or obovate, entire or serrulate, obtuse, hairy on both surfaces; heads *ca* 12 mm across, yellow, on hairy peduncles; achenes *ca* 1.5 mm long, scarcely quadrangular, with a few scattered hairs; pappus 2–seriate: outer row, a setulose– laciniate cup; inner row of white, barbellate hairs.

Fl.-Fr.: Aug. – Nov.

Alirajpur, Burhanpur, Dhar, Indore, Khandwa.

Pulicaria foliolosa DC., Prodr. 5: 480.1834; Hook.f., Fl. Brit. India 3: 298. 1881; Shukla in Verma *et al.,* Fl. Madhya Pradesh 1: 603. 1993; S. Kumar in Hajra *et al.,* Fl. India 13: 35. 1995. Samvatsar, Fl. West. Trib. Madhya Pradesh 164. 1996.

Annual, erect much branched herbs, 20–40 cm high, stem and branches pubescent; leaves 2–6 × 0.6-1.8 cm, linear–oblong or oblanceolate, obtuse at apex, hairy on both surfaces; heads *ca* 7 mm across, glandular woolly; achenes *ca* 0.7 mm long, oblong, terete, sparsely hairy; pappus 1–2.5 mm long, biseriate, outer minute toothed; inner longer, slender white.

Fl.-Fr.: Feb. – March

Dharl,Indore, Khandwa.

Pulicaria wightiana (DC.) Clarke, in Comp. Ind. 128. 1876; *Poloa wightiana* DC. Arch. Bot. (Paris) 2: 514. 1833; *Callistephus wightianus* DC. Prodr. 5: 275. 1836; Hook.f., Fl. Brit. India 3: 298. 1881; Shukla in Verma *et al.,* Fl. Madhya Pradesh 1: 605. 1993; Samvatsar, Fl. West. Trib. Madhya Pradesh 165. 1996.

Erect herbs; leaves 1.5-4 × 0.3-1.2 cm, sessile, oblong, usually serrate, pubescent on both surfaces; heads bright yellow, solitary; involucral bracts elliptic-oblong to elliptic-obovate, pubescent; achenes 1 mm long, terete, oblong or obscurely quadrangular, brown, shining; outer row of pappus a setulose-lanciniate cup, inner row of barbellate hairs.

Fl.-Fr.: Sept. -Oct.

Common along water streams, in rocky and shady places.

SCLEROCARPUS Jacq. W. 11; MP 1; SA 1

Sclerocarpus africanus Jacq., Icon. Pl. Rar. 1: 17, t. 176. 1781; Hook.f., Fl. Brit. India 3: 305. 1881; Shukla in Verma *et al.,* Fl. Madhya Pradesh 1: 605. 1993; Samvatsar, Fl. West. Trib. Madhya Pradesh 165. 1996.

Annual, hispid herbs; leaves ovate, 2.5-7 × 1-3.5 cm, remotely serrate, 3-nerved, acute or acuminate; heads yellow, terminal or leaf-opposed, sessile, supported by 2-3 floral leaves; achenes obovoid, smooth, curved; pappus absent.

Fl.-Fr.: Aug. -Nov.

Very common forest ground cover, all over the area.

SONCHUS L. W. 131; MP 3; SA 3

Key to Species

1. Outer involucral bracts densely clothed with purplish gland-tipped hairs; achenes transversely rugose; perennials .... **S. brachyotus**
1. Outer involucral bracts glabrous or nearly so; achenes not rugose, muriculate or smooth; annuals
   2. Cauline leaves with spreading, acute auricles; achenes transversely muriculate .... **S. oleraceous**
   2. Cauline leaves with rouded, appressed auricles; achenes smooth .... **S. asper**

Sonchus asper (L.) Hill. Herb. Brit. 1: 47. 1769; *Sonchus oleraceus* var. *asper* L., Sp. Pl. 794. 1753; Hook.f., Fl. Brit. India 3: 414. 1881; Shukla in Verma *et al.,* Fl. Madhya Pradesh 2: 610.1993; Mamgain & Rao in Hajra *et al.,* Fl. India 12: 318. 1995; Samvatsar, Fl. West. Trib. Madhya Pradesh 165. 1996.

Annual, glabrous herbs, 15–70 cm high; lower leaves 12–20 cm oblanceolate or spathulate, lobed or undivided, middle leaves often runcinate–pinnatifid, some times entire and basal leaves similar, dentate or sub-spinose at margins, auricles well developed, rounded at margins; uppermost leaves lanceolate entire, sub-spinose, dentate; heads *ca* 15 mm across, with glandular peduncles; achenes 2.5 –3 mm long, obovate, narrowly winged, 3–ribbed, on each side, brown; pappus *ca* 8 mm long, silky white.

Fl.-Fr.: Dec. – April

Common weed of wastelands and fields, all over the area.

Sonchus brachyotus DC., Prodr. 7: 186. 1838; *Sonchus arvensis auct. non* L. 1753; Hook.f., Fl. Brit. India 3: 414. 1881; Shukla in Verma *et al.,* Fl. Madhya Pradesh 2: 610.1993; Mamgain & Rau in Hajra *et al.,* Fl. India 12: 320.1995; Samvatsar, Fl. West. Trib. Madhya Pradesh 165. 1996.

Perennial, erect herbs, 10 –100 cm high, with thick tap roots; leaves often radical, 5–15 cm long; middle leaves oblong or spathulate–oblong, entire or lyrate, pinnati–lobed, with amplexicaul base, auricles well developed, acutely dentate at margins; upper leaves linear; heads cylindric, *ca* 15 mm across, on hairy or hispid peduncles; achenes *ca* 3 mm long oblong, oblanceolate, transversely rugose, 5– ribbed, brown; pappus *ca* 10 mm long, silky white.

Fl.-Fr.: Dec. April

Common in moist and shady places in the entire area.

Sonchus oleraceus L., Sp. Pl. 794. 1753; Hook.f., Fl. Brit. India 3: 414. 1881; Shukla in Verma *et al.,* Fl. Madhya Pradesh 2: 610.1993; Mamgain & Rao in Hajra *et al.,* Fl. India 12: 321. 1995; Samvatsar, Fl. West. Trib. Madhya Pradesh 165. 1996.

Annual, erect, branched, 20-70 cm high herbs; cauline leaves sessile, amplexicaul 10–20 × 4–8 cm, ovate-oblong, with acute auricles, margins runcinate; upper leaves lanceolate; heads *ca* 1 cm across, in panicles, peduncles hairy, glandular or eglandular; achenes *ca* 3 mm long, oblanceolate, compressed, rugose, 3– ribbed, minutely muricate between the ribs; pappus *ca* 6 mm long silky white.

Fl.-Fr.: Feb. –March

Common in wastelands, fields, along irrigation canals, all over the area.

SPHAERANTHUS L. W. 34; MP 2; SA 1

Sphaeranthus indicus L., Sp. Pl. 927. 1753; Hook.f., Fl. Brit. India 3: 275. 1881; Shukla in Verma *et al.,* Fl. Madhya Pradesh 1: 611. 1993; S. Kumar in Hajra *et al.,* Fl. India 13: 160.1995. Samvatsar, Fl. West. Trib. Madhya Pradesh 166. 1996. *Gorakh mundi*

Annual, procumbent, ascending, tomentose, aromatic herbs, 30–60 cm high; leaves sessile, 2.5 – 5 × 1.5 – 2 cm obovate – oblong. acute to rounded at tip, mucronate, margins dentate, glandular–villous on both surfaces; heads *ca* 10 mm across, globose or ellipsoid, purplish; achenes *ca* 1 mm long, glandular hairy; pappus absent.

Fl.-Fr.: Nov. -March

Common in waste lands, fields, roadsides in moist ground.

SYNEDRELLA Gaertn. Monotypic

Synedrella nodiflora (L.) Gaertn., Fruct. Sem. 2: 456. t. 171. f. 7. 1791; *Verbesina nodiflora* L., Cent. Pl. 28. 1755; Hook.f., Fl. Brit. India 3: 308. 1881; Shukla in Verma *et al.,* Fl. Madhya Pradesh 1: 613. 1993; Chowdhery in Hajra *et al.,* Fl. India 12: 413. 1995.

Annual, erect dichotomously branched herbs, 40–90 cm high; leaves opposite, petioled 3–8 × 1– 4.5 cm ovate, lanceolate, margins serrate dentate, appressed-pilose on both surfaces; heads *ca* 5 mm across, sessile axillary; achenes of disc florets 3–4.5 mm long, black, compressed; pappus *ca* 6 mm long paleae, lanceolate.

Fl.-Fr.: Nov. -Dec.

Rare, **Dhar**, as forest undergrowth, at Mandu plateau.

TRICHOLEPIS DC. W. 17; MP 3; SA 3

Key to Species

1. Pappus absent .... **T. amplexicaulis**
1. Pappus present
   2. Palea of the receptacle reaching above the pappus; stem sparsely leafy .... **T. glaberrima**
   2. Palea of the receptacle shorter than pappus; stem prominently leafy .... **T. radicans**

Tricholepis amplexicaulis C.B. Clarke Comp. Ind. 240.1876; Hook.f., Fl. Brit. India 3: 381. 1881; Shirodkar, *et al.,* in N.P. Singh *et al.,* Fl. Maharashtra 2:245. 2001.

Annual, erect, stout, glabrous herb; stems rigid; branches angled, ribbed, quite smooth, shining; leaves linear-oblong, 7.5-12.5 × 1.2-5. 7 cm, quite glabrous, pale when dry, half amplexicaul at base; heads 1.8-2.5 cm across; peduncles short, stout, angular, bracteate; achenes pale brown, compressed, *ca* 5 mm long, perfectly smooth, glossy; pappus absent or of a few bristles.

Fl.-Fr.: Oct. -Feb.

Burhanpur, Khandwa in open ground.

Tricholepis glaberrima DC. Prodr. 6: 564, 1838; Hook.f., Fl. Brit. India 3: 381. 1881; Samvatsar, Fl. West. Trib. Madhya Pradesh 166. 1996; Khanna *et al.,* Suppl. Fl. Madhya Pradesh 1: 143.2001.

Erect glabrous herbs; stems and branches angled and ribbed; leaves 2-14 × 0.5-4 cm, sessile, linear oblong or lanceolate, acute entire, spinous, serrate; heads bright purple, solitary, terminal; involucral bracts bristle pointed; achenes oblong, faintly ribbed; pappus shorter than the achenes, copious, yellowish-brown, rigid, sub-paleaceous.

Fl.-Fr.: Dec. -Feb.

In moist and shady places, Alirajpur, Dhar, Indore, Jhabua.

Tricholepis radicans (Roxb.) DC. Prodr. 6: 584. 1838; *Carduus radicans* Roxb. Fl. Ind. 3: 408. 1832; Hook.f., Fl. Brit. India 3: 381. 1881; Samvatsar, Fl. West. Trib. Madhya Pradesh 166. 1996; Khanna *et al.,* Suppl. Fl. Madhya Pradesh 1: 143.2001.

Erect or decumbent herbs; stem and branches angled, glabrous or minutely scaberulous; leaves 2.5-5 × 0.5-2 cm, oblong, toothed or serrate, glabrous, punctate, teeth as well as apex bristle pointed; heads purple, woolly pubescent; pappus yellowish, shorter than achenes, the setae flattened, many; achenes oblong, smooth, striate.

Fl.-Fr.: Sept. -Nov.

In dry, rockyor sandy, open habitat, **Alirajpur** only, Narmada bank, at Shakarja.

TRIDAX L. W. 29; MP 1; SA 1

Tridax procumbens L. Sp. Pl. 900.1753; Hook.f., Fl. Brit. India 3: 311. 1881; Shukla in Verma *et al.,* Fl. Madhya Pradesh 1: 614. 1993; Samvatsar, Fl. West. Trib. Madhya Pradesh 167. 1996. *Ghogouldi*

Procumbent or erect, hairy herbs; leaves 2-6 × 0.5-2.5 cm, ovate, ovate-lanceolate, elliptic or elliptic-lanceolate, coarsely serrate; heads solitary, yellow or pale yellow; outer involucral bracts herbaceous, very hairy; achenes silky pilose; pappus bristles plumose, aristate.

Fl.-Fr.: Almost throughout the year.

Very common along roads, dry habitat, all over the area.

XANTHIUM L. W. 11; MP 1; SA 1

Xanthium strumarium L. Sp. Pl. 987. 1753; Hook.f., Fl. Brit. India 3: 303. 1881; Grierson in Dassan & Fosb. Rev. Hand. Fl. Ceylon 1: 209. 1990; Shukla in Verma *et al.,* Fl. Madhya Pradesh 1: 619. 1993; Samvatsar, Fl. West. Trib. Madhya Pradesh 167. 1996. *Gokhru*

Erect, simple or branched, aromatic, robust herbs; leaves 2.5-12 × 2-12 cm, ovate or acuminate, dentate, hispidulous, glandular; male and female heads separate; males globular, green, fascicled; female at the base, solitary, involucre ovoid-oblong, apex with 2 spiny hooks, 2-celled, florets 2, enclosed within with two hooked stout spines, two celled, florets two, achenes enclosed in the involucres, glabrous, compressed.

Fl.-Fr.: Jan. -Feb.

Common everywhere in open ground, around villages.

ZINNIA L. W. 22; MP 1; SA1

Zinnia elegans Jacq. Coll. 3: 152. 1789; Bailey 1001. 1949; Naik Fl. Osmanabad 190.1979; Samvatsar, Fl. West. Trib. Madhya Pradesh 167. 1996; Khanna *et al.,* Suppl. Fl. Madhya Pradesh 1: 145. 2001.

An annual, hispid, 30-80 cm tall, erect herb; leaves opposite, ovate to elliptic, more or less clasping, 3-nerved; heads white, rose, orange, yellow, or red, 5-10 cm across; achenes 5-7 mm long, oblanceolate to cuneate.

Fl.-Fr.: Sept. -Dec.

Along village roadsides, during rainy season, as an escape, not wild.

CAMPANULACEAE 88/2385; 3/9; 3/4

Key to Genera

1. Flowers zygomorphic; anthers united .... LOBELIA
1. Flowers actinomorphic; anthers free, not epipetalous
   2. Capsule dehiscing at base or sides, below calyx teeth; stigma lobed; corolla divided less than halfway down .... CAMPANULA
   2. Capsule dehiscing loculicidally, within calyx teeth; stigma capitate; corolla divided nearly to the base .... WAHLENBERGIA

CAMPANULA L. W. 440; MP 3; SA 1

Campanula dimorphantha Schweinf., Beitr. Fl. Aethiop. 140.1867; *C. canescens* Wallich *ex* DC., Monogr. Campan. 292. 1830 & Prodr. 7: 473. 1839, *non* Roth 1827; C.B. Clarke in Hook.f., Fl. Brit. India 3: 439. 1881; *C. benthamii* Wall. ex Kitam., Fl. Afgh. 377. 1960; *C. wallichii* Babu, J. Bomb. Nat. Hist. Soc. 65: 808. 1969; Mukherjee in Verma *et al.,* Fl. Madhya Pradesh 1: 621. 1993; Haridasan & Mukherjee in Fasc. Fl. India 22: 41. f. 8. 1996; Samvatsar, Fl. West. Trib. Madhya Pradesh 168. 1996.

Annual, hairy herbs, up to 40 cm high; stems many, arising from rootstock; leaves sessile, linear-oblong or linear-oblanceolate, 3-6.8 × 0.5-1 cm, margin crenate or serrate, apex subacute, attenuate at base; flowers lilac-pink or greyish- purple, usually subspicate on branches of narrow panicles, dimorphic (one form without corolla or stamens); capsules hemispheric, 3-5 mm across; seeds many, ellipsoid, slightly compressed.

Fl.-Fr.: Jan. -March

Alirajpur, Burhanpur, Dhar, Indore.

LOBELIA L. W. 414; MP 3; SA 2

Key to Species

1. All anthers penicillate at apex; seeds trigonous; leaves crenate, base round .... **L. alsinoides**
1. Only 2 anthers penicillate at apex; seeds ellipsoid; leaves with an acute, cuneate base .... **L. heyneana**

Lobelia alsinoides Lam., Encycl. 3: 588. 1791; *L. trigona* Roxb., Fl. Ind.) 2: 111. 1824; C.B. Clarke in Hook.f., Fl. Brit. India 3: 423. 1881; *L. teriminalis* C.B. Clarke in Hook.f., Fl. Brit. India 3: 424. 1881; Haridasan & Mukherjee in Fasc. Fl. India 19: 42. 1988; Mukherjee in Verma *et al.,* Fl. Madhya Pradesh 1: 622. 1993; Samvatsar, Fl. West. Trib. Madhya Pradesh 168. 1996.

Annual, branched,herbs, up to 20 cm high; leaves broadly ovate, 0.6-1.7 × 0.5-1.4 cm, margin crenate-dentate, apex obtuse or subacute, base rounded or subtruncate; flowers white, solitary, axillary, with violet tinge, pedicels angular, bracts conspicuous; capsules ellipsoid to obovoid, 2-4 × 1.5-2 mm; seeds many, trogonous, ellipsoid, narrowed at both ends.

Fl.-Fr.: Sept. –Jan.

Common, in moist and shady places, all over.

Lobelia heyneana Schult., Syst. Veg. 5: 50.1819, *non* Spreng., 1825; *L. trialata* Buch. -Ham. ex D. Don, Prodr. Fl. Nepal. 157. 1825; *L. zeylanica* Moon, Cat. 14. 1824. *non* L., 1753; C.B. Clarke in Hook.f., Fl. Brit. India 3: 425. 1881, *incl.* var. *walkeri;* C.B. Clarke in Hook.f., Fl. Brit. India 3: 425. 1881; Haridasan & Mukherjee in Fasc. Fl. India 19: 51. 1988; Mukherjee in Verma *et al.,* Fl. Madhya Pradesh 1: 623. 1993; Samvatsar, Fl. West. Trib. Madhya Pradesh 168. 1996.

Annual, branched, erect, herbs, up to 25 cm high; leaves ovate or elliptic, 1-3.5 × 0.5-2.2 cm, margin crenate-serrate, apex subacute or obtuse, narrowed at base; flowers bluish- violet,solitary, axillary; pedicels 3-8 mm long; capsules ellipsoid, 3-5 × 1.5-2 mm; seeds many, ellipsoid, compressed.

Fl.-Fr.: Aug. –Dec.

Alirajpur, Dhar, Indore, Khandwa.

WAHLENBERGIA Schrad. ex Roth W. 263; MP 3; SA 1

Wahlenbergia erecta (Roth ex Schult.) Tuyn, Fl. Males. 6: 113 1960*; Dentella erecta* Roth ex Schult., Syst. Veg. 5: 25 1819*; Cephalostigma schimperi* Hochst. ex A. Rich., Tent. Fl. Abyss. 2: 2. 1850*;* C.B. Clarke in Hook.f., Fl. Brit. India 3: 425. 1881; Mukherjee in Verma *et al.,* Fl. Madhya Pradesh 1: 624. 1993; Samvatsar, Fl. West. Trib. Madhya Pradesh 169. 1996.

Annual,erect herbs, 10-35 cm high; leaves oblong or elliptic, 0.5-2.2 cm, margin thickened, undulate, shallowly and remotely dentate; flowers axillary and terminal; capsules obconic to bell-shaped, 5-nerved, 3-valved; seeds numerous, compressed.

Fl.-Fr.: Nov. – Feb.

Rare, **Alirajpur** only, Chatliya pani (Kattiwada).

PLUMBAGINACEAE 24/635; 2/2; 2/2

Key to Genera

1. Calyx not glandular; leaves perfoliate .... DYEROPHYTUM
1. Calyx glandular; leaves not perfoliate .... PLUMBAGO

DYEROPHYTUM Kuntze W. 3; MP 1; SA 1

Dyerophytum indicum (Gibs. & Wight) Kuntze, Revis. Gen. Pl. 3:394. 1891; *Vogelia indica* Gibs. & Wight in Calc. J. Nat. Hist. Soc. 7:17. 1847; C.B. Clarke in Hook.f. Fl. Brit. India 3:481. 1882; Anand Kumar in Verma *et al.,* Fl. Madhya Pradesh 1:625. 1993.

Shrubs up to 2 m tall; stem and branches scurfy; leaves perfoliate, glaucous, covered in scurfy scales, lower larger, 14 × 8 cm, ovate, margin undulate, upper smaller, up to 6 × 5 cm, orbicular; flowers reddish-pink, in dense, paniculate spikes; calyx 5-ribbed, tube winged between the ribs; corolla tube 2-2.5 cm long; stamens included; style densely hairy at base; capsules 6-8 mm long; seeds smooth, pear-shaped.

Fl.-Fr.: Dec. -Feb.

Rare, **Indore** only, in open places, roadsides, Hatod-Banedia road.

PLUMBAGO L. 17; W. 17; MP 2; SA 1

Plumbago zeylanica L., Sp. Pl. 151. 1753; C.B. Clarke in Hook.f., Fl. Brit. India 3: 480.1882; Kumar in Verma *et al.,* Fl. Madhya Pradesh 1: 626. 1993; Samvatsar, Fl. West. Trib. Madhya Pradesh 169. 1996. *Chitra* **(Plate 6, Fig. 6)**

Undershrubs, up to 1.5 m high; leaves membranous, elliptic, 3.5-6.5 × 1-3.2 cm, margin entire, apex acute or subacuminate, narrowed at base into a short petiole; flowers white, in elongate, terminal spikes; capsules oblong, circumscissile near base, enclosed in persistent calyx; seeds oblong.

Fl.-Fr.: Sept. – April

Common along roadsides of villages, as scattered undershrubs, in all districts (uprooted for the medicinal roots).

PRIMULACEAE 68/2788; 6/7; 3/4

Key to Genera

1. Leaves opposite or alternate; fruit capsular; herbs .... ANAGALLIS
1. Leaves alternate; fruit baccate; shrubs or stragglers
   2. Petals free, imbricate; berry 0.3-0.5 cm across .... EMBELIA
   2. Petals united, valvate; berry 0.8-1cm across .... ARDISIA

ANAGALLIS L. 34; MP 2; SA 2

Key to Species

1. Leaves opposite; corolla blue, longer than calyx; filaments villous .... **A. arvensis**
1. Leaves alternate; corolla white or tinged with pink, shorter than calyx; filamens glabrous .... **A. pumila**

Anagallis arvensis L., Sp. Pl. 148. 1753; *Anagallis arvensis* L. var. *coerulea* (Schreb.) Gren. & Godr., Fl. France 2: 467. 1852; Hook.f., Fl. Brit. India 3: 506. 1882; Samvatsar, Fl. West. Trib. Madhya Pradesh 169. 1996; Ramesh Kumar in Mudgal *et al.,* Fl. Madhya Pradesh 2: 1. 1997. (This is Scarlet pimpernel of hill stations) **(Plate 6, Fig. 7)**

Annual erect, branched herbs, up to 30 cm high; leaves sessile, ovate, 1-2.5 × 0.7-1.2 cm, gland-dotted, apex acute or subacute, rounded at base; flowers blue, solitary, axillary; capsules 4-5 mm across, operculately dehiscent; seeds many, minute, trigonous.

Fl.-Fr.: Oct. – March

Common along irrigation canals in fields, in all districts, not encountered outside fields.

Anagallis pumila Sw., Prodr. Veg. Ind. Occ. 40.1788; *Centunculus tenellus* Duby, Prodr. 8: 72. 1844; Hook.f., Fl. Brit. India 3: 506. 1882; Samvatsar in Fl. West. Trib. Madhya Pradesh 169. 1996; Ramesh Kumar in Mudgal *et al.,* Fl. Madhya Pradesh 2: 3. 1997.

Annual, branched, herbs, up to 18 cm high; leaves subsessile, obovate-elliptic, apex acute, narrowed at base; flowers white, solitary, axillary; capsules globose, membranous, operculately dehiscent; seeds many, trigonous.

Fl.-Fr.: Sept. – Nov.

In wet, open, hard soil, along flowing water, Alirajpur, Dhar, Indore.

ARDISIA Swartz W. 789; MP 1; SA1

Ardisia solanacea Roxb., Pl. Coromandel 1: 27. 1795; *A. humilis auct. non* Vahl, 1794; C.B. Clarke in Hook.f., Fl. Brit. India 3 : 529. 1822, *p.p. excl. syn.* Samvatsar, Fl. West. Trib. Madhya Pradesh 170.1996; U.S. Vaish in Mudgal *et al.,* Fl. Madhya Pradesh 2: 8. 1997.

Shrubs or trees, up to 3 m high; leaves crowded near apex of branches, obovate or olanceolate, 7.5-16.5 × 2.8-7.2 cm subcoriaceous, succulent, veins obscure, margin entire, apex subacuminate or caudate-acuminate, narrowed at base into 1-2 cm long petiole; flowers pink, in axillary, corymbose racemes; fruits depressed globose, 5-9 mm across, black when ripe; seed solitary, globose, striate, excavated at base.

Fl.-Fr.: March – Dec.

Along banks of streams, Barwani, Khargone.

EMBELIA Burm.f. W. 134; MP 1; SA 1

Embelia basaal (Roem. & Schult.) A. DC. in Trans. Linn. Soc. London 17: 131. 1834; *Ardisia basaal* Roem. & Schult., Syst. Veg. 4:517. 1819; U.S. Vaish in Mudgal *et al.,* Fl. Madhya Pradesh 2: 9. 1997. *Baberang* **(Plate 6, Fig. 8)**

Deciduous shrubs, up to 2 m high; leaves broadly elliptic or obovate, 4.2-12.2 × 2-5 cm, gland-dotted, rusty pupescent on the nerves beneath, margin crenate, apex sub-acute or subacuminate, base rounded or subacute; petioles 1-2 cm long; flowers in axillary, rusty puberulous racemes; capsules pink, subglobose with persistent style; seed solitary, globose, excavated at base.

Fl.-Fr.: July – Feb.

Rare, Barwani, Khandwa, Khargone.

SAPOTACEAE 60/1343/; 4/5; 2/2

Key to Genera

1. Sepals 4; corolla fleshy, rugose when dry; stamens 16-20 in 2 whorls .... MADHUCA
1. Sepals 6; corolla not fleshy; stamens 6, alternating with 6 staminodes .... MANILKARA

MADHUCA J.F. Gmelin W. 114; MP 1; SA 1

Madhuca longifolia (J. Koenig ex L.) J.F. Mac Br. var. **latifolia** (Roxb.) Chev., Rep. Bot. Appl. 23: 149. 1943; *Bassia longifolia* J. Koenig ex L., Mant. Pl. 2 (App.): 563. 1771; *Bassia latifolia* Roxb., Pl. Coromandel 1: 20.1795; C.B. Clarke in Hook.f., Fl. Brit. India 3: 544. 1882; *Madhuca latifolia* (Roxb.) J.F. MacBr., Contr. Gray Herb. n.s., 53:18. 1918; Samvatsar, Fl. West. Trib. Madhya Pradesh 170.1996; Ram Lal in Mudgal *et al.,* Fl. Madhya Pradesh 2: 13. 1997. *Mahua*

Trees, up to 25 m high; bark black; young parts pubescent; leaves clustered at the apex of branches, elliptic or elliptic-oblong, 7-20.5 × 3-10.5 cm, coriaceous, pubescent or tomentose when young, apex subacuminate, base acute or rounded; petioles 2-4 cm long, pubescent when young; stipules subulate, pubescent, caducous; flowers fleshy cream-coloured, clustered at the apex of branches; berries ovoid, 2-4.5 cm long, succulent, 1-4-seeded; seeds ellipsoid.

Fl.-Fr.: March-Aug.

Common in all districts, but abundant in Alirajpur and Jhabua (**Kamdhenu** tree of Bhils of that area).

MANILKARA Adans. W. 81; MP 1; SA1

Manilkara hexandra (Roxb.) Dubard in Ann. Mus. Col. Marseille ser. 3, 23: 9. t. 2. 1915; *Mimusops hexandra* Roxb., Pl. Corom. 1: 16. t. 15. 1795; C.B. Clarke in Hook.f., Fl. Brit. India 3: 549. 1882; Samvatsar, Fl. West. Trib. Madhya Pradesh 171. 1996; Ram Lal in Mudgal *et al.,* Fl. Madhya Pradesh 2: 14. 1997. *Khirni*

Trees, up to 18 m high; bark greyish black, furrowed; leaves obovate or oblong, 5-10.5 × 2.2-5.5 cm, coriaceous, nerves prominent, apex obtuse or emarginate, base cuneate or subtruncate; petioles 1-1.5 cm long; flowers white, axillary, solitary or in fascicles of 2-6; berries ellipsoid, 0.7-1.3 cm long, yellow when ripe; seed solitary, ovoid-ellipsoid, shining, brown.

Fl.-Fr.: November–May

Alirajpur, Burhanpur, Dhar, Indore.

EBENACEAE 3/751; 1/6; 1/3

DIOSPYROS L. W. 725; MP 6; SA 3

Key to Species

1. Un-armed trees; male flowers 3-12; stamens 12-16; staminodes in female flowers 8-10; ovary densely hairy; styles 2 .... **D. melanoxylon**
1. Armed trees (spines casual in **D. chloroxylon**) branches reduced to spines; male flowers 2-6; stamens 16 in 2 whorls; staminodes in female lowers 4-8; ovary glabrous; styles 4
   2. Staminodes 4; berry 1.5-2 cm across; anther tips awned; leaves thick, chartaceous ....**D. montana**
   2. Staminodes 8; berry up tp 1 cm across; anther tips acute; leaves thin, coriaceous...........**D. chloroxylon**

Diospyros chloroxylon Roxb., Pl. Coromandel 1:38. 1795; *D. capitata* Wight, Icon. Pl. Ind. Orient. 4:t. 1224. 1848; C.B. Clarke in Hook.f., Fl. Brit. India 3:560.1882; Hanfi, in Mudgal *et al.,* Fl. Madhya Pradesh 2:18. 1997.

Trees, branchlets often modified into thorns; leaves 2-3 × 1-2 cm, alternate, elliptic or oblong, apex apiculate, rounded at base; male flowers in axillary, short, 6-7 flowered clusters; female flowers solitary; calyx 4-lobed, cupular; corolla white, 4-lobed, urceolate to campanulate, larger in female flowers; stamens 16, 2-seriate; berries globose, glabrous, *ca* 1.2 cm across; fruiting calyx flat; seeds 2-3.

*Fl.-Fr.:*July-Dec.

Barwani, Indore, Khargone.

Diospyros melanoxylon Roxb., Pl. Coromandel 1:36. 1795; C.B. Clarke in Hook.f., Fl. Brit. India 3:564. 1882; Samvatsar, Fl. West. Trib. Madhya Pradesh 171. 1996; Hanfi in Mudgal *et al.,* Fl. Madhya Pradesh 2:19. 1997. *Tendu*

Medium sized, deciduous trees, up to 10 m high, twigs softly pubescent; leaves 4-20 × 3-10 cm, ovate-elliptic; apex acute, base rouded or cuneate, margin entire, softly tomentose on both sides when young, pubescent when mature; male flowers in axillary, simple or branched, 3-15 flowered, drooping cymes; female flowers solitary; calyx lobes 5, acute in males, cordate in females, broadly 2-winged; corolla lobes 5, densely hairy; stamens 10; berries 2.5-3.5 cm across, globose or ovoid, fruiting calyx flat, with thick, undulate lobes; seeds 2-8.

*Fl.-Fr.:*Feb. -April

Very common in all districts. Perhaps number 2 in forest revenue after Sagwan.

Diospyros montana Roxb., Pl. Coromandel 1:37. 1795; C.B. Clarke in Hook.f., Fl. Brit. India 3:555. 1882; Samvatsar, Fl. West. Trib. Madhya Pradesh 171. 1996; Hanfi in Mudgal *et al.,* Fl. Madhya Pradesh 2:21. 1997. *Bis-tendu*

Shrubs or trees, with pubescent branches; leaves 3-10 × 2-5 cm, ovate-elliptic, apex acute, cuneate to cordate at base, pubescent beneath, margin entire, veins distinct; male flowers in axillary, short, usually 3-flowered cymes; female flowers solitary; calyx lobes 4, rounded; corolla lobes 5, ovate, acute; stamens 15; berries 2.5-3.5 cm across, calyx persistent with reflexed, brown lobes; seeds 2-8.

*Fl.-Fr.:*March-Aug.

Alirajpur, Barwani, Khandwa, Khargone.

OLEACEAE 25/688; 3/13; 3/5

Key to Genera

1. Fruit baccate; plants normally perennial twiners .... JASMINUM
1. Fruit capsular; small or medium trees
   2. Capsules bipartite, compressed; seeds not winged, erect; leaves simple .... NYCTANTHES
   2. Capsules pyriform, woody; seeds winged, pendulous; leaves compound .... SCHREBERA

JASMINUM L. W. 198; MP 11; SA 3

Key to Species

1. Leaves compound; 3-foliolate; calyx lobes shorter than its tube; corolla 3-3.5 cm long................ **J. flexile**
1. Leaves simple
   2. Erect, large shrubs; leaves up to 18 cm long and 10 cm broad, petiole 1-2.5 cm long; corolla lobes usually 9, tube 1 cm long, lobes 1-1.5 cm long .... **J. arborescens**
   2. Twiners; leaves up to 8 cm long and 3 cm broad, petiole 1 cm long; corolla lobes usually 7, tube 2.5 cm long, lobes 1.5 cm long . ... **J. coarctatum**

Jasminum arborescens Roxb., Fl. Ind. 1:94. 1820; *J. arborescens* var. *montana* (Roth) C.B. Clarke, in Hook.f., Fl. Brit. India 3:594. 1882; *J. montanum* Roth, Nov. Pl. Sp. 5. 1821; *J. arboreum* Schult., Mant. 1:90.1822; Khanna & Srivastava in Mudgal *et al.,* Fl. Madhya Pradesh 2:27. 1997. *Jangli chameli*

Suberect, 2-3 m high shrubs, branchlets scandent, densely hairy; leaves simple, opposite, 4-9 × 2.5-6 cm, ovate or orbicular, apex acuminate, base rounded, tomentose when young, glabrous with age; flowers white, fragrant, in trichotomous, lax, pubescent cymes; bracts linear; calyx lobes 5-6, linear; corolla lobes usually 9, lanceolate, as long as or longer than the tube; tube 1.5 cm long; berries oblong or ellipsoid, 1-1.5 cm long, black at maturity.

*Fl.-Fr.:*March-Oct.

Rare, **Indore** only, Nand Lal ghati.

Jasminum coarctatum Roxb., Fl. Ind. 1:91 1820; *J. rottlerianum* Wall. ex DC. Prodr. 8:305 1844; C.B. Clarke in Hook.f. Fl. Brit. India 3:593. 1882 (including vars. *thwaitsii* C.B. Cl. loc cit, var. *glabrior* C.B. Cl. loc cit.); Khanna & Srivastava in Mudgal *et al.,* Fl. Madhya Pradesh 2:33. 1997.

Large, climbing, fulvous hairy shrubs; leaves 8-11 × 4-5 cm, elliptic, apex acute to acuminate, cordate to rounded at base, margin entire, lateral nerves 4 pairs; flowers fascicled in terminal, dense, bracteate cymes; corolla 6-lobed; berries about 1 cm across, black at maturity.

Fl.-Fr.: Nov. -July

Rare, as forest undergrowth in **Indore**, Nahar jhabua forest.

Jasminum flexile Vahl, Symb. Bot. 3:1. 1704; C.B. Clarke in Hook.f. Fl. Brit. India 3:601. 1882; Khanna & Srivastava in Mudgal *et al.,* Fl. Madhya Pradesh 2:30.1997. *Malti*

Scandent, glabrous shrubs; leaves 3-foliolate; leaflets 3-6 × 2-3 cm, apex acuminate; flowers in axillary or terminal, dichasial cymes; calyx lobes longer than the tube; berries 0.8-1 cm long.

Fl.-Fr.: Oct. -Dec.

Indore only, Nand Lal ghati.

NYCTANTHES L. W. 2; MP 1; SA 1

Nyctanthes arbor-tristis L., Sp. Pl. 6. 1753; C.B. Clarke in Hook.f., Fl. Brit. India 3:603. 1882; Samvatsar, Fl. West. Trib. Madhya Pradesh 172. 1996; Khanna & Srivastava in Mudgal *et al.,* Fl. Madhya Pradesh 2:35. 1997. *Sirali*

Large shrubs, 3-4 m high, with 4-angled branches; leaves ovate or ovate-lanceolate, 4-11 × 2-7 cm, magin serrate, apex acute, base truncate, very scabrid, coriaceous, covered with stiff-white hairs beneath; flowers very gragrant, white with an orange tube, sessile, in terminal, trichotomous cymes; calyx lobes 5, truncate; corolla lobes 5-7, 7-10 mm long; capsules suborbicular, compressed, 1.2-2 × 1.2-2 cm, 2-celled, separating into 2, 1-seeded parts.

Fl.-Fr.: Aug. -Nov.

Very common on hill slopes, in all districts.

SCHREBERA Roxb. W. 8; MP 1; SA 1

Schrebera swietenioides Roxb., Pl. Coromandel, 2:101. 1799; C.B. Clarke in Hook.f., Fl. Brit. India 3:604. 1882; Samvatsar, Fl. West. Trib. Madhya Pradesh 172. 1996; Khanna & Srivastava in Mudgal *et al.,* Fl. Madhya Pradesh 2:37.1997. *Mokha*

Moderate sized trees, about 10-15 m tall; leaflets 7-9, opposite, 6-12 × 2-5 cm, ovate, acute or acuminate, obliquely attenuate at base, entire, coriaceous; flowers white, in trichotomous, axillary and terminal, paniculate cymes; calyx irregularly toothed, teeth ciliate; corolla with brown, glandular dots inside, lobes undulate, obtuse, tube *ca* 1.5 cm long, pubescent at base, inside; capsules *ca* 5 × 3 cm, drooping, verrucose, pear-shaped; seeds winged.

*Fl.-Fr.:*Oct. -Dec.

Alirajpur, Barwani, Dhar, Indore, Khandwa, Khargone.

SALVADORACEAE 3/8; 1/3; 1/2

SALVADORA L. W. 3; MP 3; SA 2

Key to Species

1. Leaves ovate; berries reddish-brown .... **S. persica**
1. Leaves lanceolate; berries white .... **S. alii**

Salvadora alii Rajput & Syeda, Pak. J. Bot. 42:63.2010; Shaikh Mujaffar *et al.,* Rheedea 22 (2): 80-82.2012. *Mishwak*

Evergreen shrubs or small trees, up to 3 m tall; branches drooping; leaves thick, 4-6 × 1-3 cm, apex mucronate, base cuneate, lanceolate, margin entire, glabrous, coriaceous, 3-nerved, petioles 1-2.5 cm long; stipules ovate, caducous; inflorescence axillary and terminal, lax, 5-20 cm long, panicles; flowers 2-3 mm long on slender pedicels; bracts acute, ovate, caducous; sepals, petals and stamens 4; lobes of sepals rounded; petals persistent, lobes inflexed, oblong, obtuse, deeply cleft; stamens exserted; berries 5-5.5 mm across, white when ripe; seed one, globose, smooth.

*Fl.-Fr.:*Jan. -April

Rare, **Burhanpur** only, Jainabad.

Salvadora persica L., Sp. Pl. 122. 1753; C.B. Clarke in Hook.f., Fl. Brit. India 3:619. 1882; Samvatsar, Fl. West. Trib. Madhya Pradesh 172. 1996; Khanna in Mudgal *et al.*, Fl. Madhya Pradesh 2:39. 1997. *Pilu*

Large or medium, evergreen trees; branches drooping; leaves thick, 3-6 × 2-3.5 cm, elliptic-oblong or ovate, rounded to acute at base, coriaceous; flowers greenish-yellow, in axillary and terminal, lax panicles; bracts ovate, caducous; sepals, petals and stamens 4; calyx copular, lobes rounded; corolla very thin, lobes oblong, acute, deeply cleft; stamens exserted; drupes smooth, supported by the persistent, yellow cup of calyx, red when ripe; seeds brown.

*Fl.-Fr.:*Nov. -Feb.

Common in hedges of fields, open ground, along river banks, all over the area.

APOCYNACEAE 410/5556; 28/41; 23/30

Key to Genera

1. Armed, shrubby plants; carpels completely united; ovary 2-celled; petals overlapping to the right; fruit baccate; latex milky . ... CARISSA
1. Unarmed plants
   2. Fruit drupaceous, 2-seeded; anthers included, free; petals overlapping to left; leaves whorled .... RAUVOLFIA
   2. Fruit follicular
      3. Seeds without coma, black, muriculate; flowers white, 1-3,in leaf axils; petals overlapping to left; annual herbs .... CATHARANTHUS
      3. Seeds with coma
         4. Anthers connivent around stigma, but not fusing with it to form a gynostegium
            5. Corona absent; petals overlapping to right
               6. Trees; coma apical; flowers white in terminal, corymbose cymes; follicles 30-45 cm long, lenticellate .... HOLARRHENA
               6. Woody twiners; anthers united in a cone around stigma; coma apical
                  7. Inflorescence paniculate, its branches rusty-tomentose; leaves not gland- dotted; anthers included; follicles slender, 0.4 cm across ....ICHNOCARPUS
                  7. Inflorescence dichasial cyme, branches not as above; leaves pellucid; anthers exserted; follicles 2-2.5 cm across .... VALLARIS
            5. Corona present, corolline, white or orange, fimbriate; coma basal; petals overlapping to left; follicles free or completely united, lenticellete; shrubs or small trees .... WRIGHTIA
         4. Anthers connivent and fused with stigma, forming gynostegium; coma apical; pollen masses 10 in all (5 pairs)
            8. Pollen granular, transferred through spathulate or spoon-shaped translators; filaments free; corona coralline

9. Corolla 0.5-2 cm long; follicle round in cross-section
    10. Flowers in dense, subsessile cymes; corolla valvate; follicles 7-12 cm long; coronal scales fleshy .... HEMIDESMUS
    10. Flowers in lax, dichotomous, peduncled cymes; petals imbricate; corona of 5, clavate scales; follicles 3-angled in cross-section .... CRYPTOLEPIS
9. Corolla 5-8 cm long, purple, campanulate; flowers in trichotomous cymes; follicle 3-winged .... CRYPTOSTEGIA

8. Pollen waxy, transferred through pollinia; filaments united, forming a tube
    11. Corona double
        12. Anthers without membranous appendages; pollinia erect; pollen pellucid-dotted; corolline corona in sinuses of corolla lobes, staminal corona annular; flowers greenish-yellow; latex watery .... LEPTADENIA
        12. Anthers with membranous appendages; pollinia pendulous; pollen not pellucid-dotted
            13. Plants leafless (leaves caducous), profusely branched; flowers fragrant, in sessile umbels; both coronal whorls staminal, outer of 10 lobes, inner of 5 lobes; follicles 10-12 cm long........ SARCOSTEMMA
            13. Leaves well developed
                14. Follicles softly echinate; flowers greenish-yellow, small, in peduncled, corymbose cymes; leaves deeply cordate with wide sinuses........ PERGULARIA
                14. Follicles smooth; flowers white with purple veins, large, drooping, in umbellate cymes; leaf base rounded, blade linear-lanceolate .... OXYSTELMA
    11. Corona single
        15. Corona corolline; anthers with membranous appendages
            16. Pollinia erect; flowers yellow in sub-sessile umbels; corona adnate to corolla tube; follicles pointed, usually single .... GYMNEMA
            16. Pollinia pendulous; flowers blueish-purple, in axillary, few flowered cymes; corona annular, 10-lobed; stigma 5-winged; follicles rounded at apex, paired .... HOLOSTEMMA
        15. Corona staminal
            17. Shrubs; pollinia pendulous; flowers white or purple; follicles thick, broadly ovate; plants with cottony bloom .... CALOTROPIS
            17. Twiners
                18. Tips of petals united at apex; pollen pellucid-dotted; anthers without membranous appendages; pollinia erect .... CEROPEGIA
                18. Tips of petals free; pollen not pellucid-dotted; anthers with membranous appendages
                    19. Pollinia pendulous; flowers greenish-yellow; follicles slender .... PENTATROPIS
                    19. Pollinia erect or horizontal; corolla imbricate; corona adnate to staminal column
                        20. Coronal scales with ligular appendages; flowers 10-20, in sub-sessile umbellate cymes; follicles green .... TELOSMA

20. Coronal scales unappendaged
   21. Pollinia horizontal; flowers in axillary umbels, greenish-yellow,with a purple centre; follicle narrow at both ends ....TYLOPHORA
   21. Pollinia erect; flowers greenish-yellow
      22. Follicles 5-10 cm long, glabrous; flowers in umbellate cymes; corolla rotate; stigma dome-shaped ....DREGEA
      22. Follicles 2-3 cm long, velvety; flowers in cymose panicles; corolla campanulate; stigma cone-shaped .... MARSDENIA

CALOTROPIS R. BR.W. 3; MP 2; SA 2

Key to Species

1. Corolla lobes white or uniformly purplish; corona lobes shorter than the staminal column, with 2, obtuse auricles below the apex .... **C. gigantea**
1. Corolla lobes white with purple tips; corona lobes equalling or exceeding the staminal column, auricles absent ....**C. procera**

Calotropis gigantea (L.) R. Br. in W. T. Aiton, Hortus Kew. Ed. 2.2: 78. 1811; *Asclepias gigantea* L., Sp. Pl. 214. 1753; Hook.f., Fl. Brit. India 4: 17. 1883; Samvatsar, Fl. West. Trib. Madhya Pradesh 178. 1996; Ram Lal in Mudgal *et al.,* Fl. Madhya Pradesh 2: 67. 1997. *Aak*

Undershrubs or shrubs; leaves elliptic, ovate to obovate, 10-20 × 8-14 cm, acute at apex, cordate or slightly arcuate, clasping at base, thick, glabrous or appressedly white floccose; flowers purple to white, in umbelliform cymes, 3.7-5 cm across, not fragrant; follicles recurved, 7-10 × 3-4 cm; seeds broadly ovoid, *ca* 6 mm long, flat, minutely tomentose; coma white, silky.

Fl.-Fr.: March – May

Common in all districts.

Calotropis procera (Ait.) R. Br. in Ait. Hort. Kew. ed. 2, 2: 78. 1811; *Asclepias procera* Ait. Hort. Kew. ed. 1, 1: 305, 1789; Hook.f., Fl. Brit. India 4: 18. 1883; Samvatsar, Fl. West. Trib. Madhya Pradesh178. 1996; Ram Lal in Mudgal *et al.,* Fl. Madhya Pradesh 2: 67. 1997. *Aak*

An erect shrub, covered with white-tomentum; leaves 10-15 × 6-9 cm ovate or oblong-obovate, acute at the apex; flowers pinkish-purple or white, in lateral, umbellate cymes; follicles in pairs; seeds comose.

Fl.-Fr.: March – May

Common in all districts.

CARISSA L. W . 8; MP 6; SA2

Key to Species

1. Ovules 4 in each cell; berry 1.2-1.5 cm across; leaves more than 3 cm long .... **C. carandas**
1. Ovule 1 in each cell; berry 2-seeded; leaves up to 3 cm long.. **C. spinarum**

Carissa carandas L. Mant. Pl. 1: 52 1767*;* Hook. f, Fl. Brit. India 3: 631. 1882; *Carissa congesta* Wight, Icon. Pl. Ind. Orient. t. 1284. 1848; Samvatsar, Fl. West. Trib. Madhya Pradesh 174. 1996; Ram Lal in Mudgal *et al.,* Fl. Madhya Pradesh 2:48. 1997. *Karonda*

Shrubs, profusely branched, 2-2.5 m high; spines straight, often forked at tip, 1-2.5 cm long; leaves usually oblong, sometimes oblong-elliptic, 3-8 × 2-4 cm, obtuse or emarginate at apex, rounded or subcordate at base; flowers in puberulous, corymbose cymes; berries ellipsoid, 1.2-1.5 cm long, red, turning purple or black on ripening, 4-seeded.

Fl.-Fr.: May - Oct.

Common in all districts, in scrub.

Carissa spinarum L., Mant. Pl. 559. 1771; Hook. f, Fl. Brit. India 3: 631. 1882, *pp;* Ram Lal in Mudgal *et al.,* Fl. Madhya Pradesh 2: 51. 1997. *Karonda*

Shrubs, bushy, evergreen; branches diffuse, erect or slightly scandent; spines simple or forked, straight, 2-3.5 cm long; leaves broadly ovate or elliptic, 1.5-4 × 1.2-2.5 cm, acute or apiculate at apex, acute or rounded at base; flowers in terminal or axillary, often branched, pubescent, sessile or shortly pedunculate cymes; berries subglobose, 0.7-1 cm across, dark purple on ripening.

Fl.-Fr.: Jan. - April

Common everywhere.

CATHARANTHUS G. Don W. 8; MP; 1 SA 1

Catharanthus pusillus (Murr.) G. Don, Gen. Hist. 4: 95. 1837-1838; *Vinca pusilla* Murr. in Comment. Soc. Regiae Sci. Gotting. 3: 66. 1773; Hook. *f*, Fl. Brit. India 3: 640.1882. Samvatsar, Fl. West. Trib. Madhya Pradesh 174. 1996; Ram Lal in Mudgal *et al.,* Fl. Madhya Pradesh 2: 52. 1997.

Herbs, 10-25 cm tall; leaves 1.5-4.5 × 0.3-2.5 cm, elliptic-lanceolate, acuminate; flowers white; 1-3 in the axils; follicles linear, 3-5 cm long, straight, glabrous; seeds black, muriculate.

Fl.-Fr.: Aug. -Dec.

Common weed of fields, not found outside fields.

CEROPEGIA L. W. 217; MP 5; SA 3

Key to species

1. Leaves fleshy; glabrous annuals; flowers greenish-purple, 3-5 in umbellate cymes; corolla tube less than 2.5 cm long; follicles 9-10 cm long ....**C. bulbosa**

1. Leaves membranous
   2. Flowers bright yellow; corolla lobes equal to or longer than the tube; stem glabrous ....**C. odorata**
   2. Flowers light green with purple stripes; corolla lobes shorter than the tube; stem hirsute . ....**C. hirsuta**

Ceropegia bulbosa Roxb., Pl. Coromandel 1: 11. t. 7. 1795; Hook.f., Fl. Brit. India 4: 67. 1883; Ansari, Fasc. Fl. India 16: 10.1984; Samvatsar, Fl. West. Trib. Madhya Pradesh 178. 1996; Ram Lal in Mudgal *et al.,* Fl. Madhya Pradesh 2: 69. 1997. *Gaker kand*

Twining or trailing, slender, glabrous herbs; roots tuberous; leaves ovate, 3-6 cm long, slightly cordate at base, fleshy, glabrous; flowers in pedunculate, umbelliform cymes; corolla pale brown outside, reddish- brown inside, purple, striate; follicles glabrous, *ca* 10 cm long, terete, tapering to a slender point.

*Fl.-Fr.:*Aug. - Oct.

Alirajpur, Burhanpur, Indore, Khandwa, in hedges, twining round adjacent vegetation.

Ceropegia hirsuta Wight & Arnott in Wight, Contr. Bot. India 30.1834; Hook.f., Fl. Brit. India 4: 71. 1883; Ansari, Fasc. Fl. India 16: 16. 1984; Ram Lal in Mudgal *et al.,* Fl. Madhya Pradesh 2: 69. 1997. *Bosia kand*

Hispid twiners; roots tuberous; leaves variable, ovate, elliptic- lanceolate, up to 10 × 3 cm, acute to obtuse at apex, rounded at base, pilose; flowers in few-flowered cymes; peduncles 1.2-2.5 cm long, stout, hispid; corolla light green and variously striped with purple; follicles up to 9 cm long.

*Fl.-Fr.:*July- Oct.

Indore, Khandwa, among grasses and twining if support available.

Ceropegia odorata Hook.f., Fl. Brit. India 4:75.1883; Sabnis & Bedi, in Kew Bull. 25:59, f. 1-6. 1971; Ansari, in Fasci. Fl. India 16:26.1984; Jagtap & N.P. Singh in Fascic. Fl. India 24:234. 1999; Jagtap & S.K. Das in Fl. Maharashtra State 2:354.2001. *Kundher*

Twining herbs; leaves 6-7 × 0.5-1 cm, petiolate, linear to lanceolate, acuminate; flowers bright yellow, fragrant, 3-10, in umbellate cymes; peduncles hirsute; corolla inflated near the base; corona 2-seriate, outer of 5 entire lobes, inner linear or elongated, triangular, erect, divergent at apex; follicles *ca* 10 cm long, tapering towards apex; seeds 4x2 mm, brown, coma silky, *ca* 3 cm long.

*Fl.-Fr.:*Aug. -Oct.

Rare, **Burhanpur** only, Nepanagar.

CRYPTOLEPIS R. Br.W. 42; MP 1; SA 1

Cryptolepis dubia (Burm.f.) M.R. Almeida in Fl. Maharashtra 3A: 221 2001; *Periploca dubia* Burm.f., Fl. Indica 70.1768*; Cryptolepis buchanani* Roem. & Schult., Syst. Veg. 4: 409. 1819; Hook.f., Fl. Brit. India 4: 5. 1883; Samvatsar, Fl. West. Trib. Madhya Pradesh 176. 1996; Ram Lal in Mudgal *et al.,* Fl. Madhya Pradesh 2: 71. 1997. *Kala patha*

Large, twining shrubs; branches white, terete, becoming black with age; leaves oblong, ovate or elliptic, 8-16 × 2.5-6 cm, apiculate or acuminate at apex, usually acute at base, shining above, glaucous beneath, coriaceous; flowers greenish-yellow, in axillary, short, pedunculate cymes; coronal scales 5, fleshy, clavate; follicles lanceolate, 5-10 × 1.2-1.8 cm, stout, straight, tapering.

Fl.-Fr.: May – Feb.

Common in all districts.

CRYPTOSTEGIA R. Br.W. 2; MP 1; SA 1

Cryptostegia grandiflora Roxb. ex R. Br. in Bot. Reg. 5: t. 435. 1820; Hook.f., Fl. Brit. India 4: 6. 1883; Samvatsar, Fl. West. Trib. Madhya Pradesh 176. 1996; Ram Lal in Mudgal *et al.,* Fl. Madhya Pradesh 2: 73. 1997. *Rubber dudhi*

Large, glabrous twiners; bark grey; sap milky; leaves elliptic or oblong, 7-11 × 3.5-5 cm, shortly acuminate at apex, rounded or truncate at base, coriaceous, shining, glabrous; flowers light- purple, fragrant, showy, up to 5 cm across, in terminal, 3-5- flowered cymes; peduncles up to 4 cm long; follicles paired, ovoid-oblong, 8-12 × 2.5 cm, straight.

Fl.-Fr.: June – Sept.

Common in towns and villages, not encountered in forests.

DREGEA E. Meyer W. 12; MP 1; SA 1

Dregea volubilis (L.f.) Benth. ex Hook.f., FI. Brit. India 4: 46. 1883; *Asclepias volubilis* L.f., Suppl. PI. 170.1782; *Wattakaka volubilis* (L.f.) Stapf in Curtis, Bot. Mag. 148. sub. t. 8976. 1923; Samvatsar, Fl. West. Trib. Madhya Pradesh 180.1996; Ram Lal in Mudgal *et al.,* Fl. Madhya Pradesh 2: 89. 1997 **(Plate 7, Fig. 1)**

Large climbers, woody; leaves ovate, 10-14 × 5-10 cm, acuminate at apex, rounded to cordate at base, hoary with curled pubescence; flowers in axillary, drooping, densely flowered, umbelliform cymes; peduncles 2.5-7 cm long; follicles usually paired, divaricate, broadly lanceolate, 7-10 cm long, reddish brown, tapering to a blunt point; seeds broadly ovate with white margin; coma *ca* 3.7 cm long, white.

Fl.-Fr.: March - Nov.

Common in hedges of fields, also casually on roadside trees.

GYMNEMA R. Br.W. 51; MP 2; SA 1

Gymnema sylvestre (Retz.) R. Br. ex Schult. in Roem. & Schult., Syst. Veg. 6: 57. 1820; *Periploca sylvestris* Retz., Observe. Bot. 2:15. 1781; Hook.f., Fl. Brit. India 4: 29. 1883; Samvatsar, Fl. West. Trib. Madhya Pradesh 178. 1996; Ram Lal in Mudgal *et al.,* Fl. Madhya Pradesh 2: 75. 1997. *Gudmar*

Large twiners; young stems and branches terete, pubescent; leaves broadly elliptic-obovate or oblong, 3-6 cm long, shortly acuminate at apex, rounded to cordate at base, sparsely to densely pubescent; flowers in umbelliform cymes; follicles lanceolate, 5-7.5 cm long, glabrous; seeds nearly ovoid or oblong, flat, *ca* 1.2 cm long, broadly margined.

Fl.-Fr.: Oct. – Feb.

Alirajpur, Barwani, Burhanpur, Khargone.

HEMIDESMUS R. Br.W. 2; MP 1; SA 1

Hemidesmus indicus (L.) R. Br. ex Schult., Syst. Veg. 6:126. 1820; *Periploca indica* L., Sp. Pl. 211. 1753; Hook.f., Fl. Brit. India 4:5. 1883; Samvatsar, Fl. West. Trib. Madhya Pradesh 177. 1996; Ram Lal in Mudgal *et al.,* Fl. Madhya Pradesh 2:75. 1997. *Anant mool*

Shrubby twiners with slender branches and thick roots (with age); leaves variable, 5-10 cm long, oblong elliptic, elliptic-linear, apiculate at apex, obtuse at base, dark green with white streaks above, petioles up to 4 mm long; flowers in 5-15 flowered, fasciculate, subsessile cymes; pedicels with many imbricating bracts; corolla light-brown above and deep brown to purple inside, 4-5 mm long, tube very short; follicles thin, tapering towards apex, 10-16 cm long; seeds black, coma *ca* 2.5 cm long.

*Fl.-Fr.:*Aug. -Jan.

Common on forest floor, twining over supporting plants.

HOLARRHENA R. Br.W. 5; MP 1; SA 1

Holarrhena pubescens Wallich ex G. Don, Gen. Hist. 4: 78. 1837; *Holarrhena antidysenterica* (Roth) Wall. ex A. DC., Prodr. 8: 413. 1844; Hook.f., Fl. Brit. India 3: 644. 1882; Samvatsar, Fl. West. Trib. Madhya Pradesh 174. 1996; Ram Lal in Mudgal *et al.,* Fl. Madhya Pradesh 2:53. 1997. *Kuda*

Shrubs or small trees; leaves broadly ovate to elliptic-oblong, 10-20 × 6-9 cm, acuminate at apex, rounded or tapering at base, glabrous or more or less tomentose especially beneath; flowers fragrant, in terminal, many-flowered, 7.5-15 cm across cymes; follicles 20-40 × 0.5-0.8 cm, usually marked with narrow, white lenticels; seeds linear, *ca* 1.5 cm long; coma brown, apical, about twice as long as the seeds.

Fl.-Fr.: Feb. -June

Common forest denizen, in all districts.

HOLOSTEMMA R. Br. Monotypic

Holostemma ada-kodien Schult., Syst. Veg. 6: 95. 1820; Hook.f., Fl. Brit. India 4:21. 1883; *H. annularis* (Roxb.) K. Schum., Nat. Pflanzenfam. 4(2): 250.1895; *Asclepias annularis* Roxb., Hort. Bengal. 20.1814; Ram Lal in Mudgal *et al.,* Fl. Madhya Pradesh 2:78. 1997. *Palkhe vel*

Large twiners; stems glabrous, shining; leaves oblong or ovate, 7.5-15.5 cm long, cuspidate at apex, cordate with deep sinus at base, glabrous above, puberulous or sparsely pubescent beneath; flowers in umbelliform or subracemose cymes; corona single, annular, fleshy; filaments of anthers connate to a 10-winged column; style slender, 5-winged at apex; follicles 10-12 cm long; seeds winged, *ca* 6 mm long, thin; coma 0.6-2.5 cm long.

Fl.-Fr.: June – Aug.

Rare, **Khandwa** only, Kalibhit and Sunder dev forest.

ICHNOCARPUS R. Br.W. 3; MP 1; SA 1

Ichnocarpus frutescens (L.) W. T. Aiton, Hortus Kew. 2: 69 1811; *Apocynum frutescens* L., Sp. Pl. 213. 1753; Hook.f., Fl. Brit. India 3: 669. 1882; Samvatsar, Fl. West. Trib. Madhya Pradesh 174. 1996; Ram Lal in Mudgal *et al.,* Fl. Madhya Pradesh 2: 54. 1997. *Dhimar bel*

Twining, evergreen shrubs; young branches, inflorescence and undersurface of leaves rusty tomentose; leaves variable, lanceolate to ovate, 4-8 × 2-4 cm, acute or subacuminate at apex, obtuse or rounded at base, pale beneath; flowers in axillary and terminal, rusty pubescent panicles, *ca* 8 mm across; branches short, 3-flowered; follicles 5-15 cm long, straight or somewhat curved, pubescent when young; seeds 1.2-1.75 cm long; coma as long as seeds.

Fl.-Fr.: Aug. - Feb.

Barwani, Indore, Jhabua, Khandwa, Khargone.

LEPTADENIA R. Br.W. 5; MP 2; SA 1

Leptadenia reticulata (Retz.) Wight & Arnott in Wight, Contr. Bot. India 47. 1834; *Cynanchum reticulatum* Retz., Observ. Bot. 2:15. 1781; Hook.f., Fl. Brit. India 4: 63. 1883; Samvatsar, Fl. West. Trib. Madhya Pradesh 179. 1996; Ram Lal in Mudgal *et al.,* Fl. Madhya Pradesh 2: 80.1997. *Dodi*

Twining, leafy, much branched shrubs; branches hoary pubescent; leaves ovate or lanceolate, 4-9 × 2.5-6 cm, acute or acuminate at apex, rounded to slightly cordate at base, pubescent; petioles 1.2-1.8 cm long; flowers pale-green, in lateral or subaxillary, many-flowered, umbelliform cymes; follicles usually solitary, 5-8.5 cm long, beak curved, minutely white-dotted; seeds ovoid or oblong, flat; coma *ca* 3 cm long.

*Fl.-Fr.:*July – April

Common in forest, as well as along village roads.

MARSDENIA R. Br.W. 151; MP 1; SA 1

Marsdenia tenacissima (Roxb.) Moon, Cat. Pl. Ceylon 21. 1824; *Asclepias tenacissima* Roxb., Pl. Coromandel 3:35. 1815; Hook.f., Fl. Brit. India 4: 35. 1883; Samvatsar, Fl. West. Trib. Madhya Pradesh 179. 1996; Ram Lal in Mudgal *et al.,* Fl. Madhya Pradesh 2: 80.1997. *Dudhia bela*

Twiners; young parts and inflorescence densely softly pubescent; sap milky; leaves broadly ovate to orbicular, 7.5-15 × 7.5-10 cm, abruptly acuminate at apex, cordate, truncate or rounded at base, glandular above; petioles 5-7 cm long, softly pubescent; flowers greenish-yellow, *ca* 1`cm across, in corymbose cymes; peduncles inter-petiolar; follicles ovoid- lanceolate, 10-15 cm long, wrinkled, horny; seeds ovoid or oblong, *ca* 1.2 cm long.

Fl.-Fr.: April- June

Common twiner on roadside trees, both in cities and villages.

OXYSTELMA R. Br.W. 2; MP 1; SA 1

Oxystelma esculentum (L. f) Sm., Cycl. 25: (unnumbered) 1813; *Periploca esculenta* L.f., Suppl. Pl. 168. 1781; *Oxystelma secamone* H. Karst., Deut. Fl. 1031. 1883; Hook.f., Fl. Brit. India 4: 17. 1883; *Sarcostemma esculentum* (L.f.) R.W. Holm, Ann. Missouri Bot. Gard. 37:482. 1950; Samvatsar, Fl. West. Trib. Madhya Pradesh 179. 1996; Ram Lal in Mudgal *et al.,* Fl. Madhya Pradesh 2: 82. 1997. *Dudhia lata*

Much branched, slender twiners; branches terete; sap milky; leaves linear or lanceolate, 7-14 × 0.5-0.7 ern, acuminate at apex, rounded or cordate at base; flowers white or pink, with purple veins, in axillary, lax, racemose or umbelliform cymes; peduncles *ca* 6 cm long; bracts deciduous; corolla lobes overlapping to right in bud; follicles ovoid or oblong-lanceolate, 4-7 cm long; seeds ovoid; coma silky, dull white.

Fl.-Fr.: Sept. - Dec.

Alirajpur, Burhanpur, Dhar, Khandwa.

PENTATROPIS Wight & Arn. W. 5; MP 1; SA 1

Pentatropis nivalis (J.F. Gmelin) D.V. Field & J.R.I. Wood, Kew Bull. 38: 215; 1984; *Asclepias nivalis* J.F. Gmelin, Syst. Nat. ed. 13.2(1): 444. 1791; *Pentatropis spiralis* Decne. in Ann. Sci. Nat. ser. 2. 9: 327. 1838; *non Asclepias spiralis* Forssk., 1775; Hook.f., Fl. Brit. India 4: 19. 1883; Samvatsar, Fl. West. Trib. Madhya Pradesh 179. 1996; Ram Lal in Mudgal *et al.,* Fl. Madhya Pradesh 2:83. 1997.

A slender, twining shrub with tuberous roots; stem slender, wiry, terete, glabrous; leaves 2.4-2.8 × 1.3-1.8 cm, fleshy, ovate-oblong, acute or obtuse, base rounded or cordate, glabrous; petiole very short; flowers greenish-yellow, in 5-7 flowered umbellate cymes; follicles 2.5-5 cm, lanceolate, glabrous; seeds *ca* 4 mm long, ovate, flat, crenate at the lower ends.

Fl. -Fr. : Jan. -Aug.

Alirajpur, Burhanpur, Dhar, Indore, Khandwa.

PERGULARIA L.W. 14; MP 1; SA 1

Pergularia daemia (Forssk.) Chiov, Res. Sc. Miss. Stefan. Paoli, Somal. Ital. 1: 115. 1916; *Asclepias daemia* Forssk. Fl. Aegypt. -Arab. 51. 1775; *Daemia extensa* (Jacq.) R. Br. in Ait. Hort. Kew. ed. 2, 2: 76. 1811; *Pergularia extensa* N. E. Br. in Dyer, Fl. Cap. 4: 758. 1908; Hook.f., Fl. Brit. India 4: 21. 1883; Samvatsar, Fl. West. Trib. Madhya Pradesh 180.1996; Ram Lal in Mudgal *et al.,* Fl. Madhya Pradesh 2:84. 1997. *Sadovani*

A perennial climber with milky latex; stem terete, clothed with short, stiff, pubescent, hairs; leaves 11-12.5 × 6-7 cm, broadly ovate, with a deep sinus, acuminate, cordate at the base, pubescent, margin ciliate; flowers yellowish-green with pink tinge, in peduncled corymbose cymes or racemes; follicles 5-7 cm long reflexed, lanceolate, covered with long, soft spines; seeds ovate, densely pubescent; coma 3-5 cm long.

Fl. -Fr. : Dec. -March

Very common along village roads, all over the area.

RAUVOLFIA L. W. 77; MP 1; SA 1

Rauvolfia tetraphylla L. Sp. Pl. 208. 1753; *R. canescens* L., Sp. Pl. (ed. 2) 303. 1762; Rani & K.M. Mathew in Mathew, Fl. Tamilnadu Carnatic 2:917. 1983.

Shrubs; leaves in whorls of 4; elliptic-obovate, 5-9 × 3-4 cm, acute at both ends, thinly pubescent above, densely pubescent beneath; flowers white, in axillary or terminal, 5-7 flowered corymbs; drupes connate up to the top, 0.7 cm across.

*Fl.-Fr.:*Aug. -Oct.

Rare, deep within the forest, **Khandwa** only, Chandgarh range.

SARCOSTEMMA R. Br.W. 20; MP 1; SA 1

Sarcostemma acidum (Roxb.) Voigt, Hort. Suburb. Calcutt. 542. 1845; *Asclepias acida* Roxb., Fl. Ind. 2: 31. 1832; *Sarcostemma brevistigma* Wight & Amott in Wight, Contr. Bot. India 59. 1834; Hook.f., Fl. Brit. India 4: 26. 1883; Samvatsar, Fl. West. Trib. Madhya Pradesh 180.1996; Ram Lal in Mudgal *et al.,* Fl. Madhya Pradesh 2: 85. 1997. *Samar bel.*

Trailing or erect shrubs, with caducous leaves; stems stout; flowers fragrant, waxy, greenish-white, in terminal, umbelliform cymes; follicles lanceolate, 10-12 cm long; seeds glabrous.

Fl.-Fr.: July- Jan.

Alirajpur, Barwani, Dhar, Indore, Khandwa, Khargone, it rarely sets fruit in our area.

TELOSMA Coville W. 10; MP 1; SA 1

Key to Species

1. Petioles *ca* 0.6 cm long; peduncles intrapetiolar; pedicels 1.5 cm long; calyx *ca* 7 mm long; corolla tube 10 mm long .... **T. cordata**
1. Petioles 1.2-2 cm long; peduncles interpetiolar; pedicels 1.2-2 cm long; calyx *ca* 5 mm long; corolla tube *ca* 6 mm long .... **T. pallida**

Telosma cordata (Burm.f.) Merr., Philipp. J. Sci. 19(3): 372. 1921; *Asclepias cordata* Burm.f., Fl. Indica 72. 1768; *Telosma minor* (Andrews) W.G. Craib, Bull. Misc. Inform. Kew 1911(10): 418. 1911; *Pergularia minor* Andrews, Bot. Repos. pl 184. 1801; Hook.f., Fl. Brit. India 4:38. 1883; *Telosma odoratissima* (Lour.) Coville, Contr. U.S. Natl. Herb. 9:384. 1905; *Cynanchum odoratissimum* Lour., Fl. Cochinch. 166. 1790; K.M. Matthew & M. Rani, Fl. Tamilnadu Carnatic 2:959. 1983; *Oxystelma ovatum* P.T. Li & S.Z. Huang, Guihaia 5(2): 81, f. 3. 1985.

Twining, sweet scented perennials; leaves chartaceous, apex acuminate, base cordate, basal lobes imbricate; petiole about 0.6 cm long; inflorescence a condensed, intrapetiolar umbel; peduncle 1.5 cm long; bracts 1 cm long; bracteoles 0.5 cm long; pedicels 1.5 cm long; calyx about 7 mm long, imbricate, glandular, margin ciliate; corolla greenish-yellow, 1.5 cm across, lobes twisted to right, tube 1 cm long; corona single, staminal, forked into short,outer lobe and long, incurved inner lobes; follicles linear-lanceolate, 12-15 cm long, tip curved.

*Fl.-Fr.:*Aug. -Oct.

Khandwa only, at Pandhana.

Note:- This species was recently collected by Dr. Shakun Mishra (Personal communication). It is native to China & Indo-China. Appears to be an escape from cultivation.

Telosma pallida (Roxb.) W.G. Craib, Kew Bull. 1911: 418. 1911; *Asclepias pallida* Roxb., Fl. Ind. 2: 48. 1832. *Pergularia pallida* Wight & Arn., Wight, Contr. Bot. India 42. 1834; Hook.f., Fl. Brit. India 4: 38. 1883; Ram Lal in Mudgal *et al.,* Fl. Madhya Pradesh 2: 86. 1997.

Twining shrubs; leaves ovate, 5-12.5 cm long, acuminate at apex, cordate at base, membranous, glabrous or pubescent when young. flowers in axillary, umbelliform cymes; peduncles interpetiolar; pedicels *ca* 1.2 cm long; corolla greenish- yellow, tube *ca* 6 mm long; follicles 7.5-10 cm long, tapering to a blunt point, longitudinally ribbed, glabrous; seeds ovoid, flat, pale brown, broadly margined.

Fl.-Fr.: May - July

Burhanpur, Indore, Khandwa.

TYLOPHORA R. Br.W. 88; MP 2; SA 2

Key to Species

1. Plants erect; leaves 2-4 cm long; peduncles 2-3 mm long; pedicels 5-10 mm long; follicles 3.5-5 cm long .... **T. fasciculata**
1. Plants creeping; leaves more than 4 cm long; peduncles 1-5 cm long; pedicels 2-5 cm long; follicles *ca* 5 cm long .... **T. rotundifolia**

Tylophora fasciculata Buclr-Ham. *ex* Wight, Contr. Bot. India 50.1834; Hook.f., Fl. Brit. India 4: 40.1883; Ram Lal in Mudgal *et al.,* Fl. Madhya Pradesh 2: 87. 1997.

Erect, slender shrubs; roots fascicled; leaves ovate or lanceolate, 1-4 cm long, acute or shortly acuminate at apex, rounded or cordate at base, coriaceous, glabrous above and puberulous on the veins beneath; flowers in lateral, umbelliform or paniculate cymes; corolla purplish-brown, lobes lanceolate, 2.5-3 mm long, pilose inside; follicle usually solitary, acute, ovoid, 3.5-5 cm long,longitudinally striate, glabrous.

Fl.-Fr.: April- May

Burhanpur, Indore.

Tylophora rotundifolia Buch. -Ham. ex Wight, Contr. Bot. India 50.1834; Hook.f., Fl. Brit. India 4: 43. 1883; Ram Lal in Mudgal *et al.,* Fl. Madhya Pradesh 2: 88. 1997.

Perennial, creeping herbs; leaves closely appressed to ground, rotundate or orbicular, 4-8 cm across, rounded to obtuse or sometimes mucronate at apex, rounded to slightly cordate at base; flowers in, lateral or terminal, umbelliform or paniculate cymes; corolla pale greenish- purple to greenish- yellow; follicles broadly ovoid, *ca* 5 cm long.

Fl.-Fr.: May - Sept.

Burhanpur, Dhar, Indore, Khandwa.

VALLARIS Burm.f. W. 3; MP 1; SA 1

Vallaris solanacea (Roth) Kuntze, Rev. Gen. Pl. 2: 417. 1891; *Peltanthera solanacea* Roth, Nov. Pl. Sp. 132. 1821; *Vallaris heynei* Spreng., Syst. Veg. 1: 635. 1824; Clarke in Hook.f., Fl. Brit. India 3:650.1882; Samvatsar, Fl. West. Trib. Madhya Pradesh 175. 1996; Ram Lal in Mudgal *et al.,* Fl. Madhya Pradesh 2:60.1997. *Dhimar vel.*

Woody climbers; leaves 5-12 × 2.5-3 cm, elliptic or ovate- oblong, glabrous, glands in axil of petiole; flowers white, 3-10, in axillary, lax cymes; follicles cylindric, gradually tapering, 20 × 2.5 cm; seeds 1 cm, yellowish gray with a tuft of silky white coma.

Fl.-Fr.: Dec. -April

Alirajpur, Dhar, Indore, Khandwa.

WRIGHTIA R. Br.W. 32; MP 2; SA 2

Key to Species

1. Follicles connate at the apex only, smooth; corona white, fimbriate; leaves glabrous beneath .... **W. tinctoria**
1. Follicles connate throughout their length, rough due to lenticels; corona coloured, entire or lobed; leaves tomentose on both sides .... **W. arborea**

Wrightia arborea (Dennst.) Mabberly in Taxon, 26: 533. 1977; *Periploca arborea* Dennst. Schlussel Hortus Malabar. 13.23.25. 1818; *Wrightia tomentosa* (Roxb.) Roem. & Schult. Syst. Veg. 4: 414. 1819; *Nerium tomentosum* (Roem. & Schult.) Roxb. Fl. Ind. 2:6. 1824; Clarke in Hook.f., Fl. Brit. India 3:65. 1882; Samvatsar, Fl. West. Trib. Madhya Pradesh 175. 1996; Ram Lal in Mudgal *et al.,* Fl. Madhya Pradesh 2:62. 1997. *Dharauli*

Small trees; bark smooth, yellowish-grey; branchlets pubescent; leaves 3.5-1 × 2-5 cm, elliptic-oblong, acuminate, tomentose on both surfaces; flowers in short, dense, erect, terminal, corymbose, tomentose cymes; follicles cylindric, with a groove on each side at the junction of the carpels, rough with white tubercles.

Fl.-Fr.: April-June

Alirajour, Barwani, Burhanpur, Khandwa, Khargone.

Wrightia tinctoria R. Br. var. **rothii** (G. Don) Ngan in Ann. Miss. Bot. Gard. 52: 144. 1965; *W. tinctoria* R. Br. var. *rothii* (G. Don) Clarke in Hook.f., Fl. Brit. India 3: 653. 1882; Samvatsar, Fl. West. Trib. Madhya Pradesh 176. 1996; Ram Lal in Mudgal *et al.,* Fl. Madhya Pradesh 2: 62. 1997. *Kali dudhi*

Deciduous, large shrubs; bark whitish-grey; leaves 4-11 × 2-5 cm, elliptic-oblong to elliptic-lanceolate or obovate-oblong; base rounded, apex acuminate, rarely rounded or retuse, subentire, pubescent on both the sides; flowers white, in lax, terminal, dichotomous cymes; follicles up to 45 cm long, pendulous, paired, deep green; seeds 1.2-1.5 cm long, linear, glabrous, brown; coma 3-5 cm long.

Fl.-Fr.: Dec. -June

Very common in open ground.

LOGANIACEAE 15/351; 4/5; 3/3

Key to Genera

1. Fruits baccate; flowers in axillary cymes; trees .... STRYCHNOS
1. Fruits capsular; herbs
    2. Flowers in axillary or terminal unilateral spikes; leaves 5-15 cm long; capsule warty .... SPIGELIA
    2. Flowers in polychasial cymes; leaves 2.5-6 cm lonf; capsule 2-horned .... MITREOLA

MITREOLA L. W. 7; MP 1; SA 1

Mitreola petiolata (J.F. Gmelin) Torr. & A. Gray, Fl. N. Amer. 2:45. 1841; *Cynoctonum petiolatum* J.F. Gmelin, Syst. Nat. 2:443. 1791; *Ophiorrhiza mitreola* L. Sp. Pl. 150.1753; *Mitreola oldenlandioides* Wall. ex DC., Prodr. 9:9. 1845; C.B. Clarke in Hook.f., Fl. Brit. India 4:79. 1883; *Cynoctonum mitreola* (L.) Britton, Torrey Bot. Club 5:258. 1894; Samvatsar, Fl. West. Trib. Madhya Pradesh 181. 1996; Srivastava in Mudgal *et al.,* Fl. Madhya Pradesh 2:91. 1997.

Annual, 20-30 cm high herbs; leaves 3-5 × 1.5 3 cm, ovate or ovate-oblong, narrow at both ends, membranous; flowers in terminal and axillary cymes, initially multichasial, gradually ending into a helicoid cyme, pedicela short; bracts deltoid; calyx lobes linear, membranous, corolla white, 1-2 mm long, lobes pubescent inside, at the base; capsules wedge-shaped, deeply emarginated, 4-5 mm across, 2-horned, horns incurved; seeds ellipsoid, trigonous with a ventral cavity.

*Fl.-Fr.:*Sept. -Dec.

Common in wet and shady habitats all over the area.

SPIGELIA L. W. 89; MP 1; SA 1

Spigelia anthelmia L., Sp. Pl. 149. 1753; Oomachan & Shrivastava in J. Bombay Nat. Hist. Soc. 84:730.1987; Srivastava in Mudgal *et al.,* Fl. Madhya Pradesh 2: 93. 1997.

Annual, erect, glabrous herbs, up to 50 cm high; leaves in a pseudo whorl of 4, ovate-oblong to ovate-lanceolate, apex acuminate, base cuneate, often decurrent, margin entire, scabrous above, glabrous beneath; stipules deltoid; flowers subsessile, in cincinni (helicoid), in the axils of upper leaves, appearing to be terminal; bracts lanceolate; sepals almost free to the base, ovate-lanceolate; petals purplish to white; capsules 4-5 × 5-6 mm, tuberculate; seeds ovoid or obliquely ellipsoid.

*Fl.-Fr.:*Aug. -Sept.

In waste land, wet soil, Burhanpur, Khandwa. Appears to be a recent addition.

STRYCHNOS L. W. 168; MP 2; SA 1

Strychnos potatorum L.f., Suppl. Pl. 148. 1782; C.B. Clarke in Hook.f., Fl. Brit. India 4:90.1883; Samvatsar, Fl. West. Trib. Madhya Pradesh 181. 1996; Srivastava in Mudgal *et al.,* Fl. Madhya Pradesh 2:95. 1997. *Nirmali*

Moderate sized tree, up to 3 m tall; stem covered with black, thick, rectangular bark; leaves 5-10 × 3-5 cm, elliptic or ovate, acute at both ends, leathery; flowers fragrant, in dense, glabrous cymes; calyx puberulous, *ca* 1.5 mm long; corolla whitish, *ca* 4 mm across; berries *ca* 1.5 cm across, dark blue to black, thin-walled; seeds 1-2, lenticular.

*Fl.-Fr.:*Jan. -April

Not common, Alirajpur, Barwani, Dhar, Khargone.

GENTIANACEAE 96/1682; 6/12/; 5/8

Key to Genera

1. Flowers zygomorphic; ovary 1-celled
   2. Stamen 1; stigma hardly lobed; flowers sessile; dwarf herbs .... HOPPEA
   2. Stamens 4; stigma 2-lobed; flowers pedicellate, pink; larger herbs . ... CANSCORA
1. Flowers acinomorphic
   3. Stigma 2-lobed; anthers twisted; ovary 1-celled; flowers pink .... CENTAURIUM
   3. Stigma capitate; anthers straight; ovary 1-2 celled; flowers white or blue or purple
      4. Sepals 5; flowers white, in sessile, axillary clusters; ovary 1-celled .... ENICOSTEMA
      4. Sepals 4; flowers blue-purple, in pedunculate cymes; ovary 2-celled ....EXACUM

CANSCORA Lam. W. 11; MP 3; SA 2

Key to Species

1. Stem not winged, angular; pedicels filiform; calyx striate or keeled; flowers pinkish-red; long stamen not appendaged; stigma glabrous .... **C. diffusa**
1. Stem as well as calyx and pedicels winged; flowers white; long stamen appendaged; stigma hispidulous .... **C. alata**

Canscora alata (Roth) Wall., Numer. List 4363. 1831; *Exacum alatum* Roth, Syst. Veg. 3:159. 1818; *Canscora decussata* (Roxb.) Schult. & Schult. F., Mant. 3:229. 1827; *Pladera decussata* Roxb., Fl. Ind. 1:418. 1820; C.B. Clarke, Hook.f., Fl. Brit. India 4:104. 1883; S. Garg, Indian Gent. 1. 1988; Samvatsar, Fl. West. Trib. Madhya Pradesh 182. 1996; Srivastava in Mudgal *et al.,* Fl. Madhya Pradesh 2:99. 1997.

Erect, up to 15 cm high, annual herbs with a 4-winged stem; branches decussate; leaves sessile, 1.5-3 × 0.6-1.2 cm, oblong, lanceolate, apex acute, base subcordate; flowers white, in leafy, *ca* 1.5 cm long cymes; calyx winged, lobes *ca* 2 mm long, ovate-lanceolate with acute apex; corolla irregular, pedicels winged, lobes oblong, up to 1.2 cm long, tube about 1 cm long; long stamen with, circular, membranous ring below the anther; capsules linear-oblong, 6-8 mm long; seeds angular.

*Fl.-Fr.:*Oct. -Nov.

Common along roadsides, forest paths, all over the area.

Canscora diffusa (Vahl) R. Br. ex Roem. &Schult., Syst. Veg. 3:301. 1818; *Gentiana diffusa* Vahl, Symb. Bot. 3:47. 1794; C.B. Clarke, in Hook.f., Fl. Brit. India 4:103. 1883; S. Garg, Indian Gent. 1. 1988; Samvatsar, Fl. West. Trib. Madhya Pradesh 182. 1996; Srivastava in Mudgal *et al.,* Fl. Madhya Pradesh 2:100.1997. *Agya* **(Plate 7, Fig. 2)**

Annual, erect, up to 20 cm high herbs with 4-angled stems (not winged); leaves sessile, 1-3 × 0.5-1.5 cm, ovate-elliptic, narrowed at both ends; flowers pinkish-red, in terminal, lax, diffuse, paniculate cymes, pedicels filiform; calyx 5 mm long, lobes 4; corolla lobes 4, tube as long as the calyx lobes; one stamen longer than the other 3, not appendaged; capsules oblong,compressed, up to 6-8 mm long; seeds angular, minutely rugose.

*Fl.-Fr.:*Oct. -March

Very common on moist, rocky places. Along banks of streams, all over the area.

CENTAURIUM Hill. W. 31; MP 1; SA 1

Centaurium centaurioides (Roxb.) S.R. Rao & Hemadri, J. Bombay Nat. Hist. Soc. 67:357. 1970; *Chironia centaurioides* Roxb., Fl. Ind. 1:584. 1832; *Exacum roxburghii* G. Don, Gen. Hist. 4:206. 1838; *Centaurium meyeri* (Bunge) Druce, Rep. Bot. Exch. Club, Brit. Isles 1916:613. 1919; *Erythraea meyeri* Bunge, Ledeb. Fl. Alt. 1:220.1829; C.B. Clarke, in Hook.f., Fl. Brit. India 4:102. 1883; S. Garg, Indian Gent. 3. 1988; Samvatsar, Fl. West. Trib. Madhya Pradesh 182. 1996; Srivastava in Mudgal *et al.,* Fl. Madhya Pradesh 2:101. 1997.

Erect, 10-20 cm high herbs, branched from the base; stem 4-angled; radical leaves rosulate, laceolate-oblong, apex obtuse; upper cauline leaves linear, 1-4 × 0.5-1 cm, apex acute to subobtuse; flowers pedicellate, in each fork, pinkish-white, in terminal, dichotomous cymes; calyx 5-6 mm long, segments linear, shorter than corolla tube; corolla *ca* 1 cm long, lobes 3-4 mm long; capsules lanceolate-oblong, up to 1 cm long, glabrous; seeds numerous, round, brown, faintly reticulate.

*Fl.-Fr.:*Dec. -March

Common in moist river beds, margins of moist fields, in all districts.

ENICOSTEMA Blume W. 3; MP 1; SA 1

Enicostema hyssopifolium (Willd.) Verd., Bothalia 7:462. 1961; *Exacum hyssopifolim* Willd., Sp. Pl. 1:640.1798; *Gentiana verticillata* L., Syst. Nat. ed. 12.200.1767; *Gentiana axillaris* Lam., Tabl. Encycl. 1(2):487. 1793; *Enicostema axillare* (Poir. ex Lam.) A. Raynal, Adansonia 9:75. 1969; C.B. Clarke in Hook.f., Fl. Brit. India 4:101. 1883; S. Garg. Indian Gent. 5. 1988; Samvatsar, Fl. West. Trib. Madhya Pradesh 183. 1996; Srivastava in Mudgal *et al.,* Fl. Madya Pradesh 2:102. 1997. *Kadvi nai*

Perennial, 10-35 cm high herbs with angular, glabrous branches; leaves opposie, decussate, 1-7 × 0.5-1.2 cm, sessile, apex subacute, base cuneate, linear or linear-oblong, 3-neved; flowers white, sessile, in axillary verticillate clusters; calyx cupular, lobes 5, unequal; corolla salver-shaped, lobes 5, twisted, equal, about 7 mm long; stamens 5, included, all fertile with 2-fid scales at the base of filaments; capsules unilocular, placentation parietal, placentas intruded in the chamber, 3.5 mm long, septicidally 2-valved; seeds orbicular, minute, brown, reticulate.

*Fl.-Fr.:*July-Oct.

Common along banks of streams, moist ground, all over the area.

EXACUM L. W. 70; MP 5; SA 2

Key to Species

1. Plants more than 30 cm high; flowers 2-2.5 cm across; leaves 4-15 cm long, more than 2 cm broad . ... **E. tetragonum**
1. Plants up to 30 cm high, flowers up to 1.3 cm across; leaves 1.5-4.5 cm long, less than 2 cm broad . ... **E. pedunculatum**

Exacum pedunculatum L., Sp. Pl. 112. 1753; C.B. Clarke in Hook.f., Fl. Brit. India 4:97. 1883; S. Garg, Indian Gent. 7. 1988; Samvatsar, F. West. Trib. Madhya Pradesh 183. 1996; Srivastava in Mudgal *et al.,* Fl. Madhya Pradesh 2:104. 1997.

Erect herbs up to 25 cm high; stem 4-angled; leaves sessile, 1.5-4.5 × 0.5-1 cm, lanceolate, apex obtuse, base cuneate, 5-nerved; flowers blue, in axillary and terminal, much branched cymes; calyx lobes ovate, margin winged; corolla lobes elliptic, acute, about 1 cm long; capsules glabrous, smooth, shining, about 4 mm across; seeds brown, minute.

*Fl.-Fr.:*Dec. -Feb.

Along margin of streams, throughout the area.

Exacum tetragonum Roxb., Fl. Ind. 1:413. 1820; C.B. Clarke in Hook.f., Fl. Brit. India 4L95. 1883; S. Garg, Indian Gent. 8. 1988; Srivastava in Mudgal *et al.,* Fl. Madhya Pradesh 2:104. 1997.

Annual, erect, up to 50 cm high herbs; stem 4-angled; leaves subsessile, 4-13 × 1-5 cm, ovate, elliptic or lanceolate, apex acute or obtuse, base cuneate; flowers in terminal, paniculate cymes; calyx lobes ovate, 0.5-1 cm long, acuminate, winged on the back; corolla white, with violet-blue patches towards the edges, lobes lanceolate, 1-2 cm long; capsules globose, 0.8-1.3 cm long, smooth, shining, tipped with remains of style.

*Fl.-Fr.:*Sept. -Feb.

Rare, among grass along forest paths, **Khandwa** only, Mortakka.

HOPPEA Willd. W. 2; MP 2; SA 2

Key to Species

1. Stem 4-angled; corolla included within calyx . ... **H. dichotoma**
1. Stem 4-winged; corolla exserted from calyx .... **H. fastigiata**

Hoppea dichotoma Willd., Neue Schriften Ges. Naturf. Freunde Berlin 3:435. 1801; C.B. Clarke in Hook.f., Fl. Brit. India 4:100.1883; S. Garg, Indian Gent. 24. 1988; Samvatsar, Fl. West. Trib. Madhya Pradesh 183. 1996; Srivastava in Mudgal *et al.,* Fl. Madhya Pradesh 2:107. 1997.

Small, annual herbs, up to 10 cm high; stem 4-angled; leaves glabrous, 3-6 × 1-4 cm, 3-nerved, ovate-elliptic or elliptic-lanceolate, apex acute, upper leaves gradually passing into bracts; flowers tetramerous, in leafy, paniculate cymes, pedicels short; calyx *ca* 3 mm long, lobes 4, ribbed, veins prolonged into an apiculate tip; corolla included, greenish-yellow, lobes 4, deltoid, *ca* 2 mm long; capsules *ca* 2 mm long, oblong, 2-valved, shorter than calyx; seeds ovoid, red or brown.

*Fl.-Fr.:*Oct. -Feb.

Common in wet places, everywhere.

Hoppea fastigiata (Griseb.) C.B. Clarke in Hook.f., Fl. Brit. India 4: 100.1883; *Cicendia fastigiata* Griseb., Gen. Sp. Gent. 158. 1839 and in DC. Prodr. 9:62. 1845; S. Garg, Indian Gent. 24. 1988; Samvatsar, Fl. West. Trib. Madhya Pradesh 183. 1996; Srivastava in Mudgal *et al.,* Fl. Madhya Pradesh 2:107. 1997.

Annual, 10-15 cm high herbs; stem narrowly 4-winged; leaves 5-8 × 5-6 cm, broadly ovate to rotundate, apex obtuse, base rounded; flowers in axillary and terminal cymes, about 2 cm long; calyx lobes 4, lanceolate; corolla white, lobes 4, exserted from the calyx; capsules oblong, *ca* 3 mm across.

*Fl.-Fr.:*Sept. -Dec.

Same habitat as above, Alirajpur, Burhanpur, Khandwa.

MENYANTHACEAE 7/55; 1/2; 1/2

NYMPHOIDES Seguier W. 34; MP 2; SA 2

Key to Species

1. Corolla lobes densely fimbriate on the inner surface, eglandular; style heterostylous; seeds smooth .... **N. indica**
1. Corolla lobes not fimbriate, glandular; style homostylous; seeds tuberculate .... **N. hydrophylla**

Nymphoides hydrophylla (Lour.) Kuntze, Revis. Ge. Pl. 2:429. 1891; M*enyanthes hydrophyllum* Lour., Fl. Cochinch. 105. 1790; *Limnanthemum hydrophyllum* (Lour.) Griseb., Gen. Sp. Gent. 348. 1839; *L. cristatum* (Roxb.) Griseb., Gen. Sp. Gent. 342. 1839; C.B. Clarke in Hook.f., Fl. Brit. India 4:131. 1883; *Nymphoides cristata* (Roxb.) Kuntze, Rev. Gen. Pl. 429. 1891; Srivastava in Mudgal *et al.,* Fl. Madhya Pradesh 2:110.1997. *Kumudini*

Floating herbs with numerous stems originating from a slender rhizome; leaves *ca* 10 × 8 cm, orbicular , or orbicular-oblong, deeply cordate wih an acute sinus, margin sinuate, fleshy, smooth above, verrucose beneath, petoles short; flowers densely fascicled at nodes; sepals 5-7, narrowly lanceolate or oblong; petals white, rotate, with a yellow base, lobes linear-oblong; capsules ovoid or oblong, *ca* 5 mm long, 5-10 seeded; seeds tuberculate.

*Fl.-Fr.:*June-Sept.

In ponds, Barwani, Burhanpur, Indore, Khandwa.

Nymphoides indica (L.) Kuntze, Revis. Gen. Pl. 2:429. 1891; *Menyanthes indica* L., Sp. Pl. 145. 1753; *Limnanthemum indicum* (L.) Griseb., Gen. Sp. Gent. 343. 1839; C.B. Clarke in Hook.f., Fl. Brit. India 4:131. 1883; Samvatsar, Fl. West. Trib. Mdhya Pradesh 184. 1996; Srivastava in Mudgal *et al.,* Fl. Madhya Pradesh 2:110.1997. *Bara chuli* **(Plate 7, Fig. 3)**

Perennial, aquatic herbs with a thick rhizome; leaves floating, thick, petiolate, 7-20 × 5-20 cm, orbicular, peltate, base deeply cordate, margin coarsely crenate; flowers fascicled at nodes; sepals ovate-lanceolate, acute; petal white, with a yellow throat, lobes ovate-lanceolate, eglandlar, densely fimbriate; capsules globose, 6-8 mm across, obovoid, dehiscing irregularly; seeds yellow, globose, smooth.

*Fl.-Fr.:*Oct. -Dec.

Occasional in ponds, Barwani, Dhar, Khargone.

BORAGINACEAE 155/2685; 8/27; 8/21

Key to Genera

1. Trees or shrubs
  2. Styles 2 or 4; flowers white
    3. Styles 4; calyx accrescent .... CORDIA
    3. Styles 2; calyx not accrescent .... EHRETIA
  2. Style 1; flowers pink .... ROTULA
1. Herbs or low shrubs
  4. Style gynobasic; nutlets glochidiate .... CYNOGLOSSUM
  4. Style terminal or sub-terminal; nutlets not glochidiate
    5. Ovary deeply 4-lobed; style sub-terminal; anthers connivent in a cone; fruiting calyx enlarged .... TRICHODESMA
    5. Ovary entire or slightly 4-lobed; style terminal; anthers not connivent; fruiting calyx not enlarged
      6. Styles 2; leaves with dentate margin . ... COLDENIA
      6. Style 1, dilated above in a ring below the stigma; leaves with entire margin
        7. Anthers free; bracts present; style minute; calyx 5-partite; corolla up to 0.25 cm long; stamens inserted above the middle of corolla tube; depressed hairy undershrubs .... EUPLOCA
        7. Anthers united or free; other combination not as above .... HELIOTROPIUM

COLDENIA L. Monotypic

Coldenia procumbens L., Sp. Pl. 125, 1753; Clarke in Hook.f., Fl. Brit. India 4: 144, 1883; Samvatsar, Fl. West. Trib. Madhya Pradesh 187. 1996; Khanna in Mudgal *et al.,* Fl. Madhya Pradesh 2: 117. 1997. *Tripankhi*

A prostrate, deep-rooted, white-hairy herb; leaves 1-2 × 0.3-1 cm, obovate-oblong, crenate-crisped to sub-pinnatifid, oblique and cuneate at base; flowers pale yellow, axillary, solitary; drupes pubescent, pyramidal, 4-lobed, separating into 4, 1-seeded pyrenes.

Fl.-Fr.: Feb. -May

Common in dry ditches, particularly on rocky grounds, but more frequent in ponds when the water recedes.

CORDIA L.W. 405; MP 5; SA 4

Key to Species

1. Leaves sub-opposite, peninerved; filaments glabrous at base .... **C. sinensis**
1. Leaves alternate or occasionally sub-opposite, 3-5 nerved at base; filaments villous or glabrous at base
    2. Mature leaves densely tomentose below; calyx ribbed .... **C. macleodii**
    2. Mature leaves glabrous or softly, rusty-hairy below; calyx not ribbed
        3. Leaves essentially ovate to broadly orbicular; lower surface pubescent or with soft, rusty hairs . ...**C. monoica**
        3. Leaves essentially elliptic, lower surface glabrous, base rounded or obtuse .... **C. dichotoma**

Cordia dichotoma G. Forst., Prodr. 18. 1786; Santapau in Rec. Bot. Surv. India 16(1):185. 1953; *C. indica* Lam. Encycl. 7: 49. 1806; *C. obliqua* Willd. Phytogr. 4. 1794; *C. myxa* auct. plur. non L., 1753; Clarke in Hook.f., Fl. Brit. India 4: 136, 1883 (excl. var.); Samvatsar, Fl. West. Trib. Madhya Pradesh 186. 1996; Khanna in Mudgal *et al.,* Fl. Madhya Pradesh 2: 118. 1997. *Lasora*

A moderate sized, deciduous tree; bark greyish brown, with shallow-fissures; leaves 6-12 × 4.5-7 cm, oblong-lanceolate, orbicular, obtuse, semi-cordate or cuneate at the 3-nerved base; margin entire, glabrescent on both the surfaces, coriaceous; flowers white, in large, lax, terminal and axillary peduncled cymes; drupes ovoid, 1. 1-2.5 cm long, apiculate, yellow, with a pinkish tinge, mucilaginous.

Fl.-Fr.: Mar. -July

Common in almost all villages, also noted wild in forests and along road-sides.

Cordia macleodii (Griff.) Hook.f. & Thomson in J. Linn. Soc. London (Bot.) 2: 128. 1858; *Hemigymnia macleodii* Griff., Calc. J. Nat. Hist. 3: 363. 1843; C.B. Clarke in Hook.f., Fl. Brit. India 4: 139. 1883; Samvatsar, Fl. West. Trib. Madhya Pradesh 186. 1996; Khanna in V. Mudgal *et al.,* Fl. Madhya Pradesh 2: 119. 1997. *Telia sag* **(Plate 7, Fig. 4)**

Polygamous trees; leaves elliptic or ovate, 8-18 × 7-17 cm, cordate or rounded at base, obtuse at apex, fulvous-tomentose below; flowers white, in terminal and axillary, dense, tomentose, paniculate cymes; drupes ovoid, supported by cupular, tomentose, ribbed-calyx, non-edible.

Fl.-Fr.: March - May

In mixed dry deciduous forests.

Cordia monoica Roxb., Pl. Coromandel 1: 43. t. 58. 1776; C.B. Clarke in Hook.f., Fl. Brit. India 4: 137. 1883; Khanna in V. Mudgal *et al.,* Fl. Madhya Pradesh 2: 120.1997.

Trees; branches densely puberulentl leaves ovate or elliptic, 6-10 × 3-6 cm, rounded at base, acute or obtuse at apex; flowers white, in terminal, paniculate cymes; drupes ovoid, orange at maturity, longitudinally striated.

Fl.-Fr.: July – Oct.

Khandwa, Kalibhit forest.

Cordia sinensis Lam., Tabl. Encycl. 1: 423. 1792*; Cornus gharaf* Forssk. Fl. Aegypt. -Arab 95. 1775; *Cordia rothii* Roem. & Schult., Syst. Veg. 4: 798. 1819; *Cordia gharaf* (Forssk.) Ehrenb. ex Asch. Sitzungsber. Ges. Naturf. Freunde Berlin 46 1879; C.B. Clarke in Hook.f., Fl. Brit. India 4: 138. 1883; Samvatsar, Fl. West. Trib. Madhya Pradesh 186. 1996; Khanna in V. Mudgal *et al.,* Fl. Madhya Pradesh 2: 119. 1997. *Chooti gondi*

A small, 5-6 m tall tree; bark greyish -brown; leaves sub opposite, oblong-lanceolate, 4-10 × 1.2-2.5 cm, rounded at apex; flowers white, in large, lax, terminal and axillary peduncled cymes; drupes orange-yellow when ripe, ovoid, 0.6-1 cm long, apiculate, mucilaginous.

*Fl.-Fr.:*Dec. – June

Alirajpur, Burhanpur, Dhar, Indore.

CYNOGLOSSUM L. W. 83; MP 3; SA 2

Key to Species

1. Flowers 6-8 mm across; style 2 mm long .... **C. zeylanicum**
1. Flowers 3-4 mm across; style 5 mm long .... **C. lanceolatum**

Cynoglossum lanceolatum Forssk., Fl. Aegypt. -Arab. 41. 1775; C.B. Clarke in Hook.f., Fl. Brit. India 4: 156. 1883; Samvatsar, Fl. West. Trib. Madhya Pradesh 188. 1996; Khanna in V. Mudgal *et al.,* Fl. Madhya Pradesh 2: 122. 1997.

Herbs, up to 50 cm high; branches covered with bulbous-based hairs; leaves elliptic-oblong or lanceolate, hispid, 5-10.5 × 2-4 cm, rounded at base, acute at apex; flowers white with a bluish or purple eye, in axillary or terminal, paniculate cincinni; nutlets 4, glochidiate.

Fl.-Fr.: Sept. – May

In mixed forests, all districts.

Cynoglossum zeylanicum (Vahl ex Hornem.) Thunb. ex Lehm Neue Schr. Naturf. Ges. Pl. Hort. Halle 3(2): 20, 1817; *Anchusa zeylanica* Vahl ex Hornem. Enum Hafn. 3. 1807; Clarke in Hook.f., Fl. Brit. India 4:157. 1883; Kazmi. in J. Arn. Arb. 52: 344, 1971; Samvatsar, Fl. West. Trib. Madhya Pradesh 188. 1996; Khanna in Mudgal *et al.,* Fl. Madhya Pradesh 2: 123. 1997.

Erect, much branched, strigosely hairy undershrub; leaves radical and cauline; radical large, broadly ovate, oblong, long petiolate; cauline hairy; flowers pale-blue or purplish-white, first capitate and then in 10-15 cm long, terminal and axillary, unilateral cymes; nutlets glochidate with distinct margins.

Fl.-Fr.: Aug. -Dec.

Alirajpur, Barwani, Dhar, Indore, Khargone.

EHRETIA L. W. 65; MP 2; SA 2

Key to Species

1. Leaves glabrous beneath, 1.5-14 × 1-8 cm .... **E. laevis**
1. Leaves persistently hairy beneath, 4-8 × 3.5-6 cm .... **E. acuminata**

Ehretia acuminata R. Br., Prodr. Fl. Nov. Holl. 497 1810; C.B. Clarke in Hook.f., Fl. Brit. India 4: 141. 1883; Khanna in Mudgal *et al.,* Fl. Madhya Pradesh 2: 125. 1997.

Deciduous trees; leaves coriaceous, elliptic or ovate-oblong, 8-14 × 3.5-5 cm, glabrous or minutely hispidulous above; flowers white, in large, pubescent panicles; drupes orange, ellipsoid, 4-seeded.

Fl.-Fr.: March-July

Alirajpur, Burhanpur, Dhar, Indore.

Ehretia laevis Roxb. Pl. Cor. 1: 42. t. 56, 1796; C.B. Clarke in Hook.f., Fl. Brit. India 4: 141, 1883; *E. laevis* var. *floribunda* (Benth.) Brandis, Forest Fl. N. W. India 340.1874; Samvatsar, Fl. West. Trib. Madhya Pradesh 187. 1996; Khanna in Mudgal *et al.,* Fl. Madhya Pradesh 2: 126. 1997. *Datranga*

A small tree; leaves, coriaceous, obovate-oblong or elliptic-oblong, 1.5-14 × 1-8 cm, glabrous; flowers white, in terminal and axillary, lax cymes; drupes sub- globose, black and wrinkled when dry.

Fl.-Fr.: Dec. -May

Alirajpur, Barwani, Burhanpur, Dhar, Indore, Khargone.

EUPLOCA Nutt. W. 21; MP 1; SA 1

Euploca rariflora (Stocks) Diane Hilger, Bot. Jahrb. Syst. 125:48. 2003; *Heliotropium rariflorum* Stocks, Hooker's J. Bot. Kew Gard. Misc. 4:174. 1852; C.B. Clarke in Hook.f., Fl. Brit. India 4:152. 1883; Samvatsar, Fl. West. Trib. Madhya Pradesh 189. 1996; Khanna in Mudgal *et al.,* Fl. Madhya Pradesh 2:130.1997.

Erect, stout, branched undershrubs; leaves 0.4-3 × 0.2-0.4 cm, linear, apex subacute, margin revolute, hairy; flowers in 2-8 cm long, terminal or extra-axillary, bracteate spikes; calyx divided up to the base; corolla white; drupes 4-lobed, *ca* 0.25 cm across; nutlets 4, appressed hairy.

*Fl.-Fr.:*Feb. -April

In waste places, Dhar, Indore.

HELIOTROPIUM L. W. 156; MP 7; SA 7

Key to Species

1. Anthers united by their prolonged connectives; bracts present
    2. Stigmatic ring not distinct on style; flowers with narrow bracts .... **H. strigosum**
    2. Stigmatic ring distinct on style; flowers white, with conspicuous bracts

3. Leaves 0.3-1.3 × 0.2-0.4 .... **H. marifolium**

3. Leaves 2.5-5 × 0.3-0.5 cm .... **H. zeylanicum**

1. Anthers free; bracts present or absent

4. Bracts present; style minute; leaves 1.5-2 × 0.5-1.5 cm; plants villous .... **H. supinum**

4. Bracts absent; style hardly any; stamens inserted below the middle of corolla tube

5. Fruit turban-shaped (mitriform), 4-beaked; stigma obtuse at apex, frilled at base; flowers pale-violet .... **H. indicum**

5. Fruit of 2-4 nutlets, not beaked; stigma conical; flowers white; calyx persistent even after the nutlets are shed

6. Corolla lobes with intermediate teeth; leaves 3-7 cm long, with tubercled hairs ......**H. ellipticum**

6. Corolla lobes without intermediate teeth; leaves 0.5-2 cm long, with silky hairs . **H. ovalifolium**

Heliotropium ellipticum Ledeb. in Eichw., Pl. Nov. 10. t. 4. 1831; *H. eichwaldi* Steud. Nomencl. ed. 2, 1: 744, 1840; Clarke in Hook.f., Fl. Brit. India 4: 149, 1883; Samvatsar, Fl. West. Trib. Madhya Pradesh 189. 1996; Khanna in Mudgal *et al.,* Fl. Madhya Pradesh 2: 128. 1997.

Annual herbs, 30-80 cm tall, erect, softly hairy, stout herbs; leaves 3-7 × 1-2.5 cm, elliptic-oblong or obovate, hairy; flowers white, 2-ranked, ebracteate, in unilateral spikes; nutlets 4, ellipsoid, minutely verrucose and finely puberulous.

Fl.-Fr.: March-Aug.

Alirajpur, Burhanpur, Dhar, Khandwa.

Heliotropium indicum L. Sp. Pl. 130, 1753; *Tiaridium indicum* Lehm., Pl. Asperif. Nucif. 14. 1818; Clarke in Hook.f., Fl. Brit. India 4: 152, 1883; Samvatsar, Fl. West. Trib. Madhya Pradesh 189. 1996; Khanna in Mudgal *et al.,* Fl. Madhya Pradesh 2: 128. 1997. *Hathi sund*

Erect herbs; branches white hairy; leaves 4-10 × 2-6 cm ovate or ovate-oblong, sparsely hairy above, pilose beneath, petiole partially winged; flowers pale-violet in 2-rows; fruit deeply 2-fid, each lobe compressed, produced into a short beak and containing 1-seeded nutlets.

Fl.-Fr.: Dec. -Mar.

Dhar, Khandwa.

Heliotropium marifolium J. Koenig ex Retz. Observ. Bot. 2:8. 1781; *H. scabrum* Lehm., Pl. Asperif. Nucif. 70.1818; *Cynoglossum marifolium* Roxb., Fl. Ind. 2:8. 1824; Clarke in Hook.f., Fl. Brit. India 4:152. 1883; Samvatsar, Fl. West. Trib. Madhya Pradesh 189. 1996; Khanna in Mudgal *et al.,* Fl. Madhya Pradesh 2: 129, 1997.

Prostrate or diffuse herbs with radially spreading appressed hairy branches; leaves 0.3-1.3 × 0.2-0.4 cm,linear, sessile or subsessile, appressed hairy; flowers white, small in terminal and axillary, unilateral spikes; nutlets 4, subglobose hairy.

Fl.-Fr.: Sept. -Dec.

Alirajpur, Burhanpur, Dhar, Khandwa.

Heliotropium ovalifolium Forssk. Fl. Aegypt.-Arab. 38, 1775; *H. coromandelianum* Retz., Observ. Bot. 2: 9. 1819; Clarke in Hook.f., Fl. Brit. India 4: 150, 1883; Samvatsar, Fl. West.

Trib. Madhya Pradesh 189. 1997; Khanna in Mudgal *et al.,* Fl. Madhya Pradesh 2: 129. 1997.

Erect, decumbent or prostrate, perennial herbs; leaves alternate to subopposite 0.7-3.5 × 0.3-2 cm, petiolate, margins subentire; flowers in forked, scorpioid spikes; nutlets 4, usually subglobose, densely hairy.

Fl.-Fr.: Almost throughout the year

Common in drying ditches and in fields, throughout the area.

Heliotropium strigosum Willd. Sp. Pl. 1: 743, 1798; Clarke in Hook.f., Fl. Brit. India 4: 151, 1883; Samvatsar, Fl. West. Trib. Madhya Pradesh 189. 1996. Khanna in Mudgal *et al.,* Fl. Madhya Pradesh 2: 130.1997.

Decumbent or prostrate, appressedly hairy herbs; leaves 1.2-2.5 cm long, shortly petiolate, margins slightly revolute, entire; flowers pedicellate or subsessile, in axillary or terminal spikes; nutlets 4, appressedly hairy, with rounded back, surrounded by the enlarged calyx.

Fl.-Fr.: Aug. -Oct.

Alirajpur, Dhar, Indore.

Heliotropium supinum L. Sp. Pl. 130, 1753; *H. malabaricum* Retz., Observ. Bot. 4: 24. 1786; Clarke in Hook.f., Fl. Brit. India 4: 149. 1883; *H. supinum* L. var. *malabarica* (Retz.) Clarke loc. cit. Samvatsar, Fl. West. Trib. Madhya Pradesh 189. 1996. Khanna in Mudgal *et al.,* Fl. Madhya Pradesh 2: 131. 1997.

Prostrate, spreading, much branched, villous herbs; leaves 1.5-2 × 0.5-1.5 cm, elliptic-obovate, ovate or suborbicular, entire or crenate; flowers in simple or forked, villous,scorpioid spikes; nutlets 1-4, rounded on the back, margined, obscurely tuberculate, brown.

Fl.-Fr.: Alomost throughout the year

Common in drying, moist ground along margins of ponds throughout the area.

Heliotropium zeylanicum (Burm.f.) Lam., Encycl. 3: 94. 1789; *H. curasavicum* L. var. *zeylanicum* Burm.f., Fl. Ind. 41. t. 16. f. 2. 1768; *H. paniculatum* R. Br., Prodr. 494. 1810; C.B. Clarke in Hook.f., Fl. Brit. India 4: 151. 1883; Khanna in Mudgal *et al.,* Fl. Madhya Pradesh 2: 131. 1997.

Annual, erect herbs; leaves subsessile, linear-lanceolate, 1.2-2.5 × 0.1-0.3 cm, entire, sparsely hairy bebeath; flowers white, in simple or forked, bracteate, scorpioid spikes; nutlets lobed, compressed.

Fl.-Fr.: Aug. -Sept.

Khandwa only, Chandgarh range.

ROTULA Lour. Monotypic

Rotula aquatica Lour. Fl. Cochinch. 121, 1790; *Rhabdia lycioides* C. Mart. Nov. Gen. Sp. Pl. 2: 136. 1827; Clarke in Hook.f., Fl. Brit. India 4: 415, 1883; Samvatsar, Fl. West. Trib. Madhya Pradesh 187. 1996; Khanna in Mudgal *et al.,* Fl. Madhya Pradesh 2: 134. 1997. *Sawariya*

A shrub; branches virgate; leaves alternate or fascicled, 0.5-3 × 0.2-0.6 cm,obovate-ellipttic or spathulate, glabrous,or nearly so; flowers pink, solitary or 2-3, in axillary, condensed cymes; drupes sub- globose, reddish-brown or orange.

Fl.-Fr.: Oct. -Dec.

Alirajpur, Burhanpur, Dhar, Indore, Khandwa

TRICHODESMA R.Br.W. 17; MP 4; SA 3

Key to Species

1. Fruiting calyx lobes ovate or rounded at base; anthers exserted; corolla tube equal to lobes; lobes acuminate, without spots, not recurved .... **T. zeylanicum**

1. Fruiting calyx lobes cordate or hastate at base

  2. Staminal cone exserted, anthers permanently joined, not separable; corolla tube as long as lobes; lobes acuminate, with spots, recurved .... **T. indicum**

  2. Stamnal cone inserted; corolla tube longer than lobes; lobes shortly acuminate, without spots, not recurved .... **T. inaequale**

Trichodesma inaequale Edgew., J. Asiat. Soc. Bengal 21:175. 1852; C.B. Clarke in Hook.f., Fl. Brit. India 4:153. 1883; Khanna in Mudgal *et al.,* Fl. Madhya Pradesh 2:137. 1997. *Audhi jhar*

Diffuse, hispid herbs; leaves 3-8 × 1-3 cm, ovate elliptic, semiamplexicaule at base; flowers violet, solitary, axillary; anthers included, backs of connectives penicillate with straight, glistening, white, opaque hairs; fruit ovoid, smooth, of 4 nutlets.

*Fl.-Fr.:*July-Oct.

Alirajpur, Burhanpur, Dhar, Indore, Khandwa, along road-sides and waste ground.

Trichodesma indicum (L.) Lehm., Pl. Asperif. Nucif. 193. 1818; *Borago indica* L. Sp. Pl. 137. 1753; *Trichodesma amplexicaule* Roth, Nov. Sp. Pl. 104. 1821; C.B. Clarke in Hook.f. Fl. Brit. India 4:153. 1883; Samvatsar, Fl. West. Trib. Madhya Pradesh 190.1996; Khanna in Mudgal *et al.,* Fl. Madhya Pradesh 2:135. 1997. *Ghammara*

Annual, 15-60 cm tall herbs; branches sparsely or densely hairy with bulbous-based, stiff hairs; leaves 3-8 × 1.2-2 cm, ovate-oblong to ovate-lanceolate, hairy on both surfaces, apex acute or obtuse, base auriculate; flowers pale blue, up to 1.8 cm long, solitary or in few flowered terminal cymes; calyx hispid, deeply divided; nutlets 4, *ca* 0.5 cm long, ovoid, smooth on outer face, rugosely pitted on inner face.

Fl.-Fr.: Oct. -Feb.

Common in all districts, as forest undergrowth, open grassland, along road-sides, on slopes of ravines.

Trichodesma zeylanicum (Burm.f.) R. Br. Prodr. Fl. Nov. Holl. 496. 1810; *Borago zeylanica* Burm.f., Fl. Indica 41. 1768; C.B. Clarke in Hook.f., Fl. Brit. India 4:154. 1883; Bannerjee & Pramanik in Bull. bot. Surv. India 17:120 (1975) 1978; Samvatsar, Fl. West. Trib. Madhya Pradesh 190.1996; Khanna in Mudgal *et al.,* Fl. Madhya Pradesh 2:137. 1997. *Jalsirsa*

An erect or diffuse, 20-50 cm tall herb; stem hispid, terete, branched with long, bulbous-based, stiff hairs; leaves 5-10 × 1-3 cm, oblong, lanceolate, acute, base cuneate, prickly hairy on both surfaces; flowers blue or violet, in terminal, panicled cymes; nutlets 4, 0.4-0.5 cm long, grey, dotted brown on dorsal surface, spotted or rugose on inner face.

Fl.-Fr.: Aug. -Oct.

Alirajpur, Dhar, Indore, Khandwa, on margins of fields and open, hilly areas.

CONVOLVULACEAE 67/1296; 12/46; 11/39

Key to Genera

1. Parasitic plants; stem yellow; leaves absent or reduced to scales .... CUSCUTA
1. Autotrophs
   2. Pollen not spinulose
      3. Styles 2; capsule 4-valved
         4. Stigmas 4; petals twisted; stamens included; prostrate herbs .... EVOLVULUS
         4. Stigmas 2; petals imbricate; stamens exserted; erect herbs .... CRESSA
      3. Style 1; stigmas 2, bilobed or capitate
         5. Fruit indehiscent; calyx accrescent, falling with fruit; flowers in panicles; seed 1 .... PORANA
         5. Fruit dehiscent, 4 or 8 valved; seeds 4
            6. Capsule 8-valved; stem and leaves with stellate trichomes; stigma ellipsoid-oblong; leaf base truncate or rounded .... JACQUEMONTIA
            6. Capsule 4-valved
               7. Capsule irregularly dehiscing or by valves; sepals unequal; leaf blade reniform or ovate-lanceolate; prostrate herbs or twiners .... MERREMIA
               7. Capsule dehiscing by valves only; sepals more or less equal; leaf blade usually hastate or sagittate; herbs or twiners .... CONVOLVULUS
   2. Pollen spinulose; style 1; stigma bi-lobed or capitate
      8. Fruit indehiscent, woody or fleshy twiners
         9. Fruit fleshy; corolla salver-shaped, creamy-white; stigma linear; leaves broadly orbicular, with 2 purple glands at the base .... RIVEA
         9. Fruit woody; corolla funnel-shaped, rose-purple; stigma bi-lobed; corolla hairy outside; leaves broadly ovate, gland absent .... ARGYREIA
      8. Fruit dehiscent; corolla funnel-shaped; herbaceous or woody twiners
         10. Calyx in fruit enlarged, completely enclosing the capsule; flowers purple; leaves dotted with black glands on the lower surface .... STICTOCARDIA
         10. Calyx not so enlarged; flowers white, yellow or purple; leaves without glands.......... IPOMOEA

ARGYREIA Lour . W. 28; MP 5; SA 4

Key to Species

1. Peduncles more than 20 cm long; sepals woody .... **A. nervosa**

1. Peduncles less than 20 cm long; sepals herbaceous
   2. Flowers in lax, corymbose cymes .... **A. strigosa**
   2. Flowers in capitate or sub-capitate cymes; peduncles more than 5 cm long; sepals more than 1 cm long
      3. Sepals nearly equal, ovate-lanceolate, glabrous inside . ... **A. kleiniana**
      3. Sepals unequal, elliptic-lanceolate, sparsely hairy inside . ... **A. sericea**

Argyreia kleinina (Roem. & Schult.) Raizada, Indian For. 92:302. 1966; *Ipomoea kleiniana* Roem. & Schult., Syst. Veg. 4:789. 1819; *Argyreia bracteata* Choisy, Convolv. Orient. 39. 1834; C.B. Clarke in Hook.f., Fl. Brit. India 4:188. 1883; Srivastava in Mudgal *et al.,* Fl. Madhya Pradesh 2:145. 1997.

Perennial twiners; leaves 6-15 × 5-14 cm, ovate, apex acuminate, base cordate, sparsely hairy aove, densely silky beneath; petioles 3-8 cm long; flowers red with purple throat, in sub-capitate cymes, on 8-10 cm long peduncles; bracts prominent, silky villous, persistent; sepals 1-1.5 cm long, glabrous inside, hairy outside; corolla campanulate, *ca* 4.5 cm long; berry yellowish-brown, *ca* 1.5 cm across.

*Fl.-Fr.:*Oct. -Dec.

Dhar only, at Mandu plateau.

Argyreia nervosa (Burm.f.) Bojer, Hortus Maurit. 224. 1837; *Convolvulus nervosus* Burm.f., Fl. Ind. 48. t. 20. f. 1768; *Lettsomia nervosa* (Burm.f.) Roxb., Fl. Ind. ed. 1820 2:78. 1824; *Argyreia speciosa* Sweet, Hort. Brit. ed. 2.373. 1830; C.B. Clarke in Hook.f., Fl. Brit. India 4:185. 1883; Samvatsar, Fl. West. Trib. Madhya Pradesh 191-192. 1996; Srivastava in Mudgal *et al.* Fl. Madhya Pradesh 2:145. 1997. *Samundar- ka- pat*

White or grey, tomentose, woody twiners; leaves 10-15 cm long, broadly ovate, white tomentose beneath, glabrous above; flowers red-purple, in axillary, capitate, 20-25 cm long, pedunculate cymes; sepals ovate, *ca* 2 cm long, white tomentose; corolla 6-7 cm long, pubescent outside on tube; berry dry, sub- globose, 1-1.5 cm across; seeds light or dark brown, glabrous.

Fl.-Fr.: Aug. -Dec.

Dhar and Indore, occasional in forest, sometimes as an escape.

Argyreia sericea Dalz., Bombay Flora 169. 1861; C.B. Clarke in Hook.f., Fl. Brit. India 4:188. 1883; Samvatsar, Fl. West. Trib. Madhya Pradesh 192. 1996; Srivastava in Mudgal *et al.,* Fl. Madhya Pradesh 2:145. 1997.

Appressed, silky hairy twiners; leaves 4-12 × 6-11 cm, broadly ovate, acute to acuminate at apex, cordate at base, densely silky beneath, hispid or strigose above; flowers pink, in axillary, capitate cymes; peduncles 6-16 cm long; bracts persistent; calyx segments unequal, elliptic-lanceolate, 1.5-2.5 cm long, densely hairy outside; corolla *ca* 5-5.5 cm long; berries 1-1.5 cm across, globose, glabrous, orange-red when ripe.

Fl.-Fr.: Sept. -Dec.

Barwani, Dhar, Indore, Khandwa, Khargone.

Argyreia strigosa Roberty in Candollea XIV:44. 1952; *Ipomoea strigosa* Roth, Nov,Pl. Sp. 113. 1821; *Lettsomia setosa* Roxb., Hort. Bengal. 13 (1814), *nomen* Fl. Ind. ed. (Carey & Wallich) 11:80.1824; C.B. Clarke in Hook.f., Fl. Brit. India 4:194. 1883; Samvatsar, Fl. West. Trib. Madhya Pradesh 192. 1996; Srivastava in Mudgal *et al.,* Fl. Madhya Pradesh 2:146. 1997. *Bagh choura*

Stout, twining shrubs, strigosely hairy on younger parts; leaves 6-14 × 4-13 cm, acute or acuminate at apex, cordate at base, glabrous above, appressed hairy beneath; flowers rose, in lax corymbose cymes; peduncles 12-20 cm long; bracts ovate, persistent, appressedly stigose; sepals densely appressed hairy outside, longer than the bracts; corolla mauve, densely setose; berries *ca* 1 cm across, red when ripe.

Fl.-Fr.: Sept.-Dec.

Common in all districts.

CONVOLVULUS L. W. 72; MP 4; SA 4

Key to Species

1. Leaves sagittate-hastate at base, glabrous on both sides; corolla 4-5 times longer than calyx .... **C. arvensis**
1. Leaves not as above, villous on both sides; corolla equal to or slightly longer than calyx
    2. Flowers sessile or subsessile, 1-5, in leaf axils; plants prostrate .... **C. prostratus**
    2. Flowers in comparatively long peduncles
        3. Peduncles 4 cm long; sepals acuminate; corolla 1.2-1.4 cm; capsule 3 mm long .... **C. deserti**
        3. Peduncle 2-3 cm long; sepals acute; corolla 0.7-0.8 cm; capsule 5 mm long . ... **C. rottlerianus**

Convolvulus arvensis L. Sp. Pl. ed. 1: 153, 1753; Clarke in Hook.f., Fl. Brit. India 4: 219. 1883; Samvatsar, Fl. West. Trib. Madhya Pradesh 192. 1996; Srivastava in Mudgal *et al.,* in Fl. Madhya Pradesh 2:147. 1997. *Chand mul*

Creeping or twining, perennial herbs; leaves 2-5.5 × 1-2 cm; flowers pale pink, bracteate, axillary, solitary or 2-3 on long peduncles; capsule 0.4-0.6 × 0.3-0.4 cm, ovoid-globose, glabrous, with persistent calyx and style base, 4-seeded; seeds subtrigonous, dark brown, black or reddish brown.

Fl.-Fr.: Nov. -March

Very common as a weed of fields and wastelands, throughout the area.

Convolvulus deserti Hochst. & Steud. ex Steud., Nom. ed. 2(1): 408. 1841 (nom. nud.); Baker & Rendle in Fl. Trop. Africa 4(2):164. 1905; Bhandari in Fl. Indian desert 220.1978; Samvatsar, Fl. West. Trib. Madhya Pradesh 192. 1996.

Suffruticose herbs; stem with spreading hairs; leaves 3-6 × 0.2-1 cm, linear-oblong, subsessile; flowers pink, axillary, 1-3, on slender, hairy, long peduncles; bracts narrow, linear with ferrugineous hairs; capsules about 3 mm long; seeds 2-4, dark brown.

Fl.-Fr.: Sept. -Dec.

Burhanpur, Dhar, Jhabua. This species is not mentioned in the Plant List.

Convolvulus prostratus Forssk., Fl. Aegypt-Arab. 203. 1775; *C. microphyllus* Sieb. ex Spreng., Syst. Veg. 1:611. 1825; Samvatsar, Fl. West. Trib. Madhya Pradesh 192. 1996; Srivastava in Mudgal *et al.,* Fl. Madhya Pradesh2:147. 1997. *Shankha pushpi*

Perennial, prostrate herbs; leaves 1-3 × 0.5-0.8 cm, linear to oblong sessile, villous; bracts acute; bracteoles filiform; sepals villous; flowers pinkish-white on lateral shoots, in 2-7 flowered, short, pedunculate fascicles; ovary seated on a cup-shaped disc; capsules 4-seeded; seeds brown-black.

Fl.-Fr.: Dec. -Feb.

Burhanpur, Indore, Khandwa, in open ground.

Convolvulus rottlerianus Choisy in Mem. Soc. Phys. Genev. 6:477. 1833; C.B. Clarke in Hook.f., Fl. Brit. India 4: 219. 1883; Bhandari in Fl. Indian Desert 221. 1978.

An erect, 20-30 cm high, hardly branched perennial, with leafy, silky-villous stem; leaves 2.5-5 × 3-6 cm, linear, acute; flowers 1-3, on slender peduncles in lax racemes, up to 3.5 cm long; bracts 2, beneath each pedicel; corolla 1 cm long, capsules 5 mm across; seeds puberulent.

Fl.-Fr.: Oct. -Dec.

Alirajpur, Dhar, Indore.

CRESSA L. W. 3; MP 1; SA1

Cressa cretica L. Sp. Pl. 223. 1753; C.B. Clarke in Fl. Brit. India 4:225. 1883; Bhandari Fl. Indian Desert 223. 1978.

A dwarf perennial, up to 25 cm high; stem slender, with many, densely leaved, very hairy branches; leaves 3-7 × 1.5-3.5 cm, base rounded or narrow; flowers white or pink, solitary terminal or in short, terminal spikes; bracts leafy; corolla 4 mm long, deflexed; capsules 4-5 mm long, pointed with style bases; seeds rarely produced, brown, usually 1 per capsule.

Fl.-Fr.: Nov. -March

Burhanpur, Shahpur and Daryapur.

CUSCUTA L. W. 171; MP 3; SA 3

Key to Species

1. Style 1; calyx divided almost to the base; stem comparatively thick .... **C. reflexa**
1. Styles 2; calyx divided more than half way down; stem thread-like
    2. Scales absent; calyx divided more than half way down .... **C. hyalina**
    2. Scales present; calyx divided less than half way down .... **C. chinensis**

Cuscuta chinensis Lam. Encycl. 2:229. 1786; C.B. Clarke in Fl. Brit. India 4:226. 1883; Samvatsar, Fl. West. Trib. Madhya Pradesh 201. 1996; Srivastava in Mudgal *et al.,* Fl. Madhya Pradesh 2:150.1997. *Amar bel*

Small, yellow herbs with thread-like stem, forming tangled masses over surrounding plants; flowers white, solitary or in few flowered fascicles; peduncles short; corolla throat with scales; capsules globose, 2 mm long; seeds glabrous.

Fl.-Fr.: June-Aug.

Rare, on *Senna tora* (normally), Alirajpur, Burhanpur, Dhar, Indore, Khandwa.

Cuscuta hyalina Roth, Nov. Pl. Sp. 100.1821, non Wight, 1848; C.B. Clarke in Hook.f. Fl. Brit. India 4:226. 1883; Samvatsar in Fl. West. Trib. Madhya Pradesh 201. 1996; Srivastava in Mudgal *et al.,* Fl. Madhya Pradesh 2:151. 1997. *Niradhari*

Stems yellow, closely twined, very slender; flowers in fascicles on shortly pedunculate cymes or racemes; calyx about 3 mm long, divided more than halfway down; scales in throat of corolla absent; capsules *ca* 2.5 mm across.

*Fl.-Fr.:*July-Sept.

Alirajpur, Burhanpur, Dhar, Khandwa.

Cuscuta reflexa Roxb., Pl. Coromandel 2:3. T. 104. 1799; C.B. Clarke in Hook.f. Fl. Brit. India 4:225. 1883; Samvatsar, Fl. West. Trib. Madhya Pradesh 203. 1996; Srivastava in Mudgal *et al.,* Fl. Madhya Pradesh 2:151. 1997. *Amar bel*

A large, parasitic twiner; stem yellow, branched, forming close tangled masses on trees; flowers in subglobose, small clusters or racemes, whitish-yellow; bracts scarious; calyx lobes fleshy, unequal; corolla 5-5. 7 mm long; capsules globose; seeds glabrous, black.

Fl.-Fr.: Dec. -Jan.

Very common in all districts.

EVOLVULUS L. W. 100; MP 2; SA 2

Key to Species

1. Plants not rooting at nodes; leaves borne above ground, lanceolate, hairy; flowers blue; capsule 2-4 valved; seeds usually 4 .... **E. alsinoides**
1. Plants rooting at nodes; leaves appressed to ground, suborbicular, glabrous; flowers white; capsule 4-valved; seeds usually 2 .... **E. nummularius**

Evolvulus alsinoides (L.) L. Sp. Pl. ed. 2.392. 1762; C.B. Clarke in Hook.f. Fl. Brit. India 4:220.1883; *Convolvulus alsinoides* L. Sp. Pl. 157. 1753; Samvatsar, Fl. West. Trib. Madhya Pradesh 194. 1996; Srivastava in Mudgal *et al.,* Fl. Madhya Pradesh 2:153. 1997. *Shankha phuli*

A small, prostrate (normally) herb, covered with silky indumentum; leaves elliptic to linear oblong, 1-2.4 × 0.2-2 cm, appressed hairy on both surfaces; flowers blue, 1-5 on a peduncle; peduncles filiform, 0.6-4 cm long; capsules 3-4 mm long, 4-valved and 4-seeded.

Fl.-Fr.: July –March

Common in all districts, in open ground and on hill slopes.

Evolvulus nummularius (L.) L., Sp. Pl. ed. 2.391. 1762; *Convolvulus nummularius* L., Sp. Pl. 157. 1753; *Volvulopsis nummularia* Roberty, Candollea 14:28. 1952; Srivastava in Mudgal *et al.,* Fl. Madhya Pradesh 2:153. 1997.

Perennial, prostrate, herbs, rooting at nodes; leaves ovate to orbicular, 4-10 × 7-9 mm; flowers white, solitary or paired in leaf axils; pedicels 2-6 mm long; corolla 5-7 mm wide; capsules globose; seeds brown to black.

Fl.-Fr.: Aug. -Feb.

Burhanpur only, as ground cover in waste lands, Shahi kila and Hina garden.

IPOMOEA Jacq. W. 449; MP 19; SA 15

Key to Species

1. Sepals distinctly awned; corolla salver-shaped or hypocrateriform; capsules glabrous; annuals
    2. Corolla red
        3. Corolla up to 4-5 cm long, salver-shaped; leaves 3-5 lobed .... **I. hederifolia**
        3. Corolla 2-3 cm long, hypocrateriform; leaves pinnately segmented (not compound), segments filiform .... **I. quamoclit**
    2. Corolla white or purple, salver-shaped, up to 6 cm long; outer sepals 5-10 mm long; stem muricate .... **I. turbinata**
1. Sepals not awned; corolla funnel-shaped
    4. Leaves entire, rarely 3-lobed
        5. Outer sepals hastate, larger than the inner, accrescent; seeds villous, fringed on the margin with soft, white hairs; stem smooth .... **I. biflora**
        5. Outer sepals not hastate
            6. Flowers in sessile or nearly sessile clusters; bracts pilose; capsule hairy . ... **I. eriocarpa**
            6. Flowers in pedunculate cymes
                7. Calyx hispid
                    8. Outer bracts several; corolla pink or mauve, less than 2.5 cm long; leaves densely woolly beneath .... **I. arachnosperma**
                    8. Outer bracts few; corolla more than 2.5 cm long; leaves pubescent beneath, 3-lobed .... **I. nil**
                7. Calyx glabrous or finely pubescent
                    9. Leaves hastate; sepals with glands
                        10. Flowers 7-10 cm long; outer selpals with a gland on either side; seeds hairy .... **I. carnea**
                        10. Flowers 5-6 cm long; sepals glandular within; seeds glabrous .... **I. aquatica**
                    9. Leaves not hastate; sepals not glandular
                        11. Sepals acute; corolla campanulate; flowers creamy-yellow; pedicels thickened at apex; seeds softly pubescent .... **I. obscura**
                        11. Sepals obtuse; corolla infundibuliform; flowers white, pink or purple; pedicels not thickened; seeds densely tomentose .... **I. marginata**

4. Leaves digitately lobed or deeply emarginate; capsule glabrous; seeds minutely hairy or tomentose; annuals or perennials
   12. Leaves deeply emarginate; flowers usually solitary, pink or purple; seeds densely brownish-tomentose; perennials.. **I. pes-caprae**
   12. Leaves digitately 7-9 lobed
      13. Plants covered with long, yellow, bristly trichomes; leaves without pseudostipules; flowers rose, subsessile; seeds grey-tomentose; annuals . ... **I. pes-tigridis**
      13. Plants glabrescent; leaves with pseudostipules; flowers yellow or white, pedicellate; stem smooth or muricate; perennials
         14. Flowers purple or white, 3-6 cm long, in sub-umbellate cymes; leaves ovate-orbicular, deeply 5-lobed; stem muricate; seeds densely short tomentose .... **I. cairica**
         14. Flowers yellow, 5-10 cm long, in dichasial cymes; leaves 5-foliolate; seeds villous, margin woolly ....**I. tuberculata**

Ipomoea alba L., Sp. Pl., 1:161. 1753; *I. bonabox* L., Sp. Pl. (ed. 2) 1: 228. 1762; *Calonyction speciosum* Choisy, Mem. Soc. Phys. Geneve 6:441. 1834; *C. bona-box* (L.) Bojer, Hortus Maurit. 227. 1837; C.B. Clarke in Hook.f., Fl. Brit. India 4:197. 1883; Samvatsar, Fl. West. Trib. Madhya Pradesh 195. 1996; Srivastava in Mudgal *et al.,* Fl. Madhya Pradesh 2:157. 1997.

A large, woody twiner with warty stem; leaves 4-12 × 4-10 cm, entire or 3-5 angled; flowers white, 7-12 cm long, in axillary and terminal cymes; capsules ovoid, apiculate, 4-seeded; seeds dark brown.

Fl.-Fr.: Oct. -Dec.

On hedges of fields, Burhanpur, Dhar,Indore, Khandwa

Ipomoea aquatica Forssk., Fl. Aegypt. -Arab. 44. 1775; *Convolvulus repens* Vahl, Symb. Bot. 1:17. 1790, non L. 1753; *Ipomoea repens* (Vahl) Roth, Nov. Pl. Sp. 110.1821; C.B. Clarke in Hook.f. Fl. Brit. India 4:210.1883; Samvatsar, Fl. West. Trib. Madhya Pradesh 195. 1996; Srivastava in Mudgal *et al.,* Fl Madhya Pradesh 2:157. 1997. *Kalmi sag*

Floating and creeping, aquatic and marshy perennials; stem hollow, rooting at nodes; leaves variable, 4-8 × 2-5 cm, ovate to oblong, lanceolate or linear, mostly hastate, apex acute to obtuse; flowers white to purple, solitary or in, 1-5 flowered, pedunculate cymes, pedicels 2-7 cm long; sepals ovate-oblong, 6-7 mm long, obtuse or subacute, glabrous; corolla 2-6 cm long; capsules ovoid, 7-8 mm long; seeds densely soft-pilose.

*Fl.-Fr.:*Most part of the year

Common in aquatic or semi-aquatic habitat, in all districts.

Ipomoea arachnosperma Welw., Apont. 588. 1859; *Ipomoea dichroa* Choisy in DC. Prodr. 9:364. 1845; *Convolvulus dichrous* Roem. & Schult., Syst. Veg. 4:263. 1819, *nom. illegit.,* C.B. Clarke in Hook.f. Fl. Brit. India 4:213. 1883 *p.p;* Khanna in Mudgal *et al.,* Fl. Madhya Pradesh 2:159. 1997.

Twining annuals clothed with bulbous-based, patent, bristly hairs; leaves broadly ovate, 7-12 × 5-10 cm, acuminate at apex, cordate at base, green and thinly hairy above, densely cobweby-tomentose beneath; petioles up to 5 cm long; flowers pink to mauve, in lax or dense, axillary bracteate cymes; peduncles 3-8 cm long; sepals 8-10 × 1.5-2 mm, setose; petals 14-16 mm long; capsules pubescent, globose, 8 mm across; seeds clothed with velvety pubescence mixed with long, cottony hairs.

*Fl.-Fr.:*Aug. -Nov.

Rare, in waste land, **Indore**, Hatod village.

Ipomoea biflora (L.) Pers.,Syn. Pl. 1:183. 1805; *Convolvulus biflorus* L., Sp. Pl. (ed. 2) App. 1668. 1763; *Convolvulus sinensis* Desr., Encycl., 3(2): 557. 1789; *Ipomoea sinensis* (Desr.) Choisy, Mem. Soc. Phys. Geneve 6:459. 1834; *Ipomoea calycina* Benth. ex C.B. Clarke in Fl. Brit. India 4(10):201. 1883; Samvatsar, Fl. West. Trib. Madhya Pradesh 197. 1996; Srivatsava in Mudgal *et al.,* Fl. Madhya Pradesh 2:165. 1997.

Twining herbs covered with long, spreading hairs; leaves cordate with a wide sinus at base, 6-8 × 2.5-4.5 cm; flowers white, solitary or lax in few flowered cymes; sepals unequal, outer in fruit sagittate with obtuse auricles; capsules ovoid, pointed, glabrous; seeds oblong, villous.

Fl.-Fr.: Oct. -Dec.

Common in all districts.

Ipomoea cairica (L.) Sweet, Hort. Brit. 287. 1826; *Convolvulus cairicus* L., Syst. Nat. ed. 10.922 1759; *Ipomoea palmata* Forssk., Fl. Aegypt. -Arab. 43. 1775; C.B. Clarke in Hook.f., Fl. Brit. India 4:214. 1883; Samvatsar, Fl. West. Trib. Madhya Pradesh 195. 1996; Srivastava in Mudgal *et al.,* Fl. Madhya Pradesh 2:158. 1997. *Railway creeper*

Perennial twiners; stem smooth or muricate; leaves palmately divided into 5-7 lobes, 4-8 cm long; flowers rose-purple, in axillary, sub-umbellate cymes; capsules sub-globose, 1-1.5 cm across, 4-seeded, glabrous; seeds densely short tomentose. (seed set is very poor).

Fl.-Fr.: Throughout the year.

Common in hedges, as ornamental or escape (not truely wild).

Ipomoea carnea Jacq., Enum. Syst. Pl. 13. 1760; *I. fistulosa* Mart. ex Choisy in DC. Prodr. 9:349. 1845; *I. carnea* Jacq. subsp. *fistulosa* (Mart. ex Choisy) Austin in Taxon 26:237. 1977; *I. crassicaulis* (Benth.) B.L. Robinson, Proc. Amer. Acad. 51:530.1916; Samvatsar, Fl. West. Trib. Madhya Pradesh 196. 1996; Srivastava in Mudgal *et al.,* Fl. Madhya Pradesh 2:158. 1997. *Besharam*

Erect, ascending or scandent shrubs; leaves ovate or lanceolate, 4-20 × 3-10 cm, apex acuminate, base shallowly cordate or truncate; flowers pink to purple, in many flowered cymes; corolla with hairy, mid-petaline bands; capsules globose, up to 1.5 cm long; seeds covered with long, comose, brown trichomes.

Fl.-Fr.: Throughout the year.

Very common in all districts, in hedges of fields, roadsides and in moist ground.

Ipomoea eriocarpa R. Br. Prodr. 484. 1810; *Convolvulus hispidus* Vahl, Symb. Bot. 3:29. 1794; *Ipomoea hispida* (Vahl) Roem. & Schult. Syst. Veg. 4:238. 1819; *I. sindica* Stapf, Kew Bull. 1894: 346. 1804; C.B. Clarke in Hook.f. Fl. Brit. India 4:204. 1883; Samvatsar, Fl. West.

Trib. Madhya Pradesh 196. 1996; Srivastava in Mudgal *et al.,* Fl. Madhya Pradesh 2:161. 1997.

Slender, trailing or twining herbs, stem hairy; leaves 4-7 × 1.3-2.5 cm, broadly ovate or ovate-lanceolate, base cordate; flowers pinkish-purple in axillary, pedunculate cymes or congested heads; ovary bristly hairy, seated on an annular disc; capsules 4-5 mm in diameter, hairy, beaked, globose; seeds glabrous, blackish-brown.

Fl.-Fr.: Aug. -Dec.

Common as a weed of fields, twining on hedges, monsoon crops.

Ipomoea hederifolia L., Syst. Nat. ed 10.925. 1759; *Ipomoea coccinia auct. non* L., 1753; *I. angulata* Lam., Tabl. Encycl. 1:464. 1793; *Quamoclit coccinia auct. non* Moench 1794; *Ipomoea phoenicea* Roxb., Fl. Ind. 2:92. 1824; *Q. phoenicea* (Roxb.) Choisy, in Mem. Soc. Phys. Hist. Nat. Geneve 6:433. 1834; *Quamoclit angulata* (Lam.) Bojer, Hortus Maurit. 224. 1837; C.B. Clarke in Hook.f. Fl. Brit. India 4:199. 1883; Samvatsar, Fl. West. Trib. Madhya Pradesh 196. 1996; Srivastava in Mudgal *et al.,* Fl. Madhya Pradesh 2:161. 1997. *Sunder bel*

Annual twiners; stem glabrous or sparsely pubescent; leaves 2.5-8 × 1.5-8.5 cm, ovate-cordate, 3-lobed; flowers red, in long, pedunculate, forked, axillary, cymes; capsules 6-8 mm long, ovoid; seeds 4, ellipsoid, black, microscopically reticulate.

Fl.-Fr.: Sept.-Jan.

Common in all districts on hedges, low shrubs, forest roads.

Ipomoea marginata (Desr.) Verdc., Kew Bull. 42(3):658. 1987; *Convolvulus marginatus* Desr., Encycl. 3(2):558. 1792; *Ipomoea sepiaria* J. Koenig ex Roxb. Fl. Ind. 2:90.1824; C.B. Clarke in Hook.f. Fl. Brit. India 4:209.1883; Srivastava in Mudgal *et al.,* Fl. Madhya Pradesh 2:164. 1997.

Perennial twiners; stem slender, glabrous or covered with erect, pilose hairs; leaves 3.5-7.5 × 3-7 cm, acute to acuminate at apex, lobes rounded at base; flowers white, pink or pale purple, in few to many flowered cymes; peduncles 2.5-20 cm long; capsules globose; seeds 3-4 mm long, densely tomentose.

Fl.-Fr.: Aug. -Jan.

Common in all districts, along drains and in wastelands.

Ipomoea nil (L.) Roth, Catal. Bot. 1:36. 1797; *Convolvulus nil* L., Sp. Pl. ed. 2.219. 1762; *Ipomoea hederacea auct. non* Jacq. 1787; C.B. Clarke in Hook.f. Fl. Brit. India 4:199. 1883; Samvatsar, Fl. West. Trib. Madhya Pradesh 196. 1996; Srivastava in Mudgal *et al.,* Fl. Madhya Pradesh 2:162. 1997. *Kala dona,*

Annual twiners; stem slender, clothed with long spreading, brownish hairs; leaves 7-11 × 4-7 cm, ovate or cordate, 3-lobed, pubescent on both surfaces; petioles long; flowers blue, turning bright purple when dry, generally solitary axillary or a few in pedunculate cymes; capsules ovoid, beaked, 8-10 mm across; seeds often 4, rarely 6.

Fl.-Fr.: Sept. -Dec.

Alirajpur, Burhanpur, Dhar, Indore, Khandwa, Khargone.

Ipomoea obscura (L.) Ker-Gawl.,Bot. Reg. 3:t. 239. 1817; *Convolvulus obscurus* L., Sp. Pl. ed. 2.220.1762; C.B. Clarke in Hook.f. Fl. Brit. India 4:207. 1883; Samvatsar, Fl. West. Trib. Madhya Pradesh 196. 1996; Srivastava in Mudgal *et al.,* Fl. Madhya Pradesh 2:162. 1997.

Perennial twiners; leaves 2-7 × 1.5-6 cm, ovate-cordate, apex acuminate, base 7-9 nerved; flowers creamy-yellow on 1-2 flowered, axillary peduncles; capsules ovoid, glabrous, 5-10 mm across; seeds deep-brown, velvety.

Fl.-Fr.: July-Dec.

Common on hedges in all districts.

Ipomoea pes-caprae (L.) R. Br., Narr. Exped. Zaire 477. 1818; *Convolvulus pes-caprae* L. Sp. Pl. 1:159. 1753; *Ipomoea biloba* Forssk., Fl. Aegypt.-Arab. 44. 1775; C.B. Clarke in Hook.f. Fl. Brit. India 4:200. 1883; Bhandari in Fl. Indian Desert 229. 1978.

A perennial, glabrous plant, rooting at nodes or trailing or twining; leaves 3-10 × 3-10 cm, firm, sub-orbicular, emarginate, midrib with 2 glands near the base on the back-side; petioles 10-15 cm long; peduncles stout, 3-12 cm long; corolla pink or purple, 5-6 cm long capsules 1-1.2 cm long; seeds black, densely brownish, tomentose-villous.

Fl.-Fr.: Aug. -March

Khandwa only, banks of Narmada, Mortakka.

Ipomoea pes-tigridis L., Sp. Pl. 162. 1753; *I. hepaticifolia* L., Sp. Pl. 161. 1753; C.B. Clarkei Hook.f. Fl. Brit. India 4:204. 1883; Samvatsar, Fl. West. Trib. Madhya Pradesh 196. 1996; Srivastava in Mudgal *et al.,* Fl. Madhya Pradesh 2:162. 1997. *Paanch patri*

Annual herbs; branches hispid, prostrate, spreading or twining over small bushes; leaves variable, 3-8 cm in diameter, entire or palmately 5-7 lobed, lobes ovate to acuminate, hirsute on both sides; petioles long, slender, pubescent; flowers pinkish-white, in long peduncled heads; sepals villous; capsules membranous, globose, glabrous, hidden in the calyx; seeds grey pubescent.

Fl.-Fr.: Aug. -Nov.

Very common in all districts as forest undergrowth.

Ipomoea quamoclit L., Sp. Pl. 159. 1753; *Quamoclit vulgaris* Choisy, Mem. Soc. Phys. Geneve 6:434. 1833; C.B. Clarke in Hook.f. Fl. Brit. India 4:199. 1883; Samvatsar, Fl. West. Trib. Madhya Pradesh 197. 1996; Srivastava in Mudgal *et al.,* Fl. Madhya Pradesh 2:164. 1997. *Ganesh bel*

Slender twining annual herbs; leaves 3-5 cm long, pinnatipartite, segments linear, oblong, opposite or alternate; flowers red, solitary or on 2-5 flowered cymes; pedicels 1-2 cm long; sepals obtuse with a mucro; corolla 2-3 cm long; capsules ovoid, beaked, 6-8 mm long; seeds black.

Fl.-Fr.: July-Nov.

Alirajpur, Burhanpur, Dhar, Indore, Khandwa, on hedges or an escape from cultivation.

Ipomoea tuberculata Ker Gawl., in Edwards, Bot. Reg. 1. t. 86(Feb.) 1816; *I. dasysperma* Jacq., Eclog. 1:132, t. 89 1811-1816; C.B. Clarke in Hook.f. Fl. Brit. India 4:215. 1883; Samvatsar, Fl. West. Trib. Madhya Pradesh 197. 1996.

Perennial twiners with hirsute stem (glabrous, vide FBI); leaves palmately 5-foliolate; lealets lanceolate, lobes unequal, 3-8 cm long, acute at both ends; petioles 2.5 cm long; peduncles axillary, 4-5 cm long; flowers in dichasial cymes, yellow, with a purple centre, 3.5-4 cm long; capsules 6-11 mm long; seeds trigonous, brown, 5 mm long with appressed pubescence, occasionally with long trichomes on the angles.

Fl.-Fr.: Nov. -Jan.

Alirajpur, at Kattiwada, near forest rest house.

Ipomoea turbinata Lag. Gen. Sp. Pl. 10.1816; *Convolvulus muricatus* L., Mant. Pl. 44. 1767; *Ipomoea muricata* (L.) Jacq., Pl. Hort. Schoebr. 3:40. T. 323. 1798 (1803), *non* Cav.,1799; *Calonyction muricatum* (L.) G. Don, Gen. Hist. 4:264. 1837-38. C.B. Clarke in Hook.f. Fl. Brit. India 4:197. 1883; Gunn in Brittonia 24:163. 1972; Samvatsar, Fl. West. Trib. Madhya Pradesh 197. 1996; Srivastava in Mudgal *et al.,* Fl. Madhya Pradesh 2:165. 1997.

Slender, twining, annual herbs; stem often prostrate, muricate; leaves broadly ovate to orbicular, apex acuminate, base cordate; flowers purplish, in many flowered cymes; peduncles often muricate; capsules ovoid, 1.5-2 cm long; seeds 4, black.

Fl.-Fr.: Sept. -Dec.

Alirajpur, Burhanpur, Dhar, Indore, Khandwa, in wastelands and hedges of fields.

JACQUEMONTIA Choisy W. 78; MP 1; SA 1

Jacquemontia paniculata (Burm.f.) Hall.f., in Engl., Bot. Jahrb. Syst. 16:541. 1893; *Ipomoea paniculata* Burm.f., Fl. Ind. 50. t. 21. f. 3. 1768, *non* Desr., 1792; C.B. Clarke in Hook.f. Fl. Brit. India 4:220.1883; Srivastava in Mudgal *et al.,* Fl. Madhya Pradesh 2:166. 1997.

Twiners; stem pubescent with white tomentum; leaves 4-10 × 2-4 cm, ovate, apex acuminate, base cordate; flowers white, pink or blue, in axillary, few to several flowered cymose clusters; corolla *ca* 8 mm long; capsules globose, 4-5 mm across; seeds brown to black.

Fl.-Fr.: Sept. -Nov.

Burhanpur, Indore on hedges of fields.

MERREMIA Dennst. W. 62; MP 5; SA 5

Key to Species

1. Leaves palmate compound, 5-foliolate; flowers white
   2. Sepals 1.8-2.5 cm long, calyx accrescent; corolla 3-4.5 cm long, with a purple centre; seeds black .... **M. dissecta**
   2. Sepals up to 2 cm long, not accrescent; corolla 2-3 cm long, uniformly white; seeds brown .... **M. aegyptia**
1. Leaves simple, entire, crenate or atmost 3-lobed; flowers solitary or in lax cymes
   3. Prostrate plants; peduncles very short or absent; flowers yellow; leaves reniform . ...**M. emarginata**
   3. Twining plants; peduncles long; flowers yellow, cream-coloured or purplish; inner sepals obtuse

4. Leaves 3-lobed, ovate; flowers yellow .... **M. hederacea**

4. Leaves not lobed, linear or lanceolate; flowers white, cream-coloured or purplish ...... **M. tridentata**

Merremia aegyptia (L.) Urban, Symb. Antill. 4:505. 1910; *Ipomoea aegyptia* L., Sp. Pl. 162. 1753; *Convolvulus pentaphyllus* L., Sp. Pl. ed. 2.223. 1762; *Ipomoea pentaphylla* (L.) Jacq., Collecteana 2:297. (1788) 1789; C.B. Clarke in Hook.f. Fl. Brit. India 4:202. 1883; *Merremia pentaphylla* (L.) Hall.f., in Engl., Bot. Jahrb. Syst. 16:552. 1893; Samvatsar, Fl. West. Trib. Madhya Pradesh 199. 1996; Srivastava in Mudgal *et al.,* Fl. Madhya Pradesh 2:167. 1997.

Annual, slender, hairy twiners; leaves palmately compound, 5-foliolate; leaflets elliptic, acute to acuminate, entire or dentate, about 2 cm long; flowers white, solitary or 2-3 in cymes; sepals hirsute; corolla 2-3 cm long; capsules 1-2 cm across, papery, 4-seeded; seeds brown, glabrous.

Fl.-Fr.: Sept. -Nov.

Common in all districts on hedges, along roadsides, moist ground.

Merremia dissecta (Jacq.) Hall.f., Bot. Jahrb. Syst. 16:552. 1893; *Convolvulus dissectus* Jacq.,Obs. Bot. 2:4. t. 28. 1767; C.B. Clarke in Hook.f. Fl. Brit. India 4:214. 1883; Bhandari in Fl. Indian Desert 234 . 1978.

A large slender, perennial twiner; stem hirsute with yellowish hairs; leaves palmately divided nearly to the base, into 5-7 pinnately lobed segments, coarsely dentate; flowers white with a purple centre, in axillary, 1-4 flowered peduncles; pedicels about 2 cm long; corolla 2-3.5 cm long; capsules 1-2 cm across, partially surrounded by the accrescent calyx; seeds black, glabrous.

Fl.-Fr.: Almost throughout the year

Khandwa only, Tapalchal (Kalibhit) and Awlia forest.

Merremia emarginata (Burm.f.) Hall.f., in Engl., Bot. Jahrb. Syst. 16:552. 1893; *Evolvulus emarginatus* Burm.f., Fl. Ind. 77. t. 30.1768; *Convolvulus reniformis* Roxb., Fl. Ind. 2:67. 1824; *Ipomoea reniformis* (Roxb.) Choisy in Mem. Soc. Phys. Geneve 6:446. 1833; C.B. Clarke in Hook.f. Fl. Brit. India 4:206. 1883; Samvatsar, Fl. West. Trib. Madhya Pradesh 199. 1996; Srivastava in Mudgal *et al.,* Fl. Madhya Pradesh 2:168. 1997. *Musa kani*

Prostate herbs; stem softly hairy, rooting at nodes; leaves 0.6-3 × 1-3.7 cm, reniform; flowers yellow, solitary on short, axillary peduncles; capsules subglobose, 5-6 mm long, glabrous; seeds brown, glabrous.

Fl.-Fr.: Aug. -Feb.

Common in all districts as weed of fields, in moist places, wastelands.

Merremia hederacea (Burm.f.) Hall.f., in Engl., Bot. Jahrb. Syst. 18:118. 1894; *Evolvulus hederaceous* Burm.f., Fl. Ind. 77. t. 30. f. 2. 1768; *Ipomoea chryseides* Ker-Gawl., Bot. Reg. t. 270.1818; C.B. Clarke in Hook.f. Fl. Brit. India 4:168. 1883; Srivastava in Mudgal *et al.,* Fl. Madhya Pradesh 2:168. 1997.

Twiners with a smooth stem; leaves 3- lobed, 1-4 × 1-3.5 cm, apex obtuse, base broadly cordate; flowers yellow, solitary or in cymes; corolla 7-11 mm long; capsules conical to depressed globose; seeds blackish, 2-5 mm long.

Fl.-Fr.: Sept. -Nov.

Indore only, banks of Yeshwant Sagar pond.

Merremia tridentata (L.) Hall.f., in Engl., Bot. Jahrb. Syst. 16:552. 1893; *Convolvulus tridentatus* L., Sp. Pl. 157. 1753; *Ipomoea tridentata* (L.) Roth in Roem., Arch. Bot. 1(2):38. 1798; C.B. Clarke in Hook.f. Fl. Brit. India 4:205. 1883; *Xenostegia tridentata* (L.) Austin & Stapels in Brittonia 32:533. 1980; Samvatsar, Fl. West. Trib. Madhya Pradesh 199. 1996; Srivastava in Mudgal *et al.,* Fl. Madhya Pradesh 2:169. 1997.

Prostate, slender herbs; leaves 1.2-3 × 0.2-0.8 cm, linear, linear-oblong or obovate oblong; flowers yellow, solitary, axillary or a few in cymes; capsules globose to ovoid, glabrous; seeds 4, blackish, smooth.

Fl.-Fr.: July-Dec.

Common all over in wastelands, river beds, plains.

PORANA Burm.f. W. 3; MP 1; SA 1

Porana paniculata Roxb., Pl. Coromandel 3:31. t. 235. 1815; C.B. Clarke in Hook.f. Fl. Brit. India 4:222. 1883; Samvatsar, Fl. West. Trib. Madhya Pradesh 199. 1996; Srivastava in Mudgal *et al.*, Fl. Madhya Pradesh 2:173. 1997 **(Plate 7, Fig. 5).** *Safed bel*

A large twiner with greyish tomentum; leaves 4-10 cm long, ovate, base cordate, pubescent on both surfaces; flowers white, 3-6 mm long, in terminal, drooping panicles; fruits indehiscent; outer 3 sepals enlarged in fruit, like a wing; seeds brownish, ovoid.

Fl.-Fr.: Oct. -Jan.

Dhar, Indore, on hedges of fields, probably an escape from cultivation.

RIVEA Choisy Monotypic

Rivea hypocrateriformis (Desr.) Choisy in Mem. Soc. Phys. Hist. Nat. Geneve 6:408. 1834; *Convolvulus hypocrateriformis* Desr. in Lam. Encycl. 3; 561. 1792; C.B. Clarke in Hook.f. Fl. Brit. India 4:184. 1883; Samvatsar, Fl. West. Trib. Madhya Pradesh 201. 1996; Srivastava in Mudgal *et al.,* Fl. Madhya Pradesh 2:174. 1997. *Phang*

Large twining shrubs; leaves 5-11 × 6-11 cm, broadly ovate to orbicular, shortly acuminate, base cordate with a pair of purplish glands at the base of lamina, lower surface white, pubescent; flowers white, clove-scented, pedunculate, axillary, 1-3 together; bracts deciduous; corolla hypocrateriform, 5-8 cm long; capsules globose or ovoid, mucronate, glabrous; seeds dark-brown.

Fl.-Fr.: Aug. -Nov.

Common on hedges, open ground, roadsides, throughout the area.

STICTOCARDIA Hall.f. W. 7; MP 1; SA 1

Stictocardia tiliifolia (Desr.) Hall. f., Bot. Jahrb. Syst. 18:159. 1894; *Convolvulus tiliifolius* Desr., Encycl. 3(2):544. 1789; *Ipomoea campanulata* L. Sp. Pl. 160.1753; *Argyreia tiliifolia* (Desr.) Wight Icon. Pl. 4(2):12. 1848; *Stictocardia campanulata* Merr., Philipp. J. Sci. 9:133-134. 1914; C.B. Clarke in Hook.f. Fl. Brit. India 4:211. 1883; Samvatsar, Fl. West. Trib. Madhya Pradesh 201. 1996.

A large twiner; stem striate, younger parts pubescent; leaves 6-15 × 5-15 cm, broadly ovate to orbicular, base cordate, apex shortly acuminate, lower surface black-dotted; flowers purple, in corymbosely branched, many flowered cymes; capsules 1-2 cm across, surrounded by the accrescent calyx; seeds obovoid, grayish-brown, 8-9 mm long.

Fl.-Fr.: Oct. -March

Rare, only **Alirajpur**, Phulmal (Chhaktala) area, along streams.

SOLANACEAE 115/2678; 6/16; 6/14

Key to Genera

1. Fruits capsular
 2. Flowers in terminal panicles, 4-5 cm long; fruits enclosed by calyx; plants glandular-pubescent; capsule smooth, 2-celled .... NICOTIANA
 2. Flowers axillary, solitary, more than 5 cm long; fruits not enclosed; plants eglandular; capsule prickly, 4-celled .... DATURA
1. Fruits baccate
 3. Anthers connivent around style; calyx slightly enlarged .... SOLANUM
 3. Anthers not connivent around style
  4. Flowers in clusters, sessile; berry enclosed by the enlarged calyx; stem and leaves stellately tomentose .... WITHANIA
  4. Flowers axillary, solitary, pedicellate; stem and leaves glabrous
   5. Flowers yellow; sepals completely united, basal auricles absent; berry orange- yellow, enclosed in the inflated,10-ribbed calyx .... PHYSALIS
   5. Flowers blue or white; sepals free beyond the middle, each with 2 basal auricles; berry brown, exposed .... NICANDRA

DATURA L. W. 13; MP 4; SA 4

Key to Species

1. Plants tomentose, hairs erect, glandular; capsules deflexed .... **D. innoxia**
1. Plants glabrous or very sparsely pubescent; capsule erect or deflexed
 2. Capsule deflexed; flowers more than 15 cm long; plants with a purple tinge; capsule with a large, flattened calyx base, dehiscence irregular .... **D. metel**
 2. Capsule erect; flowers up to 10 cm long; calyx base small; plants green
  3. Capsule 4-7 cm long, broadly ovate, spines less than 2.5 cm .... **D. stramonium**
  3. Capsule up to 8 cm long, elongated, spines 2.5 cm long .... **D. ferox**

Datura ferox L., Demonstr. Pl. 6. 1753; *D. laevis* Bertol. Fl. Ital. 4:780.1841; Naik in Fl. Marathwada 2:606. 1998; K. Madhusudan Rao in N.P. Singh *et al.,* Fl. Maharashtra 2:495. 2001.

Stout, annual herbs, about 80 cm high; stem and branches appressed pubescent; leaves 7-12 × 7-10 cm, broadly ovate, apex acte or acuminate, base oblique, lamina entire, coarsely lobed, petioles pubescent, 5-7 cm long; flowers solitary, from the forks of branches, peduncle short, up to 1.5 cm long; calyx 5-angled, tube narrowed towards apex, 2.5 -3 cm long, teeth 4-6 mm long; corolla white, 4-6 cm long, tube widened upwards, lobes very short; capsules erect, ovoid to ellipsoid, 3-3.5 cm long, spiny; spines broad-based, fewer, compared to other species, half as long as the capsule; seeds flat., brown.

Fl.-Fr.: Sept. -Dec.

Rare, confined to **Burhanpur**, at Tidgaon and Jainabad.

Datura innoxia Mill., Gard. Dict. ed. 8 5. 1768; *D. meteloides* DC. ex Dunal, Prodr. 13 (1):544. 1852; C.B. Clarke in Hook.f., Fl. Brit. India 4:243. 1883; Samvatsar, Fl. West. Trib. Madhya Pradesh 203. 1996; Srivastava & Dixit in Mudgal *et al.,* Fl. Madhya Pradesh 2:179. 1997. *Dhatura*

Erect, 90-100 cm tall undershrubs; leaves 5-18 × 2-10 cm, pubescent, elliptic-lanceolate or ovate, base oblique, margin sinuate-dentate; flowers white, axillary, solitary, in forks, pedicels 2-3 cm long, recurved in fruit; calyx reflexed in fruit; corolla *ca* 15 cm long, including the 10-12 long tube and 3 cm long limb, teeth linear, 10; capsules globose, 4-5 cm across, minutely grey-tomentose, green, covered with sharply pointed, 7-10 mm long spines, dehiscence irregular; seeds sub-reniform, brown, smooth.

*Fl.-Fr.:*Almost throughout the year

Common on refuse dumps, around villages, all over the area.

Datura metel L., Sp. Pl. 179. 1753; C.B. Clarke, in Hook.f., Fl. Brit. India 4:243. 1883; Samvatsar, Fl. West. Trib. Madhya Pradesh 203. 1996; Srivastava & Dixit in Mudgal *et al.,* Fl. Madhya Pradesh 2:179. 1997. *Kala dhatura*

Erect, branched, undershrubs; stem often muriculate, tinged with purple, as are the petioles, midrib and and other veins; leaves 10-18 × 7-10 cm, elliptic-ovate, base oblique, irregularly sinuate, dentate; flowers white to purplish, especially at the top edge of petals, solitary, axillary, in the forks, pedicels bent in fruit; calyx tubular, *ca* 5 cm long; corolla tube 10-15 cm long, limb 4 cm long, teeth 5, lanceolate, about 1 cm long; capsules 4 × 3.5 cm, drooping, spines short, stout; seeds sub-reniform, pale brown.

Fl-Fr.: July-March

Occasional, in the same habitat, in all districts, easily identified by the purple tinge of the plant.

Datura stramonium L., Sp. Pl. 179. 1753; C.B. Clarke, in Hook.f., Fl. Brit. India 4:242. 1883; Samvatsar, Fl. West. Trib. Madhya Pradesh 204. 1996; Srivastava & Dixit in Mudgal *et al.,* Fl. Madhya Pradesh 2:180.1997.

Stout, annual herbs; leaves 8-15 × 5-10 cm, sinuate, lanceolate, apex acute, narrowed at base; petioles up to 5 cm long; flowers white, solitary, axillary, borne in the forks, pedicels 0.5-1 cm long; calyx 3-3.5 cm long, 5-ribbed, teeth 5-6 mm long; corolla up to 8 cm long, lobes acuminate; capsules globose, 4-5 cm across, minutely tomentose, covered with numerous, sharp spines, dehiscing by 4 valves; seeds ovoid-reniform, black.

*Fl.-Fr.:*Sept. -Dec.

Not common, Alirajpur, Burhanpur, Dhar, Indore, in the same habitat as all the above species.

NICANDRA Adans. W. 2; MP 1; SA 1

Nicandra physalodes (L.) Gaertn., Fruct. Sem. Pl. 2:237. 1791; *Atropa physalodes* L., Sp. Pl. 181. 1753; C.B. Clarke in Hook.f., Fl. Brit. India 4:240.1883; Srivastava & Dixit in Mudgal *et al.,* Fl. Madhya Pradesh 2:183. 1997.

Erect, much branched, annual herbs, about 1 m tall; stem hollow, glabrous; leaves 5-15 × 2-7 cm, oblong, ovate-lanceolate, apex acute, base oblique and decurrent, petioles winged in the upper part, 5-10 cm long; flowers pale blue, 1.5-2.5 cm across, solitary, axillary, erect initially, later becoming decurved, pedicels short; calyx 1-1.5 cm long, enlarging up to 3 cm in fruiting, sepals ovate-lanceolate, becoming cordate-sagittate later; corolla 2-4 cm long, pubescent outside, lobes ovate-orbicular; stamens 5, included; berries globose, 10-15 mm across, brown, enclosed in the accrescent calyx; seeds numerous, flat, brown.

*Fl.-Fr.:*Jan. -March

As forest undergrowth, Burhanpur, Indore.

NICOTIANA L. W. 55; MP 2; SA 1

Nicotiana plumbaginifolia Viv., Elench. Pl. Dinegro 26. 1802; C.B. Clarke, in Hook.f., Fl. Brit. India 4:246. 1883; Samvatsar, Fl. West. Trib. Madhya Pradesh 204. 1996; Srivastava & Dixit in Mudgal *et al.,* Fl. Madhya Pradesh 2:185. 1997. *Jangli tamakhu*

Erect, viscous-pubescent, 15-40 cm tall herbs; basal leaves in rosette, obovate, 7-12 cm long, narrowed at base into a winged petiole, margin entire, obtuse or acute, glandular white-dotted beneath, cauline leaves sessile, obovate to elliptic, 5-10 × 3-6 cm, semi-amplexicaul at base, upper leaves passing into bracts, petioles 2-5 cm long, gradually becoming shorter above in the flowering zone; flowers pinkish-white, in lax, terminal panicles, pedicels 5-10 mm long; calyx about 1 cm long, 10-ribbed, glandular hairy, slightly enlarging in fruit, teeth unequal; corolla 2.5-3 cm long, glandular-hairy outside; capsules up to 1 cm long, ovoid, conical, pale brown, enclosed in accrescent calyx; seeds numerous, rugose, pale brown.

Fl.-Fr.: Nov. -March

Common weed of waste places, shady and wet soil, wet fields etc., in all districts.

PHYSALIS L. W. 124; MP 2; SA 1

Physalis angulata L., Sp. Pl. 183. 1753; C.B. Clarke, in Hook.f., Fl. Brit. India 4:238. 1883; Samvatsar, Fl. West. Trib. Madhya Pradesh 204. 1996; Srivastava & Dixit in Mudgal *et al.,* Fl. Madhya Pradesh 2:189. 1997. *Ban phutka*

Erect, much branched, 20-40 cm tall, annual herbs; stem angular, often rooting from lower nodes; leaves 5-10 × 3-6 cm, ovate or elliptic, apex acuminate, base cuneate, unequal sided, petioles 2-4 cm long; flowers solitary, axillary or terminal, 1-1.5 cm long, pedicels slender; calyx 4-5 mm long, divided half way down, teeth triangular-ovate, ciliate; corolla pale yellow, with 5, faint brown spots inside, ciliate; anthers yellow, blue or violet; berries 6-7 × 5-6 mm, ovoid-globose, orange-yellow, enclosed in the accrescent calyx; seeds many, compressed.

*Fl.-Fr.:*Oct. -Feb.

Common weed in all districts.

SOLANUM L. W. 1199; MP 8; SA 6

Key to Species

1. Unarmed plants; stem angular; flowers white
   2. Berries red, 6-10 mm across .... **S. villosulum**
   2. Berries black, 4-7 (8) mm across . ... **S. americanum**
1. Armed plants; berries 0.8-2.5 cm across
   3. Annuals; spines straight, narrow, up to 1 cm long; flowers bisexual; berry 1.5-2 cm across . ... **S. virginianum**
   3. Perennials; spines hooked with a broad base; flowers unisexual or bisexual
      4. Flowers unisexual; berries 1.8-2.5 cm across .... **S. incanum**
      4. Flowers bisexual; berries 0.8-1.3 cm across; plants stellately hairy .... **S. anguivi**

Solanum americanum Miller,Gard. Dict. ed. 8, no 5 1768; *S. nigrum* L. Sp. Pl. 1:186. 1753; Edmonds, Kew Bull. 27:103. 1972; Hepper, Rev. Handbk. Fl. Ceylon 6:368. 1987; Edmonds, J.M. & Chweya, J.A. Black Nightshades. *Solanum nigrum* L. and Related Species IPGRI Italy. 1997.

Plants moderately pilose, eglandular; stem angular with dentate ridges; leaves 3-6 (11) × 1.4-6. 6 cm, ovate-lanceolate, margin entire to sinuate; inflorescence an umbellate, 6-10 flowered cyme; corolla white; berries black, 4-7 mm across.

Fl.-Fr.: Nov. -Jan.

Burhanpur only, Jainabad area

Solanum anguivi Lam.,Tabl. Encycl. 2:28. 1794; *S. indicum* auct. non. oblique L. 1753; C.B. Clarke in Hook.f., Fl. Brit. India 4:234. 1883; Sant. in J. Bombay Hist. nat. Soc. 47:653. 1948; Deb, J. Econ. Taxon. Bot 1:47. 1980; K. Madhusudan Rao in Fl. Maharashtra 2:503. 2001. (*S. violaceum* Ortega, Nov. Pl. Descr. Dec. 56. 1798, is a separate species)

Much branched, prickly undershrubs, about 1 m tall, prickles stout with a broad base; branches covered with minute, stellate hairs; leaves 3-8 × 2-5 cm, ovate, cordate, cuneate or truncate at base, lobed sinuately into ovate, acute lobes, midrib prickly, pubescent above, densely so beneath, hairs simple and stellate mixed, petioles 1.5-2.5 cm long, prickly; flowers in extra axillary, racemose cymes, peduncles short; calyx up to 3 mm long, teeth triangular, acute; corolla purple, 6-8 mm long, outside covered with purple, stellate hairs; berries yellow, globose, glabrous; seeds flat, 3-4 mm across, pitted, brown.

*Fl.-Fr.:*July-Dec.

Common in waste places, all over the area.

Solanum incanum L., Sp. Pl. 188. 1753; *S. sanctum* L., Sp. Pl (ed. 2) 269. 1762; C.B. Clarke in Hook.f., Fl. Brit. India 4:236. 1883; Deb in J. Econ. Taxon. Bot. 1:49. 1980; Samvatsar, Fl. West. Trib. Madhya Pradesh. 106; Srivastava & Dixit in Mudgal *et al.,* Fl. Madhya Pradesh 2:191. 1997. *Jangli bhata*

Erect, stellately fulvous hairy, armed shrubs, up to 1 m high; leaves 6-10 × 3-6 cm, ovate-elliptic, stellately pubescent, cuneate at base, shallowly, sinuately lobed, midrib sparsely prickly, petioles 2-4 cm long; flowers polygamous, in lateral cymes, one flower perfect, others male, peduncles short, pedicels 5-8 mm long; calyx and corolla similar to *S. anguivi* Lam; berries 2-2.5 across, globose, yellow; seeds flat, 3-4 mm across.

*Fl.-Fr.:*Oct. -Jan.

Common waste land weed in the area.

Solanum villosum Mill. Gard. Dict. (ed. 8) no. 2 1768; *S. miniatum* Bernh. ex Willd., Enum. Pl. 1:236. 1809; *S. nigrum* var. *humile* (Bernh. ex Willd.) C.Y. Wu & S.C. Huang Acta. Phytotax. Sin. 16 (2):72. 1978; Naik, Fl. Marathwada 2:617. 1998. *Makoy*

Annual, bushy herbs, 40-60 cm high, branches thinly pubescent, with eglandular hairs; leaves 3-6 × 1.3-3 cm, ovate, base narrow, irregularly sinuate-dentate, apex acute, petioles 1.5 2 cm; flowers in extra-axillary, racemiform, 4-6 flowered cymes, peduncles short, pedicels 1-2 cm long; calyx 2-3 mm long, teeth aute, strongly reflexed in fruit; corolla white, about 6 mm across, acute; berries globose, 6-7 mm across, orange-red, shining; seeds 2 mm long, pitted.

*Fl.-Fr.:*June-Feb.

Common weed in waste land everywhere.

Solanum virginianum L., Sp. Pl. 187. 1753; *S. xanthocarpum* Schrad. & H. Wendl. Sert. Hannov. 1:8, pl. 2 1785; C.B. Clarke, in Hook.f., Fl. Brit. India 4:236. 1883; Samvatsar, Fl. West. Trib. Madhya Pradesh 206. 1996; Srivastava & Dixit in Mudgal *et al.,* Fl. Madhya Pradesh 2:195. 1997. *Bhui ringni*

Diffuse to ascending annual or perennial, prickly herbs; stem and leaves clothed with stellate hairs and prickles; leaves 7-10 × 3-6 cm, ovate or elliptic, pinnately lobed, lobes acute; flowers blue or deep purple, in extra axillary, few-flowered cymes; berries globose, yellow or white with green stripes; seeds subreniform, 2-2.5 mm across.

Fl.-Fr.: Feb. -June

Very common in, open, waste ground.

WITHANIA Pauq. W. 8; MP 1; SA 1

Withania somnifera (L.) Dunal, Prodr. 13 (1): 453. 1852; *Physalis somnifera* L., Sp. Pl. 182. 1753; C.B. Clarke in Hook.f., Fl. Brit. India 4: 239. 1883; Samvatsar, Fl. West. Trib. Madhya Pradesh 205. 1996; Srivastava & Dixit in Mudgal *et al.*, Fl. Madhya Pradesh 2:196. 1997. *Asgandh*

Erect, 1-15 m tall, perennials, covered with stellate tomentum; leaves 7-9 × 3-5 cm, oblong, ovate, apex subacute, oblique at base, pubescent on both sides, base cuneate; flowers greenish, in axillary clusters of umbelliform cymes; berries 4-5 mm across, globose, enclosed in the accrescent calyx, red when ripe; seeds brown, glabrous, wrinkled.

*Fl.-Fr.:*Aug. -Dec.

Common weed of waste places, everywhere.

LINDERNIACEAE 16/64; 3/21; 2/9

Key to Genera

1. Calyx 3-5 winged . ... TORENIA
1. Calyx not winged .... LINDERNIA

LINDERNIA All. W. 85; MP 16; SA 8

Key to Species

1. Leaves palmately 3-5 nerved
   2. Leaves 3-nerved; capsule ellipsoid, longer than calyx; fertile stamens 2 .... **L. parviflora**
   2. Leaves 5-nerved; capsule ovoid, as long as calyx; fertile stamens 4 .... **L. procumbens**
1. Leaves pinnately nerved
   3. Capsule shorter or equal to calyx
      4. Calyx divided halfway down or less . ... **L. crustacea**
      4. Calyx divided almost to the base; flowers axillary, solitary
         5. Hispid herbs .... **L. pusilla**
         5. Glabrous herbs .... **L. multiflora**
   3. Capsule much longer than calyx
      6. Leaves sharply, closely spinulose-serrate; fertile stamens 2 . ... **L. ciliata**
      6. Leaves sub-entire or shallowly serrate
         7. Stamens 2 . ... **L. antipoda**
         7. Stamens 4 .... **L. anagallis**

Lindernia anagallis (Burm.f.) Pennell, J. Arnold Arbor. 24: 252. 1943; *Ruellia anagallis* Burm.f., Fl. Indica 135. 1768; *Gratiola cordifolia* Colsm., Orodr. Descr. Gratiol. 15. 1793; *Vandellia pedunculata* Benth., Scroph. Ind. 37. 1835; *V. cordifolia* (Colsm.) G. Don, Gen. Hist. 4:549. 1838; Hook.f., Fl. Brit. India 4:282. 1884; *Lindernia cordifolia* (Colsm.) Merr., Enum. Phillip. Fl. Pl. 3:437. 1923; *Vandellia anagallis* (Burm.f.) T. Yamaz., J. Jap. Bot.

30(6): 176. 1955; Samvatsar, Fl. West. Trib. Madhya Pradesh 211. 1996; Singh in Mudgal *et al.,* Fl. Madhya Pradesh 2:223. 1997.

Annual or perennial, decumbent herbs, rooting at lower nodes; stem 4-angled; leaves 1-3 × 1-2 cm, somewhat variable in shape, oblong,elliptic, ovate or obovate-lanceolate, apex acute, base truncate-cordate, margin obscurely serrate, sessile or shortly petiolate, petioles 3-5 mm long; flowers solitary axillary or in terminal racemes; pedicels 2-3 cm long; calyx divided almost to the base, *ca 4* mm long, segments acuminate; corolla pink, blue or violet, up to 1 cm long; capsules 1-1.5 cm long, linear, cylindrical, beaked with a persistent style; seeds ellipsoid.

*Fl.-Fr.:*Aug. -Dec.

Rare, in moist and shady places, **Alirajpur**, Aamkhut area.

Lindernia antipoda (L.) Alston, Handb. Fl. Ceylon 6 (suppl.): 214. 1931; *Ruellia antipoda* L., Sp. Pl. 635. 1753; *Gratiola veronicifolia* Retz., Observ. Bot. 4:8. 1786; *Bonnaya veronicifolia* (Retz.) Spreng., Syst. Veg. 1:41. 1824; *Lindernia veronicifolia* (Retz.) F. Muell., Fragm. Pl. Filip. 6:101. 1867; Hook.f., Fl. Brit. India 4:285. 1884; Samvatsar, Fl. West. Trib. Madhya Pradesh 211. 1996; Singh in Mudgal *et al.,* Fl. Madhya Pradesh 2:224. 1997.

Annual or perennial, prostrate, glabrous herbs; branches 4-angled, ascending, up to15 cm long; leaves 1-5 × 0.3-0.6 cm, subsessile, obovate or lanceolate-oblong, acute or obtuse at apex, margin subentire or serrulate; flowers bluish-purple, solitary, axillary or in axillary or terminal, leafy racemes; pedicels 5-10 mm long; calyx *ca* 5 mm long, deeply divided, segments serrate in the upper part; corolla blue-violet, *ca 8* mm long; staminodes 2; capsules 10-12 mm long, linear, striate; seeds many, minute, light brown, rugulose.

*Fl.-Fr.:*Aug. -Dec.

Rare, in moist place, Alirajpur, Burhanpur, Khandwa.

Lindernia ciliata (Colsm.) Pennell, Brittonia 2:182. 1936; *Gratiola ciliata* Colsm., Prodr. Descr. Gratiol. 14. 1793; *Bonnaya brachiata* Link & Otto, Icon. Pl. Select. 2:25. 1821; Hook.f., Fl. Brit. India 4:284. 1884; Samvatsar, Fl. West. Trib. Madhya Pradesh 212. 1996; Singh in Mudgal *et al.,* Fl. Madhya Pradesh 2:225. 1997.

Much branched, erect, annual, up to 15 cm high, herbs; stem 4-angled, glabrous, lower branches decumbent; leaves sessile, 1-3 × 0.3-0.8 cm, broadly oblong or obovate, obtuse or subacute at apex, penninerved, margin densely serrate with bristle-tipped teeth; flowers in terminal, lax, up to 12 cm long racemes; pedicels slender, 3-4 mm long; bracts 4-5 mm long, linear, subulate; calyx *ca* 5 mm long, divided almost to the base, segments linear, subulate, ciliate, aristate; corolla white, pink or red, *ca* 8 mm long; capsules 10-12 mm long, linear or oblong, acute; seeds black, subquadrate, rugose, truncate at both ends.

*Fl.-Fr.:*Aug. -Nov.

Common in moist places, in all districts.

Lindernia crustacea (L.) F. Muell., Syst. Census. Austral. Pl. 97. 1882; *Capraria crustacea* L., Mant. Pl. 1:87. 1767; *Vandellia crustacea* Benth., Scroph. Ind. 35. 1835; Hook.f., Fl. Brit. India 4:279. 1884; Samvatsar, Fl. West. Trib. Madhya Pradesh 212. 1996; Singh in Mudgal *et al.,* Fl. Madhya Pradesh 2:225. 1997.

Diffuse, annual, herbs, sometimes rooting at nodes; stem 4-angled, pubescent on the angles; leaves 1-2 × 0.5-1.5 cm, ovoid, deltoid, obtuse at apex, rounded to cordate at base, margin entire or serrate, penninerved, petioles 4-8 mm long; flowers axillary, solitary and sub-racemose; pedicels 5-10 mm long; calyx segments 3.5-4 mm long, lobed only up to ¼ th its length, teeth deltoid, acute; corolla purple, pink, yellow-red in the throat, 8-12 mm long; capsules oblong-ovoid, 2-3 mm long, truncate; seeds foveolate, subquadrate, pale yellow, granulate, with a minute tail at one end.

*Fl.-Fr.:*July-Dec.

Very common throughout the area, in moist places.

Lindernia multiflora (Roxb.) Mukerjee, J. Indian Bot. Soc. 24:131. 1945; *Torenia multiflora* Roxb., Fl. Ind. 3:96. 1832; *Vandellia multiflora* G. Don, Gen. Hist. 4:549. 1837; Hook.f., Fl. Brit. India 4:280.1884; Samvatsar, Fl. West. Trib. Madhya Pradesh 212. 1996; Singh in Mudgal *et al.,* Fl. Madhya Pradesh 2:227. 1997.

Erect, simple or branched, 5-15 cm high, herbs, branching from the base; stem acutely 4-angled, ciliately ribbed; leaves 1-2.5 × 0.2-1.2 cm, sessile or subsessile, obovate-oblong, apex obtuse, margin entire; flowers white or pale-purple, solitary, axillary or in terminal, glandular-pubescent racemes; pedicels 5-10 mm long; bracts minute, subulate; calyx *ca* 1.5 mm long, divided almost to the base; corolla white, bluish white or purple, 3-4 mm long; capsules orbicular, 3-4 mm long, apex apiculate; seeds oblong, pale brown, rugulose.

*Fl.-Fr.:*Aug. -Dec.

Rare, in wet and shady places, Alirajpur, Khandwa.

Lindernia parviflora (Roxb.) Haines, Bot. Bihar Orissa 4:635. 1922; *Gratiola parviflora* Roxb., Pl. Coromandel 3: 3. 1811; *Ilysanthes parviflora* (Roxb.) Benth., Prodr. 10: 419. 1846; Hook.f., Fl. Brit. India 4: 283. 1884; Samvatsar, Fl. West. Trib. Madhya Pradesh 212. 1996; Singh in Mudgal *et al.,* Fl. Madhya Pradesh 2: 229. 1997.

Annual, erect or decumbent, simple to diffusely branched herbs, up to 15 cm high; stems quadrangular, glabrous, rooting at lower nodes; leaves 0.5-1.5 × 0.2-0.8 cm, sessile, ovate-lanceolate, apex subacute, margin entire, 3-nerved; flowers white, axillary, solitary or in terminal racemes; pedicels 5-10 mm long; calyx segments linear-lanceolate, *ca* 3 mm long, 1-nerved; corolla 5-6 mm long; staminodes 2, lobed or dentate; capsules 3-5 mm long, ellipsoid, striate; seeds minute, granular, pale brown.

*Fl.-Fr.:*Oct. -March

Common weed of moist and shady places, all over the area.

Lindernia procumbens (Krock.) Philcox, Taxon 14:30.1965; *Anagalloides procumbens* Krock., Fl. Siles. 2(1):398. 1790; *Lindernia pyxidaria* L., Mant. Pl. 2:252. 1771; *Vandellia erecta* Benth., Scroph. Ind. 36. 1835; Hook.f., Fl. Brit. India 4:281. 1884; Samvatsar, Fl. West. Trib. Madhya Pradesh 212. 1996; Singh in Mudgal *et al.,* Fl. Madhya Pradesh 2:229. 1997.

Annual, erect, glabrous, 10-20 cm high herbs; stem angular, branched from the base; leaves sessile, 10-25 mm long, elliptic or oblong, apex obtuse, base cuneate, margin entire or crenulate, 3-5 nerved; flowers axillary, solitary, pedicels long; calyx deeply divided, sepals linear or lanceolate, about 2 mm long, 3-nerved; corolla white to pale purple, 5-8 mm long; capsules ellipsoid, 4 mm long; seeds rugose.

*Fl.-Fr.:*Aug. -Dec.

In damp and shady places, Alirajpur, Burhanpur, Khandwa.

Lindernia pusilla (Willd.) Bold., Zakfl. Java 165. 1916; *Gratiola pusilla* Willd., Sp. Pl. 1:105. 1797; *Vandellia scabra* Benth., Scroph. Ind. 36. 1835; Hook.f., Fl. Brit. India 4:281. 1884; *V. pusilla* Merr., Philipp. J. Sci., C 7:246. 1912; Samvatsar, Fl. West. Trib. Madhya Pradesh 212. 1996; A. Godbole & V. Prasad in N.P. Singh *et al.,* Fl. Maharashtra 2:541.2001.

Prostrate, hirsute, annual herbs; stems and branches densely to sparsely pubescent with long hairs; leaves 0.8-2 × 0.5-1.5 cm, sessile, ovate to suborbicular, hairy on both sides, margins crenate; flowers white, axillary, solitary, sometimes in pairs; calyx *ca* 3 mm long, hipid; corolla violet with yellow centre, 7 mm across, tube 5 mm long, lobes 5, upper lip entire, shorter than lower lip; capsules 3 mm long, equalling the fruiting calyx.

*Fl.-Fr.:*Aug. -Oct.

Rare, in damp, shady places in forest, **Alirajpur**, Kattiwada, Kadwal.

TORENIA L. W. 14; MP 3; SA 1

Torenia indica Saldanha in Bull. Bot. Surv. India 8(2):126. 1966; Hook.f. Fl. Brit. India 4:276. 1884; Singh in Mudgal *et al.,* Fl. Madhya Pradesh 2:242. 1997.

Annual erect herbs, simple or trichotomously branched; stem 4-angled, rooting at nodes; leaves 2-4 cm long, ovate-lanceolate; petioles up to 1.5 cm long; flowers purple-violet, axillary, whorled or umbellate, at the tips of branches; pedicels deflexed; calyx 2-lipped, upper 3-toothed, lower 2-toothed, winged; wings ciliate, base cordate; corolla 1.5 cm long, tube equalling calyx; capsules oblong-ellipsoid.

Fl.-Fr.: Sept. -Dec.

In moist soil, where water is leaking from a source (pipe or tap), **Indore** only, Kishanpura pond (Chhota Betma).

PLANTAGINACEAE 120/1614; 12/23; 11/14

Key to Genera

1. Stamens 2
    2. Erect or decumbent glandular plants; flowers white, in axillary racemes . ... VERONICA
    2. Prostrate, creeping, eglandular, minute plants; flowers greenish-yellow, solitary, axillary . ... MICROCARPAEA
1. Stamens 4
    3. Fertile stamens 2 or all with 1 imperfect anther cell; fruiting calyx unequal .... ADENOSMA
    3. All or atleast 2 stamens fertile, with both anther cells perfect; fruiting calyx not as above
        4. Staminodes present; dehiscence of capsule loculicidal; plants of marshy habitat ...... DOPATRIUM
        4. Staminodes absent

5. All leaves alternate; flowers yellow; corolla 2-lipped; corolla spurred ....NANORRHINUM
5. All leaves opposite
6. Flowers tetramerous, actinomorphic .... SCOPARIA
6. Flowers pentamerous, zygomorphic
7. Plants heterophyllous, habitat aquatic or marshy; dehiscence of capsule both loculicidal and septicidal ....LIMNOPHILA
7. Plants not heterophyllous, habitat terrestrial; dehiscence of capsule either loculicidal or septicidal
8. Plants eglandular; bracteoles present
9. All sepals similar, not accrescent; anther cells not stipitate . ...BACOPA
9. Three sepals larger than others, accrescent; anther cells stipitate ....MECARDONIA
8. Plants glandular pubescent; anther cells stipitate, 2-celled
10. Flowers yellow; capsule 2-valved, loculicidal; bracteoles absent .. LINDENBERGIA
10. Flowers deep violet; casule 4-valved; dehiscence septicidal; bracteoles present.. STEMODIA

ADENOSMA R. Br. W. 26; MP 2; SA 1

Adenosma ovatum (Benth.) Benth., in Hook.f., Fl. Brit. India 4:263. 1884; Samvatsar, Fl. West. Trib. Madhya Pradesh 206; Singh, in Mudgal *et al.,* Fl. Madhya Pradesh 2:201. 1997.

Erect, branched, glabrous herbs; leaves 3-5 cm long, apex obtuse, base narrowed, margin crenate-serrate, petioles *ca* 5 mm long; flowers axillary, sessile, bluish; calyx bilabiate, 3 outer sepals large, cordate, 2 inner sepals smaller, lanceolate; capsules lanceolate; seeds numerous.

*Fl.-Fr.:*Sept. -Nov.

Rare, in moist habitat, Alirajpur, Dhar.

BACOPA Aubl. W. 68; MP 3; SA 1

Bacopa monnieri (L.) Wettst., Nat. Pflanzenfam. 4(3b.):77. 1891; *Lysimachia monnieri* L., Cent. Pl. II 9. 1756; *Herpestis monnieri* (L.) Kunth, Nov. Gen. Sp. (quarto ed.) 2:366. 1818; Hook. f, Fl. Brit. India 4:207. 1884; Samvatsar, Fl. West. Trib. Madhya Pradesh 208. 1996; Singh in Mudgal *et al.,* Fl. Madhya Pradesh 2:207. 1997. *Nir bramhi*

Creeping, sub-succulent, marshy glandular punctate herbs, rooting at nodes; branches spreading radially, up to 30 cm long; leaves fleshy, sessile, 1.5-2 × 0.5-0.7 cm, opposite, decussate, obovate-oblong, or spathulate, apex obtuse or emarginate, base cuneate, margin entire; flowers axillary, solitary, pedicels up to 1 cm long, in flowers, elongating to 2 cm, when fruiting; bracteoles 2, linear, *ca* 5 mm long; sepals unequal, 3 outer, larger ones broadly ovate-oblong, the 2 inner sepals narrow, lanceolate; petals purplish or pink, campanulate, *ca* 1 cm long; capsules ovoid, acute, glabrous, 4-5 × 3-4 mm; seeds striate, minute, brown.

*Fl.-Fr.:*June-Oct.

Common in marshy habitat, where sewage water openly flows and emits a foetid smell.

DOPATRIUM Buch.-Ham. W. 14; MP 2; SA 3

Key to Species

1. Capsule linear-oblong; fruiting pedicels reflexed; bracts and calyx lobes acuminate .... **D. nudicaule**
1. Capsule globose; fruiting pedicels erect
    2. Leaves 1.3-2 cm long, obovate-oblong; flowers dark pink; stigma lamellate; seeds tuberculate .... **D. junceum**
    2. Leaves 3.2-5.5 cm long, lanceolate; flowers blue-violet; stigma triangular- rhomboid; seeds longitudinally ribbed, transversely rugose between ribs .... **D. lobelioides**

Dopatrium junceum (Roxb.) Buch.-Ham. ex Benth., Edwards' Bot. Reg. 21 : t. 1770 [4] 1835; *Gratiola juncea* Roxb., Pl. Coromandel 2:16. 1799; Hook.f., Fl. Brit. India 4:274. 1884; Samvatsar, Fl. West. Trib. Madhya Pradesh 209. 1996; Singh in Mudgal *et al.,* Fl. Madhya Pradesh 2:212. 1997.

Fleshy, glabrous, annual herbs, up to 20 cm high; leaves 1.3-2 × 0.2-0.5 cm, opposite, sessile, obovate or oblong, apex acute or obtuse, margin entire; flowers dark pink, in slender racemes of distant, opposite pairs, lower flowers sessile, upper on long, filiform pedicels; calyx 5-partite, 1-2 mm long, segments oblong; corolla *ca* 4 mm long, glandular-pubescent; capsules globose, 2.5-3.5 mm long, fruiting pedicels erect; seeds ellipsoid, tuberculate, blackish-brown.

*Fl.-Fr.:*Aug. -Nov.

In moist places, Alirajpur, Burhanpur, Dhar, Indore, Khandwa.

Dopatrium lobelioides (Retz.) Benth., Scroph. Ind. 31. 1835; *Gratiola lobelioides* Retz., Observ. Bot. 4:7. 1786; *Gratiola hyssopioides* Roxb., Pl. Coromandel 2:16. 1799; Hook.f., Fl. Brit. India 4: 274. 1884; A. Godbole & V. Prasad in N.P. Singh *et al.,* Fl. Maharashtra 2: 527. 2001.

Small, erect herbs, up to 15 cm high, stems slender, fleshy; leaves opposite, 2.5-3 × 0.7 cm, apex acute; flowers solitary, axillary, white or faintly violet; bracteoles absent; calyx 5-lobed; corolla bilabiate, upper lip 2-lobed, lower lip 3-lobed; capsules globose, *ca* 4 mm long, apex obtuse, loculicidal, fruiting pedicels horizontally spreading; seeds many.

*Fl.-Fr.:*Nov. -Jan.

Rare, on moist soil, along the margins of water bodies, **Khandwa**.

Dopatrium nudicaule (Willd.) Benth., Scroph. Ind. 31. 1835 & in Prodr. 10.407. 1846; *Gratiola nudicaulis* Willd., Neue Schriften Ges. Naturf. Freunde Berlin 4:193. 1803; Hook.f., Fl. Birt. India 4:213. 1884; Samvatsar, Fl. West. Trib. Madhya Pradesh 209. 1996; Singh in Mudgal *et al.,* Fl. Madhya Pradesh 2:213. 1997.

Very slender, erect herbs, up to 15 cm high; leaves broadly oblong or ovate-oblong, 0.5-1.5 × 0.2-0.6 cm, apex obtuse, base semi-amplexicaul; flowers solitary, axillary or paired; sepals linear-lanceolate; petals dull blue, throat minutely hirsute; capsules linear-oblong, *ca* 3 mm long, bluntly pointed at apex, with a transparent wall; seeds cuneiform, rugose, black.

*Fl.-Fr.:*Sept. -Dec.

Rare, in marshy places at **Alirajpur**, Aamkhut forest.

LIMNOPHILA R. Br.W. 18; MP 7; SA 1

Limnophila indica (L.) Druce, Rep. Bot. Exch. Club Soc. Brit. Isles 3:420.1913, publ. 1914; *Hottonia indica* L., Sp. Pl. ed. 2 208. 1762; *Limnophila gratioloides* R. Br., Prodr. Fl. Nov. Holl. 442. 1810; Hook. f, Fl. Brit. India 4:271. 1884; Samvatsar, Fl. West. Trib. Madhya Pradesh 209. 1996; Singh in Mudgal *et al.,* Fl. Madhya Pradesh 2:219. 1997. *Kuttra*

Perennial, marshy herbs, inhabiting ponds and muddy places, stems up to 15 cm high, striate and glandular in the upper part, rooting at lower nodes; aerial leaves whorled, 1.5-3.5 × 0.5-1.5 cm, variously dissected, uppermost undissected with crenate serrate margin, submerged leaves in whorls of 6-12, pinnatisect, lobes flattened or capillary; flowers axillary, solitary or in terminal racemes, pedicels glandular; bracteoles 2, linear, 3-4 mm long; sepals up to 6 mm long, lobes linear, 2-3 mm long; petals white, purple or bluish, up to 1 cm long; capsules ellipsoid, or globose, 3.5-5 mm long; seeds black, truncate or rounded at both ends.

*Fl.-Fr.:*Oct. -Feb.

Common in marshy and aquatic habitat, all over the area.

LINDENBERGIA Lehm. W. 4; MP 1; SA 1

Lindenbergia indica Vatke, Oesterr. Bot. Z. 25:10.1875; *L. urticifolia* Lehm., Index Seminum (HBG) 1829: 8. 1829; *L. abyssinica* Hochst. ex Benth., Prodr. 10:377. 1846; Hook.f., Fl. Brit. India 4:221. 1884; Samvatsar, Fl. West. Trib. Madhya Pradesh 209. 1996; Singh in Mudgal *et al.,* Fl. Madhya Pradesh 2:221. 1997. *Bhit chatti*

Annual, erect, glandular hairy herbs, about 20 cm high; leaves 1.5-7 × 1.5-5 cm, ovate-elliptic, apex acute to subobtuse, base rounded or acute, margin crenate-serrate, patently glandular hairy on both sides, lower leaves petiolate, upper sessile; flowers yellow, in axllary or terminal, leafy racemes; sepals 3-6 mm long, glandular hairy, ciliate; petals slightly glandular hairy, upper lip hooded, reflexed at the tip, palate gibbous; capsules oblong, brown, compressed, beaked, 4 × 2 mm; seeds many, dark brown, ellipsoid.

*Fl.-Fr.:*July-March

Common on old walls, waste places, throughout the area.

MECARDONIA Ruiz & Pavon W. 10; MP 1; SA 1

Mecardonia procumbens (Mill.) Small, Fl. S. E. U.S. 1065. 1903; *Erinus procumbens* Mill., Gard. Dict. ed. 8 6. 1768; *Lindernia dianthera* Sw., Prodr. 62. 1788; *Bacopa procumbens* (Mill.) Greenm., Publ. Field Columbian Mus. Bot. Ser. 2:261. 1907; Samvatsar, Fl. West. Trib. Madhya Pradesh 213. 1996; Singh in Mudgal *et al.,* Fl. Madhya Pradesh 2:207. 1997 (as *Bacopa procumbens*).

Annual, prostrate, decumbent, glabrous herbs; stem 4-angled, branches rooting from lower nodes; leaves 1.5-3 × 0.5-1.2 cm, opposite, decussate, ovate-lanceolate, crenate-serrate, glandular beneath, sessile or subsessile; flowers yellow, solitary, axillary, pedicels 2-4 cm long; bracteoles 2, linear; sepals unequal, one or both lateral sepals with or without basal, bent appendage; petals with a hairy throat; capsules cylindrical *ca* 6 mm long, 2-valved, dehiscence septicidal; seeds reticulate.

*Fl.-Fr.:*Oct. -May

Alirajpur, Burhanpur, Dhar, Indore, Khandwa, in moist and shady places.

MICROCARPAEA R. Br. Monotypic

Microcarpaea minima (K. D. Koenig ex Retz.) Merr., Phlipp. J. Sci. 7(2):100.1912; *Paederota minima* K. D. Koenig ex Retz., Observ. Bot. 5:10.1789; *Microcarpaea muscosa* R. Br., Prodr. 435. 1810; Hook.f., Fl. Brit. India 4:287. 1884; Samvatsar, Fl. West. Trib. Madhya Pradesh 213. 1996; Singh in Mudgal *et al.,* Fl. Madhya Pradesh 2:233. 1997.

Very small, slender, creeping annuals, stem diffusely branched; leaves opposite, sessile, entire, 2-3 mm long; flowers greenish, very small, solitary, axillary, sessile, ebracteate; sepals tubular, 5-angled or ribbed, 5-fid; petals 5-lobed, sublabiate, tube short; capsules ovoid to ellipsoid, loculicidally 2-valved; seeds ovoid.

*Fl.-Fr.:*Sept. -Nov.

Rare, on wet soil, in shady places, Alirajpur, Dhar.

NANORRHINUM Betsche W. 3; MP 2; SA 2

Key to Species

1. Leaves 2.8-4 × 0.7-1.3 cm, hastate to cordate; seeds echinulate, pale brown .... **N. ramosissimum**
1. Leaves 0.3-4.2 × 0.3-3 cm, triangular-rhomboid; seeds rugose, brown-black .... **N. incanum**

Nanorrhinum incanum (Wall.) Betsche, Courier Forschungsinst Senckenberg 71:132. 1984; *Linaria incana* Wall., Pl. Asiat. Rar. 2:43. 1831; *Kickxia incana* (Wall.) Pennell, Acad. Nat. Sci. Philadelphia Monogr. 5:59. 1943; Hook.f., Fl. Brit. India 4:251. 1883; Singh in Mudgal *et al.,* Fl. Madhya Pradesh 2:214. 1997.

Perennial, prostrate, villous or hirsute herbs, branched from the woody root-stock; leaves variable, 1.2 × 0.8 cm, lower ovate or sub-hastate, lobed, upper ones lanceolate, sagittate; flowers solitary axillary, pedicels often longer than leaves; sepals lanceolate; petals *ca* 7 mm long, upper lip purplish, palate purple spotted; spur straight or slightly curved forward, much shorter than the corolla tube; capsules *ca* 3 mm across, subglobose, dehiscing irregularly; seeds angular, rugose.

*Fl.-Fr.:*Sept. -Dec.

Rare, on hilly slopes, amidst boulders, Burhanpur, Indore.

Nanorrhinum ramosissimum (Wall.) Betsche, Courier Forschungsinst Seckenberg 71:132. 1984; *Linaria ramosissima* Wall., Pl. Asiat. Rar. 2:43. 1831; Hook.f., Fl. Brit. India 4:251. 1883; *Kickxia ramosissima* (Wall.) Janch., Oesterr. Bot. Z. 82:152. 1933; Samvatsar, Fl. West. Trib. Madhya Pradesh 209. 1996; Singh in Mudgal *et al.,* Fl. Madhya Pradesh 2:214. 1997.

Slender, perennial herbs, trailing or twining; stems much branched from the root-stock; branches wiry, 24-45 cm long; leaves variable, glabrous, lower ones deltoid, 3-5 lobed, base hastate, upper ones usually lanceolate, apiculate at apex, base sagittate; flowers yellow, solitary, axillary or passing into terminal racemes in favourable habitat, pedicels filiform, 1-2 cm long; sepals linear-lanceolate; petals up to 1 cm long, spur curved upwards; capsules 2.5-4 mm long, opening on one side by a valve and other side, by a subapical pore; seeds many, echinulate.

*Fl.-Fr.:*Oct. -Jan.

Common member of wall flora, rock cuttings, in ravines as lithophytes, crevices of old wells, throughout the area.

SCOPARIA L. W. 10; MP 1; SA 1

Scoparia dulcis L., Sp. Pl. 116. 1753; Hook.f., Fl. Brit. India 4:289. 1884; Samvatsar, Fl. West. Trib. Madhya Pradesh 213. 1996; Singh in Mudgal *et al.,* Fl. Madhya Pradesh 2:236. 1997.

Erect herbs, up to 80 cm high, stems 4-6 angled, ribbed, glabrous; leaves opposite or in threes, rhomboid-elliptic, 1.2-3.5 × 0.6-1.4 cm, apex obtuse or subacute, base acute, narrowed into a petiole, margin serrate, glandular punctate below; flowers white, 1-3 in axillary whorls, pedicels 8-10 mm long; sepals 4, acute, entire, 3-5 nerved; petals *ca* 4 mm across, pilose, throat hairy; capsules globose, 2-3 mm across; seeds ovoid or obovoid.

*Fl.-Fr.:*Almost throughout the year.

Alirajpur, Barwani, Dhar, Indore, Khandwa, Khargone.

STEMODIA L. W. 36; MP 1; SA 1

Stemodia viscosa Roxb., Pl. Coromandel 2:33. 1802 & Fl. Ind. 3:94. 1832; Hook.f., Fl. Brit. India 4:265. 1884; Samvatsar, Fl. Madhya Pradesh 215. 1996; Singh in Mudgal *et al.,* Fl. Madhya Pradesh 2:238. 1997. *Nak chunni*

Perennial, simple or branched, viscid-pubescent, herbs, up to 40 cm high; branches 4-angled, ascending, viscid; leaves 1.5-2 × 0.2-0.4 cm, glandular hairy, sessile,oblong to ovate-oblong, apex acute, base amplexicaul, scabrid above and strigose along the nerves beneath, margin entire or serrulate; flowers bluish-purple or violet, solitary, axillary or in few-flowered terminal racemes, pedicels 6-9 mm long, glandular-strigose; bracteoles 2, *ca* 3 mm long, linear-subulate; sepals 3-5 mm long, minutely glandular-pubescent, ciliate; petals *ca* 1 cm long, upper lip broadly orbicular, emarginate, lower lip *ca* 7 mm across, throat pilose; capsules ovoid-oblong, glabrous, acuminate, 4-6 × 2 mm, 4-valved, enclosed in the calyx; seeds oblong, cuneiform, ends truncate.

*Fl.-Fr.:*Oct. -April

Common winter weed in fields, open, wet ground, in all districts.

VERONICA L. W. 198; MP 1; SA 1

Veronica anagallis-aquatica L., Sp. Pl. 12. 1753; Hook.f., Fl. Brit. India 4:293. 1884; Samvatsar, Fl. West. Trib. Madhya Pradesh 216. 1996; Singh in Mudgal *et al.,* Fl. Madhya Pradesh 2:245. 1997.

Perennial, erect or decumbent herbs, about 15-30 cm high; stems creeping, rooting at the nodes, glandular-pubescent above; leaves 1.2-7 × 0.5-2.5 cm, oblong, linear-oblong or lanceolate, apex subacute or obtuse, base subcordate, margin entire, often serrate or crenate; flowers pinkish-white, in axillary, lax slender, 7-10 cm long, racemes, pedicels filiform, 8-10 mm long; bracts linear-lanceolate; sepals 2-3 mm long; oblong-lanceolate, subacute; petals 4-lobed, *ca* 4 mm across; capsules orbicular-oblong, 2-3 × 3-4 mm, 2-grooved, glabrous, enclosed in the calyx; seeds oblong, cuneate, smooth.

*Fl.-Fr.:*Nov. -Feb.

Common in wet and moist places, all over the area.

PHRYMACEAE 10/199; 3/3; 3/3

Key to Genera

1. Perfect stamens 2; corolla not bilabiate; flowers bluish-white; calyx 3-4 lobed, persistent; capsules 0.2 cm long; annuals . ... GLOSSOSTIGMA
1. Perfect stamens 4
2. Corolla bilabiate, lobes unequal; flowers white, throat violet, solitary, axillary; capsules 0.5 cm long; perennials . ... MIMULUS
2. Corolla not bilabiate; flowers pale blue or white in scape-like racemes; capsules 0.3-0.4 cm long; annuals .... MAZUS

GLOSSOSTIGMA Wight & Arn. W. 6; MP 1; SA 1

Glossostigma diandrum (L.) Kuntze, Revis. Gen. Pl. 2:461. 1891; *Limosella diandra* L., Mant. Pl 1:252. 1771; *Microcarpaea spathulata* Hook., Bot. Misc. 2:101. 1831; *Glossostigma spathulatum* (Hook.) Wight & Arn., Nova Acta Phys.-Med. Acad. Caes. Leop.-Nat. Carol. Cur. 18:355. 1836; Hook.f., Fl. Brit. India 4:288. 1884; Singh in Mudgal *et al.,* Fl. Madhya Pradesh 2:213. 1997.

Creeping herbs, rooting at nodes; leaves opposite, 5-8 mm long, narrowed into a petiole of variable length, linear-spathulate; flowers pedicellate, ebracteate, axillary; calyx 3-lobed; corolla bluish or pinkish-white, lobes 5, subequal; stamens 2; capsules 2 mm across, globose, as long as or shorter than the calyx and enclosed in it.

*Fl.-Fr.:*March-Sept.

Rare, confined to **Khandwa**, in ponds and aquatic habitat, at Chira Khadan.

MAZUS Lour.W. 27; MP 1; SA 1

Mazus pumilus (Burm.f.) Steenis, Nova Guinea n.s., 9:31. 1958; *Lobelia pumila* Burm.f., Fl. Indica 186. 1768; *Lindernia japonica* Thunb., Syst. Veg. (ed. 14) 56. 1784; *Mazus japonicus* (Thunb.) Kuntze, Revis. Gen. Pl. 2:462. 1891; *M. rugosus* Lour., Fl. Cochinch. 2:385. 1790; Hook.f., Fl. Brit. India 4:259. 1884; Singh in Mudgal *et al.,* Fl. Madhya Pradesh 2:233. 1997.

Erect or trailing, annual, non-stoloniferous herbs; radical leaves in rosette, oblanceolate or obovate to spathulate, 2-7 cm long, petioles short, margin crenate-dentate; flowers in leafless or leafy, pubescent, scape-like, glandular racemes; calyx lobes ovate-lanceolate, *ca* 4 mm long, spreading in fruit; corolla white or light blue, 6-13 mm long, ridges of lower lip brown, spotted and bearded wth club-shaped hairs; capsules subglobose, 3-4 mm long, included in the calyx; seeds minute, pale-yellow.

*Fl.-Fr.:*Nov. -April

Rare, in moist, shady places, along drains or irrigation canals, **Khandwa** only, at Nimar nursery area.

MIMULUS L.W. 155; MP 1; SA 1

Mimulus strictus Benth., Numer. List 3918. 1831 and Scroph. India 28. 1835; Hook.f., Fl. Brit. India 4:259. 1884; Singh in Mudgal *et al.,* Fl. Madhya Pradesh 2:234. 1997.

Perennial, erect, 25-60 cm high, branched herbs; stem 4-angled, striate; leaves sessile, radical and cauline, oblong, 3-8 × 0.5-2.5 cm, apex obtuse or acute, base narrowed, margin dentate, lamina gland-dotted beneath; flowers solitary, axillary, with stout, 1-4 cm long pedicels; calyx 5-6 mm long; corolla white, pink or pale-blue, 12-13 mm long; capsules *ca* 6.5 mm long, included in the calyx; seeds oblong-ellipsoid, smooth.

*Fl.-Fr.:*Aug. -April

Rare, as forest undergrowth in moist places at **Indore** only, Yeshwant nagar.

SCROPHULARIACEAE 76/1576; 4/6; 3/3

Key to Genera

1. All leaves alternate; flowers yellow; corolla sub-equally 5-lobed; glandular plants .... VERBASCUM
1. All leaves opposite
   2. Plants glandular pubescent; anthers 1-celled; flowers white; leaves pinnatifid ...... JAMESBRITTENIA
   2 . Plants eglandular; one cell of anther perfect, other empty; flowers rose-purple; leaves cut into filiform segments (our area) .... PARASOPUBIA

JAMESBRITTENIA Kuntze W. 83; MP 1; SA 1

Jamesbrittenia dissecta (Delile) Kuntze, Revis. Gen. Pl. 2:261. 1891; *Capraria dissecta* Delile, Descr. Egypte Hist. Nat. 239. 1813; *Sutera glandulosa* Roth, Nov. Pl. Sp. 291. 1821; *S. dissecta* (Delile) Walp., Repert. Bot. Syst. 3:271. 1844; Hook.f., Fl. Brit. India 4:258. 1884; Samvatsar, Fl. West. Trib. Madhya Pradesh 215. 1996; Singh in Mudgal *et al.,* Fl. Madhya Pradesh 2:241. 1997.

Annual, erect or spreading, glandular-viscid herbs, up to 30 cm tall; leaves 1.5-2 × 0.7-1 cm, pinnatisect, segments ovate, ciliate, dentate, glandular on both surfaces; flowers white, solitary, axillary or in terminal racemes, sessile or subsessile; calyx 3-4 mm long, segments linear-oblong, gland-dotted; corolla tubular, 4 mm long; capsules ellipsoid-oblong, *ca* 4 mm long, viscid-pubescent; seeds minute, yellow.

*Fl.-Fr.:*Nov. -March

Common along edges of drying ponds, wet soil, all over the area.

PARASOPUBIA H.P. Hofm. & Eb. Fisch. Monotypic

Parasopubia delphinifolia (L.) H.P. Hofm. & Eb. Fisch., Bot. Jahrb. Syst. 125 (3):357.2004; *Gerardia delphinifolia* L., Cent. Pl. II 21. 1756; *Rophanthera delphinifolia* (L.) Raf., New. Fl. 2:58. 1836; *Sopubia delphinifolia* (L.) G. Don, Gen. Hist. 4:560.1838; Hook.f., Fl. Brit. India 4:302. 1884; Samvatsar, Fl. West. Trib. Madhya Pradesh 213. 1996; Singh in Mudgal *et al.,* Fl. Madhya Pradesh 2:236. 1997.

Erect, annual, much branched herbs, up to 60 cm tall; branches 4-angled, grooved, glabrous, often spotted with purple; leaves pinnatisect, 2.5-5 cm long, segments scabrid, filiform or linear, passing into simple, floral bracts; flowers rose or purple, in racemes, solitary, axillary in scarcity of water, pedicels 3-6 mm long; bracteoles filiform, 1-1.5 mm long; calyx 10-nerved, *ca* 1 cm long, teeth linear-subulate; corolla ventricose, up to 3.5 cm long; capsules oblong, 5-8 mm long; seeds oblong, longitudinally striate.

*Fl.-Fr.:*Sept. -Dec.

Common in wet places, in all districts.

VERBASCUM L. W. 116; MP 2; SA 1

Verbascum chinense (L.) Santapau, Fl. Purandhar & in Rec. Bot. Surv. India 16:177. 1967; *Scrophularia chinensis* L., Mant. Pl. 2:250.1771; *Celsia coromandeliana* Vahl, 1794, *nom. illeg; Verbascum celsioides* Benth., Prodr. 10:229. 1846; Hook.f., Fl. Brit. India 4:251. 1883; Samvatsar, Fl. West. Trib. Madhya Pradesh 216. 1996; Singh in Mudgal *et al.,* Fl. Madhya Pradesh 2:244. 1997. *Gidad tamakhu*

Glandular pubescent, annual herbs, up to 60-70 cm tall; basal leaves in rosettes, 5-9 × 3-3.5 cm, apex obtuse, serrate-dentate, lyrate, pinnatifid; upper ones smaller, ovate or oblong, acute at apex, serrate-dentate, sessile; flowers yellow, in lax, simple or branched, 20-25 cm tall, glandular pubescent, terminal racemes; calyx lobes oblong or ovate, acute, pubescent; corolla 12-13 mm across, lobes unequal; capsules 5-6 mm across, subglobose or ovoid, glandular; seeds oblong, truncate.

*Fl.-Fr.:*Nov. -April

Common on moist, shady places and in gravelly locations, all over the area.

OROBANCHACEAE 89/1613; 6/13; 6/10

Key to Genera

1. Corolla salver-shaped; fertile anthers with 1 cell aborted; calyx ribbed
    2. Corolla tube straight; inflorescence spicate; flowers crowded; lower leaves broad, dentate upper narrow, entire .... BUCHNERA
    2. Corolla tube apically curved; inflorescence laxly spicate; all leaves narrow and entire, sometimes reduced to scales . ... STRIGA
1 . Corolla not salver-shaped; fertile anthers 2-celled
    3. Calyx campanulate or nearly so, covering the corolla tube and inflated after flowering ...... ALECTRA
    3. Calyx spathaceous

4. Leaves absent or hardly up to 1.5 mm long .... OROBANCHE
4. Leaves well developed
   5. Leaves 5-10 cm long; flowers usually solitary; pedicels very long (10-15) cm .... AEGINETIA
   5. Leaves 1-5 cm long; flowers in spike-like racemes; pedicels short ................CENTRANTHERA

AEGINETIA L. W. 6; MP 1; SA 1

Aeginetia indica L., Sp. Pl. 632. 1753; Hook.f., Fl. Brit. India 4:320.1884; Singh in Mudgal *et al.,* Fl. Madhya Pradesh 2:247. 1997.

Perennial, erect, 10-30 cm high herbs; stolons subterranean, giving out succulent, fibrous roots terminating in haustoria; scapes slender, purple-brown or violet, covered by scales at the base; flowers purple, solitary; calyx 1.5-3 cm long, ending in a beak at the posterior end; corolla obscurely bilabiate, dark purple, tubular, curved, 2-4 cm long, tube partly enclosed in calyx, lobes *ca* 0.6 cm long; stamens 2, included, anthers 1-lobed, spurred; stigma peltate; capsules dehiscent, 1.5-2 cm long, faintly 2-valved; seeds brown, pitted.

*Fl.-Fr.:*Sept. -Feb.

Moist, shady places in bamboo and mixed, forests, Indore, Khandwa.

ALECTRA Thunb. W. 34; MP 3; SA2

Key to Species/sub species

1. Inflorescence unbranched; flowers sessile; filaments glabrous .... **A. parasitica** subsp. **chitrakutensis**
1. Inflorescence branched; flowers pedicellate; filaments hairy .... **A. thompsonii**

Alectra parasitica subsp. **chitrakutensis** (M.A. Rau) K.K. Khanna & An. Kumar, J. Econ. Taxon. Bot. 31:537.2007; *A. parasitica* var. *chitrakutensis* M.A. Rau, Bull. Bot. Surv. India 3:25. 1961; Singh in Mudgal *et al.,* Fl. Madhya Pradesh 2:201. 1997.

Small, parasitic, rhizomatous herbs, underground part 2-3 cm thick, orange-yellow, with many axillary buds; leaves linear-oblong, *ca* 6 mm long, apex obtuse; flowers sessile, yellow with purple streaks, in terminal racemes; bracts linear; bracteoles slender; calyx 8-ribbed, lobes deltoid; corolla *ca* 5 mm across; capsules globose, smooth, about 5 mm across, dehiscence loculicidal; seeds cuneiform.

*Fl.-Fr.:*Aug. -Oct.

Rare, along stream banks, **Burhanpur** only, Tukaithad area.

Alectra thompsonii Hook.f., Fl. Brit. India 4:297. 1884; Kambley & Pradhan in Bull. Bot. Surv. India 26:131. f. 1-9 (1984) 1985; Singh in Mudgal *et al.,* Fl. Madhya Pradesh 2:202. 1997.

Annual, parasitic, 10-40 cm high herbs; rhizome orange-red; leaves reduced to 3-6 mm long scales, longest oblong, obtuse at apex; flowers in lax racemes; pedicels 5-12 mm long; bracts alternate, minute; bracteoles 2; calyx 5-7 mm long, hemispheric in fruit; corolla yellow, 6-8 mm long, with red nerves; anthers purple; stigma oblong-clavate, green; capsules globose, broader than long, as long as calyx; seeds linear, numerous, cuneiform.

*Fl.-Fr.:*July-Nov.

Rare, in shady places on *Vitex negundo* in **Burhanpur**, Tukaithad area.

BUCHNERA L. W. 138; MP 1; SA 1

Buchnera hispida Buch.-Ham. ex D.Don, Prodr. Fl. Nepal 91. 1825; Hook.f., Fl. Brit. India 4:298. 1884; Samvatsar, Fl. West. Trib. Madhya Pradesh 208. 1996; Singh in Mudgal *et al.,* Fl. Madhya Pradesh 2:209. 1997.

Erect, hispid, semiparasitic, 15-30 cm high, annual herbs; lower leaves in a rosette, upper opposite or alternate, 2.5-6.5 cm, sessile, margin dentate; flowers in terminal spikes; bracts linear or lanceolate, glabrous or pilose; bracteoles 2, *ca* 3.5 mm long; calyx 10-ribbed, 4-8 mm long; corolla rose-purple, 10-15 mm long, lobes subequal; capsules 5-7 mm long, enclosed in calyx; seeds reticulate.

*Fl.-Fr.:*Sept. -Nov.

In moist rocky places, among grass, Alirajpur, Burhanpur, Dhar, Indore, Khandwa.

CENTRANTHERA R. Br.W. 12; MP 2; SA 1

Centranthera indica (L.) Gamble, Fl. Presi. Madras 2:683. 1957, BSI rep; *Rhinanthus indicus* L., Sp. Pl. 603. 1753; *Centranthera procumbens* Benth. in DC., Prodr. 10:525. 1846; Hook.f., Fl. Brit. India 4:301. 1884; Samvatsar, Fl. West. Trib. Madhya Pradesh 208. 1996; Singh in Mudgal *et al.,* Fl. Madhya Pradesh 2:210.1997.

Erect, scabrid, 10-20 cm high, annual herbs; leaves 1-5 × 0.2-0.5 cm, sessile, oblong, apex subacute, margin sub-entire or shallowly serrate, ciliate, upper leaves gradually passing into bracts; flowers in terminal, spiciform racemes; calyx 8-15 hispid, spathaceous, mm long; corolla 15-30 mm long, brownish –purple; capsules about 1 cm long, oblong.

*Fl.-Fr.:*Sept. -Nov.

In moist and shady localities, Alirajpur, Dhar, Indore.

OROBANCHE L. W. 119; MP 2; SA 1

Orobanche cernua Loefl., Iter Hispan. 152. 1758; *O. nicotianae* Wight, Ill. Ind. Bot. 2:t. 158. 1850; Hook.f., Fl. Brit. India 4:326. 1884; Samvatsar, Fl. West. Trib. Madhya Pradesh 216. 1996; Singh in Mudgal *et al.,* Fl. Madhya Pradesh 2:248. 1997. *Bhoopada*

Erect, simple or branched, 10-30 cm high, pale-brown to bluish, annual, herbaceous total parasites; leaves reduced to ovate or acuminate scales; flowers lilac, sessile, in terminal, dense, unbranched, bracteate spikes; bracts 1-1.5 cm long, bracteoles absent; calyx 10-12 mm long, lobes 2-fid up to middle; corolla 12-18 mm long; capsules ellipsoid, 8-12 mm long, oblong, brown; seeds many, ovoid, brown.

*Fl.-Fr.:*Jan. -March

Common parasite of Solanaceous cultivated plants, like Brinjal, Tomato, Chillies, Potato, in all districts.

STRIGA Lour.W. 26; MP 4; SA 4

Key to Species

1. Calyx 15 ribbed, 0.9-1.5 cm long, lobes as long as its tube; corolla tube up to 2 cm long....**S. angustifolia**
1. Calyx 5-10 ribbed less than 0.8 cm long
   2. Calyx 10 ribbed; corolla tube curved near apex, limb bright yellow ....**S. asiatica**
   2. Calyx 5-ribbed
      3. Leaves reduced to subulate scales; flowers purple . ...**S. gesnerioides**
      3. Leaves developed; flowers white to pink ....**S. densiflora**

Striga angustifolia (D.Don) C.J. Saldanha, Bull. Bot. Surv. India 5:70.1963; *Buchnera angustifolia* D.Don, Prodr. Fl. Nepal 81. 1825; *B. euphrasioides* Vahl, Symb. Bot. 3:81. 1794; *Striga euphrasioides* Benth., Companion Bot. Mag. 1:364. 1836; Hook.f., Fl. Brit. India 4:299. 1884; Samvatsar, Fl. West. Trib. Madhya Pradesh 215. 1996; Singh in Mudgal *et al.,* Fl. Madhya Pradesh 2:239. 1997. *Agia*

Annual, erect, simple or branched, up to 30 cm high herbs; leaves mostly alternate, sessile, *ca* 1.5 cm long, narrowly oblong to lanceolate, entire or 1-2 dentate; flowers white, axillary, solitary, often forming terminal spikes; upper bracts linear to subulate, lower bracts foliaceous; bracteoles linear, 3-6 mm long; calyx 6-9 mm long, 15-nerved, 5-lobed, each lobe terminating into 3 ribs; corolla white, with a pale pink throat, 10-18 mm long, abruptly incurved at or above the middle; capsules ovoid, *ca* 7 mm long; seeds obovoid, black, striate.

*Fl.-Fr.:*Aug. -Dec.

In grasslands, waste places, all over the area.

Striga asiatica (L.) Kuntze, Revis. Gen. Pl. 2:466. 1891; *Buchnera asiatica* L., Sp. Pl. 630.1753; *Striga lutea* Lour., Fl. Cochinch 1:22. 1790; Hook.f., Fl. Brit. India 4:299. 1884; Singh in Mudgal *et al.,* Fl. Madhya Pradesh 2:239. 1997.

Erect, branched, 10-20 cm high, annual herbs; leaves sessile, hispid or strigose, 1-3 × 0.1-0.2 cm, linear, subacute at apex; flowers axillary, solitary or in lax, terminal, leafy spikes; bracteoles linear; calyx 5-6 mm long, strigose, 10-15 ribbed, 1 rib reaching the apex of each lobe, others ending the sinus; corolla yellow or white, 10-12 mm long; capsules hispid, 1-4 mm long, grooved, loculicidal; seeds dark brown.

*Fl.-Fr.:*Sept. -Dec.

Hill slopes, forest floor, Burhanpur, Indore, Khandwa.

Striga densiflora (Benth.) Benth., Companion Bot. Mag. 1:363. 1836; *Buchnera densiflora* Benth. Scroph. Ind. 41. 1835; Singh in Mudgal *et al.,* Fl. Madhya Pradesh 2:240.1997.

Simple or branched, 15-20 cm high, scabrid, annual herbs; leaves linear, 1-4 × 0.1 cm, apex acute, margin crustaceous-glandular, passing into bracts; flowers in terminal, lax, elongated, bracteate spikes; bracteoles acute, linear; calyx 4-6 mm long, scabrid, 5-nerved, teeth lanceolate, ciliate; corolla white, 10-15 mm long, tube *ca* 7 mm long, throat hairy; capsules 2-4 mm long, obovoid, 2-valved, enclosed in the calyx; seeds minute, striate, dark brown.

*Fl.-Fr.:*Aug. -Dec.

Rare, among grass, **Indore** only, Harsola area.

Striga gesnerioides (Willd.) Vatke, Oesterr. Bot. Z. 25:11. 1875; *Buchnera gesnerioides* Willd., Sp. Pl. 3:338.1800; *B. orobanchoides* R.Br. Voy. Abyss. App. 4:64.1814; *Striga orobanchoides* (R. Br.) Benth., Companion Bot. Mag. 1:361.1836; Hook.f., Fl. Brit. India 4:299.1884; Samvatsar, Fl. West. Trib. Madhya Pradesh 215.1996; Singh in Mudgal *et al.,* Fl. Madhya Pradesh 2:240.1997.

Erect, branched, 10-25 cm high, annual herbs with purple stem; rootstock woody; leaves reduced to scales, 4-7 × 1.5-2 mm, purple, upper ones passing into bracts; flowers sessile, in terminal lax or compact spikes; bracts acute, lanceolate; calyx 2-3 mm long, hispid, 4-5 ribbed, lobes ciliate, lanceolate; corolla purple, tube incurved above the middle; capsules *ca* 5 mm long, ellipsoid, tipped with stylar base; seeds minute, striate, cuneate.

*Fl.-Fr.:*Aug. -Feb.

Common in waste land, among hedges, grasslands, throughout the area.

LENTIBULARIACEAE 4/312; 1/11; 1/5

UTRICULARIA L. W. 218; MP 11; SA 5

Key to Species

1. Peduncles with whorls of spongy floats; flowers yellow .... **U. stellaris**
1. Peduncles without such floats; style persistent
    2. Style equalling or exceeding the capsules . ... **U. aurea**
    2. Style less than half length of the capsules
        3. Scales and bracts attached just above the middle of the peduncles; flowers violet .... **U. caerulea**
        3. Scales and bracts near the base of the peduncles; flowers yellow
            4. Corolla with short, stipitate glands; capsules dehiscing by valves; seeds lenticular, with irregular wing . ... **U. gibba** subsp. **exoleta**
            4. Corolla without glands; capsules circumscissile; seeds 5-6 angled .... **U. australis**

Utricularia aurea Lour., Fl. Cochinch., 26. 1790; *U. fasciculata* Roxb., Fl. Ind. 1:143. 1820; C.B. Clarke in Hook.f., Fl. Brit. India 4:329. 1884; Samvatsar, Fl. West. Trib. Madhya Pradesh 217. 1996; Srivastava in Mudgal *et al.,* Fl. Madhya Pradesh 2:251. 1997.

Aquatic herbs with much branched stolons; leaves whorled, 3-6 cm long, numerous, multifid, ultimate segments capillary; traps near the base of each pinna or at the point of bifurcation; scapes erect, racemose, on 20-22 cm long, without any floating vesicles; scales always absent; flowers yellow, pedicels sharply deflexed in fruit; bracts basifixed, orbicular; bracteoles absent; calyx lobes ovate, subequal; capsules globose, beaked, *ca* 4 mm across; seeds lenticular, prismatic, 5-angled, reticulate, minutely winged.

*Fl.-Fr.:*Oct. -Jan.

Alirajpur, Dhar, Indore, Jhabua, in short lived or permanent ponds.

Utricularia australis R. Br. Prodr. Fl. Nov. Holl. 430.1810; C.B. Clarke in Hook.f., Fl. Brit. India 4:329. 1884; Srivastava in Mudgal *et al.,* Fl. Madhya Pradesh 2:252. 1997.

Submerged aquatics with filiform, branched stolons, *ca* 35 cm long; leaves many, *ca* 2 cm long, primary segments dichotomously branched; traps numerous, 1-2 mm across, lateral on the leaf segments just above the bifurcation point, mouth lateral; flowers yellow, in *ca* 12 cm long, up to 10 flowered racemes; scales 2-3, in upper half of the scape; bracts orbicular, basifixed; bracteoles absent; calyx *ca* 3 mm long, lobes ovate or ovate-oblong; corolla *ca* 12 mm long, obscurely 3-lobed, upper lip longer than the upper lip of calyx, lower lip suborbicular; spur curved under lower lip; seeds prismatic, angled, narrowly winged.

*Fl.-Fr.:*March-April

Rare, in shallow waters, confined to **Indore**, Banedia pond, beyond Hatod.

Utricularia caerulea L., Sp. Pl. 1:18. 1753; *U. racemosa* Wall. ex Walp., Nov. Actorum. Acad. Caes. Leop. -Carol. Nat. Cur. 19. (suppl.)1:401. 1843; *U. roseopurpurea* Stapf ex Gamble Fl. Madras 2:983. 1924; C.B. Clarke in Hook.f., Fl Brit. India 4:333. 1884; Srivastava in Mudgal *et al.,* Fl. Madhya Pradesh 2:253. 1997.

Terrestrial herbs, up to 15 cm tall; leaves simple, linear-spathulate, 3-nerved, 5×2 mm; traps minute, ovoid-obovoid; scapes erect or scandent, 5-7 flowered, filiform, glabrous; scales and bracts medifixed; bracteoles smaller; calyx lobes subequal, ovate, obtuse; corolla cream to purple, 3x1 mm across, upper lip oblong or obovate, 2-lobed, lower lip 5 mm long; spur curved upwards, 4-7 mm long; capsules globose, papillose, 2.5× 2 mm, with persistent calyx, shorter than capsule; seeds oblong-ellipsod, papillose, reticulate.

*Fl.-Fr.:*Sept. -April

Rare, in muddy places, shade of surrounding vegetation, **Indore** only, Banedia pond beyond Hatod.

U. gibba L. subsp. **exoleta** (R. Br.) P. Taylor, Mitt. Bot. Staatssamml. Munchen 4:101. 1961; *U. exoleta* R. Br., Prodr. Fl. Nov, Holl. 430.1810; C.B. Clarke in Hook.f., Fl. Brit. India 4:329. 1884; Srivastava in Mudgal *et al.,* Fl. Madhya Pradesh 2:253. 1997.

Aquatic, free floating herbs; stolons much branched, filiform; leaves numerous, capillary, alternate on the stolons, dichotomously branched; traps ovate to reniform, mouth lateral, upper lip with 2, branched appendages; scapes erect, slender, 2-4 flowered; scale 1, near the middle of scape; scale and bract basifixed; bracteoles absent; flowers pedicellate, yellow, often with pink striations; calyx lobes 2, subequal, obtuse, persistent, thick; corolla 8×3 mm across, upper lip broadly ovate, longer than the upper lobe of calyx, lower lip 4-5 mm long; spur *ca* 3 mm long, facing downwards; capsules globose, *ca* 0.8 mm across; seeds lenticular, winged, testa smooth.

*Fl.-Fr.:*Sept. -April

Indore, rare, in shallow, stagnant water, Lekoda pond, Ujjain road.

Utricularia stellaris L.f., Suppl. Pl. 80.1782; C.B. Clarke in Hook.f., Fl. Brit. India 4:328. 1884; Samvatsar, Fl. West. Trib. Madhya Pradesh 217. 1996; Srivastava in Mudgal *et al.,* Fl. Madhya Pradesh 2:255. 1997.

Slender, free-floating or submerged aquatic herbs, except the inflorescence; stolons much branched, terete, filiform; leaves whorled, auricled at base, repeatedly branched into filiform segments, auricles with ciliated segments; traps numerous, mouth lateral, upper lip with 2, branched appendages; scapes erect, *ca* 10 cm long, 4-12 flowered, with a whorl of ellipsoid, white, spongy, 5-7 floats, above the middle; floats with filiform, external pinnules at their apices, pedicels deflexed in fruit; bracts basifixed; scales and bracteoles absent; calyx lobes unequal, obtuse, accrescent, upper lobe rounded, lower lobe truncate or emarginated; corolla yellow, upper lip broadly ovate, larger than upper calyx lobe, lower lip up to 1.5 cm long; spur 4 mm long, blunt, curved under the lower lip; capsules ellipsoid, globose, 6×4 mm, 4-6 angled; seeds prismatic, 4-6 angled, narrowly winged, testa glochidiate.

Fl.-Fr.:June-Oct.

Alirajpur, Barwani, Burhanpur, Dhar, Indore, Khandwa.

GESNERIACEAE 164/3122; 4/4; 1/1

DIDYMOCARPUS Wallich W. 87; MP 1; SA 1

Didymocarpus pygmaeus C.B. C.B. Clarke, Monogr. Phan. 5:82. 1883; C.B. Clarke in Hook.f., Fl. Brit. India 4:345. 1884; Samvatsar, Fl. West. Trib. Madhya Pradesh 217. 1996; Singh in Mudgal *et al.,* Fl. Madhya Pradesh 2:258. 1997.

Annual, small pubescent herbs, 3-8 cm tall; leaf usually solitary, 2.5-5 cm long, elliptic or oblong, apex acute, base cordate, membranous, sessile, secondary nerves 6-9; flowers white, 1-4 in axillary cymes, but often solitary due to reduction, pedicels up to 1 cm long; calyx lobes linear, 2-5 mm long; corolla tubular, 2-5 mm long; capsules linear, up to 1-1.5 cm long, 2-valved; seeds ellipsoid.

Fl.-Fr.:July-Nov.

On damp walls, slopes of hills when moisture is available, Alirajpur, Dhar, Indore.

BIGNONIACEAE 86/852; 5/6; 4/5

Key to Genera

1. Stamens 5; seeds winged all around, except base; capsule 6-9 cm broad; leaves 2-3 pinnate .... OROXYLUM
1. Stamens 4; seeds winged at two ends; capsule up to 2 cm broad; leaves unipinnate
    2. Calyx spathaceous; flowers white; capsule flat, falcately curved .... DOLICHANDRONE
    2. Calyx not spathaceous; capsule cylindric; flowers white or purple
        3. Capsule lenticellate, septum prominently recessed to accommodate the seed (not wings); seeds trigonous; ovules 2-seriate; flowers yellow or dull purple .... STEREOSPERMUM
        3. Capsule tubercled; septum faintly recessed; seeds flat; ovules many seriate; flowers white .... RADERMACHERA

DOLICHANDRONE (Fenzl) Seem. 10; MP 1; SA1

Dolichandrone falcata (Wall. ex DC.) Seem., J. Bot. 8:381. 1870; *Spathodea falcata* Wall. ex DC., Prodr. 9:206. 1845; *Dolichandrone lawii* Seem., J. Bot. 8:380.1870; C.B. Clarke in Hook.f., Fl. Brit. India 4:380.1884; Samvatsar, Fl. West. Trib. Madhya Pradesh 218. 1996; Khanna & Ram Saran in Mudgal *et al.,* Fl. Madhya Pradesh 2:264. 1997. *Meda singi*

Deciduous, up to 8 m high trees; young parts pubescent or shortly villous; leaves 2-5 × 1.5-4 cm, unipinnate; leaflets ovate, sub-orbicular or obovate, apex abruptly acuminate, base cuneate or rounded, margin entire; flowers white, in terminal, few flowered racemes, pedicels *ca* 1 cm long; calyx stellately pubescent; corolla *ca* 2.5 cm long, lobes obovate-oblong with undulate, crisped margin, tube narrow at base, enlarging upwards; capsules flat, falcately curved, 30-40 cm long, turning black, when mature; seeds 4-angled, winged.

*Fl.-Fr.:*March-June

Common on the fringe of forests, village roads, waste land, in all districts. The ripe fruits of last year remain on the trees, when they bloom next year.

OROXYLUM Venten. Monotypic

Oroxylum indicum (L.) Kurz, Forest Fl. Burma, 2:237. 1877; *Bignonia indica* L., Sp. Pl. 625. 1753; *B. pentandra* Lour., Fl. Cochinch., 379. 1790; *Spathodea indica* (L.) Pers., Syn. Pl. 2:173. 1807; *Calosanthes indica* (L.) Blume, Bijdr. Fl. Ned. Ind. 760.1826; C.B. Clarke in Hook. f, Fl. Brit. India 4:378. 1884; Samvatsar, Fl. West. Trib. Madhya Pradesh 218. 1996; Khanna & Ram Saran in Mudgal *et al.,* Fl. Madya Pradesh 2:268. 1997. *Tetu* **(Plate 7,Fig. 6)**

Trees, 5-15 m tall; leaves bipinnate compound, borne often at the end of twigs, petioles long, leaflets 8-18 × 4-12 cm, ovate or ovate-elliptic, apex acuminate, base rounded or cordate, margin entire or coarsely serrate-dentate; flowers dull-white (deep maroon, Fl. M.P.), in terminal, erect, racemes borne singly on tubercles, peduncles *ca* 40 cm long; calyx dirty violet, 3-5 cm long, turning woody in fruit; corolla 5-7 cm long, glandular inside; disc shallowly 5-lobed; capsules flat, hanging, 50-75 × 6-9 cm, tapering at both ends; seeds *ca* 6x 3 cm including wings.

*Fl.-Fr.:*June-March

Rare, Alirajpur, Burhanpur.

RADERMACHERA Zoll. & Moritzi W. 17; MP 1; SA 1

Radermachera xylocarpa (Roxb.) Roxb. ex K. Schum., Nat. Pflanzenfam. 4(3b) :243. 1895; *Bignonia xylocarpa* Roxb., Fl. Ind. ed 1832 3:108. 1832; *Stereospermum xylocarpum* (Roxb.) Bentham & Hook.f., Gen. Pl. 2:1047. 1876; C.B. Clarke in Hook.f., Fl. Brit. India 4:383. 1884; Samvatsar, Fl. Madhya Pradesh 218. 1996; Khanna & Ram Saran in Mudgal *et al.,* Fl. Madhya Pradesh 2:270.1997. *Garud vriksha*

Large, deciduous trees, 15-20 m tall; leaves bi to tripinnate, 50-60 cm long; leaflets 2-4 pairs and an odd one, leaflets 5-10 × 3-5 cm, elliptic-oblong, apex acute to acuminate, base rounded, margin entire or coarsely serrate, glabrous; flowers white with a yellow tinge, fragrant, in tomentose, corymbose panicles; calyx campanulate *ca* 1 cm long, pubescent outside, lobes 3-5;

corolla *ca* 3 cm long, glabrous, lobes crisp, rounded; disc cupular, fleshy; capsules *ca* 1 m long, woody, pericarp densely tubercled; seeds flat, with a membranous wing on either side.

*Fl.-Fr.:*April-Feb.

Alirajpur, Dhar, on hilly slopes.

STEREOSPERMUM Cham. W. 23; MP 2; SA 2

Key to Species

1. Capsule obscurely 3-ribbed; filaments woolly at base; flowers yellow; leaflets long, caudate-acuminate . ... **S. chelonoides**
1. Capsule 4-ribbed; filaments not woolly at base; flowers dull purple; leaflets shortly caudate-acuminate .... **S. tetragonum**

Stereospermum chelonoides (L.f.) DC., Biblioth. Universelle Geneve n.s., 17:124. 1838; *Bignonia chelonoides* L.f., Suppl. Pl. 282. 1782; *Stereospermum suaveolens* (Roxb.) DC., Biblioth. Universelle Geneve n.s., 17:124. 1838; *Bignonia suaveolens* Roxb., Fl. Ind. ed. 1832 3:104. 1832; *Tecoma suaveolens* (Roxb.) G. Don, Gen. Hist. 4:224. 1838; C.B. Clarke in Fl. Brit. India 4:382. 1884; Samvatsar, Fl. West. Trib. Madhya Pradesh 219. 1996; Khanna & Ram Saran in Mudgal *et al.,* Fl. Madhya Pradesh 2:272. 1997. *Padar* **(Plate 7, Fig. 7)**

Deciduous trees, up to 20 m tall; leaves unipinnate, 30-50 cm long, leaflets 3-4 pairs, 5-15 × 3-7 cm, obovate-oblong, apex shortly acuminate, base cuneate, margin entire, glandular pubescent; flowers fragrant, yellow, *ca* 2.5 cm across, in drooping, terminal, viscidly pubescent panicles; calyx shortly 5-lobed; corolla 2-3 cm long, tube pink; capsules 20-50 cm long, obscurely 3-ribbed, pericarp rough with white tubercles, septum (axis) thick, corky, with recesses to accommodate the nucleus of seeds, not wings; seeds furrowed in the middle, *ca* 3 cm long, including wings, with 2 papery wings on either side of the nucleus.

*Fl.-Fr.:*March-Dec.

Occasional on hill slopes, Alirajpur, Burhanpur, Indore, Khandwa.

Stereospermum tetragonum DC., Biblioth. Universelle Geneve n.s., 17: 124. 1838; *Bignonia tetragona* Wall. ex DC., Prodr. 9:211. 1845; *Dipterosperma personatum* Hassk., Flora 25 (2 beibl.) :28. 1842; *Stereospermum personatum* (Hassk.) Chatterjee, Bull. Bot. Soc. Bengal 2:90.1948; *Bignonia colais* Buch.-Ham. ex Dillwyn, Hortus Malab. 28. 1839; *Stereospermum colais* (Buch. -Ham. ex Dillwyn) Mabb., Taxon, 27:553. 1978; C.B. Clarke in Hook.f., Fl. Brit. India 4:382. 1884; Samvatsar, Fl. West. Trib. Madhya Pradesh 219. 1996; Khanna & Ram Saran in Mudgal *et al.,* Fl. Madhya Pradesh 2:272. 1997. *Chhoti padar*

Deciduous, 10-15 m tall trees; leaves unipinnate, 20-30 cm long, leaflets 3-6 pairs, 4-12 × 2-5 cm, oblong, elliptic-oblong, apex caudate-acuminate, base cuneate, glabrous, margin entire; flowers pale yellow, fragrant, in *ca* 40 cm long, drooping, glandular hairy, trichotomous drooping panicles, peduncles 2-7 cm long, calyx purple, 6-8 mm long; corolla lobes crisped, tube 4-6 mm long; capsules up to 50 cm long, bluntly 4-ribbed, pericarp rough with lenticels; seeds *ca* 2 cm long, including wings.

*Fl.-Fr.:*May-Aug.

Rare, on hill slopes, Alirajpur, Khandwa.

MARTYNIACEAE 5/21; 1/1; 1/1

MARTYNIA L. W. 4; MP1; SA1

Martynia annua L., Sp. Pl. 618. 1753; *M. diandra* Gloxin, Observe. Bot. Univ. Arg. 14. 1785; C.B. Clarke in Hook.f., Fl. Brit. India 4:386. 1884; Samvatsar, Fl. Wesst. Trib. Madhya Pradesh. 219. 1996; Khanna in Mudgal *et al.,* Fl. Madhya Pradesh 2:278. 1997. *Bagh nakhi* **(Plate 7, Fig. 8)**

Annual, rough, tall herbs, all parts velvety pubescent; roots thick; stem fistular; leaves very large, compared to the size of the plant, 18-20 × 16-18 cm, opposite, broadly ovate, apex acute, base cordate with 5-7 palmate nerves, margin dentate, petioles 15-20 cm long, glandular hairy on both surfaces; flowers ill smelling, pink, shaded with dark purple, in axillary or often terminal, drooping, glandular racemes; calyx and corolla glandular pubescent; drupes ovoid-green initially, turning black at maturity, with 2 prominent, large hooks.

*Fl.-Fr.:*Aug. -Sept.

A common weed on rubbish heaps, waste places, along roadsides, particularly of villages, during rains.

ACANTHACEAE 242/3947; 23/72; 20/48

Key to Genera

1. Seeds not supported on retinaculi
    2. Stem reduced; bracts spinous pointed, rigid, acuminate; peduncle covered with scales .... ELYTRARIA
    2. Stem well developed; bracts not rigid, apex obtuse; peduncle without scales .... NELSONIA
1. Seeds supported on rigid, hooked retinaculi
    3. Corolla with a 3-lobed lower lip, upper absent; bracteoles 8-10, tips spiny; leaves in whorls .... BLEPHARIS
    3. Corolla 2-lipped or subequally 5-lobed; bracteoles 2-many, tips not spiny
        4. Stamens 2
            5. Flowers surrounded by many, reduced spine- tipped cladodes .... HAPLANTHUS
            5. Flowers not surrounded as above
                6. Anther cells placed at different levels, not parallel
                    7. Flowers in spikes
                        8. Inflorescence 1-sided .... RUNGIA
                        8. Inflorescence not 1-sided; flowers white or rose-purple; anthers spurred at base or not .... JUSTICIA
                    7. Flowers in cymules (clustered)

9. Cymules in axillary or terminal panicles . ... PERISTROPHE
9. Cymules in sessile, axillary clusters, containing many abortive flowers .....DICLIPTERA

6. Anther cells parallel
10. Flowers large, in spikes or corymbose cymes
11. Corolla twisted in bud .... ERANTHEMUM
11. Corolla imbricate in bud ....ECBOLIUM
10. Flowers comparatively small, in sympodial cymes ................................ANDROGRAPHIS

4. Stamens 4
12. Staminodes present
13. Sepals 4, in 2 opposite decussate pairs, lateral pair very large, leafy ....BARLERIA
13. Sepals 5
12. Staminodes absent
14. Corolla 2-lipped
15. Bracts and bracteoles similar, ending in spiny tips; inflorescence congested in a head (our area) ....LEPIDAGATHIS
15. Bracts and bracteoles dissimilar, bracteoles modified as spines in some species ....HYGROPHILA
14. Corolla sub-equally 5-lobed
16. Inflorescence 1-sided; flowers white; capsule 4-seeded (our area) ....ASYSTASIA
16. Inflorescence not as above
17. Ovules 2 per cell; flowers white, pink-purple or blue
18. Anther cells spurred at base; corolla tube ventricose
19. Bracteoles small, subulate; stigma linear; bracts absent; branches rooting at nodes; flowers blue .... DYSCHORISTE
19. Bracteoles large, ovate; stigma unequally 2-lobed; bracts present; large twiners; flowers white .... PETALIDIUM
18. Anther cells not spurred at base; bracts present, viscid; bracteoles absent; leaves bristly above; shrubs . ... STROBILANTHES
17. Ovules more than 2 per cell
20. Capsule with a solid, clavate base; bracteoles foliaceous ....RUELLIA
20. Capsule seed bearing throughout; bracteoles narrow or absent ....HEMIGRAPHIS

ANDROGRAPHIS Wall. ex Nees W. 32; MP 2; SA 2

Key to Species

1. Inflorescence terminal, paniculate ....**A. paniculata**
1. Inflorescence axillary, branched, longer than the leaves . ... **A. longipedunculata**

Andrographis longipedunculata (Sreem.) L.H. Cramer in Kew Bull. 51:555. 1996; *Neesiella longipedunculata* Sreem., Phytologia 15:271. 1967, *nom. illeg; Erianthera longipedunculata* (Sreem.) M.R. Almeida, Fl. Maharashtra 4A: 41.2003; Karthik. *et al.,* Fl. Pl. India

1:3.2009; Gnanasekaran, G. *et al.,* Taxonomy and distribution of *Andrographis longipedunculata* (Acanthaceae) Rheedea 27(2):159-164.2017.

Herbs, *ca* 80 cm high; stems 4-angled, nodes swollen, densely glandular-villous; leaves 2-7.5 × 0.5-2.5 cm, sessile, ovate-lanceolate, apex obtuse or acute, base obtuse, margin entire, lateral veins 5-7 pairs, with intermixed, long, glandular hairs on both surfaces; inflorescence axillary, racemose, 4-10 cm long, often branched and longer than the subtending leaf, flowers distant on the peduncle; bracts 2, bracteoles 2; calyx 5-lobed, lobes subequal; corolla 2-lipped, white, 1-1.3 cm across, tube ventricose, upper lip 2-lobed, lower 3-lobed; stamens 2; capsules 8-10 × 3-4 mm, with a median longitudinal groove; seeds 4, deeply pitted, echinate, yellowish-brown.

*Fl.-Fr.:*July-March

Common as forest undergrowth in all districts.

Note- According to Gnanasekaran, *loc cit*, this taxon is identified as *Indoneesiella echioides* (L.) Sreem. in most of the Indian floras and after thoroughly going through literally hundreds of herbarium specimens in various herbaria, this is now identified correctly as above. This is a part of the study of *Andrographis* in India by Gnanasekaran and associates.

Andrographis paniculata (Burm.f.) Nees, Pl. Asiat. Rar. 3:116. 1832; *Justicia paniculata* Burm.f., Fl. Ind. 9. 1768; C.B. Clarke in Hook.f., Fl. Brit. India 4:501. 1884; Khanna *et al.,* in Mudgal *et al.,* Fl. Madhya Pradesh 2:287. 1997. *Kalmegh*

Erect, much branched, glabrous, 40-60 cm high herbs; leaves 5-8 × 1-2 cm, linear-obovate or lanceolate, apex acutr, base tapering, petioles short; flowers white to purplish, in axillary and terminal, paniculate racemes; bracts opposite; bracteoles 2, small or lacking; calyx lobes pubescent; corolla 2-lipped, glandular pubescent, lower lip with deep purple or pink markings inside, at base; capsules *ca* 1.5 cm long, linear-oblong; seeds subquadrate, rugosely pitted, yellow-brown.

*Fl.-Fr.:*Sept. -April

Barwani, Burhanpur, Khandwa, Khargone.

ASYSTASIA Blume W. 54; MP 2; SA 2

Ket to Species

1. Flowers in racemes; capsules about 1 cm long .... **A. gangetica**
1. Flowers in spikes; capsules about 1.5-2 cm long .... **A. mysorensis**

Asystasia gangetica (L.) T. Anderson, Enum. Pl. Zeyl. 235. 1860; *Justicia gangetica* L., Cent. Pl. II 3-4. 1756; *Asystasia coromandeliana* Nees, Pl. Asiat. Rar. 3:89. 1832; C.B. Clarke in Hook.f., Fl. Brit. India 4:493. 1884; Khanna *et al.,* Fl. Madhya Pradesh 2:288. 1997.

Suffruticose, erect or straggling herbs or undershrubs; leaves 3-5 × 2-3 cm, elliptic-ovate, apex acute to acuminate, base truncate or rounded, margin entire or dentate; flowers in terminal or axillary, lax racemes; bracts lanceolate, 2-4 mm long, shorter than the calyx; bracteoles small, lanceolate; calyx pubescent, divided nearly to the base; corolla white with purplish streaks, lobes orbicular, pubescent outside; capsules *ca* 1 cm long; seeds 4, tuberculate, *ca* 5 mm across.

*Fl.-Fr.:*March-April

Rare, Burhanpur, Indore.

Asystasia mysorensis (Roth) T. Anderson, J. Linn. Soc. Bot. 9:524. 1867; *Ruellia mysorensis* Roth, Nov. Pl. Sp. 303. 1821; *Asystasia lawiana* Dalzell in Hooker's J. Bot. Kew Gard. Misc. 4:344. 1852; C.B. Clarke in Hook.f., Fl. Brit. India 4:496. 1884; Samvatsar, Fl. West. Trib. Madhya Pradesh 221. 1996; Khanna *et al.,* in Mudgal *et al.,* Fl. Madhya Pradesh 2:489. 1997.

Annual, 30-40 cm high herbs; stems 4-angled, grooved; leaves petiolate, 4-10 × 3-5 cm, elliptic-oblong, apex acute, base acute, slightly hairy on the nerves on upper side, more densely so beneath; flowers white, in terminal, 2-5 cm long spikes; bracts lanceolate, pubescent, acute; bracteoles smaller, lanceolate; capsules 1.5-2 cm long, glandular-pubescent, clavate, with a solid base, enlarged part constricted in middle between seeds, 4-seeded; seeds obliquely ovoid, flat, faces tuberculate and with hexagonal areolae along edges.

*Fl.-Fr.:*Sept. -Nov.

In moist and shady places, Alirajpur, Burhanpur, Dhar, Indore.

BARLERIA L. W. 253; MP 7; SA 6

Key to Species

1. Armed plants
    2. Flowers yellow; leaves 6-11 cm long .... **B. prionitis**
    2. Flowers white; leaves 4-7 cm long .... **B. acanthoides**
1. Unarmed plants
    3. Flowers blue; corolla less than 8 cm long; seeds silky hairy
        4. Outer sepals spiny and ciliate; leaves 4-15x1-5 cm .... **B. critata**
        4. Outer sepals neither spiny nor ciliate; leaves 8-15x4-8 cm . ... **B. montana**
    3. Flowers white or pink; corolla more than 8 cm long
        5. Flowers white; calyx and seeds pubescent .... **B. lawii**
        5. Flowers pink; calyx and seeds glabrous .... **B. gibsonii**

Barleria acanthoides Vahl, Symb. Bot. 1:47. 1790; C.B. Clarke in Hook.f., Fl. Brit. India 4:484. 1884; Bhandari, Fl. Indian Desert 265. 1995.

Prickly branched undershrubs, up to 30 cm high; stems densely canescent; leaves pubescent, 2.5-6 × 1-2 cm, dark green, oblong to obovate, apex mucronate, base narrowed; petiole up to 7 mm long, interpetiolar spines simple, 2-4; racemes 1-8 flowered, unilateral (scorpioid), often condensed; bracts 15-17 mm long, toothed, foliaceous or with pinnatifid spines; calyx divided almost to the base, outer sepals subequal, 12-15 mm long, broadly ovate, reticulately veined, rarely pectinately toothed on the margins, inner sepals smaller; corolla white, 6-7 cm long, pubescent outside, tube cylindric, lobes ovate; capsules 1-1.2 cm long, ellipsoid, glabrous, attenuated at both ends; seeds 4, compressed, silky hairy.

*Fl.-Fr.:*Sept. -Nov.

Rare, **Burhanpur** only, Jainabad and Shahpur.

Barleria cristata L., Sp. Pl. 636. 1753; Hook.f., Fl. Brit. India 4:488. 1884; Samvatsar, Fl. West. Trib. Madhya Pradesh 221. 1996; Khanna *et al.,* in Mudgal *et al.,* Fl. Madhya Pradesh 2:290.1997. *Koranta*

Perennial, up to 70 cm high shrubs, stems 4-angled; leaves tomentose, 4-15 × 1.5-3 cm, elliptic, or oblong-lanceolate, acute to acuminate at apex, base tapering; flowers purple, in axillary and terminal, short, capituliform spikes; bracts 1-1.5 cm long, membranous, spinous-denticulate; calyx deeply divided, 2 outer segments white, bristle-tipped on margin, 2 inner narrower, lanceolate, ciliate; corolla pubescent outside, *ca* 8 mm long, ovate-oblong; capsules ellipsoid, 1-1.5 cm long, 4-seeded; seeds *ca* 4 mm across, orbicular, silky hairy.

*Fl.-Fr.:*Sept. -March

Occurs as forest undergrowth, along roadsides in forest, Alirajpur, Barwani, Dhar, Indore, Khandwa, Khargone.

Barleria gibsonii Dalzell, Hooker's J. Bot. Kew Gard. Misc. 2:339. 1850; Hook.f., Fl. Brit. India 4:487. 1884; Samvatsar, Fl. West. Trib. Madhya Pradesh 221. 1996; Khanna *et al.,* in Mudgal *et al.,* Fl. Madhya Pradesh 2:291. 1997.

Suffruticose, glabrous undershrubs, *ca* 1.5 m high; leaves glabrous, 4-12 × 2-6 cm, elliptic, apex acuminate, base rounded or subacute; flowers pink, solitary, axillary, passing into terminal spikes; bracts foliaceous, linear-lanceolate; bracteoles linear; segments of calyx unequal, outer pair ovate-oblong, inner pair linear or lanceolate; corolla lobes obovate, hirsute outside, tube white with blue base; capsules oblong, acuminate, 4-seeded; seeds *ca* 6 × 5 mm, black, glabrous.

*Fl.-Fr.:*Sept. -Nov.

As forest undergrowth, Alirajpur, Dhar, Indore, Khandwa.

Barleria lawii T. Anderson, J. Linn. Soc. Bot. 9:492. 1867; *B. beddomei* T. Anderson ex Beddome, Icon. Pl. Ind. Orient. 1: t. 62. 1874; Hook.f., Fl. Brit. India 4:486. 1884; Samvatsar, Fl. West. Trib. Madhya Pradesh 222. 1996; Khanna *et al.,* in Mudgal *et al.,* Fl. Madhya Pradesh 2:291. 1997.

Perennial, up to 1 m high undershrubs; branches terete, appressed pubescent; leaves 4-14 × 1.5-3.5 cm, elliptic or elliptic-lanceolate, glabrous and lineolate above, hairy on nerves beneath, apex acuminate, base narrowed; flowers in axillary and terminal, leafy spikes; bracts linear-oblong, *ca* 1 cm long; bracteoles hairy, linear-lanceolate; outer calyx segments elliptic-ovate, inner lanceolate; corolla white, 6-9 mm long, pubescent; capsules obovoid, compressed, *ca* 2 cm long; seeds 5-6 mm long and nearly as broad, silky hairy.

*Fl.-Fr.:*Oct. -Dec.

Common along streams, also as forest undergrowth in Alirajpur, Barwani, Dhar, Indore, Khandwa, Khargone.

Barleria montana Nees, Pl. Asiat. Rar. 3:92. 1832; Hook.f., Fl. Brit. India 4:487. 1884; Samvatsar, Fl. West. Trib. Madhya Pradesh 222. 1996; Khanna *et al.,* in Mudgal *et al.,* Fl. Madhya Pradesh 2:292. 1997. **(Plate 8, Fig. 1)**

Undershrubs, 60-80 cm high; leaves petiolate, 8-15 × 4-8 cm, elliptic, apex acuminate, base attenuate and decurrent; flowers blue, axillary solitary and running into terminal spikes; bracts shorter than calyx; outer calyx segments ovate, up to 3 cm long, 6-nerved, nerves diverging from the base of midrib, inner linear or linear-lanceolate; corolla blue or purple, lobes obovate, obtuse; capsules *ca* 2 cm long, beaked above; seeds 4, densely clothed with yellowish, silky hairs.

*Fl.-Fr.:*Sept. -Oct.

Common forest undergrowth in all districts.

Barleria prionitis L., Sp. Pl. 636. 1753; Hook.f., Fl. Brit. India 4:482. 1884; Samvatsar, Fl. West. Trib. Madhya Pradesh 222. 1996; Khanna *et al.,* in Mudgal *et al.,* Fl. Madhya Pradesh 2;295. 1997. *Katsaraiya*

Perennial, armed shrubs, about 1 m high; leaves 6-10 × 2-5 cm, ovate to elliptic-lanceolate, apex acute or obtuse, base tapering, entire, bristle-tipped, with 3-4, divaricate, white thorns in axils; flowers orange-yellow, solitary or terminal,in leafy and spiny spikes upwards; bracts 1-2 cm long, spine-tipped; bracteoles bristle-tipped; calyx segments broadly lanceolate, acuminate, outer pair spine-tipped; corolla glabrous or puberulous; capsules ovoid, about 1 cm long, with a long, tapering solid beak; seeds 2, hairy.

*Fl.-Fr.:*Oct. -Jan.

Common along village roads, in forest clearings, throughout the area.

BLEPHARIS A. L. Juss. W. 126; MP 3; SA 2

Key to Species

1. Leaves petiolate, up to 12-14 cm long, glabrous beneath; bracteoles 8; flowers diry white with purple dots; seeds echinate, glabrous .... **B. maderaspatensis**
1. Leaves sessile, up to 2.5 cm long, oblong,thick, hairy beneath; bracteoles 10; flowers blue; seeds smooth, hairy .... **B. integrifolia**

Blepharis integrifolia (L.f.) E. May & Drege ex Schinz, Vierteljahrsschr. Naturf. Ges. Zurich 60.416. 1915; *Acanthus integrifolius* L.f., Suppl. Pl. 294. 1782; *A. repens* Vahl, Symb. Bot. 2:76. 1791; *Blepharis molluginifolia* Pers., Syn. Pl. 2:180.1806; *B. repens* (Vahl) Roth, Nov. Pl. Sp. 321. 1821; Hook.f., Fl. Brit. India 4:479. 1884; Samvatsar, Fl. West. Trib. Madhya Pradesh 222. 1996; Khanna *et al.,* in Mudgal *et al.,* Fl. Madhya Pradesh 2:300.1997.

Annual, prostrate herbs, stems diffuse, wiry; leaves in whorls of 4, oblong or obovate, 1-2.7 × 0.2-0.6 cm, apex mucronate, margin entire, slightly succulent; flowers violet or blue, sessile, solitary axillary; bracteoles 8- 10, in opposite pairs, 2-6 mm long, bristly, innermost pair oblong-spathulate, all with spinulose apex; calyx divided to the base, sepals unequal, margin membranous; corolla pubescent, 8-10 mm long; larger anther lobes fertile, smaller obsolete; capsules ovoid or ellipsoid, 5-6 mm long, shining; seeds ellipsoid, 3-4 mm long, appressedly hairy.

*Fl.-Fr.:*Sept. -Oct.

Common in open places, grasslands, all over the area.

Blepharis maderaspatensis (L.) B. Heyne ex Roth, Nov. Pl. Sp. 320.1821; *Acanthus maderaspatensis* L., Sp. Pl. 639. 1753; *Blepharis boerhavifolia* Pers., Syn. Pl. 2:180.1806; C.B. Clarke in Hook.f., Fl. Brit. India 4:478. 1884; Samvatsar, Fl. West. Trib. Madhya Pradesh 222. 1996; Khanna *et al.,* in Mudgal *et al.,* Fl. Madhya Pradesh 2:299. 1997. *Dudhia choti*

Perennial, diffuse or prostrate herbs; stems wiry; leaves in whorls of 4, 2 larger and 2 smaller at each node, 3-9 × 1.5-3.5 cm, elliptic, ovate-lanceolate or oblanceolate, apex acute, base cuneate, margin ciliate; flowers axillary solitary or 2-3 together; bracteoles 8, in opposite pairs, upper with bristles on the margin; calyx segments lanceolate, acute; corolla pale blue with a yellow spot insde the lower lip; capsules broadly ellipsoid, *ca* 7 mm long, brown; seeds 2, echinate with obtuse spines.

*Fl.-Fr.:*Oct. -Nov.

Common, throughout the area.

DICLIPTERA A. L. Juss. W. 223; MP 3; SA 1

Dicliptera verticillata (Forssk.) C. Chr., Dansk. Bot. Ark. 4(3): 11. 1922; *Dianthera verticillata* Forssk., Fl. Aegypt. -Arab. 9. 1775; *Dicliptera micranthus* Nees, Pl. Asiat. Rar. 3:112. 1832; C.B. Clarke in Hook.f., Fl. Brit. India 4:533. 1885; Samvatsar, Fl. West. Trib. Madhya Pradesh 223. 1996; Khanna *et al.,* in Mudgal *et al.,* Fl. Madhya Pradesh 2:304. 1997.

A much branched, 30-60 cm tall, annual herb; stem and branches angular; leaves 4-10 × 1.5-3 cm, elliptic, oblong acute, opposite, nearly glabrous; flowers reddish-purple, sessile, in axillary custers; bracts oblong, pubescent outside, cuspidate, margins scarious, ciliolate; bracteoles linear-lanceolate, ciliate; capsule ellipsoid, hairy at the top; seeds tuberculate, brownish-yellow.

*Fl.-Fr.:*Sept. -Dec.

Common in shady places, hedges, throughout the area.

DYSCHORISTE Nees W. 102; MP 1; SA 1

Dyschoriste erecta (Burm.f.) Kuntze, Revis. Gen. Pl. 2:485. 1891; *Ruellia erecta* Burm.f., Fl. Indica 135. 1768; *Dyschoriste depressa* (L.) Nees, Pl. Asiat. Rar. 3:81. 1832; C.B. Clarke in Hook.f., Fl. Brit. India 4:410.1884; Samvatsar, Fl. West. Trib. Madhya Pradesh 224. 1996; Khanna *et al.,* in Mudgal *et al.,* Fl. Madhya Pradesh 2:310.1997. [*Dyschoriste nagchana* (Nees) Bennet, 1983, *Dipteracanthus nagchana* Nees, 1832, is a separate species]

Prostrate, spreading perennials; branches rooting at the nodes; leaves 1.8-4 × 1.2-2.5 cm, ovate, acute toothed; flowers blue, solitary axillary or in a few flowered cymes; corolla truncate at the throat, limb oblong, spreading; capsule oblong, stipitate, glabrous; seeds 4, compressed, hairy when wetted.

*Fl.-Fr.:*March-June

Common weed of fields, near water-logged soil, in shade of hedges of fields, forest undergrowth in moist soil, all over the area.

ECBOLIUM Kurz W. 21; MP 1; SA 1

Ecbolium viride (Forssk.) Alston, Handb. Fl. Ceylon 6(Suppl.): 229. 1931; *Justicia viridis* Forssk., Fl. Aegypt. -Arab. 5. 1775; *Ecbolium linneatum* Kurz, J. Asiat. Soc. Bengal, pt. 2, Nat. Hist. 40 (2):75. 1871; C.B. Clarke in Hook.f., Fl. Brit. India 4:544. 1885; Khanna *et al.,* in Mudgal *et al.,* Fl. Madhya Pradesh 2:311. 1997.

Undershrubs; leaves 10 × 6 cm, obovate or elliptic-ovate, apex acute, base attenuate; flowers in axillary and terminal, leafy spikes; bracts obtuse; bracteoles narrow; calyx segments acuminate, sub-equal; corolla bluish-green, lower lip with two, wing-like lateral lobes; capsule compressed, beaked, *ca* 1.5 cm long, 2-seeded; seeds orbicular.

*Fl.-Fr.:*Almost throughout the year

Burhanpur, Indore.

ELYTRARIA Michx. W. 21; MP 1; SA 1

Elytraria acaulis (L.f.) Lindau, Nat. Pflanzenfam. Nachtr. 1:304. 1897; *Justicia acaulis* L.f., Suppl. Pl. 84. 1781; *Elytraria crenata* Vahl. Enum. Pl. 1:106. 1804; C.B. Clarke in Hook.f., Fl. Brit. India 4:394. 1884; Samvatsar, Fl. West. Trib Madhya Pradesh 224. 1996; Khanna *et al.,* in Mudgal *et al.,* Fl Madhya Pradesh 2:312. 1997. *Dharia*

Scapigerous, fibrous-rooted herbs; leaves oblanceolate to spathulate, 6-25 × 3-9 cm, margins crenate; flowers exceeding the bracts, white or pale blue, in simple or branched, 8-30 cm long spikes; scapes many, clothed with small, rigid, ovate, spinous pointed bracts pubescent inside and ciliate margined; capsules ovoid, 6-8 mm long; seeds ovoid, flattened.

*Fl.-Fr.:*Aug. -Nov.

Alirajpur, Barwani, Dhar, Indore, Khandwa, Khargone.

ERANTHEMUM L. W. 22; MP 3; SA 3

Key to Species

1. Bracts green, nerves many .... **E. purpurascens**
1. Bracts white with green nerves
   2. Leaves 10-20 cm long; capsule 1 cm long .... **E. roseum**
   2. Leaves 5-10 cm long; capsule 1.2-1.6 cm long .... **E. pulchellum**

Eranthemum pulchellum Andrews, Bot. Repos. 2:t. 88. 1800; *Justicia nervosa* Vahl, Enum. Pl. 1:164. 1804; *Eranthemum nervosum* (Vahl) R. Br. ex Roem. & Schult., Syst. Veg. 1:174. 1817; C.B. Clarke in Hook.f., Fl. Brit. India 4:418. 1884; *Daedalacanthus nervosus* (Vahl) T. Anderson, J. Linn. Soc., Bot. 9:487. 1967; Samvatsar, Fl. West. Trib. Madhya Pradesh 224. 1996; Khanna *et al.,* in Mudgal *et al.,* Fl. Madhya Pradesh 2:313. 1997.

Erect perennial herbs ot undershrubs; stem 4-angled; leaves 5-15 × 2-8 cm, ovate-elliptic, acuminate at apex, attenuate at base, margin entire or crenulate; flowers in terminal, closed panicles composed of short, uninterrupted spikes; bracts foliaceous, white with dark-green nerves, elliptic-ovate; bracteoles lanceolate, subacuminate; calyx 9-10 mm long, densely pubescent, lobes lanceolate, unequal; corolla 2.5-3 cm long, lobes 7-8 mm long, rounded; capsules *ca* 1.5 cm long, glabrous; seeds hairy.

*Fl.-Fr.:*March-May

Alirajpur, Dhar, Jhabua.

Eranthemum purpurascens Wight ex Nees, Pl. Asiat. Rar. 3:106. 1832; *Daedalacanthus purpurascens* (Wight ex Nees) T. Anderson, J. Linn. Soc., Bot. 9:488. 1867; C.B. Clake in Hook.f., Fl. Brit. India 4:313. 1884; Khanna *et al.,* in Mudgal *et al.,* Fl. Madhya Pradesh 2:313. 1997. *Dash mooli* **(Plate 8, Fig. 2)**

Erect, stout, herbs or undershrubs, *ca* 1 m tall; stem 4-angled; leaves 13-18 × 6-8 cm, ovate or ovate –elliptic, apex acuminate, base cuneate; flowers in terminal or axillary, 3-8 cm long, dense spikes; bracts ovate-lanceolate, about 2.5 cm long, glandular pubescent; bracteoles linear-lanceolate, glandular pubescent; calyx up to 9 mm long, pubescent, white; corolla bluish-purple, 3-5 cm long, lobes ovate to obovate; capsules glabrous, *ca* 1.5 cm long; seeds about 4 mm long, densely hygroscopically hairy.

*Fl.-Fr.:*Oct. -April

Burhanpur, Indore, Khandwa, Khargone.

Eranthemum roseum (Vahl) R. Br., Prodr. Fl. Nov. Holl. 477. 1810; *Justicia rosea* Vahl, Enum. Pl. 1:165. 1804; *Daedalacanthus roseus* (Vahl) T. Anderson, J. Linn. Soc., Bot. 9:488. 1867; C.B. Clake in Hook.f., Fl. Brit. India 4:419. 1884; Samvatsar, Fl. West. Trib. Madhya Pradesh 225. 1996; Khanna *et al.,* in Mudgal *et al.,* Fl. Madhya Pradesh 2:314. 1997.

Erect, 50-100 cm tall herbs with a 4-angled stem; leaves 10-20 × 4-7 cm, oblong-lanceolate, entire, faintly crenate, apex acuminate, base cuneate; flowers blue, showy, turning pinkish later, in terminal and axillary, linear, 7-15 cm long spikes; bracts white with green veins, *ca* 1.5 cm long, glandular strigose; bracteoles ciliate, linear, pubescent; calyx scarious, white, pubescent; capsules *ca* 1 cm long, clavate, pointed; seeds 3-4 mm across, hygroscopically hairy, very densely so on the margin.

*Fl.-Fr.:*Nov. -Jan.

Common as forest undergrowth in all districts.

HAPLANTHUS Nees

(Both species described below are unresolved according to The Plant List)

Key to Species

1. Stem glandular hairy; cladodes less than 3 cm long, apex not spiny; bracteoles shorter than calyx; capsules pubescent .... **H. tentaculatus**

1. Stem glabrous; cladodes more than 3 cm long, apex with 2, sharp spines; bracteoles longer than calyx; capsules glabrous .... **H. verticillaris**

Haplanthus tentaculatus Nees, Prodr. 11:513. 1847; *H. nilgherrensis* Wight, Icon. Pl. Ind. Orient. 4:t. 1556. 1850; *Bremekampia tentaculata* (L.) Sreem., Bull. Bot. Surv. India 6:324. 1965; *Ruellia tentaculata* L., Cent. Pl. II 22. 1756; C.B. Clarke in Hook.f., Fl. Brit. India

4:506. 1884; Samvatsar, F. West. Trib. Madhya Pradesh 225. 1996; Khanna *et al.,* in Mudgal *et al.,* Fl. Madhya Pradesh 2:316. 1997.

Slender herbs; stems quadrangular in the upper part, glandular pubescent throughout; leaves 2-14 × 1.5-7 cm, ovate, acuminate; axillary spines (cladodes) shortly, densely villous, with few long hairs; flowers sessile, in between the cladodes; bracteoles subulate, shorter than the calyx; calyx segments linear-subulate, pubescent; corolla blue or lilac; capsules *ca* 1 cm long, hairy.

*Fl.-Fr.:*Dec. -Jan.

Common as forest undergrowth, all over the area.

Haplanthus verticillaris Nees, Prodr. 11:513. 1847; C.B. Clarke in Hook.f., Fl. Brit. India 4:506. 1884; *Dianthera verticillaris* (Nees) Stehle & M. Stehle, Fl. Agron. Antill. Franc. 1:106. 1957; Samvatsar, Fl. West. Trib. Madhya Pradesh 225. 1996; Khanna *et al.,* in Mudgal *et al.,* Fl. Madhya Pradesh 2:317. 1997. *Kastula*

Erect annual herbs, about a meter tall; leaves pubescent, 5-12 × 2-7 cm, ovate-oblong, apex acute or acuminate; cladodes up to 4 cm long, 4-angled, 10-15 at each node, with 2 spinous teeth at the apex; flowers sessile, arising in between he cladodes; bracteoles longer than the calyx; segments of calyx 4 mm long, lanceolate, pointed, glandular pubescent; corolla lobes, violet, tube white, *ca* 1.5 cm long; capsules grooved, glabrous, *ca* 1.5 cm long, green with white horns; seeds compressed, brownish-yellow, hygroscopically densely hairy.

*Fl.-Fr.:*Nov. -March

Common throughout the area in moist, shady places like gullies in the forest.

HEMIGRAPHIS Nees W. 22; MP 4; SA 4

Key to Species

1. Filaments of longer stamens glabrous above; plants densely strigose; flowers in involucrate heads; corolla tube ventricose; petals clothed in brown hairs; capsule glabrous, grooved, 6-8 seeded .... **H. urens**
1. Filaments of longer stamens densely bearded above; plants not strigose; corolla funnel-shaped; capsule pubescent, 6-12 seeded
   2. Creeping plants; flowers in 2-6 flowered spikes; capsule up to 12-seeded .... **H. hirta**
   2. Erect plants
      3. Flowers terminating axillary branches; plants not viscid; capsule 6-seeded .... **H. latebrosa**
      3. Flowers in dense, axillary and terminal heads; plants viscid; capsule 6-8 seeded .... **H. rupestris**

Hemigraphis hirta (Vahl) T. Anderson, J. Linn. Soc., Bot. 9:462. 1867; *Ruellia hirta* Vahl, Symb. Bot. 3:84. 1794; C.B. Clarke in Hook.f., Fl. Brit. India 4:422. 1884; Khanna *et al.,* in Mudgal *et al.,* Fl. Madhya Pradesh 2:318. 1997.

Annual, prostrate, much branched, hirsute-villous herbs; leaves 1-2 × 0.5-1.3 cm, broadly ovate or elliptic, apex acute or obtuse, margin crenate; flowers violet, in condensed, 2-6 flowered heads; bracts wooly, elliptic, 10-15 mm long; bracteoles absent; calyx segments linear, *ca* 8 mm long; corolla fugacious, 10-15 mm long; capsules glabrous, linear-cylindrical, *ca* 1 cm long, about 12 seeded; seeds brownish.

*Fl.-Fr.:*April-July

Barwani, Burhanpur, Khargone.

Hemigraphis latebrosa (Roth) Nees, Prodr. 11:723. 1847; *Ruellia latebrosa* Roth, Nov. Pl. Sp. 303. 1821; C.B. Clarke in Hook.f., Fl. Brit. India 4:423. 1884; Samvatsar, Fl. West. Trib. Madhya Pradesh 226. 1996; Khanna *et al.,* in Mudgal *et al.,* Fl. Madhya Pradesh 2:318. 1997.

Diffuse, annual herbs, viscidly pubescent, *ca 50* cm high; leaves 2-7 × 1-4, dimorphic, ovate-oblong, apex acuminate, base cuneate or truncate, leaves of side branches smaller in overall dimensions; flowers blue, usually borne on amall axillary branches; bracts ovate or elliptic, petaloid; sepals unequal, 6-8 mm long, pubescent; corolla 1.5 2 cm long, tube yellowish, limb blue, bearded inside; capsules *ca* 6 mm long, 6-seeded, brown, pubescent; seeds covered with white hairs, 1-1.5 mm across.

Fl.-Fr.: Dec. -Jan.

Common forest undergrowth in all districts.

Hemigraphis rupestris T. Anderson, J. Linn. Soc., Bot. 9:462. 1867; *H. latebrosa* (B. Heyne ex Roth) Nees, var. *rupestris* (B. Heyne ex T. Anderson) C.B. Clarke in Hook.f., Fl. Brit. India 4:424. 1884; Khanna *et al.,* in Mudgal *et al.,* Fl. Madhya Pradessh 2:319. 1997.

Perennial, erect or decumbent, aromatic, glandular-pubescent herbs; stem 4-angled; leaves 4-10 × 1-6 cm, ovate-elliptic or lanceolate, apex acute to acuminate, base rounded or cuneate; flowers blue, sessile, in terminal or axillary spikes; bracts foliaceous, imbricate, ovate-lanceolate, 5-12 mm long; calyx *ca* 5 mm long, glandular pubescent, segments ciliate; corolla *ca* 1 cm long, lobes ovate, tube yellow; capsules *ca* 5 mm long, beaked, 6-seeded.

*Fl.-Fr.:*Nov. -April

Common, same localities as above.

Hemigraphis urens (Roth) J.R.I. Wood, Edinburgh J. Bot. 51 (1):108. 1994; *Ruellia urens* Roth, Nov. Pl. Sp. 302. 1821; *Gantelbua urens* (B. Heyne ex Roxb.) Bremek., Verh. Kon. Ned. Akad. Wentensch. Afd. Natuurk., Sect. 2 41(1) : 148. 1944; C.B. Clarke in Hook.f., Fl. Brit. India 4:422. 1884; Samvatsar, Fl. West. Trib. Madhya Pradesh 225. 1996; Khanna *et al.,* in Mudgal *et al.,* Fl. Madhya Pradesh 2:315. 1997. (as *Gantelbua urens*)

Procumbent or diffuse, strigosely hairy herbs; leaves 1.5-4 × 1-2 cm, spathulate-oblong, hispid, apex obtuse; flowers blue, in terminal, many flowered spikes, densely covered with white and yellowish-brown hairs; bracts spathulate, pubescent, ovate-oblong, 2-3 cm long; calyx *ca* 1.2 cm long, densely pubescent, ciliate; corolla *ca* 1.5 cm long, tube slender, lobes unequal; capsules *ca* 1.5 cm long, glabrous. 6-8 seeded; seeds ellipsoid or orbicular, yellowish-brown.

*Fl.-Fr.:*Jan. -March

Common in dry ditches, outskirts of village roads, all over the area.

HYGROPHILA R. Br. emend Heine W. 46; MP 5; SA 3

Key to Species

1. Armed hispid plants; leaves in whorls of 6, up to 15 cm long; flowers in congested axillary whorls; bracteoles absent; sepals 4; seeds 4-6 .... **H. auriculata**

1. Unarmed shrubs; leaves opposite; flowers in axillary clusters or spikes

  2. Fertile stamens 2; capsules 20 or more seeded, pubescent .... **H. polysperma**

  2. Fertile stamens 4; seeds 8-15

    3. Leaves more than 8 cm long; plants not rooting at nodes .... **H. phlomoides**

    3. Leaves up to 3 cm long; plants rooting at lower nodes .... **H. serpyllum**

Hygrophylla auriculata (Schumach.) Heine, Kew Bull. 16:172. 1963; *Barleria auriculata* Schumach., Beskr. Guin. Pl. 285. 1827; *B. longifolia* L. Cent. Pl. II 22. 1756; *Hygrophila spinosa* T. Anderson, J. Proc. Linn. Soc., Bot. 7:22. 1804; *Asteracantha longifolia* Nees, Pl. Asiat. Rar. 3:90.1832; C.B. Clarke in Hook.f., Fl. Brit. India 4:408. 1884; *H. schulii* M.R. Almeida & S. M. Almeida, J. Bombay Nat. Hist. Soc. 83 (Suppl.):221. 1987; Samvatsar, Fl. Madhya Pradesh 226. 1996; Khanna *et al.,* in Mudgal etal, Fl. Madhya Pradesh 2:321. 1997. *Tal makhana* (seeds)

Annual or perennial herbs, about 1 m tall; stem stout, nodes thickened with axillary whorls of 1.5-3 cm long, hispid, yellow spines (thorns); leaves 5-15 × 2-3 cm, 6 in a whorl, subsessile, oblong or linear-lanceolate; flowers blue, in dense clusters in the axils of leaves, usually 8 in a whorl; bracts 2-seriate, up to 3 cm long, linear-lanceolate, ciliate; bracteoles linear-lanceolate, smaller; calyx 4-partite, hairy, segments linear-lanceolate; corolla glabrous, tube *ca* 1.2 cm long, widening abruptly; capsules *ca* 8 mm long, pointed, 4-8 seeded; seeds orbicular *ca* 3 mm across.

*Fl.-Fr.:*Aug. -Dec.

Very common in roadside ditches, moist places, in all districts.

(**Note:** K. Subramanyum, Aquatic Angiosperms, 1961, has thoroughly explained the morphology of spines in this species. The axillary buds of two opposite species are transformed into thorns,rather than branches. These two thorns, at their base on either side bear two more leaves, so that there are 4 smaller leaves. These smaller leaves in their axils bear more thorns, shorter than the first 2. At the base of these 4 smaller thorns, 2 more leaves develop and their axillary buds later develop into flowers, so that there are 3 whorlsof leaves, arising one above the other, as do the floral leaves in a flower, the lowermost whorl consisting of 2 big,opposite leaves, each developing 3 thorns in their axils, 6 leaves and 4 flowers. As the uppermost or innermost whorl of small leaves bear flowers in their axils, they can be regarded as bracts. All these structures are sessile and crowded due to very short internodes to form clusters. Thus, 2 larger and 4 smaller thorns at each node represent modified axillary branches of the first and second order).

Hygrophila phlomoides Nees, Pl. Asiat. Rar. 3:80.1832; C.B. Clarke in Hook.f., Fl. Brit. India 4:408. 1884; Samvatsar, Fl. West. Trib. Madhya Pradesh 227. 1996; Khanna *et al.,* in Mudgal *et al.,* Fl. Madhya Pradesh 2:322. 1997.

Stout, erect herbs; stem striate, glandular pubescent; leaves 7-15 cm long, sessile, linear-oblong, pubescent; flowers purple, in dense, villous, axillary clusters, one sepal larger than the others in fruit; *ca* 12 mm long; corolla *ca* 12 mm long; capsules equal in length to calyx and corolla, 10-12 seeded; seeds ovoid, abruptly acuminate, compressed, margined.

*Fl.-Fr.:*Oct. -Dec.

Not common, in moist localities, Burhanpur, Dhar.

Hygrophila polysperma (Roxb.) T. Anderson, J. Linn. Soc., Bot. 9:456. 1867; *Justicia polysperma* Roxb., Fl. Ind. ed. 1820 1:120.1820; *Hemiadelphis polysperma* (Roxb.) Nees, Pl. Asiat. Rar. 3:80.1832; C.B. Clarke in Hook.f., Fl. Brit. India4:406. 1884; Samvatsar, Fl. West. Trib. Madhya Pradesh 226. 1996; Khanna *et al.,* in Mudgal *et al.,* Fl. Madhya Pradesh 2:324. 1997.

Erect or procumbent, pubescent herbs; stem rooting at lower nodes; leaves 1-2.5 × 0.5-1 cm, subsessile, acute or obtuse at apex, tapering ar base; flowers blue, in terminal, dense spikes; bracts foliaceous hairy, elliptic-lanceolate; bracteoles linear; calyx 5-partite , segments scarious, linear-lanceolate, margin ciliate; corolla hairy, *ca* 7 mm long; capsules linear-oblong, *ca* 7 mm long, brown, pubescent; seeds 20-30, orbicular.

*Fl.-Fr.:*Dec. -Jan.

Common in moist places all over the area.

Hygrophila serpyllum T. Anderson, J. Linn. Soc., Bot. 9:456. 1867; *Ruellia polysperma* Roth. Nov. Pl. sp. 305. 1821; *Physichilus serpyllum* Nees, Compan. Bot. Mag. 2:311. 1837; *Adenosma polysperma* Spreng., Syst. Veg. 2:829. 1825; C.B. Clarke in Hook.f., Fl. Brit. India 4:406. 1884; Samvatsar, Fl. West. Trib. Madhya Pradesh 227. 1996; Khanna *et al.,* in Mudgal *et al.,* Fl. Madhya Pradesh 2:325. 1997.

Annual, procumbent, pubescent herbs; stem often rooting at the nodes; leaves 1-4 × 0.5-2 cm, elliptic-lanceolate, apex acute, subsessile, pubescent; flowers bright purple, in dense, leafy spikes; bracts and bracteoles pubescent, elliptic-lanceolate, 5-8 mm long; calyx 5-partite, segments densely ciliate; corolla throat white, bearded; capsules linear-oblong, 5-8 mm long, glabrous, pointed; seeds many, ovoid-orbicular.

*Fl.-Fr.:*Jan. -April

Common in moist places in all districts.

JUSTICIA L. W. 652; MP 11; SA 10

Key to Species

1. Sepals 5; bracts longer than calyx; undershrubs
    2. Bracts scarious with green nerves; flowers white with red-purple dots; capsules *ca* 1 cm long .... **J. betonica**
    2. Bracts not scarious; flowers uniformly white; capsules 2-3 cm long .... **J. adhatoda**
1. Sepals 4
    3. Spikes cylindric; bracts and bracteoles equal to or longer than sepals; margins scarious or not
        4. Bracts and bracteoles very slender or linear
            5. Bracts and bracteoles very slender with long, spreading white, bulbous-based hairs .... **J. segoviaensis**

5. Bracts and bracteoles linear, pectinately white-ciliate, scabrous and hirsute .... **J. procumbens**
4. Bracts and bracteoles elliptic to ovate-lanceolate; plants diffuse
6. Bracts 2.5 mm or more broad; leaves up to 2.5 cm long, main nerves about 6 pairs; plants pale when dry .... **J. japonica**
6. Bracts much less than 2.5 mm broad; leaves up to 5 cm long, main nerves about 4 pairs; plants black when dry .... **J. mollissima**
3. Spikes linear; bracts and bracteoles shorter than sepals, midrib green and margins scarious
7. Bracts, bracteoles and sepals obtuse at apex .... **J. quinqueangularis**
7. Bracts, bracteoles and sepals acute at apex
8. Bracts and bracteoles hardly 1-1.2 mm long, ovate or lanceolate .... **J. diffusa**
8. Bracts and bracteoles nearly 2.5 mm long, lanceolate
9. Plants prostrate, rooting at nodes; spikes short; leaves up to 1.2 cm long, orbicular to ovate .... **J. prostrata**
9. Plants erect; spikes long, leaves up to 7.5 cm long, linear to lanceolate .... **J. vahlii**

Justicia adhatoda L., Sp. Pl. 15. 1753; *Adhatoda vasica* Nees, Pl. Asiat. Rar. 3:103. 1832; *Adhatoda zeylanica* Medik., Hist. & Commentat. Acad. Elect. Sci. Theod.-Palat. 6:393. 1790; C.B. Clarke in Hook.f., Fl. Brit. India 4:540.1885; Samvatsar, Fl. West. Trib. Madhya Prsdesh 221. 1996; Khanna *et al.,* in Mudgal *et al.,* Fl. Madhya Pradesh 2:284. 1997. *Adusa*

Perennial, evergreen, shrubs, up to 1-1.5 m tall; leaves 6-15 × 2-7 cm, elliptic-lanceolate or ovate-lanceolate, apex acuminate, base cuneate or decurrent, margin subentire or undulate-crenulate; flowers white, in leafy spikes; bracts foliaceous, elliptic-obovate to ovate, apiculate, ciliate; upper lip of corolla curved, notched at apex, lower lip with rose-purple lines, throat villous; capsules clavate, 2-3 cm long, dark brown, 4-seeded; seeds sub-orbicular, rugose.

*Fl.-Fr.:*Jan. -April

Common along stream and river banks, also planted as an absolutely ungrazable hedge.

Justicia betonica L., Sp. Pl. 15. 1753; *Adhatoda betonica* (L.) Nees, Pl. Asiat. Rar. 3:103. 1832; *Justicia trinervia* Vahl, Enum. Pl. 1:156. 1804; C.B. Clarke in Hook.f., Fl. Brit. India 4:525. 1885; Samvatsar, Fl. West. Trib. Madhya Pradesh 228. 1996; Khanna *et al.,* in Mudgal *et al.,* Fl. Madhya Pradesh 2:328. 1997. *Mokandar*

Undershrubs, up to 1 m tall; branches swollen above nodes; leaves glabrous, 4-10 × 2-4 cm, ovate-lanceolate, apex acuminate, base cuneate, margin entire or crenate; flowers red with purple dots, in terminal, simple or branched spikes; bracts foliaceous, acuminate, white with green nerves, up to 1.5 cm long; bracteoles more or less similar; calyx 5-partite, segments linear-lanceolate, subequal; corolla *ca* 1.5 cm long; capsules clavate, *ca* 1.5 cm long, pubescent, beaked; seeds 4, suborbicular, *ca* 4 mm across.

*Fl.-Fr.:*Aug. -April

Rare, **Dhar** only, planted as hedge for flower beds and now escape.

Justicia diffusa Willd., Sp. Pl. 1:87. 1797; *Rostellaria diffusa* (Willd.) Nees, Pl. Asiat. Rar. 3:100.1832; *Rostellularia diffusa* (Willd.) Nees, Prodr. 11:37. 1847; C.B. Clarke in Hook.f.,

Fl. Brit. India 4:538. 1885; Samvatsar, Fl. West. Trib. Madhya Pradesh 228. 1996; Khanna *et al.,* in Mudgal *et al.,* Fl. Madhya Pradesh 2:328. 1997.

Small, decumbent herbs, about 30 cm tall, much branched; leaves 1.5-3.5 × 0.8-1.5 cm, broadly elliptic, apex acute or acuminate, base rounded or cuneate, margin entire, petioles 5-10 mm long; flowers pink, in axillary and terminal narrow spikes, up to 15 cm long; bracts ovate, acuminate, 1-1.5 mm long, margins scarious; bracteoles similar to bracts but narrower; calyx 4-partite to base, 2.5 3 cm long, segments acute with scarious margins; corolla 3.5-4 mm long, lobes short, obtuse; capsules glabrous, elliptic-oblong, 4-5 mm long, pointed and grooved on back; seeds 4, orbicular, 1 mm across, subconcentrically rugose.

*Fl.-Fr.:*Sept. -March

Common in shady places throughout the area.

Justicia japonica Thunb., Syst. Veg. ed. 14 63. 1784; *J. simplex* D. Don, Prodr. Fl. Nepal. 118. 1825; C.B. Clarke in Hook.f., Fl. Brit. India 4:539. 1885; *Rostellularia japonica* (Thunb.) J.L. Ellis, Bull. Bot. Surv. India 22(1-4):196. 1980 [Flora Madhya Pradesh, 2:332 (1997) has treated this species and *J. mollissima* (Nees) Y.F. Deng & T.F. Daniel, 2011, under *J. simplex* D. Don]

Slender herbs, 15-45 cm tall, branching from the base; branches grooved, hairy; leaves 3-7 × 1.5-2.5 cm, ovate to elliptic, apex and base acute, petioles 4-5 mm long; flowers pink, in dense, axillary and terminal, cylindric, 2-6 cm long spikes; bracts 3-4 mm long, acute, hairy; bracteoles equal to bracts in length, but narrower, all with scarious margins, with many jointed hairs; calyx 4-5 mm long, hairy, divided almost to the base, segments with scarious, densely ciliate margins; corolla 6-8 mm long, tube as long as the 2-lipped limb; capsules 4-6 mm long, acute and pubescent at the apex, stalk solid; seeds orbicular, 1 mm across, concentrically rugose.

*Fl.-Fr.:*Aug. -Oct.

Common in all districts on gravelly slopes of hills and near the banks of streams.

Justicia mollissima (Nees) Y.F. Deng & T.F. Daniel, Fl. China 19:451-452.2011; *Rostellaria mollissima* Nees, Pl. Asiat. Rar. 3:101. 1832; *Rostellularia mollissima* (Nees) Nees, Prodr. 11:373. 1847; *Justicia procumbens* var. *latispica* C.B. Clarke in Fl. Brit. India 4:539. 1885; Samvatsar, Fl. West. Trib. Madhya Pradesh 228. 1996; Khanna *et al.,* in Mudgal *et al.,* Fl. Madhya Pradesh 2:329. 1997.

Diffuse herbs or undershrubs, stem covered with stiff, strigose scattered hairs; leaves 4-5 × 2-2.5 cm, elliptic or ovate, main nerves 4 pairs, petioles 2-3 mm long; flowers pale lilac, in terminal and axillary, cylindrical, 3-6 cm long spikes; bracts lanceolate, 6-7 mm long, strongly ciliate along scarious margins; bracteoles similar, but narrower; calyx 4-partite, segments divided almost to the base, sepala pectinately ciliate, about equalling in length to the bracts; corolla 10-12 mm long, pubescent outside, tube 7-8 mm long, limb 2-lipped, lobes rounded, short, palate with 2 white ridges; capsules oblong, 5-6 mm long, acute, glabrous; seeds 4, ovoid, tubercled, brown.

*Fl.-Fr.:*Sept. -Feb.

Rare, confined to **Indore** only, Nehru Park (Indore city) as a weed.

Justicia procumbens L., Sp. Pl. 15. 1753; *Rostellaria procumbens* Nees, Pl. Asiat. Rar. 3:101. 1832; *Rostellularia procumbens* (L.) Nees, Prodr. 11:371. 1847; C.B. Clarke in Hook.f., Fl. Brit. India 4:539. 1885; Khanna *et al.,* in Mudgal *et al.,* Fl. Mdhya Pradesh 2:330.1997.

Annual, erect or diffuse herbs, rooting at lower nodes; branches zigzag; leaves petiolate, glabrous or pubescent, 1.5-3.5 × 0.5-1 cm, elliptic, ovate or lanceolate, apex acute or obtuse, base acute; flowers pale-purple, in 1.5-3 cm long, terminal, slender spikes; bracts and bracteoles pubescent, linear-lanceolate, margins scarious; calyx 4-partite, segments linear-lanceolate, margins ciliate; corolla pubescent outside, upper lip broadly ovate, notched above, lower lip shortly 3-lobed, tube as long as the limb; capsules *ca* 5 mm long, stalked, beaked, hairy at the tip; seeds small, tuberculate.

*Fl.-Fr.:*Dec. -Jan.

Alirajpur, Burhanpur, Dhar, Indore, Khandwa.

Justicia prostrata (C.B. Clarke) Gamble, Fl Madras 2: 1081. 1924; *J. prostrata* Willd var. prostrata C.B. Clarke in Hook.f., Fl. Brit. India 4: 538.1885; *Rostellularia prostrata* (C.B. Clarke) Majundar in Bull. Bot. Ser. Bengal 25: 75. 1971; Samvatsar, Fl. West. Trib. Madhya Pradesh 230.1996; Khanna et al., Fl. Madhya Pradesh 2: 331.1997.

Annual, slender, prostrate herbs with spreading branches; leaves 0.7-3 × 0.5-1.5 cm, ovate-elliptic or suborbicular, apex acute, base rounded, margins entire; flowers purple, in axillary and terminal, erect, 3-8 cm long spikes; bracts and bracteoles lanceolate, pubescent, shorter than calyx; calyx 4-partite, segments lanceolate, scarious, ciliate along midrib and margin; corolla *ca* 7 mm long; capsules *ca* 3 mm long, oblong-ellipsoid, grooved, beaked, pubescent; seeds black, tuberculate.

*Fl.-Fr.:*Oct. -March

Not common, Alirajpur, Barwani, Khargone.

Justicia quinqueangularis K. D. Koenig ex Roxb., Fl. Ind. 1:134. 1820; *J. peploides* (Nees) T. Anderson, J. Linn. Soc., Bot. 9: 511. 1867; *Rostellaria peploides* Nees, Pl. Asiat. Rar. 3:101. 1832; *Rostellularia quinqueangularis* (K. D. Koenig ex Roxb.) Nees, Prodr. 11: 375. 1847; C.B. Clarke in Hook.f., Fl. Brit. India 4:538. 1885; Samvatsar, Fl. West. Trib. Madhya Pradesh 230.1996; Khanna *et al.,* in Mudgal *et al.,* Fl. Madhya Pradesh 2:331. 1997.

Annual, erect, ascending, glabrous herbs about 50 cm tall; stems 4-5 angled; leaves 2.5-7.5 × 0.3-1 cm, pairs distantly placed, ovate-oblong, linear or linear-lanceolate, apex obtuse or subacute, base rounded or cuneate; flowers in slender, white with purplish blotch, in terminal, lax, elongated or short spikes; bracts and bracteoles linear, glabrous, 3-4 mm long; calyx 4-partite nearly to the base, segments 4-5 mm long, linear, equal, glabrous; capsules ellipsoid or oblong, stalked, *ca* 8 mm long, pointed; seeds 4, reticulately rugose.

*Fl.-Fr.:*Aug. -Nov.

Common in moist, shady places all over the area.

Justicia segoviaensis P. Soumya & Sunojk., Phytotaxa 238(2): 222.2018; *Rostellaria crinita* Nees, Pl. Asiat. Rar. 3:101. 1832; *Rostellularia crinita* (Nees) Nees, Prodr. 11:373. 1847; Wall. ex C.B. Clarke in Fl. Brit. India 4:536. 1885; *Justicia neesii* Ramamoorthy, Fl. Hassan

Dist. 551. 1976; Khanna *et al.,* in Mudgal *et al.,* Fl. Madhya Pradesh 2:330.1997. [*J. micrantha* (Oerst.) V.A.W. Graham, Kew Bull. 43:603. 1988, is a later homonym *of J. micrantha* Wall. ex C.B. Clarke, 1885; *Chaetothylopsis micrantha* Oerst., Vidensk. Meddel. Naturhist. Foren. Kjobenhavn 1854:163. 1855, therefore, P. Soumya and Sunojk., have made the above combination, which is accepted here]

Subshrubby perennials, about 60 cm tall; leaves 2.5-6 × 1.5-3 cm, broadly elliptic to obovate, apex acute, base acute to attenuate, margin entire, petioles up to 2 cm long; spikes termina and in the forks of branches, up to 2 cm long; bracts and bracteoles 4-5 mm long, margins scarious; calyx lobes unequal, scarious; corolla pink, up to 6 mm long; capsules 4 mm long, oblong; seeds *ca* 1 mm across, smooth or papillose.

*Fl.-Fr.:*Aug. -March

Common on hilly slopes in all districts.

Justicia vahlii Roth, Nov. Pl. Sp. 14. 1821; *Rostellaria vahlii* (Roth) Nees, Pl. Asiat. Rar. 3:102. 1832; *Rostellularia vahlii* (Roth) Nees, Prodr. 11:376. 1847; *J. diffusa* var. *vahlii* (Roth) C.B. Clarke in Hook.f., Fl. Brit. India 4:538. 1885; Samvatsar, Fl. West. Trib. Madhya Pradesh 230.1996; Khanna *et al.,* in Mudgal *et al.,* Fl. Madhya Pradesh 2:333. 1997.

Erect or procumbent herbs, up to 30 cm tall, stem scabrid on the angles, rooting at lower nodes; leaves 3-8 × 0.5-2 cm, apex acute to acuminate, base cuneate, margin entire, petioles 5-10 mm long, scabrid; flowers pink, in axillary and terminal, up to 3 cm long spikes; bracts and bracteoles lanceolate, *ca* 2 mm long, shorter than calyx, scarious and ciliate on the margin; calyx 4-partite, *ca* 3 mm long, scarious, ciliate; corolla *ca* 5 mm long, throat with dark pink markings; capsules beaked, oblong-ellipsoid, 3-4 mm long, pubescent; seeds slightly flattened, prominently tuberculate.

*Fl.-Fr.:*Oct. -March

Rare, along streams, **Alirajpur** only, Aamkhut forest.

LEPIDAGATHIS Willd. W. 71; MP 7; SA 2

Key to Species

1. Calyx 5-partite; leaves 3-nerved; main stem well developed, erect; flowers aggregated near the base as well as in axils of lower leaves .... **L. trinervis**
1. Calyx 4-partite; leaves peninerved; main stem hardly any, lateral branches spreading on all sides; flowers aggregated in a congested bunch at the base, very rarely in axils of upper leaves .... **L. cristata**

Lepidagathis cristata Willd., Sp. Pl. 3:400; 1800; *Acanthus cristata* Rottler ex Nees, Prodr. 11:256. 1847; C.B. Clarke in Hook.f., Fl. Brit. India 4:516. 1885; Samvatsar, Fl. West. Trib. Madhya Pradesh 231. 1996; Khanna *et al.,* in Mudgal *et al.,* Fl. Madhya Pradesh 2:335. 1997.

Perennial, procumbent undershrubs, branching from the base; branches spreading radially, almost winged; leaves 2-4 × 0.5-0.8 cm, sessile, ovate-oblong, apex acute or obtuse, base rounded, margins spinous- serrate; flowers pale pink, in globose clusters at the base of stems;

bracts ovate, acuminate, 8-9 mm long, spinous tipped, haity; bracteoles similar to bracts but membranous; calyx 4-partite, divided almost to the base, 6-8 mm long, lower sepals often 2-fid, all ciliate and softly hairy, spinous pointed; corolla with brown dots, 10-12 mm long, tube narrow, limb 2-lipped; capsules 4-5 mm long, glabrous, acute; seeds ovoid-oblong, 2-3 mm long, densely clothed with hygroscopic hairs.

*Fl.-Fr.:*Oct. -Dec.

Common at the base of gravelly hills all over the area.

Lepidagathis trinervis Nees, Pl. Asiat. Rar. 3:96. 1832; C.B. Clarke in Hook.f., Fl. Brit. India 4:517. 1885; Samvatsar, Fl. West. Trib. Mdhya Pradesh 231. 1996; Khanna *et al.,* in Mudgal *et al.,* Fl. Madhya Pradesh 2:339. 1997.

Procumbent or diffuse herbs up to 25 cm high, with a woody root-stock; branches slender, 4-angled; leaves sessile, 2.5-6 × 0.5-1 cm, linear or linear-lanceolate, apex acute, base tapering, margins entire, scabrid; flowers pink to purple, in axillary spikes crowded at the base of stem into a single head, spikes 1.5-2 cm long; bracts ovate-elliptic, spinous-cuspidate; bracteoles oblong-lanceolate, aristate; calyx segments 5, spinous tipped, 8-10 mm long, densely hairy on the back; corolla throat with transverse streaks and mottled with yellow, 6-9 mm long, densely white-hairy; capsules compressed, *ca* 6 mm long, beaked; seeds 2, 2.5-3 mm long, oblong, rounded at apex clothed with long, mucilaginous, hygroscopic hairs.

*Fl.-Fr.:*Sept. -Feb.

Common on dry rocky or sandy grounds, in all districts.

NELSONIA R. BR.W. 3; MP 1; SA 1

Nelsonia canescens (Lam.) Spreng., Syst. Veg. 1:42. 1824; *Justicia canescens* Lam., Tabl. Encycl. 1:41. 1791; *Nelsonia campestris* R. Br. Prodr. F. Nov. Holl. 481. 1810; C.B. Clarke in Fl. Brit. India 4:394. 1884; Khanna *et al.,* in Mudgal *et al.,* Fl. Madhya Pradesh 2:339. 1997. *Chhoti aandhi*

Perennial, erect or diffuse herbs, up to 50 cm tall; young parts villous; roots tuberous, fleshy; lower leaves large, 10-20 cm long, petiolate, upper small, subsessile, 1-5 cm long, elliptic-oblong to obovate, canescent; flowers pink or blue, in axillary and terminal, 1-5 cm long, grey woolly spikes; bracts ovate-obovate, 4-6 mm long; bracteoles represented by 2 opposite tuft of hairs; calyx segments 5-6 mm long, glandular ciliate; corolla 7-8 mm long; capsules ovoid, conical, *ca* 6 mm long, glabrous, 8-12 seeded; seeds broadly ovoid, brown, granular.

*Fl.-Fr.:*April-June

Rare, **Khandwa** only, Kalibhit area.

PERISTROPHE Nees W. 20; MP 2; SA 1

Peristrophe bicalyculata (Retz.) Nees, Pl. Asiat. Rar. 3:113. 1832; *Dianthera bicalyculata* Retz., Kongl. Vetensk. Acad. Handl. 1775:297. 1775; *Justicia bicalyculata* (Retz.) Vahl, Symb. Bot. 2:13. 1791; C.B. Clarke in Fl. Brit. India 4:554. 1885; Samvatsar, Fl. West. Trib. Madhya Pradesh 231. 1996; Khanna *et al.,* in Mudgal *et al.,* Fl. Madhya Pradesh 2:345. 1997. *Atrilal*

Annual, erect, much branched hispid, herbs about 1 m tall; stem and branches 6-angled; leaves 4-8 × 2-3.5 cm, ovate or ovate-lanceolate, apex acute to acuminate, base rounded, margins subentire or crenulated, ciliate, petioles 6-12 mm long; flowers rosy or rosy-pink, in axillary and terminal, lax, divaricate, paniuculate trichotomous cymes; bracts 2, linear, very unequal, 6-10 mm long; bracteoles 4, similar to bracts, but shorter, subequal; calyx 5-partite almost to the base, segments 3-4 mm long; corolla 10-12 mm long, pubescent outside, deeply 2-lipped, lobes acute, short; capsules oblong, 6-7 mm long, pubescent; seeds 4, orbicular, 2-3 mm across, papillose and rugose.

*Fl.-Fr.:*Sept. -Feb.

Common on dry rocky or sandy ground, all over the area.

PETALIDIUM Nees W. 34; MP 1; SA 1

Petalidium barlerioides (Roth) Nees, Pl. Asiat. Rar. 3:82. 1832; *Ruellia barlerioides* Roth, Nov. Pl. Sp. 310.1821; *R. bracteata* Roxb., Fl. Ind. ed. 1832 3:47. 1832; C.B. Clarke in Fl. Brit. India 4:416. 1884; Khanna *et al.,* in Mudgal *et al.,* Fl. Madhya Pradesh 2:345. 1997.

Erect or scandent undershrubs, 30-100 cm tall; stem subterete, glabrous; leaves on the main stem broadly ovate 5-10 × 3-6 cm, apex acuminate, base rounded, margin crenate-serrate, glabrous above, pubescent on the nerves beneath, petioles 1-2 cm long, upper leaves smaller; flowers white, solitary, axillary, crowded at the ends of branchlets, sessile or shortly pedicellate; bracts absent; bracteoles foliaceous, ovate-elliptic, up to 2.5 cm long; calyx segments glandular hairy, 5-partite to the base, 8-10 mm long, lanceolate-subulate; corolla 3-3.5 cm long, pubescent outside, strongly nerved, tube bent upwards, lobes 10-12 mm long, margin irregularly crenate; capsules clavate, 1.5 -2 cm long, glabrous, pointed; seeds 2, orbicular, 4-5 mm across, softly villous.

*Fl.-Fr.:*Jan. -May

Rare, Barwani, Khandwa, Khargone.

RUELLIA L. W. 269; MP 3; SA 2

Key to Species

1. Plants prostrate or decumbent, up to 50 cm tall; leaves 2-8 cm long; bracteoles 1-3 cm long .... **R. prostrata**
1. Plants erect, up to 60 tall; leaves 2-6 cm long; bracteoles up to 1 cm long .... **R. patula**

Ruellia patula Jacq., Misc. Austriac. 2:358. 1779; *Dipteracanthus patulus* (Jacq.) Nees, Pl. Asiat. Rar. 3:82. 1832; C.B. Clarke in Hook.f., Fl. Brit. India 4:412. 1884; Samvatsar, Fl. West. Trib. Madhya Pradesh 223. 1996; Khanna *et al.,* in Mudgal *et al.,* Fl. Madhya Pradesh 2:307. 1997 (as *Dipteracanthus patulus*).

Erect, branched, hoary pubescent woody herbs, up to 50 cm high; internodes long; leaves 3.5-4.5 × 1-2-1-7 cm, ovate or ovate-elliptic, apex obtuse, base rounded or subacute, margin entire, hirsute, petioles 6-15 mm long; flowers solitary or 2-3 together, pale purplish, subsessile; bracteoles foliaceous, elliptic, 10-12 mm long, pubescent, margin ciliate; calyx 5-6 mm long, divided up to 2/3 its length, densely pubescent, teeth acute, linear-lanceolate; corolla 2.5-3 cm long, lobes pubescent outside, tube narrow in the lower part, lobes ovate, 3.5-4 mm long; capsules clavate, 12-15 mm long, glabrous, base solid, *ca* 12 seeded; seeds 2 (at maturity), suborbicular, flat, glabrous with hygroscopic hairs along margins.

*Fl.-Fr.:*Sept. -Oct.

Not common, on wet, rocky soil, Alirajpur, Dhar, Indore, Khandwa.

Ruellia prostrata Poir., Encycl. 6:349. 1804; *Dipteracanthus prostratus* (Poir.) Nees, Pl. Asit. Rar. 3:81. 1832; *D. dejectus* Nees, Pl. Asiat. Rar. 3:82. 1832; *D. prostrata* var. *dejecta* (Nees) C.B. Clarke in Hook.f., Fl. Brit. India 4:412. 1884; Samvatsar, Fl. Madhya Pradesh 223. 1996; Khanna *et al.,* in Mudgal *et al.,* Fl. Madhya Pradesh 2:309. 1997.

Procumbent or often straggling undershrubs, up to 40 cm long; stems much branched, terete at internodes; leaves 1.2-6 × 1.0.8-3 cm, ovate or elliptic, acute to acuminate at apex, acute at base, margin entire, ciliate, petioles 5-15 mm long; flowers violet-blue, sessile, axillary, usually solitary; bracteoles elliptic, 8-20 mm long, pubescent, ciliate on the margins; calyx 6-8 mm long, divided ¾ th way down, teeth acute, linear-subulate; corolla 3-3.2 cm long, ventricose part 2-3 cm long, twice the length of tube; capsules clavate, *ca* 2 cm long, brown, pubescent; seeds 16-20, orbicular, 2.5-3 mm across, glabrous with hygroscopic hairs along margins.

*Fl.-Fr.:*Aug. -Nov.

Under shade of hedges, Alirajpur, Dhar, Indore, Khandwa.

RUNGIA Nees W. 23; MP 3; SA 3

Key to Species

1. Bracts dimorphic; spikes distinctly unilateral; flowers 3-5 mm long; upper lip of corolla acute; seeds yellow, minutely echinulate .... **R. pectinata**

1. Bracts uniform; spikes imperfectly unilateral

2. Erect plants; stem densely pubescent; capsules obovoid .... **R. elegans**

2. Diffuse or decumbent plants; stem glabrous; capsules ovoid-oblong . ... **R. repens**

Rungia elegans Dalz. & Gibs., Bombay Flora, 196. 1861; *Diapedium elegans* (Dalz. & Gibs.) Kuntze, Revis. Gen. Pl. 2:485. 1891; *Rungia repens* (L.) Nees var. *villosa* Nees, Prodr. 11:472. 1847; C.B. Clarke in Fl. Brit. India 4:549. 1885; Khanna *et al.,* in Mudgal *et al.,* Fl. Madhya Pradesh 2:351. 1997.

Erect, 15-45 cm tall, annual herbs; stem angular, covered with soft, white hairs; leaves 2-10 × 1.2-5 cm, sessile or shortly petiolate, elliptic-lanceolate, apex acute to acuminate, base rounded, margin entire; flowers blue, sessile, in terminal, imperfectly unilateral spikes, 2-5 cm long; bracts broadly elliptic, 8-10 mm long, hairy with broad, scrious margins, rounded and cuspidate at apex; bracteoles linear-lanceolate, 4-5 mm long, with scarious margins; calyx 5-partite, divided nearly to the base, 4-5 mm long, sepals linear-subulate; corolla 8-10 mm long, pubescent outside, 2-lipped, lobes obtuse; capsules obovoid, 4-5 mm long, pubescent on the margins, mucronate; seeds 4, orbicular, 1-1.5 mm across, brown, glabrous.

*Fl.-Fr.:*Sept. -Jan.

Rare, in wastelands at **Jhabua** only, in Meghnagar forest.

Rungia pectinata (L.) Nees, Prodr. 11:470.1847; *Justicia pectinata* L., Cent. Pl. II 3. 1756; *Rungia parviflora* Nees, Pl. Asiat. Rar. 3:110.1832; *R. parviflora* var. *pectinata* (L.) C.B. Clarke in Fl. Brit. India 4:550.1885; Samvatsar, Fl. West. Trib. Madhya Pradesh 232. 1996; Khanna *et al.,* in Mudgal *et al.,* Fl. Madhya Pradesh 2:352. 1997.

Annual, procumbent, much branched herbs; branches rooting at the lower nodes; leaves 1.5-5.5 × 0.5-2.5 cm, acute at both ends, elliptic or lanceolate, sessile or petiolate; flowers light blue, in terminal and axillary, one-sided, subsessile, up to 2 cm long, dense spikes; barren bracts in 2 rows, lanceolate, 4 mm long, cuspidate with scarious margins; fertile bracts orbicular, 2.5-3 mm across, cuspidate with scarious margins; bracteoles elliptic-oblong, 2-2.5 mm long ciliate with scrious margins, apex 2-fid; calyx segments linear-lanceolate, margins ciliate; corolla 2-lipped, pubescent outside, lobes obtuse, 2 rows of darker patches present on the lower lip; capsules 2.5 mm long, hairy; seeds 2-4, orbicular, rugose, yellowish-brown.

*Fl.-Fr.:*Aug. -Feb.

Common weed of shady places and a herbaceous member of forest undergrowth, in all districts.

Rungia repens (L.) Nees, Pl. Asiat. Rar. 3:110.1832; *Justicia repens* L., Sp. Pl. 15. 1753; *Dicliptera repens* (L.) Juss., Ann. Mus. Hist. Nat. 9:269. 1807; *Diapedium repens* (L.) Kuntze, Revis. Gen. Pl. 2:485. 1891; C.B. Clarke in Hook.f., Fl. Brit. India 4:549. 1885; Samvatsar, Fl. West. Trib. Madhya Pradesh 232. 1996; Khanna *et al.,* in Mudgal *et al.,* Fl. Madhya Pradesh 2:352. 1997.

Annual, diffuse or procumbent herbs; leaves subsessile 4-8 × 1-2 cm, oblong or elliptic-oblong, apex acute or obtuse, base cuneate; flowers in terminal 1-3 cm long, pubescent spikes; bracts ovate-orbicular, foliaceous, pubescent, margin scarious; bracteoles lanceolate, ciliate; calyx 4 mm long, segmentslanceolate, pubescent; corolla rosy or pink with deep red or purple spots in the throat of lower lip; capsules 4-5 mm long, compressed, pubescent; seeds 4, sub-orbicular, brown, glabrous.

*Fl.-Fr.:*Sept. Dec.

Very common in open ground, shady slopes of hills, near ponds and streams, throughout the area.

STROBILANTHES Blume W. 167; MP 1; SA 1

Strobilanthes callosa Nees, Pl. Asit. Rar. 3:85. 1832; *Carvia callosa* (Nees) Bremek., Vehr. Con. Ned. Akad. Wetensch., Afd. Natuurk., Sect. 2 41(1):187. 1944; C.B. Clarke in Hook.f., Fl. Brit. India 4:451. 1884; Samvatsar, Fl. West. Trib. Madhya Pradesh 223. 1996; Khanna *et al.,* in Mudgal *et al.,* Fl. Madhya Pradesh 2:300.1997 (as *Carvia callosa*). *Karvi*

Perennial, up to 2m tall shrubs; leaves 10-20 × 3.5-7 cm, one of each pair smaller than its counterpart, elliptic-lanceolate, hairy on nerves beneath, apex acute or acuminate, base tapering into wing-like, expanded petiole, margin crenate, ciliate; floers bluish-purple in axillary, simple or branched, pedunculate, subtetragonal spikes; bracts violet to whitish-green, boat-shaped, glabrous or white hispid; calyx divided almost up to the base, pubescent, persistent; corolla *ca* 3.5

cm long, lobes orbicular, tube pale white, cylindrical; capsules *ca* 2 cm long, obovoid, acute, narrowed at base; seeds 2, broadly ovate, acute, densely clothed with white, appressed hairs, except on the nucleus.

*Fl.-Fr.:*Oct. -Dec.

Alirajpr, Burhanpur, Dhar, Indore, Khandwa.

Note- The plants flower at an interval of 2-5 years. Due to root suckers, we find a group of 40-50 plants, surrounding the mother plant. The region of inflorescence is very sticky.

VERBENACEAE 34/1035; 6/7; 6/7

Key to Genera

1. Trees; flowers white, in elongated racemes; stamens included; stigma dilated; ovary 4-locular; drupes of 2 pyrenes .... CITHAREXYLUM
1. Herbs or shrubs; flowers in spikes
    2. Stamens 2; peduncle elongated; flowers deep blue; fruits splitting into 2 pyrenes .... STACHYTARPHETA
    2. Stamens 4; peduncle very short
        3. Calyx truncate; corolla sub-equally 5-lobed; fruits fleshy; shrubs .... LANTANA
        3. Calyx 2-4 lobed; corolla 2-lipped; fruits dry
            4. Spikes elongated during flowering; flowers purple; stamens didynamous; leaves whorled .... ALOYSIA
            4. Spikes not so elongated; flowers white or lilac
                5. Shrubs or under-shrubs; leaves not fleshy, more than 5 cm long .... LIPPIA
                5. Herbs; leaves fleshy, less than 5 cm long .... PHYLA

ALOYSIA Palau W. 44; MP 1; SA 1

Aloysia citriodora Palau, Parte Pract. Bot. 1:708. 1784; *Verbena citriodora* (Palau) Cav., Descr. Pl. 68. 1801; *Lippia citriodora* (Palau) Kunth, Nov. Gen. Sp. 2:269. 1818; *Aloysia triphylla* (L'Herit.) Britton, Bot. Porto Rico 6:140.1925; H. Moldenke & A. Moldenke in Rev. Handb. Fl. Ceylon 4:232. 1983; Saamvatsar, Fl. West. Trib. Madhya Pradesh 233. 1996.

Aromatic creeping shrubs; leaves lemon scented, 2-7 × 0.5-3 cm, oblong, caducous, whorled in groups of 3-4 at each node, petioles short; inflorescence spicate, flowers purple-white, fragra-grant, sessile; calyx tubular, purple; corolla hypocrateriform, 2-lipped; stamens 4, didynamous; fruit dry.

*Fl.-Fr.:*Sept. -Nov.

Rare, along Narmada bank, **Dhar** (Dharampuri bet). The occurrence of this plant in a remote area is surprising. It is an alien species naturalized in that area, but how and when is a mystery.

CITHAREXYLUM L. W. 114; MP 1; SA 1

Citharexylum spinosum L., Sp. Pl. 625. 1753; *C. subserratum* Sw. Prodr. Veg. Ind. Occ. 91. 1788; Srivastava in Mudgal *et al.,* Fl. Madhya Pradesh 2:368. 1997.

Small, glabrous trees; leaves petiolate, 7-16 × 4-6 cm, ovate-elliptic or obovate, remotely, shallowly serrate towards the apex; flowers white, fragrant, in long, drooping racemes; calyx shortly 5-lobed; drupes succulent.

*Fl.-Fr.:*July-Oct.

Rare, Burhanpur, Indore, Khandwa.

LANTANA L. W. 129; MP 2; SA2

Key to Species

1. Plants armed with recurved spines; peduncle much reduced, flowers in various colours; leaves not lemon-scented .... **L. × aculeata**

1. Plants not armed; peduncle elongated; flowers white or lilac; leaves lemon-scented .... **L. indica**

Lantana × aculeata L., Sp. Pl. 627. 1753; *L. camara* L. var. *aculeata* (L.) Moldenke, Torreya 34:9. 1934; Hook.f., Fl. Brit. India 4:562. 1885; Samvatsar, Fl. West. Trib. Madhya Pradesh 235. 1996; Srivastava in Mudgal *et al.,* Fl. Madhya Pradesh 2:379. 1997. *Rai muniya*

Prickly, about 3 m high shrubs; leaves 2-9 × 2-6 cm, ovate, shortly acuminate at apex, base cuneate or rounded, crenate-serrate, scabrous above; flowers of various mixed colours, ranging from red, yellow to light pink, in capitates, pedunculate spikes; bracts inconspicuous, about as long as the corolla tube; calyx very small, inconspicuous, *ca* 3 mm long; limb of corolla 6-8 mm wide, length of tube *ca* 1 cm; drubs globose, 2-3 mm across, shining black.

*Fl.-Fr.:*All the year round

Very common throughout the area and serious menace to the local plants.

Lantana indica Roxb., Fl. Ind. ed. 1832 3:89. 1832; Hook.f., Fl. Brit. India 4:562. 1885; Samvatsar, Fl. West. Trib. Madhya Pradesh 235. 1996; Srivastava in Mudgal *et al.,* Fl. Madhya Pradesh 2:379. 1997.

Strongly aromatic, unarmed, straggling shrubs; leaves 1.5-8 × 0.5-3.5 cm, ovate or lanceolate, apex acute, crenate-serrate, softly hairy above; flowers white, pink to pale purple, in pedunculate, capitates spikes, peduncles 3-5 cm long; bracts ovate or elliptic, 6-8 mm long; calyx *ca* 1.5 mm long, pubescent outside; corolla *ca* 8 mm long, hairy outside, limb *ca* 6 mm wide; drupes 3-6 mm across, purple, fruiting calyx thin.

*Fl.-Fr.:*Throughout the year

In scrub forest, along stream banks, Alirajpur, Burhanpur, Dhar, Indore, Khandwa.

LIPPIA L. W. 191; MP 1; SA 1

Lippia javanica (Burm.f.) Spreng., Syst. Veg. 2:752. 1825; *Verbena javanica* Burm.f., Fl. Indica 12. 1768; Hook.f., Fl. Brit. India 4:563. 1885; Srivastava in Mudgal *et al.,* Fl. Madhya Pradesh 2:380.1997.

Erect shrubs; leaves ternately whorled, elliptic-oblong, pubescent below, 5-8 × 2-3.5, apex acume, base rounded; margin crenate-serrate, petioles up to 1 cm long; flowers pink, in axillary spikes; bracts ovate-lanceolate, *ca* 3 mm long; calyx lobes lanceolate; corolla 5-6 mm long; drupes subglobose.

*Fl.-Fr.:*July-Jan.

Rare, **Khandwa** only, at Mandhata.

PHYLA Lour.W. 13; MP 1; SA 1

Phyla nodiflora (L.) Greene, Pittonia 4:46. 1899; *Verbena nodiflora* L., Sp. Pl. 20.1753; Hook.f., Fl. Brit. India 4:563. 1885; Srivastava in Mudgal *et al.,* Fl. Madhya Pradesh 2:382. 1997.

Prostrate herbs; leaves obovate-spathulate, 1-2.5 × 0.5-1 cm, apex obtuse, base cuneate, decurrent into the petiole, serrate above the middle; flowers in pedunculate spikes; bracts broadly ovate, concave; bracteoles absent; calyx cup-like, lobes 2, acute; corolla purplish-white, 2-2.5 mm across, tube *ca* 1.5 mm long; drupes globose, *ca* 1 mm across.

*Fl.-Fr.:*Throughout the year

Common in moist places, all over the area.

STACHYTARPHETA Vahl W. 130; MP 1; SA 1

Stachytarpheta jamaicensis (L.) Vahl, Enum. Pl. Obs. 1:206. 1804; *Verbena jamaicensis* L., Sp. Pl. 19. 1753; Hook.f., Fl. Brit. India 4:564. 1885 *p.p;* Samvatsar, Fl. West. Trib. Madhya Pradesh 235. 1996; Srivastava in Mudgal *et al.,* Fl. Madhya Pradesh 2:386. 1997.

Herbs or undershrubs, 20-60 cm high, stems dichotomously branched; leaves 4-8 × 2-4 cm, elliptic, obtuse or acute, cuneate and long decurrent at base margin serrate; petioles slender, winged; flowers deep blue in slender, nearly continuous spikes; calyx *ca* 4 mm long, 4-toothed; corolla lobes *ca* 5 mm long, tube curved, hairy; fruits oblong, ribbed, splitting into 2 pyrenes.

*Fl.-Fr.:*Aug. -Nov.

Along forest paths, not common, Alirajpur, Barwani. Indore, Khargone.

LAMIACEAE 245/7886; 30/78; 24/54

Key to Genera

1. Leaves digitately 3-5 foliolate .... VITEX
1. Leaves simple
    2. Fruit drupaceous
        3. Flowers actinomorphic
            4. Inflorescence terminal; white, 4-6 merous; calyx inflated in fruit; stamens 4-6 .... TECTONA
            4. Inflorescence axillary or supra-axillary; flowers actinomorphic, lilac-purple, pentamerous; calyx not inflated; stamens 4 .... CALLICARPA
        3. Flowers zygomorphic, white, yellow or purple, pentamerous; corolla 2-lipped
            5. Drupe composed of one 4-loculate and two, 2-loculate pyrenes; stamens included

6. Corolla tube large, ventricose .... GMELINA
6. Corolla tube short, cylindrical .... PREMNA

5. Drupe composed of pyrenes; stamens exserted

7. Drupe composed of 4 or many 1-loculate pyrenes; flowers white; leaves opposite........ CLERODENDRUM
7. Drupe not splitting into 4 pyrenes at maturity; flowers blue; leaves whorled .... ROTHECA

2. Fruit schizocarpic

8. Calyx lobes or teeth 6-10

9. Flowers white; upper lip of corolla as long as or half as long as lower lip; calyx teeth not spinescent .... LEUCAS
9. Flowers scarlet-red; upper lip of corolla twice as long as the lower lip; calyx teeth spinescent........ LEONOTIS

8. Calyx lobes or teeth 5 or less

10. Calyx 2-lipped

11. Upper lip of calyx entire, lower lobed

12. Fruiting calyx deflexed

13. Lower lip of corolla bent downward, not concave . ... OCIMUM
13. Lower lip of corolla not bent downward, concave .... ANISOCHILUS

12. Fruiting calyx not deflexed; stamens 2; anther lobes separated by extended connective .... SALVIA

11. Both upper and lower lips of calyx lobed

14. Two teeth of lower lip of calyx awned; stigma clavate or sub-clavate .... ORTHOSIPHON
14. Teeth of lower lip of calyx not awned; stigma not clavate

15. Calyx 15-nerved; upper lip of corolla notched or 2-fid .... NEPETA
15. Calyx not 15-nerved; upper lip of corolla 4-fid, tube decurved; lower boat-shaped, narrowed into a neck .... PLECTRANTHUS

10. Calyx not lipped

16. Corolla subequally 4-lobed

17. Corolla 2-lipped

18. Upper lip of corolla 3-lobed, lower flat .... POGOSTEMON
18. Upper lip of corolla entire, lower 3-lobed

20. Calyx 13-15 nerved

19. Calyx 13-nerved, actinomorphic, revolute; petiole obscure . ... MICROMERIA
19. Calyx 15-nerved, zygomorphic, not revolute; petiole well developed .... CLINOPODIUM

20. Calyx 5-10 nerved

21. Lower lip of corolla 3-lobed, midlobe large, 2-fid or emarginate ...ANISOMELES
21. Lower lip of corolla 3-fid, midlobe entire, obcordate; leaves pinnatipartite .... LEONURUS

17. Corolla not 2-lipped, 4-lobed; calyx 5-partite, lobes plumose; stamens included .... COLEBROOKIA

16. Corolla 5-lobed

22. Lower lip of corolla 3-partite; leaves divided into linear segments (our area) .... LAVANDULA

22. Lower lip of corolla entire, upper lip 4-lobed

23. Lower lip of corolla boat-shaped, not saccate or deflexed .... ISODON

23. Lower lip of corolla not boat-shaped, saccate or deflexed .... HYPTIS

ANISOCHILUS Wall. W. 17; MP 2; SA 1

Anisochilus carnosus (L.f.) Wall. Pl. Asiat. Rar. 2:18. 1830; *Lavandula carnosa* L.f., Suppl. Pl. 273. 1782; Hook.f., Fl. Brit. India 4:627. 1885; Samvatsar, Fl. West. Trib. Madhya Pradesh . 237. 1996; Mukherjee in Mudgal *et al.,* Fl. Madhya Pradesh. 2:399. 1997.

Annual, 15-40 cm high herbs; leaves fleshy, hairy and gland-dotted on both surfaces, 4-6 × 2.5-3 cm, ovate or oblong, apex obtuse, base sub-cordate or rounded, margin crenate; petiole up to 6 cm long; flowers pinkish-purple, in terminal, paniculate, 1.5 2 cm long heads or spikes; bracts ovate, truncate at base; calyx acute, uo to 2,5 mm long; corolla up to 1.5 cm long, tube 4-9 mm long; nutlets brown, when ripe.

*Fl.-Fr.:*Sept. -Oct.

In the forest in rocky areas, on shingled roofs of houses, Alirajpur, Barwani, Burhanpur, Khargone.

ANISOMELES R. Br.W. 5; MP 2; SA 2

Key to Species

1. Plants densely villous all over; flowers lilac to pale blue; leaves 3-8 cm long .... **A. malabarica**

1. Plants not villous, only hairy; flowers white or pink to purple, in dense, sessile, axillary whorls; leaves 5-12 cm long .... **A. indica**

Anisomeles indica Kuntze, Revis. Gen. Pl. 2:512. 1891; *Nepeta indica* L. Sp. Pl. 571. 1753; *Anisomeles ovata* W. T. Aiton, Hortus Kew. 3:364. 1811, *nom. illeg;* Hook.f., Fl. Brit. India 4:672. 1885; Samvatsar, Fl. West. Trib. Madhya Pradesh 238. 1996; Mukherjee in Mudgal *et al.,* Fl. Madhya Pradesh 2:400.1997.

Perennial, acutely 4-angled, densely villous herbs; leaves 3-10 × 3-8 cm, apex acute, base truncate, rounded or sub-cordate; petioles 1-7 cm long; flowers bluish-purple, *ca* 2 cm long in axillary verticillasters; calyx hisute, teeth as long as the tube, about 6 mm long; corolla *ca* 2 cm long; nutlets *ca* 2.5 mm long, black, smooth, shining.

*Fl.-Fr.:*Aug. -Feb.

Common along water courses and in hedges of fields.

Anisomeles malabarica (L.) R. Br. ex Sims, Bot. Mag. 46:t. 2071. 1819; *Nepeta malabarica* L., Mant. Pl. 2:566. 1771; Hook.f., Fl. Brit. India 4:673. 1885; Mukherjee in Rec. Bot. Surv. India 14(1) 153. 1940; Kamble in N.P. Singh *et al.,* Fl. Maharashtra 2:716.2001.

Undershrubs, about a m high, stems and branches densely tomentose, with white, woolly hairs; leaves 3.5-10 × 3.5 cm, lanceolate to oblong-lanceolate, tomentose on both surfaces, apex acute, base rounded or narrowed, margin crenate-serrate; flowers pinkish, fragrant, in dense whorls of terminal spikes; calyx 0.8 cm long, teeth equaling the tube; corolla *ca* 2 cm long; nutlets *ca* 0.2 cm long, polished brown.

*Fl.-Fr.:*Oct. -Jan.

Rare, in hilly areas, **Alirajpur** only, Bakhat garh area.

CALLICARPA L. W. 185; MP 1; SA 1

Callicarpa macrophylla Vahl, Symb. Bot. 3:13. 1794**;** Hook.f., Fl. Brit. India 4:568. 1885; Srivastava in Mudgal *et al.,* Fl. Madhya Pradesh 2:365. 1997.

Shrubs; leaves 12-25 × 5-11 cm, elliptic-oblong to lanceolate, apex acuminate, base cuneate, margin crenate-dentate, teeth gland-tipped; petioles 4-12 mm long; flowers lilac to purple, in axillary, solitary or often numerous cymes; pedicels *ca* 2mm long; calyx campanulate, 1.5-2 mm long; corolla *ca* 3 cm long, lobes 4, ovate; drupes white, globose, *ca* 2 cm across.

*Fl.-Fr.:*Oct. -Jan.

Rare, Barwani, Khargone.

CLERODENDRUM L. W. 308; MP 4; SA 3

Key to Species

1. Leaves whorled, 6-9 cm long; flowers in axillary, 3-7 flowered cymes .... **C. indicum**
1. Leaves opposite, decussate
    2. Leaves 3-5.5 cm long; flowers in dichotomous panicles; calyx marginally enlarged, up to 1 cm long; drupes black when mature .... **C. phlomidis**
    2. Leaves 8-24 cm long; flowers in dichotomous or trichotomous panicles; calyx enlarged in fruit, 1-1.5 cm long; drupes bluish-black when mature .... **C. infortunatum**

Clerodendrum indicum (L.) Kuntze, Revis. Gen. Pl. 2:586. 1891; *Siphonanthus indica* L., Sp. Pl. 109. 1753; *Clerodendrum siphonanthus* R. Br. in Hortus Kew. 4:65. 1812, *nom. illeg;* C.B. Clarke in Hook.f., Fl. Brit. India 4:595. 1885; Srivastava in Mudgal *et al.,* Fl. Madhya Pradesh 2:370.1997.

Erect, woody herbs; leaves 6-19 × 2-4 cm, lanceolate to oblong; apex acute or sub-acuminate; base acute; flowers in axillary, 3-7 flowered cymes; pedicels *ca* 2 cm long; calyx 5-fid, *ca* 1.5 cm long, glandular-punctate; corolla white, *ca* 2 cm long; drupes 4-lobed, turning reddish-black, fruiting calyx accrescent, red, fleshy.

*Fl.-Fr.:*Sept. -Oct.

On sandy soil, along river beds, Alirajpur, Dhar, Khandwa.

Clerodendrum infortunatum L., Sp. Pl. 637. 1753; *C. viscosum* Vent., Jard. Malmaison 1:t. 25. 1803; Santapau in Bull. Bot. Surv. India 3:14. 1961; Samvatsar, Fl. West. Trib. Madhya Pradesh 234. 1996; Srivastava in Mudgal *et al.,* Fl. Madhya Pradesh 2:374. 1997.

Shrubs, uo to 2m high; leaves 9-20 × 4-15 cm, elliptic to ovate, apex acute, base obtuse or cordate, margin denticulate or serrate; flowers white in terminal and in many, opposite, few-flowered, divaricate cymes; calyx 1-1.5 cm long; corolla *ca* 2 cm long; drupes bluish-black at maturity, fleshy and enclosed in accrescent fruiting calyx.

*Fl.-Fr.:*Feb. -April

In hilly area, rare, **Dhar** only, below Neelkanth Mandir hill.

Clerodendrum phlomidis L.f., Suppl. Pl. 292. 1782; *C. multiflorum* (Burm.f.) Kuntze, Revis. Gen. Pl. 3:526. 1891, *nom. illeg; Volkameria multiflora* Burm.f., Fl. Ind. 137. 1768; C.B. Clarke in Hook.f., Fl. Brit. India 4:590.1885; Samvatsar, Fl. West. Trib. Madhya Pradesh 233. 1996; Srivastava in Mudgal *et al.,* Fl. Madhya Pradesh 2:371. 1997. *Arni*

Small trees with pubescent branches; leaves 3-6 × 2.5- 4 cm, ovate-rhomboid, apex acute to shortly acuminate, acute or sub-cordate at base, margin crenate-serrate; flowers in axillary and terminal, dichotomous cymes; calyx 9-10 mm long, lobes sub-acuminate; corolla white to yellowish-white, *ca* 2 cm long; drupes *ca* 1 cm long, wrinkled, black when mature.

*Fl.-Fr.:*Oct. -Feb.

Common in villages, waste lands, all over the area.

CLINOPODIUM L. W. 148; MP 1; SA 1

Clinopodium capitellatum (Benth.) Kuntze, Revis. Gen. Pl. 2:515. 1891; *Micromeria capitellata* Benth., Prodr. 12:218. 1848; Hook.f., Fl. Brit. India 4:649. 1885; Samvatsar, Fl. West. Trib. Madhya Pradesh 241. 1996; Mukherjee in Mudgal *et al.,* Fl. Madhya Pradesh 2:419. 1997.

Slender, aromatic, 30-40 cm high annual herbs with a woody root-stock; leaves 1-3 cm long, ovate or ovate-oblong, apex obtuse, base rounded or cuneate, margin entire or crenate-serrate; flowers violet in axillary verticillasters, forming a spicate raceme, lower verticillasters pedunculate, upper gradually sessile; calyx villous, teeth as long as the tube, throat with a ring of hairs; corolla 3-4 mm long, hairy outside, tube as long as the calyx or a little more longer; nutlets minute, elliptic-oblong, smooth, brown.

*Fl.-Fr.:*Feb. -April

Very common in moist places, especially below bridges, in all districts.

COLEBROOKIA J. E. Smith Monotypic

Colebrookea oppositifolia Sm., Exot. Bot. 2:111. 1806; Hook.f., Fl. Brit. India 4:401. 1885; Samvatsar, Fl. West. Trib. Madhya Pradesh 238. 1996; Mukherjee in Mudgal *et al.,* Fl. Madhya Pradesh 2:401. 1997. *Gilhari ki poonchh*

Soft, woolly, 1-2 m high shrubs; branches stout, grooved; leaves white tomentose, 10-20 × 4-8 cm, oblong-lanceolate, apex acute or acuminate, base acute, margin crenate-serrulate; petioles

2-4 cm long; verticillasters in axillary and terminal, paniculate, 5-10 cm long spikes; calyx *ca* 3 mm long, 5-toothed, plumose with hairs elongating up to 6 mm long in fruit; corolla white, very small; nutlets often solitary.

*Fl.-Fr.:*Jan. -March

In ravines of Burhanpur and Dhar.

GMELINA L. W. 38; MP 2; SA 2

Key to Species

1. Trees; leaves large, ovate-cordate, 10-25 cm long; branches unarmed; flowers erect; drupes 2-2.5 cm long . ... **G. arborea**
1. Shrubs, often straggling; leaves ovate to obovate, 2-9 cm long; branches thorny; flowers pendulous; drupes *ca* 1.8 cm long.......... **G. asiatica**

Gmelina arborea Roxb., Hort. Bengal. 46. 1814; C.B. Clarke in Hook.f., Fl. Brit. India 4:581. 1885; Samvatsar, Fl. West. Trib. Madhya Pradesh 234. 1996; Srivastava in Mudgal *et al.,* Fl. Madhya Pradesh 2:376. 1997. *Siwan*

Deciduous, 8-15 m high trees with a corky, smooth, yellowish bark; leaves 8-16 × 5-12 cm, ovate, apex acuminate, densely tomentose above when young, fulvous tomentose beneath, base cuneate or cordate with 2 glands on the upper surface, near the junction of lamina and petiole; flowers in axillary and termina, cymose panicles; calyx *ca* 5 mm long, 5-toothed, fulvous tomentose outside; corolla brownish-yellow, ventricose, *ca* 4 cm long, bilabiate, upper lip 2-lobed, lower 3-lobed; drupes orange-yellow, obovoid-pyriform, 2-2.5 cm long.

*Fl.-Fr.:*Feb. -April

Common in forests, along village roads, throughout the area.

Gmelina asiatica L., Sp. Pl. 626. 1753; *G. lobata* Gaertn. Fruct. Sem. Pl. 1:268. 1788; *G. parvifolia* Roxb., Pl. Coromandel 2:32. 1802; C.B. Clarke in Hook.f., Fl. Brit. India 4:582. 1885; Samvatsar, Fl. West. Trib. Madhya Pradesh . 234. 1996; Khanna *et al.,* Suppl. Fl. Madha Pradesh 153.2001.

Much branched, about 3 m high, deciduous shrubs armed with axillary thorns; leaves borne when the plant becomes old, 1-3.5 × 0.4-2.5 cm, apex obtuse, base acute, margins obscurely lobed; flowers sulphur yellow, in terminal, pubescent, paniculate racemes; drupes yellow, *ca* 1.8 cm long, obovoid-pyriform.

Fl.-Fr.: Nov. -March

Rare, **Dhar** Narmada bank, Dharampuri bet, only 2 shrubs are present on the top of hill. This species is not found anywhere in M.P. Prasanna in N.P. Singh *et al.,* Fl. Maharashtra 2:692.2001 has described it as frequent in deciduous forests, in 5 districts and he further says that it is grown in gardens at Aurangabad, Beed and Nanded gardens. It is not known as to how and when it entered M.P.

HYPTIS Jacq. W. 295; MP 1; SA 1

Hyptis suaveolens (L.) Poit., Ann. Mus. Hist. Nat. 7:472. 1806; *Ballota suaveolens* L., Syst. Nat. ed. 10, 2:1100.1759; Hook.f., Fl. Brit. India 4:630.1885; Samvatsar, Fl. West. Trib. Madhya Pradesh 238. 1996; Mukherjee in Mudgal *et al.,* Fl. Madhya Pradesh 2:403. 1997. *Dona*

Rigid, strongly aromatic, annual herbs about 1 m high; stems and branches obtusely 4-angled, patently hirsute; leaves 2.5-10 × 1.5-8 cm, broadly ovate or sub-orbicular, apex acuminate, base acute, margin serrulate, thinly hairy above, densely hairy below; petioles 1.5-4 cm long; flowers bluish-violet, in contracted, 1-5 fascicled, secund cymes, paniculate in late stage; calyx glandular and hispid outside, striate , 10-ribbed; corolla small, 0.6 cm long; nutlets blackish-brown, rugose, notched at the apex and with a white mark on either side of the pointed base.

*Fl.-Fr.:*Aug. -Feb.

Very common along roads, waste lands, banks of drains, in all districts.

ISODON (Benth.) Schrader & Spach W. 107; MP 4; SA 3

Key to Species

1. Fruiting calyx distinctly 2-lipped, subsequently 5-toothed; leaves rugose .... **I. rugosus**
1. Fruiting calyx not 2-lipped; leaves not rugose
   2. Corolla white with purple spots, tube straight .... **I. lophanthoides**
   2. Corolla pale violet, tube deflexed . ...**I. coetsa**

Isodon coetsa (Buch. -Ham. ex D. Don) Kudo, Mem. Fac. Sci. Taihoku Imp. Univ. 2:131. 1929; *Plectranthus coetsa* Buch.-Ham. ex D. Don, Prodr. Fl. Nepal. 117. 1825; *P. japonicus* (Burm.f.) Koidz., Bot. Mag. (Tokyo) 43:380.1929; *Scutellaria japonica* Burm.f., Fl. Ind. 130.1768; Hook.f., Fl. Brit. India 4:619. 1885; Santapau & Wagh, Bull. Bot. Surv. India 5:107. 1964; Mukherjee in Mudgal *et al.,* Fl. Madhya Pradesh 2:426. 1997. [As *Plectranthus japonicus* (Burm.f.) Koidz.]

Strongly aromatic, branched undershrubs, about 1.5 m high; leaves ovate, 3-10 × 1.5-4.5 cm, margin serrate, base cuneate, hoary beneath; petioles much shorter than the length of leaves; flowers pale violet, in axillary and terminal cymose panicles; pedicels pubescent, slender, deflexed; calyx *ca* 2 mm long in flower, slightly longer in fruit, 10-nerved, 5-toothed, hoary pubescent, lips indistinct, tube curved above the middle; corolla *ca* 5 mm long, pubescent outside, upper lip short, erect, tube deflexed, much shorter than the lower lip; nutlets brown, *ca* 1 mm across, oblong or subglobose.

Fl.-Fr.: Oct. -Dec.

Rare, **Khandwa** only, Kalibhit range.

Isodon lophanthoides (Buch.-Ham. ex D. Don) H. Hara in J. Jap. Bot. 60:235. 1985; *Hyssopus lophanthoides* Buch.-Ham. ex D. Don, Prodr. Fl. Nepal 110.1825; *Plectranthus stocksii* Hook.f., Fl. Brit. India 4:618. 1885; Samvatsar. Fl. West. Trib. Madhya Pradesh 243. 1996. (As *Plectranthus striatus* Benth.)

Slender, perennial herbs, up to 1 m high, stems 4-angled, sulcate; leaves 2-8 × 0.8-3.0 cm, oblong-ovate, chartaceous, upper surface olive-green, lower surface greenish, dotted with brown glands, apex acuminate, base cuneate or cuneate-rounded, margin crenate; flowers white to rosy, with purple spots on the limbs; calyx *ca* 0.2 cm long, dotted with red-brown glands, fruiting calyx longer, 0.4-0.5 cm long; corolla *ca* 0.7 cm long, limb 2-lipped, upper lip reflexed, 4-fid, lower broadly ovate; nutlets about 1 mm long, brown, smooth.

*Fl.-Fr.:*Sept. -Nov.

In moist places along hills, Alirajpur, Burhanpur, Dhar.

Isodon rugosus (Wall. ex Benth.) Codd in Taxon 17:239. 1968; *Plectranthus rugosus* Wall. ex Benth. Pl. Asiat. Rar. 2:17. 1830; Hook.f., Fl. Brit. India 4:620.1885; Mukherjee in Mudgal *et al.,* Fl. Madhya Pradesh 2:428. 1997. (As *Plectranthus rugosus*)

Erect undershrubs up to 1.5 m high, branches hoary with stellate hairs; leaves 2-4 × 1-2 cm, ovate or elliptic, apex obtuse, base sub-cordate, margin crenate, slightly thick, rugose, finely reticulate above, white-tomentose below; petioles short; flowers in axillary, lax, paniculate, leafy cymes; calyx 4-5 mm long in fruit, subequally 5-toothed or very shortly 2-lipped; corolla white with purple or pink spots, about 4 mm long, lower lip longer, tube very short, declinate, usually gibbous at base; nutlets oblong, triquetrous, dark brown.

*Fl.-Fr.:*Aug. -Nov.

Rare on hill slopes, **Burhanpur** only, at Jambu pani.

LAVANDULA L.W. 47; MP 1; SA 1

Lavandula bipinnata (Roth) Kuntze, Revis. Gen. Pl. 2:521. 1891; *Bysteropogon bipinnatus* Roth, Nov. Pl. Sp. 255. 1821; *Lavandula burmanni* Benth., Labiat. Gen. Sp. 151. 1833, *nom. illeg;* Hook.f., Fl. Brit. India 4:631. 1885; Samvatsar, Fl. West. Trib. Madhya Pradesh 238. 1996; Mukherjee in Mudgal *et al.,* Fl. Madhya Pradesh 2:405. 1997. **(Plate 8, Fig. 3)**

Erect, herbs, up to 1 m high; stems 4-angled, pubescent; leaves 3-10 cm long, pinnatisect, segments linear, entire or dentate; flowers blue in simple or branched in axillary and terminal verticillasters; bracts glandular pubescent, persistent, each subtending a flower; calyx 4-6 mm long, hoary pubescent, glandular, teeth acute with penicillate tips; corolla *ca* 12 mm long, tube slender below, slightly dilated above; nutlets ellipsoid, black, with a white , large areole on the dorsal side, very mucilaginous when wet.

*Fl.-Fr.:*Oct. -Feb.

Along drains, slopes of hills, forest roads, all over the area.

LEONOTIS (Pers.) R. BR.W. 9; MP 1; SA 1

Leonotis nepetifolia (L.) R. Br., Hortus Kew. 3:409. 1811; *Phlomis nepetifolia* L., Sp. Pl. 586. 1753; Hook.f., Fl. Brit. India 4:691. 1885; Samvatsar, Fl. West. Trib. Madhya Pradesh 239. 1996; Mukherjee in Mudgal *et al.,* Fl. Madhya Pradesh 2:406. 1997. *Bhimal gutta*

Erect, annual, 1.5-2 m high herbs; stems stout, quadriangular, puberulous; leaves membranous, 9-15 × 4-5 cm, ovate, apex acute, base cuneate, coarsely crenate-serrate, teeth blunt, gland-dotted, long petiolate; flowers orange-red, in axillary, globose, 3-7 cm across heads, borne in acropetal succession on the stem; bracts deflexed, spine-tipped, pedicels up to 2 mm long; calyx up to 2 cm long, teeth spine-tipped, unequal, upper one *ca* 8 mm long, mouth oblique; corolla *ca* 2-3 cm long, glabrous below and densely villous above as well as on the long upper lip, tube hairy inside, with 3 parallel rings of white hairs; nutlets *ca* 4 mm long, obliquely truncate.

*Fl.-Fr.:*Oct. -Nov.

Along village roads, waste places, Alirajpur, Dhar, Indore, Khandwa.

LEONURUS L. W. 24; MP 1; SA 1

Leonurus sibiricus L., Sp. Pl. 584. 1753; Hook.f., Fl. Brit. India 4:678. 1885; Mukherjee in Mudgal *et al.,* Fl. Madhya Pradesh 2:407. 1997.

Annual, 60-70 cm high herbs; stems 4-angled, grooved, densely pubescent; leaves membranous, deeply lobed, palmatisect, apex acute to acuminate, base truncate, lobes 2-6 × 0.5-1 cm, irregularly serrate, upper leaves narrow, less divided; petioles up to 5 cm long; verticillasters dense; bracteoles spinescent; calyx up to 5 mm long, teeth spinescent; corolla pink *ca* 1 cm long, annual ring present inside; nutlets up to 2 mm long, smooth, triquetrous.

*Fl.-Fr.:*Jan. -March

As a weed in waste lands, **Indore** only, Rala mandal road.

LEUCAS R. Br. 111; MP 16; SA 13

Key to Species

1. Calyx mouth oblique
   2. Calyx mouth much produced below .... **L. urticifolia**
   2. Calyx mouth produced above
      3. Calyx tube bent in the middle, curving upwards or downwards; teeth long and slender
         4. Calyx tube 'S'-shaped; bracts spinous-tipped; lower lip of corolla twice as long as the upper, tube not annulate within .... **L. martinicensis**
         4. Calyx tube abruptly deflexed; bracts foliaceous; lower lip of corolla about equal to the upper, tube annulate within .... **L. nutans**
      3. Calyx tube not bent in the middle, erect
         5. Calyx mouth villous within; flowers in terminal whorls
            6. Whorls very large, about 5 cm across; bracts foliaceous .... **L. cephalotus**
            6. Whorls smaller, less than 5 cm across; bracts linear to filiform
               7. Corolla up to 2 cm long; bracts linear, spinous tipped . ... **L. zeylanica**
               7. Corolla up to 0.6 cm long; bracts filiform, not spinous tipped .... **L. clarkei**
         5. Calyx mouth not villous within; corolla tube annulate in the middle
            8. Calyx smooth below, ribbed and hispid above .... **L. aspera**
            8. Calyx smooth throughout . ... **L. lavandulifolia**
1. Calyx mouth straight or nearly so; teeth erect; whorls 6 or more flowered

9. Perennials with a stout root-stock; calyx annulate within; leaf margin entire . ... **L. longifolia**
9. Annuals
    10. Whorls up to 4-flowered ....**L. biflora**
    10. Whorls 6-many flowered
        11. Whorls up to12 flowered; leaves 3-8 cm long, base decurrent .... **L. wightiana**
        11. Whorls up to 6 flowered; plants hairy, tomentose or hispid
            12. Calyx teeth long, unequal; erect plants .... **L. stricta**
            12. Calyx teeth very short; straggling; weak plants .... **L. decemdentata**

Leucas aspera (Willd.) Link, Enum. Hort. Berol. Alt. 2:113. 1822; *Phlomis aspera* Willd., Enum. Pl. 621. 1809; Hook.f., Fl. Brit. India 4:690.1885; Samvatsar, Fl. West. Trib. Madhya Pradesh 240.1996; Mukherjee in Mudgal *et al.,* Fl. Madhya Pradesh 2:409. 1997. **(Plate 8, Fig. 4)**

Erect, annual herbs, up to 50 cm high; branches many, hispid or scabrid; leaves 3-8 × 0.5-1 cm, linear or narrowly oblong-lanceolate, apex obtuse, base cuneate, margin minutely crenate; petioles very short; verticillasters dense, terminal or axillary; bracts linear, ciliate, sub-spinulose; calyx 8-12 mm long, curved, smooth and glabrous below, ribbed and scabrid or hispid above, teeth spinulose, mouth very oblique, produced on the upper side; corolla white, *ca* 1 cm long; nutlets *ca* 2.5 mm long, smooth, brown.

Fl. -Fr: Aug. -March

Common weed of roadsides, waste lands, along streams, everywhere.

Leucas biflora (Vahl) Sm., Cycl. 20(2):7. 1812; *Phlomis biflora* Vahl,Symb. Bot. 3:77. 1794; *Leucas procumbens* Thwaites, Enum. Pl. Zeyl. 240.1860; Hook.f., Fl. Brit. India 4:683. 1885, *excl. syn;* Samvatsar, Fl. West. Trib. Madhya Pradesh 240.1996; Mukherjee in Mudgal *et al.,* Fl. Madhya Pradesh 2:409. 1997.

Procumbent, slender, diffuse herbs, up to 60 cm long, spreading from a woody rootstock; leaves 1.5-5 × 0.5 2 cm, ovate, elliptic or oblong, apex obtuse, base rounded, margin coarsely serrate or crenate; petioles up to 1 cm long; verticillasters axillary, 1-4 flowered; bracts setaceous; calyx ribbed 7-9 mm long, hispid outside, teeth subequal, *ca* 3 mm long; corolla tube annulate inside; nutlets *ca* 1.5 mm long, trigonous, brown, apex truncate, irregularly prominently tuberculate.

Fl.-Fr.: -Sept. -Feb.

Common in gullies, roadsides, along river banks, waste lands, forest undergrowth, in all districts.

Leucas cephalotus (Roth) Spreng., Syst. Veg. 2:743. 1825; *Phlomis cephalotes* Roth, Nov. Pl. Sp. 262. 1821; Hook.f., Fl. Brit. India 4:689. 1885; Samvatsar, Fl. West. Trib. Madhya Pradesh 240.1996; Mukherjee in Mudgal *et al.,* Fl. Madhya Pradesh 2:410.1997. *Dhupri sag*

Annual, erect, 40-70 cm high herbs, stems and branches clothed with spreading hairs; leaves 3-10 × 1.5-2 cm, elliptic, lanceolate or ovate-lanceolate, apex acute, base cuneate, margin crenate-serrate, glandular-punctate below, petioles short; verticillasters terminal, in many flowered, globose heads, about 3.5 cm across; bracts lanceolate, 1.5-2.5 cm long, awned, numerous, imbricating, covering the heads; calyx tubular, curved, glabrous below, pubescent above, mouth oblique, produced on the upper side; corolla up to 2.5 cm long; nutlets *ca* 3 cm long, brown, smooth.

*Fl.-Fr.:*Sept. -Dec.

Common along roadsides, waste places, weed of fields, all over the area.

Leucas clarkei Hook.f., Fl. Brit. India 4:688. 1885; Samvatsar, Fl. West. Trib. Madhya Pradesh 240.1996; Mukherjee in Mudgal *et al.,* Fl. Madhya Pradesh 2:411. 1997.

Annual, 10-50 cm high herbs; branches with deflexed hairs; leaves 3-6 × 0.3-0.8 cm, linear or linear-lanceolate, apex obtuse, base narrowed, blade entire, hispid above, scabrid, hirsute beneath; verticillasters terminal, whorls in subglobose, 2-3 cm across heads; bracts as long as the calyx tube, ciliate; calyx *ca* 6 mm long, upper 3 teeth much longer, acicular, mouth slightly oblique; lower lip of corolla about 1 cm long; nutlets *ca* 2 mm long, smooth, apex rounded.

*Fl.-Fr.:*Aug. -Dec.

Rare as a weed in fields, **Alirajpur** only, at Kadwal.

Leucas decemdentata (Willd.) Sm., Cycl. 20 (2):6. 1812; *Phlomis decemdentata* Willd., Sp. Pl. 3:124. 1800; *Leucas mollissima* Wallich ex Benth., Pl. Asiat. Rar. 1:62. 1830; Hook.f., Fl. Brit. India 4:682. 1885; Samvatsar, Fl. West. Trib. Madhya Pradesh 240.1996; Mukherjee in Mudgal *et al.,* Fl. Madhya Pradesh 2:414. 1997.

Straggling, hairy tomentose herbs, about 1 m high; leaves 2-8 × 1-4 cm, ovate or ovate-lanceolate, apex acute, base cuneate, margin serrate, strigosely villous above, white tomentose below, petioles up to 2 cm long; verticillasters axillary, whorls distant, many flowered; bracts minute, villous; calyx 6-8 mm long, straight, teeth minute; corolla annulate inside; nutlets *ca* 2 mm long, oblong, obliquely truncate, dark brown.

*Fl.-Fr.:*Oct. -Feb.

Rare in moist places at Alirajpur, Indore.

Leucas lavandulifolia Sm., Cycl. 20(2):2. 1812; *Phlomis linifolia* Roth, Nov. Pl. Sp. 260.1821; *Leucas linifolia* (Roth) Spreng., Syst. Veg. 2:743. 1825; Hook.f., Fl. Brit. India 4:690.1885; Samvatsar, Fl. West. Trib. Madhya Pradesh 240.1996; Mukherjee in Mudgal *et al.,* Fl. Madhya Pradesh 2:413. 1997.

Erect, slender, 30-40 cm high, annual herbs; leaves 2.5-10 × 0.5-1.5 cm, linear or linear-lanceolate, apex sub-acute, base tapering, margin remotely crenate-serrate; petioles very short or obsolete; verticillasters axillary and terminal, in the apical region of the plant; pedicels short; bracts linear; calyx *ca* 9 mm long, mouth pubescent inside, very oblique, tube curved; corolla , *ca* 12 mm long, upper lip woolly, tube about 7 mm long; nutlets oblong, *ca* 2 mm long.

*Fl.-Fr.:*Aug. -Dec.

Common in waste places, forest, sandy river beds, all over the area.

Leucas longifolia Benth., Labiat. Gen. Spec. 744. 1835; Hook.f., Fl. Brit. India 4:684. 1885; Mukherjee in Mudgal *et al.,* Fl. Madhya Pradesh 2:413. 1997.

Annual, erect, rigid, 20-50 cm high herbs; leaves 3-6 cm long, sessile, narrowly linear, apex obtuse, base narrow, margin entire; verticillasters axillary, few flowered; bracts subulate; calyx *ca* 6 mm long, hispid, teeth 10, very small, mouth wide , truncate, villous inside; corolla *ca* 15 mm long, upper lip very short, lower 3-lobed, much longer and wide, annulate within; nutlets *ca* 2.5 mm long, top rounded.

*Fl.-Fr.:*Aug. -Nov.

In waste land, Burhanpur, Indore, Khandwa.

Leucas martinicensis (Jacq.) R. Br., Hortus Kew. 3:409. 1811; *Clinopodium martinicense* Jacq., Enum. Syst. Pl. 25. 1760; *Phlomis martinicensis* (Jacq.) Sw.,Prodr. Veg. Ind. Occ. 88. 1788; Hook.f., Fl. Brit India 4:688. 1885; Samvatsar, Fl. West. Trib. Madhya Pradesh 240.1996; Mukherjee in Fl. Madhya Pradesh 2:414. 1997.

Erect, stout, simple or branched annuals, up to 1 m high; stems and branches hirsute, 4-angled and 4-grooved; leaves 3-8 × 1.5-4 cm, ovate, oblong or broadly lanceolate, apex obtuse, base cuneate, margin widely and deeply crenate; verticillasters 2-3 cm across, axillary, distant, globose, many flowered; bracts linear, 6-9 mm long with spinescent tips and slender awns; calyx sigmoid, *ca* 1 cm long and further longer in fruits, teeth 10, with spinescent tips, uppermost tooth *ca* 5 mm long, mouth oblique; corolla *ca* 1 cm long, upper lip erect, villous, lower lip spreading, midlobe 2-fid, lateral lobes oblong, rounded; nutlets shining black, obovoid-oblong, obliquely truncate at apex.

*Fl.-Fr.:*Aug. -Dec.

In gullies, open fields, shady places, near habitations, Alirajpur, Barwani, Dhar, Indore, Khandwa, Khargone.

Leucas nutans (Roth) Spreng.,Syst. Veg. 2:743. 1825; *Phlomis nutans* Roth, Nov. Pl. Sp. 263. 1821; Hook.f., Fl. Brit. India 4:688. 1885; Samvatsar, Fl. West. Trib. Madhya Pradesh 240.1996; Mukherjee in Mudgal *et al.,* Fl. Madhya Pradesh 2:415. 1997.

Annual, 10-12 cm high herbs; branches straggling, laxly hairy; leaves 4-5 × 0.5-2 cm, sessile or shortly petioled, oblong-lanceolate, apex obtuse, base truncate or narrowed, margin closely crenate-serrate, usually with long hairs on the nerves beneath; verticillasters terminal, many flowered, often in the upper axils; bracts foliaceous, lanceolate, almost equalling the calyx or shorter; calyx 8-10 mm long, falcately curved in the middle, strongly ribbed above, teeth subequal, spinous tipped, mouth oblique; corolla *ca* 1.2 cm long, lower lip longer than the upper, tube slender, annulate near the base; nutlets brown, smooth, oblong-ellipsoid, *ca* 3 mm long.

*Fl.-Fr.:*July-Nov.

Common amidst grasses, edges of fields, roadsides, throughout the area.

Leucas stricta Benth., Pl. Asiat. Rar. 1:61. 1830; Hook.f., Fl. Brit. India 4:688. 1885; Samvatsar, Fl. West. Trib. Madhya Pradesh 241. 1996; Mukherjee in Mudgal *et al.,* Fl. Madhya Pradesh 2:416. 1997.

Annual, hispid hairy herbs, about 35 cm high; branches often diffuse; leaves 2.5-5 × 1.5-2 cm,glandular beneath, hispid, linear or linear-oblong, apex obtuse, base narrow, margin entire, subsessile; verticillasters usually terminal and solitary on the branches or sometimes in 2 tiers with several leaves close to the inflorescence; bracts filiform, ciliate; calyx *ca* 1 cm long, teeth sometimes as long as the tube, mouth oblique, densely villous inside, tube nearly straight; corolla 1.5-2 cm long, tube annulate within, lower lip much longer than the upper; nutlets 0.25-0.3 cm long, brown, smooth, rounded at top.

*Fl.-Fr.:*Oct. -Jan.

Rare, weed of fields, Alirajpur, Indore, Khandwa.

Leucas urticifolia (Vahl) Sm., Cycl. 20(2):4. 1812; *Phlomis urticifolia* Vahl, Symb. Bot. 3:76. 1794; *Leucas urticifolia* (Vahl) R. Br., Prodr. 504. 1810; Hook.f., Fl. Brit. India 4:680.1885; Samvatsar, Fl. West. Trib. Madhya Pradesh 241. 1996; Mukherjee in Mudgal *et al.,* Fl. Madhya Pradesh 2:416. 1997.

Annual, erect, about 1 m high, much branched herbs; stems finely tomentose; leaves 3-5 cm long, ovate or often lanceolate, apex subacute, base cuneate, margin coarsely serrate; petioles 0.5-1.5 cm long; verticillasters axillary, densely flowered, globose whorls, 1.5-2.5 cm across; bracts almost equalling the calyx, ciliate; calyx 8-12 mm long, membranous, villous outside, teeth spinulose, mouth oblique, split above, produced on the lower side; corolla tube 0.5 cm long, upper lip villous outside, lower 3-lobed; nutlets *ca* 0.2 cm long, obovoid, obtuse, triquetrous, brown, smooth.

*Fl.-Fr.:*Oct. -Jan.

Common throughout the area.

Leucas wightiana Wall. ex Benth., Pl. Asiat. Rar. 1:60.1830; Hook.f., Fl. Brit. India 4:690.1885; Naik in Fl. Marathwada 2:721. 1998; Prasanna in N.P. Singh *et al.,* Fl. Maharashtra 2:736.2001.

Erect, much branched, pubescent and scaberulous herbs; leaves 3-7 × 0.3-0.5 cm, linear or linear-lanceolate, apex obtuse or subacute, margin entire or very distantly and obscurely serrate, hispid on both surfaces; verticillasters in axillary and terminal whorls, close above, often forming a cylindric spike; bracts linear, villous and ciliate; calyx 5-6 mm long, prominently nerved, hispid throughout from bulbous based hairs, mouth oblique, teeth triangular; corolla small, upper lip woolly, shorter than the lower; nutlets 1.5-2 mm long, oblong, dark brown.

*Fl.-Fr.:*Oct. -Dec.

Rare on gravelly soil along stream banks, **Indore** only, Mhow area.

Leucas zeylanica (L.) W. T. Aiton, Hortus Kew. 3:409. 1811; *Phlomis zeylanica* L., Sp. Pl. 586. 1753; Hook.f., Fl. Brit. India 4:689. 1885; Samvatsar, Fl. West. Trib. Madhya Pradesh 241. 1996; Mukherjee in Mudgal *et al.,* Fl. Madhya Pradesh 2:417. 1997.

Annual, erect herbs, 15-60 cm high; stems acutely 4-angled, appressed hairy; leaves 1.5-6 × 0.2-0.5 cm, linear-lanceolate to lanceolate, apex subacute, base tapering, margin entire, densely hairy below; petioles 1-4 mm long; verticillasters 2-3, usually combined into a terminal head; bracts linear, subulate, 4-7 mm long, ciliate; calyx 7-8 toothed, 10-11 nerved, mouth oblique; corolla up to 12 mm long, upper lip up to 4 mm long, lower lip spathulate, up to 11 mm long, annulate inside, near the middle; nutlets oblong-cuneate, *ca* 3 mm long, tip truncate, 3-angled, shining black.

*Fl.-Fr.:*Aug. -Dec.

Rare, along water streams and roadsides in the forest, Alirajpur, Burhanpur, Dhar, Indore, Khandwa.

MICROMERIA Benth. W. 77; M P 1; SA 1

Micromeria biflora (Buch.-Ham. ex D. Don) Benth., Labiat. Gen. Spec. 378. 1834; *Thymus biflorus* Buch.-Ham. ex D.Don, Prodr. Fl. Nepal 112. 1825; Hook.f., Fl. Brit. India 4:650.1885; Mukherjee in Mudgal *et al.,* Fl. Madhya Pradesh 2:418. 1997.

Perennial, up to 40 cm high herbs, branches from a stout, woody, rootstock; leaves up to 6 mm long, ovate or oblong, apex sub-acute, margin entire, revolute, subsessile; verticillasters axillary, 2-4 flowered or flowers pedicellate, even solitary sometimes; calyx hirsute, teeth setaceous, subulate, tube strongly nerved; corolla white or pink, upper lip notched or 2-lobed, lower lip 3-lobed, spreading.

*Fl.-Fr.:*Nov. -April

Rare, along irrigation channels, **Indore** only, Gautampura, along banks of Chambal river.

NEPETA L. W. 251; M P 2; SA 2

Key to Species

1. Calyx 4-5 mm long; nutlets clearly depressed tuberculate; peduncles filiform .... **N. graciliflora**
1. Calyx 3.5-4 mm long; nutlets inconspicuously depressed tuberculate; peduncles not filiform .... **N. hindostana**

Nepeta graciliflora Benth., Pl. Asiat. Rar. 1:65. 1830; Hook.f., Fl. Brit. India 4:663. 1885; Mukherjee in Mudgal *et al.,* Fl. Madhya Pradesh 2:420. 1997.

Slender, erect or spreading herbs, branching from the base; leaves 1-7 cm long, deltoid-ovate or oblong-ovate, apex acute, base cordate to truncate, coarsely serrate or crenate, petioles as long as the blade; verticillasters axillary and terminal, few flowered, in loose panicles; flowers long pedicellate, bracts minute, setaceous; calyx *ca* 4 mm long, teeth aristate, shorter than the tube, mouth slightly oblique; corolla lilac or rosy, 6-8 mm long; nutlets *ca* 1 mm long, dark brown, granulate.

*Fl.-Fr.:*Jan. -March

In moist places, **Burhanpur** only, Asirgad forest.

Nepeta hindostana (B. Heyne ex Roth) Haines, Bot. Bihar Orissa 2:781. 1961 (BSI reprint); *Glechoma hindostana* B. Heyne ex Roth, Nov. Pl. Sp. 258. 1821; *Nepeta ruderalis* Buch. -Ham. ex Benth., Pl. Asiat. Rar. 1:64. 1830; Hook.f., Fl. Brit. India 4:661. 1885; Mukherjee in Mudgal *et al.,* Fl. Madhya Pradesh 2:420.1997.

Annual, erect, 10-60 cm high herbs, stems grooved, obtusely angled; leaves 3-4 cm × 2 -2.5 cm, broadly ovate or orbicular, apex obtyse, base cordate, margins crenate, petioles 1-2 cm long; verticillasters dense, 1-1.5 cm across, usually in pedunculate panicles or interrupted spikes, bracts setaceous; calyx *ca* 4 mm long, ribbed, slightly curved, mouth oblique; corolla blue-purple, *ca* 6 mm long, upper lip flat, 2-fid, lower lip 2-fid with large lateral lobes and emarginate midlobe; nutlets oblong, *ca* 1 mm long, brown with white dots.

*Fl.-Fr.:*Jan. -April

Roadsides and river banks, Barwani, Burhanpur, Khargone.

OCIMUM L. W. 66; M P 4; SA 3

Key to Species

1. Shrubby; pedicels shorter than the calyx; corolla yellowish; leaves larger, 8-12 cm long; nutlets not mucilaginous when wetted . ... **O. gratissimum**

1. Herbaceous; pedicels longer than the calyx; corolla white; leaves 2-8 cm long; nutlets mucilaginous when wetted

  2. Fruiting calyx 6-8 mm long, glabrous or sparsely hairy on the upper lip or at the base; corolla 8-15 mm long; nutlets *ca* 2 mm long .... **O. basilicum**

  2. Fruiting calyx 3-5 mm long, densely villous or pilose; corolla 5-6 mm long; nutlets 1-1.5 mm long . ... **O. americanum**

Ocimum americanum L., Cent. Pl. 1:15. 1755; *O. canum* Sims in Bot. Mag. 51: t. 2452. 1823; Hook.f., Fl. Brit. India 4:607. 1885; Samvatsar, Fl. West. Trib. Madhya Pradesh 242. 1996. *Ban tulsi*

Annual, erect, 30-60 cm high herbs, stems and branches striate, subquadrangular; leaves 2.0-3.5 × 1-1.5 cm, ovate or ovate –oblong, acute at both ends, margin entire or serrulate, gland-dotted beneath; flowers white, 5-6 in a whorl, in spiciform, 5-10 cm long racemes, bracts 0.5 cm long, lanceolate, ciliate with white hairs; pedicels shorter than calyx; calyx tubular, 2-lipped, 4-6 mm long, upper lip broader than long, entire, lower lip unequally 4-toothed, the 2 central teeth longer than the upper lip; corolla 0.5-0.6 cm long, upper lip 4-toothed, lower lip longer than the upper; nutlets ellipsoid, 1-1.5 mm long, black, mucilaginous when wetted.

*Fl.-Fr.:*July-Dec.

Common in waste places, near fields, forest and village paths all over the area.

Ocimum basilicum L., Sp. Pl. 597. 1753; *O. thyrsiflorum* L., Mant. Pl. 1:84. 1767; Hook.f., Fl. Brit. India 4:608. 1885; Samvatsar, Fl. West. Trib. Madhya Pradesh 242. 1996; Naik, Fl. Marathwada 2:723. 1998. *Marwah*

Erect, 25-50 cm high, branched herbs; nodes densely hairy; leaves punctate, 2.5-6 × 1-3 cm, ovate or ovate-lanceolate, margin entire or crenate-serrate, petioles 1-2 cm long; flowers white, in thyrsoid, 10-15 cm long racemes; whorls normally 6-flowered; bracts *ca* 3 mm long, ciliate, acuminate, with long, white hairs; pedicels 2-3 mm long; calyx deflexed, accrescent, 5-6 mm long in fruit, upper lip 3-ribbed, middle 2 teeth of the lower lip aristate and longer than the upper lip; corolla 6-8 mm long, lower lip often longer than the 4-lobed upper lip; nutlets black, ellipsoid, *ca* 2 mm long, faintly rugose or pitted.

*Fl.-Fr.:*Throughout the year

Common along village roads, all over the area. (*Ocimum basilicum* L. and *O. americanum* L. are not conspecific as given in Fl. Madhya Pradesh 2:421. 1997).

Ocimum gratissimum L., Sp. Pl., 1197. 1353; Hook.f., Fl. Brit. India 4:608. 1885; Samvatsar, Fl. West. Trib. Madhya Pradesh 242. 1996; Mukherjee in Mudgal *et al.,* Fl. Madhya Pradesh 2;422. 1997. *Ram tulsi*

Shrubby, much branched, perennials from a woody base or rootstock; leaves 8-12 × 3-6 cm, broadly ovate or elliptic-lanceolate; apex acute, base cuneate, margin crenate-serrate or coarsely dentate; petioles 2.5-3.5 cm long; flowers in 10-15 cm long, thyrsoid racemes; whorls normally 6-flowered; bracts sessile or subsessile, longer than the calyx; pedicels shorter than the calyx; calyx about 1 cm long, glandular, upper lip orbicular, longer than the lower; central teeth of lower lip short, subulate; corolla 4-5 mm long, yellowish, lower lip longer than the 4-lobed, upper lip; nutlets subglobose, 1-1.5 mm long, rugose, brownish.

*Fl.-Fr.:*Nov. -March

Confined to Burhanpur, Dhar, Indore (abundant at Mandu plateau, Dhar).

ORTHOSIPHON Benth. W. 40; MP 3; SA 3

Key to Species

1. Petiole 1-3 cm long .... **O. thymiflorus**
1. Petiole less than or up to 0.4 cm long
   2. All leaves sessile or sub-sessile; corolla tube twice longer than calyx tube .... **O. rubicundus**
   2. All leaves, except the uppermost petiolate; corolla tube as long as calyx tube .... **O. pallidus**

Orthosiphon pallidus Royle ex Benth., in Hooker's J. Bot. Kew Gard. Misc. 3:370.1833 & Labiat. Gen. Spec. 708. 1835; Hook.f., Fl. Brit. India 4:613. 1885; Samvatsar, Fl. West. Trib. Madhya Pradesh 242. 1996; Mukherjee in Mudgal *et al.,* Fl. Madhya Pradesh 2:424. 1997.

Perennial, diffusely branched, 10-40 cm high herbs, branches many from the woody rootstock; leaves petiolate, gland-dotted, 2-8 × 1-4 cm, ovate, often acute at apex. Base cuneate, margin dentate or crenate, entire at base; verticillasters in short racemes with 6 flowers at each node; calyx *ca* 3 mm long, the 2 lowest teeth with long, upcurved awns, throat naked; corolla white to pale lilac, upper lip shorter than the larger, suborbicular lower lip; nutlets compressed, yellowish-brown.

*Fl.-Fr.:*July-Oct.

Common in all districts, on gravelly soil and wastelands.

Orthosiphon rubicundus (D. Don) Benth., Pl. Asiat. Rar. 2:14. 1830; *Plectranthus rubicundus* D. Don, Prodr. Fl. Nepal. 116. 1825; Hook.f., Fl. Brit. India 4:614. 1885; Mukherjee in Mudgal *et al.,* Fl. Madhya Pradesh 2:425. 1997.

Perennial, 15-40 cm high herbs with a stout rootstock; roots tuberous; leaves sessile or subsessile, 3-15 × 1-5 cm, ovate or ovate-lanceolate, apex acute, base cuneate, margin coarsely dentate; verticillasters in long, slender racemes with distantly placed flowering whorls, bracts minute; calyx 4-6 mm long, throat naked, lower teeth subulate; corolla white to pinkish, twice longer (or less) than the calyx; nutlets compressed, suborbicular punctulate, smooth.

*Fl.-Fr.:*April-Aug.

Rare, along forest roads, Burhanpur, Khandwa.

Orthosiphon thymiflorus (Roth) Sleesen in Reinwardtia 5:42. 1959; *Ocimum thymiflorum* Roth, Nov. Pl. Sp. 269. 1821; *Orthosiphon glabratus* Benth. in Pl. Asiat. Rar. 2:14. 1830; *O. tomentosus* Benth., Pl. Asiat. Rar. 2:14. 1830; Hook.f., Fl. Brit. India 4:613. 1885; Mukherjee in Mudgal *et al.,* Fl. Madhya Pradesh 2:425. 1997.

Erect, *ca* 30 cm high herbs; leaves 3-8 cm long, ovate, apex acute, base truncate; verticillasters in terminal racemes, 5-6 flowered at each node, bracts ovate; corolla white, upper lip suborbicular, lower broad; nutlets1.5 mm long, compressed, brown, minutely tuberculate, enclosed in the calyx.

*Fl.-Fr.:*Aug. -Oct.

Rare, **Khandwa** only, at Singot and Khalwa.

PLECTRANTHUS L'Herit. W. 325; MP 2; SA 2

Key to Species

1. Roots fleshy; stem densely hirsute, with orange oil glands; leaves 2-8 × 1.5-4.5 cm; flowers up to 2 cm long; fruiting calyx 0.8 cm long; filaments united at base .... **P. barbatus**

1. Roots not fleshy; stem fleshy below, puberulous, without orange oil glands; leaves 3-11 × 2-6 cm flowers 1-1.2 cm long; fruiting calyx 1-1.2 cm long; filaments free . ... **P. mollis**

Plectranthus barbatus Andr., Bot. Repos. 10 :t. 594. 1810; *Coleus barbatus* (Andr.) Benth. ex G. Don, Hort. Brit. 483. 1830; Hook.f., in Fl. Brit. India4:625. 1885, exl. syn; Mukherjee in Mudgal *et al.,* Fl. Madhya Pradesh 2:402. 1997 as *Coleus barbatus* (Andr.) Benth. 1830.

Perennial, densely hirsute, glandular herbs; roots tuberous; leaves 2-8 × 1.5-4.5 cm, densely pubescent; inflorescence 10-25 cm long, rachis glandular; flowers bluish *ca* 2 cm long; filaments fused at base in a sheath around style; seeds smooth, dark brown.

Fl.-Fr.: Nov.-Feb.

Rare, on meadows, **Indore** only, Sitala Mata falls area.

Plectranthus mollis (Aiton) Spreng., Syst. Veg. 2:690.1825; *Ocimum molle* Aiton, Hort. Kew. 2:322. 1789; *Plectranthus incanus* Link, Enum. Hort. Berol. Alt. 2:120.1822; Hook.f., Fl. Brit. India 4:621. 1885; Mukherjee in Mudgal *et al.,* Fl. Madhya Pradesh 2:428. 1997.

Erect, robust, annual herbs about 1 m high; stems pubescent, fleshy below; leaves 3-11 × 2-6 cm, ovate-cordate, apex acute or acuminate, margin crenate, base deeply cordate; petioles 3-10 mm long; flowers pale lilac to light blue, in slender, 10-25 cm long racemes, paniculately branched below, pedicels 3-4 mm long in flower, 6-8 mm long in fruit; bracts small, ovate, cuneate; calyx 4-5 mm long in flowers, 1 cm long in fruits, broadly campanulate, upper lip short, rounded, reflexed, lower with 4, acutely pointed teeth; corolla *ca* 1 cm long, lower lip sub-hemispheric, boat-shaped; nutlets *ca* 2 mm across, subglobose with purplish-brown spots.

*Fl.-Fr.:*Aug. -March

Rare, along forest clearings, undergrowth, **Indore** only, at Khurda-Khurdi.

POGOSTEMON Desf. W. 94; MP 4; SA 4

Key to Species

1. Opposite pairs of leaves very unequal .... **P. paniculatus**
1. Opposite pairs of leaves not very unequal
   2. Stem glabrous, polished; calyx teeth shorter than the tube . ... **P. parviflorus**
   2. Stem hairy, pubescent or villous
      3. Calyx teeth shorter than the tube, 5-nerved; bracts large, foliaceous; apices of sepals not penicillate; stem with short hairs .... **P. benghalensis**
      3. Calyx teeth as long as or longer than the tube, 15 nerved; bracts small; apices of leaves penicillate; stem with long, spreading hairs . . .. **P. purpurascens**

Pogostemon benghalensis (Burm.f.) Kuntze, Revis. Gen. Pl. 2:529. 1891; *Origanum benghalense* Burm.f., Fl. Indica 128. 1768; Hook.f., Fl. Brit. India 4:632. 1885; Samvatsar, Fl. West. Trib. Madhya Pradesh 243. 1996; Mukherjee in Mudgal *et al.,* Fl. Madhya Pradesh 2:431. 1997.

Aromatic bushy shrubs, 1-2 m high, branches dark purple; young parts hoary pubescent or villous; leaves 5-10 × 3-7 cm, ovate, apex acute, margin doubly serrate, base rounded or cuneate; flowers in dense, hairy terminal and axillary spikes forming panicles; bracts pubescent, foliaceous; bracteoles linear; calyx pubescent, teeth lanceolate; stamens bearded with purple hairs; nutlets broadly ellipsoid, reddish-brown.

*Fl.-Fr.:*Dec. -Jan.

On hilly slopes, Alirajpur, Barwani, Dhar, Khandwa, Khargone.

(*Pogostemon pletrantoides* Desf. is a different species, not conspecific with the above)

Pogostemon paniculatus (Willd.) Benth., Pl. Asiat. Rar. 1:30.1830; *Elscholtzia paniculata* Willd. Sp. Pl. 3:59. 1800; Hook.f., Fl. Brit. India 4:631. 1885; Samvatsar, Fl. West. Trib. Madhya Pradesh 243. 1996; Mukherjee in Mudgal *et al.,* Fl. Madhya Pradesh 2:433. 1997.

Erect, villous undershrubs, 1-2 m high; leaves pubescent, 7-10 × 5-8 cm, opposite, pairs unequal, apex acute, base cuneate, irregularly serrate, membranous; petioles 1-3 cm long; verticillasters in axillary and terminal, 5-10 cm long, laxly paniculate spikes; bracts imbricating, membranous, pubescent, about as long as calyx; calyx *ca* 5 mm long, teeth lanceolate; corolla white, as long as the calyx; nutlets minute, brownish-black, smooth, shining.

*Fl.-Fr.:*Sept. -Dec.

Rare, in the forest at **Alirajpur** only, Aamkhut forest.

Pogostemon parviflorus Benth., Pl. Asiat. Rar. 1:31. 1830; Hook.f., Fl. Brit. India 4:632. 1885; Samvatsar, Fl. West. Trib. Madhya Pradesh 243. 1996.

Shrubs about 1 m high, stem and branches pubescent, obtusely 4-angled; leaves 5-10 × 2-5 cm, ovate, apex acute, base cuneate, pubescent below, especially on the nerves, petioles long; flowers white-purple in dense spikes forming panicles; bracts pubescent, ciliate; nutlets ellipsoid, brownish.

*Fl.-Fr.:*Nov. -Jan.

Rare as forest undergrowth in **Alirajpur** only, way to Katarkheda

Pogostemon purpurascens Dalzell in Hooker's J. Bot. Kew Gard. Misc. 2:337. 1850; Hook.f., Fl. Brit. India 4:632. 1885; Samvatsar, Fl. West. Trib. Madhya Pradesh 243. 1996; Mukherjee in Fl. Madhya Pradesh 2:434. 1997.

Erect, pubescent herbs or shrubs, stems furrowed; leaves 6-15 × 4-6 cm, ovate, apex acute, base cuneate, margin doubly serrate, petioles 2-4 cm long; verticillasters globose in long, pedunculate, 6-16 cm long, spikes, lower whorls sparsely arranged; bracts oblong-lanceolate, *ca* 5 mm long, penicillate at the tip; calyx *ca* 5 mm long, pubescent, teeth lanceolate, as long as the 5-angled tube; corolla white with purple upper lip; nutlets ovoid, smooth, shining black.

*Fl.-Fr.:*Nov. -Jan.

Not common, confined to hilly slopes in **Alirajpur** only, Bhabhra-Kattiwada road.

PREMNA L. W. 104; MP 3; SA 1

Premna barbata Wall. ex Schauer, Prodr. 11:636. 1847; Hook.f., Fl. Brit. India 4:579. 1885; Srivastava in Mudgal *et al.,* Fl. Madhya Pradesh 2:383. 1997.

Shrubs about 1.5 m high, branched from the base; leaves 7-18 × 5-7 cm, oblong or ovate, apex acuminate, base subcordate or cuneate, margin dentate; flowers in terminal, more or less dense corymbs; bracts linear, caducous; calyx 4-toothed, teeth oblong, obtuse; corolla *ca* 2.5 mm long, bilabiate; drupes globose or obovoid, *ca* 2 mm across, brown at maturity, obscurely verrucose.

*Fl.-Fr.:*July-Aug.

Rare, Barwani, Khargone.

ROTHECA Raf. W. 33; MP 1; SA 1

Rotheca serrata (L.) Steane & Mabb. Novon 8:206. 1998; *Volkameria serrata* L., Mant. Pl. 1:90.1767; *Clerodendrum serratum* (L.) Moon, Cat. Pl. Ceylon 46. 1824; C.B. Clarke in Hook.f., Fl. Brit. India 4:592. 1885; Samvatsar, Fl. West. Trib. Madhya Pradesh 234. 1996; Srivastava in Mudgal *et al.,* Fl. Madhya Pradesh 2:372. 1997. *Bharangi* **(Plate 8, Fig. 5)**

Shrubs, up to 3 m high, branches 4-angled; leaves opposite or in whorls of 3, ovate or elliptic, 10-18 × 4-6 cm, subsessile, apex acute, base subcordate or rounded, margin serrate; flowers blue, in terminal, pyramidal, up to 30 cm long panicles; calyx campanulate; corolla bilabiate, unequally 5-lobed, 7-9 mm long; drupes subglobose, dark purple or black at maturity, 2-4 lobed, *ca* 7 mm long, fruiting calyx cup-like, somewhat accrescent.

Fl.-Fr.: July-Nov.

Forest clearings, paths, open areas, Alirajpur, Burhanpur, Dhar, Indore, Khandwa.

SALVIA L. W. 986; MP 2; SA 2

Key to Species

1. Upper lip of calyx 3-toothed; leaves 2-4 cm long .... **S. aegyptiaca**
1. Upper lip of calyx tube entire; leaves 6-10 cm long .... **S. plebeia**

Salvia aegyptiaca L., Sp. Pl. 23. 1753; Hook.f., Fl. Brit. India 4:656. 1885; Samvatsar, Fl. West. Trib. Madhya Pradesh 244. 1996; Mukherjee in Mudgal *et al.,* Fl. Madhya Pradesh 2:436. 1997.

Straggling undershrubs with rigid, hairy, 4-angled branches; leaves linear-lanceolate, 0.5-4 × 0.3-0.8 cm, apex acute, margin crenate, subsessile or shortly petiolate; verticillasters 2-6 flowered; calyx glandular, hairy; corolla white, 4-5 mm long; nutlets ellipsoid, bluish-black, smooth.

*Fl.-Fr.:*Aug. -Dec.

Rare in forest, along forest paths, Shakarja, **Alirajpur** only.

Salvia plebeia R. Br., Prodr. Fl. Nov. Holl. 501. 1810; Hook.f., Fl. Brit. India 4:655. 1885; Samvatsar, F. West. Trib. Madhya Pradesh 244. 1996; Mukherjee in Mudgal *et al.,* Fl. Madhya Pradesh 2: 437. 1997.

Annual, erect, up to 30 cm high herbs; stems glandular, pubescent, obtusely 4-angled, grooved; leaves oblong-lanceolate, 4-14 × 1.5-4 cm, apex acute, base narrowed or subcordate, margin irregularly crenate; verticillasters in 5-25 cm long panicles with 4-6 flowers at each node; bracts small, narrow; fruiting calyx *ca* 3 mm long, upper lip shorter than the lower, reflexed, lower lip obtusely 3-toothed; corolla white or lilac, *ca* 5 mm long; upper lip oblong-obtuse, lower lip 3-lobed, middle lobe much larger, tube exserted, annulate inside; nutlets ovoid, brown, smooth, mucilaginous when wet.

*Fl.-Fr.:*Sept. -Feb.

Common in moist and shady places, Alirajpur, Barwani, Burhanpur, Dhar, Indore, Khargone.

TECTONA L.f. W. 3; MP 1; SA 1

Tectona grandis L.f., Suppl. Pl. 151. 1782; Hook.f., Fl. Brit. India 4:570.1885; Samvatsar, Fl. West. Trib. Madhya Pradesh 236. 1996; Srivastava in Mudgal *et al.,* Fl. Madhya Pradesh 2:387. 1997. *Sagwan*

Large trees, up to 20 m high, young parts tomentose; young branches stout, 4-angled; leaves opposite, ovate-elliptic, 12-5- × 6-30 cm, stellately tomentose beneath, margin entire or repand-denticulate; flowers actinomorphic, in dichotomously branched, 20-40 cm long, tomentose panicles; calyx *ca* 4 mm long, teeth often reflexed; corolla white, erect or reflexed, tube cylindrical, *ca* 2 mm long; drupes enclosed in 4-angled, inflated calyx.

*Fl.-Fr.:*Aug. -Dec.

Very common in all districts because of extensive plantation.

VITEX L. W. 223; MP 3; SA 3

Key to Species

1. Leaves sessile, 1-2 or 3 foliolate; panicles 3-12 cm long .... **V. trifolia**
1. Leaves petiolate, always 3-5 foliolate
   2. Inflorescence axillary; panicles up to 20 cm long; leaves 3-foliolate; petiole winged on young shoots .... **V. peduncularis**
   2. Inflorescence terminal; panicles up to 30 cm long; leaves 3-5 foliolate; petiole not winged .... **V. negundo**

Vitex negundo L., Sp. Pl. 638. 1753; Hook.f., Fl. Brit. India 4:583. 1885; Samvatsar, Fl. West. Trib. Madhya Pradesh 236. 1996; Srivastava in Mudgal *et al.,* Fl. Madhya Pradesh 2:390.1997. *Samhalu*

Shrubs, up to 2.5 m high; leaves 3-5 foliolate (normally 3), oblanceolate, 2-12 × 2-3 cm, apex acuminate, base acute, white tomentose beneath; flowers blue in terminal and upper axillary, about 20 cm long, cymose panicles; calyx *ca* 3 mm long, 5-toothed, white-spotted; corolla tube *ca* 5 mm long; drupes globose, *ca* 4-5 mm long, black at maturity.

*Fl.-Fr.:*Sept. -April

Common in waste places, hedges, in all districts.

Vitex peduncularis Wall. ex Schauer, Prodr. 11:687. 1847; Hook.f., Fl. Brit. India 4:587. 1885; Samvatsar, Fl. West. Trib. Madhya Pradesh 236. 1996; Srivastava in Mudgal *et al.,* Fl. Madhya Pradesh 2:390.1997.

Large shrubs or small trees, about 6-7 m high; leaves 3-foliolate; leaflets oblanceolate, coriaceous, 9-15 × 2-4 cm, apex shortly acuminate; petioles winged; flowers in dichasial cymes, arranged in axillary, up to 25 cm long, racemose panicles; calyx *ca* 4 mm long, yellow-spotted, 5-toothed; corolla cream-coloured with bluish tinge, tube 2 mm long; drupes obovoid, red, *ca* 7 mm across.

*Fl.-Fr.:*April-Aug.

Rare in hilly areas of Mandu plateau, **Dhar**.

Vitex trifolia L., Sp. Pl. 638. 1753; Hook.f., Fl. Brit. India 4:583. 1885; Samvatsar, Fl. West. Trib. Madhya Pradesh 236. 1996; Prasanna in N.P. Singh *et al.,* Fl. Maharashtra 2:701.2001.

Woody shrubs, up to 3 m high, young parts greyish-tomentose; leaflets 1-3, terminal leaflet subsessile, lateral ones sessile, 3-7 × 1-3 cm, apex obtuse to subacute, base cuneate-attenuate, white tomentose beneath, glabrous above; flowers white, in terminal, tomentose panicles about 12-13 cm long; drupes 5 × 4 mm, globose, fruiting calyx about 5 × 5 mm.

*Fl.-Fr.:*Most part of the year

Common hedge plant in villages.

VOLKAMERIA L. W. 11; MP 1; SA 1

Volkameria inermis L. Sp. Pl. 637. 1753; *Clerodendrum inerme* (L.) Gaertn., Fruct. Sem. Pl. 1:271. 1768; Hook.f., Fl. Brit. India 4:589. 1885; Samvatsar, Fl. West. Trib. Madhya Pradesh 233. 1996; Srivastava in Mudgal *et al.,* Fl. Madhya Pradesh 2:371. 1997.

Shrubs, about 3 m high; leaves glabrous, 3-5 × 1.5-2.5 cm, elliptic or narrowly lanceolate, apex obtuse, base acute; flowers white in 3-flowered cymes; calyx campanulate, 5-toothed, *ca* 4 mm long; corolla hypocrateriform, lobes obovate, tube slender, 2-3 cm long; drupes 4-lobed, obovate, deep purple at maturity, base enclosed in the enlarged, persistent calyx.

*Fl.-Fr.:*July-April

Common hedge plant all over the area, not wild.

NYCTAGINACEAE 34/450; 1/4; 1/6

BOERHAVIA L. W. 105; MP 4; SA 6

Key to Species

1. Anthocarp ribbed, not glandular; plants erect or subscandent (diffuse in absence of support)
    2. Inflorescence of 3-8 flowered in axillary umbels; pedicels capillary; ribs of anthocarp straight . ... **B. chinensis**
    2. Inflorescence a widely branched, terminal panicle; pedicels not capillary; ribs of anthocarp undulate . ... **B. erecta**
1. Anthocarp glandular and ribbed
    3. Flowers pedicellate, in axillary or terminal pedunculate cymes
        4. Flowers in terminal, diffuse panicles .... **B. procumbens**
        4. Flowers in axillary, pedunculate cymes .... **B. repens**
    3. Flowers sessile or subsessile, in axillary or terminal umbels
        5. Perianth white; anthocarp sparsely glandular on the ribs, glands crowded at the crown .... **B. plumbaginea**
        5. Perianth pink; anthocarp densely glandular on the ribs, less so at the crown .... **B. diffusa**

Boerhavia chinensis (L.) Aschers & Schweinf., Beitr. Fl. Aethiop. 1:167. 1867; *Valeriana chinensis* L. Sp. Pl. 33. 1753; *Boerhavia repanda* Willd., Sp. Pl. 1:22. 1797; *Commicarpus chinensis* (L.) Hiemerl in Engl. & Prantl, Nat. Pflanzenfam. ed. 2.16C: 117.1934; Samvatsar, Fl. West. Trib. Madhya Pradesh 245. 1996; Mukherjee in Mudgal *et al.,* Fl. Madhya Pradesh 2:444.1997.

Diffuse or subscandent shrubs; leaves 2.5-6 × 1-4.5 cm, acute or acuminate at apex, truncate to cordate at base, margin lobed or sinuate, petiole channeled; flowers pink, in axillary, 3-8 flowered umbels, mostly from the axil of one leaf; pedicels slender, 7-12 mm long; perianth infundibuliform; anthocarps ribbed, uniformly covered with wart-like glands.

*Fl.-Fr.:*Sept. -Dec.

Common in hedges throughout the area.

Boerhavia diffusa L., Sp. Pl. 3.1753; *B. repens* var. *diffusa* (L.) Hook.f., Fl. Brit. India 4:709: 1885; Samvatsar, Fl. West. Trib. Madhya Pradesh 244.1996; Mukherjee in Mudgal *et al.,* Fl. Madhya Pradesh 2:441.1997. *Punarnava*

Perennial, diffuse, straggling herbs; stem woody below; leaves 0.5-3 × 0.5- 2.5 cm, broadly suborbicular, unequal, margin sinuate to repand, whitish tomentose beneath, obtuse or subacute at apex, roundish at base; flowers pink, sessile, in umbellate, terminal, diffuse panicles, exceeding the foliage; bracts and bracteoles ovate, acuminate, 1.5-2 mm long; anthocarp 1-3 mm long, 5-ribbed, glandular hairy.

*Fl.-Fr.:*Almost throughout the year

Very common in all districts.

Boerhavia erecta L. Sp. Pl. 3. 1753; Nair in Bull. Bot. Surv India 9:283.1967; Moorthy in N.P. Singh *et al.,* Fl. Maharashtra 2:772.2001.

Shrubs with a purplish stem when young; leaves 1-4 × 0.5-3 cm, linear-lanceolate to ovate, apex apiculate, base rounded or acute, margin wavy, lower surface whitish; flowers purplish, in groups of 1-3, in much branched, umbelliform clusters; anthocarps turbinate with a truncate crown, glands absent.

*Fl.-Fr.:*Oct. -Feb.

Rare, **Burhanpur** only, Asirgad forest.

Boerhavia plumbaginea Cav., Icon 2:7. 1793; *B. verticillata* Poir., Encycl. 5:56. 1804; *B. repanda* Roxb., Tracts St. Helena 302.1816; Samvatsar, Fl. West. Trib. Madhya Pradesh 244. 1996.

Diffuse herbs with long branches; leaves thick, ovate to suborbicular, apex obtuse, base cordate; flowers white, in few flowered whorls on long peduncles; pedicels filiform; glands stalked, on apex of the anthocarp.

*Fl.-Fr.:*Oct. -Dec.

Common in all districts.

Boerhavia procumbens Banks ex Roxb., Fl. Ind. 1:148. 1820; *B. repens* var. *procumbens* (Banks ex Roxb.) Hook.f., Fl. Brit. India 4:709. 1885; Mukherjee in Mudgal *et al.,* Fl. Madhya Pradesh 2:441. 1997.

Perennial, diffuse, straggling herbs; stems woody below; leaves 1-5 × 0.3-3.5 cm, ovate or oblong to sub-cordate, cuspidate ot obtuse, sinuate to repand; flowers in terminal, diffuse panicles, much exceeding the foliage; bracts and bracteoles ovate, 1.5-2 mm long; perianth pinkish, campanulate; anthocarps turbinate to broadly clavate, *ca* 3 mm long, glandular, papillose.

*Fl.-Fr.:*Sept. -June

Burhanpur, Khandwa, along roadsides of villages on gravelly soil.

Boerhavia repens L. Sp. Pl. 3. 1753; Hook.f., Fl. Brit. India 4:709. 1885, excl. var. *diffusa* & *procumbens*; Mukherjee in Mudgal *et al.,* Fl. Madhya Pradesh 2:442. 1997. *Shwet punarnava*

Perennial, diffuse, straggling jherbs; branches pinkish; larger leaves *ca* 5 × 4 cm, sinuate to repand; flowers pink, in axillary cymes, peduncles solitary, rarely exceeding the leaves; perianth tube glandular, constricted above the ovary; anthocarps papillose, turbinate to narrowly clavate.

Fl.-Fr.: Almost throughout the year

Barwani, Burhanpur, Indore, Khandwa, Khargone.

AMARANTHACEAE 178/2052; 14/30; 12/28

Key to Genera

1. Perianth not dry or scarious; leaves fleshy; anthers 2-celled
    2. Stigmas 5; fruit with yellow glands; plants aromatic; seeds reddish-brown ....DYSPHANIA
    2. Stigmas 2; fruit without glands; plants not aromatic; seeds blackish-brown .... CHENOPODIUM
1. Perianth, bracts and bracteoles scarious and dry
    3. Leaves alternate
        4. Perianth scarious, pink when young; fruit many seeded ....CELOSIA
        4. Perianth not scarious; fruit 1-seeded
            5. Flowers bisexual
                6. Filaments united at base; inflorescence of simple or compound spikes; seeds exarillate .... AERVA
                6. Filaments free; flowers bright pink to purple .... DIGERA
            5. Flowers unisexual, ceamy-white or greenish, in simple or branched spikes or clusters ....AMARANTHUS
    3. Leaves opposite
        7. Fertile flowers mixed with many imperfect florets modified as bristles ....PUPALIA
        7. Fertile flowers not mixed with modified florets
            8. Staminodes absent
                9. Stamens 2; flowers in axillary clusters; bracteoles without a dorsal crest . ...NOTHOSAERVA
                9. Stamens 2-5; flowers in terminal, globose or oblong spike; bracteoles with a dorsal crest ....GOMPHRENA
            8. Staminodes present; stamens 2-5
                10. Stigma 1, capitate ....ALTERNANTHERA
                10. Stigmas 2-lobed or 2-branched
                    11. Staminodes fringed or long fimbriate; midvein of bracteole spiny, utricle indehiscent... ACHYRANTHES
                    11. Staminodes triangular; midvein not as above; utricle dehiscing by lid.... TRICHURIELLA

ACHYRANTHES L. W. 12; MP 3; SA 3

Key to Species

1. Staminodes with long, fimbriate scales; wings of the bracteoles adnate to spines; spines 0.2-0.4 cm long . ...**A. aspera**
1. Staminodes simple; wings of bracteoles free; spines 0.4-0.5 cm long .... **A. bidentata**

Achyranthes aspera

Key to Varieties

1. Plants tomentose; leaves thick .... var. **aspera**
1. Plants subglabrous; leaves membranous, more or less glabrous .... var. **porphyristachya**

Achyranthes aspera L. var. **aspera** Sp. Pl. 204. 1753; Hook.f., Fl. Brit. India 4:730.1885; Samvatsar, Fl. West. Trib. Madhya Pradesh 247. 1996; Mukherjee in Mudgal *et al.,* Fl. Madhya Pradesh 2 :448. 1997. *Lat jeera*

Erect, pubescent, tomentose herbs; leaves thick, 3-15 × 1-8 cm, obovate or oblanceolate, acute to acuminate at apex, base narrowed, never silvery-silky beneath; flowers greenish, in erect, terminal spikes with spinous- tipped perianth lobes; utricle indehiscent; seeds cylindrical, smooth.

Fl.-Fr.: Aug. -Feb.

Very common all over.

Achyranthes aspera var. **porphyristachya** (Wallich ex Moq.) Hook.f., Fl. Brit. India 4:730. 1885; *A. porphyristachya* Wallich ex Moq. in DC., Prodr. 13(2): 316. 1849; Mukherjee in Mudgal *et al.,* Fl. Madhya Pradesh 2:448. 1997.

Differs from the typical variety in subglabrous stems, membranous, more or less subglabrous leaves.

*Fl.-Fr.:*Aug. -Feb.

Rare, waste lands, **Khandwa** only, at Mandhata.

Achyranthes bidentata Blume, Bijdr. 545. 1826; Hook.f., Fl. Brit. India 4:730.1885; Samvatsar, Fl. West. Trib. Madhya Pradesh 347. 1996; Mukherjee in Mudgal *et al.,* Fl. Madhya Pradesh 2:450.1997.

Erect or straggling, much branched, perennial herbs; leaves 5-15 × 2-7 cm, elliptic-oblong to broadly ovate, apex acuminate, base narrowed; petioles 3-20 mm long; flowers greenish, in initially dense but finally elongating, about 20 cm long spikes; peduncles 1-4 cm long; bracts narrowly lanceolate, membranous; 3-5 mm long; bracteoles 3.5-5.5 mm long, almost free from the spine; pseudostaminodes subquadrate; utricles 2-3 mm long; seeds smooth.

*Fl.-Fr.:*Sept. -Dec.

Rare, **Alirajpur** only, near water fall at Kattiwada.

AERVA Forssk. W. 20; MP 3; SA 3

Key to Species

1. Flowers unisexual; seeds round .... **A. javanica**
1. Flowers bisexual and bisexual and female both; seeds reniform
    2. Leaves more than 5 cm long; spikes axillary and terminal, often paniculate .... **A. sanguinolenta**
    2. Leaves 1-5 cm long; spikes axillary, not paniculate... **A. lanata**

Aerva javanica (Burm.f.) Juss. ex Schult., Syst. Veg. 5:565. 1819; *Iresine javanica* Burm.f., Fl. Indica 212. t. 65. f. 2. 1768; Hook.f., Fl. Brit. India 4:727. 1885; *Aerva tomentosa* Forssk., Fl. Aegypt. -Arab. 122, 170.1775; Samvatsar, Fl. West. Trib. Madhya Pradesh 247. 1996; Mukherjee in Mudgal *et al.,* Fl. Madhya Pradesh 2:451. 1997.

A much branched, erect, tomentose perennial herb or undershrub; leaves 5-13 × 1.5-2.5 cm, linear-oblong or oblong-spathulate, sessile, tomentose on both surfaces; flowers unisexual, in terminal, cylindrical, woolly, paniculate spikes; outer tepals ovate in male and obovate spathulate in female flowers, inner tepals smaller; utricles ovoid-orbicular, compressed; seeds lenticular, shining, brown-black.

Fl.-Fr.: Aug. -Jan.

Alirajpur and Dhar.

Aerva lanata (L.) Juss., Ann. Mus. Hist. Nat. 11:131. 1808; *Achyranthes lanata* L., Sp. Pl. 204. 1753; Hook.f., Fl. Brit. India 4:728. 1885; Samvatsar, Fl. West. Trib. Madhya Pradesh 247. 1996; Mukherjee in Mudgal *et al.,* Fl. Madhya Pradesh 2:451. 1997. *Kapuri*

An erect, suberect or diffuse, 15-40 cm tall, woolly, tomentose herb; stems and branches covered with soft, woolly hairs; leaves 1.2-4 × 1-2.5 cm, elliptic or elliptic-obovate, white woolly beneath, tapering and unequal sided at base; flowers greenish-white, in dense, axillary spikes; utricle broadly ovoid; seeds smooth, black.

*Fl.-Fr.:*Sept. -April

Very common all over.

Aerva sanguinolenta (L.) Blume, Bijdr. 547. 1826; *Achyranthes sanguinolenta* L., Sp. Pl. ed. 2.294. 1762; *Achyranthes scandens* Roxb., Fl. Ind. 1:676. 1820; Hook.f., Fl. Brit. India 4:727. 1885; Samvatsar, Fl. West. Trib. Madhya Pradesh 247. 1996; Mukherjee in Mudgal *et al.,* Fl. Madhya Pradesh 2:452. 1997.

Erect or straggling perennial herbs or undershrubs; stem terete, appressedly or patently hairy towards the apex; leaves 1.5-8 × 0.5-4 cm, membranous, alternate or rarely opposite, elliptic-lanceolate or lanceolate, usually tapering at both ends; flowers in densely tomentose, axillary or terminal spikes; panicles 1-5 cm long or terminal ones up to 10 cm long; bracts and bracteoles ovate, mucronate; tepals densely white-villous; pseudostaminodes shorter than stamens; utricles thin, rupturing irregularly, more or less transeversely.

*Fl.-Fr.:*Throughout the year

Alirajpur, Barwani, Burhanpur, Dhar, Khandwa, Khargone.

ALTERNANTHERA Forssk. W. 137; MP 5; SA 5

Key to Species

1. Heads prickly; tepals extremely dissimilar . ... **A. pungens**
1. Heads not prickly; tepals similar
    2. Plants of aquatic habitat; peduncles 2-5 cm long .... **A. philoxeroides**
    2. Plants of terrestrial habitat

3. Tepals 1-nerved; leaves linear-oblong; petioles short .... **A. sessilis**
3. Tepals 3-nerved
    4. Pseudostaminodes much shorter than the stamens; plants prostrate .... **A. paronychioides**
    4. Pseudostaminodes as long as stamens; plants erect, ascending .... **A. bettzickiana**

Alternanthera bettzickiana (Regel) G. Nicolson, Ill. Dict. Gard. 1:59. 1884; *Telanthera bettzickiana* Regel, Index Seminum (LE) 1862:28. 1862; *Achyranthes bettzickiana* (Regel) Standl. N. Amer. Fl. 21:138. 1917; Samvatsar, Fl. West. Trib. Madhya Pradesh 248. 1996; Mukherjee in Mudgal *et al.,* Fl. Madhya Pradesh 2:454. 1997.

Erect or ascending, bushy perennial herbs, stem with a purple tinge on nodes; leaves narrowly or more broadly elliptic to oblanceolate, apex aute, 2-6 × 1-3 cm; spikes axillary, sessile, usually solitary, globose or ovoid; tepals white, lanceolate, mucronate with excurrent midrib; utricles enclosed within persistent tepals.

*Fl.-Fr.:*Almost throughout the year

Very common and aggressive weed, occupying all possible niches.

Alternanthera paronychioides A. St. Hill. Voy. Distr. Diam. 2(2):43. 1833; Dutta & Mitra in Ind. For. 87:304 f. 2, 1961; Townsend in Dassan. & Fosb., Rev. Handb. Fl. Ceylon 1:50; 1980; Samvatsar, Fl. West. Trib. Madhya Pradesh 248. 1996; Mukherjee in Mudgal *et al.,* Fl. Madhya Pradesh 2:454. 1997.

Prostrate herbs, rooting at nodes; leaves 0.5-1 × 0.4-0.9 cm, obovate-spathulate, apex obtuse, base cuneate; flowers spicate, axillary, solitary or 2-3 together, shining white; bracts and bracteoles scarious; stamens 5, all fertile; utricles narrowly winged, orbicular-obcordate, enclosed in persistent perianth 1-seeded.

Fl.-Fr.: Throughout the year

Common in all districts.

Alternanthera philoxeroides (Mart.) Griseb. in Abh. Koningl. Ges. Wiss. Gottingen 24:36. 1879; *Bucholzia philoxeroides* Mart., Beitr. Amarantac. 107. 1825; Mukherjee in Mudgal Fl. Madhya Pradesh 2:454. 1997.

Perennial, floating herbs, stems *ca* 1.5 cm thick, rooting at nodes; leaves variable, obovate-spathulate or linear-lanceolate to oblong, up to 10 × 2 cm, apex acute, base cuneate; flowers in axillary, solitary, globose, pedunculate heads, ultimate terminal head, smaller, sessile; peduncles 2-5 cm long; bracts and bracteoles persistent; tepals white; utricle obcordate with thick margin; seeds discoid, brown.

*Fl.-Fr.:*Aug. -Oct.

Rare, Indore and Khandwa (In Indore, the surface of tank near Khajrana temple is so full that water is not visible).

Alternanthera pungens Kunth, Nov. Gen. Sp. 2:206. 1818; *Achyranthes repens* L., Sp Pl. 205. 1753; *Illecebrum pungens* (Kunth) Spreng. Syst. Veg. 1:820.1824; Samvatsar, Fl. West. Trib. Madhya Pradesh 248. 1996; Mukherjee in Mudgal *et al.,* Fl. Madhya Pradesh 2: 455. 1997.

Prostrate, decumbent, widely spreading herbs; stems zigzag, hairy; leaves punctuate, 1.5-4 cm long, obovate or suborbicular, apex mucronate, base tapering into a petiole; flowers whitish, sessile, in small, axillary spikes; bracts and bracteoles scarious, glabrous, white, very sharp pointed; utricles *ca* 2 mm long; seeds discoid, brown.

*Fl.-Fr.:*Sept. -Jan.

Common along road sides of villages, waste places, throughout the area (Essentially a species of dry ground).

Alternanthera sessilis (L.) R. Br. ex DC. Cat. Pl. Horti Monsp. 77. 1813; *Gomphrena sessilis* L., Sp. Pl. 225. 1753; Hook.f., Fl. Brit. India 4:732. 1885; Samvatsar, Fl. West. Trib. Madhya Pradesh 249. 1996; Mukherjee in Mudgal *et al.,* Fl. Madhya Pradesh 2:455. 1997. Guroo oblique

Prostrate, spreading, polymorphic herbs; stems rooting at nodes; internodes shortening in length upwards on the branch; leaves 4-6 × 1-3 cm, glabrous, variable, obtuse at apex, cuneate at base, linear-lanceolate to oblong, ovate or obovate-spathulate; flowers white or tinged pink, in axillary, spicate, sessile heads; bracts and bracteoles scarious; tepals ovateelliptic, equal in length; stamens 5, two filaments without anthers; utricles obcordate, 2-2.5 mm long, margin thick; seeds discoid, brown.

*Fl.-Fr.:*Throughout the year

Common in moist habitat, all over the area.

AMARANTHUS L. W. 105; MP 9; SA 9

Key to Species

1. Stamens and tepals 5
    2. Leaf axils with spines .... **A. spinosus**
    2. Leaf axils without spines
        3. Bracts and bracteoles spine-tipped, 1.5-2 times longer than the tepals
            4. Bracts ovate, mucronate; tepals obovate, acuminate .... **A. caudatus**
            4. Bracts and tepals ovate-acuminate .... **A. cruentus**
        3. Bracts and bracteoles not spine-tipped, much shorter than tepals
            5. Short, diffuse herbs; flowers in axillary clusters .... **A. polygonoides**
            5. Stout, tall herbs; flowers both in axillary clusters and interrupted spikes or panicles .... **A. dubius**
1. Stamens and tepals less than 5
    6. Stamens 2; tepals 2 .... **A. tenuifolius**
    6. Stamens 3; tepals 3
        7. Bracts and bracteoles as long as or longer than tepals .... **A. tricolor**
        7. Bracts and bracteoles shorter than tepals
            8. Flowers in axillary and terminal clusters; panicle spicate; utricle indehiscent; utricle corrugated, acute at apex, scarcely exceeding perianth .... **A. viridis**
            8. Flowers in axillary clusters only; utricle circumscissile .... **A. roxburghianus**

Amaranthus caudatus L., Sp. Pl. 990.1753; Hook.f., Fl. Brit India 4:719. 1885; Samvatsar, Fl. West. Trib. Madhya Pradesh 249. 1996; Mukherjee in Mudgal *et al.,* Fl. Madhya Pradesh 2:457. 1997. *Ramdane*

Annual, erect herbs, much branched in the upper portion; leaves broadly ovate or rhomboid-ovate, apex mucronate, base cuneate, 3-12 × 2-6 cm; petiole up to 8 cm long, but shorter than the lamina; flowers in terminal and axillary, drooping spikes; tepals of male flowers oblong, those of female obovate to spathulate; utricles ovoid-globose, 2-2.5 mm across, circumscissile; seeds compressed, shining black.

*Fl.-Fr.:*Aug. -Oct.

Common weed of fields, perhaps an escape, all over the area.

Amaranthus cruentus L., Syst. Nat. ed. 10 2:1269. 1759; *A. paniculatus* L., Sp. Pl. ed. 2 1406. 1763; *A. hybridus* L. subsp. *cruentus* (L.) Thell. Fl. Adv. Montpellier 205. 1912; *A. paniculatus* var. *cruentus* (L.) Moq. Prodr. 13 (2):257. 1849; Hook.f., Fl. Brit. India 4:718. 1885; Lakshminarasimhan *et al.,* Fl. Maharashtra 2:785. 2001.

Annual, up to 30 cm tall, stout herbs; leaves 2.5-11 × 1-4 cm, elliptic-lanceolate, apex acute, base shortly decurrent, hairy on both surfaces; petioles 2-15 cm long; flowers in clusters, arranged along axillary and terminal panicles; bracts and bracteoles long aristate; tepals green, more often red-purple; utricles urceolate, circumscissile; seed globose, dark-brown, shining.

*Fl.-Fr.:*Sept. -Dec.

Occurs as an escape, Khandwa, Khargone.

Amaranthus dubius Mart. ex Thell., Fl. Adv. Montpellier 203. 1912; *A. tristis* Willd. Hist. Amaranth 21. 1790; Townsend in Rev. Handb. Fl. Ceylon 1:11. 1980; Samvatsar, Fl. West. Trib. Madhya Pradesh 249. 1996; Naik, Fl. Marathwada 2:746. 1998.

Annual, erect herbs; leaves 1.5-6 × 0.7-4 cm, ovate or rhomboid; flowers greenish, in axillary clusters and spikes; bracts and bracteoles deltoid-ovate; utricles smooth, dehiscing irregularly.

*Fl.-Fr.:*Aug. -Dec.

Alirajpur, Burhanpur, Dhar, Khandwa.

Amaranthus polygonoides L. Fl. Jamaica Pug. 2:27. 1759; Naik, Fl. Marathwada 2:747. 1998; Lakshaminarasimhan *et al.,* in N.P. Singh *et al.,* Fl. Maharashtra 2:786.2001.

Diffuse, annual, 15-40 cm tall herbs; leaves lanceolate to rhomboid, 1-3 × 0.5-1.5 cm, narrowed at base and tapering into petioles, apex shortly mucronate, green with ash-coloured, V-shaped band on the upper surface; flowers in dense, axillary clusters; bracts and bracteoles shortly cuspidate; utricles ovoid, faintly rugulose; seed discoid, black, shining.

*Fl.-Fr.:*Aug. -Oct.

Gregarious in waste land, Burhanpur, Indore, Khandwa.

Amaranthus roxburghianus H. W. Kung Fl. III Nord Chine 4:19. 1935; *A. polygonoides* Roxb. Fl. Ind. 3:602. 1832, *nom. illeg.* non L. 1759; *A. blitum* var. *angustifolius* Moq., Prodr. 13(2):

263. 1849; *A. polygamous* sensu Hook.f., Fl. Brit. India 4:719. 1885; Naik, Fl. Marathwada 2:748. 1998.

Prostrate or diffuse, 20-30 cm long, annual herbs; stems striate; leaves ovate-oblong or spathulate, 1-3 × 0.6-2 cm, base cuneate, emarginated at obtuse apex; petioles up to 2 cm long; flowers in dense, axillary clusters; bracts and bracteoles ovate, acute; utricles ovoid, 1.5 mm long, with a broad, seedless top, circumscissile; seed globose, black.

*Fl.-Fr.:*July –Dec.

Rare, **Indore** only, in waste land, old walls, weed of fields etc.

Amaranthus spinosus L. Sp. Pl. 991. 1753; Hook.f., Fl. Brit. India 4:718. 1885; Samvatsar, Fl. West Tribal Madhya Pradesh 250.1996; Mukherjee in Mudgal *et al.,* Fl. Madhya Pradesh 2:458. 1997. *Kante choulai*

An erect or diffuse, 30-60 cm tall, spiny herb; stems glabrous, red or green, axillary thorns (not spines), straight, sharp, 1-1.5 cm long; leaves 6-11 cm long, ovate or lanceolate, apex mucronate, base cuneate; petioles up to 9 cm long; flowers green, in terminal and axillary spikes or panicles; bracts and bracteoles scaly, ovate, lanceolate; utricles ovoid, longer than broad; seed dark-brown or black, lenticular, shining.

Fl.-Fr.: Aug. -Dec.

Common along roads of villages, weed of fields, open and waste places, everywhere.

Amaranthus tenuifolius Willd., Sp. Pl. 4:381. 1805; Hook.f., Fl. Brit. India 4:722. 1885; Mukherjee in Mudgal *et al.,* Fl. Madhya Pradesh 2:459. 1997.

Annual, prostrate, branched glabrous herbs; leaves glabrous, 0.5-2 × 0.2-0.5 cm, oblong or oblong-linear, apex obtuse to emarginated, base cuneate, margin entire; flowers in axillary, cymose clusters; bracts hyaline on margin; tepals 2; utricles ellipsoid, compressed, hyaline, with longitudinal ribs.

Fl.-Fr.: March-June

Burhanpur, Indore, Khandwa.

Amaranthus tricolor L., Sp. Pl. 989. 1753; *A. tristis* L., Sp. Pl. 989. 1753; *A. mangostanus* L., Cent. Pl. I. 32. 1755; *A. gangeticus* L., Syst. Nat. ed. 10 2:1268. 1759; Hook.f., Fl. Brit. India 4:719. 1885; Samvatsar, Fl. West. Trib. Madhya Pradesh 250.1996; Mukherjee in Mudgal *et al.,* Fl. Madhya Pradesh 2:459. 1997.

Annual, decumbent or prostrate herbs; stems glabrous, much branched; leaves 3.5-4.5 × 1-3 cm, apex obtuse or acute, cuneate to attenuate at base, broadly ovate or rhomboid-ovate or broadly elliptic to lanceolate-oblong; petioles up to 8 cm long; flowers green, in axillary clusters or sometimes in terminal spikes, male and female flowers intermixed; bracts and bracteoles deltoid-ovate; tepals 3, awned; utricles 2.25-2.75 mm long, circumscissile; seed lenticular, black or brown, shining.

Fl.-Fr.: Dec. -Jan.

Common as a weed in fields, in waste lands, sometimes cultivated also.

Amaranthus viridis L., Sp. Pl. ed. 2 1405. 1763; *A. gracilis* Desf. Tab. Ecole Bot. 43. 1804; Hook.f., Fl. Brit. India 4:720.1885; Samvatsar, Fl. West. Trib. Madhya Pradesh 250.1996; Mukherjee in Mudgal *et al.,* Fl. Madhya Pradesh 2:460.1997.

Erect, much branched, 20-50 cm tall, annual herbs, branches grooved; leaves 3-8 × 2-5 cm, ovate, glabrous, apex obtuse or acute, base tapering; petioles chanelled, 2-5 cm long; flowers yellowish-green, at length reddish-brown, in axillary and terminal spikes; bracts and bracteoles deltoid-ovate to lanceolate-ovate; bracteoles shorter than the tepals; tepals 3, rarely 4; utricles obovoid, compressed, rugose, indehiscent or rupturing irregularly; seed reticulate, dark brown to black.

Fl.-Fr.: Aug. -Dec.

Common in waste lands and as a weed in fields.

CELOSIA L. W. 51; MP 1; SA 1

Celosia argentea L., Sp. Pl. 205. 1753; Hook.f., Fl. Brit. India 4:714. 1885; *C. cristata* L., Sp. Pl. 205. 1753; Hook.f., Fl. Brit. India 4:715. 1885; *C. coccinia* L., Sp. Pl. ed 2 297. 1762; Samvatsar, Fl. West. Trib. Madhya Pradesh 250.1996; Mukherjee in Mudgal *et al.,* Fl. Madhya Prasesh 2:461. 1997. *Safed murga*

An erect, glabrous, annual herb; stems and branches initially terete then striate; leaves 5-12 × 0.5-3 cm, linear-lanceolate or ovate-lanceolate, glabrous, apex acute, base tapering; the spikes terminal, 5-12 cm long, pink at the apex, gradually fading to white as the flowers mature; utricle ellipsoid, yellowish-green, muricate; seed black, shining, many per utricle.

Fl.-Fr.: Oct. -March

Very common everywhere. The fields are full of this species, giving the impression that it has been sown as a crop.

CHENOPODIUM L. W. 150; MP 2; SA 1

Chenopodium murale L., Sp. Pl. 219. 1753; Hook.f., Fl. Brit. India 5:4. 1886; Samvatsar, Fl. West. Trib. Madhya Pradesh 253. 1996; Mukherjee in Mudgal *et al.,* Fl. Madhya Pradesh 2:472. 1997. *Bathua*

Annual, erect, 15-80 cm tall herbs; leaves 2-7 × 1.5-5 cm, ovate-rhomboid-deltoid; apex acute, base cuneate, margin irregularly coarsely dentate-serrate; petioles of lower leaves 2-5 cm long; flowers in axillary and terminal, 1-4 cm long panicles; tepals connate at base; utricles depressed globose, rugose, dark chocolate in colour; seed horizontal, black.

Fl.-Fr.: Nov.-Feb.

Common in moist fields as a weed.

DIGERA Forssk. Monotypic

Digera muricata (L.) Mart., Nova Acta Phys. -Med. Acad. Caes. Leop. -Carol. Nat. Cur. 13:285. 1826; *Achyranthes muricata* L., Sp. Pl. ed. 2 295. 1762; *A. alternifolia* L., Mant. Pl. 50.1767; *Digera arvensis* Forssk., Fl. Aegypt. -Arab. 65. 1775; Hook.f., Fl. Brit. India 4:717. 1885;

Samvatsar, Fl. West. Trib. Madhya Pradesh 250.1996; Mukherjee in Mudgal *et al.,* Fl. Madhya Pradesh 2:462. 1997. *Lat madhuriya*

Erect or straggling, glabrous, 20-80 cm tall herbs; leaves broadly ovate, 1.5-7 × 0.8-3 cm, apex acute or acuminate, base narrowed; petioles up to 3 cm long; flowers in axillary, 15-30 cm long racemes, with carmine or red tinge; bracts persistent; 2 outer tepals 3-4 mm long, 5-7 nerved, 3 inner tepals shorter, narrow, 1-2 nerved; fruits subglobose, 2-fid at apex, 2-2.5 mm long, style persistent; seed one.

Fl.-Fr.: Most part of the year when moisture is available.

Common weed of fields and waste lands throughout the area.

DYSPHANIA Forssk. W. 36; MP 1; SA 1

Dysphania ambrosoides (L.) Mosyakin & Clemants, Ukrayins Bot. Zhurn. 59:382.2002; *Chenopodium ambrosoides* L., Sp. Pl. 219. 1753; Hook.f., Fl. Brit. India 5:4. 1886; Samvatsar, Fl. West. Trib. Madhya Pradesh 253. 1996; Mukherjee in Mudgal *et al.,* Fl. Madhya Pradesh 2:472. 1997. *Chandan bathua*

Strongly scented, much branched, annual herbs; leaves lanceolate-oblong to elliptic, apex obtuse or subacute, base cuneate, 4-10 × 1-3 cm, gland-dotted; flowers green, in axillary, clustered spikes; tepals 4-5, united below, ovate-deltoid; stigmas 3-5; utricles with yellow glands, enclosed in the tepals; seed horizontal or vertical, orbicular, smooth, reddish-brown.

Fl.-Fr.: Nov. -Feb.

Common weed of irrigated fields, especially along bunds and water channels.

GOMPHRENA L. W. 133; MP 1; SA 1

Gomphrena celosioides Mart., Nova Acta Phys. -Med. Acad. Caes. Leop. -Carol. Nat. Cur. 13:301. 1826; Samvatsar, Fl. West. Trib. Madhya Pradesh 251. 1996; Mukherjee in Mudgal *et al.,* Fl. Madhya Pradesh 2: 463. 1997. (*G. serrata* L. is a separate species)

Annual, prostrate or erect herbs, clothed with white, shaggy hairs; leaves subsessile, 1.5-5 × 0.5-1,5 cm, spathulate or oblong-lanceolate, apex acute, base narrowed to a poorly demarcated petiole; flowers greenish-white, in sessile, terminal, subglobose spikes, up to 4 cm long; bracts deltoid-ovate, 2.5 4 mm long, bracteoles 5-6 mm long; tepals 4-5 mm long, outer 3 more or less flat, inner 2 sigmoid; pseudostaminodes absent; utricles pyriform; seed shining brown.

*Fl.-Fr.:*July-April

Common weed of waste places, roadsides all over the area.

NOTHOSAERVA Wight Monotypic

Nothosaerva brachiata (L.) Wight, Icon. Pl. Ind. Orient. 6:1. 1853; *Achyranthes brachiata* L., Mant. Pl. 50.1767; *Illecebrum brachiatum* (L.) L., Mant. Pl. 213. 1771; *Pseudanthus brachiatus* (L.) Wight, Icon. Pl. Ind. Orient. 5:3. 1852; Hook.f., Fl. Brit. India 4:726. 1885; Samvatsar, Fl. West. Trib. Madhya Pradesh 251. 1996; Mukherjee Mudgal *et al.,* Fl. Madhya Pradesh 2:465. 1997.

Annual, 10-30 cm tall herbs; branches many, opposite, spreading from the base upwards; leaves 1-5 × 0.5-2,5 cm, ovate, elliptic or ovate-lanceolate; spikes axillary, linear-cylindric, whitish, solitary or clustered in leaf axils; peduncles pilose; bracts hyaline, concave, nerveless, bracteoles minute; utricles globose or oblong, smooth; seed lenticular, smooth, shining black.

*Fl.-Fr.:*Sept. -May

In moist ground, near ponds or puddles, Alirajpur, Barwani, Khandwa, Khargone.

PUPALIA A. L. Juss. W. 3; MP 1; SA 1

Pupalia lappacea (L.) Juss., Ann. Mus. Natl. Hist. Nat. 2:132. 1803; *Achyranthes lappacea* L., Sp. Pl. 204. 1753; *A. atropurpurea* Lam., Encycl. 1:546. 1785; *Pupalia atropurpurea* (Lam.) Moq., Prodr. 13 (2):331. 1849; Hook.f., Fl. Brit. India 4: 723-724. 1885; Samvatsar, Fl. West. Trib. Madhya Pradesh 251. 1996; Mukherjee in Mudgal *et al.,* Fl. Madhya Pradesh 2:468. 1997. *Chit chita*

Annual or perennial, up to 1 m high herbs; stems 4-angled, much branched and swollen at nodes; branches opposite, pilose to tomentose; leaves 1.5-9 × 1-5 cm, ovate-lanceolate to oblong or orbicular, apex acuminate, base cuneate, petioles 2-2.5 cm long; flowers in spikes, dimorphic, perfect flowers very few, imperfect ones reduced to awns with long, hooked bristles; outer 2 tepals 3-5 mm long, inner 3 terminating in 5-18, 3-7 mm long setae; utricles 2-2.5 mm long; seed dark brown, shining.

*Fl.-Fr.:*Nov. -Feb.

Common in waste places, roadsides, open ground. All over.

TRICHURIELLA Bennet Monotypic

Trichuriella monsoniae (L.f.) Bennet in Indian J. For. 8:86. 1985; *Illecebrum monsoniae* L.f. Suppl. Pl. 161. 1781; Mukherjee in Mudgal *et al.,* Fl. Madhya Pradesh 2:469. 1997.

Slender, up to 40 cm high, simple or branched herbs; leaves mostly crowded on lower internodes, opposite or whorled in the upper part, linear-filiform, about 2.5 cm long; flowers in *ca* 4 cm long, terminal spikes; bracts lanceolate, *ca* 2 mm long, lanate, hyaline; bracteoles smaller; tepals *ca* 2.5 mm long, hyaline, curving towards the apex of spike; utricles shining, reddish-brown.

*Fl.-Fr.:*Aug. -May

Rare, along dried river beds, Burhanpur, Khandwa.

POLYGONACEAE 59/1384; 4/15; 3/9

Key to Genera

1\. Perianth lobes 6, inner 3 accrescent in fruit; stigma fimbriate .... RUMEX

1\. Perianth lobes 3-5, not accrescent in fruit; stigmas capitates or capitellate

  2\. Flowers axillary, solitary or in clusters; ochreas hyaline .... POLYGONUM

  2\. Flowers terminal, in simple or panicled, spiciform racemes or heads; ochreas membranous or chartaceous .... PERSICARIA

PERSICARIA (L.) Mill. W. 66; MP 8; SA 6

Key to Species

1. Flowers in cymosely disposed spurious heads; stem with recurved prickles on nodes; styles 2; nut biconvex .... **P. dichotoma**
1. Flowers in pseudo-spikes, racemes or panicles
   2. Nuts 3-angled; bracts closely set; perianth not glandular
      3. Stems and peduncles with recurved prickles .... **P. strigosa**
      3. Stems and peduncles without prickles; cilia of ochrea longer than the tube
         4. Stems and leaves glabrescent, stout .... **P. barbata**
         4. Stems and leaves strigose hairy; leaves pubescent-hispid .... **P. stagnina**
   2. Nuts biconvex; ochreas without limb
      5. Ochreas and bracts glabrous .... **P. glabra**
      5. Ochreas and bracts variously hairy or ciliate .... **P. pulchra**

Persicaria barbata

Key to Varieties

1. Stems appressed hairy; leaves acute at base ... . var. **barbata**
1. Stems glabrous; leaves rounded or subcordate at base .... var. **gracilis**

Persicaria barbata (L.) H. Hara, var. **barbata** Fl. East. Himal. 1:70.1966; *Polygonum barbatum* L., Sp. Pl. 362. 1753; Samvatsar, Fl. West. Trib. Madhya Pradesh 254. 1996; Srivastava *et al.,* in Mudgal *et al.,* Fl. Madhya Pradesh 2:478. 1997. *Aesa*

Perennial, erect or prostrate herbs, about 1 m high; stem appressed hairy; leaves linear-lanceolate, 5-20 cm long, apex acuminate, base acute, ochreae long ciliate; pseudospikes in panicles; perianth white or pinkish; nutlets ovoid-ellipsoid, 3-angled, enclosed within the persistent perianth.

*Fl.-Fr.:*Mostly throughout the year in the vicinity of water.

Common in the area.

Persicaria barbata (L.) H. Hara var. **gracilis** (Danser) H. Hara, Enum. Fl. Pl. Nepal 175. 1982; *P. barbata* subsp. *gracilis* (Danser) Sojak, Preslia 46:152. 1974; *Polygonum barbatum* subsp. *gracile* Danser, Bull. Jard. Bot. Buitenzorg, ser, 3, 8:146, f. 2. 1927; Samvatsar, Fl. West. Trib. Madhya Pradesh 254. 1996; Srivastava *et al.,* in Mudgal *et al.,* Fl. Madhya Pradesh 2:480.1997.

This variety differs from the typical variety in having a glabrous stem and leaves rounded or subcordate at base; other details are similar to the typical variety.

*Fl.-Fr.:*Dec. -April

Rare, in marshy places at Alirajpur and Burhanpur.

Persicaria dichotoma (Blume) Masam.,Sci. Rep. Kanazava Univ. 2(2):82. 1954; *Polygonum dichotomum* Blume, Bijdr. Fl. Ned. Ind. 11:529. 1826; *P. pedunculare* Wall. ex Meisn., Pl. Asiat. Rar. 3:58. 1832; *P. strigosum* R. Br. var. *pedunculare* (Meisn.) Steward, Contr. Gray Herb. 88:91. 1930; Srivastava *et al.,* in Mudgal *et al.,* Fl. Madhya Pradesh 2:480.1997.

Trailing herbs up to 50 cm long, stems ascending with recurved prickles on nodes; leaves softly ciliate, 2.5-8 × 0.5-1 cm, elliptic-oblong or linear-oblong, apex acuminate, base cuneate, margins prickly, ochreae 1.5 cm long, obliquely truncate; flowers in oblong or globose heads; peduncles glandular hairy; perianth white, with a pink tinge; nutlets biconvex, enclosed within the perianth.

*Fl.-Fr.:*July-Nov.

Rare, along streams, **Khandwa** only, Golia area.

Persicaria glabra (Willd.) M. Gomez, Anales Inst. Segunda Ensen. 2:278. 1896; *Polygonum glabrum* Willd., Sp. Pl. 2:447. 1799; Hook.f., Fl. Brit. India 5:34. 1886; Samvatsar, Fl. West. Trib. Madhya Pradesh 254. 1996; Srivastava *et al.,* in Mudgal *et al.,* Fl. Madhya Pradesh 2:480.1997.

An erect, semiaquatic, 60-150 cm tall herb; old shoots red in colour; leaves gland-dotted, 10-20 cm long, linear-lanceolate, apex acuminate, base tapering, ochreae membranous, eciliate, sheathing the stem; flowers pink, in erect, slender, terminal, 4-8 cm long racemes; bracts oblong; tepals oblong, obtuse; nutlets broadly ovoid or suborbicular, biconvex, black, polished.

*Fl.-Fr.:*Sept. -March

Very common, forming pure patches in stagnant, foul smelling puddles along with *Bacopa monnieri* (L.) Wettst.

Persicaria pulchra (Blume) Sojak, Preslia 46:154. 1974; *Polygonum pulchrum* Blume Bijdr. 530.1826; Hook.f., Fl. Brit. India 5:30.1886; Srivastava *et al.,* in Mudgal *et al.,* Fl. Madhya Pradesh 2:487. 1997.

Annual, erect, up to 1 m high herbs; leaves 10-12 × 2.5-3 cm, elliptic-ovate to lanceolate, apex acuminate, silky tomentose below, ochreae 1.2-12.5 cm long, mouth ciliate; flowers in paniculate racemes; peduncles strigose; bracts orbicular, hispid; nutlets biconvex, *ca* 3 mm across.

*Fl.-Fr.:*Oct-March

Rare, along bank of streams, **Khandwa** only, Mortakka, banks of Narmada.

Persicaria stagnina (Buch. -Ham. ex Meisn.) Qaiser, Fl. Pakistan 205. 44.2001; *Polygonum stagninum* Buch. -Ham. ex Meisn., Pl. Asiat. Rar. 3:56. 1832; Naik, Fl. Marathwada 2:761. 1998; Jadhav in Sharma *et al.,* Fl. Maharashtra 2:804.2001.

Erect or ascending, annual or perennial herbs; stems 40-90 cm long, thickened at nodes; leaves subsessile, oblong-lanceolate, 5-20 × 1.5-3 cm, acute at both ends, ochreae with long, ciliate mouth; flowers in terminal, up to 5 cm long spikes, rachis and peduncles strigose hairy, bracts 2.5-3 mm long, ciliate; nutlets trigonous, black, smooth.

*Fl.-Fr.:*Sept. -Jan.

In marshes, along streams, Burhanpur, Indore.

Persicaria strigosa (R.Br.) Nakai, Rigakkawai 24:299. 1926; *Polygonum strigosum* R.Br. Prodr. Fl. Nov. Holl. 420.1810; Srivastava *et al.,* in Mudgal *et al.,* Fl. Madhya Pradesh 2:488. 1997.

Annual, creeping herbs; stems 30 80 cm long, angular, prickly; leaves 3-10 × 0.6-2.5 cm, linear-oblong, apex acute or acuminate, base truncate or cordate, margin strigose, ochreae ciliate, *ca* 2 cm long; flowers in paniculate racemes, peduncles glandular-hispid, prickly, bracts strigose, ovate; perianth pink, 2-2.5 mm long; nutlets black.

*Fl.-Fr.:*July-Nov.

Rare, Barwani, Khargone.

POLYGONUM L. W. 217; MP 1; SA 1

Polygonum plebeium R. Br. Prodr. Fl. Nov. Holl. 420.1810; Hook.f., Fl. Brit. India 5:27. 1886; Samvatsar, Fl. West. Trib. Madhya Pradesh 254. 1996; Srivastava *et al.,* in Mudgal *et al.,* Fl. Madhya Pradesh 2:484. 1997.

Small, prostrate, annual herbs; stemspurplish-green, white hairy; leaves 0.3-3 cm long, elliptic-oblong, oblong, linear, obovate or orbicular (depending upon the level of moisture, if it is more, the leaves are broader, away from moisture, they start becoming narrower), apex acuminate or obtuse, base narrow, ochreae hyaline, fimbriate; flowers axillary, solitary or 2-3 together; perianth pink, *ca* 2 mm long, divided nearly to the base; nutlets 3-angled, smooth, shining.

*Fl.-Fr.:*Nov. -May

Abundant in drying ponds, all over the area.

RUMEX L. W. 152; MP 2; SA 1

Rumex dentatus L., Mant. Pl. 2:226. 1771; Fl. Brit. India 5:59. 1886; Samvatsar, Fl. West. Trib. Madhya Pradesh 255. 1996; Srivastava *et al.,* in Mudgal *et al.,* Fl. Madhya Pradesh 2:489. 1997. *Lal bibi*

An erect, deep-rooted, 30-60 cm tall, annual herb; roots red in colour; stem and branches ribbed, red; leaves 6-10 × 2-2.5, oblong, apex obtuse, base rounded, cordate, radical leaves much longer than the cauline; flowers greenish-yellow, pedicellate in leafy or leafless, verticillate clusters; stigmas fimbriate; nutlets trigonous, reddish-brown, 2.5 mm long.

*Fl.-Fr.:*Dec. -June

Very common on the edges of ponds, ditches, throughout the area.

ARISTOLOCHIACEAE 8/ 624; 1/2; 1/2

ARISTOLOCHIA L. 485; MP 2; SA 2

Key to Species

1. Leaves broadly ovate to reniform, base cordate; flowers solitary; prostrate plants; capsules 1-2 cm long . ... **A. bracteolata**
1. Leaves elliptic-ovate, or panduriform, cuneate, subtruncate or subcordate at base; capsules 3-3.5 cm long .... **A. indica**

Aristolochia bracteolata Lam., Encycl. 1:258. 1783; *A. bracteata* Retz., Observ. Bot. 5:29. 1788; Hook.f., Fl. Brit. India 5:75. 1886; Samvatsar, Fl. West. Trib. Madhya Pradesh 255. 1996; Dixit in Mudgal *et al.,* Fl. Madhya Pradesh 2:493. 1997. *Kidamar*

Slender, prostrate herbs; leaves 2-7 cm long and as much broad, reniform, base cordate, with a wide, shallow sinus; flowers solitary, axillary, dark purple; capsules 1-3 cm long, oblong-ellipsoid; seeds deltoid with a slightly cordate base.

Fl.-Fr.: Aug. -Nov.

Weed of fields common in Alirajpur, Barwani, Burhanpur, Dhar, Khandwa, Khargone.

Aristolochia indica L., Sp. Pl. 960.1753; Fl. Brit. India 5:75. 1886; Samvatsar, Fl. West. Trib. Madhya Pradesh 255. 1996; Dixit in Mudgal *et al.,* Fl. Madhya Pradesh 2:493. 1997. *Ishar mool* **(Plate 8, Fig. 6)**

Perennial, twining herbs; stems woody below, stems and branches grooved; leaves punctuate, 5-12 × 2-5 cm, ovate -oblong to lanceolate, glabrous; flowers greenish-white in axillary racemes, limb purple or brown, tube cylindrical, inflated at base; capsules oblong, subglobose, 3-3.5 cm long, 6-valved; seeds 3-angled, many, flattened, winged.

*Fl.-Fr.:*Sept. -Nov.

In hedges of fields and open ground, Alirajpur, Burhanpur, Khandwa.

PIPERACEAE 13/2658; 2/2; 1/1

PEPEROMIA Ruiz & Pavon W. 1161; MP 1; SA 1

Peperomia pellucida (L.) Kunth, Nov. Gen. Sp. 1:64. 1816; *Piper pellucidum* L., Sp. Pl. 30.1753; Samvatsar, Fl. West. Trib. Madhya Pradesh 255. 1996; Hanfi in Mudgal *et al.,* Fl. Madhya Pradesh 2:496. 1997.

Delicate, subfleshy, diffuse herbs; stems glabrous, pellucid; leaves alternate, 1.5-3 × 1.5-2.5 cm, broadly ovate, thin, sub-truncate or cordate, 5-7 nerved at base, petioles 1-2 cm long, decurrent; flowers axillary, terminal or leaf-opposed, in 2-2.5 cm long spikes; ripe nuts 0.8 mm long, globose, pointed at apex, brown-black.

*Fl.-Fr.:*Aug. -Oct.

In moist and shady habitats, sometimes along sewage canals, Alirajpur, Barwani, Dhar, Indore, Khandwa, Khargone.

LAURACEAE 66/2978; 3/3; 1/1

CASSYTHA L. W. 19; MP 1; SA 1

Cassytha filiformis L., Sp. Pl. 35. 1753; Hook.f., Fl. Brit. India 5:188. 1886; Hanfi in Mudgal *et al.,* Fl. Madhya Pradesh 2:499. 1997. *Amar bel*

Leafless, parasitic herbs, forming tangled masses over adjacent vegetation; leaves represented by minute scales or altogether absent; flowers in lax, divaricate spikes; bracteoles small, ciliate; perianth double the length of bracteoles, lobes 6, in 2 whorls, the outer 3 rounded, ciliate, inner ones oblong, acute, concave, tube short; drupes globose, white, smooth.

*Fl.-Fr.:*Most of the year

In waste lands and open foresta, Barwani, Indore, Khargone.

LORANTHACEAE 79/886; 2/4; 2/2

Key to Genera

1. Flowers actinomorphic; corolla 5-lobed, tube not split on one side .... DENDROPHTHOE
1. Flowers zygomorphic; corolla 4-lobed, tube split on one side .... LORANTHUS

DENDROPHTHOE Mart. W. 59; MP 1; SA 1

Dendrophthoe falcata (L.f.) Ettinsh., Denkcher. Kaiserl. Akad. Wiss. Wien. Math.-Naturwiss. Kl. 32:52. 1872; *Loranthus falcatus* L.f. 1782; Samvatsar, Fl. West. Trib. Madhya Pradesh 256. 1996; Srivastava in Mudgal *et al.,* Fl. Madhya Pradesh 2:507. 1997. *Bandha*

Partial parasitic, dichotomously branched shrubs, branches often horizontal; leaves elliptic-oblong, obovate or ovate-lanceolate, entire or undulate, leathery, coriacious; flowers greenish-white, in 4-5 cm long, axillary racemes; bracts minute, ovate; corolla curved, trumpet-shaped, tube narrow, lobes 5, reflexed; drupes ovoid-oblong, 6-7 mm long, black at maturity.

*Fl.-Fr.:*July-Oct.

Very common in forest, affecting a variety of hosts.

Note: var. *coccinia* (Talb.) Santapau is not accepted in the Plant List. Fresh leaves have a much lower temperature than that of the host leaves. *Loranthus longiflorus* Desr. given as a synonym of this species is recognized as a separate species.

LORANTHUS Jacq. W. 10; MP 3; SA 2

Loranthus cordifolius Wall., Fl. Ind. 2:222. 1824, unresolved; *Scurrula cordifolia* G. Don, Gen. Hist. 3:421. 1834, Hook.f., Fl. Brit. India 5:209. 1886; Srivastava in Mudgal *et al.,* Fl. Madhya Pradesh 2:510.1997.

Herbs or undershrubs, young parts covered with rusty or white scurfy tomentum; leaves 3-7 × 2-5 cm, orbicular or broadly ovate, apex obtuse or acute, base subcordate, lower surface white pubescent; flowers in the axils of fallen leaves, densely tomentose; bracts ovate, obtuse; corolla brown; fruits clavate, rusty tomentose.

*Fl.-Fr.:*Oct. -March

Rare, **Burhanpur** only, Jambupani.

SANTALACEAE 42 / 992; 3/3; 2/2

Key to Genera

1. Leaves absent (our area); flowers unisexual; fruit baccate .... VISCUM
1. Leaves well developed; flowers bisexual; fruit drupaceous .... SANTALUM

SANTALUM L. W. 12; MP 1; SA 1

Santalum album L., Sp. Pl. 349. 1753; Hook.f., Fl. Brit. India 5:231. 1886; Samvatsar, Fl. West. Trib. Madhya Pradesh 257. 1996; Srivastava in Mudgal *et al.,* Fl. Madhya Pradesh 2:516. 1997. *Chandan*

Evergreen, small trees, initially parasitic on roots; leaves opposite, lanceolate or elliptic, 3-7 × 1.5-3 cm, apex acute, base rounded, subcoriaceous; petioles 1-1.5 cm long; inflorescence a trichotomous panicle; flowers pentamerous; perianth purplish-brown inside, tube campanulate; disc of 5, fleshy scales; drupes *ca* 1 cm across, black.

*Fl.-Fr.:*Sept. -Dec.

Very common in the hedges of fields, not at any other place, a casual tree may be present in a bunglow.

VISCUM L. W. 74; MP 1; SA 1

Viscum articulatum Burm.f., Fl. Indica 311. 1768; *V. nepalense* Spreng., Syst. Veg. ed. 16 (Sprengel) 4 (2, Cur. Post.): 47. 1827; Hook.f., Fl. Brit. India 5:226. 1886; Samvatsar, Fl. West. Trib. Madhya Pradesh 256. 1996; Srivastava in Mudgal *et al.,* Fl. Madhya Pradesh 2:511. 1997. *Badu* **(Plate 8, Fig. 7)**

Pendulous, dichotomously branched, green shrubs; stems jointed, flattened, readily disarticulating; leaves absent or minute, scale-like; flowers minute, sessile, solitary or in fascicles at the nodes, central flowers female, lateral flowers usually male; bracts copular; tepals reflexed in male flowers, and erect in female; berries yellowish-green, globose, 1-seeded.

*Fl.-Fr.:*June-Aug.

Rare, found as a parasite commonly on Mahua, Tendu, Sadad, Ber etc, Alirajpur, Barwani, Dhar, Indore Khandwa.

EUPHORBIACEAE 228 / 6547; 11/42; 9/38

Key to Genera

1. Perianth totally absent; flowers in cyathia; glands of involucres free or united in a ring .... EUPHORBIA
1. Perianth of atleasr 1-whorl, sepals present
    2. Perianth 2-seriate, composed of 5 sepals and 5 petals, both in male and female flowers
        3. Stamens numerous, free, inflexed; sepals imbricate; leaf base with 2 glands .... CROTON
        3. Stamens up to 15, united at base, straight; sepals imbricate or valvate; leaves with palmate venation
            4. Sepals valvate; plants densely stellate hairy; stamens 5-15, united .... CHROZOPHORA
            4. Sepals imbricate; plants glabrous; stamens 8-10 (our area), those of inner whorl united .... JATROPHA
                5. Climbing shrubs with stinging hairs; stamens 3, united .... TRAGIA
                5. Erect shrubs or trees without stinging hairs; stamens more than 3

6. Female flowers subtended by large, leafy bracts; anthers vermiculiform .... ACALYPHA
6. Female flowers subtended by small bracts; anthers not vermiculiform
   7. Seeds mottled; fruits without glands; sepals in male flowers imbricate; shrubs .... BALIOSPERMUM
   7. Seeds not mottled; fruits glandular or not; sepals in male flowers valvate; trees or shrubs
      8. Filaments connate; shrubs of riverine habitat . ... HOMONOIA
      8. Filaments free; trees of dry habitat .... MALLOTUS

ACALYPHA L. W. 454; MP 4; SA 5

Key to Species

1. Undershrubs, male spikes axillary, slender, dense; female spikes terminal, capitate .... **A. alnifolia**
1. Herbs; both male and female flowers in axillary spikes, male upper ones and a few in number
   2. Bracts shorter than the capsules .... **A. lanceolata**
   2. Bracts longer or as long as the capsules
      3. Bracts as long as the capsule; capsule teeth with gland-tipped hairs; leaves 5-nerved .... **A. malabarica**
      3. Bracts longer than the capsule; capsule teeth without gland-tipped hairs; leaves 3-5 nerved
         4. Leaves 3-nerved; bracts dentate or crenate; capsule hispid .... **A. indica**
         4. Leaves 5-nerved; bracts pectinate with subulate, hispid teeth; capsule glabrous .... **A. ciliata**

Acalypha alnifolia Klein ex Willd., Sp. Pl. 4:525. 1805; Hook.f., Fl. Brit. India 5:415. 1887; Gamble 2:930.1957 Rep. ed; Samvatsar, Fl. West. Trib. Madhya Pradesh 258; 1996.

Small, eglandular shrubs; leaves ovate, acute, 2.5 cm long, tomentose beneath; male flowers in axillary spikes, female in, terminal, capitate spikes; bracts many, imbricating, dentate in female spikes, enveloping the capsules; capsules hispid; seeds globose.

*Fl.-Fr.:*Aug. -Dec.

Rare, in wet places, near streams, **Dhar** only, Dharampuri bet (Narmada river).

Acalypha ciliata Forssk., Fl. Aegypt. -Arab. 162. 1775; Hook.f., Fl. Brit. India 5:417. 1887; Samvatsar, Fl. West. Trib. Madhya Pradesh 258. 1996; Shukla in Mudgal *et al.,* Fl. Madhya Pradesh 2:520.1997.

Annual, 30-80 cm high herbs; leaves 4-8 × 2-6 cm, ovate, acuminate, finely serrate, hairy on veins beneath, base rounded; petioles often exceeding the leaf blades; flowers greenish in axillary, androgynous spikes; male flowers at the top, females crowded at the base, subtended by large, fimbriate bracts; capsules enclosed in the bracts; seeds reddish-brown, globose-ovoid, smooth.

*Fl.-Fr.:*July-Nov.

Common in waste lands, along foot-paths and edges of fields, gardens, all over.

Acalypha indica L., Sp. Pl. 1003. 1753; Hook.f., Fl. Brit. India 5:416. 1887; Samvatsar, Fl. West. Trib. Madhya Pradesh 258. 1996; Shukla in Mudgal *et al.,* Fl. Madhya Pradesh 2:521. 1997. *Kuppi*

Erect, annual, 30-70 cm high herbs; stems angular; leaves 5-9 × 1.5-3 cm, ovate or obovate, apex acute, margin serrate, base cuneate; petioles very long; stipules subulate; flowers in 4-6 cm long, axillary spikes; male flowers in inclusters, near apex, females scattered, 3-5 surrounded by reniform, cup-like, leafy, truncate bracts, toothed at margins; capsules hispid, greyish-white; seeds pale brown, glabrous, ovoid.

*Fl.-Fr.:*June-Sept.

Very common weed of gardens, fields, old buildings, vacant plots etc., all over.

Acalypha lanceolata Willd., Sp. Pl. 4:524.1805; *A. fallax* Mull. Arg., Linnaea 34:43.1865; Hook.f., Fl. Brit. India 5:416.1887; Londhe in Sharma *et al.,* Fl. Maharashtra 2:856.2001.

Annual, erect, up to 40 cm tall herbs; leaves 1-5 × 0.4 cm, ovate-lanceolate, apex subacute, base acute, margins crenate-dentate; spikes axillary; male flowers yellow-red in colour, females green; capsules 1.5-2 mm long, hispid with gland-based hairs; seeds ovoid, acute.

*Fl.-Fr.:*July-Oct.

Rare, **Burhanpur** only, Asirgad and Dhulkot hills.

Acalypha malabarica Mull. Arg., Linnaea 34:42. 1865; Hook.f., Fl. Brit. India 5:416. 1887; Samvatsar, Fl. West. Trib. Madhya Pradesh 258. 1996; Shukla in Mudgal *et al.,* Fl. Madhya Pradesh 2:521. 1997.

Annual, 20-60 cm tall, tomentose herbs; leaves 3-8 × 2-6 cm, ovate-deltoid, crenate-serrate, long-petioled; flowers in *ca* 5 cm long spikes; male flowers few, terminal, ebracteate, females below, bracteates; capsules glabrous, 2-3 mm long; seeds brown, ovoid, smooth.

*Fl.-Fr.:*Aug. -Dec.

As forest undergrowth, Alirajpur, Barwani, Dhar, Indore.

BALIOSPERMUM Blume W. 5; MP 1; SA 1

Baliospermum solanifolium (Burm.) Suresh, Interpret. Rheede's Hort. Malab. 106. 1988; *Croton solanifolius* Burm., Fl. Malab. 6. 1769; *Baliospermum axillare* Blume, Bijdr. 604. 1826; *Jatropha montana* Willd.,Sp. Pl. 4:563. 1805; *B. montanum* (Willd.) Mull. Arg. Prodr. 15(2):1125. 1866; Hook.f., Fl. Brit. India 5:461. 1887; Samvatsar, Fl. West. Trib. Madhya Pradesh 259. 1996; Shukla in Mudgal *et al.,* Fl. Madhya Pradesh 2:528. 1997. *Jamal gota*

Stout undershrubs; leaves variable in shape and size, upper leaves small, lanceolate, peninerved, lower large, ovate, palmately 3-5 lobed, 20-25 cm long, base rounded or acute, with 2 glands; petioles 5-12 cm long; stipuls glandular; flowers small, greenish, in axillary racemes, all male flowers above, females few, situated below; capsules 8-15 mm long, obovoid, 3-lobed; cocci 3; seeds smooth, ellipsoid, mottled, light black.

*Fl.-Fr.:*Sept. -March

Common forest undergrowth in all districts.

CHROZOPHORA A. L. Juss. W. 9; MP 2; SA 2

Key to Species

1. Leaves eglandular at base; stellate hairs stipitate; prostrate herbs . ... **C. plicata**
1. Leaves biglandular at base; stellate hairs sessile; erect herbs . ...**C. rottleri**

Chrozophora plicata (Vahl) A. Juss. ex Spreng., Syst. Veg. 3:850.1826; *Croton plicatus* Vahl, Symb. Bot. 1:78. 1790; *Chrozophora prostrata* Dalz. in Dalz. & Gibs., Bombay Fl. 233. 1861; Samvatsar, Fl. West. Trib. Madhya Pradesh 259. 1996; Shukla in Mudgal *et al.,* Fl. Madhya Pradesh 2:635. 1997.

Prostrate, stellately woolly herbs, branches many, spreading from the base; leaves 4-9 × 3-9 cm, clothed with stellate tomentum on both the surfaces, broadly ovate or orbicular, often obtusely lobed, base cordate; flowers in axillary racemes, male flowers at the top, females below; bracts linear; capsule stellately woolly and with silvery scales; seeds rugose or tubercled.

*Fl.-Fr.:*Nov. -March

Common in waste lands throughout the area.

Chrozophora rottleri (Geiseler) A. Juss. ex Spreng., Syst. Veg. 3:850.1826; *Croton rottleri* Geiseler, Croton Monogr. 54. 1807; Samvatsar, Fl. West. Trib. Madhya Pradesh 260.1996; Shukla in Mudgal *et al.,* Fl. Madhya Pradesh 2:536. 1997.

Erect, 30-45 cm tall stellately hairy herbs; leaves 2.5-6 × 0.7-1.5 cm, variable in shape and size, ovate or ovate-lanceolate, acute or acuminate, base cuneate, margin serrate or denticulate, sparsely, stellately hairy on both the surfaces; flowers greenish-yellow, in *ca* 2 cm long axillary racemes; peduncles densely, stellately hairy; male flowers above and females below; capsules 5-7 mm long; seeds globose, ash-coloured.

*Fl.-Fr.:*Feb. -April

Common weed of waste places and fields, all over the area (Both species are part of the Ruderal flora).

CROTON L. W. 1205; MP 1; SA 1

Croton bonplandianus Baill., Adansonia 4:339. 1864; *C. sparciflorus* Morong, Ann. New York Acad. Sci. 7:221. 1893; Samvatsar, Fl. West. Trib. Madhya Pradesh 260.1996; Shukla in Mudgal *et al.,* Fl. Madhya Pradesh 2:539. 1997.

Much branched, up to 100 cm tall, stellate-tomentose, perennial herbs; leaves 3.5-8 × 1.5-3 cm, lanceolate, crenate-serrate, apex acuminate, base obtuse-subtruncate; flowers greenish-white, in 5-8 cm long racemes; male flowers above, female flowers fascicled in the axils of minute bracts on the lower side; male outer tepals obovate, minute; inner linear, oblong; stamens 10-15; female tepals lanceolate; ovary densely stellate-hairy; capsules 5-6 mm long, 3-angled, stellate hairy; seeds oblong, *ca* 4 mm long, shining with white caruncle.

Fl.-Fr.: Most of the year

Common as a roadside weed in plains, occasional in the forest.

EUPHORBIA L. W. 2046; MP 22; SA 21

Key to Species

1. Glands of involucre united in a ring; bracts of male flowers forming a tube round the female flower . ... **E. umbellata**
1. Glands of involucres free; bracts of male flowers also free
    2. Plants succulent with caducous or deciduous leaves
        3. Plants armed with stipular spines
            4. Main stem highly reduced, branches forming the main plant; leaves oblong-ovate, caducous 5-7 cm long .... **E. caducifolia**
            4. Main stem well developed, forming the main plant; leaves ovate-oblong, spathulate
                5. Stem 3-5 lobed
                    6. Stem 5-angled, angles spiral; leaves 7-15 cm long .... **E. neriifolia**
                    6. Stem sharply 3-angled or winged; leaves 0.9-1.5 cm long.......... **E. antiquorum**
                5. Stem terete .... **E. nivulia**
        3. Plants unarmed; stem cylindrical, shining .... **E. tirucalli**
    2. Plants not succulent
        7. Leaves radical, fleshy; root tuberous; cyme branched once only.......... **E. fusiformis**
        7. Leaves not radical
            8. Leaves more than 1 cm long
                9. Leaves alternate below and opposite in the floral region, exstipulate
                    10. Involucre with a single large gland on one side .... **E. heterophylla**
                    10. Involucre with 4-5 well developed glands along the upper margin
                        11. Floral leaves slightly broader than lower leaves; seeds ellipsoid with tuberculate testa .... **E. dracunculoides**
                        11. Floral leaves very broad and very different from the lower leaves
                            12. Bracts free; capsules hairy; seeds bluish-grey .... **E. rothiana**
                            12. Bracts connate at the base; capsules glabrous; seeds yellowish-brown .... **E. perbracteata**
                9. Leaves all opposite, stipulate or when exstipulate, the opposite pair of leaves joined by an inter-petiolar line
                    13. Floral leaves distinctly imbricating, concealing involucres
                        14. Capsules glabrous .... **E. pycnostegia**
                        14. Capsules hairy
                            15. Limb of glands pectinately laciniate.......... **E. cristata**
                            15. Limb of glands entire or serrulate, but not laciniate .... **E. elegans**
                    13. Floral leaves neither distinctly imbricating nor concealing the involucres
                        16. Plants nearly glabrous; cymes not capituliform; glands with distinct limbs .... **E. hypericifolia**
                        16. Plants hispid; cymes capituliform; limb of glands obsolete .... **E. hirta**

8. Leaves less than 1 cm long
   17. Glands of involucre with prominent petaloid limb .... **E. rosea**
   17. Glands of involucre without a petaloid limb
      18. Leaves subquadrate or obliquely ovate, apex retuse .... **E. serpens**
      18. Leaves ovate-oblong, elliptic oblong or obliquely oblong, apex obtuse or rounded
         19. Stem glabrous or nearly so; capsules glabrous all over .... **E. heyneana**
         19. Stem pubescent or puberulous; capsules pubescent either partly or completely
            20. Capsules pubescent at the angles of cocci, glabrous elsewhere .... **E. prostrata**
            20. Capsules pubescent all over; seeds furrowed .... **E. thymifolia**

Euphorbia antiquorum L., Sp. Pl. 450.1753; Hook.f., Fl. Brit. India 5:255. 1887; Samvatsar, Fl. West. Trib. Madhya Pradesh 261. 1996; Shukla in Mudgal *et al.*, Fl. Madhya Pradesh 2:545. 1997. *Tredhara*

Much branched shrubs or small trees; branches curving upwards, stout, fleshy, jointed, 3-angled, the angles winged; wings repand-sinuate, stipular spines borne on small tubercles, about 5 mm long; leaves deciduous, 1-1.5 cm long, subsessile, obovate-oblong or spathulate; cyathia in branched pedunculate cymes; involucres 3-nate; central flowers sessile, lateral stalked; cocci glabrous, compressed.

*Fl.-Fr.:*July-Sept.

Common as hedge plant in villages, throughout the area.

Euphorbia caducifolia Haines in Indian For. 40:154. 1914 & Bot. Bihar Orissa 1:147. 1961 (BSI Rep.); Samvatsar, Fl. West. Trib. Madhya Pradesh 262. 1996; Shukla in Mudgal *et al.*, Fl. Madhya Pradesh 2:545. 1997.

Fleshy, dendroid, 1.5- 3 m tall shrubs; branches numerous arising from the base, tuberculate, tubercles spirally arranged at the apex, in 5 spirals, each provided with broadly elliptic and black areoles, bearing paired, stipular spines; leaves at the apex of areoles, suborbicular, up to 9 × 4 cm, caducous; cyathia in triads, in the axils of fallen leaves; bracts 2, pink; involucral lobes cuneate, fimbriate, gland tumid; capsules acutely 3-angled, about 8 mm across; seeds *ca* 3 mm across, smooth, deep brown.

*Fl.-Fr.:*Jan. -April

On barren rocks, hilly areas, Burhanpur, Indore, Jhabua.

Euphorbia cristata B. Heyne ex Roth, Nov. Pl. Sp. 226. 1821; Hook.f., Fl. Brit. India 5:247. 1887; *Chamaecyce cristata* (B. Heyne ex Roth) Webster in J. Arnold Arb. 48:424. 1967; Samvatsar, Fl. West. Trib. Madhya Pradesh 262. 1996; Shukla in Mudgal *et al.*, Fl. Madhya Pradesh 2:546. 1997.

Annual, prostrate or ascending herbs, clothed with long, flexuous hairs; leaves opposite, obliquely ovate-cordate, 0.7-1 × 0.6-1 cm, apex obtuse, margin serrulate, hairy; involucres subsessile, hairy, in the axils of uppermost, distichously imbricate leaves; lobes linear-lanceolate; glands with pectinate and fimbriate limbs; capsules white, transversely furrowed.

*Fl.-Fr.:*Oct. -Jan.

Rare, on rocky, exposed places, **Alirajpur** only, Kattiwada forest.

Euphorbia dracunculoides Lam., Encycl. 2:428.1788; Hook.f., Fl. Brit. India 5:262.1887; Samvatsar, Fl. West. Trib. Madhya Pradesh 262.1996; Shukla in Mudgal *et al.,* Fl. Madhya Pradesh 2:547. 1997. *Chagulputputi*

Annual, much branched, 20-40 cm tall herbs; leaves 1.5-5 × 0.3-0.5 cm, sessile, linear to linear-lanceolate, glabrous, apex acute; involucres solitary, caMPanulate, subsessile, glabrous outside and hairy inside, lobes ciliolate, glands semi-lunate; capsules 3-4 mm long; seeds ellipsoid, 2-2.5 mm long, white, tuberculate.

*Fl.-Fr.:*Oct. -March

Common weed of wheat fields, all over the area.

Euphorbia elegans Spreng., Syst. Veg. 3:794. 1826; Hook.f., Fl. Brit. India 5:246. 1887; Shukla in Mudgal *et al.,* Fl. Madhya Pradesh 2:547. 1997.

Annual, dichotomously branched herbs; leaves 1-2 × 0.8-1.3 cm, ovate or orbicular, apex obtuse, margin serrate; bracts (floral leaves) broadly elliptic orobovate, much imbricating and and forming short spikes; involucres top-shaped, lobes rosy, obovate, cuneate; capsules 3-4 mm across; seeds about 2 mm long, 4-angled, parallel –grooved.

*Fl.-Fr.:*Sept. -March

Occasional, Barwani, Khandwa, Khargone.

Euphorbia fusiformis Buch. -Ham. ex D. Don, Prodr. Fl. Nepal 62, 1825; Hook.f., Fl. Brit. India 5:257. 1887; *E. acaulis* Roxb., Fl. Ind. 2:472. 1832; Londhe in Mudgal *et al.,* Fl. Maharashtra 2:880.2001.

Leaves in rosettes; aerial stem greatly reduced; root swollen, fusiform; leaves decurrent at base, fleshy, apex obtuse, 15-20 x-7-10 cm, obovate or oblanceolate. (Flowers not seen, description based upon vegetative characters).

Rare, **Alirajpur** only, on hard, open ground, in groups of 15-20, showing plants of all ages, Kattiwada.

Euphorbia heterophylla L., Sp. Pl. 453. 1753; *E. geniculata* Ort., Nov. Pl. Desr. Dec. 18. 1797; Hook.f., Fl. Brit. India 5:266. 1887; Samvatsar, Fl. West. Trib. Madhya Pradesh 262. 1996; Shukla in Mudgal *et al.,* Fl. Madhya Pradesh 2:549. 1997.

Erect, 30-60 cm high, annual herbs; stems fistular, branched in the upper part; leaves 3-10 × 1-5 cm, alternate below, opposite in the flowering region, broadly elliptic or obovate, upper part forming a rosette, margins entire or coarsely dentate, apex acute, base cuneate; stipular glands 3-5; cyathia in dense, corymbose cymes; involucral lobes 5, ovate, fimbriate; capsules 3- lobed, *ca* 5 mm across, glabrous; seeds black, obovoid, obscurely trigonous, truncate at one end.

*Fl.-Fr.:*Rainy season

Common weed of fields and roadsides.

Euphorbia heyneana Spreng., Syst. Veg. 3:791. 1826; *E. microphylla* B. Heyne ex Roth, Nov. Pl. Sp. 229. 1821 (Illegi.); *E. orbiculata* Miq., Fl. Ind. Bat. 1 (2):421. 1859 (Illegi.), non Humb., Bonpl., Kunth, 1817; Samvatsar, Fl. West. Trib. Madhya Pradesh 262. 1996; Shukla in Mudgal *et al.,* Fl. Madhya Pradesh 2:549. 1997.

Annual, prostrate, dichotomously branched glabrous herbs; leaves 2-8 × 1.5-5 mm, obliquely oblong, base entire, serrate above, rounded or retuse at apex; involucres axillary, solitary, stalked, broadly campanulate, lobes deltoid, acute; glands shortly stipitate; capsules obtusely keeled, about 2 mm across; seeds obtusely triangular, sides concave, smooth, brown.

*Fl.-Fr.:*Sept. -Oct.

Common on rocky habitat, throughout the area.

Euphorbia hirta L., Sp. Pl. 454. 1753; *E. pilulifera* L., Sp. Pl. 454. 1753; Hook.f., Fl. Brit. India 5:250.1887; Samvatsar, Fl. West. Trib. Madhya Pradesh 263. 1996; Shukla in Mudgal *et al.,* Fl. Madhya Pradesh 2:550.1997. *Dudhiya*

Decumbent, ascending, hispid annual herbs, branching from the root stock; branches green or pinkish, full of latex; leaves ovate- oblong to elliptic, serrulate, 1.5-2.5 × 0.5-1.5 cm, apex sub-obtuse, base oblique, rounded or cuneate; stipules subulate; cyathia axillary, in pedunculate, globose cymes; glands red, with or without limb; capsules depressed globose, pubescent, 3-lobed, about 5 mm long; seeds 4-angled, slightly transversely rugose, reddish-brown.

*Fl.-Fr.:*Throughout the year

Very common weed of the fields, along roadsides, waste and moist places, all over the area.

Euphorbia hypericifolia L., Sp. Pl. 454. 1753; Hook.f., Fl. Brit. India 5:249. 1887; Shukla in Mudgal *et al.,* Fl. Madhya Pradesh 2:550.1997. *Hakshardana* (*E. parviflora* L. is a separate species and not a synonym of this species:The Plant List).

Annual, erect or ascending herbs, up to 40 cm high; leaves 1-2.5 × 0.5-1.3 cm, broadly oblong, apex obtuse or round, base oblique, glabrous above, puberulent beneath; cymes axillary, pedunculate; involucres cup-like, lobes deltoid, ciliate; glands white, orbicular; capsules pubescent, *ca* 2 mm across; seeds ellipsoid, 4-angled, slightly transversely rugose.

*Fl.-Fr.:*Sept. -Nov.

Common, Barwani, Burhanpur, Indore, Khandwa, Khargone.

Euphorbia neriifolia L., Sp. Pl. 451. 1753; *E. ligularia* Roxb. ex Buch. -Ham., Trans. Linn. Soc. London 14:285. 1825; Samvatsar, Fl. West. Trib. Madhya Pradesh 263. 1996; Shukla in Mudgal *et al.,* Fl. Madhya Pradesh 2:551. 1997. *Patton ki send*

Deciduous, 1.5-2 m tall shrubs or a small trees, young branches 5-angled with short, stipular spines arising from thick tubercles, arranged in 5, irregular rows; leaves 7-18 × 3-7 cm, sessile, fleshy, crowded at the ends of branches, obovate, oblong, apex retuse or obtuse, midrib prominent below; involucres yellowish, lateral ones shortly pedicillate, bisexual, central ones male, sessile; lobes large, cordate, fimbriate; glands transversely oblong; capsules *ca* 1 cm across; seeds *ca* 5 mm across.

Fl.-Fr.: Dec. -March

Planted as live hedge of fields and houses in villages, occurs wild also.

Euphorbia nivulia Buch. -Ham. in Trans. Linn. Soc. London 14:286. 1825; Hook.f., Fl. Brit. India 5:255. 1887; Samvatsar, Fl. West. Trib. Madhya Pradesh 263 . 1996; Shukla in Mudgal *et al.,* Fl. Madhya Pradesh 2:552. 1997. *Kanta thuar*

Shrubs or small trees, up to 5 m high; branches fleshy, terete with pairs of spines on tubercles; leaves 10-15 × 5-6 cm, obovate, fleshy, apex retuse; involucres in crowded cymes at the ends of branches; lobes large, ovate, fimbriate; glands oblong; capsules *ca* 6 mm across, glabrous; seeds globose, *ca* 2 mm across, greenish-black, dorsally grooved.

Fl.-Fr.: Dec. -Feb.

Common in hedges of fields everywhere.

Euphorbia perbracteata Gage in Kew Bull. 1914:238. 1914; Shukla in Mudgal *et al.,* Fl. Madhya Pradesh 2:552. 1997.

Annual, 20-40 cm high herbs; leaves 1.5-5.5 × 0.3-0.5 cm, linear or linear-lanceolate, apex obtuse or apiculate, base oblique; cyathia umbellate; bracta foliaceous, often connate at base; involucres campanulate, lobes 5, sub-deltoid; glands 4, broadly oblong, truncate; capsules 4-5 mm long, 3-lobed; seeds 2-3 mm long, pale yellowish-brown, smooth.

*Fl.-Fr.:*Jan. -April

Rare, in fallow fields, **Barwani** only, as a weed.

Euphorbia prostrata Aiton, Hort. Kew. 2:139. 1789; Hook.f., Fl. Brit. India 5:266. 1887; Londhe in N.P. Singh *et al.,* Fl. Maharashtra 2:884.2001. (*E. chamaecyce* L. and *E. prostrata* Aiton are not conspecific: The Plant List)

Prostrate or erect, weak, perennial herbs; leaves opposite, oblong or elliptic-obovate, 5-9 × 3-4 cm, oblique at base, apex apiculate; cyathia 1-3, axillary on short, peduncles; involucres campanulate, 5-lobed, glands rosy; capsules 2.5 mm long with spreading hairs on the keel; seeds brown, rugulose or 3-4 furrowed.

Fl.-Fr.: Almost throughout the year

Common on hill slopes, rock crevices, all over the area.

Euphorbia pycnostegia Boiss., Cent. Euphorb. 9. 1860; Hook.f., Fl. Brit. India 5:246. 1887; Shukla in Mudgal *et al.,* Fl. Madhya Pradesh 2:553. 1997

Annual, erect, dichotomously branched, 30-50 cm high herbs; leaves 1.5-2 × 1-1.3 cm, oblong, obtuse or suacute at apex, finely serrulate; involucres turbinate, lobes sub-orbicular, fimbriate; glands oblong; capsules *ca* 2 mm across, cocci obtusely keeled; seeds 4-angled or not, smooth or tuberculate.

*Fl.-Fr.:*Sept. -Dec.

Rare, **Indore** only, Simrol grassland.

Euphorbia rosea Retz., Obser. Bot. 4:26. 1786; Hook.f., Fl. Brit. India 5:251. 1887; Samvatsar, Fl. West. Trib. Madhya Pradesh 263. 1996; Shukla in Mudgal *et al.,* Fl. Madhya Pradesh 2:554. 1997.

Perennial, prostrate herbs; stems many, flexuous; leaves 0.5-1.2 × 0.4-0.6 cm, opposite, coriaceous, obliquely obovate or linear-spathulate, apex round, base unequal sided; stipules deltoid; involucres solitary, in small, leafy, axillary and terminal cymes, lobes triangular, ciliate; glands small, transversely oblong, limbs of glands rosy; capsules glabrous, 2-2.5 mm long; seeds 1-1.5 mm long, transversely rugose.

*Fl.-Fr.:*Aug. -Sept.

On rocky, sandy soil, Alirajpur, Dhar, Indore.

Euphorbia rothiana Spreng., Syst. Veg. 3:796. 1826; *E. laeta* B. Heyne ex Roth, Nov. Pl. Sp. 230.1821, *nom. Illeg*; Rau in Bull. Bot. Surv. India 19:71. 1969; Hook.f., Fl. Brit. India 5:263. 1887; Samvatsar, Fl. West. Trib. Madhya Pradesh 263. 1996; Shukla in Mudgal *et al.,* Fl. Madhya Pradesh 2:550.1997.

Erect, 30-60 cm high, annual herbs; stems cylindric, densely leafy, upper branchlets umbellate; leaves alternate below and 4-whorled beneath the crown of branches, 4-8 × 0.5-1.5 cm, apex acute or mucronate, linear-oblong, lanceolate, glabrous; bracts (floral leaves) opposite beneath the dichotomous forks, broadly ovate or orbicular, acute, mucronate at apex, rounded at base; involucres solitary in the upper axils and terminal, solitary, lobes sub-quadrate; glands semilunate, 2-horned; capsules glabrous, deeply 3-lobed, 5-6 mm across; seeds oblong, 2-3 mm long, 4-angled, blueish-grey, grooved down on one side, smooth.

*Fl.-Fr.:*Jan. -July

Common weed of fields and gardens, in the area.

Euphorbia serpens Kunth, Nov. Gen. Sp. 2:52. 1817; *Anisophyllum serpens* (Kunth) Klotzsch & Garcke, Abhandl. Akad. Berl. Phys. 1859:23. 1860; Hook.f., Fl. Brit. India 5:252. 1887, p.p; Shukla in Mudgal *et al.,* Fl. Madhya Pradesh 2:555. 1997.

Annual or perennial, prostrate herbs; stems profusely branched, rooting at nodes; leaves entire, 2-6 × 1-5 mm, opposite, sub-quadrate or obliquely ovate, apex retuse, base cordate or sub-cordate; stipules membranous, incised; cyathia axillary, solitary; involucres broadly caMPanulate, lobes deltoid; glands 4; capsules ovoid, 1.5 mm long, 3-angled; seeds 4-angled, smooth.

*Fl.-Fr.:*Aug. -Dec.

Rare, **Barwani** only, weed of fields.

Euphorbia thymifolia L., Sp. Pl. 454. 1753; Hook.f., Fl. Brit. India 5:252. 1887; Samvatsar, Fl. West. Trib. Madhya Pradesh 264. 1996; Shukla in Mudgal *et al.,* Fl. Madhya Pradesh 2:555. 1997.

Much branched prostrate, hispidly pubescent, annual herbs; stems often pinkish-red in colour, 20-30 cm long; leaves opposite, 7-9 × 3-5 mm, elliptic-oblong, crenate, obtuse, base obliquely sub-quadrate; stipules fimbriate; involucres campanulate, subsessile, axillary, solitary or more than 2, in axillary cymes; lobes short, ciliate; glands minute or absent; capsules *ca* 1.2 mm long, obtusely keeled, clothed with appressed, stiff hairs; seeds 4-angled, brownish with 4-5 transverse furrows.

*Fl.-Fr.:*Throughout the year

A common weed of fields, roadsides, waste lands, moist walls etc.

Euphorbia tirucalli L., Sp. Pl. 452. 1753; Hook.f., Fl. Brit. India 5:254. 1887; Samvatsar, Fl. West. Trib. Madhya Pradesh 264. 1996; Shukla in Mudgal *et al.,* Fl. Madhya Pradesh 2:555. 1997. *Thor*

Shrubs or small trees with a bush-like crown of irregular, leafless (with age)branches; cylindrical, fleshy, green, dichotomously forked; leaves 7-10 mm long, elliptic-oblong, caducous; involucres shortly stalked, having peltate and punctate glands in the forks of the younger branches; flowers very rarely.

*Fl.-Fr.:*Aug. -Sept., on very rare occasions

Common hedge plant in all villages, not wild, planted.

Euphorbia umbellata (Pax) Bruyns, Euphorbia world 3:5. 2007; *Synadenium umbellatum* Pax, Bot. Jahrb. Syst. 19:125. 1894; *Synadenium grantii* Hook.f., Bot. Mag. 83:t. 5633. 1867; Samvatsar, Fl. West. Trib. Madhya Pradesh 270.1996.

Erect, 2-4 m tall, bushy shrubs; leaves fleshy, red when young, green with age, scattered, 8-10 cm long, obovate-spathulate, apex acute, base cuneate; cymes dichotomously branched, both axillary and terminal; involucres bisexual or entirely male, 5-6 mm across, saucer-shaped with an entire, rim-like gland; males in 5 fascicles with or without a single female flower in the centre; capsules red, trigonous, 6-8 mm across; seeds ellipsoid, dull brown, not well developed.

*Fl.-Fr.:*Nov.-Feb.

A recent introduction in the area, Alirajpur, Burhanpur, Dhar, Indore, planted for live hedge.

HOMONOIA Lour. W. 3; MP 1; SA 1

Homonoia riparia Lour., Fl. Cochinch. 637. 1790; Hook.f., Fl. Brit. India 5:455. 1887; Shukla in Mudgal *et al.,* Fl. Madhya Pradesh 2:562. 1997. *Surra*

Perennial, branched, 2-3 m high shrubs; leaves 10-20 × 1-2 cm, linear to lanceolate, apex shortly acuminate, base acute; male flowers *ca* 5 mm across; tepals 3, broadly elliptic, *ca* 3 mm long; female flowers *ca* 4 mm across, tepals 5, *ca* 2 mm long; capsules globose, tomentose, *ca* 4 mm across; seeds yellowish-brown, *ca* 1 mm across.

*Fl.-Fr.:*Jan. -April

Barwani, Khandwa, Khargone, along banks of river beds, drains.

JATROPHA L. W. 190; MP 3; SA 3

Key to Species

1. Leaves palmately 3-5 lobed; petioles with stalked glands; stipules a series of stalked glands; petals free, red; undershrubs .... **J. gossypifolia**
1. Leaves entire or 3-5 angled or lobed; petioles eglandular; stipules absent or minute; petals united or free; shrubs or small trees
   2. Petals free; stamens 8, all united in the lower half; capsule about 1 cm across .... **J. nana**
   2. Petals united; stamens 10, 5 inner connate into a central column; capsule 2.5 -3 cm across .... **J. curcas**

Jatropha curcas L., Sp. Pl. 1006. 1753; Hook.f., Fl. Brit. India 5:383. 1887; Samvatsar, Fl. West. Trib. Madhya Pradesh 264. 1996; Shuklain Mudgal *et al.,* Fl. Madhya Pradesh 2:563. 1997. *Ratan jyote*

Much branched, 2-3 m high, glabrous shrubs; branches green with scars of fallen leaves; leaves 9-15 × 4-13 cm, broadly ovate, palmately 3-5 lobed, apex acute, base cordate; petioles 7-12 cm long, glabrous; stipules deciduous; flowers greenish-yellow, in 6-8 cm long, axillary and terminal paniculate cymes; perianth campanulate, 5-10 lobed, outer tepals equal, ovate or obovate *ca* 4 mm long, sericeous at base inside; stamens 10, in 2 whorls of 5 each; ovary *ca* 2 mm across; capsules globose, 2.5-3 cm across, rugose, black when dry.

*Fl.-Fr.:*Nov. -Feb.

Commonly planted as hedge and also naturalized on waste lands near villages.

Jatropha gossypifolia L. var. **elegans** Mull. Arg., in DC,Prodr. 15(2):1087. 1866; Hook.f., Fl. Brit. India 5:383. 1887; Samvatsar, Fl. West. Trib. Madhya Pradesh 264. 1996; Shukla in Mudgal *et al.,* Fl. Madhya Pradesh 2:263. 1997. *Raja-rani*

Glabrous, 1-1.5 m high shrubs; stems deep reddish-brown; leaves palmately 3-5 lobed, cordate, lobes elliptic, oblong orobovate, margin with gland-tipped hairs; petioles glandular, 7-8 cm long, reddish-purple, lamina shining brown; flowers dark red or crimson in glandular, corymbose, terminal cymes; outer tepals ovate, *ca* 4 mm long, obtuse, ciliate; stamens 8, 5+3, filaments basally connate; inner 3 tepals longer; capsules broadly oblong, *ca* 2 cm across, truncate at the ends; seeds greyish-red, oblong.

*Fl.-Fr.:*Aug. -March

Common in waste lands near villages, also along roadsides, all over the area.

Jatropha nana Dalzell & A. Gibson Bombay Fl. 229. 1861; Hook.f., Fl. Brit. India 5:382. 1887; Samvatsar, Fl. West. Trib. Madhya Pradesh 264. 1996; Khanna *et al.,* Suppl. Fl. Madhya Pradesh 160.2001.

Glabrous, 30-60 cm high undershrubs; stems smooth, terete; leaves entire or 3-lobed, broadly ovate, 7-15 cm long, margin entire; petioles up to 4 cm long; flowers pedicellate, in few flowered, terminal, paniculate cymes; bracts lanceolate, acute; capsules about 1 cm long, obovoid-oblong, flattened at the top, slightly 6-lobed.

*Fl.-Fr.:*Oct. -Jan.

Rare, along water places, river banks, **Alirajpur** only, Shakarja. It is difficult to accout for this species in such a remote area as Shakarja. Reported as endemic to Maharashtra, but now the distribution extends to MP.

MALLOTUS Lour. W. 122; MP 1; SA 1

Mallotus philippensis (Lam.) Mull. Arg. in Linnaea 34:196. 1865; *Croton philippense* Lam., Encycl. 2;206. 1786; Hook.f., Fl. Brit. India 5:442. 1887; Samvatsar, Fl. West. Trib. Madhya Pradesh 266. 1996; Shukla in Mudgal *et al.,* Fl. Madhya Pradesh 2:566. 1997. *Kamala*

Evergreen trees up to 8 m high; leaves ovate-lanceolate, 8-20 × 3-10 cm, acute to acuminate at apex, round or cuneate at base, entire or closely dentate; flowers whitish-yellow; male flowers clustered in 10-20 cm long, erect, terminal spikes, tepals 4, *ca* 3 mm long; stamens numerous; female flowers distant, in 4-6 cm long spikes, tepals 3-4, thick; ovary reddish, glandular; capsules globose, 8-12 mm across, 3-lobed, covered with red granules; seeds 3-4 mm across, smooth.

*Fl.-Fr.:*Jan. -April

In forests, along roadsides, Alirajpur, Barwani, Dhar, Indore, Khandwa, Khargone.

TRAGIA L. W. 152; MP 2; SA 2

Key to Species

1. Leaves 6-10 × 3.5-5 cm; petiole 2.5 cm; tepals in female flowers 6 .... **T. involucrata**
1. Leaves 2.5-5 cm; petiole 0.9-1cm; tepals in female flowers 5 .... **T. plukenetii**

Tragia involucrata L., Sp. Pl. 980.1753; Hook.f., Fl. Brit. India 5:465. 1888; Samvatsar, Fl. West. Trib. Madhya Pradesh 270.1996; Shukla in Mudgal *et al.,* Fl. Madhya Pradesh 2:580.1997. *Bichhu buti*

Perennial, stout, 30-70 cm high twiners with thick root-stock; branches puberulous, villous as well as with scattered, pungent hairs; leaves 5-9 × 3-4 cm, apex acuminate, base rounded, ovateoblong or elliptic, margin serrate, hairy; flowers in androgynous racemes; male tepals 1 mm long; female tepals imbricate, 1.5 mm long; ovary *ca* 1.5 mm across; styles united below; capsules 3-lobed, 8-9 mm long, hispid.

*Fl.-Fr.:*Sept. -Jan.

In waste places, hedges, bunds of fields, Alirajpur, Dhar, Indore.

Tragia plukenetii Radcl. -Sm. In Kew Bull. 37:688. 1983; *T. involucrata* var. *cannabina* Mull. Arg. in DC., Prodr. 15(2):944. 1866; Shukla in Mudgal *et al.,* Fl. Madhya Pradesh 2:581. 1997. *Chur churi*

Perennial, erect or twining shrubs, up to 1.5 m high; leaves 2-6 × 1-2 cm, palmatifid or 3-5 lobed, apex acute, base truncate, 5-nerved; flowers greenish, in racemes; male flowers *ca* 2 mm across, tepals up to 1 mm long; female flowers *ca* 2 mm across, tepals 2 mm long; capsules 3-lobed, 7-8 mm across; seeds globose, *ca* 3 mm across.

*Fl.-Fr.:*Almost throughout the year

Burhanpur, Khandwa.

PHYLLANTHACEAE 58/2099; 9/26; 5/15

Key to Genera

1. Perianth 2-seriate, sepals and petals 5 each in both types of flowers; stamens 5, united on a gynophore; trees
   2. Ovary 2-celled; fruit drupaceous; evergreen trees... BRIDELIA
   2. Ovary 3-celled; fruit capsular; deciduous trees ....CLEISTANTHUS
1. Perianth 1–seriate
   3. Pistillode absent; filaments united.. ......PHYLLANTHUS

3. Pistillode present
    4. Flowers in axillary clusters; stamens 5; ovary 3-celled; fruit capsular, white .... FLUEGGEA
    4. Flowers in spikes; stamens 2-5; ovary 1-celled; fruit red, drupaceous .... ANTIDESMA

ANTIDESMA L. W. 102; MP 3; SA 1

Antidesma acidum Retz., Observ. Bot. 5:30.1788; *A. diandrum* (Roxb.) B. Heyne ex Roth, Nov. Pl. Sp. 369. 1821; *Stilago diandra* Roxb., Pl. Coromandel 2:t. 166. 1802; Shukla in Mudgal *et al.,* Fl. Madhya Pradesh 2:525. 1997. *Ban musari*

Small, deciduous, 4-8 m tall trees; leaves 4-12 × 3-7 cm, obovate-elliptic to oblong, glabrous, apex acute or acuminate, base rounded or cuneate; flowers in *ca* 8 cm long, spikes at the ends of branchlets; bracts ovate, acute; male flowers pedicellate, about 0.7 mm across, tepals 5, cup-shaped, ovate, acute, basally connate; stamens 2, horseshoe-shaped; pistillodes absent; female flowers 0.7 mm across, tepals deeply lobed; drupes purple-red, 4-5 mm long.

Fl.-Fr.: July-Oct.

Barwani, Khargone.

BRIDELIA Willd. W. 49; MP 2; SA 1

Bridelia retusa (L.) A. Juss., Euphorb. Gen. 109. 1824; *Clutia retusa* L., Sp. Pl. 1475. 1753; Hook.f., Fl. Brit. India 5:268. 1887; Samvatsar, Fl. West. Trib. Madhya Pradesh 259. 1996; Shukla in Mudgal *et al.,* Fl. Madhya Pradesh 2:533. 1997. *Kasai*

Deciduous, 4-7 m high trees, spinous when young, leaves 6-15 × 3-6 cm, elliptic-oblong, apex obtuse, base rounded, finely tomentose beneath, secondary nerves close set, up to 20 pairs; flowers in axillary clusters or in long, axillary or terminal spikes, male and female flowers are intermixed; male outer tepals deltoid-ovate, 2-3 mm long, fleshy, inner tepals greenish-white, short; stamens connate below; female tepals spathulate; drupes 0.8 mm across, supported by the persistent calyx, ripening black.

*Fl.-Fr.:*Aug. -Dec.

Common in forest, all districts.

CLEISTANTHUS Hook.f. W. 133; MP 1; SA 1

Cleistanthus collinus (Roxb.) Benth. ex Hook.f., Fl. Brit. India 5:274. 1887; *Clutia collina* Roxb., Pl. Coromandel 2:37. 1802; Samvatsar, Fl. West. Trib. Madhya Pradesh 260.1996; Shukla in Mudgal *et al.,* Fl. Madhya Pradesh 2:537. 1997. *Garari*

Small trees, up to 6 m high; leaves × 2-7 cm, distichous, orbicular, obovate or elliptic, apex acute, base round; flowers greenish, in axillary villous clusters; sepals ovate-lanceolate; corolla fleshy, incurved; ovary 3-locular, styles 3, shortly 2-fid; disc annular, with an inner fimbriate corona; capsules sessile, globose, 1.5- 2 cm across, dark brown; seeds globose, reddish-brown, smooth, shining.

*Fl.-Fr.:*March-Sept.

Barwani, Dhar, Khandwa, Khargone.

FLUEGGEA Willd. W. 16; MP 2; SA 2

Key to Species

1. Armed plants; pistillodes 2, free; leaves to 1.5 × 1cm .... **F. leucopyrus**
1. Unarmed plants; pistillodes 3, united below; leaves to 2.5 × 1.5 cm.................................... ......**F. virosa**

Flueggea leucopyrus Willd., Sp. Pl. 4:757. 1806; *Securinega leucopyrus* (Willd.) Mull. Arg., Prodr. 15 (2):451. 1866; Samvatsar, Fl. West. Trib. Madhya Pradesh 269. 1996; Shukla in Mudgal *et al.,* Fl. Madhya Pradesh 2:578. 1997. *Jondle*

Small, 2.5-3 m high shrubs; branchlets arrested, stiff, thorn-like, bearing leaves and flowers; leaves 1-2 × 0.6-1.5 cm, obovate, obcordate or rotundate; flowers in axillary clusters; bracts minute; male flowers numerous, females 1-5; tepals in male flowers up to 0.7 mm long, stamens 5, pistillodes 2; tepals in female flowers up to 0.8 mm long; capsules white, *ca* 5 mm across.

*Fl.-Fr.:*Sept. -Dec.

Forest undergrowth and in hedges of fields, Burhanpur, Dhar, Indore.

Flueggea virosa (Roxb. ex Willd.) Royle, Ill. Bot. Himal. Mts. 328. 1836; *Phyllanthus virosus* Roxb. ex Willd., Sp. Pl. 4:578. 1805; *Flueggea microcarpa* Blume, Bijdr. 580.1826; *Securinega virosa* (Roxb. ex Willd.) Baill., Adansonia 6:334. 1866; Hook.f., Fl. Brit. India 5:328. 1887; Samvatsar, Fl. West. Trib. Madhya Pradesh 269. 1996; Shukla in Mudgal *et al.,* Fl. Madhya Pradesh 2:578. 1997. *Chirgodi*

Unarmed, up to 3 m high shrubs; leaves 3-7 × 2-4 cm, obovate, apex emarginate, base cuneate; flowers in axillary fascicles; male flowers *ca* 3.5 mm across, tepals *ca* 1.5 mm long, stamens 5, pistillodes 3, connate below; female flowers *ca* 2 mm across, tepals up to 1 mm long; capsules white, 4 mm across; seeds trigonous, brown, minutely punctate.

*Fl.-Fr.:*Aug. -Nov.

Common in forest, all over the area.

PHYLLANTHUS L. W. 936; MP 11; SA 10

Key to Species

1. Ovary 5-10 celled; ovules superposed; stamens 5; fruit fleshy, purple; seeds 8-16 .... **P. reticulatus**
1. Ovary 3-celled; ovules 2 per cell, juxtaposed; stamens 3; fruit fleshy or capsular; seeds 6 or less
    2. Fruits fleshy, 2-3 cm across; trees .... **P. emblica**
    2. Fruits capsular, less than 2 cm across, herbs or shrubs
        3. Shrubs; stem grooved; seeds smooth ....**P. lawii**
        3. Herbs or small shrubs
            4. Fruits densely echinate .... **P. urinaria**
            4. Fruits smooth
                5. Stipules peltate

6. Anthers erect; styles 2-lobed; disc of female flowers consisting of free glands .... **P. maderaspatensis**
6. Anthers didymous; styles 3, bi-fid; disc of female flowers shortly cup-shaped ... **P. virgatus**
5. Stipules subulate
7. Stem angled or winged
8. Stem sharply angled, scabrid above; leaves 5-20 × 2-5 mm; capsule 2-2.5 mm across .... **P. debilis**
8. Stem wiged, scabrid all over; leaves 4-15 × 2-10 mm; capsule 3-4 mm ....**P. scabrifolius**
7. Stem terete; seeds regularly ribbed
9. Disc gland in male flowers minute; seeds with slender ribs, erect and comparatively larger plants .... **P. fraternus**
9. Disc gland in male flowers pedicelled; seeds with strong ribs; prostrate and small plants ....**P. amarus**

Phyllanthus amarus Schum. & Thonn., Kongel Danske Vidensk. Selsk. Skr., Naturvidensk. Math. Afd. 4:195. 1829; Hook.f., Fl. Brit. India 5:298. 1887; Samvatsar, Fl. West. Trib. Madhya Pradesh 267. 1996; Shukla in Mudgal *et al.,* Fl. Madhya Pradesh 2:572. 1997. *Jarmala*

Annual, erect, up to 40 cm high herbs, branchlets scabridulous on the angles; leaves 5-14 × 2-5 mm, oblong-elliptic, obtuse at both ends; stipules lanceolate-subulate; flowers yellowish; male tepals 5, *ca* 0.6 mm long, stamens connate, 3; female flowers solitary, tepals 5, 1-1.4 mm long; capsules depressed globose, smooth, *ca* 2 mm across; seeds 3-angled with longitudinal ribs on the back.

*Fl.-Fr.:*Aug. -Dec.

In moist, open places, Alirajpur, Barwani, Burhanpur, Khandwa, Khargone.

Phyllanthus debilis Klein ex Willd., Sp. Pl. 4:582. 1805; Hook.f., Fl. Brit. India 5:299. 1887; Samvatsar, Fl. West. Trib. Madhya Pradesh 267. 1996.

Erect, 15-50 cm high, annual herbs; stems sharply angled, angles scabrid in the upper region; leaves 0.5-2 × 0.2-0.5 cm, acute, elliptic; stipules lanceolate-acuminate; capsules sub-globose, 2-2.5 mm across; seeds trigonous with longitudinal and cross striations and hyaline, hygroscopic setae.

*Fl.-Fr.:*Aug. -March

Wet and shady places, Alirajpur, Indore.

Phyllanthus emblica L., Sp. Pl. 982. 1753; *Emblica officinalis* Gaertn., Fruct. Sem. Pl. 2:122. t. 108. 1790; Hook.f., Fl. Brit India 5:289. 1887; Samvatsar, Fl. West. Trib. Madhya Pradesh 267. 1996; Shukla in Mudgal *et al.*, Fl. Madhya Pradesh 2:572. 1997. *Aaonla*

Deciduous trees, up to 8 m high, bark peeling off in irregular pieces; branches finely tomentose; leaves close set, distichous, 9-14 mm long, linear –oblong, glabrous; stipules ovate, 1-1.5 mm long; flowers greenish-yellow, in axillary clusters from lower portions of the branchlets; male tepals 6, 1-1.2 mm long, stamens 3, filaments connate; female tepals 6, 1.2-1.5 mm long, valvate; styles connate at base, twice 2-fid; drupes globose, 2-3 cm across, obscurely 6-lobed; seeds *ca* 6 mm long, dark brown, 3-angled.

*Fl.-Fr.:*Feb. -April

Common member of forest flora, everywhere.

Phyllanthus fraternus G. L Webster, Contri. Gray Herb. 176:53. 1955; (*P. niruri* auct. non) L. 1753; Hook.f., Fl. Brit. India 5:298. 1887; Samvatsar, Fl. West. Trib. Madhya Pradesh 267. 1996. *Bhui aaonli*

Erect, annual herbs, stems scaberulous on angles; leaves 0.5-1.5 cm long, oblong with obtuse or rounded base; stipules lanceolate-subulate, 1 mm long; male flowers 1-3 together, disc lobes 6, glandular; female flowers solitary; capsules depressed globose, 2-2.5 mm across; seeds brown, longitudinally ribbed on the back, disc shallowly 6-9 lobed.

*Fl.-Fr.:*July-Nov.

Very common in cultivated fields, waste places, all over the area.

(Note:- *P. amarus* and *P. fraternus* are two separate species and not synonymous as given in Fl. M.P. 2:572. 1997)

Phyllanthus lawii J. Graham, Cat. Pl. Bombay 181. 1839; Hook.f., Fl. Brit. India 5:290.1887; Samvatsar, Fl. West. Trib. MP 267. 1996; Shukla in Mudgal *et al.,* Fl. Madhya Pradesh 2:572. 1997.

Shrubs, up to 1.5 m high, stems rigid; leaves distichous, crowded,linear- oblong, 3-8 mm long; stipules persistent; male flowers solitary or a few together, tepals 6, 2-3 mm long, stamens 3, filaments connate below in a short column; capsules globose, smooth, *ca* 5 mm across; seeds *ca* 2 mm long, 3-angled, rounded on back.

*Fl.-Fr.:*Sept. -Oct.

Along and in river beds, Alirajpur, Dhar.

Phyllanthus maderaspatensis L., Sp. Pl. 982. 1753; Hook.f., Fl Brit. India 5:292. 1887; Samvatsar, Fl. West. Trib. Madhya Pradesh 267. 1996; Shukla in Mudgal *et al.,* Fl. MP 2:573. 1997. *Kanocha*

Annual, erect, up to 60 cm high herbs, branches angular, often sulcate; leaves 1.5-4 × 0.5-1 cm, oblong or obovate, apex mucronate, base acute; stipules lanceolate, acuminate, *ca* 2 mm long; male tepals 6, obovate, 0.7-1 mm long, stamens 3, filaments connate at base; female flowers solitary, tepals 6, obovate-spathulate; capsules globose, 2-3 mm across; seeds 3-angled, dark brown, vertically muriculate.

*Fl.-Fr.:*Aug. -Feb.

Common weed of fields and waste lands throughout the area.

Phyllanthus reticulatus Poir., Encycl. 5:298. 1804; *Kirganelia reticulata* (Poir.) Baill. Etude Euphorb. 613. 1858; Hook.f., Fl. Brit. India 5:288. 1887; Samvatsar, Fl. West. Trib. Madhya Pradesh 267. 1996; Shukla in Mudgal *et al.,* Fl. Madhya Pradesh 2:573. 1997. *Kamoi*

Straggling, 1.5-2.5 m high shrubs, branches pinkish; leaves 1.5-2.5 × 0.6-1,5 cm, elliptic-oblong, apex obtuse, base rounded or cuneate; stipules lanceolate, 1-1.5 mm long; male tepals 5, unequal, up to 2.3 mm long, imbricate, outer smaller, oblong-obovate, inner sub-orbicular; female tepals as in males, styles 3, shortly 2-fid; berries smooth, globose, purple-black; seeds 3-angled, finely granulate.

*Fl.-Fr.:*April-Dec.

Common in hedges of fields in villages, roadsides, in all districts.

Phyllanthus scabrifolius Hook.f., Fl. Brit. India 5:299. 1887; Londhe in Sharma *et al.,* Fl. Maharashtra 2:901.2001.

Annual, 15-25 cm high herbs; leaves hispidly hairy, 1-1.2 × 0.6 cm, elliptic or obovate, apex rounded; flowers green, pedicels short; capsules 0.3 cm across, depressed-globose; seeds trigonous, with parallel ribs on the back.

*Fl.-Fr.:*Sept. -Dec.

Rare, **Burhanpur** only, Asirgad hills.

Phyllanthus urinaria L., Sp. Pl. 982.1753; Hook.f., Fl. Brit. India 5:293.1887; Samvatsar, Fl. West. Trib. Madhya Pradesh 269.1996; Shukla in Mudgal *et al.,* Fl. Madhya Pradesh 2:574.1997.

Erect, glabrous, 10-30 cm high, annual herbs; leaves 0.5-1.5 × 0.2-0.6 cm, elliptic-oblong or obovate, margin and base oblique; stipules minute, filiform; bracts linear, 1 mm long; male tepals 6, 0.6-0.8 mm long, stamens 3, filaments connate; female flowers solitary, tepals 6, *ca* 0.7 mm long; capsules 2-3 mm across, minutely echinate; seeds trigonous, transversely ribbed.

*Fl.-Fr.:*Aug. -Oct.

As forest undergrowth in wet areas, in all districts.

Phyllanthus virgatus G. Forster, Fl. Ins. Austr. 65. 1786; *P. simplex* Retz., Observ. Bot. 5:29. 1789; Samvatsar, Fl. West. Trib. Madhya Pradesh 269. 1996; Shukla in Mudgal *et al.,* Fl. Madhya Pradesh 2:575. 1997.

Annual or perennial (when moisture available), 20-30 cm high herbs, branches from a woody base, compressed, often brick-red; leaves distichous, 1-3 × 0.3-0.6 cm, obtuse-apiculate at apex, rounded at base, linear-oblong; stipules peltate, sub-sagittate, 1-1.2 mm long; flowers usually solitary, greenish-yellow; male tepals 6, *ca* 0.8 mm long, stamens 3, free; female tepals 6, ovate-oblong, *ca* 0.7 mm long; capsules 3-lobed, 2-2.5 mm across; seeds 1.2 mm long, trigonous, dark brown, finely tuberculate.

*Fl.-Fr.:*Sept. -Dec.

Common weed of fields, along banks of streams, moist, waste places, all over the area.

PUTRANJIVACEAE 3/216; 1/1; 1/1

PUTRANJIVA Wall. W. 4; MP 1; SA 1

Putranjiva roxburghii Wall., Tent. Fl. Nepal. 61. 1826; Hook.f., Fl. Brit. India 5:336. 1887; *Drypetes roxburghii* (Wall.) Hurus., J. Fac. Sci. Univ. Tokyo sect. 3, Bot. 6:335. 1954; Samvatsar, Fl. West. Trib. Madhya Pradesh 260.1996; Shukla in Mudgal *et al.,* Fl. Madhya Pradesh 2:541. 1997. *Putrajiva*

Evergreen, 10-14 m high tree with pendulous branches; leaves 4-10 × 2-3 cm, ovate-oblong, acute at apex, obliquely obtuse or rounded at base, serrulate; male flowers in axillary clusters, pedicels short, tepals 5, *ca* 2 mm long; disc absent, stamens 3, connate at base; female flowers 1-3 in an axil, tepals 5, up to 2 mm long; ovary 3-locular, finely silky, stigmas 3; drupes ellipsoid, *ca* 1.5 cm long, white-tomentose, stone hard, rugose; seeds crustaceous.

*Fl.-Fr.:*March-May

Rare in forest, Alirajpur, Dhar, Indore.

URTICACEAE 52/1303; 8/9; 6/ 6

Key to Genera

1. Plants with stinging hairs; female flowers without staminodes, their perianth lobes connate; leaves alternate; stipules intra-petiolar; achenes strongly oblique, stipitate; stigma subulate or filiform
   2. Female perianth with 3 lobes fused, fourth lobe absent or reduced to a bristle; stinging hairs more than5 mm long .... GIRARDINIA
   2. Female perianth lobes 4, dorsal lobe longest, ventral smallest, never bristle-like; stinging hairs much less than 5 mm .... LAPORTEA
1. Plants without stinging hairs; female flowers with or without staminodes
   3. Female perianth tubular; stigma not penicillate
      4. Male perianth 2-3 fid; stamens two.......... NEODISTEMON
      4. Male perianth 4-5 partite; stamens 4-5.......... GONOSTEGIA
   3. Female perianth not tubular; stigma penicillate
      5. Flowers in clusters on fleshy receptacles; leaves alternate .... ELATOSTEMA
      5. Flowers in branched cymes .... PILEA

ELATOSTEMA J. R. & G. Forst. W 525; MP 2; SA 1

Elatostema cuneatum Wight, Icon. Pl. Ind. Orient. 6: t. 2094, f. 3 1853; Hook.f., Fl. Brit. India 5:568. 1888; Srivastava in Mudgal *et al.,* Fl. Madhya Pradesh 2:586. 1997.

Annual, erect, 15-20 cm high herbs, rooting near the base; leaves sub-sessile, 1-4 × 0.5-1.5 cm, hatchet-shaped, apex ronded or subacute, base falcately cuneate, coarsely crenate above the middle; stipules ovate; female flowers in axillary, solitary, sessile heads; flowers green or white; male flowers not seen; tepals 3; achenes ellipsoid, *ca* 0.5 mm long, ribbed.

*Fl.-Fr.:*Aug. -Oct.

Rare, in moist places, **Burhanpur** only, Asirgad area.

GIRARDINIA Gaud. W. 3; MP 1; SA 1

Girardinia diversifolia (Link) Friss in Kew Bull. 36:145. 1981; *Urtica diversifolia* Link, Enum. Hort. Berol. Alt. 2:385. 1822, non Blume, 1825; Hook.f., Fl. Brit. India 5:551. 1888; Srivastava in Fl. Madhya Pradesh 2:588. 1997. *Bichhu buti*

Perennial, 50-70 cm high herbs; leaves 7-13 × 8-12 cm, broadly ovate or sub-orbicular, apex acuminate, base cordate or truncate, margin falcately dentate or 3-7 lobed, with appressed, stiging hairs; stipules foliaceous, ovate, 9-15 mm long; male flowers in simple or paniculate spikes, tepals 4, stamens 4, pistillode globose; female flowers in paniculate spikes, tepals 3; achenes compressed, obliquely ovoid.

*Fl.-Fr.:*July-Dec.

Rare, **Khandwa** only, Kalibhit forest.

GONOSTEGIA Turcz. W. 4; MP 1; SA 1

Gonostegia pentandra (Roxb.) Benn., Pl. Jav. Rar. 64. 1838; *Urtica pentandra* Roxb., Fl. Ind. ed. 1832 3:583. 1832; Hook.f., Fl. Brit. India 5:583. 1888; Srivastava in Mudgal *et al.,* Fl. Madhya Pradesh 2:594. 1997, as *Pouzolzia pentandra* (Roxb.) Benn.

[The species, *Pouzolzia pentandra* (Roxb.) Benn. is transferred to *Gonostegia pentandra* (Roxb.) Benn. according to The Plant List, hence the citation has changed as above, description and other details remain unchanged]

Perennial, erect herbs up to 90 cm high; leaves 2-7 × 0.5-1 cm, opposite, oblong-lanceolate, apex acute or acuminate, base cordate or shortly petiolate, 3-nerved; upper leaves smaller and mostly alternate, linear-oblong; flowers white, in clusters borne in up to 45 cm long, terminal spikes; male flowers pedicellate, ciliate, perianth 5-partite; female perianth with 2-3 lateral wings; achenes ovoid, shining black.

*Fl.-Fr.:*Aug. -Oct.

Rare, along streams, Barwani, Dhar, Indore, Khargone.

LAPORTEA Gaud. W. 25; MP 1; SA 1

Laportea interrupta (L.) Chew, in Gard. Bull. Singapore 21(2):200-201. 1965; *Urtica interrupta* L., Sp. Pl. 2:985. 1753; *Fleurya interrupta* (L.) Gaudich., Voy. Uraine 497. 1830; Hook.f., Fl. Brit. India 5:548. 1888; Samvatsar, Fl. West. Trib. Madhya Pradesh 273. 1996; Srivastava in Mudgal *et al.,* Fl. Madhya Pradesh 2:589. 1997. *Lal bichhua*

Annual, 10-40 cm high, densely hairy herbs with scattered stinging hairs; leaves 3.5-9.5 × 2.5-7.5 cm, broadly ovate, crenate-serrate along margins, clothed with stinging hairs, apex acuminate, base cordate to sub-truncate; flowers greenish-white, in much aggregated, interrupted, up to 30 cm long, cymose spikes or panicles longer than the petioles; male tepals 4, ovate; female tepals stalked, shortly 4-lobed or toothed, keeled on one side; keels decurrent on the stipes; achenes compressed, ovoid.

*Fl.-Fr.:*Aug. -Nov.

Common in moist and shady, waste places, all over the area.

NEODISTEMON Babu & A.N. Henry Monotypic

Neodistemon indicum (Wedd.) Babu & A.N. Henry in Taxon 19:651. 1970; *Distemon indicum* Wedd., in Monogr. Fam. Urtic. 551. 1856; Hook.f., Fl. Brit. India 5:588. 1888; Srivastava in Mudgal *et al.,* Fl. Madhya Pradesh 2:590.1997.

Annual, pubescent herbs; leaves 3-6 × 2-3.5 cm, broadly ovate, apex acuminate, base cuneate, margin serrate, cystoliths punctate; petioles 1-2.5 cm long; flowers in interrupted, axillary, pseudo-spikes; bracts ovate, acuminate; male flowers minute, tipped with hooked hairs; female flowers strigose; achenes beaked, *ca* 2 mm long.

*Fl.-Fr.:*Nov. -Dec.

Rare, **Khandwa** only, Kalibhit forest.

PILEA Lindley W. 279; MP 1; SA 1

Pilea microphylla (L.) Liebm., Mexic. Neldeagt. Pl. V, 2:302. 1851; *Parietaria microphylla* L., Syst. Nat. ed. 10.2:1308. 1759; Hook.f., Fl. Brit. India 5:551. 1888; Srivastava in Mudgal *et al.,* Fl. Madhya Pradesh 2:593. 1997.

Annual, prostrate or decumbent-ascending, delicate herbs; leaves 1-6 × 1-5 mm, apex subacute, base acute, obovate-spathulate, membranous, sessile with shining glands on the lower surface; flowers in axillary, sessile cymes; male tepals 4, ovate-rounded; female tepals 3; achenes 0.3 mm long, smooth, brown.

*Fl.-Fr.:*Dec. -Jan.

In moist places, Burhanpur, Indore, Khandwa.

ULMACEAE 8/64; 1/1; 1/1

HOLOPTELEA Planch. Monotypic

Holoptelea integrifolia Planch., in Ann. Sci. Nat. Bot. III. 10:259. 1848; *Ulmus integrifolia* Roxb. Pl. Coromandel 1:56. 1796; Hook.f., Fl. Brit. India 5:481. 1888; Samvatsar, Fl. West. Trib. Madhya Pradesh 273. 1996; Khanna in Mudgal *et al.,* Fl. Madhya Pradesh 2:599. 1997. *Chirol*

Large, deciduous, 10-15 m high trees; leaves 5-15 × 2.5-9 cm, ovate to elliptic, abruptly acuminate at apex, rounded or sub-cordate at base; petioles 1-1.5 cm long; flowers green, sessile, in fascicles,on leafless branches, male and bisexual mixed; tepals 4, free, stamens 7-8; ovary stalked; samaras oblique-elliptic to sub-orbicular, 1.5-2.5 cm across, tip emarginate, crowned by the persistent styles.

Fl.-Fr.: Feb. -May

Common along roadsides, around villages, all over the area.

CANNABACEAE 8/10; 2/4; 1/3

TREMA Lour. W. 13; MP 3; SA 3

Key to Species

1. Leaves 7-10 cm long; drupes globose, not compressed; testa pitted .... **T. polytoria**
1. Leaves more than 10 cm long
2. Leaves 7-12 cm long; petiole 1 cm long; drupes black, sub-compressed .... **T. orientalis**
2. Leaves 7-15 cm long; petiole 0.7 cm long; drupes blackish-purple, compressed . ...**T. tomentosa**

Trema orientalis (L.) Blume, Mus. Bot. 2:62. 1856; *Celtis orientalis* L., Sp. Pl. 1044. 1753; Hook.f., Fl. Brit. India 5:484. 1753; Samvatsar, Fl. West. Trib. Madhya Pradesh 273. 1996; Khanna in Mudgal *et al.,* Fl. Madhya Pradesh 2:600.1997. *Gio*

Large evergreen shrubs or small, up to 8 m high trees; leaves 7-12 × 3-5 cm, ovate, apex acuminate, margin crenate-serrate, scabrid above, soft, tomentose beneath, base unequal; petioles up to 1 cm long; flowers unisexual, 5-merous, in lax, axillary cymes; tepals 5, hirsute, stamens 5, pistillode present; drupes sub-compressed, globose, black, 3-5 mm across.

*Fl.-Fr.:*Sept. -Oct.

Common member of Teak forests, all districts.

Trema politoria (Planch.) Blume, Mus. Bot. 2:58. 1856; *Sponia politoria* Planch., Ann. Sci. Nat. 10:326. 1848; Hook.f., Fl. Brit. India 5:484. 1888; Khanna in Mudgal *et al.,* Fl. Madhya Pradesh 2:600.1997.

Small, 4-6 m high trees, branches pubescent and scabrid; leaves 4.5-12 × 1.5-4.5 cm, oblong or ovate-lanceolate, apex acute or acuminate, blade equal sided, scabrid on both the surfaces, base rounded or cordate, margin serrulate; petioles up to 1 cm long; flowers unisexual, 5-merous, bracteate in axillary cymes; tepals hispid, stamens 5, pistillode present; drupes ovoid, black.

*Fl.-Fr.:*April-Sept.

Frequent, Barwani, Burhanpur, Khargone.

Trema tomentosa (Roxb.) H. Hara, Fl. E. Himalaya 19. 1971; *Celtis tomentosa* Roxb. Fl. Ind. ed. 1832 2:66. 1832; Hook.f., Fl. Brit. India 5:484. 1888; Khanna in Mudgal *et al.,* Fl. Madhya Pradesh 2:603. 1997.

Velvety pubescent , 5-6 m high, small trees; leaves 8-15 × 4-7 cm, broadly ovate to ovate-elliptic, apex acute to acuminate-caudate, base cordate, blades asymmetrical, strongly scabrid above, tomentose beneath, margin serrate; stipules linear-lanceolate; flowers in large, coMPact cymes; male and female flowers either on the same branch or on different vegetative branches; bracts ovate, acute; tepals 4-5, elliptic, pistillode present in male flowers; drupes ovoid, black at maturity.

*Fl.-Fr.:*Jan. -Dec.

Rare, Burhanpur, Khandwa, Khargone.

MORACEAE 40/1217; 4/25; 1/13

FICUS L. W. 841; MP 18; SA 13

Key to Species

1. Leaves caudate acuminate
   2. Acumen half as long as blade; stipules minute; figs dark purple at maturity .... **F. religiosa**
   2. Acumen not so long; stipules larger
      3. Figs stalked; leaves 8-20 × 5-15 cm, 5-7 nerved at base .... **F. arnottiana**
      3. Figs sessile, paired, 1-2.5 cm long; leafbase 5-nerved; receptacle black at maturity .... **F. rumphii**

1. Leaves not caudate acuminate
  4. Leaves scabrid or hispid above
    5. Leaves opposite; receptacle 3-bracteate at base .... **F. hispida**
    5. Leaves alternate; receptacle ebracteate .... **F. exasperata**
  4. Leaves neither scabrid or hispid above
    6. Receptacle sessile
      7. Aerial roots always present
        8. Figs solitary, up to 2 cm across, orange-yellow; leaves 10-22 × 6-15 cm .... **F. drupacea**
        8. Figs paired
          9. Leaves 4-12 × 2-6 cm; receptacles purple-black at maturity .... **F. microcarpa**
          9. Leaves 12-30 × 7-12 cm; figs red or orange yellow at maturity
            10. Non stragglers; aerial roots pillar-like; leaves 10-30 × 7-20 cm; figs red .... **F. benghalensis**
            10. Epiphytic stragglers; aerial roots absent; leaves 5-15 × 2-7 cm; figs orange-yellow .... **F. tinctoria**
      7. Aerial roots absent; leaves coriaceous, 5-18 × 3-12 cm; figs 1-1.5 cm across, purple at maturity .... **F. amplissima**
    6. Receptacle shortly or prominently stalked
      11. Figs cauliflorous .... **F. racemosa**
      11. Figs not cauliflorous, axillary
        12. Lateral veins 9-16 pairs; figs 0.5-0.8 cm across, black or purple at maturity .... **F. caulocarpa**
        12. Lateral veins 5-11 pairs; figs 0.8-1.2 cm across, white, flushed with red at maturity .... **F. virens**

Ficus amplissima Sm., Cycl. 14: 1. 1810; *Ficus tsiela* Roxb., Hort. Bengal. 66. 1814; King in Hook.f., Fl. Brit. India 5:515. 1888; Khanna *et al.,* in Mudgal *et al.,* Fl. Madhya Pradesh 2:610.1997. *Pipli*

Deciduous, spreading trees, aerial roots absent; 7-15 × 4-9 cm, spirally arranged, ovate or elliptic-lanceolate, apex acute orobtusely cuspidate, base rounded or cuneate, cystoliths confined to upper surface; receptacles axillary and on the twigs below leaves, 1-1.5 cm across, spherical, glabrous, purple when ripe.

*Fl.-Fr.:*March-Oct.

On roadsides, but not common, Alirajpur, Dhar, Indore, Khandwa.

Ficus arnottiana (Miq.) Miq., Ann. Mus. Bot. Lugduno-Batavi 3:287. 1867; *Urostigma arnottiana* Miq., London J. Bot. 6:564. 1847 (unresolved); King in Hook.f., Fl. Brit. India 5:513. 1888; Samvatsar, Fl. West. Trib. Madhya Pradesh 271. 1996; Khanna *et al.,* in Mudgal *et al.,* Fl. Madhya Pradesh 2:610.1997. *Paras peepal*

Trees, up to 18 m high, aerial roots absent; leaves 8-20 × 5-15 cm, subcoriaceous, broadly ovate, apex shortly caudate, acuminate, base cordate, margin entire, petioles long; receptacles in axillary pairs or clustered on tubercles, sessile or subsessile, globose, basal bracts 3, ripening red to purple.

*Fl.-Fr.:*March-June

Common along banks of streams, ghat section, throughout the area.

Ficus benghalensis L., Sp. Pl. 1059. 1753; *Urostigma benghalense* (L.) Gasp., Nov. Gen. Fic. 7. 1844; King in Hook.f., Fl. Brit. India 5:499. 1888; Samvatsar, Fl. West. Trib. Madhya Pradesh 271. 1996; Khanna *et al.,* in Mudgal *et al.,* Fl. Madhya Pradesh 2:611. 1997. *Bargad*

Very large evergreen trees with pillar-like strong, prop. roots; leaves 10-20 × 6-12 cm, elliptic to ovate, apex obtuse, base rounded, margin entire, cystilith confined to lower surface; petioles stout; stipules up to 2 cm long; receptacles axillary, paired, 1-2 cm across, ripening red, basal bracts 3.

*Fl.-Fr.:*April-July

Common on roadsides, near temples, in the whole area.

Ficus caulocarpa (Miq.) Miq., Ann. Mus. Bot. Lugduno-Batavi 3:268. 1867; *F. infectoria* var. *caulocarpa* (Miq.) King, Ann. Roy. Bot. Gard. (Calcutta) 1:63. 1887; *F. stipulosa* (Miq.) Miq., Ann. Mus. Bot. Lugduno-Batavi 3:287. 1867; Corner in Dass. & Fosberg Rev. Handb. Fl. Celon 3:240.1981; Samvatsar, Fl. West. Trib. Madhya Pradesh 271. 1996; Khanna *et al.,* in Suppl. Fl. Madhya Pradesh 162.2001.

Large trees, aerial roots present; leaves 5-2- × 3-9 cm, elliptic to oblong, apex acuminate, base rounded to narrowly cuneate; receptacles 2-8 in the axils of fallen leaves; black or purple when ripe.

*Fl.-Fr.:*Dec. -March

Rare, **Alirajpur** only, Aamkhut forest.

Ficus drupacea Thunb., Ficus 11. 1786; *F. drupacea* Thunb. var. *pubescens* (Roth) Corner in Gard Bull. Singapore 17:381. 1960; *F. mysorensis* B. Heyne ex Roth in Roem. & Schult., Syst. Veg. 1:508. 1817; *F. mysorensis* var. *pubescens* Roth ex Roem. & Schult., Syst. Veg. 1:508. 1817; King in Hook.f., Fl. Brit. India 5:500.1888; Khanna *et al.,* in Mudgal *et al.,* Fl. Madhya Pradesh 2:615. 1997.

Large trees; branchlets and petioles brown hairy; aerial roots present; leaves 10-20 × 6-10 cm, subdistichous to spiral, elliptic-ovate or obovate, apex shortly acuminate, base truncate-cordate, cystoliths on both surfaces, basal nerves 3; petioles 2-3 cm long; receptacles paired, axillary, sessile, *ca* 2.5 cm long, ripening yellow to red, densely tomentose when young.

*Fl.-Fr.:*Feb. -May

Rare, **Khandwa** only, Chandgarh range.

Ficus exasperata Vahl, Enum. Pl. 2:197. 1805; *F. asperrima* Roxb., Fl. Ind. 1832 3:554. 1832; King in Hook.f., Fl. Brit. India 5:522. 1888; Samvatsar, Fl. West. Trib. Madhya Pradesh 271. 1996; Khanna *et al.,* in Mudgal *et al.,* Fl. Madhya Pradesh 2:615. 1997.

Small, 5-8 m high trees;aerial roots absent; leaves 10-15 × 6-10 cm, ovate or oblong-elliptic, apex acute to acuminate, base cuneate, suboriaceous, scabrid above, tomentose beneath; petioles up to 1 cm long; receptacles axillary, solitary, pedunculate, globose, ripening yellow, basal bracts 2-3.

*Fl.-Fr.:*Jan. -March

Common in forest, in all districts.

Ficus hispida L.f., Suppl. Pl. 442. 1782; *F. oppositifolia* Roxb., Sp. Pl. 4:1151. 1806; King in Hook.f., Fl. Brit. India 5:522. 1888; Samvatsar, Fl. West. Trib. Madhya Pradesh 272. 1996; Khanna *et al.,* in Mudgal *et al.,* Fl. Madhya Pradesh 2:616. 1997. *Bhui gular*

Large shrubs or small trees, about 5 m high; branches hollow; aerial roots absent; leaves opposite, decussate, 10-25 × 5-10 cm, obovate, oblong, abruptly acuminate, base rounded or truncate, coriaceous, scabrid, hipid beneath; petioles 2-4 cm long; receptacles caulocarpus and also on leafless, hanging twigs, depressed globose, 1-2.5 long, ripening yellow.

*Fl.-Fr.:*April-July

Abundant along river and stream banks all over the area.

Ficus microcarpa L.f., Suppl. Pl 442. 1782; King in Hook.f., Fl. Brit. India 5:511. 1888; Samvatsar, Fl. West. Trib. Madhya Pradesh 272. 1996; Khanna *et al.,* in Mudgal *et al.,* Fl. Madhya Pradesh 2:618. 1997. **(Plate 8, Fig. 8)**

Large trees, about 10 m high, aerial roots forming prop roots; leaves 4-12 × 1.5-6 cm, elliptic-obovate to lanceolate, acute or shortly acuminate at apex, cuneate at base, margin entire or undulate; receptacles axillary, paired, sessile, about 1 cm across, black on ripening.

*Fl.-Fr.:*Oct. -Dec.

Rare along banks of rivers and streams, Dhar, Indore, Khandwa.

Ficus racemosa L., Sp. Pl. 1060.1753; *F. glomerata* Roxb., Pl. Coromandel 2:13. 1799; King in Hook.f., Fl. Brit. India 5:535. 1888; Samvatsar, Fl. West. Trib. Madhya Pradesh 272. 1996; Khanna *et al.,* in Mudgal *et al.,* Fl. Madhya Pradesh 2:620.1997. *Gular* **(Plate 9, Fig. 1)**

Large, up to 10 m high trees; aerial roots absent; leaves 7-15 × 3-6 cm, ovate-oblong, apex acute or obtuse, base rounded or acute, subcoriaceous; receptacles cauliflorus in large clusters on the main trunk and branches, pedunculate, pyriform, 2-3 cm across, basal bracts 3, ripening reddish-purple.

*Fl.-Fr.:*Feb. -June

Very common in forests along banks of streams and rivers, also in villages, all over the area.

Ficus religiosa L., Sp. Pl. 1059. 1753; King in Hook.f., Fl. Brit. India 5:513. 1888; Samvatsar, Fl. West. Trib. Madhya Pradesh 272. 1996; Khanna *et al.,* in Mudgal *et al.,* Fl. Madhya Pradesh 2:620.1997. *Peepal*

Large, deciduous trees, up to 20 m high; aerial roots absent; leaves 8-15 × 6-13 cm, broadly ovate, caudate-acuminate at apex, acumen 7-9 cm long, base truncate, coriaceous, margin undulate; petioles 4-8 cm long; receptacles axillary, sessile, paired about 1 cm across, pink to purple on ripening.

*Fl.-Fr.:*March-June

Common roadside tree, also found as an epiphyte over other trees, all over the area.

Ficus rumphii Blume, Bijdr. 437. 1825; *F. cordifolia* Roxb., Fl. Ind. ed. 1832 3:548. 1832, nom. illeg; King in Hook.f., Fl. Brit. India 5:512. 1888; Khanna *et al.,* in Mudgal *et al.,* Fl. Madhya Pradesh 2:621. 1997.

Deciduous, up to 10 m high trees; aerial roots absent; leaves 10-14 × 7-9 cm, ovate or ovate-oblong, apex acuminate, base truncate or cordate; receptacles axillary, paired, sessile, black on ripening.

*Fl.-Fr.:*March-July

Rare, roadsides, **Burhanpur** only, Madni area.

Ficus tinctoria G. Forst., Fl. Ins. Austr. 76. 1786; King in Hook.f., Fl. Brit. India 5:497. 1888; Khanna *et al.,* in Mudgal *et al.,* Fl. Madhya Pradesh 2:624. 1997. *Kama*

Large shrubs or trees, erect or straggling, epiphytic when young, aerial roots present; leaves 5-15 × 2-7 cm, elliptic- lanceolate or sub-rhomboid, apex acute or acuminate, base cuneate, usually unequal, petioles 0.4-1 cm long; receptacles axillary, solitary or paired, pedunculate, 8-12 mm across, orange or golden yellow when mature.

Fl.-Fr.: July-Dec.

Rare, Burhanpur, Khandwa.

Ficus virens Aiton, Hort. Kew. 3:451. 1789; *F. infectoria* Roxb., Fl. Ind. ed. 1832 3:551. 1832, nom. illeg; King in Hook.f., Fl. Brit. India 5:515. 1888; Samvatsar, Fl. West. Trib. Madhya Pradesh 272. 1996; Khanna *et al.,* in Mudgal *et al.,* Fl. Madhya Pradesh 2:625. 1997. *Phedhari*

Deciduous, 6-10 m high trees; aerial roots present; leaves5-15 × 3.5-9 cm, oblong-ovate, oblong-elliptic, abruptly acuminate at apex, rounded, truncate or cuneate at base; petioles 4-6 cm long; receptacles axillary, paired, sessile or subsessile, globose, white to cream-coloured on ripening.

*Fl.-Fr.:*May-July

Common along roadsides and in forests, in the whole area.

SALICACEAE 54/1269; 3/5; 3/4

Key to Genera

1. Flowers bisexual; fruit capsular; leaves gland-dotted; seeds arillate .... CASEARIA
1. Flowers unisexual; fruit baccate or capsular; leaves not gland-dotted; seeds exarillate
   2. Fruit capsular; perianth absent; stigmas 4; seeds with long, deciduous hairs .... SALIX
   2. Fruit baccate; perianth of 1 whorl present; stigmas 4-11; seeds not hairy ....FLACOURTIA

CASEARIA Jacq. W. 217; MP 2; SA 2

Key to Species

1. Stipules long; leaves 18-20 cm long; petiole 0.6-0.8 cm long; all young parts including inflorescence with brown pubescence . ... **C. tomentosa**
1. Stipules minute; leaves 10-20 cm long; petiole 0.6-1.5 cm long; plants glabrous................. **C. graveolens**

Casearia graveolens Dalz., Hooker's J. Bot. Kew Gard. Misc. 4:107. 1852; Clarke in Hook.f., Fl. Brit. India 2:592. 1879; Mukherjee in Verma *et al.*, Fl. Madhya Pradesh 1:164. 1993; Samvatsar, Fl. West. Trib. Madhya Pradesh 49. 1996. *Girchi*

Shrubs or small trees; leaves 10-20 × 4-10 cm, broadly elliptic, obtuse or shortly acuminate, crenulate; petioles 0.5-1.5 cm long; flowers green, numerous in leafless, axillary clusters; calyx deeply 5-lobed, gland-dotted; stamens 6-8, alternating with linear-oblong, villous staminodes; capsules 1.5-1.8 cm long, deep yellow, sub-globose; seeds arillate.

*Fl.-Fr.:*May-June

Alirajpur, Barwani, Burhanpur, Dhar, Jhabua, Khandwa, Khargone.

Casearia tomentosa Roxb., Fl. Ind. 2:421. 1824; *C. elliptica* Willd., Sp. Pl. 2:628. 1799, *nom. illeg*; Clarke in Hook.f., Fl. Brit. India 2:593. 1879; Mukherjee in Verma *et al.*, Fl. Madhya Pradesh 1:163. 1993; Samvatsar, Fl. West. Trib. Madhya Pradesh 48. 1996.

Small trees, up to 8 m high; leaves 8-20 × 4-6 cm, oblong-ovate, apex acute, base rounded to sub-cordate, all young parts including inflorescence covered with brown pubescence; flowers in dense, axillary clusters; sepals pubescent; stamens 6-10, alternating with fleshy staminodes; capsules 6-angular, ellipsoid, ribbed; seeds yellow embedded in scarlet aril.

*Fl.-Fr.:*Feb. -May

Common in open ground, forest undergrowth, throughout the area.

FLACOURTIA Commers. ex L'Herit W. 22; MP 2; SA1

Flacourtia indica (Burm.f.) Merr., Interpr. Herb. Amboin. 377. 1917; *Gmelina indica* Burm.f., Fl. Ind. 132. 1768; *Flacourtia ramontchi* L'Her. Strip. Nov. 59. 1786; *F. sepiaria* Roxb., Pl. Coromandel 68. 1796; Hook.f. & Thoms. in Hook.f., Fl. Brit. India 1:193. 1872; Mukherjee in Verma *et al.,* Fl. Madhya Pradesh 1:164. 1993; Samvatsar, Fl. West. Trib. Madhya Pradesh 49. 1996. *Bilanga* **(Plate 9, Fig. 2)**

Trees, about 8 m high, armed with thorns; leaves 1.5-8.5 × 1.5-6 cm, elliptic, ovate or oblong, margin crenate-serrate; flowers unisexual; male flowers in short branched clustered racemes, sometimes borne on thorns, calyx pubescent; stamens numerous; female flowers on short branches, solitary or in pairs, sepals villous, ovary glabrous, stigmas 5-10; berries globose; seeds ovoid.

*Fl.-Fr.:*April-May

Common in scrub forest, in all districts.

SALIX L. W. 552; MP 1; SA 1

Salix tetrasperma Roxb., Pl. Coromandel 1:66, pl. 97. 1795; Hook.f. Fl. Brit. India 5:626. 1888; Khanna in Mudgal *et al.,* Fl. Madhya Pradesh 2:632. 1997. *Banjari*

Deciduous, up to 6 m high, small trees, young parts glabrous or silvery hairy; leaves 5-12 × 1.5-2 cm, lanceolate or elliptic-lanceolate, apex acuminate, base cuneate or rounded, glandular-serrate; stipules ovate, caducous; flowers in axillary or terminal, hairy catkins on dwarf shoots; peduncles 3-12 cm long; male flowers sessile; bracts 1-flowered, densely white hairy; stamens 4-

12, free, villous, exserted; female flowers shortly pedicellate; capsules about 3 mm long, valves recurved.

*Fl.-Fr.:*Oct. -Feb.

Along streams, Indore, Khandwa, Khargone.

CERATOPHYLLACEAE 1/4;1/2; 1/1

CERATOPHYLLUM L. W. 4; MP 2; SA 1

Ceratophyllum demersum L., Sp. Pl. 992. 1753; Hook.f., Fl. Brit. India 5:639. 1888; Samvatsar, Fl. West. Trib. Madhya Pradesh 274. 1996; Mishra *et al.,* in Mudgal *et al.,* Fl. Madhya Pradesh 2:633. 1997.

Submerged herbs up to 2 m long, internodes 1-3 cm long; leaves 4-9 in a whorl, 1-3.5 cm long, dichotomously branched, with 2-4, filiform, minutely dentate segments; flowers axillary, solitary, sessile; male tepals 9-12, greenish-white, linear; female tepals as in male flowers; nutlets laterally compressed, *ca* 4.5 mm long, blackish, minutely tuberculate, apical spines 5-9 mm long, basal spines 5-7 mm long.

*Fl.-Fr.:*Oct. -Feb.

Common in stagnant water in all districts.

ARTIFICIAL KEY TO FAMILIES OF MONOCOTYLEDONS

Characters pertain to study area

1. Perianth well developed, usually in 2 series of 3 each, both series coralline or sepaloid or absent
    2. Trees with palm-like or pinnatisect, plicately folded lamina or leaflets; both series sepaloid; fruit drupaceous with fibrous pericarp.......... ARECACEAE
    2. Habit various; leaves not palm-like or pinnatisect; atleast 1 series of perianth petaloid
        3. Ovary inferior
            4. Fertile stamens 1-2; flowers zygomorphic, bisexual; anthers 2-celled
                5. Stamens and styles united in a column; median petal forming a lip (labellum); stamens 1 or 2; pollen grains present in pollinia; fruit capsular; seeds numerous, very minute, incompletely developed when dispersed .... ORCHIDACEAE
                5. Stamens and styles separate; median petal not modified as labellum; stamen 1; anthers 2-celled; pollen grains not in pollinia; seeds comparatively large
                    6. Leaves spiral, borne on a curving stem; sheath closed; spikes terminating a leafy stem .... COSTACEAE
                    6. Leaves distichous; sheaths open; flowers borne directly from the rhizome (our area) .... ZINGIBERACEAE
            4. Fertile stamens usually 3-6
                7. Aquatic plants; perianth 2-seriate or altogether absent; flowers unisexual, (bisexual in **Ottelia**) enclosed in persistent, peduncled spathes (absent in **Najas graminea**); stamens 1-3; placentation lamellar; seeds numerous .... HYDROCHARITACEAE

7. Terrestrial plants; placentation not lamellar
   8. Plants not scapigerous
      9. Fertile stamens 5; aerial stem absent or present; stolens present; inflorescence terminal, pendulous, compound spadix; fruit baccate .... MUSACEAE
      9. Fertile stamens usually 6; flowers unisexual or bisexual; twiners or erect herbs; leaves entire or pedately divided; bracteoles divided into filiform segments or herbaceous; fruit capsular, 3-winged or baccate, 6-ribbed; seeds winged or not .... DIOSCOREACEAE
   8. Plants scapigerous; fertile stamens 6; flowers in spathaceous umbels or condensed racemes
      10. Leaves plicate or not; flowers in condensed racemes or 1-flowered umbels..HYPOXIDACEAE
      10. Leaves not plicate; flowers in spathaceous umbels surrounded by an involucre of membranous bract or bracts .... AMARYLLIDACEAE

3. Ovary superior
   11. Flowers bisexual; stamens 6
      12. Both perianth series petaloid
         13. Climbers; leaves with 3-5 radiating, basal nerves, venation reticulate; fruit baccate.. SMILACACEAE
         13. Erect plants
            14. Terrestrials; inflorescence not subtended by a spathe-like leaf sheath
               15. Leaves reduced to scales or well developed; fruits baccate or capsular; lateral axillary branches modified as cladodes or not; main stem well developed or bulbous; roots tuberous, fasciculated; scramblers or scapigerous plants ....ASPARAGACEAE
               15. Leaves developed; fruits capsular
                  16. Scape absent; flowers in terminal racemes; leaf base sheathing; underground stem corm or rhizome; aerial stem developed; styles 3; leaf-tip climbers or erect plants . ...COLCHICACEAE
                  16. Scape present; flowers borne on terminal scapes; aerial stem hardly any; underground stem modified as bulb or rhizomatous; leaves in rosette; style 1 . ... XANTHORRHOEACEAE
            14. Aquatics; inflorescence subtended by a spathe-like leaf sheath; plants free-floating or rooted in mud .... PONTEDERIACEAE
      12. Outer perianth series sepaloid, inner petaloid; staminodes present or absent; carpels united; placentation axile; fruits capsular ..................................COMMELINACEAE
   11. Flowers unisexual; aquatic plants; carpels free; placentation parietal or basal; fruits achenial .... ALISMATACEAE

1. Perianth reduced, scale-like or absent
   17. Flowers subtended by dry bracts or glumes
      18. Flowers unisexual, in globose heads, surrounded by involucres; ovary 3-celled; plants of marshy habitat . ... ERIOCAULACEAE
      18. Flowers bisexual, rarely unisexual; borne in spikelets; ovary 1- celled, 1-ovuled

19. Stem solid, triangular in cross-section; leaf sheaths closed; nodes condensed; glume 1, subtending one or more spikelets; fruits biconvex or triangular nut ....CYPERACEAE
19. Stem hollow; normally round in cross-section; leaf sheaths open; nodes spaced; glumes 2, subtending one or more spikelets; fruits caryopsis .... POACEAE

17. Flowers not subtended by dry bracts or glumes
20. Spathe present; flowers unisexual .... ARACEAE
20. Spathe absent; flowers unisexual or bisexual
21. Flowers unisexual, females below and males above, in long, terminal spikes; perianth absent; bracteoles long, filiform .... TYPHACEAE
21. Flowers bisexual
22. Plants aromatic; flowers borne in a compact spike in the axil of a pseudospathe; tepals 3+3 stamens 3+3, all fertile; bracteoles not filiform............................... ACORACEAE
22. Plants not aromatic
23. Latex present; perianth segments 1-3; stamens 6 or more; carpels free; placentation basal; fruit follicular .... APONOGETONACEAE
23. Latex absent; perianth segments and stamens 4; carpels united; ovary unilocular; fruit of drupelets; stipules present .... POTAMOGETONACEAE

HYDROCHARITACEAE 16/133; 5/7; 5/7

Key to Genera

1. Flowers bisexual; stem condensed; leaves mostly radical, broad and petiolate; stamens 6-15; spathe winged .... OTTELIA
1. Flowers unisexual
2. Leaves whorled, 0.5-2 cm long; stamens 3; style lobed .... HYDRILLA
2. Leaves not whorled
3. Leaves dentate or toothed, sheaths auriculate; perianth hyaline or absent; stamen 1; stigmas 2-4, slender .... NAJAS
3. Leaves with entire margin
4. Stamens 2; stem well developed; leaves cauline, alternate and opposite; spathe of male flower 2-fid .... NECHAMANDRA
4. Stamens 1-3; stem condensed; leaves radical; spathe of male flowers 3-lobed .... VALLISNERIA

HYDRILLA L. C. Rich. Monotypic

Hydrilla verticillata (L.f.) Royle, Ill. Bot. Himal. Mts. t. 376. 1839; *Serpicula verticillata* L.f., Suppl. Pl. 416. 1782; Hook.f., Fl. Brit. India 6:659. 1888; Samvatsar, Fl. West. Trib. Madhya Pradesh 276. 1996; Misra *et al.,* in N.P. Singh *et al.,* Fl. Madhya Pradesh 3:3.2001.

Submerged, free floating herbs, often with fibrous roots at the nodes; leaves 3-8 in a whorl, oblong-linear, 1-3 × 0.1-0.3 cm, serrate-dentate; male flowers white, solitary, axillary, pedicillate; perianth of 3 sepals and 3 petals; female spathes axillary, sessile; fruits echinate, 5-7 mm long.

Fl.-Fr.: Sept. -Nov.

Common in ponds, ditches, rivers, in all districts.

NAJAS L. W. 38; MP 3; SA 3

Key to Species

1. Plants dioecious; leaves coarsely dentate or almost pinnatifid . ...**N. marina**
1. Plants monoecious; leaves finely dentate with very small teeth
   2. Flowers of both sexes not enclosed in spathes .... **N. graminea**
   2. Male flowers enclosed in a spathe; leaves with 10-17 marginal teeth on each side; areoles in 25-30 longitudinal rows .... **N. indica**

Najas graminea Delile, Descr. Egypte Hist. Nat. 2: 282. 1813; Hook.f., Fl. Brit. India 6: 569. 1893; Samvatsar, Fl. West. Trib. Madhya Pradesh 300.1996; Srivastava *et al.,* in Singh *et al.,* Fl. Madhya Pradesh 3: 210.2001.

Annual, submerged, plumose herbs; leaves narrowly linear, with numerous oblique spinules on each side, 2.4-3. 4 × 0.6-1 mm; leaf-sheaths long auricled; flowers solitary or 2-4-together; achenes ellipsoid-oblong; areoles minute, subquadrate or polyhedral, irregularly disposed or in distinct longitudinal lines.

Fl.-Fr.: Aug. - Sept.

Common in tanks, ponds and water logged areas, in all districts.

Najas indica (Willd.) Cham., Linnaea 4:501. 1829; *Caulina indica* Willd., Mem. Acad. Roy. Soc. Hist. (Berlin) 1798:89. 1801; *Najas minor auct. non* All. 1785; Hook.f., Fl. Brit. India 6:569. 1893; Srivastava *et al.,* in N.P. Singh *et al.,* Fl. Madhya Pradesh 3:210.2001.

Annual, aquatic, submerged herbs; leaves 2-5 × 0.8-1.5 mm, linear, flat or sub-terete, spinulose dentate; auricles of sheath small and rounded; male flowers: spathe *ca* 3 mm long, anthers subsessile; female flowers: spathe absent or 2-3 mm long; achenes ellipsoid; seeds tetragonal with areoles in longitudinal rows.

*Fl.-Fr.:*Aug. -Sept.

Common in tanks, associated with *Sagittaria* spp. and other aquatic species, in all districts.

Najas marina L., Sp. Pl. 1015. 1753; *N. major* All., Fl. Pedem. 2;221. 1785; Hook.f., Fl. Brit. India 6:569. 1893; Srivastava *et al.,* in N.P. Singh *et al.,* Fl. Madhya Pradesh 3:21.2001.

Aquatic, dioecious, annual herbs; stem stout, fragile; leaves sub-ternately whorled, linear, dentate, with 4-8 teeth on either side and a few on the back; sheaths with rounded sides; outer perianth or spathe present in the male flowers; anthers 4-celled; achenes oblong-ellipsoid.

Fl.-Fr.: Sept. -Feb.

Common in ponds, ditches, in all districts.

NECHAMANDRA Planch. Monotypic

Nechamandra alternifolia (Roxb. ex Wight) Thw., Enum. Pl. Zeyl. 332. 1864; *Vallisneria alternifolia* Roxb. ex Wight Bot. Misc. 2:344. 1831; *Lagarosiphon alternifolia* (Roxb.)

Druce, in Rep. Bot. Exch. Cl. Brit. Is. 1916:630.1917; Samvatsar, Fl. West. Trib. Madhya Pradesh 276. 1996; Misra *et al.,* in N.P. Singh *et al.,* Fl. Madhya Pradesh 3:4.2001 [as *Lagarosiphon alternifolia* Roxb. ex Wight Thw.].

Annual, diocious, submerged herbs; stem filiform; leaves alternate and opposite, sessile, oblong-lanceolate, apex acute; 3-8 cm long; male flowers many, crowded in a sessile, 2-fid spathe; spathe of female flowers tubular, 2-fid; fruit a many seeded utricle; seeds oblong.

Fl.-Fr.: Aug. -Feb.

Burhanpur, Dhar, Indore, Khandwa.

OTTELIA Pers. W. 19; MP 1; SA 1

Ottelia alismoides (L.) Pers., Syn. Pl. 1:400.1805; *Stratiotes alismoides* L., Sp. Pl. 535. 1753; Hook.f., Fl. Brit. India 6:662. 1888; Samvatsar, Fl. West. Trib. Madhya Pradesh 277. 1996; Misra *et al.,* in N.P. Singh *et al.,* Fl. Madhya Pradesh 3:6.2001.

Submerged or partly floating, aquatic herbs; leaves variable, on 8-50 cm long, sheathing petioles, broadly ovate or ovate, reniform, with a cordate or truncate base, obtuse or rounded, margins undulate; flowers bisexual, floating, white, solitary; spathe 3 cm long, 5-toothed, 5-winged; wings membranous; fruits oblong, enclosed in the spathe; seeds many, oblong, minute.

Fl.-Fr.: Oct. -April

Common in ponds, ditches and water logged areas in all districts.

VALLISNERIA L. W. 14 ;MP 1; SA 1

Vallisneria spiralis L., Sp. Pl. 1015. 1753; Hook.f., Fl. Brit. India 6:660.1888; Samvatsar, Fl. West. Trib. Madhya Pradesh 277. 1996.

A submerged, dioecious, tufted herb, rooted in the mud; leaves linear with a sheathing base, obtuse, entire or dentate; male flowers many in a shortly peduncled, 3-lobed spathe; female flowers on long peduncles, which become coiled after anthesis; seeds oblong to fusiform, embedded in a gelatinous mass.

Flora of Madhya Pradesh 3 : 6.2001, reports *V. natans* (Lour.) Hara in J. Jap. Bot. 49: 136. 1974. This makes 2 species in MP.

Fl.-Fr.: Nov. -Jan.

Common in muddy bottom of ponds and ditches, in all districts.

ORCHIDACEAE 899/27801; 25/49; 12/17

Key to Genera

1. Epiphytic orchids; pollen waxy; labellum spurred
   2. Column with a distinct foot
      3. Leaves flat; midlobe of lamellum entire, lateral lobes not not forming a tube around the column .... AERIDES
      3. Leaves terete; midlobe of lamellum 2-fid, lateral lobes forming a tube round the column.......... PAPILIONANTHE
   2. Column with an indistinct foot or foot absent

4. Spur distant from the base of labellum ....RHYNCHOSTYLIS
4. Spur at the base of labellum; flowers lateral; leaves coriaceous
5. Flowers more than 2 cm across; pollinia 2-4 ... . VANDA
5. Flowers up to 2 cm across; pollinia 2 or unequally 2-partite . ... ACAMPE

1. Terrestrial orchids
6. Plants leafy during flowering time only; sepals and petals glabrous
7. Leaf solitary, orbicular or broadly ovate, broader than long; anther cells distinct; pollinia granular; root tuberous .... NERVILIA
7. Leaves 2 or more, linear, oblong-lanceolate, longer than broad; pollinia 2, waxy; root not tuberous . ...EULOPHIA
6. Plants leafy throughout life
8. Labellum spurred at base, spur 1; pollinia granular
9. Spur cylindrical; lateral sepal free, not forming a mentum; petals and dorsal sepal forming a hood
10. Bracts foliaceous; flowers more or less 5 cm across; anther cells adnate to column.. PECTEILIS
10. Bracts not foliaceous; flowers 2-3 cm across; ovaries and capsules making an angle with the peduncle .... HABENARIA
9. Spur globose, ellipsoid, scrotiform or conical; pollinia 2; ovaries and capsules erect, parallel to the peduncle .... PERISTYLUS
8. Labellum saccate at base
11. Inflorescence terminal, glandular-pubescent; pollinia granular; anther cells distinct...ZEUXINE
11. Inflorescence lateral, curved, not glandular; pollinia waxy ....GEODORUM

ACAMPE Lindl. W. 8; MP1; SA 1

Acampe praemorsa (Roxb.) Blatt. & Mc Cann, J. Bombay Nat. Hist. Soc. 35:495. 1932; *Epidendrum praemorsum* Roxb., Pl. Coromandel 34. 1795; *Acampe wightiana* (Lindl. ex Wight) Lindl., Fol. Orchid. 4:2. 1853; *Vanda wightiana* Lindl. ex Wight, Icon. Pl. Ind. Orient. 5: t. 1670.1851; *Saccopetalum wightianum* (Lindl. ex Wight) Hook.f., Fl. Brit. India 6:62. 1890, *nom. illeg*; Samvatsar, Fl. West. Trib. Madhya Pradesh 278. 1996; Khanna in N.P. Singh *et al.,* Fl. Madhya Pradesh 3:14.2001.

Epiphytic herbs; stems stout, sheathed with leaf bases; leaves 9-10 × 1-3 cm, thick, coriaceous, chanelled, emarginate with 2, unequal, rounded lobes; flowers yellow, with red bands, in supra axillary, corymbose panicles; peduncles stout; sepals and petals more or less similar, petals creamy yellow with pinkish stripes; labellum 3-lobed, saccate, fleshy, midlobe ligulate, reflexed, lateral lobes erect, small; column with 2, erect, terminal horns; capsules, longitudinally ribbed, 4-6 cm long.

Fl.-Fr.: Aug. -Dec.

Common on Mango and Mahua trees.

AERIDES Lour. W. 31; MP 3; SA1

Aerides maculosa Lindl., Edward's Bot. Reg. 31: 58. 1845; *Saccolobium speciosum* Wight, Icon. Pl. Ind. Orient. 5:t. 1674. 1851; Hook.f., Fl. Brit. India 6:45. 1890; Samvatsar, Fl. West. Trib. Madhya Pradesh 278. 1996; Khanna in N.P. Singh *et al.,* Fl. Madhya Pradesh 3:16.2001.

Epiphytes; leaves chanelled, linear-oblong, 13-23 × 2-3 cm, mottled with purple, apical lobes 2, unequal; flowers pedicellate, in branched, up to 24 cm long racemes; sepals subequal; petals pinkish-white; labellum straight, 3-lobed, midlobe finely crisped on margin, lateral lobes smaller, spur hook-like; column pink; capsules 3-5 cm long, strongly ribbed.

Fl.-Fr.: May-Sept.

On trees like *Acacia catechu*, Mango, *Terminalia* etc, Alirajpur, Dhar, Khandwa.

EULOPHIA R. Br. ex Lindl. W. 201; MP 2; SA 2

Key to Species

1. Flowers appearing long before leaves; racemes 5-12 cm long; capsules up to 2 cm long ....**E. dabia**
1. Flowers and leaves appearing together, racemes up to 60 cm long; capsules up to 3.5 cm long .... **E. epidendraea**

Eulophia dabia (D. Don) Hochr., Bull. New York Bot. Gard. 6:270.1910; *Bletia dabia* D. Don, Prodr. Fl. Nepal. 30.1825; *Eulophia ramentacea* (Roxb.) Lindl., Gen. Sp. Orchid. Pl. 185. 1837; Hook.f., Fl. Brit. India 6:4. 1890; Sharma *et al.,* in Fl. Maharashtra 3:28. 1996.

Perennial herbs; pseudobulbs oblong-orbicular; leaves plicate, oblong-lanceolate to linear, 20-30 cm long; flowers in lax racemes, appearing before leaves; scapes up to 50 cm long; bracts persistent; pedicels including ovary 10-15 mm long; sepals greenish or brownish-yellow, 5-nerved; spur 2-3 mm long; petals yellow, shorter than sepals, lip broader than long, side lobes large, midlobe orbicular, 3 mm long; capsules 2.5-3.5 cm long, ribbed.

Fl.-Fr.: Nov. -March

Along streams and pastures, amidst grasses, Indore, Khandwa.

Eulophia epidendraea (J. Koenig ex Retz.) C.E. C. Fisch., Gamble, Fl. Pres. Madras 1434. 1928 (3:1003. 1957, Rep. ed.); *Serapis epidendraea* J. Koenig ex Retz. Observ. Bot. 6:65. 1791; Hook.f., Fl. Brit. India 6:1. 1890; Sharma *et al.,* in Fl. Maharashtra 3:28. 1996.

Perennial herbs; pseudobulbs up to 10 cm long; leaves distichous, sheathing at base, 15-35 cm long, linear-lanceolate; scape up to 60 cm long, appearing with the leaves; flowers pale greenish, tinged with maroon; capsules up to 3.5 cm long, drooping, mauve with 3 pale green,broad bands.

Fl.-Fr.: Nov. -March

Rare, **Khandwa** only, at Sunder Dev.

GEODORUM G. Jackson W. 12; MP 1; SA 1

Geodorum densiflorum (Lam.) Schltr., Repert. Spec. Nov. Regni Veg. Beih. 4:259. 1919; *Limodorum densiflorum* Lam., Encycl. 3:516. 1792; Hook.f., Fl. Brit. India 6:16. 1890; Khanna in N.P. Singh *et al.,* Fl. Madhya Pradesh 3:37 2001.

Terrestrial, rhizomatous herbs; stems short, sheathed; pseudobulbs tuberous, ovoid-conical; leaves 2-3, elliptic-lanceolate, up to 24 cm long, acuminate at apex; flowers in *ca* 30, cm long racemes, pedicels short, racemes sharply decurved terminally; scape thick, arising from the pseudobulb; bracts leafy; lateral sepals adnate to the dorsal sepal forming a hood; petals yellowish-white; labellum flagellate, narrowed to a claw; capsules fusiform, pendulous, 3-4 cm long, conspicuously 6-ribbed.

Fl.-Fr.: July-Oct.

Burhanpur, Khandwa. Rapidly becoming rare as plants of any size are uprooted and sold in the market as an aphrodisiac.

HABENARIA Willd. W. 837; MP 9; SA 5

Key to Species

1. Plants less than 25 cm tall; leaves in rosette; flowers white or yellow
    2. Leaves ovate to sub-orbicular; capsules strongly ribbed, straight .... **H. grandifloriformis**
    2. Leaves oblong-lanceolate or oblong-elliptic; capsules not strongly ribbed, fusiform, turgid
        3. Leaves oblong-lanceolate; flowers white; petals and sepals equal in length . ...**H. plantaginea**
        3. Leaves oblong-elliptic; flowers yellow; petals smaller than sepals . ...**H. marginata**
1. Plants more than 35 cm tall; leaves cauline; flowers greenish-white; petals entire; leaves obovate-lanceolate or ovate-oblong
    4. Leaves obovate-lanceolate; bracts ovate-lanceolate; capsules decurved . ...**H. furcifera**
    4. Leaves ovate-oblong; bracts lanceolate; capsules straight ....**H. digitata**

Habenaria digitata Lindl., Gen. Sp. Orchid. Pl. 307. 1835; *H. trinervia* Wight, Icon. Pl. Ind. Orient. 5:t. 1701. 1851; Hook.f., Fl. Brit. India 6:134. 1890; *H. graveolens* Duthie, Ann. Roy. Bot. Gard. (Calc.) 9:177. 1906; Khanna in N.P. Singh *et al.,* Fl. Madhya Pradesh 3:42.2001.

Terrestrial, 30-45 cm tall herbs; tubers 1-2; leaves elliptic-lanceolate or ovate-oblong, acute at apex, 3-6 cm long, base sheathing; flowers many, in 8-10 cm long spikes, pedicels short; bracts leafy; sepals unequal, green; petals green, 2-partite, upper segment filiform, linear, lower filiform, lying on the labellum; labellum 3-partite up to the base, fleshy, spur green; column green, oblong; capsules ribbed.

Fl.-Fr.: July-Oct.

Rare, in humus soil and moist, rocky places, **Khandwa** only, Kalibhit forest.

Habenaria furcifera Lindl., Gen. Sp. Orchid. Pl. 319. 1835; *H. modesta* Dalzell, Hooker's J. Bot. Kew. Gard. Misc. 2:262. 1850; *H. ovalifolia* Wight, Icon. Pl. Ind. Orient. 5:t. 1708.

1851; Hook.f., Fl. Brit. India 6:149. 1890; Samvatsar, Fl. West. Trib. Madhya Pradesh 278. 1996; Khanna in N.P. Singh *et al.,* Fl. Madhya Pradesh 3:44.2001.

An erect, 40-60 cm tall herb, with ovoid-ellipsoid tubers; leaves 13-18 cm long, oblanceolate, or obovate; flowers small, shortly stalked, in 20-40 cm long, stout, bracteate spikes; sepals unequal; petals greenish-yellow; labellum 3-partite up to base, midlobe broad, lateral lobes filiform, longer than the midlobe, spur slender; capsules 1-1.5 cm long, turgid, fusiform with strong ribs.

Fl.-Fr.: Aug. -Nov.

Occasional as forest undergrowth, Alirajpur, Khandwa.

Habenaria grandifloriformis Blatt. & Mc Cann in J. Bombay Nat. Hist. Soc. 36:17. 1932; *H. grandiflora* Lindl. ex Dalzell & Gibson, 1861, *nom. illeg;* Hook.f., Fl. Brit. India 6:136. 1890; Samvatsar, Fl. West. Trib. Madhya Pradesh 278. 1996; Khanna in N.P. Singh *et al.,* Fl. Madhya Pradesh 3:45. 2001.

Terrestrial, *ca* 20 cm high herbs; tubers 1 or 2, globose or ellipsoid; leaves 1-2, radical, touching the ground, broadly ovate to suborbicular, 2-3.5 cm long, coriaceous; flowers pedicellate, in 1-many flowered , 10-15 cm long racemes; peduncles longitudinally ribbed, bracts ovate or oblong, amplexicaul; sepals unequal, white; petals white, 2-partite; labellum white, 3-lobed, midlobe linear, lateral lobes linear to filiform; spur white, curved; column greenish-white; capsules stongly ribbed.

Fl.-Fr.: July-Nov.

Occasional as forest undergrowth, **Dhar** only, below Neelkantheshwar Mandir.

Habenaria marginata Colebr., Exot. Fl. t. 136. 1824; *H. fusifera* Hook.f., Fl. Brit. India 6:147. 1890; *H. flavescens* Hook.f., Fl. Brit. India 6:150.1890; Samvatsar, Fl. West. Trib. Madhya Pradesh 279. 1996; Khanna in N.P. Singh *et al.,* Fl. Madhya Pradesh 3:45. 2001

A 15-30 cm tall, erect herb with 1-2 tubers; leaves subradical, clustered in the lower part of the stem, 3-10 cm long, oblong, lanceolate or ovate; flowers yellow, in lax or dense, many flowered, 5-10 cm long spikes; peduncles green, bracteate; sepals pale green, unequal; petals yellow; labellum, yellow, 3-lobed, fleshy; spur fusiform; capsules 1.2-1.5 cm long, fusiform, turgid, obscurely beaked.

Fl.-Fr.: Sept. -Nov.

In wet and shady plaves, Alirajpur, Dhar, Indore, Khandwa.

Habenaria plantaginea Lindl.,Gen. Sp. Orchid. Pl. 323. 1835; *Plantaginorchis plantaginea* (Lind.) Szlach., Richardiana 4:65 2004; Hook.f., Fl. Brit. India 6:141. 1890; Khanna in N.P. Singh *et al.,* Fl. Madhya Pradesh 3:47.2001.

Terrestrial, 20 50 cm tall herbs; tubers 2, unequal; leaves radical, 4-8 cm long, oblong-lanceolate, uppermost small, bract-like; flowers white, in lax, few to many flowered racemes; bracts ovate, much shorter than the ovary; sepals white, unequal; petals white; labellum 3-lobed, long spurred, midlobe linear, lateral lobes much broader than midlobe, wing-like; spur white, curved, longer than ovary; capsules fusiform, turgid, beak slender.

Fl.-Fr.: Sept. -Nov.

Rare, moist and shady places, **Khandwa** only, Bhagpura forest.

NERVILIA Commers. & Gaud. W. 67; MP 3; SA 1

Nervilia concolor (Blume) Schltr., Bot. Jahrb. Syst. 45:404.1911; *Cordyla concolor* Blume, Bijdr. 416.1825; *Nervilia aragoana* Gaudich., Voy. Uraine 422.1829; *Pogonia concolor* Blume, Mus. Bot. 1:32.1849; Khanna in N.P. Singh *et al.,* Fl. Madhya Pradesh 3:53.2001.

Small, terrestrial herbs; tubers white, 1.5-2 cm across; leaves appearing after flowers, petiolate, broadly ovate, 9-12 × 8-11 cm, petioles 10-15 cm long; flowers drooping, in lax, few to many flowered, 10-20 cm long racemes, pedicels short; bracts linear-lanceolate, deflexed; sepals subequal, 3-nerved; petals green, similar to sepals; labellum obovate, about 12 mm long, 3-lobed, midlobe densely pubescent along nerves, lateral lobes erect, embracing the column; column about 7 mm long; capsules pendulous, *ca* o. 5 cm long.

Fl.-Fr.: July-Sept.

In moist, shady places, Alirajpur, Khandwa.

PAPILIONANTHE Schltr. W. 11; MP 1; SA 1

Papilionanthe teres (Roxb.) Schltr., Orchis 9:78. 1915; *Dendrobium teres* Roxb., Fl. Ind. ed. 1832 3:485. 1832; *Vanda teres* (Roxb.) Lind., Gen. Sp. Orchid. Pl. 217. 1883; Hook.f., Fl. Brit. India 6:49. 1890; Khanna in N.P. Singh *et al.,* Fl. Madhya Pradesh 3:59.2001.

Stem branched or unbranched, terete, subscandent; leaves few, alternate, terete, fleshy, 15-20 cm long; flowers fragrant, in 3-6 flowered, 25-30 cm long racemes; racemes borne opposite the leaf base and confined to the upper part of stem; sepals reddish, undulate; petals rose, widely spreading; labellum 3-lobed, midlobe yellow at base with red speckles, deeply 2-fid, lateral lobes forming a tube around the column; spur infundibuliform; column short, with a distinct foot; capsules narrowly fusiform.

Fl.-Fr.: Sept. -Dec.

Very rare, **Indore** only, deep within the forest, at Nahar-Jhabua range.

PECTEILIS Raf. W. 8; MP 1; SA 1

Pecteilis susannae (L.) Rafin. Fl. Tellur. 2:38. 1837; *Orchis susannae* L. Sp.p. 939. 1753; *Platanthera susannae* (L.) Lindl., Gen. SP. Orchid. Pl. 295. 1835; Hook.f., Fl. Brit. India 6:137. 1890; Khanna in N.P. Singh *et al.,* Fl. Madhya Pradesh 3:67.2001.

Herbs, about 60-100 cm high; tubers 2, ellipsoid to ovoid; leaves 5-15 × 2-7 cm, ovate-oblong or oblong-lanceolate, apex acute, sharp, often black; raceme terminal, 2-6 flowered, 15-20 cm long; flowers bracteate, white, 6 × 7 cm; pedicels short; capsules up to 5 cm long, beaked, strongly ribbed.

Fl.-Fr.: Sept. -Dec.

Rare, in shady places with high humidity, **Khandwa** only, Kalibhit forest.

PERISTYLUS Blume W. 103; MP6; SA 1

Peristylus lawii Wight, Icon. Pl. Ind. Orient. 5:t. 1695. 1851; *Habenaria lawii* (Wight) Hook.f., Fl. Brit. India 6:162. 1890; Samvatsar, Fl. West. Trib. Madhya Pradesh 279. 1996; Khanna in N.P. Singh *et al.*, Fl. Madhya Pradesh 3:62.2001.

Terrestrial, small, 15-40 cm tall herbs; tubers 1-2; stem stout, covered in sheaths; leaves 3-4, clustered about the middle of stem, 8-10 × 2-3 cm, lanceolate or oblong-lanceolate; flowers white or pale, in 8-15 cm long spikes; bracts acuminate; sepals white, subequal, conduplicate; petals greenish-white, broader than lateral sepals; labellum 3-lobed, clawed, lobes subequal; spur small; capsules fusiform, ribbed, shortly beaked.

Fl.-Fr.: July-Sept.

Rare in dense forest, **Alirajpur** only, at Kadya Pahar forest.

RHYNCHOSTYLIS Blume W. 3; MP 1; SA1

Rhynchostylis retusa (L.) Blume, Bijdr. 286. 1825; *Epidendrum retusum* L., Sp. Pl. 953. 1753; *Aerides retusa* (L.) Sw., J. Bot. (Schrader) 2:233. 1799; *Saccolabium guttatum* (Lindl.) Wall., Numer. List 7308. 1832; Hook.f., Fl. Brit. India 6:32. 1890; Khanna in N.P. Singh *et al.,* Fl. Madhya Pradesh 3:69.2001.

Epiphytic, stout herbs; roots thick, long; leaves strap-shaped 10-25 × 2-3 cm, unequally 2-lobed at apex; flowers in dense, axillary, drooping racemes; bracts persistent; dorsal sepal obtuse, lateral sepals larger than the dorsal; petals pinkish-violet or white with purple spots; labellum pink-purple, slightly 3-lobed; spur large; capsules oblong-obovate, about 2 cm long, ridged, stalked.

Fl.-Fr.: July-Dec.

Epiphytic on a number of trees, confined to **Khandwa** only, Battu Bagra and Bhagpura forest.

VANDA Jones ex R. Br.W. 60; MP 3; SA 2

Key to Species

1. Flowers 2-3 cm across, 6-10 in a raceme; labellum with ridges; spur not vermiform; capsules 7-9 cm long .... **V. tessellata**
1. Flowers 1.5 cm across, 3-10 in a raceme; labellum without ridges; spur vermiform at the tip; capsules up to 3 cm long .... **V. testacea**

Vanda tessellata (Roxb.) Hook. ex G. Don, Hort. Brit. 372. 1830; *Epidendrum tessellatum* Roxb., Pl. Coromandel 1: 34. 1795; *Vanda roxburghii* R. Br. Bot. Reg. 6: t. 506. 1820; *Cymbidium tesseloides* Roxb., Fl. Ind. ed. 1832 3:463. 1832; Hook.f., Fl. Brit. India 6: 52. 1890; Samvatsar, Fl. West. Trib. Madhya Pradesh 279. 1996; Khanna in Singh *et al.,* in N.P. Singh *et al.,* Fl. Madhya Pradesh 3: 73.2001. *Badang* **(Plate 9, Fig. 3)**

Epiphytic herbs; leaves alternate, linear, praemorse at apex, thick, coriaceous, keeled, 10-15 × 1-2 cm; flowers axillary, in *ca* 10-flowered racemes; sepals unequal; petals yellowish-green, shorter than sepals; labellum 3-lobed, yellow with violet tips; capsules clavate, 6-9 cm long.

Fl.-Fr.: Throughout the year

On roadside trees in mixed deciduous forests.

Vanda testacea (Lindl.) Rchb. f. in Gard. Chron. n.s., 8:166. 1877; *Aerides testacea* Lindl., Gen. Sp. Gen. Sp. Orchid. Pl. 238. 1833; *Vanda parviflora* Lindl., Edward's Bot. Reg. 30(Misc.): 45. 1844; Hook.f., Fl. Brit. India 6:50.1890; Khanna in N.P. Singh *et al.,* Fl. Madhya Pradesh 3:74.2001.

Epiphytes with a thick, 10-20 cm long stem; leaves alternate, linear or linear-oblong, 10-15 cm long, 3- toothed at apex, coriaceous, keeled; flowers in axillary, 3-15 cm long racemes; peduncles stout, with few sterile sheathing bracts; bracts 3-nerved; sepals unequal; petals yellow; labellum 3-lobed, spurred, reflexed; spur yellow, *ca* 2 mm long, vermiform at tip; capsules oblong-fusiform, 2-3 cm long, winged.

Fl.-Fr.: Sept. -April

Normally epiphytic on *Madhuca longifolia* var. *latifolia* and *Terminalia bellirica*, **Khandwa** only, Bhagpura forest.

ZEUXINE Lindl. W. 75 :MP 1; SA 1

Zeuxine strateumatica (L.) Schltr., Bot. Jahrb. Syst. 45:394. 1911; *Orchis strateumatica* L., Sp. Pl. 943. 1753; *Spiranthes strateumatica* (L.) Lindl., Bot. Reg. 10:t. 823. 1824; Hook.f., Fl. Brit. India 6: 106. 1890; Samvatsar, Fl. West. Trib. Madhya Pradesh 279. 1996; Khanna in N.P. Singh *et al.,* Fl. Madhya Pradesh 3: 74.2001.

Terrestrial, 20-25 cm high herbs; leaves 1.5-10 × 0.2-0.6 cm, alternate, linear or linear-lanceolate, acute at apex, sessile; flowers white,in terminal, dense, erect or slightly decurved spikes; sepals unequal; petals oblong-elliptic; labellum cymbiform, as long as lateral sepals; column very short; capsules ellipsoid, ribbed, without beak.

Fl.-Fr.: Dec. -March

In wet and marshy places, Alirajpur, Barwani, Burhanpur, Dhar, Indore, Khargone.

COSTACEAE 7/139; 1/1; 1/1

CHEILOCOSTUS C.D. Specht W. 5; MP 1; SA 1

Cheilocostus speciosus (J. Köenig) C. D. Specht, Taxon 55:159.2006; *Banksea speciosa* J. Köenig in Retz., Observ. Bot. 3:75.1783; *Costus speciosus* (J. Koenig) Sm.,Trans. Linn. Soc. London 1:249. 1791; Baker in Hook.f., Fl. Brit. India 6: 249. 1892; Samvatsar, Fl. West. Trib. Madhya Pradesh. 280.1996; Mudgal *et al.,* in N.P. Singh *et al.,* Fl. Madhya Pradesh 3:92.2001. *Keo-kand* **(Plate 9, Fig. 4)**

Herbs *ca* 1 m high; leaves oblong-lanceolate, 5-30 × 5-9 cm, borne spirally on the stem; flowers in terminal, ovoid heads; bracts 2-3 cm long, greenish to bright red, oblong-ovate; calyx tubular; corolla white; labellum white with yellow centre; capsules bright red, trigonous; seeds black with white aril.

Fl. - Fr.: Aug. – May

Alirajpur, Barwani, Burhanpur, Dhar, Indore, Khandwa, Khargone, along shady places, rivers.

ZINGIBERACEAE 50/1548; 7/18; 2/6

Key to Genera

1. Aerial stem greatly reduced; lateral staminode broad, free from the lip ....CURCUMA
1. Aerial stem well developed; lateral staminode very small, adnate to the lip; flowers in dense, compact spikes, originating directly from the rhizome (our area) .... ZINGIBER

CURCUMA L. W. 93; MP 6 ;SA 5

Key to Species

1. Spikes lateral
   2. Rhizomes white-yellow; coma bracts pink or red . ...**C. aromatica**
   2. Rhizome white, pale yellow inside; coma bracts purple .... **C. angustifolia**
1. Spike central, appearing after leaves; coma bracts purple
   3. Root –stock large; tubers sessile, pale yellow inside .... **C. amada**
   3. Root –stock small, bearing stalked tubers at the ends of fibres; tubers white inside
      4. Flowers yellow . ... **C. pseudomontana**
      4. Flowers purple .... **C. decipiens**

Curcuma amada Roxb., Asiat. Res. 11:341. 1810; Baker in Hook.f., Fl. Brit. India 6:213. 1890; Mudgal *et al.,* in N.P. Singh *et al.,* Fl. Madhya Pradesh 3:81. 2001. *Amada*

Large herbs, up to 1 m tall; rhizomes pale yellow inside, smelling of unripe mango; leaves oblong-elliptic or oblanceolate, 15-45 × 7-16 cm, acuminate, ending in a short, twisted cusp, base narrow; flowers in spikes arising from the centre of leafy tuft; bracts of coma pink, floral bracts white; calyx very thin; corolla white or pale yellow; labellum yellow, 3-lobed; capsules globose.

Fl.-Fr.: July-Oct.

Rare, in moist places between boulders, **Khandwa** only, Sunder Dev forest.

Curcuma angustifolia Roxb., Asiat. Res. 11: 338. t. 3. 1810; Baker in Hook.f., Fl. Brit. India 6: 210.1890; Samvatsar, Fl. West. Trib. Madhya Pradesh 280.1996; Mudgal *et al.,* in N.P. Singh *et al.,* Fl. Madhya Pradesh 3: 81.2001. *Tikhur*

Rhizomatous herbs; rhizome white-pale yellow inside; leaves oblong-lanceolate, subcaudate at apex, 25-50 × 6-12 cm; flowers in lateral, 8-14 cm long spikes, as long as or slightly longer than bracts; peduncles *ca* 10 cm long; coma bracts purple; corolla lobes yellow; labellum yellow, emarginate; capsules ovoid, appressed hairy; seeds reddish-brown.

Fl.-Fr.: May – Oct.

Common in all districts.

Curcuma aromatica Salisb., Parad. Lond. t. 96. 1808; Baker in Hook.f., Fl. Brit. India 6:210.1890; Mudgal *et al.,* in N.P. Singh *et al.,* Fl. Madhya Pradesh 3:82 . 2001. *Ban haldi*

Herbs; rhizome whitish-yellow, smelling like camphor, when freshly cut; roots ending in yellow, oblong tubers; leaves lanceolate, 30-60 × 15-20 cm, sometimes with a pale or deep purple band along the midrib, apex acuminate, petioles long; flowers in lateral, 25 cm long spikes; coma bracts pink, lower floral bracts green, upper with pink tips; corolla pinkish-white; labellum yellow; lateral staminodes pinkish-white; capsules globose, membranous.

Fl.-Fr.: May-Aug.

In fresh ground, in open spaces of forest, Burhanpur, Khandwa.

Curcuma decipiens Dalz, Hooker's J. Bot. Kew Gard. Misc. 2:144. 1850; Hook.f., Fl. Brit. India 6:215. 1890; Samvatsar, Fl. West. Trib. Madhya Pradesh 280 . 1996; Mudgal *et al.,* in N.P. Singh *et al.,* Fl. Madhya Pradesh 3:82 . 2001.

Rhizomes small; tubers white inside, almond-like; leaves 15-30 × 7-12 cm, acuminate, rounded to subcordate at base; petioles as long as the blades; flowers in spikes arising from the centre of the previously formed leafy tuft; coma bracts purple, numerous; floral bracts purple, saccate; corolla purple; labellum 2-fid, crisped on margin; capsules hairy, up to 1 cm long.

Fl.-Fr.: June-Sept.

Occasional in the forest, Dhar, Khandwa.

Curcuma pseudomontana J. Graham, Cat. Pl. Bombay 210 . 1839; Baker in Hook.f., Fl. Brit. India 6:214. 1890, *p.p*; Mudgal *et al.,* in N.P. Singh *et al.,* Fl. Madhya Pradesh 3:84.2001.

Aromatic herbs; rhizomes yellow inside; tubers white inside, stipitate; leaves oblong-lanceolate, 10-30 × 4-10 cm, uniformly green, acuminate, base tapering; flowers arising in spikes, 2-3 per bract, from previously formed leafy tuft; peduncles 20-25 cm long, enclosed in the leaf sheath; coma bracts white with purple tips, floral bracts greenish-white; corolla yellow; labellum yellow; capsules ovoid, 1.5 × 1.0 cm.

Fl.-Fr.: Aug. Oct.

Rare, as undergrowth in forest, **Khandwa** only, Kalibhit forest.

ZINGIBER Boehmer W. 144; MP 3; SA 1

Zingiber cernuum Dalzell, Hooker's J. Bot. Kew Gard. Misc. 4:342. 1852; Baker in Hook.f., Fl. Brit. India 6:245. 1892; Samvatsar, Fl. West. Trib. Madhya Pradesh 281. 1996; Mudgal *et al.,* in N.P. Singh *et al.,* Fl. Madhya Pradesh 3:2001. *Jangli adrak*

Perennials, stem 1-1.5 m tall; leaves oblong-lanceolate, 50-6- × 6-8 cm, glabrous, acuminate; spikes ovate, obtuse, pedunculate, arising very close to ground; peduncles up to 3 cm long; bracts greenish-yellow; labellum 3-lobed, white with red blotches; capsules deep red, 4-5 × 1.5-1.7 cm; seeds black, striated, aril white.

Fl.-Fr.: Dec. -Jan.

Occasional in wet and shady places, Alirajpur, Dhar.

MUSACEAE 2/78; 2/2; 2/2

Key to Genera

1. Plants not stoloniferous; pseudostems swollen at base; offsets absent .... ENSETE
1. Plants stoloniferous; pseudostems not swollen at base; offsets present .... MUSA

ENSETE Bruce & Horan. W. 8; MP 1; SA 1

Ensete superbum (Roxb.) Cheesman, Kew Bull. 2:100. (1947) 1948; *Musa superba* Roxb., Pl. Coromandel 3: 17. 1811; Baker in Hook.f., Fl. Brit. India 6: 261. 1892; Mudgal *et al.,* in Singh *et al.,* Fl. Madhya Pradesh 3:95. 2001.

Large, 3.0-3.6 m high herbs; trunk purple-brown, swollen at base, narrow below leaves; leaves oblong, 2-3 × 0.5-0.8 m, glabrous, acute at apex, long petiolate; midrib pale-red beneath; sheaths convolute, persistent; inflorescence of sub-terminal, stout spikes; flowers in clusters of 20-30 in each bract; bracts maroon-coloured, 15-20 cm long; berries *ca* 7.5 × 3.7 cm, elongated, trigonous, ripening dry; seeds brown, sub globose.

Fl.-Fr.: Aug. -Oct.

Alirajpur, Barwani, Burhanpur, Khandwa, Khargone, in hilly areas, although occurring in 5 districts, the plants are nowhere common.

MUSA L. W. 70; MP 1; SA 1

Musa balbisiana Colla, Mem. Reale Accad. Sci. Torino 25:384. 1820; *M. rosacea* Jacq. Pl. Hort. Schoenbr. 4:22. 1804; Baker in Hook.f., Fl. Brit. India 6:263. 1892; Samvatsar, Fl. West. Trib. Madhya Pradesh 281. 1996; Mudgal *et al.,* in N.P. Singh *et al.,* Fl. Madhya Pradesh 3: 96.2001. *Jangli kela*

A 1-3 m tall stoloniferous plant; pseudostems cylindrical; leaves spiral, linear-oblong, 0.8-1 × 0.2-0.25 m; flowers in erect or drooping spikes; bracts pale red, ovate, few-flowered; male flowers more numerous than female; berries linear-oblong, slightly incurved, many-seeded; seeds tuberculate.

Fl- Fr.:-July – Nov.

Alirajpur, Barwani, Dhar, Khargone, the seeds are angular and very hard, harder than betel nuts.

AMARYLLIDACEAE 80/2258; 2/4; 2/3

Key to Genera

1. Corona present, formed by expanded filaments .... PANCRATIUM
1. Corona absent .... CRINUM

CRINUM L. W. 106; MP 4; SA 2

Key to Species

1. Perianth funnel-shaped, erect .... **C. latifolium**
1. Perianth salver –shaped, lobes linear; leaves deeply chanelled .... **C. viviparum**

Crinum latifolium L., Sp. Pl. 291. 1753; Hook.f., Fl. Brit. India 6: 283.1892; Samvatsar, Fl. West. Trib. Madhya Pradesh 283. 1996; Dixit *et al.,* in N.P. Singh *et al.,* Fl. Madhya Pradesh 3:103.2001. *Madho parnika*

A perennial herb with bulbous stem; leaves clustered at tip of short caudex, linear-oblong or linear-lanceolate, obtuse or acute at apex, 40-50 cm long; bracts 2, ovate, tinged with pink; bracteoles 2, shorter than bracts; flowers white, with pink stripes, in 6-12-flowered umbels; perianth 10-18 cm long, hypocrateriform; lobes linear, cuspidate, cusp small, fimbriate, appendage present just inside the apex; fruits *ca* 2.5 cm across.

Fl. -Fr.: June-Aug.

Alirajpur, Burhanpur, Khandwa, in moist, shady places.

Crinum viviparum (Lam.) R. Ansari & V.J. Nair, J. Econ. Taxon. Bot. 11:205. (1987) 1988; *Amaryllis vivipara* Lam., Encycl. 1:123.1783; *Crinum defixum* Ker-Gawl., J. Sci. Arts (London) 3:105.1817; Hook.f. Fl. Brit. India 6:281.1892; Samvatsar, Fl. West. Trib. Madhya Pradesh 282.1996; Dixit *et al.,* in N.P. Singh *et al.,* Fl. Madhya Pradesh 3:102.2001 (as *C. defixum*) *Magar kanda*

A perennial herb with ovoid bulb, 6-8 cm across; leaves 10-50 cm long, linear, apex acute; scape 30-50 cm long; bracts 2, spathaceous, membranous; flowers white; fruit sub globose, 2-4 cm across, beaked; seeds 1-2, rugose.

Fl.-Fr.: Aug. -Oct.

Very common in rocky river beds and river banks in all districts.

PANCRATIUM L. W. 21; MP 1; SA 1

Pancratium triflorum Roxb., Fl. Ind. ed. 1832 2:126. 1832; *P. malabathricum* Herb., Amaryllidaceae 202. 1837; Hook.f., Fl. Brit. India 6:285. 1892; Lakshminarsimhan in Sharma *et al.,* Fl. Maharashtra 3:99. 1996.

Bulbs globose, 3-5 cm across; leaves appearing with flowers, 20-40 × 1.2-2.5 cm, linear or linear-lanceolate, apex acute; scapes slender, 15.0-22 cm long; flowers 3-8 in an umbel, fragrant; perianth tube 3-5 cm long; staminal cup with bifid teeth between short filaments; fruits trigonous, ovoid.

Fl.-Fr.: April-June

Khandwa only, Khalwa and Sunder Dev forests.

HYPOXIDACEAE 10/154; 2/3; 2/2

Key to Genera

1. Leaves linear-lanceolate or elliptic-lanceolate, plicate, petiolate; ovary and perianth bridged by a long, filiform tube; fruit indehiscent .... CURCULIGO
1. Leaves grass-like, not plicate, sessile; perianth sessile on ovary; fruit circumscissile or valvular .... HYPOXIS

CURCULIGO Gaertn. W. 15; MP2; SA 1

Curculigo orchioides Gaertn., Fruct. Sem. Pl. 1: 63. 1788; Hook.f., Fl. Brit. India 6: 279. 1892; Samvatsar, Fl. West. Trib. Madhya Pradesh 282. 1996; Dixit *et al.,* in N.P. Singh *et al.,* Fl. Madhya Pradesh 3: 107.2001. *Kali musli.*

Perennial, acaulescent, tuberous herbs; leaves very variable, linear or linear-lanceolate, narrowed at base, acuminate at apex, 10-30 × 1-2 cm, plicate; flowers appear almost at the ground level, yellow, on a beak-like hypanthium; perianth 6-lobed; lobes oblong-elliptic; fruits ovoid; seeds many, globose.

Fl.-Fr.: June-Sept.

Common as forests-undergrowth in all districts.

HYPOXIS L. W. 93 MP 1; SA 1

Hypoxis aurea Lour., Fl. Cochinch. 200.1790; Samvatsar, Fl. West. Trib. Madhya Pradesh 282. 1996; Dixit *et al.,* in N.P. Singh *et al.,* Fl. Madhya Pradesh 3:107.2001. *Bhui khajur*

Perennial herbs with sub globose root stock; leaves 10-30 × 0.8-1.2 cm, narrowly linear, sub coriaceous, acute, keeled; flowers yellow, solitary, terminal on 5-15 cm long scape; capsules oblong, opening by 3 valves; seeds black, tuberculate.

Fl.-Fr.: July-Aug.

Common in all districts, among grasses in low lying areas where water accumulates.

DIOSCOREACEAE 6/653; 2/13; 2/8

Key to Genera

1. Flowers unisexual, borne in spikes, scape absent; twiners; leaves simple or compound; fruit capsular; capsule and seeds winged .... DIOSCOREA
1. Flowers bisexual, borne on a scape, umbellate; erect plants with globose root-stock; leaves simple, palmately 3-partite, petiole up to 80 cm long; fruit baccate, 6-ribbed; seeds not winged .... TACCA

DIOSCOREA L. W. 613 ;MP 12; SA 7

Key to Species

1. Stem twining to right
    2. Stems armed, glabrous, terete; seeds winged all round . ... **D. wallichii**
    2. Stems unarmed
        3. Tubers small; lower leaves sagittate-ovate, upper cordate-ovate; male spikes unilateral, 1-3 on special leafless branches , female spikes solitary or 2; capsules 1-2 across ......**D. belophylla**
        3. Tubers very long; leaves uniform; spikes axillary on normal branches; capsules 3.5-4.5 cm across .... **D. oppositifolia**
1. Stems twining to left

4. Leaves simple; unarmed twiners; seeds winged at the base .... **D. bulbifera**
4. Leaves 3-5 foliolate
   5. Fertile stamens 6; leaves 3-foliolate . ... **D. hispida**
   5. Fertile stamens 3
      6. Stems tomentose; leaves usually 3-foliolate; staminodes and stamens as long as sepals .... **D. tomentosa**
      6. Stems glabrous; leaves 3-5 foliolate; staminodes and stamens shorter than sepals **D. pentaphylla**

Dioscorea belophylla (Prain) Voigt ex Haines, Forest Fl. Chota Nagpur 530.1910; *D. nummularia* var. *belophylla* Prain, Bengal Pl. 2:1065. 1903; Hook.f., Fl. Brit. India 6:294. 1892; Prain & Burkill in Ann. Roy. Bot. Gard. Calc. 14 (2):348. 1938; Khanna in N.P. Singh *et al.,* Fl. Madhya Pradesh 3:120.2001.

Climbers; tubers small, fusiform; stem twining to right; bulbils present; leaves simple, usually opposite, 8-10 × 3-8 cm, apex acuminate, base cordate; male spikes laxly flowered, unilaterally 1-3 nate on special leafless branches; stamens 6, all fertile; female spikes solitary or 2 together in the axils of leafy branches; capsules *ca* 2 cm long, broader than long, pendulous; seeds winged around.

Fl.-Fr.: Aug. -Dec.

Burhanpur only, in mixed forest, at Asirgad and Nepa nagar.

Dioscorea bulbifera L., Sp. Pl. 1033. 1753; Hook.f., Fl. Brit. India 6: 295. 1892; Prain & Burkill in Ann. Roy. Bot. Gard. Calc. 14 (2): 348. 1938; Samvatsar, Fl. West. Trib. Madhya Pradesh 284. 1996; Khanna in N.P. Singh *et al.,* Fl. Madhya Pradesh 3: 121.2001. *Ratalu*

Large climbers; stems twining to the left; leaves simple, alternate, broadly ovate-cordate, cordate at base, subcaudate at apex, 8-25 × 4-12 cm; leafless branches; male flowers flowers in raceme-like, axillary spikes, greenish-purple; female flowers purple, sessile, axillary or in fascicled spikes; capsules oblong, slightly widened upwards; seeds winged at base.

Fl.-Fr.: Aug. -Oct.

Common in all districts, in shady places of the forest.

Dioscorea hispida Dennst., Schlüssel Hort. Malab. 33. 1818; *D. daemona* Roxb., Fl. Ind. ed. 1832 3: 805. 1832; Hook.f., Fl. Brit. India 6: 289. 1892; Samvatsar, Fl. West. Trib. Madhya Pradesh 285. 1996; Khanna in N.P. Singh *et al.,* Fl. Madhya Pradesh 3: 124.2001. *Baichandi*

Stems hirsute, twining to the left, with recurved prickles; leaves 3-foliolate, glabrous or puberulous, middle leaflet elliptic-oblong or obovate, acute at base, acuminate at apex, 10-20 × 8-10 cm; lateral leaflets smaller, inequilateral, 5-costate; male flowering axes usually solitary; female flowers in solitary pendulous spikes; capsules brown, oblong, 3-winged, ascending; seeds winged at base.

Fl.-Fr. : July-Oct.

Alirajpur, Barwani, Burhanour, Indore, Khandwa, Khargone, in hilly places.

Dioscorea oppositifolia L., Sp. Pl. 1033, 1753; Hook.f. Fl. Brit. India 6:292. 1892; Prain & Burkill in Ann. Roy. Bot Gard. Calc. 14 (2):392. 1938; Khanna in N.P. Singh *et al.,* Fl. Madhya Pradesh 3:124.2001. *Targaria kanda*

Stem climbing to the right, unarmed; tubers cylindrical; bulbils small; leaves 6-10 × 4-8 cm, simple, opposite or alternate, elliptic-ovate, apex acuminate, base obtuse or rounded; spikes axillary, clustered; stamens 6, all fertile; capsules 15-20 × 10-15 mm, winged; seeds brown, winged around.

Fl.-Fr.: July-Nov.

Burhanpur, Khandwa, climbing over adjacent vegetation.

Dioscorea pentaphylla L., Sp. Pl. 1032. 1753; Hook.f., Fl. Brit. India 6: 289. 1892; Samvatsar, Fl. West. Trib. Madhya Pradesh; Khanna in N.P. Singh *et al.,* Fl. Madhya Pradesh 3: 126.2001. *Suwar kand*

Twining herbs with globose to pyriform tubers; slender, herbaceous twiners with recurved prickles, twining to the left; leaves alternate, 3-5-foliolate, middle leaflet elliptic-oblanceolate, oblique at base, acuminate at apex, 6-12 × 2-6 cm; lateral leaflets smaller; male flowers white, in dense, pubescent, axillary or terminal panicles borne on large leafless branches; female flowers in slender, 2-3 nate racemes; capsules oblong, ascending, pubescent when young; seeds winged only at base.

Fl.-Fr.: Sept. -Nov.

Common in all districts, in the forest.

Dioscorea tomentosa J. Koenig ex Spreng., Pl. Min. Cogn. Pug. 2:92. 1815; Roxb. & B. Heyne in Roth, Nov. Pl. Sp. 371. 1821; Hook.f., Fl. Brit. India 6:289. 1892; Samvatsar, Fl. West. Trib. Madhya Pradesh 285. 1996; Khanna in N.P. Singh *et al.,* Fl. Madhya Pradesh 3:127.2001.

Stem twining to the left, cottony white-tomentose, often prickly towards base; tubers present; leaves sometimes simple in the upper part, normally 3 (5) foliolate, elliptic obovate, 12 × 7 cm, densely white- tomentose beneath; male spikes short, 1-3 together on slender, tomentose pendulous panicles; female flowers in simpe or branched racemes; flowers white; stamens 3 fertile and staminodes 3; capsules oblong, 3-winged, white-brown tomentose; seeds winged at base.

Fl.-Fr.: Sept. -Nov.

Alirajpur and Khandwa on the banks of streams on surrounding vegetation.

Dioscorea wallichii Hook.f., Fl. Brit. India 6:295. 1892; Samvatsar, Fl. West. Trib. Madhya Pradesh 285. 1996; Khanna in N.P. Singh *et al.,* Fl. Madhya Pradesh 3:128. 2001.

Climbing shrubs with numerous, fascicled, cylindrical tubers; stem twining to the right, armed at base; leaves alternate, 10-15 × 8-10 cm, 7-9 nerved, acuminate at apex, cordate at base; flowers red-brown; male flowers in axillary, recurved, leafless branches in simple clusters or panicles; female flowering axes 1-2 together; stamens 6, all fertile; capsules stramineous with brown flecks; wings of capsule 18-20 × 15-18 mm; seeds winged around.

Fl.-Fr.: Sept. -Nov.

Alirajpur, Dhar, Indore, along water streams.

TACCA J. R. Forst. & G. Forst. W. 17 ;MP 1; SA 1

Tacca leontopetaloides (L.) Kuntze, Revis. Gen. Pl. 2: 704. 1891; *Leontice leontopetaloides* L., Sp. Pl. 313. 1753; *Tacca pinnatifida* J. R. Forst. & G. Forst., Char. Gen. Pl. 35. 1775; Hook.f., Fl. Brit. India 6: 287. 1892. Samvatsar, Fl. West. Trib. Madhya Pradesh 284. 1996; Hanfi in N.P. Singh *et al.,* Fl. Madhya Pradesh 3: 117.2001. *Barwa-Barwi* **(Plate 9, Fig. 5)**

Perennial herbs with tuberous rootstock; leaves 3-partite, each irregularly pinnatifid, glabrous, *ca* 80 cm across; flowers in drooping, *ca* 8 cm across umbels, green; bracts *ca* 3 × 1 cm; bracteoles several, *ca* 10 cm long; perianth green, *ca* 1 cm long, segments oblong; berries ovoid, with persistent perianth, *ca* 2 × 1.2 cm; seeds numerous, striate.

Fl.-Fr.: Sept. -Nov.

Alirajpur, Barwani, Dhar, Indore, Khandwa, in moist, shady places and hill slopes.

ASPARAGACEAE 128/2929; 4/9; 4/9

Key to Genera

1. Leaves reduced to spinescent scales; cladodes present; roots fleshy, fasciculated .... ASPARAGUS
1. Leaves well developed, radical; cladodes absent; flowers in scapes or racemes
  2. Root stock not bulbous, root tuberous, fasciculated ....CHLOROPHYTUM
  2. Root stock bulbous
    3. Leaves appearing after the flowers; pedicels 3-7 cm long .... DRIMIA
    3. Leaves appearing with flowers; pedicels 2-5 cm long . ... LEDEBOURIA

ASPARAGUS L. W. 211; MP 2; SA 2

Key to Species

1. Cladodes flat, more than 1 mm broad; spines up to 4 mm long .... **A. gonocladOS**
1. Cladodes sub terete, less than 0.6 mm broad; spines more than 4 mm long .... **A. racemosus**

Asparagus gonoclados Baker, J. Linn. Soc. Bot. 14:627. 1875; Hook.f., Fl. Brit. India 6:318. 1892; Dixit *et al.,* in N.P. Singh *et al.,* Fl. Madhya Pradesh 3:133.2001.

Excessively branched undershrubs, branches strongly angled, spines short; cladodes in fascicles of 2-6, *ca* 1 cm long, flat, stout, narrowed to an acuminate tip; flowers in lax cymes; pedicels jointed below the middle; berries globose, didymous.

*Fl.-Fr.:*July-Nov.

Burhanpur, Khandwa, appears to be an escape.

Asparagus racemosus Willd., Sp. Pl. 2: 152.1799; Hook.f., Fl. Brit. India 6: 316.1892; Samvatsar, Fl. West. Trib. Madhya Pradesh 287.1996; Dixit *et al.,* in Singh *et al.,* Fl. Madhya Pradesh 3: 133.2001. *Khargosh ki ghas.*

Scandent undershrubs with cladodes; roots tuberous, fasciculated; stems grey-white, woody; spines decurved; cladodes in clusters of 2-6, oblong-linear, 1-3 cm long; flowers white, small, in solitary or fascicled, simple or branched racemes; pedicels jointed above the middle; perianth-segments linear-oblong, spreading; berries scarlet when ripe, globose.

Fl.-Fr.: Sept. -Dec.

Common in all districts.

CHLOROPHYTUM Ker-Gawl. W. 196; MP 3; SA 5

Key to Species

1. Leaves longer than scape; capsules 3-winged, 1-4 seeded; flowers remotely placed on the scape .... **C. laxum**
1. Leaves shorter than scape
   2. Leaves elliptic-lanceolate, lanceolate or oblanceolate; scapes naked
      3. Leaves with serrulate margin; lobes of perianth papillose at apex; filaments papillose; scape 15-25 cm long; seeds granular . ... **C. diwanjii**
      3. Leaves with smooth margin; lobes of perianth smooth at apex; filaments smooth; scape up to 40 cm long; seeds rugose. .... **C. arundinaceum**
   2. Leaves linear or ensiform, hardly contracted to a petiole; scapes longer than the leaves
      4. Anther filaments papillose ....**C. tuberosum**
      4. Anther filaments glabrous ....**C. borivilianum**

Chlorophytum arundinaceum Baker, J. Linn. Soc. Bot. 15:323. 1876; Hook.f., Fl. Brit. India 6:333. 1892; Dixit *et al.,* in N.P. Singh *et al.,* Fl. Madhya Pradesh 3:135. 2001. *Safed musli*

Perennial herbs; roots tuberous, in clusters; leaves 15-60 × 2.5-5 cm, acute to acuminate at apex, narrowed to petiole, flowers white, in dense racemes; scape as long as the leaves; pedicels jointed in the middle; anthers about twice as long as the filaments; capsules globose, about 8 mm across, emarginated at both ends; black, flat.

*Fl.-Fr.:*Aug. -Oct.

Khandwa only, amidst rocky or gravelly soil, Sunder Dev and Golai forests

Chlorophytum borivilianum Sant. & R. R. Fern., in J. Bombay Nat. Hist. Soc. 3:898. 1955; Lakshminarsimhan in Sharma *et al.,* Fl. Maharashtra 3:125, 1996.

Perennial herbs; tubers cylindrical, fascicled, sessile; leaves radical, sessile, linear or ensiform, 20-30 × 1-1.6 cm; flowers in racemes, arranged in alternate clusters, pedicellate; bracts linear or triangular, persistent; scape terminal, 25-35 cm lon; perianth white; outer segments sepaloid; inner segments petaloid; capsules triquetrous to sulcate, obcordate, 5-6 mm across; seeds black, orbicular, discoid.

Fl. - Fr.: June – July

Barwani and Burhanpur in moist places. The occurrence of this species here extends its distribution to M.P.

Chlorophytum diwanjii Mujaffar Shaikh, Arjun Tiwari & A.N. Chandore in Phytotaxa 291 (2):159-162.2017.

Perennial, up to 40 cm tall herbs; rhizome short, 2-3 cm long, giving rise to new plants laterally; roots up to 35 cm long, lateral root tubers 3-5 × 1-1.5 cm, tuberous in the middle; leaves radical, 15-38 × 2.0-5 cm, broadly lanceolate, apex acute; scape 15-25 cm long, solitary, unbranched; flowers in clusters of 2-5, white, 2.0-2.5 cm across; bracts 5-nerved; perianth segments 3+3, apex papillose; filaments papillose; capsules 0.7-o. 9 cm long, black at maturity; seeds, flat, black.

Fl.-Fr.: June-Oct.

Khandwa, as forest undergrowth, Khalwa forest. This is an addition to the Fl. of World.

Chlorophytum laxum R. Br., Prodr. Nov. Holl. 277. 1810; Hook.f., Fl. Brit. India 6: 336. 1892; Samvatsar, Fl. Wesst. Tribal Madhya Pradesh 288. 1996; Dixit *et al.,* in N.P. Singh *et al.,* Fl. Madhya Pradesh 3: 136.2001.

Perennial herbs; root fibres ending into ellipsoidal tubers; leaves strongly recurved, flat or conduplicate, linear-lanceolate, falcate, 10-18 × 0.3-0.5 cm, sessil;. flowers in simple or forked racemes; scape slender, 3-6 cm long; perianth greenish-white; segments oblong; capsules obcordate, 3-winged; seeds angular.

Fl. - Fr.: June - Sept.

Alirajpur, Barwani, Burhanpur, Dhar, Indore, Khandwa, along forest paths.

Chlorophytum tuberosum (Roxb.) Baker, J. Linn. Soc. Bot. 15: 332. 1876; *Anthericum tuberosum* Roxb., Pl. Coromandel 2:20.1800; Hook.f., Fl. Brit. India 6: 334. 1892; Samvatsar, Fl. West. Trib. Madhya Pradesh 288. 1996; Dixit *et al.,* in N.P. Singh *et al.,* Fl. Madhya Pradesh 3: 136.2001.

Perennial, acaulescent herbs; roots tuberous, cylindrical, fleshy; levaes radical, crowded at the base, 20-50 × 2-4 cm; flowers in simple or shortly branched racemes; scape 15-90 cm long, with a small sheath; perianth white; segments elliptic, 12-15 mm long; capsules oblong, 2-lobed at apex, *ca* 1 cm long; seeds irregularly orbicular, flat, opaque.

*Fl.-Fr.:*July-Sept.

Common in all districts, among grasses and also on hill slopes (All species of *Chlorophytum* are called Safed Musli).

DRIMIA Jacq. ex Willd. W. 99; MP1 ;SA 1

Drimia indica (Roxb.) Jessop, J. South Afr. Bot. 43:272. 1977; *Scilla indica* Roxb., Fl. Ind. ed. 1832 2:147. 1832; *Urginea indica* (Roxb.) Kunth, Enum. Pl. 4:333. 1843; Samvatsar, Fl. West. Trib. Madhya Pradesh; Dixit *et al.,* in N.P. Singh *et al.,* Fl. Madhya Pradesh 3:142.2001(*Urginea indica*). *Jangli piaz*

A small geophyte with tunicated, 3-4 cm across bulb; leaves radical, 12-36 × 1-1.3 cm, linear-lanceolate, flat, succulent acute; flowers pale-brown on a scape, in few flowered racemes; pedicels 2-3 cm long, articulated; capsules ellipsoid, 1.8-2 cm long; seeds 6 mm across, black.

Fl.-Fr.: March-May

Common as forest undergrowth, also on hill slopes in all districts.

LEDEBOURIA Roth W. 53; MP 1; SA 1

Ledebouria revoluta (L.f.) Jessop, J.S. African Bot. 36:255. 1970; *Hyacinthus revolutus* L.f., Suppl. Pl. 204. 1782; *Scilla hyacinthina* (Roth) MacBr.,Contrib. Gray Herb. (n.s.) 56:14. 1918; *Scilla indica* Roxb., Fl. Ind. ed. 1832 2:147. 1831; Samvatsar, Fl. West. Trib. Madhya Pradesh 289. 1996; Dixit *et al.,* in N.P. Singh *et al.,* Fl. Madhya Pradesh 3:141.2001.

Small plants with tunicated bulb; leaves 2-4, radical, glabrous, 3.5-7.5 × 1-2.5 cm, upper surface with purple or brown spots, narrowing into sheathing petiole; flowers in a scape, greenish-purple, in many flowered racemes; bracts scarious, serrate, bidid above; capsules brown, 5 × 5 mm, globose, unequally 3-lobed; seeds obovoid, brown, wrinkled.

Fl.-Fr.: June-Aug.

Common in all districts on hill slopes, among grasses.

COLCHICACEAE 18/284; 3/4; 2/3

1. Leaftips ending into tendrils; perianth segments strongly undulate; root stock V-shaped ....GLORIOSA
1. Leaftips not ending into tendrils; perianth segments not undulate; root stock, tunicated, cormous ....IPHIGENIA

GLORIOSA L. W. 10; MP 1; SA 1

Gloriosa superba L., Sp. Pl. 305. 1753; Hook.f., Fl. Brit. India 6: 358. 1892; Samvatsar, Fl. West. Trib. Madhya Pradesh 288. 1996; Dixit *et al.,* in N.P. Singh *et al.,* Fl. Madhya Pradesh 3: 138. 2001. *Kan kasni* **(Plate 9, Fig. 6)**

Climbing, 1-1.5 m tall, scrambling herbs; rootstock poisonous, V-shaped, 2-7 cm long; leaves lanceolate or oblong-lanceolate, cordate at base, tendrillar at apex, 8-16 × 1-3 cm; flowers showy; pedicels 10-15 cm long; perianth yellow in lower half, red in upper half, finally red with age, 5-8 × 1-5 cm; segments oblong-lanceolate, undulate on margin; capsules linear-oblong, up to 5 cm long.

Fl.-Fr.: Aug. -Oct.

Common in hedges during monsoon, but becoming scarce due to over exploitation.

IPHIGENIA Kunth W. 13 MP 2; SA 2

Key to Species

1. Perianth white, segments oblanceolate .... **I. pallida**
1. Perianth purplish, segments linear-subulate .... **I. indica**

Iphigenia indica (L.) A. Gray ex Kunth, Enum. Pl. 1:213. 1833; *Melanthium indicum* L.,Mant. Pl. 226. 1771; Hook.f., Fl. Brit. India 6:357. 1892; Samvatsar, Fl. West. Trib. Madhya Pradesh 288. 1996; Dixit *et al.,* in N.P. Singh *et al.,* Fl. Madhya Pradesh 3:139.2001. *Santhal chutia*

An erect, delicate herb with sub globose, tunicated corm; leaves 10-20 × 0.5-1.5 cm, sessile, linear-lanceolate, acuminate, glabrous; flowers purple in terminal racemes; bracts linear or subulate; capsules oblong, obtuse, 3-4 grooved; seeds ellipsoid, reddish-brown.

Fl.-Fr.: July-Aug.

Alirajpur, Burhanpur, Dhar, Indore, Khandwa, on hill slopes, open grass-land.

Iphigenia pallida Baker, J. Linn. Soc. Bot. 17:451. 1879; Hook.f., Fl. Brit. India 6:357. 1892; Dixit *et al.,* in N.P. Singh *et al.,* Fl. Madhya Pradesh 3:140.2001.

Perennial herbs with globose, 1 cm across corm; sheaths pale-brown; stem leafy, erect; leaves 8-20 × 0.2-0.5 cm, linear-lanceolate; flowers solitary or in corymbs; pedicels 1-2 cm long; bracts linear, foilaceous; flowers white, perianth segments oblanceolate, much broader, many nerved; capsules oblong; seeds sub globose.

Fl.-Fr.: Aug. -Nov.

Burhanpur, Indore, among grasses and forest floor.

XANTHORRHOEACEAE 34/1236; 2/2; 2/2

Key to Genera

1. Leaves fleshy, mucilaginous, margin distantly spiny .... ALOE
1. Leaves neither fleshy nor mucilaginous, tubular; roots fleshy . ... ASPHODELUS

ALOE L. W. 558; MP 1; SA 1

Aloe vera (L.) Burm.f., Fl. Indica 83. 1768; *A. perfoliata* var. *vera* L., Sp. Pl. 1:320.1753; *A. barbadensis* Mill.,Gard. Dict. ed. 8,2 1768; Samvatsar, Fl. West. Trib. Madhya Pradesh 287. 1996; Dixit *et al.,* in N.P. Singh *et al.,* Fl. Madhya Pradesh 3: 131.2001. *Gawar patha*

Acaulescent herbs; leaves erect, crowded, lanceolate, thick, succulent, glaucous-green, 30-50 cm long, margin spiny; flowers in simple, erect racemes, the lower flowers falling off as the raceme elongates; perianth dull- red or reddish- yellow, tubular; capsules *ca* 1.5 × 1 cm, cylindrical or ellipsoid-oblong (hardly 1-2 per inflorescence).

Fl.-Fr.: Nov. -March

Common around villages, planted as live hedge or self propagated.

ASPHODELUS L. W. 19; MP 1; SA 1

Asphodelus tenuifolius Cav., Anales Ci. Nat. 3:46. 1801; *A. clavatus* Roxb., Fl. Ind. ed. 1832 2:24. 1832; Hook.f., Fl. Brit. India 6:332. 1892; Samvatsar, Fl. West. Trib. Madhya Pradesh 287. 1996; Dixit *et al.,* in N.P. Singh *et al.,* Fl. Madhya Pradesh 3: 134.2001. *Piazi*

Herbaceous,20-50 cm tall, stemless annual; leaves 6-16 cm long, hollow, acute, base sheathing, glabrous; scapes 30-40 cm, several from the base, simple or branched; flowers white; capsules 4 mm long, globose, smooth, 3-valved; seeds black, rugose, 3-gonous.

Fl.-Fr.: Dec.-March

Common as a winter weed in all districts.

SMILACACEAE 2/261; 1/1; 1/1

SMILAX L. W. 248; MP 1; SA 1

Smilax zeylanica L., Sp. Pl. 1029. 1753; *S. macrophylla* Roxb.,1832, *nom. illeg;* Hook.f., Fl. Brit. India 6:309. 1892; Dixit *et al.,* in N.P. Singh *et al.,* Fl. Madhya Pradesh 3:144.2001.

Climbing perennials; stem 4-angled or terete; branches with small prickles, often zigzag; leaves 10-30 × 5-15 cm, broadly orbicular, apex cuspidate, base rounded or sub cordate, usually 5-nerved from the base; petioles 1.5-2.5 cm long, jointed in the middle, narrowly sheathing at base but not auricled; peduncle 20-30 flowered; staminodes 3 in the female flowers; berries red; seeds biconvex.

Fl.-Fr.: Sept. -April

Khandwa only, in valleys, at Battu Bagra.

PONTEDERIACEAE 6/33; 2/3; 2/3

Key to Genera

1. Plants floating; petioles swollen; flowers sessile and zygomorphic; stamens 6, sub equal in length, filaments of larger stamens hairy .... EICCHORNIA
1. Plants not floating, rooted in mud; petioles not swollen; flowers pedicellate, actinomorphic; stamens 6, 1 much longerthan others, all filaments glabrous .... MONOCHORIA

EICHHORNIA Kunth W. 6; MP 1; SA 1

Eichhornia crassipes (Mart.) Solms, Monogr. Phan. 4:527. 1883; *Pontederia crassipes* Mart., Nov. Gen. Sp. Pl. 1:9. 1824; Samvatsar, Fl. West. Trib. Madhya Pradesh 289. 1996; Misra *et al.,* in N.P. Singh *et al.,* Fl. Madhya Pradesh 3:145. 2001. *Jal kumbhi*

Aquatic, floating herbs; stolons sub cylindric, spongy; leaves forming rosettes, broadly ovate-rhomboid, 5-15 cm long,sub cordate at base, fleshy; petioles 15-20 cm long, spongy; flowers lilac blue in dense, up to 15 cm long spikes; perianth limb 2-lipped, lobes 6; capsules ovoid-oblong, *ca* 1.5 cm long; seeds many, ovoid *ca* 1 mm long.

Fl.-Fr.: Rainy season

Common in ponds, stagnant water, rivers, streams and becoming very troublesome to eradicate.

MONOCHORIA K. Presl W. 7; MP 2; SA 2

Key to Species

1. Root-stock creeping, branched; leaf base sagittate to hastate, basal lobes divergent ....**M. hastata**
1. Root-stock sub-erect or oblique, usually short; leaf base obtuse, rounded, truncate or cordatate, basal lobes when present, broadly rounded . ...**M. vaginalis**

Monochoria hastata (L.) Solms, Monogr. Phan. 4:523. 1883; *Pontederia hastata* L., Sp. Pl. 288. 1753; *Monochoria hastifolia* C. Presl, Reliq. Haenk. 1:128. 1827; Hook.f. Fl. Brit India 6. :128. 1892; Misra *et al.,* in N.P. Singh *et al.,* Fl. Madhya Pradesh 3:146.2001.

Perennial, rhizomatous marshy herbs; leaves 7-15 × 4-12, triangular-ovate, obtuse, acute or acuminate, sagittate, hastate, many nerved; petioles of basal leaves long; racemes dense, shortly peduncled, finally deflexed; perianth1.5-2 cm long, twisted around the fruit when drying up; stamens 6, the large anther blue; stigma obscurely 3-lobed; capsules 5-6 mm across; seeds with 10, thin, longitudinal ribs and finely transversely striate.

Fl.-Fr.: June-Sept.

Indore only, along margins of canals, tanks, water –logged places, at Mhow.

Monochoria vaginalis (Burm.f.) C. Presl, Reliq. Haenk. 1:128. 1827; *Pontederia vaginalis* Burm.f., Fl. Indica 80.1768; Hook.f., Fl. Brit. India 6:363. 1892; Samvatsar, Fl. West. Trib. Madhya Pradesh 289. 1996; Misra *et al.,* in N.P. Singh *et al.,* Fl. Madhya Pradesh 3:147.2001.

Aquatic herbs, rooted in mud, about 30-40 cm tall; leaves ovate, 3-10 × 1.5-5 cm, acute to acuminate at apex, rounded to sub cordate or truncate at base; flowers blue, regular, in 6-10 flowered racemes, enclosed initially in the leaf-sheath, later deflexed; perianth campanulate, 6-partite (3+3), outer 8-10 mm long, inner10-12 mm long; stamens 6, filaments unequal in length; stigma 3-lobed; capsules with brown ribs, 8-10 mm long, oblong; seeds many ribbed.

Fl.-Fr.: Dec. -March

Burhanpur, Dhar, Indore, Khandwa, in ditches and ponds.

COMMELINACEAE 41/728; 4/24; 3/16

Key to Genera

1. Fertile stamens 6, staminodes absent; filaments bearded or hairy; leaf sheath inflated or not; capsules beaked or not . ... CYANOTIS
1. Fertile stamens 2 or 3, rest 4 or 3 staminodes
    2. Inflorescence cymose; flowers zygomorphic, enclosed in boat-shaped bracts........ COMMELINA
    2. Inflorescence paniculate or a cluster of 1-5 flowers; flowers actinomorphic, not enclosed by bracts .... MURDANNIA

COMMELINA L. W. 215; MP 10; SA 8

Key to Species

1. Spathes funnel-shaped
    2. Flowers blue, in terminal spathes
        3. Seeds pitted; cleistogamous flowers present; spathes not subtended by floral leaves .... **C. benghalensis**
        3. Seeds smooth; cleistogamous flowers absent; spathes subtended by floral leaves
            4. Capsules sub-equally 3-valved; seeds free in the cell; leaf base oblique .... **C. paludosa**

4. Capsules 2-valved; seeds adnate to cells; leaf base not oblique . ... **C. erecta**

2. Flowers blue or white in axillary spathes; capsules 2-celled

5. Flowers white; leaves 18-22 × 4.5 cm; seeds with large pits; roots fleshy . ...**C. suffruticosa**

5. Flowers blue; leaves 2.5-5 × 0.6-1.2 cm; seeds smooth; roots not fleshy .... **C. forsskalii**

1. Spathes with posterior margins free

6. Capsule 2-celled; seeds rugose; leaves 5-12 × 2.5-5 cm, shortly petiolate . ..... **C. rufipes** var. **glabrata**

6. Capsules 3-celled, posticous cell keeled; leaves sessile

7. Seeds tuberculate, reticulate, brown .... **C. diffusa**

7. Seeds smooth, dark grey . ...**C. caroliniana**

Commelina benghalensis L., Sp. Pl. 41. 1753; Hook.f., Fl. Brit. India 6: 370.1892; Samvatsar, Fl. West. Trib. Madhya Pradesh 290.1996; Khanna *et al.,* in N.P. Singh *et al.,* Fl. Madhya Pradesh 3: 154.2001. *Kan kouwa* **(Plate 9, Fig. 7)**

Annual, slender, decumbent-ascending herbs; leaves ovate, elliptic or suborbicular, 1-8 × 1-4 cm, acute-obtuse at apex, subtruncate at base; spathes 2-3 at the end of branches, terminal or leafy-opposed, subsessile, hairy outside; petals blue or lilac; capsules oblong, trigonous, unequally 3-celled; seeds 5, pitted, wrinkled.

*Fl.-Fr.:*Aug. -Jan.

Very common all over the area, in moist soil.

Commelina caroliniana Walter, Fl. Carol. 68. 1788; *C. hasskarlii* C.B. Clarke, Commelyn. Cyrtander. Bengal t. 5. 1874 & Monogr. Phan. 3:157. 1881; Hook.f. Fl. Brit. India 6:370.1892; Samvatsar, Fl. West. Trib. Madhya Pradesh 291. 1996; Khanna *et al.,* in N.P. Singh *et al.,* Fl. Madhya Pradesh 3:156.2001.

Prostrate or erect, much branched herbs; leaves 3-8 × 0.7-1.5 cm, lanceolate or ovate-lanceolate, subacute, sheath mouth ciliate; cymes pubescent, lower branches 1-2 flowered, upper 2-4 flowered; flowers dark blue; capsules quadrate, glabrous; seeds brown, without appendages, truncate at one end, rounded at the other.

*Fl.-Fr.:*Aug. -Nov.

Common weed of fields, also in wet and shady places, in all districts.

Commelina diffusa Burm.f., Fl. Indica18. 1768; Hook.f. Fl. Brit. India 6:369. 1892; Samvatsar, Fl. West. Trib. Madhya Pradesh 290.1996; Khanna *et al.,* in N.P. Singh *et al.,* Fl. Madhya Pradesh 3:154 2001.

Erect or diffuse, glabrous, annual herbs, branches diffuse; leaves 4-9 × 1.2-1.7 cm, cordate, acuminate, base sheathing; capsules 5-seeded, 5 mm long, concealed in the spathe; seeds brown, oblong, reticulate and tubercled.

Fl.-Fr.: Aug. -Oct.

Common in all districts, among grasses, moist areas and moist fields.

Commelina erecta L., Sp. Pl. 41. 1753; *C. kurzii* var. *glochidiata* K. D. Koenig ex C.B. Clarke, Hook.f., Fl. Brit. India 6:373. 1892; Samvatsar, Fl. West. Trib. Madhya Pradesh 290.1996; Khanna *et al.,* in N.P. Singh *et al.,* Fl. Madhya Pradesh 3:155. 2001.

Erect or straggling, annual herbs; leaves lanceolate, acuminate, 5-15 × 1.4-4.5 cm, sheaths 2.5 cm long, sparsely hairy, ciliolate along margins at the mouth; spathes 2-5, terminal, accompanied by 2-4 smaller leaves; flowers in bifid cymes; capsules 3-celled, 3-valved, each cell with a free seed; seeds black with a white, membranous margin.

Fl.-Fr.: Sept. -Dec.

Common in all districts in wet and shady places.

Commelina forsskalii Vahl, Enum. Pl. Obs. 2: 172. 1805; *C. falcata* Hassk., Flora 46:387. 1863; Hook.f., Fl. Brit. India 6: 371. 1892; Samvatsar, Fl. West. Trib. Madhya Pradesh 291. 1996; Khanna *et al.,* in N.P. Singh *et al.,* Fl. Madhya Pradesh 3: 155. 2001.

Herbs, rooting at lower nodes; branches with aerial chasmogamous and underground cleistogamous flowers; leaves lanceolate, 4-10 × 0.7-1.6 cm, acute at apex, cuneate at base, entire or undulate; sheaths 1-2 cm long, pubescent; spathes axillary, sparsely hairy, intermixed with long slender hairs; flowers violet, in branched axillary cymes; sepals 3; petals 3; capsules oblong, 3-seeded; seeds subglobose, smooth; cleistogamous flowers often solitary in a spathe, bisexual with 1-seeded capsule.

Fl.-Fr.: Aug. -Oct.

Common weed of fields.

Commelina paludosa Blume, Enum. Pl. Javae 1:2:. 1827; *C. obliqua* Buch. -Ham. ex D. Don, 1825, *nom. illeg., non* Vahl, 1805; Hook.f., Fl. Brit. India 6:372. 1892; Samvatsar, Fl. West. Trib. Madhya Pradesh 291. 1996; Khanna *et al.,* in N.P. Singh *et al.,* Fl. Madhya Pradesh 3:157.2001.

Diffuse, ascending, perennial herbs; stem decumbent or creeping and rooting at lower nodes; leaves obliquely lanceolate, 5-20 × 2-5 cm, scabrid; sheaths 15-30 mm long, usually red-brown, glabrous, except a line of hairs at the mouth and margin; spathes 3-8, clustered together, narrowly ovate-lanceolate, 15-30 mm long; flowers small, pale-violet or white; capsules ovoid, *ca* 6 mm across, 3-celled, 1-3 seeded; seeds smooth.

Fl.-Fr.: Oct. -Jan.

Alirajpur, Burhanpur, Dhar, Khandwa, in shady places, also as forest undergrowth.

Commelina rufipes Seub. var. **glabrata** (D. R. Hunt) Faden & D. R. Hunt, Ann. Missouri Bot Gard. 74:122. 1987; *Commelinopsis glabrata* D. R. Hunt, Kew Bull. 36:195. 1981; *C. persicariifolia* Delile, Liliac. 8:t. 472. 1815; Hook.f. Fl. Brit. India 6:372. 1892; Samvatsar, Fl. West. Trib. Madhya Pradesh 291. 1996 (*C. persicariifolia* Wight ex C.B. Clarke); Khanna *et al.,* in N.P. Singh *et al.,* Fl. Madhya Pradesh 3:156.2001. (*C. imberbis* Ehrenb. ex Hassk. is a separate species)

Diffuse herbs, rooting at lower nodes; leaves 5-15 × 2-5 cm, ovate-lanceolate or lanceolate, apex acuminate, base sub cordate, petoles short; sheaths *ca* 2 cm long; spathes 2-2.5 cm long, not auricled; flowers blue, in axillary cymes; capsules oblong, *ca* 6 mm long, 2-celled; seeds obscurely rugose.

Fl.-Fr.: Oct. -Jan.

Rare, **Khandwa** only, Kalibhit forest.

Commelina suffruticosa Blume, Catalogus 35. 1823; Hook.f., Fl. Brit. India 6:374. 1892; Samvatsar, Fl. West. Trib. Madhya Pradesh 291. 1996; Khanna *et al.,* in N.P. Singh *et al.,* Fl. Madhya Pradesh 3:157.2001.

Perennial herbs, erect, tufted, stout, branched; leaves linear-lanceolate, 7-15 × 2-5 cm, margins scabrid; sheaths *ca* 1.5 cm long, ciliate at mouth; spathes solitary or clustered together, 1.5 cm long; flowers white, 6-12 in unbranched cincinni; capsules 6-8 per spathe, 2-celled; seeds pale-yellow, about 3.5 mm long, ellipsoid with large pits.

F. -Fr. - Oct-. Dec.

Alirajpur only, in moist places, Chatliya pani, Kattiwada range.

CYANOTIS D. Don W. 52; MP 6; SA 6

Key to Species

1. Flowers sessile, enclosed in inflated leaf sheaths
   2. Capsules 3-horned at apex; seeds sub-quadrangular, pitted . ... **C. cucculata**
   2. Capsules beaked at apex; seeds cylindrical, not pitted . ... **C. axillaris**
1. Flowers pedicellate
   3. Annuals; leaves all cauline; capsules 0.2-0.4 mm long, floccosely woolly herbs .... **C. fasciculata**
   3. Perennials; leaves of two types: in rosettes on sterile shoots, cauline on fertile shoots; rosette leaves larger than cauline leaves; inflorescence terminal; capsules pilose or densely hairy
      4. Roots tuberous; capsules pilose; seeds faintly rugose .... **C. tuberosa**
      4. Roots not tuberous
         5. Capsules densely hairy at apex . ... **C. arcotensis**
         5. Capsules nearly glabrous … .... **C. cristata**

Cyanotis arcotensis Rolla S. Rao, Blumea 14 :345. 1967; Samvatsar, Fl. West. Trib. Madhya Pradesh 292; Khanna *et al.,* in N.P. Singh *et al.,* Fl. Madhya Pradesh 3:159.2001.

Annual herbs with pubescent branches; leaves linear-lanceolate, 2-5 × 0.2-0.5 cm, acute or obtuse at apex, cordate at base, glabrous above, pubescent beneath; sheaths short, pubescent; flowers purple, in terminal cymes, bracts ovate-lanceolate, falcately recurved; bracteoles ovate, cross-nerved, ciliate; filaments bearded with blue hairs at the apex; capsule 2-2.5 mm long; seeds 6, pitted.

Fl.-Fr.: Aug. -Dec.

Alirajpur only, in wet places as forest undergrowth, at Chatliya pani, Kattiwada range.

Cyanotis axillaris (L.) D. Don ex Sweet, Hort. Brit.,430.1826; *Commelina axillaris* L., Sp. Pl. 42. 1753; *Tradescantia axillaris* (L.) L.,Mant. Pl. 2:321. 1771; *Tonningia axillaris* (L.) Kuntze, Revis. Gen. Pl. 2:722. 1891; Hook.f., Fl. Brit. India 6: 388. 1892; *Amischophacelus axillaris* (L.) R.S. Rao & Kammathy, J. Linn. Soc. Bot. 59:306. 1966; Samvatsar, F. West. Trib. Madhya Pradesh 293. 1996; Khanna *et al.,* in N.P. Singh *et al.,* Fl. Madhya Pradesh 3:169.2001.

Diffuse, fleshy herbs, rooting at lower nodes; leaves linear-lanceolate or ensiform, 4-15 × 0.3-1.2 cm, acute to acuminate at apex, often falcate, sessile; sheaths scarious, 5-10 mm long, ciliate on margins; flowers pink or blue, in axillary fascicles; filaments purple bearded; capsules glabrous, oblong; seeds conical, pitted, truncate at base.

Fl.-Fr.: July-Dec.

Common in all districts, in moist places.

Cyanotis cristata (L.) D. Don, Prodr. Fl. Nepal. 46. 1825; *Commelina cristata* L. Sp. Pl . 42. 1753; Hook.f., Fl. Brit. India 6: 385. 1892; Samvatsar, Fl. West. Trib. Madhya Pradesh 292. 1996; Khanna *et al.,* in N.P. Singh *et al.,* Fl. Madhya Pradesh 3: 159.2001.

Perennial, trailing herbs; leaves ovate-oblong or oblong-oblanceolate, 5-9 × 1-2 cm, sessile, acute at apex, cordate or rounded at base; flowers in sessile or in long pedunculate cincinni; petals blue-violet; capsules oblong, trigonous, nearly glabrous; seeds black, conical, pitted.

*Fl.-Fr.:*Aug. – Sept.

Common in all districts in moist and shady places.

Cyanotis cuccullata (Roth) Kunth, Enum. Pl. 4:107. 1843; *Tradescantia cuccullata* Roth, Nov. Pl. Sp. 189. 1821; *Tonningia cuccullata* (Roth) Kuntze, Revis. Gen. Pl. 2:722. 1891; Hook.f., Fl. Brit. India 6:389. 1892; *Amischophacelus cuccullata* (Roth) R.S. Rao & Kammathy in J. Linn. Soc. Bot) 59:306. 1966; Samvatsar, Fl. West. Trib. Madhya Pradesh 293. 1996 (as *Tonningia cucculata*); Khanna *et al.,* in N.P. Singh *et al.,* Fl. Madhya Pradesh 3:170.2001.

Annual, erect or decumbent herbs; branches stout, rooting at lower nodes; Leaves 2.5-7 × 0.5-1 cm, distichous, linear-lanceolate, reflexed or spreading, sessile; sheaths inflated; flowers blue, generally 1-2 in the hollow axils of leaf sheaths; capsules ellipsoid-oblong, trigonous, 3-horned at the top with a central depression; seeds brown, pitted.

Fl.-Fr.: Aug. -Nov.

Alirajpur, Burhanpur, Dhar, Indore, Khandwa, frequently in fields and moist places.

Cyanotis fasciculata (B. Heyne ex Roth) Schult. &. Schult. f., Syst. Veg. 7: 1152. 1830; *Tradescantia fasciculata* B. Heyne ex Roth, Nov. Pl. Sp. 189. 1821; Hook.f., Fl. Brit. India 6: 387. 1892; Samvatsar, Fl. West. Trib. Madhya Pradesh 292. 1996; Khanna *et al.,* in N.P. Singh *et al.,* Fl. Madhya Pradesh 3: 159.2001.

Erect or decumbent, branched herbs; leaves linear-lanceolate or narrowed from base to the apex, 2-6 × 0.4-1.2 cm, acute or acuminate at apex; flowers in compact, axillary and terminal, 5-15 mm long cincinni; petals blue, purple or pink; capsules oblong, trigonous, *ca* 4 mm long; seeds 2, oblong, rugulose.

Fl.-Fr.: Aug. – Sept.

Common on shingled roofs of huts and amidst boulders.

Cyanotis tuberosa (Roxb.) Schult. &. Schult. f., Syst. Veg. 7;1153. 1830; *Tradescantia tuberosa* Roxb., Pl. Coromandel 2:5. 1799; Hook.f., Fl. Brit. India 6:386. 1892; Khanna *et al.,* in N.P. Singh *et al.,* Fl. Madhya Pradesh 3:160.2001.

Erect or creeping herbs with tuberous roots; stems densely pubescent, up to 1 m long, rooting at lower nodes; leaves radical and cauline, 15-25 × 2-4 cm, oblong-lanceolate, acute at apex, often purple beneath, sessile, pilose; sheaths up to 2.5 cm long; flowers in axillary and terminal, densely hairy scorpioid cymes, enclosed in foliaceous bracts; petals bluish-purple; capsules *ca* 3 mm long; seeds brown.

Fl.-Fr.: Throughout the year.

Rare, **Khandwa** only, Kalibhit range.

MURDANNIA Royle W. 53; MP 7; SA 2

Key to Species

1. Leaves 4-8 × 0.5-1.5 cm; capsule 6-seeded; seeds prominently tuberculate .... **M. nudiflora**
1. Leaves 1.5-2.5 × 0.5-1 cm; capsule 9-10 seeded; seeds faintly tubercled .... **M. spirata**

Murdannia nudiflora (L.) Brenan, Kew Bull. 7: 189. 1952; *Commelina nudiflora* L., Sp. Pl. 41. 1753; *Aneilema nudiflorum* (L.) Sweet, Hort. Brit. 430.1826; *Tradescantia malabarica* L., Sp. Pl. ed. 2 412. 1762; *Aneilema malabaricum* (L.) Merr., Philipp. J. Sci. 7:232. 1912; *Murdannia malabarica* (L.) G. Bruck., Nat. Pflanzenfam. ed. 2, 15a:173. 1930; Hook.f., Fl. Brit. India 6:378. 1892; Samvatsar, Fl. West. Trib. Madhya Pradesh 292. 1996; Khanna *et al.,* in N.P. Singh *et al.,* Fl. Madhya Pradesh 3: 164.2001.

Suberect or diffuse, ascending herbs;. leaves linear-lanceolate, 4-10 × 0.4-0.8 cm, sessile, subacute at apex, rounded or subcordate at base; flowers in terminal and axillary, lax, cymose panicles; sepals ovate; petals purple, blue or whitish; capsules subglobose, mucronate, purple-streaked, 6-seeded; seeds trigonous, tuberculate.

Fl.-Fr.: Aug. -Oct.

Common in wet soil, along roadsides in moist ground in all districts.

Murdannia spirata (L.) G. Bruck., Nat. Pflanzenfam. ed. 2, 15a:173. 1930; *Commelina spirata* L.,Mant. Pl. 2:176. 1771; *Aneilema spiratum* (L.) R. Br., Prodr. 271. 1810; Hook.f., Fl. Brit. India 6:377. 1892; Samvatsar, Fl. West. Trib. Madhya Pradesh 292. 1996; Khanna *et al.,* in N.P. Singh *et al.,* Fl. Madhya Pradesh 3:166.2001.

Erect or decumbent, ascending annual herbs; leaves 1.5-2.5-x 0.5-1 cm, sessile, ovate-lanceolate, apex acute, base rounded, margins wavy, scaberulous; flowers blue, in terminal and axillary, cymose panicles; capsules 3-angled; seeds trigonous, faintly tubercled, umbonate.

Fl.-Fr.: Aug. -Oct.

Alirajpur, Burhanpur, Dhar, Khandwa in wet places.

ARECACEAE 185/2522; 3/4; 3/4

Key to Genera

1. Leaves palmately lobed; flowers unisexual; drupe 12-15 cm across .... BORASSUS
1. Leaves pinnately compound, unipinnate, lower pinnae spinescent; berry 1-1.5 cm in diameter

2. Plants low, clumpy; petioles with long spinous teeth .... CHAMAEROPS
2. Plants tree-like, with tall trunks; petioles with short spinous teeth; lower leaf segments reduced to spines .... PHOENIX

BORASSUS L. W. 5; MP 1; SA 1

Borassus flabellifer L., Sp. Pl. 1187. 1753; Becc. & Hook.f., Hook.f., Fl. Brit. India 6:482. 1893; Samvatsar, Fl. West. Trib. Madhya Pradesh 293. 1996; Dixit in N.P. Singh *et al.,* Fl. Madhya Pradesh 3: 176.2001. *Taad*

Tall palms,trunk annulate, 20-30 m high; leaves flabelliform, 60-100 cm across; leaflets 60-70, linear-lanceolate, plicate; petioles spinous on the margin; spadices branched; branches of male spadix large, clustered; female spadix simple; drupes subglobose, with 1-3-fibrous pyrenes.

Fl.-Fr.: Jan. -April

Alirajpur, Barwani, Burhanpur, Dhar, Indore, Khandwa, Khargone. The species is actually rare, except Alirajpur, where it appears naturalized. It is planted, not indigenous.

CHAMAEROPS L. Monotypic

Chamaerops humilis L., Sp. Pl. 1187. 1753; *Phoenix humilis* (L.) Cav. 2:12. 1793; *P. humilis* Royle, Ill. Bot. Himal. Mts. 1:394. 1839; Samvatsar, Fl. West. Trib. Madhya Pradesh 294. 1996, as *Phoenix humilis* Royle; Dixit in N.P. Singh *et al., Fl.* Madhya Pradesh 3:180.2001. *Chhind* (*Phoenix loureiroi* Kunth is a separate species, not synonymous to above, as given in Fl. M.P. loc. cit) District Chhindwada is named after vernacular name of this species).

Dwarf palms, about 3 m tall; leaves 100-200 cm long; leaflets 25-40 × 0.8-1.2 cm, bases not decurrent, basal spines up to 8 cm long; fruiting peduncle up to 30 cm long, not concealed among the leaves; fruits orange, ovoid-oblong.

Fl.-Fr.: March-May

Rare, **Dhar** only, forming dense grove along river and stream banks, at Kukshi.

PHOENIX L. W. 14; M P 3; SA 2

Key to Species

1. Stem to 15 m tall; spikes more than 30 cm long ....**P. sylvestris**
1. Stem almost underground, only leaf crown visible above ground; spikes less than 30 cm long ....**P. acaulis**

Phoenix acaulis Roxb., Pl. Coromandel 3:69. 1820; Buch. -Ham. ex Roxb. Fl. Ind. 3:783. 1832; Beccari & Hook.f., Fl. Brit. India 6:426. 1892; Samvatsar, Fl. West. Trib. Madhya Pradesh 294. 1996; Dixit in N.P. Singh *et al.,* Fl. Madhya Pradesh 3:179.2001. *Chhinna*

A dwarf, almost stemless palm, difficult to distinguish from the juvenile stage of *P. sylvestris*; leaves 60-150 cm long; leaflets 25-50 cm, bases thickened and decurrent; basal spines 5-15 cm long; fruiting peduncle short, concealed among the leafbases; fruit red, black at maturity.

Fl.-Fr.: March-May

Rare, Dhar, Khandwa, along river and stream banks, the tribal people eat the apical, soft stem, it is sweet, and therefore existence endangered.

Phoenix sylvestris (L.) Roxb., Fl. Ind. 3: 787. 1832; *Elate sylvestris* L., Sp. Pl. 1189. 1753; Becc. & Hook.f. in Hook.f., Fl. Brit. India 6: 425. 1892; Dixit in N.P. Singh *et al.,* Fl. MP 3: 180.2001. *Sindole*

Tall, unbranched, (not always), dioecious palms; leaves 2-4 m long, lower leaves arching; leaflets in several fascicles, grayish- green, rigid, ensiform, canaliculated, spinous-pointed, 15-45 × 2-3 cm; petioles compressed towards the apex and bearing a few short spines at the base; male flowers white, fragrant; female inflorescence drooping; drupes orange-yellow, scattered, ellipsoid-oblong; seeds rounded at the both ends.

Fl.-Fr.: Dec. – June

Very common along stream and river banks, in all districts. Due to extensive tapping for toddy, the stem has a zigzag appearance.

TYPHACEAE 2/65; 1/1; 1/1

TYPHA L. W. 38; M P 1; SA 1

Typha angustifolia L., Sp. Pl. 971. 1753; *T. angustata* Bory & Chaub., Nour. Fl. Pelop. 4. 1838; Hook.f., Fl. Brit. India 6: 489. 1893; Samvatsar, Fl. West. Trib. Madhya Pradesh 294. 1996; Misra *et al.,* in N.P. Singh *et al.,* Fl. MP 3: 183.2001. (*T. domingensis* Pers. 1807 is a separate species). *Ghao bajra*

Perennial, aquatic or marshy herbs, 1-3 m tall; leaves mostly basal, linear, subterete above the sheath, flattened towards apex, 30-60 × 1-2 cm; inflorescence an interrupted spike; spikes reddish- brown, *ca* 35 cm long; male and female spikes separated by a considerable interval; male flowers paler and more slender, mixed with clavate- tipped pistillodes; female spikes pale brown; fruits brown, fusiform or elongate-ovoid, smooth; nutlets brown, minute, ellipsoid.

Fl.-Fr.: Aug. -Dec.

Common in pure formations along river, stream, pond margins, throughout the area.

ACORACEAE 1/2; 1/1; 1/1

ACORUS L. W. 2 ;MP 1; SA 1

Acorus calamus L., Sp. Pl. 324. 1753; Hook.f., Fl. Brit. India 6:555. 1893; Vaish in N.P. Singh *et al.,* Fl. Madhya Pradesh 3:185. 2001. *Bach*

Perennial herbs; rhizomes aromatic, up to 2 cm thick; stem erect, glabrous; leaves linear or ensiform, 60-80 × 1-2 cm; peduncles 25-30 × 1cm; spadix at an angle to the leaf; flowers compactly borne on the peduncle; tepals oblong-obovate; seeds obconical, *ca* 2 mm long.

*Fl.-Fr.:*July-Sept.

Rare, marshy places at **Khandwa** only, Sunder Dev.

ARACEAE 117/3368; 13/25; 11/19

Key to Genera

1. Free floating, thalloid or large aquatic plants; flowers minute or small
   2. Roots absent; fronds up to 1.5 mm, borne on the upper surface; anthers 1-celled .... WOLFFIA
   2. Root/s present
      3. Flowers borne in marginal clefts of fronds; fronds much larger than 1.5 mm; anthers 2-celled
         4. Root 1 ....LEMNA
         4. Roots many .... SPIRODELA
      3. Flowers not borne in tubular spathes; spadix very small; male and female flowers separated by neutral flowers; leaves 3.5-5 cm long (or much longer), densely pubescent on both surfaces.. PISTIA
1. Terrestrial plants
   5. Leaves and spadices present simultaneously
      6. Leaves compound, pedatisect or 3-5 lobed
         7. Monoecious; male and female flowering zones separated by neuters . ... TYPHONIUM
         7. Dioecious or when monoecious, male and female flowering zones contiguous . ... ARISAEMA
      6. Leaves simple
         8. Leaves linear, lanceolate or elliptic
            9. Spadices inconspicuous; lower margins of spathe united, enclosing the spadix ....CRYPTOCORYNE
            9. Spadices conspicuous; lower margins of spathe overlapping, spadix exserted ....THERIOPHONUM
         8. Leaves peltate or broadly ovate .... COLOCASIA
   5. Leaves and spadices appearing separately; spadices appendaged
      10. Lower margins of spathe united into a tube .... SAUROMATUM
      10. Lower margins of spathe not united, convolute .... AMORPHOPHALLUS

AMORPHOPHALLUS Blume ex Decne. W. 198; MP 3; SA 6

Key to Species

1. Male and female flowers contiguous, not separated by neutrals
   2. Spathe greenish-pink, campanulate; appendage irregularly lobed, dark purple . ... **A. paeonifolius**
   2. Spathe brownish-purple, ovate-lanceolate; appendage cylindrical, cream coloured ..... **A. commutatus**
1. Male and female flowers separated by neutrals
   3. Male flowers with long, projected connectives; appendix of spadix stalked, very short, 2 cm long with irregularly formed sterile structures .... **A. longiconnectivus**
   3. Male flowers connectives not as above
      4. Stigma papillose; spathe slightly longer than spadix; ovary 2-locular .... **A. margaritifer**
      4. Stigma not papillose; spathe shorter than spadix; ovary 2-3 locular

5. Ovary 3-locular; style 0.3 cm long; seeds 3; neutral flowers rhomboid shaped .... **A. konkanensis**
5. Ovary 2-locular; style 1-1.5 cm long; seeds 2; neutral flowers diamond shaped .... **A. sylvaticus**

Amorphophallus commutatus (Schott) Engl., Monogr. Phan. 2:319. 1879; *Conophallus commutatus* Schott, Bonplandia (Hannover) 7:28. 1889; Hook.f., Fl. Brit. India 6:515. 1893; Samvatsar, Fl. West. Trib. Madhya Pradesh 295. 1996. *Jangli suran*

Annual, about 75-100 cm tall herbs; corm depressed, spherical; leaf solitary, 3-partite, 60 cm across; peduncles 30-90 cm long, solid, variously spotted; spathe 5-25 × 6-10 cm, brownish-purple, rough inside, obovate, lanceolate, acute; spadix sessile, columnar, appendages longer than the spathe, cream-coloured, obtuse at apex, male flowers above and female below, berries 0.8-1 cm, deep orange.

Fl.-Fr.: May-June

Common forest undergrowth at Alirajpur, Dhar, Indore.

Amorphophallus konkanensis Hett.,S.R. Yadav & K. S. Patil, Blumea 39:289. 1994; Jaleel V. B *et al.,* Revision of **Amorphophallus** Blume ex Decne. Sect. *Rhapiophallus* (Schott) Engl, (Araceae) in India, Bangladesh J. Pl. Taxon. 18 (1):1-26.2011.

Corm depressed globose, 3-4 cm in across; petiole 20-80 cm long, with pale yellowish-green to white stripes and pinkish dots; lamina 35-80 cm diam. rachis winged; leaflets 7-20 cm long, linear-lanceolate, apex acuminate; peduncle 25-60 cm long; spathe 15-20 cm long, veins purple-brown outside, dark maroon inside; spadix much longer than spathe; female and male zones separated by broad zone of neuters; ovary normally 3-locular; style very short; seeds usually 3.

Fl.-Fr.: April-June

Rare, **Khandwa** only, along forest paths, Kalibhit range.

Amorphophallus longiconnectivus Bogner, Kew Bull. 50(2):397 1995; Sivad. & V.B. Jaleel, Rheedea 8(2): 243. 1998.

Corms 4.5-7.00 cm in diam in the vegetative phase; petiole 35-77 cm long, smooth, with dark-brownish stripes; sessile, linear-lanceolate, up to 14 cm long, apex acuminate, base decurrent on the rhachis; peduncle smooth, 60-110 cm long; sheath 10-14 cm long, completely convolute, not differentiated into basal tube and upper limb; spadix as long as or slightly shorter than the spathe; female zone 1.3-2.5 long; staminodial zone 8-9 mm long; male zone 3-4 cm long; upper sterile zone 1.3-3 cm long, usually without a stipitate appendix; ovary 2-3 locular; style very short, stigma 2-3 lobed; staminodial zone with echinate, sterile flowers; male flowers 2-3 mm long, connective elongated up to 1.0-1.5 mm; seeds 2-3.

*Fl.-Fr.:*June –Sept.

In moist, shady places, **Khandwa,** Sunder Dev, Golai forest.

Amorphophallus margaritifer (Roxb.) Kunth, Enum. Pl. 2:34 1837; *Arum margaritiferum* Roxb., Fl. Ind. 1832 3:512. 1832; 3:512 1832; *Plesmonium margaritiferum* (Roxb.) Schott, Syn. Aroid. 34. 1856; Hook.f., Fl. Brit. India 6: 518. 1893; Hett. & de Sarkar, Aroideana 19: 131. 1996.

Corms up to 15 cm diam; petiole mottled; lamina 70 cm in diam; leaflets linear to lanceolate, 3-4 cm broad, apex acute; peduncles 18-90 cm long; spathe erect, broader than long, 6-15 × 6-20 cm, pale green to yellow on both surfaces; spadix equal to or longer than the spathe, 8-18 cm long; sterile zone between male and female zones 2-4 cm long, staminodes very large; appendix absent; style very short; stigma capitate, 2-lobed; berries red, *ca* 0.8 cm long.

Fl.-Fr.: May-Aug.

Rare,**Khandwa**, Kalibhit forest.

Amorphophallus paeoniifolius (Dennst.) Nicolson, Taxon 26: 337. 1997; *Dracontium paeonii-folium* Dennst., Schlüssel Hort. Malab. 38. 1818;. *Amorphophallus campanulatus* Decne., Nouv. Ann. Mus. Hist. Nat. 3: 366. 1834; Hook.f., Fl. Brit. India 6: 513. 1893; Samvatsar, Fl. West. Trib. Madhya Pradesh 295. 1996; Vaish in N.P. Singh *et al.,* Fl. Madhya Pradesh 3: 187.2001.

Herbs, up to 1 m tall; corms dark brown, depressed globose; leaves large, radical, twice 2-fid; leaflets obovate or oblong, acute, many veined; petioles dark green, warty; spathes light red to pink, campanulate; spadix as long as spathe, stout; appendage globose, amorphous; male inflorescence turbinate; anthers pale yellow; female inflorescence sessile on the ground.

Fl.-Fr.: July-Oct.

In moist shady places near cultivated fields, in all districts.

Amorphophallus sylvaticus (Roxb.) Kunth, Enum. Pl. 3:34. 1841; *Arum sylvaticum* Roxb., Fl. Ind. ed. 1832 3:511.1832; *Synantherias sylvatica* (Roxb.) Schott, Syn. Aroid. 35.1856; Hook.f., Fl. Brit. India 6:518.1893; Samvatsar, Fl. West. Trib. Madhya Pradesh 295.1996; Vaish in N.P. Singh *et al.,* Fl. Madhya Pradesh 3:188.2001.

Corms up to 8 cm across; leaf solitary, 30-35 cm across, leaflets lanceolate, 10-20 cm long; spathes 12-15 cm long, convolute, except at the top, green outside, with a purple blotch inside; spadix stalked; male zone brown, *ca* 5 cm long, female zone up to 3 cm long; stigma yellow; neutral zone up to 2 cm long, appendage brown, cylindrical, up to 12 cm long.

Fl.-Fr.: Sept. -Nov.

Rare, in moist and shady places, Alirajpur, Burhanpur, Khandwa.

ARISAEMA Mart. W. 180; MP 1; SA 1

Arisaema tortuosum (Wall.) Schott, Melet. Bot. 17. 1832; *Arum tortuosum* Wall., Pl. Asiat. Rar. 2:10.1830; Hook.f., Fl. Brit. India 6:502. 1893; Vaish in N.P. Singh *et al.,* Fl. Madhya Pradesh 3:189.2001. *Bagh jandra*

Corms up to 5 cm across; leaflets 7-19, pedatisect, *ca* 12 × 4 cm, petioles with purple spots, up to 40 cm long, base sheathing up to ¾ th length, peduncles longer than the petioles, 40-45 cm long; spathes up to 12 cm long, basal part convolute, tubular, upper part expanded and horizontally bent, ovate, acuminate; spadix usually monoecious, appendage sigmoid at the base, tapering to a 25 cm long tail; fertile area 2.55-6.5 cm long, lowest 1-2 cm part pistillate and upper 2-5 cm part staminate; ovaries ovoid; berries 4-5 seeded.

Fl.-Fr.: June-Nov.

Very rare, **Indore** only, deep within dense forest, at Kalakund.

COLOCASIA Schott W. 8; MP 1; SA 1

Colocasia esculenta (L.) Schott in Schott & Endl., Melet. Bot. 18. 1832; *Arum esculentum* L., Sp. Pl. 965. 1753. Samvatsar, Fl. West. Trib. Madhya Pradesh 296. 1996; Vaish in N.P. Singh *et al.,* Fl. Madhya Pradesh 3: 190.2001. *Arvi*

Tall herbs with tuberous stoloniferous rhizomes; leaves radical, peltate; lamina broadly ovate, sagittate usually with tip pointing downwards and basal lobes pointing upwards, 10-30 cm long; petioles long, vaginate at base; spadix 1 or more, axillary, pedunculate, up to 10 cm long, differentiated into 4 distinct portions: basal female followed by neuteriferous, male and apical naked sterile appendage; berries oblong; seeds many.

Fl.-Fr. : Aug.-Sept.

Commonly cultivated, occurs as an escape.

CRYPTOCORYNE Fisch. ex Wydler W. 63; MP 1; SA 1

Cryptocoryne spiralis (Retz.) Fisch. ex Wydler, Linnaea 5:428. 1830; *Arum spirale* Retz. Observ. Bot. 1:30.1779; Hook.f., Fl. Brit. India 6:494. 1893; Sharma *et al.,* in F. Maharashtra 3:217. 1996; Samvatsar, Fl. West. Trib. Madhya Pradesh 297. 1996.

Rhizomatous, aquatic or marshy herbs; leaves linear-lanceolate, 6-15 cm long, sessile or subsessile; limb of spathe expanded, spirally twisted, purplish outside, dark purple and with sub-horizontal lamellations inside, margins sometimes denticulate; spathes 7.5-12.5 cm long; spadix up to 2 cm long; berries subglobose; seeds *ca* 0.5 cm long, spindle-shaped, irregularly verrucose-rugose.

Fl.-Fr.: Oct. -Dec.

Common in rocky river beds, all districts.

LEMNA L. W. 14; MP 3; SA 3

Key to Species

1. Root sheath winged at base; roots up to 3 cm long; fronds with 1 very distinct papilla apex on the upper surface; seeds with distinct, 8-26 ribs; wing of root sheath 1-. 2.5 times as long as broad .... **L. aequinoctialis**
1. Root sheath not winged at base; roots often longer than 3 cm; papilla present or absent
    2. Ovary (1) 2-7 ovulate .... **L. gibba**
    2. Ovary 1-ovulate .... **L. minor**

Lemna aequinoctialis Welw., Apont. 578. 1859; *L. paucicostata* Hegelm.,Lemnac. 138. 1868; Hook.f., Fl. Brit. India 6: 556. 1893; Samvatsar, Fl. West. Trib. Madhya Pradesh 297. 1996 (as *L. paucicostata* Hegelm.) Anand Kumar in N.P. Singh *et al.,* Fl. Madhya Pradesh 3: 198. 2001.

Annual, free-floating herbs; root solitary, root-sheath appendaged; fronds ovate-obovate, 1.5-3 × *ca* 0.5 mm, light green, asymmetrical, more or less flat on both surfaces; flowers unisexual, monoecious, in marginal cavity; spathe open; tepals absent; staminate flowers 2; pistillate flower solitary; fruits ellipsoid, asymmetric, slightly laterally compressed; seeds longitudinally ribbed.

Fl.-Fr.: Sept. -May

Common in temporary ponds and roadside ditches.

Lemna gibba L., Sp. Pl. 970.1753; Hook.f., Fl. Brit. India 6:556. 1893; Anand Kumar in Singh *et al.,* Fl. Madhya Pradesh 3:197.2001.

Free floating herbs; fronds solitary or in groups of 2-4, ovate, obovate or orbicular, up to 5 × 3.5 mm, obtuse or rounded at apex, entire, lower side flat or inflated, nerves 3-5, sometimes not distinct; floral pocket ventral to the margin of frond; ovary 1-4 ovuled; fruits laterally winged; seeds 1-4, longitudinally ribbed.

*Fl.-Fr.:*April-Oct.

In stagnant water, throughout the area.

Lemna minor L., Sp. Pl. 970.1753; Hook.f., Fl. Brit India 6:556. 1893; Anand Kumar in N.P. Singh *et al.,* Fl. Madhya Pradesh 3:198. 2001.

Small, floating herbs; fronds solitary or in groups of 2-many, round-elliptic or elliptic-ovate, up to 4.5 × 3 mm, 3-nerved; floral pocket coinciding with the margin of frond; ovule 1 in the ovary; seed smooth.

Fl.-Fr.: Rainy season

Common in ponds and ditches.

PISTIA L. Monotypic

Pistia stratiotes L., Sp. Pl. 963. 1753; Hook.f., Fl. Brit. India 6:497. 1893; Samvatsar, Fl. West. Trib. Madhya Pradesh 296. 1996; Vaish in N.P. Singh *et al.,* Fl. Madhya Pradesh 3:192.2001. *Water cabbage*

Aquatic, free-floating plants with bunch of adventitious roots; leaves 3.5-5 × 2-3.5 cm, wedge-shaped or obovate, cuneate, forming a cup, densely pubescent on both surfaces; spathe 7-7.5 mm long, tubular, boat-shaped or gibbous; spadix very small, enclosed in spathe, adnate to the back of tube of spathe; male flowers few, middle flowers neuter, forming a ring; female flower solitary; fruit ovoid, opening irregularly; seeds rugose, ovoid.

Fl.-Fr.: Oct. -Dec.

Found in still waters, Alirajpur, Dhar, Indore (in a pond at Kukshi (Dhar), the plants are very large, having a diameter of 30-50 cm).

SAUROMATUM Schott W. 9; MP 0; SA 1

Sauromatum venosum (Dryand. ex Aiton) Kunth, Enum. Pl. 3:28. 1841; *Arum venosum* Dryand. ex Aiton Hort. Kew. 3:315. 1789; *Arum guttatum* Wall., Pl. Asiat. Rar. 2:10.1830;

Sauromatum guttatum (Wall.) Schott, Endl. Melet. Bot. 17. 1832; Samvatsar, Fl. West. Trib. Madhya Pradesh 296. 1996; Vaish in N.P. Singh *et al.,* Fl. Madhya Pradesh 3:194.2001.

Perennial herbs with a depressed, globose tuber; leaf solitary, appearing after flowers, on a,60-8- cm long, purple-blotched petiole, pedately partite, segments 3-15, elliptic-obovate, or oblong, acuminate, midrib and veins impressed above, prominent beneath; peicel 10-15 cm long; spathe lanceolate-oblong, 20-50 cm long, purple on the outside, dark purple and blotched on the inner side; spadix shorter than the spathe; male zone about 2 cm above the female zone; berries ovoid, black.

Fl.-Fr.: April-June

Common on hill slopes and forest edges, in moist, shady places, all over the area.

SPIRODELA Schleid. W. 4; MP1; SA 1

Spirodela polyrrhiza (L.) Schleid., Linnaea 13: 392. 1839; *Lemna polyrrhiza* L., Sp. Pl. 970.1753; Hook.f., Fl. Brit. India 6: 557. 1893. Samvatsar, Fl. West. Trib. Madhya Pradesh 297. 1996; Anand Kumar in N.P. Singh *et al.,* Fl. Madhya Pradesh 3: 199.2001. *Bhos*

Free- floating herbs; fronds green above, purple beneath, orbicular-ovate, 1-3 × 1-2.5 mm; daughter fronds budding from near the point of root insertion in a slit in the parent thallus; flowers surrounded by a small, open spathe in a lateral slit-pouch; staminate flowers 2-3; pistillate flowers with 2 ovules; fruits slightly winged on margin.

*Fl.-Fr.:*Sept. -Oct.

Common in all districts in still water of ponds and ditches.

THERIOPHONUM Blume W. 7; MP 2; SA 1

Theriophonum dalzellii Schott, Aroid. 15. 1853; Hook.f. Fl. Brit. India 6:513. 1893; *T. uniseriatum* Blatt. & McCann, J. Bombay Nat. Hist. Soc. 35:25. 1932; Sivadasan & Nicols. in Kew Bull. 37:278. 1982; Hanfi in N.P. Singh *et al.,* Fl. Madhya Pradesh 3:195. 2001.

Corms globose, *ca* 2 cm across; leaves hastate-sagittate, 3-lobed, lobes ovate-lanceolate, apex acuminate; petioles 5-10 cm long; peduncles 7-8 cm long; spathes up to 15 cm long, tube short, cylindrical or oblong; spadix 10-12 cm long, shorter than spathe; appendage ceamy to purplish, 7-9 cm long; stalk short; female and male flowers separated by naked rachis; berry 4-5 seeded.

*Fl.-Fr.:*July-Aug.

Rare, **Khandwa**, in moist and shady places, in Kalibhit forest.

TYPHONIUM Schott W. 68; MP 1; SA 2

Key to Species

1. Corms 2-3 cm across; basal lobes of leaves short, subacute; petiole 10 -25 cm long; neuters above the female zone purple- headed, those above the male zone white; berry 1-seeded .... **T. flagelliforme**

1. Corms 0.8-1.5 cm across; basal lobes of leaves orbicular, acuminate; petioles 5-30 cm long; neuters in both, female and male zones yellow; berry 1-2 seeded .... **T. inopinatum**

Typhonium flagelliforme (Lodd.) Blume, Rumphia 1:134. 1837; *Arum flagelliforme* Lodd. Bot. Cab. 4:t. 396. 1819; Blatt., J. Bombay Nat. Hist. Soc. 35:22. 1931; Hook.f. Fl. Brit. India 6:511. 1893; *T. divaricatum* Blume 1834, *nom. illeg;* Sharma *et al.,* Fl. Maharashtra 3:223-225. 1996.

Corms 2-3 cm across; lower rudiments of sterile flowers clavate, upper ones subacute or only verruciform; leaves 5.0-11 cm long, breadth variable, basal lobes short, subacute; petioles 10-25 cm long; spadix shortly stipitate; spathe10-17 cm long; neuters above the female flowers purple-headed; those above the male flowers white; seed solitary, rugulose.

*Fl.-Fr.:*May-Oct.

Dhar, Khandwa , Khargone in moist and shady places.

Typhonium inopinatum Prain, J. Asiat. Soc. Bengal, Pt. 2,Nat. Hist. 67:301. 1898 and Prain, Bengal Plants 1107. 1903; Anand Kumar *et al.,* Rheedea 24(2):120-123.2014.

Corms 0.8-1.5 cm across; leaves 5-11 cm long; spathe *ca* 5-11 cm long, margin entire, apex acuminate; spadix up to 9 cm long; neuters yellow, both in female and male zones; berry 1-2 seeded.

*Fl.-Fr.:*June-Oct.

Rare, **Burhanpur** , Chandani, Nepanagar forest.

WOLFFIA Horkel ex Wimm. W. 11; MP2; SA 1

Wolffia arrhiza (L.) Horkel ex Wimm., Fl. Schles. ed. 3. 140.1857; *Lemna arrihza* L., Mant. Pl. 2: 294. 1771; Hook.f., Fl. Brit. India 6:557. 1893; Samvatsar, Fl. West. Trib. Madhya Pradesh 298. 1996; Anand Kumar in N.P. Singh *et al.,* Fl. Madhya Pradesh 3:199.2001.

Very minute, floating herbs; fronds almost orbicular, up to 1 × 1 mm, dorsal side flat or slightly convex, ventral side inflated; floral pocket median, infundibuliform, with cicular opening; inflorescence dorsal, composed of 1 staminate and 1 pistillate flower; anther monothecous; ovary with solitary ovule; fruit globose; seeds smooth.

Fl.-Fr.: After rainy season

Rare in old, unused wells, stagnant water in a corner, Alirajpur, Dhar, Indore, Khandwa.

ALISMATACEAE 17/120; 4/6; 3/4

Key to Genera

1. Stamens more than 6; carpels many on a globose or oblong receptacle; achenes compressed .... SAGITTARIA
1. Stamens 6
    2. Leaves 10-15 × 7-9 cm; flowers pale blue to almost white, many in a whorl inflorescence 15-24 cm long; filaments of anthers not persistent; achenes obovoid compressed .... LIMNOPHYTON
    2. Leaves 6-15 × 4-9 cm, deeply cordate; flowers white, 3 in a whorl; inflorescence longer than 60 cm; filaments of anthers persistent; achenes reniform, dorsally 4-ridged .... CALDESIA

CALDESIA Parl. W. 5; MP 1; SA 1

Caldesia oligococca (F. Muell.) Buchanan, Bot. Jahrb. Syst. 2:479. 1882; *Alisma oligococcum* F. Muell., Fragm. 1:23. 1858; Hook.f., Fl. Brit India 6:560.1893; Srivastava *et al.,* in N.P. Singh *et al.,* Fl. Madhya Pradesh 3:201.2001.

Scapigerous herbs, at times submerged; leaves ovate, 8-12 × 5-8 cm, very flaccid, base cordate withacute lobes and narrow sinus, nerves 11-17, nervules 2-3 mm apart, perpendicular to the principal nerves; petioles 40-60 cm long; panicles 25-35 cm long; peduncles as long as the petioles; lower bract 6-8 cm long; pedicels 1-1.5 cm long; petals 3; stamens 6; carpels 3-7; style ventral; achenes reniform, 2-3 mm long, spinulous warty.

Fl.-Fr.: Dec. -May

Along banks of ponds, muddy places, in shallow water, **Khandwa** only, Mortakka, along Narmada river.

LIMNOPHYTON Miq. W. 3; MP 1; SA 1

Limnophyton obtusifolium (L.) Miq., Fl. Ned. Ind. 3:243. 1856; *Sagittaria obtusifolia* L., Sp. Pl. 993. 1753; Hook.f., Fl. Brit. India 6:560.1893; Srivastava *et al.,* in N.P. Singh etal Fl. Madhya Pradesh 3:202.2001.

Aquatic, scapigerous herbs; root-stock stout with numerous long root fibres; leaves radical, broadly sagittate, 10-15 × 7-9 cm, floating, pellucid-punctate, with ovate-triangular acute, basal lobes, apical lobe 7-12 cm long; panicles pyramidal; peduncles ribbed; flowers pale-bluish whie, whorled on 15-25 cm long scape; achenes turgid, obovoid, 5 mm long, wrinkled.

Fl.-Fr.: Oct. -Jan.

In shallow water, at edges of ponds, Dhar, Indore, Jhabua.

SAGITTARIA L. W. 40; MP 2; SA 2

Key to Species and subsp.

1. Leaves broadly ovate, rounded or obtuse at apex; lower whorls of flowers bisexual in racemes .... **S. guayanensis** subsp. **lappula**
1. Leaves hastate or sagittate, acute or acuminate at apex; lower whorls of flowers female in racemes .... **S. trifolia**

Sagittaria guyanensis Kunth subsp. **lappula** (D. Don) Bogin, Mem. New YorkBot. Gard. 9:192. f. 5. 1955; *S. lappula* D. Don, Prodr. Fl. Nepal 22. 1825; Hook.f., Fl. Brit. India 6:561. 1893; Samvatsar, Fl. West. Trib. Madhya Pradesh 298. 1996; Srivastava *et al.,* in N.P. Singh *et al.,* Fl. Madhya Pradesh 3:204.2001. **(Plate 9, Fig. 8)**

Annual aquatic herbs; leaves broadly ovate, 3-10 × 2-8 cm, floating, obtuse or rounded at apex, deeply cordate at base; racemes up to 8 cm long; peduncles about 20 cm long, flaccid, floating, ribbed; flowers greenish-white; lower whorls bisexual, upper ones male; sepals 10 mm long; petals 12-18 mm long, obovate-suborbicular; achenes ellipsoid, 3-4 mm long, surrounded by broad, spiny wing; seeds brown.

Fl.-Fr.: April-Oct.

In marshy places, not common, Alirajpur, Dhar.

Sagittaria trifolia L. Sp. Pl. 993. 1753; *S. sagittifolia auct. non* L., 1753; *S. sagittifolia* L. var. *leucopetala* Miq., Ill. Fl. Archip. Ind. 2:49. 1870; Hook.f., Fl. Brit. India 6:561. 1893; *S. sagittifolia* subsp. *leucopetala* (Miq.) Hartog, Fl. Males. 5:332. 1957; Srivastava *et al.,* in N.P. Singh *et al.,* Fl. Madhya Pradesh 3:206.2001.

Erect, aquatic herbs with thick, stoloniferous rhizome; mature leaves hastate or sagittate, 5-20 cm long, acute at apex; young and floating leaves often strap-shaped or cordate-oblong, petioles trigonous, to 60 cm long; flowers white, lower female, nearly sessile; upper male with longer pedicels; stamens *ca* 24; bisexual flowers rare; claw usually purple; achenes obovate, winged.

Fl.-Fr.: Sept. -Feb.

Barwani, Burhanpur, Khandwa, along river banks.

APONOGETONACEAE 2/58; 1/2; 1/2

APONOGETON L.f. W. 57; MP 2; SA 1

Aponogeton natans (L.) Engl. & Krausse, Pflanzenr. IV, 13:11. 1906; *Saururus natans* L., Mant. Pl. 227. 1771; *Aponogeton monostachyon* L.f., Suppl. Pl. 214. 1782; Hook.f., Fl. Brit. India 6:564. 1893; Samvatsar, Fl. West. Trib. Madhya Pradesh 299. 1996; Srivastava *et al.,* in N. PSingh *et al.,* Fl. Madhya Pradesh 3:213.2001.

Aquatic, floating, scapigerous herbs; leaves 7.5-9.5 × 2-2.5 cm, floating, oblong-lanceolate, base cordate; petioles longer than the leaf blade; flowers bright pink or purple blue, in densely flowered, solitary, 7-9 cm long spike; tepals 2, obovate or suborbicular; stamens 6; anthers bluish-purple; follicles smooth, exceeding the tepals; seeds 4-8, oblong, ribbed.

Fl.-Fr.: Aug. -Dec.

In still water of ponds, Alirajpur, Burhanpur, Dhar, Indore, Khandwa.

POTAMOGETONACEAE 6/186; 2/5; 3/5

Key to Genera

1. Leaf sheaths not auriculate, deciduous, not leaving a scar on the node; male flowers solitary, pedicellate, without perianth, stamen 1(or 2); female flowers solitary, sessile, with a cupular perianth of 3 scales; fruit keeled on the lower side; keel winged; plants totally submerged ... ZANNICHELLIA

1. Leaf sheaths auriculate, not deciduous

   2. Stipular sheaths of submersed leaves free from the leaf blade or when united, then less than ½ the length of stipule; both submersed and floating leaves flat, not channeled; peduncle compact, projecting above the surface of water . ... POTAMOGETON

   2. Stipular sheaths of submersed leaves united with the leaf blade to more than 2/3 length of the stipule; all leaves submersed, channeled; peduncle not projecting above the surface of water ... STUCKENIA

POTAMOGETON L. W. 160; MP 3; SA 3

Key to Species

1. Plants with floating, petiolate leaves, up to 12 cm long; stipules 2-3 cm long ....**P. nodosus**
1. Plants with submerged, sessile leaves
   2. Leaves perfoliate, 3-4 cm long, 5-9 nervedmargin entire, spikes many flowered ....**P. perfoliatus**
   2. Leaves semi-amplexicaul, 3-6 cm long, 3-nerved, margin crisped and serrulate; spikes few flowered ....**P. crispus**

Potamogeton crispus L., Sp. Pl. 126. 1753; Hook.f., Fl. Brit. India 6: 566. 1893; Samvatsar, Fl. West. Trib. Madhya Pradesh 299. 1996; Srivastava. *et al.,* in N.P. Singh *et al.,* Fl. Madhya Pradesh 3: 214.2001.

Floating or submerged herbs; leaves narrow, crisped, finely serrulate, distichous, sessile, 3-nerved, 2-6 × 0.2-0.6 cm; spikes generally aerial, small; flowers green, *ca* 2 mm long; tepals suborbicular; drupelets oblique-obovoid, turgid, shortly beaked, rounded and obscurely 3-keeled on the back, *ca* 3 mm long.

Fl.-Fr.: Dec. -May

Common in ponds, tanks, streams and rivers.

Potamogeton nodosus Poir., Encycl. Suppl. 4: 535. 1816; *P. indicus* Roxb., 1820, *nom. illeg;* Hook.f., Fl. Brit. India 6:566. 1893; Samvatsar, Fl. West. Trib. Madhya Pradesh 299. 1996; Srivastava *et al.,* in N.P. Singh *et al.,* Fl. Madhya Pradesh 3: 216.2001. **(Plate 10, Fig. 1)**

Floating or marshy, brownish- purple herbs; leaves petiolate; lower leaves submerged, very thin with short petioles; upper ones floating, alternate or uppermost opposite, many-nerved, 3-6 × 2-3 cm; petioles 2-5 cm long; stipules scarious; spikes generally above the surface of water, erect, reddish- brown; peduncles axillary or leaf opposed; fruits 3-ribbed on the back.

Fl.-Fr.: Dec. -March

In ponds, lakes, canals and river banks, all over the area.

Potamogeton perfoliatus L., Sp. Pl. 126. 1753; Hook.f., Fl. Brit. India 6: 566. 1893; Samvatsar, Fl. West. Trib. Madhya Pradesh 300.1996; Srivastava *et al.,* in N.P. Singh *et al.,* Fl. Madhya Pradesh 3: 217.2001.

Aquatic, rhizomatous herbs, 20-40 cm high; leaves submerged, ovate-lanceolate, 2.5-5 × 0.8-1.5 cm, acute at apex, cordate-amplexicaul at base; stipules 3-6 mm long; peduncles axillary, 1-3 cm long; spikes 2-2.5 cm long, densely flowered; flowers bisexual, ebracteate; tepals green, 4, elliptic, 1.5-2.5 × 1-2 mm, clawed; drupelets obliquely ovoid, 2-3 × 1-1.5 mm, compressed, with a short curved beak at apex.

Fl. - Fr-. Dec. – June

Common in shallow waters of ponds and rivers in all districts.

STUCKENIA Borner W. 10; MP 1; SA 1

Stuckenia pectinata (L.) Borner, Fl. Deut. Volk. 713. 1912; *Potamogeton pectinatus* L., Sp. Pl. 127. 1753; Hook.f., Fl. Brit. India 6: 567. 1893; Samvatsar, Fl. West. Trib. Madhya Pradesh 299. 1996; Srivastava *et al.,* in N.P. Singh *et al.,* Fl. Madhya Pradesh 3: 217.2001.

Aquatic, grass like, much branched herbs; leaves alternate, filiform, 2-8 × 0.1-0.15 cm, flaccid, opaque with slightly thickened or inrolled margins; spikes 5-15 mm long; flowers interruptedly whorled on a spike; drupelets obovoid, oblique, hard keeled, ridged, *ca* 4 mm long.

Fl.-Fr.: Oct. -March

In ponds and water-logged places, in all districts.

ZANICHELLIA L. W. 7; MP 1; SA 1

Zanichellia palustris L., Sp. Pl. 969. 1753; Hook.f., Fl. Brit. India 6: 568. 1893; Misra *et al.,* in N.P. Singh *et al.,* Fl. Madhya Pradesh 3:218. 2001.

Perennial, rhizomatous, submerged herbs, 15-20 cm long; leaves opposite or in whorls, narrowly linear to filiform, 2-6 cm long; flowers in hyaline, axillary, cupular, deciduous spathe, enclosing 1 male and 2-5 female flowers; perianth absent in male flowers, and hyaline in female flowers; carpels 1-9, sessile, drupelets 2-5, curved and dentate on back.

Fl.-Fr.: Feb.-March

Rare, **Indore** only, Yeshwant nagar pond.

ERIOCAULACEAE 11/1206; 1/14; 1/8

ERIOCAULON L. W. 477; MP 15; SA 8

Key to Species

1. Anthers white or yellow when young; cells of seed coat transversely elongated; female petals absent .... **E. cinereum**
1. Anthers black or brown
    2. Cells seed coat isodiametric; receptacle villous .... **E. elenorae**
    2. Cells of seed coat not isodiametric, either vertically or transversely elongated
        3. Cells of seed coat vertically elongated .... **E. truncatum**
        3. Cells of seed coat transversely elongated
            4. Male sepals free; floral bracts glabrous; female petals eglandular .... **E. duthiei**
            4. Male sepals connate into a spathe; floral bracts hoary; seed surface with appendages
                5. Leaves and sheaths on peduncles turning red or purple on drying; involucral bracts shorter than floral, bracts; female petals glandular .... **E. quinquangulare**
                5. Leaves and sheaths on peduncles not as above; appendages of seeds setose; male spathe split on one side, distinctly 3- lobed, atleast in the upper half
                    6. Female sepals very unequal; middle sepal of female flower hoary towards apex .... **E. heterolepis**

6. Female sepals subequal
   7. Scapes 30-40 cm long; floral bracts scabrid or puberulous; heads 0.75 cm across .... **E. oryzoterum**
   7. Scapes 6-15 cm long; not scabrid; heads 0.3-0.5 cm across .... **E. sollyanum**

Eriocaulon cinereum R. Br., Prodr. Fl. Nov. Holl. 254. 1810; *E. sieboldianum* Siebold & Zucc. ex Steud., Syn. Pl. Glumac. 2: 272. 1855; Hook.f., Fl. Brit. India 6: 577. 1893; *E. reductum* Ruhland, Pflanzenr. IV, 30:113. 1903; Ansari & Balakr. Eriocaul. India 181. 1994; Samvatsar, Fl. West. Trib. Madhya Pradesh 301. 1996; Khanna *et al.,* in N.P. Singh *et al.,* Fl. Madhya Pradesh 3: 223.2001.

Herbs; rootstocks absent; leaves acicular, glabrous, 2-5 × 0.15-0.2 cm; heads black, hemispherical; peduncles 3-8 cm long; involucral bracts straw-coloured, obovate-oblong, *ca* 1.6 mm long, obtuse, glabrous; floral bracts black, linear-elliptic, *ca* 1.6 mm long, glabrous; male flowers: sepals black, connate into a 3-lobed spathe; petals 3; female flowers: sepals 1 or 2, black, linear; petals absent; seeds ovoid-ellipsoid, 0.3-0.35 mm long; cells of seed-coat transversely elongated, without appendages.

Fl.-Fr.: Sept. -Nov.

Along the edges of ponds, quite common all over the area.

Eriocaulon duthiei Hook.f., Fl. Brit. India 6:578. 1893; Ansari & Balakr., Eriocaul. India 95. 1994; Samvatsar, Fl. West. Trib. Madhya Pradesh 301. 1996; Khanna *et al.,* in N.P. Singh *et al.,* Fl. Madhya Pradesh 3:225. 2001.

Small herbs without root stock; leaves 1-1.5 cm long, broadly lanceolate; peduncles severa 18-14 cm long, heads grey or black, hemispherical, about 4 mm across, sheaths 2-3 cm long; involucral bracts straw-coloured; floral bracts black, oblong-lanceolate, cuspidate; sepals 2 and petals 3 in both sexes; anthers black; seeds ovoid-oblong; cells of seed coat transversely elongated, appendages setiform, arising from transverse, radial cells.

Fl.-Fr.: Sept. -Dec.

Common in marshy places, endemic to Central India, Fl. M.P. 3:225. 2001.

Eriocaulon elenorae Fyson, J. Ind. Bot. Soc. 2:316. 1921; Ansari & Balakr., Eriocaul. India 137. 1994; Samvatsar, Fl. West. Trib. Mdhya Pradesh 302 . 1996; Khanna *et al.,* in N.P. Singh *et al.,* Fl. Madhya Pradesh 3:225. 2001.

Small tufted herbs; root stock absent; leaves linear, 3-5 × 0.1 cm, glabrous; peduncles many, slender, up to 10 cm long; heads hemispheric, white, 3-4 mm across; sheaths 2-2.3 cm long; involucral bracts straw-coloured; floral bracts hyaline; male sepals black,connate into a 3-lobed spathe, petals 3, glabrous with a black gland in each; anthers black; female sepals 3, the 2 laterals shortly keeled, middle broadly keeled, petals 3, eglandular; seeds purple; cells of seedcoat isodiametric, appendages absent.

*Fl.-Fr.:*Sept. -Oct.

In marshy places, Alirajpur, Indore.

Eriocaulon heterolepis Steud., Syn. Pl. Glumac. 2:271. 1855; Hook.f., Fl. Brit. India 6:585. 1893; *E. dianae* Fyson, J. Ind. Bot. 2:259. 1921; Ansari & Balakr., Eriocaul. India 48. 1994;

Samvatsar, Fl. West. Trib. Madhya Pradesh 301. 1996; Khanna *et al.,* in N.P. Singh *et al.,* Fl. Madhya Pradesh 3:226.2001.

Herbs; root stock absent; leaves linear-lanceolate, 1.5-3 × 0.2-0.3 cm; heads grey or white, 2-3 mm across; peduncles 4-8 cm long; sheaths 1.5-2.5 cm long; involucral bracts straw-coloured; floral bracts black, acuminate; male flowers: sepals black, connate into a 3-lobed spathe, petals 3 with a black gland in each, anthers black; female flowers: sepals 3, unequal, black, 2 lateral ones conduplicate, keeled, middle sepal much narrower, not keeled, petals 3, subequal, hyaline, black gland present or absent; cells of seed coat transeversely elongated, appendages setae-like, 5-6 from transverse radial walls.

Fl.-Fr.: Oct. -Dec.

Rare, in moist places at **Khandwa** only, Kalibhit range.

Eriocaulon oryzoterum Mart., Wall. Pl. Asiat. Rar. 3:28. 1832; Hook.f., Fl. Brit. India 6:579. 1893; Ansari & Balakr., Eriocaul. India 41. 1994; Khanna *et al.,* in N.P. Singh *et al.,* Fl. Madhya Pradesh 3:228. 2001.

Herbs; root stock absent; leaves oblong-lanceolate, 8-10 × 0.7-0.8 cm; heads hemispherical, straw-coloured, 6-7 mm across; peduncles 15-20 cm long; sheaths 7-9 cm long; involucral and floral bracts straw-coloured; male flowers: sepals connate into a 3-lobed spathe, petals 3, unequal, anthers black; female flowers: sepals 3, subequal, 2 lateral ones conduplicate, strongly keeled, middle one flat, petals 3 eglandular; seeds purple; cells of seed coat transversely elongated, appendages setiform, 3-6 from transverse radial walls.

*Fl.-Fr.:*Oct. Dec.

Rare, moist places in **Khandwa** only, Kalibhit range.

Eriocaulon quinquangulare L., Sp. Pl. 87. 1753; *E. martianum* Wall. *ex* Korn., Linnaea 27. 642. 1854; Hook.f., Fl. Brit. India 6: 582. 1893; Ansari & Balakr. Eriocaul. India 104. 1994; Samvatsar, Fl. West. Trib. Madhya Pradesh 302.1996; Khanna *et al.,* in N.P. Singh *et al.,* Fl. Madhya Pradesh 3: 229.2001.

Acaulescent herbs, up to 25 cm high; leaves linear-oblong, 1.8-5 × 0.2-0.4 cm, usually changing to reddish on drying; heads globose or spherical, white or grey; peduncles 10-20 cm long; involucral bracts oblong-obovate, *ca* 1.5 mm long, obtuse; floral bracts black, oblanceolate, *ca* 2 mm long, hoary towards apex; male flowers: sepals connate into *ca* 1.2 mm long, 3-lobed spathe, hoary towards apex, black; petals 3; female flowers: sepals 3, free, spathulate, *ca* 1.5 mm long, obtuse, hoary towards apex, black; petals 3; seeds ovoid or oblong, pale purple; cells of seed-coat transversely elongated, with 2-4 appendages from transverse radial walls.

Fl.-Fr.: Sept. -Nov.

In moist places near pond, in all districts.

Eriocaulon sollyanum Royle, Ill. Bot. Himal. Mts. 409, t. 97, f. 1. 1840; *E. trilobum* Buch. - Ham. ex Korn., 27:645. 1856; Hook.f., Fl. Brit. India 6:583. 1893; Ansari & Balakr., Eriocaul. India. 46. 1994; Samvatsar, Fl. West. Trib. Madhya Pradesh 302 . 1996; Khanna *et al.,* in N.P. Singh *et al.,* Fl. Madhya Pradesh 3:233.2001.

Herbs; root stock absent; leaves linear, 2-4 × 0.2-0.3 cm; heads globose, grey or black, 4-6 mm across; peduncles 3-12 cm long; sheaths 1-2.5 cm long; involucral bracts straw-coloured; floral bracts black, subacute or subacuminate; male flowers: sepals connate into a 3-lobed spathe, hoary towards apex, petals 3, with a black gland in each, anthers black; female flowers: sepals 3, subequal, black, conduplicate, keel hoary, petals 3, hyaline with black gland in each; seeds ellipsoid; cells of seed coat transversely elongated, appendages setiform, 3-6 from transverse radial walls appearing in transverse rings on its surface.

Fl.-Fr.: Sept. -Dec.

Common in marshy, wet places in Alirajpur and Dhar.

Eriocaulon truncatum Buch. -Ham. ex Mart., Pl. Asiat. Rar. 3:29. 1832; Hook.f., Fl. Brit. India 6:578. 1893; Ansari & Balakr. Eiocaul. India 118. 1994; Samvatsar, Fl. West. Trib. Madhya Pradesh 302. 1996; Khanna *et al.,* in N.P. Singh *et al.,* Fl. Madhya Pradesh 3: 234.2001.

Small, scapigerous herbs; root stock absent; leaves, lanceolate,acute, 3-6 × 0.2-0.6 cm; peduncles many, 5-13 cm long; heads hemispheric, 4-5 mm across; sheaths 2.5-3.5 cm long; involucral bracts hyaline; floral bracts hyaline; male flowers: sepals connate into a 2-lobed spathe, black, petals 3, subequal with a black gland in each, anthers black; female flowers: sepals 2 or 3, black, petals 3 with a black gland in each; seeds oblong or ovoid; cells of seed coat rectangular, vertically elongated, appendages ribbon-like, paired and continuous from vertical walls, solitary and incomplete from transverse, radial walls.

Fl.-Fr.: Sept. -Oct.

In **Alirajpur** only, open marshy ground, Harod, Mordundiya.

CYPERACEAE 110/5784; 11/103; 14/84

Key to Genera

1. Flowers unisexual; style 3-fid; nut trigonous; glumes of female spikelets 3 .... SCLERIA
1. Flowers bisexual
    2. Glumes, atleast the lower distichous
        3. Style articulated with the nuts, 3-fid, more or less fimbriate; lower glumes distichous, upper spiral or all distichous ....................FIMBRISTYLIS
        3. Style continuous with the nuts, 2-3 fid, usually glabrous; all glumes distichous
            4. Style 2-fid
                5. Rhachilla persistent ....PYCREUS
                5. Rhachilla deciduous . ... KYLLINGA
            4. Style 3-fid
                6. Rhachilla persistent .... CYPERUS
                6. Rhachilla deciduous . ...COURTOISINA
    2. Glumes spirally arranged
        7. Hypogynous scales or bristles absent
            8. Style articulated with the nuts; nuts crowned by a button-like tumour of persistent style base....................BULBOSTYLIS

8. Style not articulated

9. Leaves represented by leaf sheaths only, lamina absent

10. Hypogynous scales or bristles absent; stigmas 3 .... SCHOENOPLECTIELLA

10. Hypogynous scales or bristles present or not; stigmas 2 or 3 . ...SCHOENOPLECTUS

7. Hypogynous bristles or scales present

9. Leaves with lamina present or leaf-sheaths alone present

11. Plants leafless; involucral bracts stem or glume-like or absent; inflorescence single, terminal spikelet; style articulated with nut, base dilated; bristles barbellate; scales absent .... ELEOCHARIS

11. Plants leafy; involucral bracts foliaceous, many, covering the inflorescence

12. Spikelets comose due to exseted hypogynous bristles; plants scaberulus ERIOPHORUM

12. Spikelets not comose, many to numerous; bristles when present, 3-6, not exserted; plants glabrous; involucral bracts 1-4, not covering the inflorescence; empty glumes 1-3, similar to the fertile glumes; style continuous with the nuts; hypogynous bristles and scales, both present

13. Stem nodeless or atmost 1-noded at base; all leaves basal; glumes glabrous or atmost ciliolate on margins; hypogynous scales 2, hyaline, tightly adhering to the nuts; bristles absent; inflorescence of capitate heads with 1-10 sessile or subsessile spikelets . ...LIPOCARPHA

13. Stems 1-5 noded; leaves mainly cauline; glumes with long cilia or with an awn or mucro

14. Leaves with a collar around stem; glumes long ciliate; hypogynous scales 3 (absent in **Fuirena cuspidata**); bristles when present, 3; rhizome not tuberiferous ....FUIRENA

14. Leaves without a collar; glumes not ciliate; hypogynous bristles 6; scales absent; rhizome emitting tuberiferous stolons .... BOLBOSCHOENUS

BOLBOSCHOENUS Palla W. 16; MP 1; SA 1

Bolboschoenus maritimus (L.) Palla, Syn. Deut. Schweiz. Fl. ed. 3, 3: 2532. 1905; *Scirpus maritimus* L., Sp. Pl. 51. 1753; C.B. Clarke in Hook.f., Fl. Brit India 6: 658. 1893; *Schoenoplectus maritimus* (L.) Lye, Blyttia 29:145. 1971; Samvatsar, Fl. West. Trib. Madhya Pradesh 319. 1996; Verma in N.P. Singh *et al.,* Fl. Madhya Pradesh 3:326.2001, as *Scirpus maritimus* L.

Perennial, rhizomatous plants, up to 1 m tall; leaves as long as the stem, sheaths long; inflorescence simple or compound; bracts 1-several, longer than the anthela; spikelets many, ovoid-oblong to subcylindric, 10-20 mm long, brown; glumes ovate, 5-7 mm long; awn curved, scabrid; perianth bristles 3-6, unequal, retrorsely scabrid; nut narrowly obovoid, smooth, beaked, brown, 2.5-3.5 mm long.

Fl.-Fr.: Oct. -Dec.

Along margins of ponds, Alirajpur, Khandwa.

BULBOSTYLIS Kunth W. 217 ;MP 2; SA 1

Bulbostylis barbata (Rottb.) C.B. Clarke in Hook.f., Fl. Brit. India 6: 651. 1893; *Scirpus barbatus* Rottb., Descr. Icon. Rar. Pl. 52. t. 1773; Samvatsar, Fl. West. Trib. Madhya Pradesh 303. 1996; Verma in N.P. Singh *et al.,* Fl. Madhya Pradesh 3: 238. 2001.

Annual, tufted herbs, 4-20 cm tall; leaves shorter to subequalling the stem, *ca* 0.3 mm broad; inflorescence capitate, 4-16 mm across; spikelets linear, polygonal, rhachilla zig-zag, 2-8 × 1.0-1.2 mm; glumes brown, laxly imbricating, acute or mucronate, glabrous or hispidulous; nuts stramineous to pale brown, smooth.

Fl.-Fr.: July-Jan.

In roadside ditches and stream beds, all over the area.

COURTOISINA Sojak W. 2; MP. 1; SA 1

Courtoisina cyperoides (Roxb.) Sojak, Cas. Nar. Mus. Odd. Prir. 148:193.1979 (pub. 1980); *Kyllinga cyperoides* Roxb., Fl. Ind. 1:182.1820; *Courtoisia cyperoides* (Roxb.) Nees, Linnaea 9:286.1834; *Cyperus pseudokyllingoides* Kukenth.,Pflanzenr. IV, 20 (101):14. 1935; Verma in N.P. Singh *et al.,* Fl. Madhya Pradesh 3:278. 2001.

Annuals, *ca* 10-50 cm high; leaves 1.5-4 mm broad; anthela compound or subcompound, bracts 3-7, foliaceous, up to 40 cm long; spikelets many together, congested in subglobose, 5-8 mm broad spikes, strongly compressed, 3-4 × 1.2-1.5 mm, 1-2 flowered; rachilla winged; glumes 2.5-3 mm long, grooved, with a short, stout mucro; keel broadly winged; stamens and stigmas 3; nuts deep brown, fusiform, 2.3-2.7mm long, puncticulate.

Fl.-Fr.: Sept. -March

In water logged areas, along streams, Burhanpur, Khandwa.

CYPERUS L. W. 699; MP. 53 (*sensu lato*); SA 32 (*sensu stricto*)

Key to Species

1. Stigmas 2; nuts biconvex
    2. Inflorescence capitate; glumes spiral ... . **C. michelianus**
    2. Inflorescence anthelate or capitate; glumes distichous
        3. Inflorescence pseudolateral cluster of spikelets; bract solitary, stem-like, erect .... **C. laevigatus**
        3. Inflorescence various, distinctly terminal; bracts several, spreading; glumes mucronate or mucronulate; perennials; glumes 5-7 nerved . ... **C. alopecuroides**
1. Stigmas 3; nuts trigonous
    4. Rhachilla of the spikelets caducous; spikelets falling off along with persistent glumes; anthela simple; spikes cylindrical
        5. Rhizomes emitting slender stolons; spikelets bearing a single nut . ... **C. paniceus**
        5. Rhizomes not emitting stolons; spikelets bearing 1-4 nuts
            6. Spikelets cylindrical, pedunculate; nuts linear-oblong .... **C. cyperoides**
            6. Spikelets narrowed at the base, sessile or subsessile; nuts oblong-ellipsoid .... **C. cyperinus**

4. Rhachilla of the spikelets persistent; glumes shedding from below upwards as they mature
   7. Stamen 1 (-2); nuts linear or narrowly oblong with almost parallel sides .... **C. castaneus**
   7. Stamens 1-3; nuts ovoid-oblong or sub-globose
      8. Inflorescence capitate
         9. Small annual herbs, glumes with a long arista, half as long as to equalling its main body .... **C. cuspidatus**
         9. Small or large, annual or perennial herbs; glumes muticous or shortly mucronate
            10. Spikelets 10-35 mm long, 3-6 mm broad; stamens 3
               11. Spikelets green; stigma *ca* 0.5 mm long .... **C. compressus**
               11. Spikelets pale-brown to whitish
                  12. Inflorescence 2-4 cm broad; glumes whitish-stramineus; stamens 3 .... **C. niveus**
                  12. Inflorescence 6-8 cm broad; glumes stramineus; stamens usually 2 .... **C. kyllingiella**
            10. Spikelets 3-8 cm long, 1-3 mm broad
               13. Stamens 3 .... **C. clarkei**
               13. Stamens 1-2
                  14. Glumes broadly obovate-suborbicular, 0.5-0.8 mm long . ... **C. difformis**
                  14. Glumes elliptic, oblong or lanceolate, more than 1 mm long; spikelets 1.5-2 mm broad; glumes 3-nerved . ... **C. pulchellus**
      8. Inflorescence anthelate with distinct rays
         15. Spikelets in well elongated spikes; rhachis several times longer than the spikelets
            16. Glumes orbicular or broadly ovate, about as long as broad
               17. Rhachis of the spike smooth
                  18. Glumes 3-5 nerved . ... **C. iria**
                  18. Glumes 5-9 nerved . ... **C. javanicus**
               17. Rhachis of the spike scabrid on the angles; glumes 7-nerved, winged . ... **C. alulatus**
            16. Glumes linear-lanceolate, ovate or oblong, about half or less broad as long
               19. Rhachis hispidulous; anthela compound; spikelets 6-12 mm long .... **C. pilosus**
               19. Rhachis glabrous
                  20. Connective of the anther produced into a crest, about as 1/3 rd to ½ as long as the anther cells .... **C. elatus**
                  20. Connective of the anther scarcely produced
                     21. Base of the glumes decurrent along the rhachilla internodes; spikelets linear to lanceolate, 0.15-0.2 mm broad; glumes close together, acute .... **C. tenuiculmis**
                     21. Base of the glumes not decurrent as above
                        22. Spikes penicillate, 2-10 mm broad; spikelets ascending .... **C. nutans**
                        22. Spikes cylindrical to flattened, 10-20 mm broad; spikelets divaricate
                           23. Spikelets subterete or slightly compressed; wings of rhachilla caducous .... **C. digitatus**
                           23. Spikelets strongly compressed; wings of rhachilla long persistent

24. Inflorescence compound or decompound; spikes sessile or pedunculate, lax with spreading spikelets; mucro of the glumes erect .... **C. exaltatus**

24. Inflorescence simple or compound; spikes subsessile, very dense with ascending spikelets; mucro of the glumes curved .... **C. imbricatus**

15. Spikelets digitate or in very short spikes; rhachis obscure or much shorter than to about as long as the spikelets

25. Spikelets digitate, 3-20 mm long

26. Glumes mucronate; leaves 1.5-4 mm broad; bracts shorter or equalling the anthela .... **C. haspan**

26. Glumes muticous or mucronulate; spikelets 2-10 together; stamens 1 or 2; nuts much shorter than the glumes; rhachilla clearly visible between the widely spreadinglumes .... **C. tenuispica**

25. Spikelets shortly spicate, 5-60 mm long

27. Anthela simple; glumes distinctly mucronate, 1-2 mm long, mucro 0.5-1.5 mm long; spikelets squarrose .... **C. squarrosus**

27. Anthela simple or compound; glumes muticous or mucronulate

28. Stem stout, 4-8 mm thick in the middle; leaves absent or solitary

29. All bracts short, rarely as long as the anthela .... **C. corymbosus**

29. Lower bracts much exceeding the anthela .... **C. pangorei**

28. Stem slender or stout, 1-1.5 thick in the middle; leaves 2-several; stolons present; bracts equal to or longer than the anthela

30. Base of stem enclosed in a hard, irregularly disintegrating coat; involucral bracts clearly separated from one another or only the lowermost obvious .... **C. bulbosus**

30. Base of stem covered with membranous scales when young, scales soon disintergrating into fibres; involucral bracts close together; stolons soft, always bearing tubers

31. Tubers covered with grey tomentum; spikelets 5-18 mm long, 8-16 flowered; glumes golden brown, with 7-9 prominent nerves evenly distributed over their whole breadth .... **C. esculentus**

31. Tubers without tomentum; spikelets 10-35 mm long, 10-20 flowered; glumes usually deep brown, rarely yellowish, with 5-7 nerves, the lateral nerves becoming less prominent, away from the keel .... **C. rotundus**

Cyperus alopecuroides Rottb., Descr. Icon. Rar. Pl. 38. 1773; *Juncellus alopecuroides* (Rottb.) C.B. Clarke in Hook.f., Fl. Brit. India 6:595. 1893; Samvatsar, Fl. West. Trib. Madhya Pradesh 304. 1996; Verma in N.P. Singh *et al.,* Fl. Madhya Pradesh 3:255. 2001.

Perennials; culms 80-140 cm tall; leaves shorter than culms, inverted 'W'-shaped in cross section; anthela compound or decompound; bracts 4-7, 2 or 3 surpassing the anthela; spikelets ovate to oblong-lanceolate, 4-7 mm long; glumes ovate-elliptic, 1.5-2.5 mm long, margins hyaline; achenes 0.8-0.9 mm long, elliptic to obovate.

*Fl.-Fr.:*Sept. -Dec.

Common in marshy places, along streams throughout the area.

Cyperus alulatus J. Kern, Reinwardtia 1:463. 1952; *C. iria auct. non* L., 1753; Samvatsar, Fl. West. Trib. Madhya Pradesh 304. 1996; Verma in N.P. Singh *et al.,* Fl. Madhya Pradesh 3:256.2001.

Tufted annuals, up to 50 cm high; leaves as long as or longer than the culms, 2-5 mm broad; anthela lax, simple or compound, 5-15 cm across; bracts foliaceous; rays up to 10 cm long; rachilla scabrid on the angles; spikelets numerous, 5-10 mm long; glumes loosely imbricating, broadly ovate or suborbicular, *ca* 1.3 mm long, keeled, muticous or mucronulate; keel 7-nerved, its wing serrate near the top; nuts about as long as the glumes, broadly stipitate.

Fl.-Fr.: Aug. -Dec.

Common in all districts in fields and waste places.

Cyperus bulbosus Vahl, Enum. Pl. Obs. 2:342. 1805; C.B. Clarke in Hook.f., Fl. Brit. India 6:611. 1893; Verma in N.P. Singh *et al.,* Fl. Madhya Pradesh 3:258. 2001.

Perennials, 10-30 cm high; rhizome bearing long, slender stolons; culm solitary, bulbously swollen at base due to a cover of stiff, and irregularly splitting brown or blackish scales; leaves usually shorter than the culms; anthela simple with up to 2 cm long divaricate rays; bracts foliaceous, 4-8; spikes cylindrical, 1-3 cm long; spikelets linear-ellipsoid, bearing a solitary nut; glumes straw-coloured, oblong-lanceolate, 9-11 nerved, with a green keel; nuts brown or blackish.

Fl.-Fr.: July-Dec.

Rare, in stagnant water, **Indore** only, grassland at Hirniya khedi.

Cyperus castaneus Willd., Sp. Pl. 1:278. 1797; C.B. Clarke in Hook.f., Fl. Brit. India 6:598. 1893; Samvatsar, Fl. West. Trib. Madhya Pradesh 304. 1996; Verma in N.P. Singh *et al.,* Fl. Madhya Pradesh 3:258. 2001.

Tufted annuals; culms 2-10 cm high; leaves filiform, about as long as the culm; anthela usually simple, congested in a single head; bracts 2-5, surpassing the heads; spikelets stellately clustered in groups of 5-30, 5-20 mm long, including mucros; glumes oblong or oblong-lanceolate, 1.2-1.5 mm long including mucro; keel strongly 3-nerved; nuts 1.6-2 mm long, oblong-obovoid.

*Fl.-Fr.:*Sept. -Oct.

In wet and shady places, Alirajpur, Burhanpur, Khandwa.

Cyperus clarkei T. Cooke, Fl. Bombay 2:873. 1908 [3: 387. 1958 (repr. ed.)]; *Mariscus clarkei* (T. Cooke) T. Koyama, J. Jap. Bot. 51:316. 1976; *M. bulbosus* C.B. Clarke in Fl. Brit. India 6:620.1893; Sharma *et al.,* in Fl. Maharashtra 3: 334. 1996.

Small sedges with a tuberous root-stock, covered with striate, pale-brown scales; leaves linear, acute, lower leaf-sheaths much inflated; spikelets many, closely packed, arranged at the apex of the culm; nuts brown, ellipsoid, trigonous.

*Fl.-Fr.:*Aug. -Oct.

Rare, wet places, **Khandwa** only, Nagchun pond vicinity.

Cyperus compressus L., Sp. Pl. 46. 1753; C.B. Clarke in Hook.f., Fl. Brit. India 6: 605. 1893; Samvatsar, Fl. West. Trib. Madhya Pradesh 305. 1996; Verma in N.P. Singh *et al.,* Fl. MP 3: 259.2001.

Annual herbs, 3-40 cm high; leaves greyish green, 1.4-4.0 mm broad; anthela simple, lax; spikelets in clusters of 3-13, linear-oblong, 10-30 × 3-4 mm; glumes pale yellowish or brownish, ovate, mucronate, several-nerved, broadly scarious-margined, 3-4.5 mm long; nuts blackish-brown, glossy, obovoid, triquetrous, apiculate, shortly stipitate.

Fl.-Fr.: July-Dec.

In wet open places, all over the area.

Cyperus corymbosus Rottb., Descr. Icon. Rar. Pl. 42. 1773; C.B. Clarke in Fl. Brit. India 6:612. 1893; Verma in N.P. Singh *et al.,* Fl. Madhya Pradesh 3:261.2001.

Rhizomatous perennials, up to 1 m high; stems spongy, 5-8 mm thick, terete below and trigonous above; leaves absent or solitary, up to 13 cm long; anthela compound, 15-20 cm across, bracts with spinulose, revolute margin; spikelets clustered, 4-8 together, linear, 6-40 × 1 mm, wings of rachilla hyaline; glumes reddish-brown, obovate-oblong, 2.5-3 mm long; nuts *ca* 1.5 mm long, apiculate.

Fl.-Fr.: Dec. -Jan.

River beds, Barwani, Burhanpur, Indore, Khandwa, Khargone.

Cyperus cuspidatus Kunth, Nov. Gen. Sp. 1:204. 1817; C.B. Clarke in Hook.f., Fl. Brit. India 6:598. 1893; Verma in N.P. Singh *et al.,* Fl. Madhya Pradesh 3:261.2001.

Annuals, culms slender, up to 20 cm high; leaves 0.5-1 mm broad; anthela simple or compound, bracts 2-4, foliaceous; rays erect or spreading; spikelets up to 22 together, linear, 4-12 × 1-1.3 mm; glumes spreading, ferrugineus, *ca* 2 mm long, aristate, arista about 1 mm long; nuts chestnut-brown, long, puncticulate.

Fl.-Fr.: Aug. -Nov.

In moist places, Burhanpur, Indore, Khandwa.

Cyperus cyperinus (Retz.) Suringar, Cyperus Mal. Arch. 154. 1898; *Kyllinga cyperina* Retz., Observ.. Bot. 6:21. 1791; *Mariscus cyperinus* (Retz.) Vahl, Enum. Pl. Obs. 2:377. 1805; C.B. Clarke, in Hook.f., Fl. Brit. India 6:621. 1893; Verma, in N.P. Singh *et al.,* Fl. Madhya Pradesh 3:263.2001.

Perennials, 10-60 cm high, with a short rhizome, clothed with brown fibres, not bearing any stolons; leaves shorter than the culms, narrowly linear, long attenuate, 4-8 mm broad; anthela simple, with well developed, pedunculate rays; bracts foliaceous, up to 30 cm long; spikes 1-3.5 cm long, sessile or borne on *ca* 3.5 cm long peduncles, bearing up to 80 spikelets; spikelets 2-4 flowered, 4-7 mm long, linear lanceolate or linear-oblong; glumes elliptic, keel 3-nerved, sides 3-4 striate; nuts brown, apiculate, slightly muricate.

Fl.-Fr.: Aug. -Oct.

Rare, confined to **Khandwa**, in moist places, at Kalibhit.

Cyperus cyperoides (L.) Kuntze, Revis. Gen. Pl. 3(2): 333. 1898; *Scirpus cyperoides* L., Mant. Pl. 2:181. 1771; *Mariscus sieberianus* Nees ex C.B. Clarke in Hook.f. Fl. Brit India 6:622. 1893; Samvatsar, Fl. West. Trib. Madhya Pradesh 305. 1996; Verma in N.P. Singh *et al.,* Fl. MP 3: 263.2001.

Perennial herbs, up to 1 m high; leaves 2.4-9.0 mm broad; spikes cylindrical, 6-35 × 6-12 mm; spikelets linear-lanceolate, flowering obliquely spreading, fruiting divaricate or the lower even deflexed, usually maturing 2 nuts, 3.5 mm long; glumes greenish, ovate-oblong; nuts yellowish when young, mature deep brown, linear-oblong, trigonous, apiculate, *ca* 2.2 × 0.4 mm.

Fl.-Fr.: Aug. -March

In open places and as forest-undergrowth, Alirajpur, Barwani, Indore, Khargone.

Cyperus difformis L., Cent. Pl. 2: 6. 1756; C.B. Clarke in Hook.f., Fl. Brit. India 6: 599. 1893; Samvatsar, Fl. West. Trib. Madhya Pradesh 305. 1996; Verma in N.P. Singh *et al.,* Fl. Madhya Pradesh 3: 263.2001.

Annual, 10-50 cm high herbs; roots reddish; stems compressed-triquetrous; leaves rather abruptly acuminate, 1.5-5.0 mm broad; spikelets numerous together, forming 5-13 mm broad, subglobose heads, linear-lanceolate, 3-6 × 1 mm; glumes hyaline with a reddish- brown band on each side of the midrib, broadly obovate-suborbicular with rotundate tips; nuts yellowish- brown, obovoid, trigonous, apiculate, sessile.

Fl.-Fr.: July-April

In moist depressions and paddy fields, all over the area.

Cyperus digitatus Roxb., Fl. Ind. 1:209. 1820; C.B. Clarke in Hook.f., Fl. Brit. India 6:618. 1893; Samvatsar, Fl. West. Trib. Madhya Pradesh 305. 1996; Verma in N.P. Singh *et al.,* Fl. Madhya Pradesh 3:264.2001.

Perennials; culms 60-150 cm tall; leaves shorter or equalling the stem; anthela compound or decompound; bracts 4-8, overtopping the anthela; spikelets linear, 5-20 mm long; glumes 1.8-2.2 mm long, narrowly elliptic, keeled on the back; nuts up to 1 mm long, oblong-ellipsoid.

Fl.-Fr.: Sept. -Dec.

Marshy and wet places, Alirajpur, Dhar, Khandwa.

Cyperus elatus L., Cent. Pl. 11 5. 1756; C.B. Clarke in Fl. Brit. India 6:618. 1893; Verma in N.P. Singh *et al.,* Fl. Madhya Pradesh 3:265. 2001.

Perennials; culms stout, up to 1.2 m high; leaves nearly as long as the stem, up to 15 mm broad; anthela compound, bracts foliaceous, several, up to 70 cm long; spikes 3-8 cm long, narrowly cylindrical, clustered at the ends of rays; spikelets densely, spirally inserted, erect, linear, *ca* 5 mm long; glumes closely imbricating, ovate, *ca* 1 mm long; nuts pale brown, ellipsoid, about half to two-third as long as the glume.

Fl.-Fr.: March-May

Rare, **Burhanpur** only, Asirgad, Dahinala.

Cyperus esculentus L., Sp. Pl. 45. 1753; C.B. Clarke in Hook.f., Fl. Brit. India 6:616. 1893; Samvatsar, Fl. West. Trib. Madhya Pradesh 305. 1996; Verma in N.P. Singh *et al.,* Fl. Madhya Pradesh 3:266.2001. *Chichoda*

Perennial, 30-90 cm tall, stoloniferous herbs; stolons covered with strongly veined scales; leaves linear, acuminate; anthela lax, compound, bracts about 4, unequal; spikelets linear-lanceolate, rachis narrowly winged; glumes boat-shaped, keeled, 3-nerved; nuts obovoid-ellipsoid.

Fl.-Fr.: Aug. -Oct.

Marshy places, along water bodies, Burhanpur, Dhar, Khandwa.

Cyperus exaltatus Retz., Observ. Bot. 5: 11. 1788; C.B. Clarke in Hook.f., Fl. Brit. India 6: 617. 1893; Samvatsar, Fl. West. Trib. Madhya Pradesh 306. 1996; Verma in N.P. Singh *et al.,* Fl. Madhya Pradesh 3: 266.2001.

Perennial, rhizomatous herbs, 1-2 m high; leaves short or long, up to 1 cm broad; anthela large, compound; spikes umbellate or corymbose; spikelets numerous, approximate or distant; glumes closely or laxly imbricating, broadly ovate, *ca* 1.2 mm long; keel green or pale; nuts yellowish, dark brown or black, very small.

Fl.-Fr.: Sept. -Feb.

Along edges of ponds and streams, in all districts.

Cyperus haspan L., Sp. Pl. 45. 1753; C.B. Clarke in Fl. Brit. India 6:600.1893; Samvatsar, Fl. West. Trib. Madhya Pradesh 306. 1996; Verma in N.P. Singh *et al.,* Fl. Madhya Pradesh 3:267.2001.

Perennials or rarely annuals, with short rhizome; culms 10-40 cm tall; leafblade elongated or reduced to lanceolate appendage of the sheath, when elongated, shorter than the stem; anthela compound to decompound, bracts 2-3, shorter or subequalling the antherela; spikelets linear-oblong or linear- lanceolate, 5-10 mm long; glumes elliptic to oblong, 1-1.5 mm long, keel 3-nerved; nuts broadly obovate, creamy-yellow, granulate, verruculose.

Fl.-Fr.: Oct. -Nov.

Confined to wet and shady places in the forest, Alirajpur, Barwani, Dhar, Khargone.

Cyperus imbricatus Retz., Observ. Bot. 5:12. 1788; *C. radiatus* Vahl, 1805, *nom. illeg;* C.B. Clarke in Hook.f., Fl. Brit. India 6:617. 1893; Verma in N.P. Singh *et al.,* Fl. Madhya Pradesh 3:267.2001.

Annuals, culms 30-90 cm tall; stem solitary, stout; leaves up to 8 mm broad; anthela dense, simple or compound; spikes subsessile, digitate, cylindrical, 2-4 cm long, bracts foliaceous, 5-7, up to 60 cm long; spikelets oblong-lanceolate, 3-5 × 1.2 mm, broadly ovate; glumes closely imbricating, broadly ovate, *ca* 1.2 mm long, shortly cuspidate; nuts black, broadly ellipsoid.

Fl.-Fr.: Aug. -Oct.

Muddy places, confined to **Khandwa** only, Mortakka, banks of Narmada.

Cyperus iria L., Sp. Pl. 45. 1753; C.B. Clarke in Fl. Brit. India 6:606. 1893; Samvatsar, Fl. West. Trib. Madhya Pradesh 306. 1996; Verma in N.P. Singh *et al.,* Fl. Madhya Pradesh 3:269.2001.

Annuals; culms 8-60 cm tall; leaves much shorter to a little longer than the stem, scabrous on the upper margins; anthela compound, bracts 3-5, lower 1 or 2 longer thant the anthela; spikelets oblong-elliptic to linear-oblong, 9-11 mm long; glumes obovate-orbicular, 1.3-1.5 mm long, rounded, often shallowly emarginate, straw coloured, keel 3-5 nerved; nuts obovate, dark brown.

Fl.-Fr.: Throughout the year.

Common weed of fields in all districts.

Cyperus javanicus Houtt., Nat. Hist. t. 88, f. 1. 1782; *C. pennatus* Lam., Tabl. Encycl. 1:144. 1791; *Mariscus javanicus* (Houtt.) Merr. & Metcalf, 1945, *nom. illeg;* Sharma *et al.,* Fl. Maharashtra 3:335. 1996; Samvatsar, Fl. West. Trib. Madhya Pradesh 306. 1996.

Perennials; culms 30-90 cm tall; leaves mostly overtopping the culms; anthela compound or decompound, lower branches much longer than the anthela; spikelets oblong-lanceolate, 4-10 mm long; rachilla broadly winged; glumes broadly ovate, 2.5-3.5 mm long; nuts 1.2-1.5 mm long.

Fl.-Fr.: Sept. -Nov.

Rare, Alirajpur, Dhar.

Cyperus kyllingiella Larridon, Pl. Ecol. Evol. 144:351.2011; *Kyllinga microcephala* Steud., Flora 25:597. 1842; *Isolepis kyllingoides* A. Rich.,Tent. Fl. Abyss. 2:502. 1850; *Scirpus kyllingoides* (A. Rich.) Boeckler, Linnaea 36:733. 1870; C.B. Clarke in Hook.f., Fl. Brit. India 6:662. 1893; *S. microcephalus* (Steud.) Dandy, Fl. Pl. Sudan 3:366. 1956; Verma in N.P. Singh *et al.,* Fl. Madhya Pradesh 3:327.2001.

Perennials; culms 10-25 cm tall, sometimes bearing stolons; leaves about 1 mm broad; anthela solitary, terminal, 6-8 mm broad, bearing 5-15 spikelts, bracts foliaceous, up to 10 cm long; spikelets many flowered, ovate-lanceolate; glumes straw coloured with purple dots, keeled, 3-5 nerved; perianth absent; stamens 2 (3); nuts brown or blackish, obovoid-oblong, trigonous to subcompressed, puncticulate.

Fl.-Fr.: Sept. -Oct.

Rare, moist, wet places, **Indore** only, at Machal village (Kishanpura pond).

Cyperus laevigatus L., Mant. Pl. 2: 179. 1771; *Juncellus laevigatus* C.B. Clarke in Hook.f., Fl. Brit. India 6:596. 1893; Verma in N.P. Singh *et al.,* Fl. Madhya Pradesh 3:270.2001.

Rhizomatous perennials; culms 10-60 cm tall, base with chestnut- red sheaths; leaves long, short orabsent, 2-3 mm broad when present; anthela capitate, bracts 2, one very short and broad, the other up to 5 cm long, appearing as if continuation of stem; spikelets 5-12 mm long, linear-oblong, straw-coloured with a red tinge, few to several in a densely congested pseudolateral head, 16-40 flowered; glumes broadly ovate, mucronulate; stigmas 2; nuts obovoid, plano-convex, compressed at right angles to the rachilla, about half to two-thirds as long as the glume.

Fl.-Fr.: July-Sept.

Rare, at moist places, Burhanpur, Khandwa.

Cyperus michelianus

Key to subspecies

1. Glumes spiral .... subsp. **michelianus**
1. Glumes distichous .... subsp. **pygmaeus**

Cyperus michelianus (L.) Del. subsp. **michelianus** Descr. Egypte Hist. Nat. 3:50.1813; *Scirpus michelianus* L., Sp. Pl. 52. 1753; C.B. Clarke in Hook.f., Fl. Brit. India 6:662. 1893; Samvatsar, Fl. West. Trib. Madhya Pradesh 306. 1996; Verma in N.P. Singh *et al.,* Fl. Madhya Pradesh 3:273.2001.

Annuals; culms 10-15 cm tall; leaves 1-2 mm broad, often longer than the culms; anthela capitate, consisting of 4-6, ovoid-subglobose, 6-15 mm broad spikes, bracts 3-8, longer than the heads; spikelets numerous, congested, 3-5 × 1-2 mm; glumes spiral, oblong-lanceolate, tightly imbricating, 1.5-2 mm long; stamens usually 2; stigmas 2 (rarely 3); nuts pale brown, plano-convex (rarely trigonous).

Fl.-Fr.: Sept. -Dec.

In wet places, not common, Alirajpur, Dhar, Khandwa.

Cyperus michelianus subsp. **pygmaeus** (Rottb.) Asch. & Graebn.,Syn. Metteleur. Fl. 2(2):273. 1904; *C. pygmaeus* Rottb., Descr. Icon. Rar. Pl. 20.1773; *Juncellus pygmaeus* (Rottb.) C.B. Clarke in Hook.f., Fl. Brit. India 6:596. 1893; Samvatsar, Fl. West. Trib. Madhya Pradesh 309. 1996; Verma in N.P. Singh *et al.,* Fl. Madhya Pradesh 3:281.2001.

Annuals; culms 5-30 cm tall; leaves shorter than to slightly longer than the culms, scabrid on the upper margin; inflorescence congested in a lobed head; spikelets broadly ovate to lanceolate, 3-5 mm long; glumes lanceolate to elliptic, 1.8-2.2 mm long, 3 or 4-nerved, acute to mucronulate; nuts oblong to ovate-oblong, plano-convex, minutely puncticulate, pale brown.

Fl.-Fr.: Sept. -Dec.

Occasional in moistplaces, Alirajpur, Burhanpur, Dhar, Indore.

Cyperus niveus Retz.,Observ. Bot. 5:12. 1788; C.B. Clarke in Hook.f., Fl. Brit. India 6:601. 1893; Verma in N.P. Singh *et al.,* Fl. Madhya Pradesh 3:274.2001.

Perennials; culms 15-80 cm tall, rhizomatous; leaves 1.5-2 mm broad; anthela capitates, 1-4 cm broad, bracts 2-3, foliaceous, up to 8 cm long; spikelets oblong-lanceolate 10-25 × 4-6 mm, rachilla not winged; glumes tightly imbricating, pale brown, ovate-lanceolate, 4-5 mm long, chartaceous; nuts triquetrous, brown, *ca* 1.5 mm long, densely puncticulate.

Fl.-Fr.: Aug. -Dec.

In wastelands, Burhanpur, Indore.

Cyperus nutans

Key to subspecies

1. Spikelets crowded into cylindric spikes; plants not stoloniferous .... subsp. **eleusinoides**
1. Spikelets remotely arranged; plants stoloniferous .... subsp. **nutans**

Cyperus nutans subsp. **eleusinoides** (Kunth) T. Koyama, Gard. Bull. Singapore 30:136. 1977; *C. eleusinoides* Kunth, Enum. Pl. 2:39. 1937; C.B. Clarke in Hook.f., Fl. Brit. India 6:608. 1893; Samvatsar, Fl. West. Trib. Madhya Pradesh 307. 1996; Verma in N.P. Singh *et al.,* Fl. Madhya Pradesh 3:275. 2001.

Perennials; culms 30-100 cm tall; leaves shorter than culms, bracts leafy, surpassing the anthela; spike-lets 8-10 mm long, less than 8-flowered, linear or narrowly elliptical; glumes lanceolate, 2-2.2 mm long; nuts 1-1.2 mm long, obovate to broadly elliptic.

Fl.-Fr.: July-Dec.

Weed of fields and waste places, common throughout the area.

Cyperus nutans subsp. **nutans** Vahl, Enum. Pl. 2:363. 1805; C.B. Clarke in Hook.f., Fl. Brit. India 6:607. 1893; Samvatsar, Fl. West. Trib. Madhya Pradesh 307. 1996; Verma in N.P. Singh *et al.,* Fl. Madhya Pradesh 3:275. 2001.

Perennials with a short, corm-like rhizome; culms 80-150 cm tall; leaves shorter than the culms, scabrous on the upper margin; anthela compound to decompound, bracts 4-6, lower 1 or 2 longer than the anthela; spikelets linear to linear-oblong, 8-16 flowered, 6-15 mm long; glumes oblong, 2-2.5 mm long, folded with keels and with a broad hyaline margin in the upper half, 3-nerved on both sides of midnerve; nuts oblong to ovate-oblong, apiculate, brown.

Fl.-Fr.: Aug. -Feb.

Common in marshy places in all districts.

Cyperus pangorei Rottb., Descr. Icon. Rar. Pl. 31. 1773; C.B. Clarke in Hook.f., Fl. Brit. India 6:612. 1893; C.B. Clarke in Hook.f., Fl. Brit. India 6:613. 1893; Samvatsar, West. Trib. MP 307. 1996; Verma in N.P. Singh *et al.,* Fl. Madhya Pradesh 3:275. 2001.

Rhizomatous perennials; culms up to 1 m tall; leaves usually reduced to subaphyllous sheaths surround-ding bases of culms; anthela compound, bracts 4-5, all longer than anthela; spikelets linear, 1-2 cm long; glumes oblong to lanceolate, 2.2-2.7 mm long, nerveless; nuts oblong-obovate, brown,minutely apiculate.

Fl.-Fr.: Sept. -Jan.

In moist places of forest, Alirajpur, Burhanpur, Dhar, Indore, Khandwa.

Cyperus paniceus (Rottb.) Boeckler, Linnaea 36:381. 1870; *Kyllinga panicea* Rottb., Descr. Icon. Rar. Pl. 15. 1773; *Mariscus paniceus* Vahl, Enum. Pl. 2:373. 1805; C.B. Clarke in Hook.f., Fl. Brit. India 6:620.1893; Samvatsar, Fl. West. Trib. Madhya Pradesh 307. 1996; Verma in N.P. Singh *et al.,* Fl. Madhya Pradesh 3:275. 2001.

Perennials; culms 30-60 cm tall; leaves shorter to more or less equalling the stem; anthela simple, bracts 3-6, surpassing the anthela; spikelets horizontally divergent, 2.5-4 mm long, bearing 4 glumes; nuts elliptic-oblong, 1.8-2.5 mm long.

Fl.-Fr.: Sept. -Dec.

Not common, wet places, Alirajpur, Indore.

Cyperus pilosus Vahl, Enum. Pl. Obs. 2:354. 1805; C.B. Clarke in Hook.f., Fl. Brit. India 6:609. 1893; Verma in N.P. Singh *et al.,* Fl. Madhya Pradesh 3:276.2001.

Rhizomatous perennials bearing long, slender stolons; culms up to 1 m tall, scabrid on the angles above; leaves 6-10 mm broad; anthela compound, 10-40 cm across, bracts 3-5, foliaceous, up to 70 cm long; spikes 3-7 together, 2-5 cm long, diverging, rachis hispidulous; spikelets divaricate or lower deflexed, 6-12 × 1.5-2 mm long, rachilla hyaline on margins; glumes yellowish to reddish-brown, 1.5-2 mm long, broadly scarious and inrolled on margins; nuts blackish, 1-1.3 mm long, apiculate, shortly stipitate.

Fl.-Fr.: July-March

Rare, muddy places, **Indore** only, banks of Sirpur pond.

Cyperus pulchellus R. Br., Prodr. Fl. Nov. Holl. 213. 1810; C.B. Clarke in Hook.f., Fl. Brit. India 6:602. 1893; Verma in N.P. Singh *et al.,* Fl. Madhya Pradesh 3:280.2001.

Perennials; culms 10-30 cm tall; leaves very narrow; anthela solitary, terminal, dense, head, globose, 5-12 mm across, bracts 3-5, foliaceous up to 10 cm long; spikelets numerous, elliptic, congested 3-5 × 1.5-2 mm, several flowered; glumes loosely imbricating, 1-2 mm long, obscurely 3-nerved; stamen 1; nuts black, oblong-ellipsoid, about half as long as the glumes.

Fl.-Fr.: Sept. -Nov.

Not common, Burhanpur, Khandwa.

Cyperus rotundus L., Sp. Pl. 45. 1753; C.B. Clarke in Hook.f., Fl. Brit. India 6: 614. 1893; Samvatsar, Fl. West. Trib. Madhya Pradesh 308. 1996; Verma in N.P. Singh *et al.,* Fl. Madhya Pradesh 3: 281.2001. *Motha*

Perennial, stoloniferous herbs, 10-40 cm high; leaves 2-5 mm broad; anthela usually simple, sometimes compound; spikelets 3-10-together, linear-lanceolate, 10-30 × 1.4-2.0 mm; glumes tightly imbricating, reddish or deep brown, broadly ovate, muticous or mucronulate, keeled, 3.0-3. 4 mm long; nuts glossy- black, broadly obovate or obovoid-ellipsoid, trigonous, 1.4-1.6 mm long.

Fl.-Fr.: July-April

In open moist places, everywhere.

Cyperus squarrosus L., Cent. Pl. 11: 6. 1756; *Mariscus squarrosus* C.B. Clarke in Hook.f., Fl. Brit. India 6:623. 1893; *C. aristatus* Rottb., 1772, *nom. illeg*; Hooper in Saldanha & Nicolson, Fl. Hassan Dist. 690.1976; Samvatsar, Fl. West. Trib. Madhya Pradesh 308. 1996; Verma in N.P. Singh *et al.,* Fl. Madhya Pradesh 3: 283.2001.

Annual, 5-15 cm high herbs; leaves flaccid, 1.4-2.4 mm broad; anthela simple, with 2-4 rays; spikelets numerous, congested, usually brownish, oblong, 6-20-flowered, 5-8 × 3-4 mm; glumes pale or brown, oblong-lanceolate, strongly keeled, mucronate, *ca* 3 mm long; nuts reddish-brown, narrowly obovoid, trigonous, apiculate, *ca* 1.2 mm long.

Fl.-Fr.: Sept. -Nov.

Common in all districts.

Cyperus tenuiculmis Boeckler, Linnaea 36:286. 1870; *C. zollingeri* sensu C.B. Clarke in Hook.f., Fl. Brit. India 6:613. 1893 *non* Steud; *C. zollingeri* Steud. var. *condensatus* Kuk.,

Engl. Pflanzenr. IV, 20 (101:) 134.1935; Kern in Steenis Fl. Males. 1, 7:608.1974; Samvatsar, Fl. West. Trib. Madhya Pradesh 308.1996.

Perennials with corm-like rhizome or knotty and often with short stolons; culms 30-80 cm tall; leaves shorter than culms; anthela simple or subcompound, bracts 2-5, lowest equalling to or longer than the anthela; spikelets linear to linear-lanceolate, 1-3 cm long; glumes ovate-elliptic, boat-shaped, 3.5-4 mm long, 7-9 nerved; nuts elliptic to obovate, 1.7-2 mm long, minutely puncticulate.

Fl.-Fr.: July-Nov.

Occasional as a weed of fields, Alirajpur, Dhar.

Cyperus tenuispica Steud.,Syn. Pl. Glumac. 2:11. 1854; *C. flavidus* sensu C.B. Clarke in Hook.f., Fl. Brit. India 6:600.1893, non Retz. 1789; Samvatsar, Fl. West. Trib. Madhya Pradesh 310.1996; Verma in N.P. Singh *et al.,* Fl. Madhya Pradesh 3:284.2001.

Annuals; culms 5-40 cm tall; leaves much shorter than to equalling the stem; anthela compound to decompound, bracts 1-3, 1 or 2 longer than the anthela; spikelets digitate in clusters of 3-9, linear-oblong; glumes ovate, 0.8-1 mm long, membranous, apex recurved, nerves inconspicuous, keel ending in a short recurved mucro; nuts orbicular-obovate, creamy-yellow, minutely tuberculate.

Fl.-Fr.: Oct. -Dec.

Common weed of fields and waste places throughout the area.

ELEOCHARIS R. Br.W. 288; MP 7; SA 7

Key to Species

1. Annuals
    2. Persistent style base not differentiated from the apex of nut by constriction; nuts 1.7-3. 9 mm, narrowly winged; style 3-fid; perianth bristles 6, shorter than the nuts; culms 5-angled; plants often stoloniferous .... **E. retroflexa**
    2. Persistent style base differentiated from the apex of nut by constriction; style 2- fid; bristles equal to nuts in length; nuts black
        3. Bristles typically 4 (4-7); nuts 0.3-0.5 mm .... **E. atropurpurea**
        3. Bristles typically 7 (4-8); nuts 0.5-1.1 mm .... **E. geniculata**
1. Perennials
    4. Culms not septate at nodes, with 8-30 blunt ridges; style 2-fid; bristles 4, shorter than the nuts .... **E. palustris**
    4. Culms septate at nodes
        5. Style 2-3 fid, base conical, dilated, articulation at rhe junction persistent; spikelets 2-3.5 × 2.5-4 mm; bristles 6-7, unequal, 1or 2 shorter, 1or 2 longer than the nuts,upper 2/3 rd. portion spinous; nuts 1. 9-2.3 mm, abruptly narrow at base, neck short ...... **E. khandwensis**
        5. Style 3-fid; bristles 6-8, retrorsely spinulose

6. Spikelets 1.5-4 cm × 6-7 mm; bristles equal in length to nuts; nuts smooth, 2.5 × 1.8 mm, apex with annulated thickening; stolons ending in corms; culms strongly ribbed .... **E. dulcis**

6. Spikelets 2.5-4 cm × 3-4.5 mm, angular; culms strongly ribbed; bristles unequal, overtopping the annuli but never exceeding the style base; nuts 2 × 1.5 mm, striated, base attenuate .... **E. lankana**

Eleocharis atropurpurea (Retz.) J. Presl & C. Presl, Reliq. Haenk. 1:196. 1828, excl. spec. cit; *Scirpus atropurpureus* Retz., Observ. Bot. 5: 14. 1788; C.B. Clarke in Hook.f., Fl. Brit. India 6: 627. 1893; Samvatsar, Fl. West. Trib. Madhya Pradesh 310.1996; Verma in N.P. Singh *et al.,* Fl. Madhya Pradesh 3: 286.2001.

Annual, slender herbs, 5-16 cm high; leaf sheaths obliquely truncate at mouth, 5-10 mm long; spikelets ovoid-oblong, many-flowered, 3-4 × 2.4 mm; glumes loosely imbricating, elliptic-oblong, 1.0-1.4 mm long; perianth bristles 4-6, glistening white, shorter than to equalling the nut; nuts glossy, brownish -black, obovoid, biconvex, smooth, long.

*Fl.-Fr.:*Aug. -Dec.

In moist open places, throughout the area.

Eleocharis dulcis (Burm.f.) Trin. ex Hensch., Vita Rumphii 186. 1833; *Andropogon dulce* Burm.f., Fl. Ind. 219. 1768; *Eleocharis plantaginea* (Retz.) Roem. & Schult., Syst. Veg. 2:150.1817; C.B. Clarke in Hook.f., Fl. Brit. India 6: 625. 1893; Samvatsar, Fl. West. Trib. Madhya Pradesh 310.1996; Verma in N.P. Singh *et al.,* Fl. Madhya Pradesh 3: 288. 2001.

Perennial, stoloniferous herbs, 0.4-1.5 m tall; leaf sheaths usually reddish, 5-20 cm long; spikelets terete, 15-50 × 3-5 mm; glumes stramineous, with red dots, ovate-oblong, 5-8 mm long; perianth bristles 6-9, yellowish, subequal or 6 well developed and 3 small and thread like; nuts glossy, yellowish- brown, obovoid, biconvex, 1.4-2.6 mm long.

Fl.-Fr.: Sept. -Dec.

In marshy places, all over the area.

Eleocharis geniculata (L.) Roem. & Schult.,Syst. Veg. 2:150.1817; *Scirpus geniculatus* L., Sp. Pl. 48. 1753; C.B. Clarke in Hook.f., Fl. Brit. India 6:627. 1893; Samvatsar, Fl. West. Trib. Madhya Pradesh 310.1996; Verma in N.P. Singh *et al.,* Fl. Madhya Pradesh 3:289.2001.

Annuals; culms 10-30 cm tall; leaf-sheath obliquely truncate at mouth; spikelets ovoid-globose to ovoid-ellipsoid, 3-7 mm long, much broader than the stem; glumes broadly elliptic to ovate, with a red tinge, membranous, 1-nerved, 1.8-2 mm long; perianth bristles 6-8, pinkish, longer than the nut, retrorsely scabrid; stigmas 2; nuts biconvex, shining dark brown or blackish, 1 mm long, depressed conical, apex rounded.

Fl.-Fr.: Aug. -Dec.

In wet and marshy places, Alirajpur, Burhanpur, Dhar, Khandwa.

Eleocharis khandwensis Mujaffar, Chandore and S.R. Yadav Nordic J. Bot. 32:710.2014.

Densely tufted rhizomatous perennials; culms up to 120 cm tall, transversely septate; stolons slender, 10-20 cm long; sheaths membranous, outer up to 8 cm long, inner 12-30 cm long;

spikelets 2.0-3.5 cm long, many flowered, rachilla persistent, glume-like; glumes imbricate, 4-5 ranked, 5.5-6.5 × 2.5-3.5 mm; hypogynous bristles 6-7, 1.5-2.5 mm long; style much dilated into a conical base; nuts biconvex, 2-2.3 mm long, dark brownish to black, epidermal cells isodiametrically hexagonal to pentagonal.

Fl.-Fr.: July-Dec.

Newly discovered species from **Khandwa**, village Bamangaon.

Eleocharis lankana T. Koyama, Bull. Natl. Sci. Mus. Tokyo n.s., 17:67. 1974 & in Rev. Handbook Fl. Ceyl. 5:260.1985.

Densely tufted rhizomatous perennials; culms 20-55 cm tall, indistinctly transversely septate; basal sheaths 2-3, uppermost sheath 8-17 cm long, obliquely truncate at mouth; spikelets cylindrical, 12-30 mm long, many flowered, bract 3-5 mm long; glumes ovate to elliptic, 5.5-6.5 mm long, keel obtuse; nuts biconvex, 2-2.2 mm long, rounded above the ringed apex, epidermal cells isodiametrically hexagonal, not pitted; hypogynous bristles 6-8, shorter than to equalling the nuts.

Fl.-Fr.: July-Dec.

Endemic to Ceylon, but now the distribution extended to M.P. (Burhanpur and Khandwa).

Eleocharis palustris (L.) Roem. & Schult.,Syst. Veg. 2:151. 1817; *Scirpus palustris* L. Sp. Pl. 47. 1753; C.B. Clarke in Hook.f., Fl. Brit. India 6:628. 1893; Verma in N.P. Singh *et al.,* Fl. Madhya Pradesh 3:289.2001.

Perennials with a long, creeping rhizome; culms 10-40 cm tall; leaf sheaths reddish-brown, upper truncate or bearing a short lamina; spikelets 7-15 × 3-5 mm, densely many flowered; glumes ovate-lanceolate, 3.5-4 mm long; perianth bristles 4, subequal, *ca* 1.5 mm long; stigmas 2; nuts glossy, biconvex, 1.2-1.7 mm long.

Fl.-Fr.: April-June

Along water drains, Burhanpur, Khandwa.

Eleocharis retroflexa (Poir.) Urban, Symb. Antill. 2:165. 1900; *Scirpus retroflexus* Poir., Encycl. 6:753. 1805; C.B. Clarke in Hook.f., Fl. Brit. India 6:629. 1893; Verma in N.P. Singh *et al.,* Fl. Madhya Pradesh 3:289.2001.

Annuals (sometimes perennial also); culms 3-10 cm or more tall; leaf sheaths lax, inner 0.5-2.5 cm long; spikelets ovoid-ellipsoid, 1.5-4 × 1.5-3 mm, 6-8 flowered, but maturing 1-4 nuts; glumes 2.5-3.2 mm long; perianth bristles 6 or less, very unequal; stigmas 3; nuts straw-coloured about 1 mm long, surface cells subquadrate in 6-10 vertical rows on each face.

Fl.-Fr.: Sept. -Dec.

Rare, near wet areas, **Burhanpur** only, Jaswadi.

ERIOPHORUM L. W. 27; MP. 1; SA 1

Eriophorum comosum (Wall.) Nees, Contr. Bot. India 110.1834; *Scirpus comosus* Wall., Roxb. Fl. Ind. 1:234. 1820; C.B. Clarke in Hook.f., Fl. Brit. India 6:664. 1893; Samvatsar, Fl.

West. Trib. Madhya Pradesh 311. 1996; Verma in N.P. Singh *et al.,* Fl. Madhya Pradesh 3:290.2001.

Rhizomatous perennials; culms 15-50 cm high; leaves equalling or exceeding the stem, sheaths glossy; anthela compound or supradecompound, up to 25 cm long, bracts up to 15 cm long, foliaceous; spikelets solitary or paired, ellipsoid, reddish-brown, 3-7 × 1-1.5 mm, comose at maturity, rachilla persistent; glumes 2.5-3 mm long, oblong-lanceolate, keeled, acute, mucronate; perianth segments 7-11 mm long in fruit; stamen1; nuts glossy, black, 2.5-2.8 mm long, compressed-trigonous.

Fl.-Fr.: Sept. -Dec.

Hanging from crevices of rocks, in all districts.

FIMBRISTYLIS Vahl W. 306; MP 17; SA 15

Key to Species

1. Style 2-fid; nut biconvex
 2. Inflorescence of single, terminal spikelet (sometimes 2-3 added)
  3. Plants leafless, short, leaves, if present, then eligulate
   4. Nuts cylindrically oblong with almost parallel sides, longer than broad, tuberculate between 5-8 vertical ridges; style base as broad as nuts . ... **F. tetragona**
   4. Nuts obovoid-ellipsoid or suborbicular, transversely undulate with 8 wavy ridges; style base narrower than nuts; spikelets erect, 1-3.5 mm wide . ... **F. acuminata**
  3. Plants leafy; leaves ligulate; glumes with distinct, 4-7 nerves, 3x3 mm, broader than long; spikelets whitish, largest 3-4.5 mm broad; stamens 3; nuts 1.5-2 mm long, reticulate .... **F. schoenoides**
 2. Inflorescence of few to many spikelets
  5. Leaves eligulate
   6. Annuals, hairy all over; glumes without scarious margins; leaves up to 7 cm long; stamen usually 1 .... **F. aestivalis**
   6. Perennials with stout rhizomes; glumes with broad, scarious margins; leaves up to 30 cm long; stamens usually 3 .... **F. cymosa**
  5. Leaves ligulate
   7. Glumes greyish tomentose; style densely fimbriate throughout; nuts smooth; perennials .... **F. ferruginea**
   7. Glumes glabrous
    8. Nuts smoothly tuberculate; spikelets solitary
     9. Nuts 0.8-0.9 mm long, with 12-16 vertical striations; spikelets up to 2 mm broad .... **F. albo-viridis**
     9. Nuts 0.4-0.5 mm long, with 4-9 vertical ridges; spikelets 1.5-2.2 mm broad . ... **F. bisumbellata**
    8. Nuts coarsely tuberculate, 0.8-1. 1 mm long with 7-9 vertical ridges . ... **F. dichotoma**
1. Style 3-fid; nuts trigonous or triquetrous
 10. Inflorescence a single spikelet .... **F. ovata**

10. Inflorescence of few to many spikelets
 11. Spikelets in clusters of 2-5 .... **F. falcata**
 11. Spikelets not clustered
  12. Stems strongly compressed; leaves ligulate . ... **F. complanata**
  12. Stems not compressed; leaves if present, eligulate
   13. Cauline leaves well developed, eligulate; stamens 1 or2; nuts with 3-10 vertical rows of cells; annuals .... **F. tenera**
   13. Cauline leaves reduced to bladeless sheaths
    14. Leaf sheaths strongly laterally compressed; nuts 0.5-1.2 mm long .... **F. littoralis**
    14. Leaf sheaths rounded; stem 5-angled; nuts 0.5-1.2 mm long.......... **F. quinquangularis**

Fimbristylis acuminata Vahl, Enum. Pl. Obs. 2: 285. 1805; C.B. Clarke in Hook.f., Fl. Brit. India 6: 631. 1893; *F. setacea auct. non* Benth., 1843: C.B. Clarke in Hook.f., Fl. Brit. India 6: 632. 1893; Samvatsar, Fl. West. Trib. Madhya Pradesh 312. 1996; Verma in N.P. Singh *et al.,* Fl. Madhya Pradesh 3: 295. 2001.

Perennials, 5-25 cm high; stems obtusely quadrangular; leaves absent; spikelets solitary, terminal, ovoid-oblong or lanceolate, 6-12 × 2-4 mm; glumes stramineous or with brown tinge, acute or mucronate, 3.5-5.0 × 3-5 mm; nuts stramineous or brownish, broadly obovoid or suborbicular, biconvex, shortly stipitate, with 5-8 brownish ridges, 1.2-1.7 × 1.0-1.5 mm.

Fl.-Fr.: Sept. -Dec.

In moist shady places, Alirajpur, Barwani, Burhanpur, Dhar, Khandwa.

Fimbristylis aestivalis Vahl, Enum. Pl. Obs. 2:288. 1805; *Scirpus aestivalis* (Vahl) Retz., Observ. Bot. 4:12. 1786; C.B. Clarke in Hook.f., Fl. Brit. India 6:637. 1893; Samvatsar, Fl. West. Trib. Madhya Pradesh 312. 1996; Verma in N.P. Singh *et al.,* Fl. Madhya Pradesh 3:296.2001.

Annual, densely tufted herbs; culms 5-20 cm high; leaves shorter than or subequalling the stem, sheaths hairy; anthela compound or decompound, up to 5 cm across, umbellate, bracts filiform; spikelets oblong-lanceolate, 2.5-6 × 1-1.5 mm, acute, bracts 1-3 or several, scabrid, equalling the anthela; glumes pale-brown, 1.5 mm long, keeled, oblong, 2-4 aristate; stamens 1-2; nuts biconvex, shortly stipitate, straw-coloured.

Fl.-Fr.: Sept. -Jan.

Common in marshy places in all districts.

Fimbristylis albo-viridis C.B. Clarke, in Hook.f., Fl. Brit. India 6:638. 1893; Verma in N.P. Singh *et al.,* Fl. Madhya Pradesh 3:296.2001.

Annuals; culms 15-30 cm high; leaves several; anthela lax, compound or decompound, bracts foliaceous; spikelets solitary, ovoid, 4-10 × 2-2.5 mm; glumes broadly ovate, muticous, 2-2.5 mm long, 3-nerved; stamens 1-2; stigmas 2; nuts straw-coloured, biconvex, 1-1.2 mm long, verruculose, surface cells arranged in 12-16 rows on each face, making them trabeculate.

Fl.-Fr.: July-Oct.

Rare, in moist locations, **Indore** only, Banedia pond.

Fimbristylis bisumbellata (Forssk.) Bubani, Dodecanthea 30.1850; *Scirpus bisumbellatus* Forssk., Fl. Aegypt. -Arab. 15. 1775; *Fimbristylis dichotoma auct. non* Vahl, 1805: C.B. Clarke in Hook.f., Fl. Brit. India 6: 635. 1893; Samvatsar, Fl. West. Trib. Madhya Pradesh 312. 1996; Verma in N.P. Singh *et al.,* Fl. Madhya Pradesh 3: 297.2001.

Annual herbs, 10-20 cm high; leaves usually falcate, sometimes pubescent beneath, 1.5-2.5 mm broad; sheaths hairy; anthela compound to supradecompound, 2-4 cm across; spikelets solitary, polygonal, ovoid-oblong or lanceolate, 3-7 × 1.0-1.4 mm; glumes yellowish- brown, sharply keeled, 1.2-1.4 mm long; nuts whitish stramineous, broadly obovoid.

Fl.-Fr.: June-Dec.

Common in moist places throughout the area.

Fimbristylis complanata (Retz.) Link, Hort. Berol. 1:292. 1827; *Scirpus complanata* Retz., Observ. Bot. 5:14. 1788; C.B. Clarke in Hook.f., Fl. Brit. India 6:646. 1893; Samvatsar, Fl. West. Trib. Madhya Pradesh 312. 1996; Verma in N.P. Singh *et al.,* Fl. Madhya Pradesh 3:298. 2001.

Annuals or perennials; culms 30-80 cm high; stem compressed below the inflorescence; leaves flat, ½ to ¾ th length of the stem, ligule a fringe of hairs; anthela of compound umbels, 2-8 cm across, bracts suberect, lowest foliaceous; spikelets solitary, oblong-lanceolate, 5-7 × 2 mm; glumes spiral, 2-2.5 mm long, keel green, 3-nerved; nuts whitish-stramineous, trigonous, 0.7-1.2 mm long, verruculose.

Fl.-Fr.: Aug. -Dec.

Common weed of waste places and fields, in all districts.

Fimbristylis cymosa R. Br., Prodr. Fl. Nov. Holl. 228. 1810; Kern, Fl. Males. 1,7:557. 1974; C.B. Clarke in Hook.f., Fl. Brit. India 6:640.1893; Sharma *et al.,* in Fl. Maharashtra 3:307. 1996.

Perennial, rhizomatous herbs; culms tufted, 10-20 cm tall, base many-leaved; leaves much shorter than the stem, abruptly acuminate, scabrid along margins; umbels compound, contracted, bracts 2-3, short, rigid; spikelets ellipsoid, 3-5 mm long, rachilla broadly winged; glumes ovate, obtuse, notched, keeled, 1-2 mm long; nuts 0.7-01 × 0.0.5 cm, shortly stipitate, black, tubercled.

Fl.-Fr.: Aug. -Dec.

Rare, in moist places, **Indore** only, Choral river.

Fimbristylis dichotoma (L.) Vahl, Enum. Pl. Obs. 2: 287. 1805, excl. var; *Scirpus dichotomus* L., Sp. Pl. 50.1753; *S. diphyllus* Retz., Observ. Bot. 5:15. 1788; *Fimbristylis diphylla* (Retz.) Vahl, Enum. Pl. Obs. 2: 289. 1805; C.B. Clarke in Hook.f., Fl. Brit. India 6: 636. 1893; Samvatsar, Fl. West. Trib. Mdhya Pradesh 312. 1996; Verma in N.P. Singh *et al.,* Fl. Madhya Pradesh 3: 298. 2001.

Annual or perennial herbs, 10-80 cm high; leaves mostly from the base, flat, linear, acute to obtuse at apex, glabrous or pubescent beneath, 1.5-5.0 mm broad; anthela simple to supra-decompound, 3-8 cm across; spikelets ovoid-globose, terete; glumes spiral, glossy chestnut, ovate-oblong, 2-3(-5) mm long, 3-nerved; nuts pale stramineous, obovoid, 0.8-1.2 × 0.6-0.8 mm.

Fl.-Fr.: Aug. -Dec.

Very common on rocky surfaces in fields and marshy places, all over the area.

Fimbristylis falcata (Vahl) Kunth, Enum. Pl. 2:239. 1837; *Scirpus falcatus* Vahl, Enum. Pl. Obs. 2:275. 1805; *Fimbristylis junciformis* (Nees) Kunth, Enum. Pl. 2:239. 1837; C.B. Clarke in Hook.f., Fl. Brit. India 6:647. 1893; Samvatsar, Fl. West. Trib. Madhya Pradesh 312. 1996; Verma in N.P. Singh *et al.,* Fl. Madhya Pradesh 3:301.2001.

Perennials; culms 10-40 cm tall, rhizome short, creeping; stems 4-5 angled; anthela simple or coMPound, up to 10 cm across, usually bearing 20-30 or sometimes only 3-5 clusters of spikelets; spikelets in clusters of 2-5, sometimes solitary, 3-6 mm long; glumes spiral, broadly ovate, 2-2.7 mm long; nuts pale-stramineous, trigonous, 0.8-1 mm long, stipe short, smooth or verruculose.

Fl.-Fr.: Oct. -Dec.

Common in wet places, all over the area.

Fimbristylis ferruginea

Key to Subspecies

1. Nut with a distinct stalk . ... subsp. **sieberiana**
1. Nut without a stalk or if present, very obscure ... . subsp. **ferruginea**

Fimbristylis ferruginea (L.) Vahl subsp. **ferruginea**, Enum. Pl. Obs. 2:291. 1805; *Scirpus ferrugineus* L., Sp. Pl. 50.1753; C.B. Clarke in Hook.f., Fl. Brit. India 6:638. 1893; Samvatsar, Fl. West. Trib. Madhya Pradesh 313. 1996; Verma in Fl. Madhya Pradesh 3:301.2001.

Tufted, rhizomatous perennials; culms 10-60 cm tall, covered in grey-brown scales; leaves almost absent, short or up to 15 cm long; anthela simple or compound, corymbose; spikelets ovate-lanceolate, 5-10 × 3-4 mm, solitary; glumes ovate-oblong, 3-4 mm long, apiculate, keeled upwards; stigmas 2; nuts white or pale yellow, strongly compressed, 1-1.5 mm long, shortly stipitate, epidermal cells in many vertical rows.

Fl.-Fr.: Aug. -Oct.

Common in open marshy places, all districts.

Fimbristylis ferruginea subsp. **sieberiana** (Kunth) Lye,Nordic J. Bot. 2:235. 1982; *F. sieberiana* Kunth, Enum. Pl. 2:237. 1837; *F. ferruginea* (L.) Vahl, var. *sieberiana* (Kunth) Boeckler, Linnaea 37; 17, 1817; Kern in Steenis, Fl. Males. 1, 7:572. 1974; Sharma *et al.,* Fl. Maharashtra 3:317. 1996; Samvatsar, Fl. West. Trib. Madhya Pradesh 313. 1996.

Perennials with a small, woody rhizome; leaves flat, narrowly linear, scabrid on margins; anthela simple or compound; spikelets ovate, obtuse, apiculate; nuts elliptic, flattened, finely striated.

Fl.-Fr.: Aug. -Dec.

In marshy places, Alirajpur, Dhar, Jhabua.

Fimbristylis littoralis Gaudich., Voy. Uraine 413. 1829; C.B. Clarke in Hook.f., Fl. Brit. India 6:644. 1893; Verma in N.P. Singh *et al.,* Fl. Madhya Pradesh 3:303.2001.

Annuals or perennials; culms 10-6- cm high, 4-5 angled; leaves several, more or less distichous, as long as the stem, grooved, upper leaves reduced to sheaths only; anthela compound or supra- decompound, bracts short, foliaceous; spikelets solitary, 2-5 × 1.5-2 mm; glumes spiral, ovate-oblong, 1-1.5 mm long, 3-nerved; nuts brown, narrowly obovoid, 0.6-0.7 mm long, shortly stipitate, verruculose, transversely lineolate.

Fl.-Fr.: Sept. -April

Along ponds and streams, Indore, Khargone.

Fimbristylis ovata (Burm.f.) J. Kern, Blumea 15:126. 1967; *Carex ovata* Burm.f., Fl. Indica 194. 1768; *Cyperus monostachyos* L., Mant. Pl. 2:180.1771; *Fimbristylis monostachyos* (L.) Hassk., Pl. Jav. Rar. 61. 1848; C.B. Clarke in Hook.f., Fl. Brit. India 6:649. 1893; Samvatsar, Fl. West. Trib. Madhya Pradesh 313. 1996; Verma in N.P. Singh *et al.,* Fl. Madhya Pradesh 3:304.2001.

Perennials; culms 10-30 cm high; leaves crowded near the base, ½ the length of the stem; spikelets solitary, terminal, ovate-lanceolate, 6-17 × 3-6 mm, rachilla winged, bracts setaceous; lowr glumes distichous, upper spiral, broadly ovate-oblong, 4-6 mm long, keeled, mucronate, 3-5 nerved; nuts yellowish-white or brownish, trigonous, 2-3 mm long, stipitate, tuberculate.

Fl.-Fr.: Aug. -Jan.

In marshy places, not common, Alirajpur, Dhar, Indore, Khandwa.

Fimbristylis quinquangularis (Vahl) Kunth, Enum. Pl. 2:229. 1837; *Scirpus quinquangularis* Vahl, Enum. Pl. Obs. 2:279. 1805; *F. miliacea* (L.) Vahl, Enum. Pl. 2:267. 1805; *Scirpus miliaceus* L., Syst. Nat. ed. 10.2: 868. 1759; *Fimbristylis mucronata* Vahl, Enum. Pl. Obs. 2:293. 1805; C.B. Clarke in Hook.f., Fl. Brit. India 6: 644. 1893; Samvatsar, Fl. West. Trib. Madhya Pradesh . 313. 1996; Verma inN.P. Singh *et al.,* Fl. Madhya Pradesh 3: 303.2001.

Annuals or perennials, 10-80 m high; leaves on sterile shoots several, 2-3 mm broad; sheaths on the flowering stems inflated; anthela compound to supra -decompound; bracts short, usually setaceous; spikelets all or nearly all solitary; glumes brown, spiral, ovate, 1-2 mm long; nuts stramineous or pale brown, trigonous, 0.4-0.7 × 0.3-0.4 mm.

Fl.-Fr.: Sept. -Nov.

In puddles and along edges of ponds, all over the area.

Fimbristylis schoenoides (Retz.) Vahl, Enum. Pl. Obs. 2: 286. 1805; *Scirpus schoenoides* Retz., Observ. Bot. 5: 14. 1789; C.B. Clarke in Hook.f., Fl. Brit. India 6: 634. 1893; Samvatsar, Fl. West. Trib. Madhya Pradesh 313. 1996; Verma in N.P. Singh *et al.,* Fl. Madhya Pradesh 3: 306.2001.

Perennial herbs with a small woody rhizome; leaves narrowly linear, ligulate, 0.4-1.0 mm broad; anthela of 2-3 spikelets or reduced to a solitary, terminal spikelet; spikelets broadly ovoid or ovoid-oblong, terete, 4-8 × 3.0-4. 4 mm; glumes spiral, glossy, stramineous or with red tinge, muticous or mucronulate, 2.4 -3.3 mm long; nuts stramineous, obovoid, biconvex, 1.2-1.6 × 0.8-1.2 mm.

Fl.-Fr.: Sept. -Oct.

Common, along roadside drains, all over the area.

Fimbristylis tenera Schult., Mant. 2: 57. 1824; C.B. Clarke in Hook.f., Fl. Brit. India 6: 642. 1893; Samvatsar, Fl. West. Trib. Madhya Pradesh 314. 1996; Verma in N.P. Singh *et al.,* Fl. Madhya Pradesh 3: 308. 2001.

Annuals, 10-30 cm tall herbs; leaves almost filiform, usually shorter than the stem; anthela simple or subcompound; spikelets ellipsoid or oblong-ellipsoid, 4-6 mm long; glumes yellowish or brown, ovate-lanceolate, acute, glabrous or glandular puberulous; keel shortly excurrent and strongly incurved at the base; nuts grayish- white, obovoid, tuberculate, *ca* 0.6 mm long.

Fl. Fr. -Oct. -Feb.

Common weed of fields and waste places.

Fimbristylis tetragona R.Br., Prodr. Fl. Nov. Holl. 226. 1810; C.B. Clarke in Hook.f., Fl. Brit. India 6:631. 1893; Samvatsar, Fl. West. Trib. Madhya Pradesh 314. 1996; Verma in N.P. Singh *et al.,* Fl. Madhya Pradesh 3:308. 2001.

Annuals or perennials; culms 10-60 cm high, tetragonous; leaves absent, sheaths 2-3; spikelets solitary, terminal, ovoid-globose, 6-18 × 4-6 mm; glumes spiral, ovate-oblong, 3-5 mm long; stamens 1-3; stigmas 2-3; nuts brown, linear-oblong with almost parallel sides, 1.5-2 mm long, stipitate, trabeculate due to the surface cells being arranged in about 9 vertical rows.

Fl.-Fr.: Oct. -Nov.

Occasional in wet places, Alirajpur, Dhar, Indore, Khandwa.

FUIRENA Rottb. W. 59; MP 5; SA 2

Key to Species

1. Hypogynous scales absent; leaves 1-ribbed . ... **F. cuspidata**
1. Hypogynous scales present; leaves 3-5 ribbed .... **F. ciliaris**

Fuirena ciliaris (L.) Roxb., Hort. Bengal. 81. 1814; *Scirpus ciliaris* L. Mant. Pl. 182. 1771; C.B. Clarke in Hook.f., Fl. Brit. India 6:666. 1893; Samvatsar, Fl. West. Trib. Madhya Pradesh 314. 1996; Verma in N.P. Singh *et al.,* Fl. Madhya Pradesh 3:310.2001.

Annuals; culms 10-40 cm gigh, pubescent all over, including leaves and glumes; leaves linear-lanceolate, 5-15 cm long, 3-5 nerved; partial panicles more or less congested at the tip; spikelets 4-12 mm long, in clusters of 3-10, ovoid-oblong; glumes 1.5-1.7 mm long, with hairy, recurved awns; perianth bristly, length variable, sometimes absent; perianth scales brown, subquadrate, as long as the nut, clawed, 3-dentate at apex, 3-nerved; nuts brownish, triquetrous, smooth or faintly transversely lineolate.

Fl.-Fr.: Sept. -Dec.

Wet and marshy places, Alirajpur, Barwani, Dhar, Khandwa.

Fuirena cuspidata (Roth) Kunth, Enum. Pl. 2:187. 1837; *Scirpus cuspidatus* Roth, Nov. Pl. Sp. 31. 1821; *Fuirena wallichiana* Kunth, Enum. Pl. 2:182. 1837; C.B. Clarke in Hook.f., Fl.

Brit. India 6:665. 1893; Samvatsar, Fl. West. Trib. Madhya Pradesh 314. 1996; Verma in N.P. Singh etal Fl. Madhya Pradesh 3:311.2001.

Perennials; culms 30-60 cm tall; leaves shorter than the stem, linear-lanceolate; spikelets 6-8 mm long, in corymbose clusters; glumes 3-3.5 mm long with red streaks, apex rounded; perianth scales absent; nuts trigonous, broadly obovate to suborbicular, *ca* 1.2 mm long, longitudinally ribbed.

Fl.-Fr.: July-Dec.

In sandy river beds, Alirajpur, Dhar, Khandwa.

KYLLINGA Rottb. W. 78; MP 3; SA 3

Key to Species

1. Annuals; midrib of glumes winged, laciniate .... **K. tenuifolia**
1. Perennials
   2. Rhizome thin, creeping; plants 5-30 cm tall; culms 3-angled; nuts brown .... **K. brevifolia**
   2. Rhizome stout, short; plants 30-100 cm tall; culms compressed; nuts black .... **K. melanosperma**

Kyllinga brevifolia Rottb., Descr. Icon. Rar. Pl. 13. 1773; *Cyperus brevifolius* (Rottb.) Hassk., Cat. Hort. Bot. Bogor. 24. 1844; C.B. Clarke in Hook.f., Fl. Brit. India 6:588. 1893; Samvatsar, Fl. West. Trib. Madhya Pradesh 316. 1996; Verma in N.P. Singh *et al.,* Fl. Madhya Pradesh 3: 256.2001.

Perennial, rhizomatous herbs, 10-80 cm high; leaves much shorter than to equalling the stem, 1.4-3. 4 mm broad; inflorescence capitate; bracts 3-4, foliaceous, spreading or deflexed; spikelets numerous, congested, elliptic-lanceolate, 1-flowered, 2.4-3. 4 × 1 mm; glumes keeled, mucronate; keel with spinules up to two third way up; nuts ultimately brown, obovoid or oblong-obovoid, 1.0-1.4 × 0.6 mm.

Fl.-Fr.: Sept. -Dec.

In open wet places throughout the area.

Kyllinga melanosperma Nees, Contr. Bot. India 91. 1834; *Cyperus melanospermus* (Nees) Suringar, Cyperus Mal. Arch. 50.1898; Verma in Singh *et al.,* Fl. Madhya Pradesh 3; 272 2001.

Perennials, up to 1 m high; rhizome short, creeping, woody, up to 5 cm long; stems few to many; leaves much shorter than the stem; anthela solitary, terminal, up to 1 cm long; bracts 3-4, foliaceous, 15 cm or more long; spikelets numerous, congested in a head, bearing 1, rarely 2 nuts; nuts black, obovoid-oblong.

Fl.-Fr.: Aug. -Nov.

Burhanpur only, Shahi Kila, Asirgad.

Kyllinga tenuifolia Steud., Syn. Pl. Glumac. 2:69. 1855; *Kyllinga triceps* Rottb., 1773, *nom. illeg; Cyperus triceps* Endl., Cat. Horti. Vindob. 1:94. 1842; C.B. Clarke in Hook.f., Fl. Brit. India 6: 587. 1893. Verma in N.P. Singh *et al.,* Fl. MP 3: 284.2001.

Annual or (rarely) perennial herbs, 5-40 cm high; leaves shorter or longer than the stems, linear; gradually acuminate, 1-3 mm wide; sheaths somewhat stout, often brownish at base; inflorescence capitate, ovoid-oblong or subcylindrical, 4-6 mm across; spikelets numerous, congested, ovate-lanceolate, 1-flowered, *ca* 2 mm long; glumes green, ovate-lanceolate, 2.0-2.4 mm long; nuts yellowish brown, *ca* 1.4 mm long.

Fl.-Fr.: Aug. - Dec.

Common in open, moist places.

LIPOCARPHA R. Br. W. 35; MP 3; SA 3

Key to Species

1. Involucral bracts strictly erect; spike 1; glumes of spikelets present; stigmas 2 .... **L. hemisphaerica**
1. Involucral bracts reflexed; spikes 1-3; glumes of spikelets absent
   2. Stigmas 3; nuts rounded-triquetrous in cross section .... **L. squarrosa**
   2. Stigmas 2; nuts sub-terete in cross section . ... **L. kernii**

Lipocarpha hemisphaerica (Roth) Goetegh., Wageningen Agric. Univ. Pap. 89(1):37. 1989; *Scirpus hemisphaericus* Roth, Nov. Pl. Sp. 29. 1821; *Hemicarpha isolepis* Nees, Edinburgh New Philos. J. 17:263. 1834; *Scirpus isolepis* (Nees) Boeckler, Linnaea 36:498. 1870; Samvatsar, Fl. West. Trib. Madhya Pradesh 318. 1996.

A small, tufted annual herb; culms setaceous, 2-12 cm tall; leaves 2-6 cm long, linear; spikelets usually single, 20-30 flowered, bracts erect; glumes minute, obovate, rhomboid, minutely mucronate; hypogynous bristles absent; a thin, hyaline, irregular scale present between nuts and glumes; nuts biconvex, obovoid.

Fl.-Fr.: Oct. -Dec.

In marshy and open moist ground, Alirajpur, Burhanpur, Khandwa.

Lipocarpha kernii (Raymond) Goetegh., Wageningen Agic. Univ. Pap. 89(1):42. 1989; *Scirpus kernii* Raymond, Naturaliste Canad. 86:230.1959; *Isolepis kernii* (Raymond) Lye, Bot. Not. 124:479. 1971; *Rikliella kernii* (Raymond) J. Raynal, Adansonia n.s. 13:155. 1973; Hooper in Sald. & Nicols. Fl. Hassan 696. 1976; Sharma in Fl. Maharashtra 3:356. 1996.

Annuals, culms up to 10 cm high; leaves filiform, canaliculated, obtuse, sheaths papery, red-purple, bracts 2-3, spreading; spikelets oblong, 5-8 mm long, obtuse; glumes 1 mm long, aristate, arista 1 mm long; nuts compressed *ca* 0.75 mm long, black with 18-20 rows of epidermal cells.

Fl.-Fr.: Oct. -Dec.

Rare, **Khandwa** only, Jaswadi.

Lipocarpha squarrosa (L.) Goetegh., Wageningen Agric. Univ. Pap. 89(1):71. 1989; *Scirpus squarrosus* L., Mant. Pl. 2:181. 1771; C.B. Clarke in Hook.f., Fl. Brit. India 6:663. 1893; *Rikliella squarrosa* (L.) J. Raynal, Adansonia n.s. 13:154. 1973; Samvatsar, Fl. West. Trib. Madhya Pradesh 319. 1996.

Annuals; culms 10-20 cm tall; leaves 2 or 3, lowest reduced to sheath; anthela a pseudolateral head with 1 or 3 spikes, bracts 1 or 2, longer than the anthela; spikes sessile, ovate-elliptic, squarrose, 3-6 mm long; glumes spirally imbricate, without awn, obscurely 2-nerved on each side, keel projecting beyond the apex of glume, into a recurved, smooth, 1.8-2.2 mm long awn; nuts obovate, 0.5-0.6 mm long, minutely puncticulate.

Fl.-Fr.: Sept. -Dec.

Common in marshy places, Alirajpur, Dhar, Khandwa.

PYCREUS P. Beauv. W. 121; MP 5; SA 5

Key to Species

1. Glumes mucronate; spikes globose; spikelets white, congested; annuals ....**P. pumilus**
1. Glumes muticous; annuals or perennials
   2. Nuts with transverse, wavy lines; glumes 1.8-3 mm long ....**P. diaphanus**
   2. Nuts puncticulate or finely reticulate
3. Glumes with a depression on either side, 1.8-2.5 mm long, blood-red; keel curved; nuts 1- 1.3 mm long .... **P. sanguinolentus**
3. Glumes without depression
   4. Glumes 1.5 -2.5 mm long ....**P. polystachyos**
   4. Glumes 0.8-1.2 mm long ....**P. flavidus**

Pycreus diaphanus (Schrad. ex Roem. & Schult.) S. S. Hooper & T. Koyama, in Jap. J. Bot. 51:316. 1976; *Cyperus diaphanus* Schrad. ex Roem. & Schult., Mant. 2:477. 1824; *C. latespicatus* Boeckeler, Flora 42:441. 1859; *Pycreus latespicatus* (Boeckeler) C.B. Clarke in Hook.f., Fl. Brit. India 6:696. 1894; Verma in N.P. Singh *et al.,* Fl. Madhya Pradesh 3:271.2001.

Tufted annual herbs, 15-20 cm tall; leaves flat, margins scabrid in the upper part; anthela simple, umbellate of 2-3 rays; spikelets spreading horizontally, 1-2 cm long; glumes reddish-brown, membranous, ovate, obliquely, patently keeled, sides nerveless; stamens and stigmas 2; nuts laterally compressed, stipitate, 1. 1 × 0.8 mm.

Fl.-Fr.: Aug. -Nov.

Rare in marshy places, Alirajpur, Dhar.

Pycreus flavidus (Retz.) T. Koyama in J. Jap. Bot. 51 : 316. 1976; *Cyperus flavidus* Retz., Observ. Bot. 5 : 13. 1788; *Pycreus capillaris* (J. Koenig ex Roxb.) Nees ex C.B. Clarke Fl. Brit. India 6:591. 1893; *Cyperus capillaris* J. Koenig ex Roxb. Fl. Ind. 1:194. 1820; Samvatsar, Fl. West. Trib. Madhya Pradesh 306. 1996; Verma in N.P. Singh *et al.,* Fl. Madhya Pradesh 3: 267.2001.

Annuals or perennials, 10-70 cm high; leaves 1-2 mm broad; lower sheaths reddish- brown; anthela simple; bracts 2-4, foliaceous; rays 1-4, up to 11 cm long; spikelets numerous, linear, 8-30 × 1.4-3.0 mm; glumes pale- brown to chestnut, muticous, hyaline on margins, 1.5-2.5 mm long; nuts dark brown, laterally compressed, biconvex, oblong-obovate to oblong-elliptic, apiculate at apex, 0.7-1.2 × 0.4-0.6 mm.

Fl.-Fr.: Aug. -Jan.

Occasional in marshy places in all districts.

Pycreus polystachyos (Rottb.) P. Beauv., Fl. Oware 2:48. 1816; *Cyperus polystachyos* Rottb., Descr. Icon. Rar. Pl. 39. 1773; C.B. Clarke in Hook.f., Fl. Brit. India 6:592. 1893; Verma in N.P. Singh *et al.,* Fl. Madhya Pradesh 3:278. 2001.

Annuals or perennials; culms 10-70 cm high; leaves greyish-green, 2-4 mm broad; anthela compound, bracts 3-6, foliaceous, up to 20 cm long; spikelets congested, 8-25 × 1.5 mm, rachilla narrowly winged glumes ovate-lanceolate, 1.5-2 mm long acute, membranous; stigma deeply 2-fid; nuts ultimately glossy-black, lenticular, linear-oblong, 1. 1.2 mm long, apiculate.

Fl.-Fr.: Aug. -Nov.

Rare, **Khandwa** only, Mortakka, river Narmada banks.

Pycreus pumilus (L.) Nees, Linnaea 9 :283. 1834; *Cyperus pumilus* L., Cent. Pl. II 6 1756; *P. nitens* (Retz.) Nees, Nov. Actorum Acad. Caes. Leop.-Carol. Nat. Cur. 19 (Suppl 1):53. 1843; *Cyperus nitens* Retz. Observ. Bot. 5:13. 1788; *Pycreus pumilus* (L.) Nees ex C.B. Clarke in Hook.f., Fl. Brit. India 6: 591. 1893; Samvatsar, F. West. Trib. Madhya Pradesh 307, 1996; Verma in Singh *et al.,* Fl. Madhya Pradesh 3: 280.2001.

Annual herbs, 5-20 cm tall; leaves 1-2 mm broad; anthela simple or subcompound, up to 18 cm long; spikelets several to many, congested together in subglobose heads, linear-oblong, 5-10 × 1-2 mm; glumes stramineous, ovate, mucronate, 1.0-1.4 mm long; nuts brownish, compressed, biconvex, oblong-obovate, truncate to obtuse and apiculate at apex, 0.5-0.8 × 0.3-0.5 mm.

Fl.-Fr.: Sept. -Dec.

Common in all districts.

Pycreus sanguinolentus (Vahl) Nees, Linnaea 9:283. 1834; *Cyperus sanguinolentus* Vahl, Enum. Pl. Obs. 2:351. 1805; Samvatsar, Fl. West. Trib. Madhya Pradesh 308. 1996; Verma in N.P. Singh *et al.,* Fl. Madhya Pradesh 3:282.2001.

Annuals or perennials; culms 30-90 cm high; rhizomes when present, short and creeping; leaves several, clustered at the base; anthela open, bracts 2-5, foliaceous, up to 20 cm long; rays when present1-3, up to 4 cm long; spikelets congested, 3-20 together, oblong-lanceolate or ovate, 5-25 × 2-3 mm;glumes imbricating, ovate, 1.5-2.5 mm long, keel green, 3-5 nerved; stigma 2-fid; nuts blackish-brown,broadly obovoid or suborbicular, 1-1.3 mm long, apiculate.

Fl.-Fr.: July-Dec.

Common as a weed, in forest, waste places in all districts.

SCHOENOPLECTIELLA Lye W; 26; MP 5; SA 5

Key to Species

1. Hypogynous scales absent; bristles present, barbellate; inflorescence apparently pseudolateral due to the solitary stem-like bract, appearing like a continuation of stem

   2. Stem sharply triquetrous . ... **S. mucronata**

2. Stem terete .... **S. juncoides**
1. Hypogynous scales or bristles absent; stigmas 3; stems sub-terete
3. Stems solid; nuts with transverse wavy lines .... **S. lateriflora**
3. Stems hollow, usually septate; glumes 3-6 mm long
4. Nuts smooth .... **S. articulata**
4. Nuts with transverse wavy lines . ... **S. roylei**

Schoenoplectiella articulata (L.) Lye, Lidia 6:20.2003; *Scirpus articulatus* L. Sp. Pl. 47. 1753; *Isolepis articulata* (L.) Nees, Contri. Bot. India 108. 1834; *Schoenoplectus articulatus* (L.) Palla,Bot. Jahrb. Syst. 10:299. 1888; C.B. Clarke in Hook.f., Fl. Brit. India 6: 656. 1893; Samvatsar, Fl. West. Trib. Mdhya Pradesh 318. 1996; Verma in N.P. Singh *et al.,* Fl. Madhya Pradesh 3: 321.2001

Annual or perennial herbs; leaves reduced to 2-3 basal sheaths; inflorescence capitate, pseudolateral, bearing 15-60 spikelets, 1.3 cm across; spikelets ovoid-oblong, terete, subacute, 8-16 × 4-6 mm; glumes broadly ovate or suborbicular, acute or mucronulate, 3-5 mm long; perianth absent; nuts black, obovoid, triquetrous, apiculate, sessile, transversely wavy, wrinkled, 1.6-2.0 mm long.

Fl.-Fr.: Sept. -April

Along edges of ponds and tanks, in all districts.

Schoenoplectiella juncoides (Roxb.) Lye, Lidia 6:25. 2003; *Scirpus juncoides* Roxb. Fl. Ind. 1:218. 1820; *Isolepis juncoides* Miq., Fl. Ned. Ind. 107. 1834; *Schoenoplectus juncoides* (Roxb.) Palla, Bot. Jahrb. Syst. 10:299. 1888; Verma in N.P. Singh *et al.,* Fl. Madhya Pradesh 3:323.2001.

Perennials; culms10-80 cm high; rhizomes when present, short and woody; leaves absent; sheaths 2-3; anthela pseudolateral, capitate, usually with 2-4 spikelets, bract solitary, erect, stem-like, 2-16 cm long; spikelets ovoid-oblong, 7-25 × 3-5.5 mm; glumes straw-coloured, 3-4 mm long, broadly ovate or suborbi-cular, broadly scarious on margins; perianth bristles 5-6, unequal, retrorsely scabrid, longest shorter or equalling the nut; stigmas 2 in lower flowers, 3 in upper flowers; nuts black, plano-convex, obovoid, 1.7-2.2 mm long, sessile, smooth or faintly, transversely wrinkled.

Fl.-Fr.: Feb. -March

Rare, in wet, muddy places, **Indore** only, Simrol.

Schoenoplectiella lateriflora (J.F. Gmelin) Lye, Lidia 6:25.2003; *Scirpus lateriflorus* J.F. Gmelin Syst. Nat. 2:127.1791; *Isolepis juncoides* Miq., Fl. Ned. Ind. 3:312.1856; *Schoenoplectus lateriflorus* (J.F. Gmelin) Lye, Bot. Not. 124:290.1971; Verma in N.P. Singh *et al.,* Fl. Madhya Pradesh 3: 324.2001.

Annual, 10-20 cm high herbs; leaves reduced to 2-3 basal sheaths; inflorescence pseudolateral, capitate, bearing 2-3 spikelets or sometimes reduced to a solitary spikelet; spikeletsovoid-oblong,subterete, 5-8 × 2.4-3.0 mm; glumes with brown-purplish dots, ovate, mucronate, 2.0-2.4 mmlong; perianth absent; nuts blackish, broadly obovoid, trigonous, sessile,transversely wavy-wrinkled, rugulose, 1.2-1.4 mm long.

Fl.-Fr.: Aug. -May

Burhanpur, Indore, Khargone, in shallow water.

Schoenoplectiella mucronata (L.) J. Jung & H.K. Choi, J. Pl. Biol. 53(3):230.2010; *Scirpus mucronatus* L. Sp. Pl. 50.1753; *Schoenoplectus mucronatus* (L.) Palla, Bot. Jahrb. Syst. 10:299. 1888; Samvatsar, Fl. West. Trib. Madhya Pradesh 319. 1996; Verma in N.P. Singh *et al.,* Fl. Madhya Pradesh 3:328. 2001.

Perennials; culms 0.5-1.5 m high; rhizomes short and creeping; leaves absent; sheaths 1-3, purplish-black, cuspidate; anthela pseudolateral, capitate, 1-2.5 cm across; bract stem-like, solitary, 1-5 cm long; spikelets ovoid-oblong, 2-18, 8-20 × 3.5-4.5 mm; glumes 3.2-4.5 mm long, mucronulate, striate; perianth bristles 5-6, unequal, retrorsely scabrid; stamens 1-3; stigmas 3; nuts brown to blackish,plano-convex, 1.7-2.2 mm long, sessile, apiculate, surface cells isodiametric.

Fl.-Fr.: Oct. -March

In marshy places, Alirajpur, Dhar, Indore.

Schoenoplectiella roylei (Nees) Lye, Lidia 6:26.2003; *Isolepis roylei* Nees, Contr. Bot. India 107. 1834; C.B. Clarke in Hook.f., Fl. Brit. India 6:657. 1893; *Scirpus roylei* (Nees) R. Parker, Fl. Gangetic Plain 3:361. 1929; Samvatsar, Fl. West. Trib. Madhya Pradesh 319; Verma in N.P. Singh *et al.,* Fl. Madhya Pradesh 3:328. 2001.

An annual, tufted, 5-30 cm high herb; culms transversely septate when dry; leaves absent; sheaths 2-3; anthela pseudolateral, capitate, 8-12 mm across; spikelets 1-1.5 cm long, yellow or brown; perianth bristles absent; glumes laxly imbricating, oblong-lanceolate, *ca* 3.5 mm long, membranous; stamens and stigmas 3; nuts blackish, obovoid, trigonous, sessile, *ca* 1.3 mm long, transversely wavy-wrinkled.

Fl.-Fr.: Aug. -Oct.

Common in marshy places, Alirajpur, Indore, Khandwa, Khargone.

SCHOENOPLECTUS Reichb. W. 46; MP 2; SA 2

Key to Species

1. All spikelets sessile and clustered; inflorescence corymbose, rays distinct; stigmas 3 .... **S. corymbosus**
1. All spikelets solitary on ray-like peduncles; hypogynous scales plumose; stigmas 2 .... **S. litoralis**

Schoenoplectus corymbosus (Roth ex Roem. & Schult.) J. Raynal, Cat. Pl. Vasc. Niger 343. 1976; *Isolepis corymbosa* Roth ex Roem. & Schult. Syst. Veg. 2:110.1817; *Scirpus brachyceras* Hochst. ex A. Rich., Tent. Fl. Abyss 2:496. 1850; C.B. Clarke in Hook.f., Fl. Brit. India 6:657. 1893; *Schoenoplectus brachyceras* (Hochst. ex A. Rich.) Lye, Bot. Not. 124:290.1971; *Schoenoplectus corymbosus* var. *brachyceras* (Hochst. ex A. Rich.) Lye, Nordic J. Bot. 3:242. 1983; Samvatsar, Fl. West. Trib. Madhya Pradesh 318. 1996; Verma in N.P. Singh *et al.,* Fl. Madhya Pradesh 3:321.2001.

Rhizomatous perennials; culms 0.5-2 m high; leaves reduced to sheaths with a very oblique, cuspidate mouth; anthela simple or compound, rays 3-10, bearing clusters of spikelets at tips;

spikelets rusty-brown, 5-8 mm long; glumes ovate, 3.5-4.5 mm long, strongly keeled; perianth bristles absent; nuts trigonous, obovoid, yellow, *ca* 2 mm long, smooth.

Fl.-Fr.: Sept. -Oct.

In marshy places, Alirajpur, Dhar, Indore, Khandwa.

Schoenoplectus litoralis (Schrad.) Palla, Bot. Jahrb. Syst. 10:299. 1888; *Scirpus litoralis* Schrad. Fl. Germ. 1:142. 1806; C.B. Clarke in Hook.f., Fl. Brit. India 6:659. 1893; Samvatsar, Fl. West. Trib. Madhya Pradesh 318. 1996; Verma in Fl. Madhya Pradesh 3:326.2001.

Perennials; culms 0.5-2 m high; leaves long or short or reduced to sheaths only; anthela lateral, compound or decompound, up to 10 cm across; bracts erect, as if continuation of stem, 2-15 cm long; spikelets mostly solitary, 10-18 mm long; glumes elliptic-oblong, *ca* 3 mm long, red-brown, notched, cuspidate, margins scarious; perianth 2-7, red-brown, bristle or scale-like, plumose segments, longer than the nut; stigmas 2; nuts plano-convex, dark-brown, *ca* 2 mm long, smooth.

Fl.-Fr.: Sept. -Dec.

Along marshy places, Alirajpur, Dhar, Indore, Khandwa.

SCLERIA Bergius W. 262; MP 11; SA 3

Key to Species

1. Disc lobes subquadrate, rounded or truncate at the top; nuts obtuse . ... **S. tessellata**
1. Disc lobes lanceolate or ovate-oblong; nuts with prominent pits in more or less vertical rows
   2. Disc lobes lanceolate, acuminate with reddish-brown striations . ... **S. stocksiana**
   2. Disc lobes ovate-oblong, acute to subacute, without reddish-brown striations .... **S. parvula**

Scleria parvula Steud., Syn. Pl. Glumac. 2:174. 1855; C.B. Clarke in Hook.f., Fl. Brit. India 6:686. 1894, Samvatsar, Fl. West. Trib. Madhya Pradesh 320.1996; Verma in N.P. Singh Fl. Madhya Pradesh 3:336.2001.

Tufted, slender annuals, up to 1 m tall; leaves linear, 5-35 × 0.2-0.6 cm, 3-nerved; anthela paniculate,terminal and 2-4 lateral, partial panicles; spikelets unisexual, 4.5-5 mm long; glumes ovate-lanceolate, brown, acute or mucronulate; disc thick, deeply 3-lobed, lobes ovate-oblong, acute to subacute; nuts glossy white or discoloured and brownish, 2.2-2.5 × 1.5-2 mm, shortly beaked, tessellate, glabrous or brown pubescent on the transverse ridges.

Fl.-Fr.: Sept. -Dec.

Rare, in hedges of fields, Alirajpur, Dhar, Indore.

Scleria stocksiana Boeckeler, Linnaea 38:474. 1874; C.B. Clarke in Hook.f., Fl. Brit. India 6:687. 1894; Verma in N.P. Singh *et al.,* Fl. Madhya Pradesh 3:337.2001.

Tufted, 20-70 cm tall, annuals or perennials; culm smooth or scabrid; leaves 3-6 mm broad, scabrid on margins and midrib; sheaths narrowly winged or not, contraligule ovate-oblong, *ca* 2 mm long; anthela paniculate, lateral partial panicles 2-nate, 1-3 cm long; spikelets bisexual and male, in clusters of 2-3; glumes ovate-lanceolate; disc deeply 3-lobed, *ca* 0.8 mm long; nuts greyish-white, ellipsoid-oblong, terete or faintly trigonous, 3-3.7 mm long, cancellate, glabrous.

Fl.-Fr.: Sept. -Nov.

Rare, **Indore** only, Simrol, Military Grassland.

Scleria tessellata Willd., Sp. Pl. 4:315. 1805; C.B. Clarke in Hook.f., Fl. Brit. India 6:686. 1894; *Scleria biflora* C.B. Clarke, J. Linn. Soc. Bot. 36:263. 1903; Samvatsar, Fl. West. Trib. Madhya Pradesh 320.1996; Verma in N.P. Singh *et al.,* Fl. Madhya Pradesh 3:340.2001.

Tufted, 20-60 cm tall annuals; leaves 2-6 mm broad; sheaths narrowly to broadly winged; contraligule rounded, hairy; inflorescence paniculate, partial panicles solitary or binate; spikelets unisexual, 5-6 mm long; glumes ovate-lanceolate, acute, midrib green; disc deeply 3-lobed, thick, lobes with a broad, truncate top and narrow base; nuts white or frequently discoloured brown, ellipsoid, terete, 2.7-3 mm long, tessellate, pitted at the base.

Fl. Fr. - Oct. -Nov.

Common on the border of fields, associated with *Apluda mutica* L. and *Themeda* spp., in all districts.

POACEAE 759/11554; 94/262; 89/239

Key to Genera

1. Trees or shrubs; culm sheaths large; leaves articulate on sheaths, veins transverse
   2. Spikelets in globose heads; pericarp fleshy or crustaceous .... DENDROCALAMUS
   2. Spikelets in panicles; pericarp thin, adnate to seed; filaments free .... BAMBUSA
1. Herbs or shrubs, culm sheaths absent; leaves not articulate on sheaths; veins parallel
   3. Spikelets 1-many flowered, breaking up at maturity above the persistent glumes; if falling with glumes, then not 2-flowered and with the lower spikelet male or barren; upper bisexual
      4. Unawned grasses
         5. Spikelets 1-flowered
            6. Inflorescence panicle
               7. Spikelets laterally compressed; glumes absent or greatly reduced; lemma and palea strongly 1-keeled, keels bristly; plants of marshy or watery places; perennials .... LEERSIA
               7. Spikelets terete; glumes well developed; lemmas and palea not keeled, lemmas glabrous; plants of drier places .... SPOROBOLUS
            6. Inflorescence of spikes or spiciform racemes
               8. Spikes 2-8, umbellate .... CYNODON
               8. Spikes or racemes solitary; spikelets sessile
                  9. Lower glume supressed, upper 5-ribbed, ribs armed with hooked pinules; spikelets twin, facing each other .... TRAGUS
                  9. Both glumes present and similar; lower glume not ribbed; spikes alternate and secund, in 2 rows, often curved .... MICROCHLOA

5. Spikelets 2- many flowered
 10. Inflorescence of spikes, heads or racemes
  11. Inflorescence of racemes .... LEPTOCHLOA
  11. Inflorescence of spikes
   12. Spike solitary
    13. Rhachis of inflorescence undulating, alternately, distichously hollowed; spikelets sunk in the hollows; lower glume absent or minute; small grasses, up to 8 cm tall .... OROPETIUM
    13. Rhachis of inflorescence straight, not hollowed; spikelets secund; lower glume well developed .... ERAGROSTIELLA
   12. Spikes more than one, in a whorl .... ELEUSINE
 10. Inflorescence of panicles
  14. Shrubs .... THYSANOLAENA
  14. Herbs
   15. Florets 2; spikelets many; glumes deciduous; upper lemma indurated......... ISACHNE
   15. Florets 3 –many
    16. Panicles spiciform .... DESMOSTACHYA
    16. Panicles lax or contracted
     17. Spikelets 3-flowered .... PHALARIS
     17. Spikelets few to many flowered .... ERAGROSTIS

4. Awned grasses
 18. Spikelets 1-flowered
  19. Lemmas muticous; glumes tapering into long, capillary awns .... PEROTIS
  19. Lemmas awned; glumes awned or muticous; awns not keeled; lemma 1-nerved
   20. Rhachilla produced beyond the spikelets
    21. Inflorescence enclosed in a leaf sheath; lemma not bearded . ... CRYPSIS
    21. Inflorescence not enclosed in a leaf sheath; lemma bearded at base . . SCHOENFELDIA
   20. Rhachilla not produced beyond the spikelets
    22. Glumes reduced or absent
     23. Glumes minute, subulate; plants rooted in mud or water .... ORYZA
     23. Glumes absent; plants floating .... HYGRORYZA
    22. Glumes well developed
     24. Awn of lemma usually with 3 setae .... ARISTIDA
     24. Awn of lemma without setae; glumes notched and long awned .... POLYPOGON
 18. Spikelets 2-many flowered
  25. Lemmas muticous; glumes aristate .... DINEBRA
  25. Lemmas awned; glumes awned or muticous
   26. Inflorescence of plumose panicles
    27. Rhachilla long, hairy .... PHRAGMITES
    27. Rhachilla glabrous .... ARUNDO

26. Inflorescence never plumose
    28. Spikelets 2-flowered; upper bisexual, lower male or barren or bisexual; glume acute or acuminate; awn of lemma between its lobes or at apex; panicles effuse or contracted .... ARUNDINELLA
    28. Spikelets 2-many flowered; upper florets often reduced; lower bisexual; upper lemma not lobed or 2-lobed, its awn straight
        29. Lemma entire or 2-lobed, awned from the back .... AVENA
        29. Lemma entire or 2 or more lobed, awned from apex or sinus
            30. Glumes toothed or notched at apex
                31. Glumes notched at apex .... ACRACHNE
                31. Glumes toothed on one margin .... TRIPOGON
            30. Glumes entire, muticous or aristate; lemmas 1-3 nerved
                32. Glumes muticous; lemmas 2- fid or entire
                    33. Spikes 1 or 2; perfect florets up to 4; empty lemmas muticous .... TETRAPOGON
                    33. Spikelets 3-many; perfect floret 1; empty lemmas awned
                        34. Keeled lemma laterally compressed; grain trigonous or subterete .... CHLORIS
                        34. Keeled lemma dorsally compressed; grain narrow .... ENTEROPOGON
                32. Both glumes aristate
                    35. Spikelets secund; inflorescence a cluster of 4-5 spikelets on a flattened, solitary axis; lemmas entire and 1 or 3 cuspidate; rhachilla produced beyond upper lemma; glumes aristate .... MELANOCENCHRIS
                    35. Spikelets not secund
                        36. Inflorescence of densely crowded, globose clusters of spikelets, continuous in a cylindrical mass or interrupted on a simple, terminal axis . ... ELYTROPHORUS
                        36. Inflorescence of umbellate spikes . ... DACTYLOCTENIUM

3. Spikelets 2-flowered, falling entire at maturity; usually upper floret bisexual and lower male or barren, all similar or more often differing in size, shape and texture, frequently dorsally compressed
    37. Spikelets unisexual, male above and female below; lemma hyaline, membranous, thinner than glumes
        38. Female spikelet completely enclosed in a metamorphosed, cylindrical or spherical, bead-like leaf sheath .... COIX
        38. Female spikelet not as above its lower glume usually with 2 or 3 small teeth .... CHIONACHNE
    37. Spikelets bisexual or male or barren and bisexual spikelets mixed in the same inflorescence arranged such that male or barren spikelet is near the bisexual spikelet or if unisexual, then the lemma of fertile floret is indurated
        39. Spikelets solitary or paired; glumes usually membranous; upper lemma papery to tough and rigid, usually awnless

40. Grasses awned
  41. Upper lemma shortly aristate ....ALLOTEROPSIS
  41. Upper lemma muticous
    42. Lower glume awned ....OPLISMENUS
    42. Lower glume muticous
      43. Lemmas muticous ....PSEUDORAPHIS
      43. Lower lemma awned
        44. Upper glume entire, acuminate or awned ....ECHINOCHLOA
        44. Upper glume emarginate or 2-lobed, aristate from the sinus ....MELINIS
40. Grasses unawned
  45. Inflorescence of panicles
    46. Spikelets usually subtended by involucres of bristles
      47. Bristles rigid; upper lemma rugose or transversely striate . ...SETARIA
      47. Bristles rather soft; upper lemma more or less smooth .... PENNISETUM
    46. Spikelets not subrended by bristles
      48. Spikelets with a ring-like callus at base ....ERIOCHLOA
      48. Spikelets without callus
        49. Spikelets not gibbous .... PANICUM
        49. Spikelets gibbous
          50. Spikelets strongly laterally compressed ....CYRTOCOCCUM
          50. Spikelets not laterally compressed, turgid ....SACCIOLEPIS
  45. Inflorescence of spiciform racemes arranged on a common axis
    51. Spikelets solitary or 2-4 clustered ....CENCHRUS
    51. Spikelets solitary, binate or trinate, not clustered
      52. Lower glume usually absent, if present, then minute, scale-like
        53. Spikelets elliptic, elliptic-oblong or lanceolate .... DIGITARIA
        53. Spikelets orbicular, oblong or ovate, secund, 2-seriate ....PASPALUM
      52. Lower glume well developed
        54. Lower glume turned away from the axis
          55. Spikelets flattened or depressed; upper lemma mucronate ....UROCHLOA
          55. Spikelets usually turgid; upper lemma acute . ... PASPALIDIUM
        54. Lower glume turned towards the rhachis ....BRACHIARIA
39. Spikelets often paired, one sessile, other pedicellate, those of each pair often dissimilar; glumes as long as and enclosing the spikelets, rigid and firmer than the lemmas; lemmas hyaline or membranous, upper usually awned
  56. Spikelets solitary; rhachis of raceme not jointed, tough, flat or filiform; glumes keeled...DIMERIA
  56. Spikelets in pairs
    57. Spikelets of each pair similar, one sessile and other pedicellate, both bisexual, rarely sessile spikelet male and pedicellate bisexual

58. Spikelets arranged in compound racemes along a central axis; upper lemma unawned or awned from the tip or from between its two terminal teeth
   59. Rhachis of raceme tough or tardily breaking up .... IMPERATA
   59. Rhachis of raceme not tough, breaking up
      60. Spikelets awned ....SPODIOPOGON
      60. Spikelets unawned ....SACCHARUM
58. Spikelets arranged in single or in several racemes, if latter, then digitate or arranged on a very short main axis; spikelets in pairs; lower glume truncate
   61. Upper lemma 2-toothed ....EULALIOPSIS
   61. Upper lemma deeply cleft
      62. Lower glume 2-keeled, depressed between keels ....EULALIA
      62. Lower glume convex on the back, at maturity
         63. Raceme solitary .... POGONATHERUM
         63. Racemes 2 or 3 ....LOPHOPOGON

57. Spikelets of each pair dissimilar; sessile bisexual, pedicellate, male or barren or completely absent, if more or less similar, then joints of raceme and the pedicel thick and swollen
   64. Joints of rhachis and pedicel of the pedicellate spikelet swollen, 3-angled, rounded, flattened
      65. Sessile spikelet with a male and a bisexual floret; upper lemma awned
         66. Racemes 1-noded, reduced to 3, heteromorphous spikelets enclosed in a boat-shaped spathe...APLUDA
         66. Racemes many noded, not enclosed in a spathe-like sheath
            67. Racemes 2 or many, more or less digitate
               68. Pedicellate spikelets developed
                  69. Keel acute, winged; rugosities or nodules absent . ...POLYTRIAS
                  69. Keel not acute, nor winged; rugosities or nodules present . ... ISCHAEMUM
               68. Pedicellate spikelet reduced to a linear, curved pedicel . ... THELEPOGON
            67. Raceme solitary, terminal; spikelets in pairs, forming a false raceme; lower and upper glume without a tuft of hairs ....SEHIMA
      65. Sessile spikelet with bisexual floret only; upper lemma unawned
         70. Spikelets similar; rhachis tough . ... HEMARTHRIA
         70. Spikelets dissimilar (spikelets solitary in **Ophiuros**); rhachis fragile
            71. Racemes covered with shaggy hairs . ... ELIONURUS
            71. Racemes glabrous
               72. Inflorescence terminal, of solitary or more racemes; pedicel jointed to internode of rhachis; pedicellate spikelet absent .... MNESITHEA
               72. Inflorescence of axillary racemes from upper leaf axils; sessile spikelets secund; pedicellate spikelet developed; spikelets of pair dissimilar; sessile spikelet globose .... HACKELOCHLOA

73. Sessile spikelets not in pairs, pair consists of 1 sessile and other pedicellate spikelet .... .ROTTBOELLIA

73. Sessile spikelet only present, pedicellate suppressed .... OPHIUROS

64. Joints rhachis and pedicel of pedicellate spikelet not swollen, narrow; inflorescence panicle

74. Spikelets not enclosed in spathe, paired or in threes

75. Spikelets dorsally compressed

76. Racemes few-noded .... SORGHUM

76. Racemes many-noded .... PSEUDOSORGHUM

75. Spikelets laterally compressed .... CHRYSOPOGON

74. Spikelets enclosed in spathes or the espatheate racemes digitate or in pairs

77. Panicles espatheate

78. All pairs of spikelets similar

79. Upper lemma awned from back below the middle; racemes digitate .... ARTHRAXON

79. Upper lemma awned from the sinus and reduced to the stalk of the awn; joints and pedicels filiform with transluscent, longitudinal furrow

80. Lower glume of sessile spikelet pitted, if not, then raceme more than 4-jointed.. BOTHRIOCHLOA

80. Lower glume if sessile spikelet not pitted; raceme 1-4 jointed .... CAPILLIPEDIUM

78. All pairs of spikelets not similar, lowest 1 or more pairs homogamous and differing from all above them; spikes solitary; callus elongate, sharp

81. Lower glume of fertile spikelet rounded on back .... HETEROPOGON

81. Lower glume of fertile spikelet with a deep groove only on the dorsal surface ....PARAHYPARRHENIA

77. Panicles spatheate; spikes usually 2 or more on a common peduncle, sheathed by spathe

82. Spikes reduced to 1 or a few nodes with or without a surrounding involucre of modified male or empty spikelets

83. Involucre of modified spikelets present

84. Involucre persistent, the rest jointed on the rhachis and falling separately .... THEMEDA

84. Involucre of modified spikelets jointed on the peduncle and falling together with the rest .... ISEILEMA

83. Involucre absent

85. Callus of spikelets not inflated .... PSEUDANTHISTIRIA

85. Callus of spikelets inflated ....APLUDA

82. Spikes many noded, not reduced, nor involucred

86. Spikes solitary in each spatheole

87. Joints and pedicels slender .... DICHANTHIUM
87. Joints and pedicels thickened upwards to a cupped apex, often with dentate, terminal appendage .....................SCHIZACHYRIUM

86. Spikes 2 in each spatheole

88. All pairs of spikelets similar; racemes terminal ....ANDROPOGON
88. Lowest pair of spikelets or 1 or both spikes homogamous; racemes compound; aromatic plants .... CYMBOPOGON

(Data on species endemic to India and occurring in the study area is provided by Arjun Tiwari, BSI Mysore)

ACRACHNE Wight & Arn. ex Chiov. W. 3; MP 1; SA 1

Acrachne racemosa (B. Heyne ex Roth) Ohwi, Bull. Tokyo Sci. Mus. 18:1. 1947; *Eleusine racemosa* B. Heyne ex Roth, Syst. Veg. 2:583. 1817; *E. verticillata* Roxb., Fl. Ind. 1:346. 1820; Hook.f., Fl. Brit. India 7:295. 1896; Bor, Grass. Burm. Ceyl. Ind. Pak. 487. 1960; Samvatsar, Fl. West. Trib. Madhya Pradesh 325. 1996; Roy in N.P. Singh *et al.,* Fl. Madhya Pradesh 3: 356.2001.

Annuals; culms 20-90 cm tall, nodes glabrous; leaves linear-lanceolate, apex acuminate, ligule membranous, ciliate; spikes 2-many, altrnate, pseudo-whorled or scsttered, on a grooved, scberulous rachis; spikelets 3-4 mm long; rachilla continuous between the lemmas; lower glume membranous, 1-nerved, keel produced into a ashort arista; upper glume similar and with an arista; lowest lemma boat-shaped, 3-nerved, apex notched or 2 toothed; palea hyaline; caryopsis rugose.

Fl.-Fr.: Aug. -Nov.

Common in wet places in all districts.

ALLOTEROPSIS K. Presl W. 5; MP 1; SA 1

Alloteropsis cimicina (L.) Stapf, Fl. Trop. Africa 9: 487. 1919; *Milium cimicinum* L., Mant. Pl. 2: 184. 1771; *Axonopus cimicinus* (L.) P. Beauv., Ess. Agrostogr. 12. 1812; Hook.f., Fl. Brit. India 7: 64. 1896; Bor, Grasses Burma Ceylon India Pakistan 276. 1960; Samvatsar, Fl. West. Trib. Madhya Pradesh 325. 1996; Roy in N.P. Singh *et al.,* Fl. Madhya Pradesh 3: 357.2001. *Sinri*

Annual herbs; culms 30-60 cm long; leaves ovate-lanceolate, 2.3-8 × 0.5-1.5 cm, acute at apex, deeply cordate at base, hirsute with long white hairs; inflorescence of 3-8, subverticillate spikes; spikelets solitary or clustered; lower glume acuminate; upper glume mucronate or shortly awned; lower lemma *ca* 3 mm long; upper lemma *ca* 2.6 mm long; caryopsis ovoid, compressed, *ca* 1.2 mm long.

Fl.-Fr. : : July – Nov.

Common in cultivated fields and on outskirt of forests in all districts.

ANDROPOGON L. W 120; MP 2; SA 2

Key to Species

1. Culms 15-100 cm long; leaf blade 5-30 cm long; spatheole 4-6 cm long; raceme 1- 2.5 cm long, rhachis fragile; sterile spikelet longer than fertile spikelet; callus pilose . ... **A. fastigiatus**
1. Culms 15-45 cm long; leaf blades 7-12 cm long; spatheole 0.5-2 cm long; racemes 2, 1.5- 2 cm long; rhachis tough; sterile spikelet shorter than fertile spikelet; callus pubescent .... **A. pumilus**

Andropogon fastigiatus Sw. Prodr. Veg. Ind. Occ. 26. 1788; *Diectomis fastigiata* Kunth, Nov. Gen. et. Sp. 1: 193. 1816; Hook., f., Fl. Brit. Ind. 7: 167. 1896; Bor, Grass. Burm. Ceyl. Ind. Pak. 135. 1960; Roy in N.P. Singh *et al.,* Fl. Madhya Pradesh 3:407.2001.

Perennials; culms 30-50 cm tall, erect or decumbent, branched, nodes glabrous; leaves narrowly linear, acuminate, scabrid; sheath loose, ligule membranous; inflorescence a solitary spike in the axil of a spathiform leaf sheath, peduncle short or long, callus bearded; sessile spikelet about 3 mm long, laterally compressed; lower glume linear, acuminate, keeled, villous; upper glume 2.5 mm long, membranous, 2-fid, awned, keel villous; lemma 2.5-3 mm long, 2-toothed, awn 3-3.5 cm long; palea minute, hyaline; pedicelled spikelet 11-nerved, hyaline with slender awn, neuter; anthers 3, brownish- yellow, 1.5 mm long; caryopsis 1.8-2 mm long, light brown, fusiform, style persistent.

Fl.-Fr. : : Sept. -Dec.

Khandwa, Kalibhit forest.

Andropogon pumilus Roxb. Fl. Ind. 1:277. 1820; Hook., f., Fl. Brit. India 7:170.1896; Bor, Grass. Burma. Ceyl. Ind. Pak. 93. 1960; Samvatsar, Fl. West. Trib. Madhya Pradesh 325. 1996; Roy in N.P. Singh *et al.,* Fl. Madhya Pradesh 3:359.2001. *Pochat* **(Plate 10, Fig. 2)**

Annuals; culms 20-50 cm tal;, nodes glabrous; leaf sheath terete, glabrous, 1-4 cm long, ligule membranous, blade flat, glabrous, 3-5 × 0.1-0.5 cm long, linear, apex acuminate; inflorescence of paired, digitate 2 cm long,racemes, enclosed in spathes; callus of spikelet obstuse; sessile spikelet linear, 3-5-4 mm long, margin inflexed, 2-nerved, 2 keeled, apex 2-toothed; upper glume membranous, 3. 8-4 mm long, 3-nerved, 1-keeled on the back, apex acuminate; lower lemma membranous, glabrous, 1-nerved, apex acute; upper lemma membranous, linear, 1-nerved, cleft at apex in to two lobes, with geniculate, 8-10 mm long awn; palea hyaline, elliptic, apex obtuse; caryopsis ovate; pedicellate spikelet ovate, awn less; lower glume membranous, scabrous, margin inflexed, 12-nerved, 2-keeled; upper glume membranous, 3-nerved, 1-keeled; lower lemma membranous, 2-nerved; palea absent.

Fl. - Fr.: Sept. -Nov.

Grows in open rocky, waste places, especially hill slopes in all districts.

APLUDA L. Monotypic

Apluda mutica L., Sp. Pl. 82. 1753; *A. varia* subsp. *mutica* Hack., Monogr. Phan. 6: 197. 1889; Hook.f., Fl. Brit. India 7: 150.1896; Bor, Grass. Burm. Ceyl. Ind. Pak. 93. 1960; Samvatsar, Fl. West. Trib. Madhya Pradesh 325. 1996; Roy in N.P. Singh *et al.,* Fl. Madhya Pradesh 3: 360.2001. *Phuli* **(Plate 10, Fig. 3)**

Perennials; culms 30-90 cm tall, slender, erect, ascending; leaves 9-40 × 0.3-1.2 cm long, linear-lanceolate, glabrous, tapering to the apex; ligule a ring of hairs; inflorescence subtended and enclosed in a peduncled spathe; racemes combined into leafy panicle; spikelets 3-nate; one sessile and two pedicellate, of which one reduced to an empty glume; sessile spikelets: lower glume linear-lanceoate, 2-keeled; upper glume boat-shaped; lower floret male; lower lemma 3.5 mm long; lower palea divided half-way down; upper floret fertile; upper lemma awned in the sinus with geniculate base; stamens 3; pedicelled spikelet male: lower glume, acuminate; palea hyaline; upper lemma linear-lanceolate, 3-nerved; palea hyaline; caryopsis oblong.

Fl.-Fr.: -Sept. -Dec.

Common in hedges, bushes, and banks of river, in all districts.

ARISTIDA L. W. 301; M P 8; SA 8

Key to Species

1. Lemmas articulated
    2. Glumes similar; spikelets 6-8 mm long; glumes unawned . ... **A. funiculata**
    2. Glumes dissimilar; spikelets 22-25 mm long; lower glume awned, upper unawned . ... **A. stocksii**
1. Lemmas not articulated
    3. Lateral awns much reduced and different in texture from the central .... **A. redacta**
    3. Lateral awns well developed
        4. Perennials; tough, robust plants; both glumes aristate
            5. Panicle contracted; lower glume only awned .... **A. setacea**
            5. Panicles effuse
                6. Awn 2 cm long .... **A. cynantha**
                6. Awn 6 cm long .... **A. hystrix**
        4. Annuals
            7. Central awn 0.5-0.8 cm long; lemma 1.7-2 cm long .... **A. cumingiana**
            7. Central awn 1-2.5 cm long; lemma 0.5-1. 1 cm long . ... **A. adscensionis**

Aristida adscensionis L., Sp. Pl. 82. 1753; *A. depressa* Retz., Observ. Bot. 4:22. 1786; Hook., f., Fl. Brit. India 7: 224. 1896; Bor, Grass. Burm. Ceyl. Ind. Pak. 407. 1960; Samvatsar, Fl. West. Trib. Madhya Pradesh 326. 1996; Roy in N.P. Singh *et al.,* Fl. Madhya Pradesh 3:362.2001. *Mulmal*

Annuals or perennials; culms 20-90 cm tall, tufted, erect ascending from a shortly decumbent base; leaves 7-30 × 0.1-2 cm long, convolute, filiform, smooth or scabridulous; ligule a ridge of short hairs; inflorescence a panicle, 7-30 cm long; spikelets erect, 6-9 mm long, one-flowered, purplish; lower involucral glume 5-8 mm long, oblong-lanceolate, acute, membranous, often purple, scabrous on keel; upper involucral glume 6-9 mm long, 2-toothed, apiculate at tip, keel smooth; lemma 6-13 mm long, laterally compressed, convolute, scabrid on the keel, callus narrowly oblong , 0.5 mm long, obtuse; awn trifid, not articulated; central awn 7-15 mm long, lateral subequal; palea 1.8 mm long, obtuse; caryopsis as long as lemma.

Fl.-Fr.: Aug. -Dec.

Common in dry land and on hill tops, waste lands, all over the area.

Aristida cumingiana Trin. & Rupr. Sp. Gram. Stipac. 141. 1842; Hook., *f.*, Fl. Brit. India 7: 224. 1896; Bor, Grass. Burm. Ceyl. Ind. Pak. 409. 1960; Roy in Singh *et al.,* Fl. Madhya Pradesh 3:362.2001.

Annuals; culms slender, 15-40 cm tall; leaves basal, filiform small; sheath loose, glabrous; ligule very minute, membranous; inflorescence a narrow panicle, loosely branched, rachis and pedicels scabrid; spikelets 3-awned, 2.0-2.5 mm long, and without articulation, central one up to 5 mm, laterals 3.0-3.5 mm, green or purple; lower glumes 2.0-2.5 mm; upper glume 3.0-3.5 mm long, pointed; callus very minute with small white hairs; lemma 2.0-2.5 mm long, linear, laterally compresed; anthers 3, yellowish to light pink; caryopsis 2 mm long, cylindric, deep pink to brownish.

Fl.-Fr. : Sept. -Dec.

Common in dry waste land and hill slopes, everywhere.

Aristida cyanantha Steud. Syn. Pl. Glumac. 1:141. 1854; Hook., f., Fl. Brit. India 2: 225. 1896; Bor, Grass. Burm. Ceyl. Ind. Pak. 409. 1960; Roy in N.P. Singh *et al.,* Fl. Madhya Pradesh 3: 363.2001.

Perennials; culms up to 1.5 m tall, stout, erect; leaves convolute, glaucous, glabrous; sheath smooth; ligule membranous; inflorescence an effuse panicle with flexuous branches; spikelets 8-13 mm long, excluding awn, usually deep purple; lower glume 6-9 mm long, smooth, awn 1-1.5 mm long; upper glume 10-12 mm long plus 2-3 mm long awn, fimbriately toothed; lemma up to 8-12 mm long with densely hairy callus, no articulation between lemma and awns; anthers 3, yellow tinged with purple; caryopsis 8-10 mm long, terete.

Fl.-Fr.: Sept. –Feb.

Rare, **Indore**, grassland at Kampel.

Aristida funiculata Trin. & Rupr. Sp. Grass. Stipac. 159. 1842; Hook., f., Fl. Brit. Ind. 7:226. 1896; Bor, Grass. Burma. Cey. Ind. Pak. 410.1960; Samvatsar, Fl. West. Trib. Madhys Pradesh 326 . 1996; Roy in N.P. Singh *et al.,* Fl. Madhya Pradesh 3:364.2001. *Safed kusli*

Annuals; culms 20-70 cm tall, tufted, erect, slender; leaves convolute, sparsely hairy on upper surface; sheaths keeled; ligules hairy; inflorescence a narrow panicle; branches capillary, usually with long white hairs in the axil; spikelets straw-coloured, 1.5-2.5 cm long; lower and upper glumes similar and equalling the spikelets; lemma ± 4 mm long, smooth, awned; articuled column 2-4.5 cm long with 3 capillary branches; central awn 3-6 cm long, lateral awns slightly shorter; anthers 3, light yellow; caryopsis black, cylindrical, 4 mm long.

Fl.-Fr. : : Aug. -Dec.

Common in open grass lands and rocky plain ground, in all districts.

Aristida hystrix L.f. Suppl. Pl. 113. 1782; Hook., f., Fl. Brit. Ind. 7:225. 1896; Bor, Grass. Burm. Cey. Ind. Pak. 410.1960; Samvatsar, Fl. West. Trib. Madhya Pradesh 326. 1996. *Lapusri*

Perennials; culums 15-60 cm tall, diffuse, ascending, sometimes proliferously branched, stiff, stout, nodes glabrous; leaf sheath glabrous, ligule a rim of hairs, 0.5 mm long; leaf blade 10-40 cm long, narrowly linear, convolute, glaucous; panicle as broad as long, lax and rigid; spikelets narrowly ovate, 1.5-1.8 cm long; lower glume membranous, narrowly ovate, 10-12 mm, 1-nerved, 1-keeled, apex aristate; callus of lemma pungent, bearded with hairs; lower lemma coriaceous, narrowly ovate, 3-nerved, 3-awned, central awn longer than lateral, awns 3-4 cm long, scabrid on margins; palea hyaline, narrowly ovate, nerveless, apex obtuse or truncate; stamens 3; ovary 3.5-4 mm long; caryopsis linear to narrowly ovate, 7-7.5 mm long.

*Fl.-Fr. :*Aug. -Nov.

Rare in dry stony places, Dhar and Khandwa.

Aristida redacta Stapf, Bull. Misc. Inform. Kew 1892: 85. 1892; Hook., f., Fl. Brit. India 7: 227. 1896; Bor, Grass. Bur. Cey. Ind. Pak. 412; Samvatsar, Fl. West. Trib. Madhya Pradesh 327. 1996; Roy in Singh *et al.,* Fl. Madhya Pradesh 3: 364.2001. *Burgi*

Annuals; culms tufted, geniculately ascending, 10-60 cm tall, nodes glabrous;. leaf sheath terete, glabrous; ligule a rim of hairs, 0.2-0.5 mm long; leaf blade 0.1-0.5 cm long, linear, narrowly convolute; panicle effuse, 10-40 cm long; spikelets linear, 2.5-3 mm, awned, callus obtuse, covered with dense white hairs; lower glume membranous, narrowly, ovate, 1-nerved, 1-keeled, keel scabrous, cleft at apex into two lobes; upper glume membranous, narrowly ovate, 1-nerved, 1-keeled, keel scabrous, cleft at apex into 2 lobes; lemma coriaceous, linear, articulated below column, twisted, 3- nerved, awned at apex, central awn longer than lateral awns, lateral awns reduced; palea hyaline, narrowly ovate, 2-nerved, apex truncate; stamens 3; ovary 3 mm long; caryopsis 4-5 mm, narrowly ovate.

Fl.-Fr.: Nov. -Feb.

Commen in open grass lands, on dry rocky places, in all districts.

Aristida setacea Retz. Observ. Bot. 4: 22. 1786; Hook.f., Fl. Brit. India 7: 235. 1896; Bor, Grass. Burm. Ceyl. Ind. Pak. 412. 1960; Samvatsar, Fl. West. Trib. Madhya Pradesh 327 . 1996; Roy in N.P. Singh *et al.,* Fl. Madhya Pradesh 3: 364.2001.

Culms tufted, stout, robust, 1 m tall, glabrous; leaf glabrous; ligule to 0.5 mm; blades oblong, to 30 cm, convolute; panicle contracted, 15-30 cm, branches erect; rachis terete, glabrous; pedicels to 5 mm; glumes linear, convolute, arista to 2.5 mm; lower glume to 1.5 cm; upper glume 1.5-2 cm, lemma to 1.2 cm; awn sessile; setae 3, mid one to 3 cm, laterals 2.5-2.8 cm, callus to 1.5 mm; caryopsis *ca* 1.5 cm long

*Fl.-Fr.:*Nov. -Feb.

Common in open grassland on dry, rocky places, everywhere.

Aristida stocksii (Hook.f.) Domin, Biblioth. Bot. 20: 338. 1915; *A. funiculata* var. *stocksii* Hook.f., Fl. Brit. India 7:227. 1896; Bor, Grass, Burm. Ceyl. Ind. Pak. 412. 1960; Sharma *et al.,* in Fl. Maharashtra 3:397 . 1996.

Annuals; culms 20-40 cm tall, tufted, erect or geniculately ascending from base, nodes glabrous; leaf sheaths terete, glabrous; leaves 3-15 × 0.15-0.25 cm, involute, narrowly linear, finely acuminate or acute at apex, ligule a rim of hairs; leaf blade linear, 5-10 cm long, hispid on upper side; panicles lax, 5-10 cm lon;. lower glume membranous, linear, 2.8-3 mm, minutely hairy all ovar dorsal side, 1-nerved, 1-keeled, apex acuminate; upper glume membranous, narrowly ovate; spikelets yellowish or straw-coloured; column of lemma 2.7-4.5 cm long; palea hyaline, narrowly ovate, 2-nerved, apex obtuse; stamens 3; caryopsis cylindric.

Fl.-Fr. :Sept. -Nov.

Rare in open grasslands, Burhanpur, Khandwa, Khargone. Endemic to India

ARTHRAXON P. Beauv. W. 23; MP 10; SA 10

Key to Species

1. Lower glume of sessile spikelet laterally 2-keeled, its margins inflexed
   2. Keels of lower glume pectinate, spinulose .... **A. echinatus**
   2. Keels of lower glume not spinulose
      3. Lower glume of sessile spikelet with keels having 2 rows of tubercles .... **A. meeboldii**
      3. Lower glume of sessile spikelet with pectinately toothed keels
         4. Lower glume flat on the back .... **A. lanceolatus**
         4. Lower glume rounded on the back .... **A. prionodes**
1. Lower glume of sessile spikelets not keeled, margins flat; spikelets laterally compressed .... **A. junnarensis**
   5. Pedicellate spoikelets not developed .... **A. nudus**
   5. Pedicellate spikelets developed
      6. Sessile spikelets strongly scabrid, spinulose
         7. Lower glume 5-9 nerved; upper glume 5-nerved .... **A. hispidus**
         7. Lower glume 5-nerved; upper glume 1-nerved .... **A. santapaui**
      6. Sessile spikelets to scaberulus
         8. Lower glume of sessile spikelets obscurely nerved, smooth below the middle .... **A. lancifolius**
         8. Lower glume of sessile spikelets 2-4 nerved .... **A. jubatus**

Arthraxon echinatus (Nees) Hochst., Flora 39:188. 1856; *Batratherum echinatum* Nees, Edinburgh New Philos. J. 18:181. 1835; *Andropogon echinatus* (Nees) B. Heyne ex Steud., Nomencl. Bot. ed. 2, 1:91. 1840; *A. echinatus* Edgew., J. Asiat. Soc. Bengal 21:182. 1853; *A. spathaceus* Hook.f., Fl. Brit. India 7:145. 1897; Bor, Grass. Burm. Ceyl. Ind. Pak. 99. 1960; Jain, J. Indian Bot. Soc. 51:173. 1972.

Perennial herbs; culms simple, 20-50 cm tall; leaves 1.5-3.5 × 0.5-1.2 cm, hairy, ovate-lanceolate; racemes 2-4, 2.5-4 cm long; sessile spikelets up to 5 mm long; lower glumes 5-7 nerved, upper glume faintly 3-5 nerved; pedicellate spikelets up to 4-4.5 mm long, linear-lanceolate; awns 4-5 mm long.

*Fl.-Fr.:*Sept. -Dec.

Rare, Burhanpur, Khandwa.

Arthraxon hispidus (Thunb.) Makino, Bot. Mag. (Tokyo) 26:214. 1912; *Phalaris hispida* Thunb., Syst. Veg. ed. 14 104. 1784; *Arthraxon ciliaris* P. Beauv., Ess. Agrostogr. 111. 1812; Hook.f., Fl. Brit. India 7:145. 1896; *A. quartinianus* (A. Rich.) Nash, N. Amer. Fl. 17:99. 1912; Bor, Grass. Burm. Ceyl. Ind. Pak. 99. 1960; Roy in N.P. Singh *et al.,* Fl. Madhya Pradesh 3:366.2001.

Densely tufted annuals; culms 30-40 cm tall; leaves ovate-lanceolate, ciliate on margins, sheaths glabrous; ligules lacerate; inflorescence of several spikes; rachis joints *ca* 2.5 mm long, ciliate, curved; spikelets solitary, *ca* 3 mm long, scabrid; upper glume similar to lower; lower floret empty; lemma usually sbsent; upper floret bisexual; lemma *ca* 2 mm long; awns up to 9 mm long from the base of lemma; caryopsis *ca* 2 mm long, linear, terete.

*Fl.-Fr.:*Sept. -Dec.

Alirajpur, Barwani, Dhar, Indore, Khargone.

Arthraxon jubatus Hack., Monogr. Phan 6: 358. 1889; Hook.f. Fl. Brit. India 7: 147. 1896; Bor, Grass. Burm. Ceyl. Ind. Pak. 100.1960; Sharma *et al.,* Fl. Maharashtra 3:399. 1996.

An annual delicate herb, hanging from rocks, 10-25 cm long, nodes densely hairy; leaf sheath compressed, 0.5-3 cm long, densely silky hairy, ligule membranous; leaf blade silky hairy, 1-4 cm long; racemes 2, 2-3 cm long; sessile spikelets narrowly ovate, 8-9 mm, awned, callus obtuse; lower glume coriaceous, narrowly ovate, 6-10 mm, margin membranous, inflexed, 5-7-nerved, 2-keeled, apex two toothed; upper glume coriaceous, narrowly ovate, 3-nerved, 1-keeled, keel scabrous, apex acuminate; lower lemma hyaline, narrowly ovate, 1-nerved, apex acute; palea absent; upper lemma membranous, narrowly ovate, 1-nerved, awned from dorsal side, awn filiform, up to 10 mm long, apex 2-toothed; palea absent; stamens 3; pedicelled spikelet 4-4.5 mm; lower glume ovate, minutely scabrous along nerves, margins inflexed, 5-7-nerved, apex acuminate; upper glume membranous, margin inflexed, 3-nerved, apex acute; lower lemma hyaline, ovate, margin inflexed, 3-nerved, apex acute; palea absent; upper lemma hyaline, glabrous, ovate, margin inflexed, faintly 1-nerved, apex acute; stamens and lodicules not well developed; caryopsis 4-4.5 mm long.

Fl.-Fr. : -Aug. -Dec.

Along water courses, Barwani, Khargone. Endemic to India.

Arthraxon junnarensis Jain & Hemadri, J. Bombay. Nat. Hist. Soc. 68: 300.1971 & J. Indian Bot. Soc. 51: 175. 1972; Sharma *et al.,* Fl. Maharashtra 3:399. 1996 (as var. *junnarensis*).

Annuals; culms terete, erect, 10-30 cm tall, nodes hairy, rooting at lower nodes; leaf sheath compressed, hairy, ligule membranous ciliate; leaf blade elliptic, hairy, 0.2-3 cm; inflorescence of 5-15 racemes; sessile spikelets elliptic, 1.4 - 1.6 mm, awned; lower glume membranous, narrowly ovate, 1.3–1.4 mm, laterally compressed, 5-nerved, apex acute; upper glume membranous, narrowly ovate, 1.4–1.5 mm, 1-nerved, 1-keeled on back, apex acute; lower lemma absent; palea absent; upper lemma membranous, narrowly ovate, 0.8-1 × 0.1-0.2 mm, 1-nerved, 1-keeled, apex acute, awned on back, awn geniculate, up to 3.5 mm long; palea absent; stamens 2; pedicelled spikelet absent or reduced; caryopsis elliptic.

Fl.-Fr. : -Sept. -Oct.

Rare, along the road sides and on old walls, **Burhanpur** only, wall of Shahi Kila. Endemic to India.

Arthraxon lanceolatus (Roxb.) Hochst., Flora 39: 188. 1856; *Andropogon lanceolatus* Roxb., Fl. Ind. ed. 1820 1:262. 1820; *Batratherum lanceolatum* Nees, Edinburgh New Philos. J. 18:181. 1835; Hook.f., Fl. Brit. India 7: 143. 1896; Bor, Grass. Burm. Ceyl. Ind. Pak. 100.1960; *Arthraxon deccanensis* S.K. Jain, J. Bombay Nat. Hist. Soc. 68:297. 1971; Samvatsar, Fl. West. Trib. Madhya Pradesh 328. 1996.

Annuals; culms, erect, 10-100 cm tall, nodes pilose; leaf sheath compressed, covered with bulbous based hairs, ligule membranous ciliate; leaf blade elliptic, ovate, margin inflexed, glabrous, 7-nerved, 2-keeled, apex two toothed; upper glume membranous, narrowly ovate, margin inflexed, ciliate, 3-nerved, apex acuminate covered with bulbous based hairs; racemes 1-7, 3-6 cm long; sessile spikeles narrowly ovate, awned; lower glume narrowly ovate, inflexed, glabrous, 5-7-nerved, 2-keeled, keels spiculate, apex acute or two toothed; upper glume membranous, narrowly ovate, margin ciliate hairy, 3-nerved, 1-keeled, apex acute or acuminate; lower lemma hyaline, narrowly ovate, apex acute; palea absent; upper lemma membranous, narrowly ovate, 1-nerved. awned, 1-keeled, apex acute; palea hyaline, margin glabrous, nerveless, apex acute; stamens 3; pedicelled spikelet narrowly elliptic; lower glume membranous, narrowly ovate; lower lemma membranous, narrowly ovate, apex acute; palea absent; upper lemma membranous, narrowly ovate, 1-nerved, apex two toohhed; palea absent; caryopsis terete, *ca* 3 mm long.

*Fl.-Fr. :*Aug. -Dec.

In grass lands and along water courses, Alirajpur, Dhar, Indore.

Arthraxon lancifolius (Trin.) Hochst., Flora 39**:**188. 1856; *Andropogon lancifolius* Trin.,Mem. Acad. Imp. Sci. Saint- Petersbourg, Ser. 6,Sci. Math. 2:271. 1832; Bor, Grass. Bur. Ceyl. Ind. Pak. 100.1960; Samvatsar, Fl. West. Trib. Madhya Pradesh 328 . 1996; Roy in N.P. Singh *et al.,* Fl. Madhya Pradesh 3**:**367.2001. *Lotna*

Annuals; culms 10-40 cm tall; leaves ovate-lanceolate, cordate at base, usually glabrous, rarely sparsely hairy; sheaths glabrous; ligules membranous; inflorescence of 2-many racemes; rachilla hairy; sessile spikelets 2.75-3.5 mm long, 1- awned; callus obtuse, hairy; lower glume linear-lanceolate, faintly 5-nerved, apically 2-keeled; keel produced into a short arista; upper glume boat-shaped, subequal to lower, 1-keeled; lower lemma hyaline; palea absent; upper lemma hyaline, cleft at apex; awn geniculate; palea absent; caryopsis cylindric, light brown.

Fl.-Fr. :-June-Dec.

Common, on old walls, eroded hills and waste lands, in all districts.

Arthraxon meeboldii Stapf, Bull. Misc. Inform. Kew 1908. 449. 1908; *A. lanceolatus* var. *meeboldii* (Stapf) Welzen, Blumea 27:285. 1981; Bor, Grass. Burm. Ceyl. Ind. Pak. 101. 1960; Samvatsar, Fl. West. Trib. Madhya Pradesh 328. 1996.

Annuals; culms terete, erect, 50-120 cm tall, rooting at lower nodes; nodes hairy; ligule membranous; leaf blade, densely hairy on both sides, apex acuminate; racemes 2-3, 6-8 cm long, densely hairy on both sides; sessile spikelets narrowly ovate, awned, callus obtuse; lower glume coriaceous, narrowly ovate, margin inflexed, coriaceous, glabrous, 5-6-nerved, 2-keeled, keels with spicules, spicules with a bunch of hairs at apex, apex two toothed; upper glume membranous, glabrous, narrowly ovate, margin ciliate hairy, 3-nerved, keeled on the dorsal side, apex acuminate; lower lemma membranous, narrowly ovate, glabrous, nerveless, apex acute; palea absent; upper lemma membranous, narrowly ovate, 1-nerved, awned, awn up to 20 mm long; pedicelled spikelet narrowly ovate, lower glume membranous, glabrous, narrowly ovate, margin inflexed, 7-9-nerved, apex acute; upper glume membranous, glabrous, ciliate hairy, 3-nerved, apex acute; palea absent; upper lemma hyaline, narrowly ovate, apex acute; palea absent; stamens 3; caryopsis terete, brown.

Fl.-Fr. :July-Dec.

In grass lands, old walls and wet places at Alirajpur, Burhanpur, Dhar, Indore, Khandwa. Endemic to India.

Arthraxon nudus (Steud.) Hochst., Flora 39:188. 1856; *Batratherum nudum* Nees ex Steud., 1854, *nom. Invalidum; Andropogon nudus* Steud., Syn. Pl. Glumac. 1:383. 1854; Bor,Grass. Burm. Ceyl. Ind. Pak. 101. 1960; Samvatsar, Fl. West. Trib. Madhya Pradesh 329. 1996.

A weak, decumbent grass, rooting at lower nodes; leaves 3.5-4 × 1.3-1.5 cm, ovate-lanceolate, base amplexicaule with tubercle-based hairs along the margin, in the basal region only; sheath 2.5-3.0 cm long; ligule membranous; inflorescence on a long peduncle of 3-5, subdigitate racemes, 4-4.5 cm long; sessile spikelets 4.5-5 mm long, laterally compressed; lower glume 4.5-5 mm long, 7-9 nerved, spinulose hispid on the nerves; spinules originating from tubercled bases; upper glume similar to lower, 3-nerved, strongly keeled; lemma 2.5-3 mm long, strongly 2-nerved; awn dorsal, geniculate, basal portion 6 mm long, dark brown, upper portion 4 mm long, barbalate; palea hyaline, 2-keeled; caryopsis narrowly oblong; pedicelled spikelets represented by 3.5-4 mm long pedicel only.

Fl.-Fr.: Oct. -Dec.

Rare, **Alirajpur** only, way to Shakarja.

Arthraxon prionodes (Steud.) Dandy, Fl. Pl. Sudan 3:399. 1956; *Andropogon prionodes* Steud., Syn. Pl. Gumac. 1:383. 1854; Bor, Grass. Burm. Ceyl. Ind. Pak. 101. 1960; Samvatsar, Fl. West. Trib. Madhya Pradesh 329. 1996; Roy in N.P. Singh *et al.,* Fl. Madhya Pradesh 3:367.2001.

Perennial grasses, culms 40-80 cm tall, leafy, rooting at lower nodes; leaves lanceolate, base cordate; ligules membranous; inflorescence of 1 to several racemes; sessile spikelets 5.5-6 mm long, awned from the sinus of upper lemma; awns 9-11 mm long; palea absent; Anthers 3, yellow; caryopsis about 2 mm long.

Fl.-Fr.: Aug. -Dec.

Common, occurs along water streams, in all districts.

Arthraxon santapaui Bor, Kew Bull. 1951:446. 1952 and Grass. Burma. Ceyl. Ind. Pak. 102. 1960

Annuals; culms terete, 20-60 cm tall, nodes glabrous; leaf sheath glabrous, hairy on one margin, ligule membranous, hairy; leaf blade flat, elliptic, 2-3 cm, hairy; inflorescence of 6-13 or more racemes; sessile spikelets narrowly ovate, awned; lower glume coriaceous, narrowly ovate, margin glabrous, 6-nerved, 1-keeled, apex acute to acuminate; upper glume membranous, narrowly ovate, 4. 8-5 mm, margin glabrous, 3-nerved, 1-keeled, apex acute to acuminate; lower lemma membranous, narrowly ovate, margin glabrous, nerveless, apex acute; palea absent; upper lemma membranous, narrowly ovate, 2-2.4 mm, margin glabrous, 1-nerved and 1-keeled, apex acute; palea absent; stamens 2; pedicelled spikelet absent or reduced; caryopsis linear.

Fl.-Fr. : -Sept. -Dec.

Along water courses, Barwani, Khargone. Endemic to India.

ARUNDINELLA Raddi W. 49; M P 5; SA 3

Key to Species

1. Upper lemma with 2 lateral bristles arising at base of awn; pedicels of spikelets with well developed bristles or setae .... **A. setosa**
1. Upper lemma not as above
   2. Upper lemma with a geniculate awn; spikelets 1.5-1.7 mm long . ... **A. pumila**
   2. Upper lemma 2-toothed at apex; spikelets 8-10 mm long .... **A. tuberculata**

Arundinella pumila (Hochst.) Steud., Syn. Pl. 1:114. 1854; *Acratherum pumilum* Hochst. ex A. Rich., Tent. Fl. Abyss. 2:414. 1850; Hook.f., Fl. Brit. India 7:71. 1896; Bor Grass. Burm. Ceyl. Ind. Pak. 410.1960; Samvatsar, Fl. West. Trib. Madhya Pradesh 330.1996; Roy in N.P. Singh *et al.,* Fl. Madhya Pradesh 3:370.2001.

Annuals; culms 15-40 cm tall, erect, leafy; nodes glabrous; leaves lanceolate, flat, glabrous; sheaths compressed, glabrous; ligule hairy; panicles effuse, large; spikelets 1.5-2 mm long; lower glume *ca* 1.5 mm long, 3. -nerved; upper glume 1.5-1.8 mm 5-nerved; lower lemma empty, *ca* 1 mm long, chartaceous; awns 3-4 mm long; stamens 3, very small; caryopsis ellipsoid.

*Fl.-Fr.:*Aug. -Nov.

Common on hill slopes and in moist, shady places all over the area.

Arundinella setosa Trin., Gram. Panic. 63. 1826; *A. capillaris* Hook., *f.,* Fl. Brit. India 7: 74. 1896; Bor, Grass. Burm. Ceyl. India. Pak. 424. 1960; Samvatsar, Fl. West. Trib. Madhya Pradesh 330.1996; Roy in N.P. Singh *et al.,* Fl. Madhya Pradesh 3:371.2001.

Perennials; culms 70-120 cm tall, erect smooth with woody root stock; leaves 15-30 × 0.5-0.9 cm long, linear, flat, acute, hirsute; ligule a ring of stiff hairs; inflorescence a panicle, 15-20 cm long; spikelets 5-6 mm long, 2-flowered, one sessile and other long pedicelled; lower glume 4 mm long, ovate, acuminate, 3-nerved, hairy; upper glume ovate, lanceolate, 5-nerved glabrous; lower lemma 3 mm long, ovate, 7-nerved, glabrous; upper lemma 5-nerved, bilobed, awned; palea 2-nerved, hyaline, margins infolded; stamens 3; upper lemma 2 mm long, ovate, 3-nerved, inrolled; lateral awn 1.5 mm long; central awn 8 mm long; palea 2.2 m long; stamens 3; caryopsis 1.5 mm long, oblong, slightly compressed, dark brown.

Fl.-Fr. : Aug. -Dec.

Rare, **Indore**, Kampel grassland.

Arundinella tuberculata Munro ex Lisboa, List Bombay Grasses 29. 1897; Bor, Grass. Burm. Ceyl. Ind. Pak. 425. 1960; Naik in Fl. Marathwada 2:991. 1998.

Annual herbs, culms 30-6- cm tall, erect, slender, sparingly branched, nodes glabrous; leaves linear-lanceolate, 2.5-11 × 0.5-0.9 cm, densely hispid, apex acute; ligule fringe of hairs; sheaths 2.5-5 cm long, densely hispid; panicles 5-20 cm long, lax, rachis angular, erect or flexuous, branches 2.5 cm long, alternating, scaberulous, pedicels up to 4 mm long, unequal, scaberulous; spikelets, ovate, 8-10 mm long, hispid, apex acuminate; lower glume herbaceous, ovate to lanceolate, up to 6 mm long, 3-nerved, acuminate; upper glume similar, 8-10 mm long, 5-nerved, apex beaked; lower lemma membranous, elliptic-oblong, 4-5 mm long, 3-nerved; palea hyaline, 2-keeled; stamens 3, anthers 0.8-1 mm long; upper lemma coriaceous to crustaceous, 2.5-3.2 mm long, scabrid with age, apex 2-toothed with 2 apical , 2.5-3 mm long setae, awn median, 12-14 mm long, geniculate; caryopsis ellipsoid, up to 3 mm long.

Fl.-Fr.: Aug. -Dec.

On gravelly soil of hill slopes, Dhar, Indore. Endemic to India.

ARUNDO L. W. 5; M P 1; SA 1

Arundo donax L., Sp. Pl. 81. 1753; Hook.f., Fl. Brit. India 7:302. 1896; Bor, Grass. Burm. Ceyl. Ind. Pak. 413. 1960; Samvatsar, Fl. West. Trib. Madhya Pradesh 330.1996; Roy in N.P. Singh *et al.,* Fl. Madhya Pradesh 3:372.2001. *Barru*

Strong perennials; culms 1.5-3.0 m tall, basal part woody, nodes glabrous; leaves amplexicaul at base, apex acuminate, 30-50 × 2-6 cm, ligule a ring of stiff hairs; panicles large, branches ascending, scaberulous; spikelets laterally compressed, 2-6-flowered, lanceolate, 1.0-1.5 cm long with awns; glumes 1 cm long; lower lemma 3-nerved; 9-10 mm long, silky hairy; upper lemma similar but shorter than lower; anthers 3, 2 mm long; caryopsis 2 mm long, oblong.

Fl.-Fr. : June-Sept.

Occasional along water courses and streams, Burhanpur, Dhar, Indore, Khandwa, Khargone.

AVENA L. W. 22; Ind. 9 M P 2; SA 1.

Avena sterilis L. subsp. **ludoviciana** (Durieu) J. M. Gillet. & Magne, Nouv. Fl. Franc. ed. 3, 352. 1873; *A. ludoviciana* Durieu, Actes Soc. Linn. Bordeaux 20:41. 1855; Bor, Grass. Burm. Ceyl. Ind. Pak. 434. 1960; Samvatsar, Fl. West. Trib. Madhya Pradesh 330 . 1996. *Jangli jai*

Annuals; culms, erect, 50-120 cm tal;, leaf sheath glabrous, 5-9 cm long, ligule membranous; leaf blade 10-25 cm; panicle 3--12 cm long; spikelets 2-3 flowered, 2-nate; florets 1 short, other long pedicelled, 1.8-2.2 cm long; lower glume coriaceous, densely villous on dorsal side, margins inflexed, 7-nerved awned; palea coriaceous, 2-nerved, 2-keeled; stamens 3; caryopsis villous all over.

Fl.-Fr. : Sept. -March

Common weed of wheat fields, Alirajpur, Burhanpur, Dhar, Khandwa, Khargone.

BAMBUSA Schreb. W. 137; M P 4; SA 1

Bambusa bambos (L.) Voss, Vilm. Blumengartn. ed. 3 1: 1189. 1895; *Arundo bambos* L., Sp. Pl. 81. 1753; *Bambos arundinacea* Retz., Observ. Bot. 5:24. 1789; *Bambusa arundinacea* Willd.,Sp. Pl. 2:245. 1799; Gamble in Hook.f., Fl. Brit. India 7: 395. 1896; Roy in N.P. Singh *et al.,* Fl. Madhya Pradesh 3:373.2001. *Katanga bans*

Perennials; culms erect, 5-20 m tall, internodes hollow drooping with thorny branches; culm-sheath 20-30 × 10-20 cm long, rounded at apex, hairy with dark brown hairs on the back, glabrous with age; leaves 6-12 × 1.2 cm long, linear, glabrous or hairy on the back, auricles with few bristles; ligule long ciliate; spikelets ± 2 cm long, lanceolate, acute, florets 3-6; rachilla hairy below the lemmas; palea glabrous, 2-keeled, the keels ciliate; lodicules 3, 2.5 mm long, fringed with a few hairs; anthers 5 mm long; caryopsis oblong, grooved on one side.

Fl.-Fr. : Feb. -April

Confined to **Khandwa,** extensively planted in forest nurseries and escape also.

BOTHRIOCHLOA Kuntze W. 34; M P 5; SA 5

Key to Species

1. Pedicellate spikelet with a transluscent median line, present also on the internodes; branches of panicles short, bearing number of spikelets
   2. Lower racemes shorter than the main rhachis .... **B. bladhii**
   2. Lower racemes longer than the main rhachis; sessile spikelets 4-6 mm long; pedicellate spikelets 5-9 mm liong; lower glume of pedicellate spikelet 15-20 nerved .... **B. jainii**
1. Pedicellate spikelet without a transluscent median line
   3. Nodes densely villous; racemes 12-15, densely ciliate; sessile spikelets 4.5-5 mm long; both sessile and pedicellate spikelets with a dorsal slit .... **B. kuntzeana**
   3. Nodes bearded; sessile spikelet awned
      4. Lower glume of pedicellate spikelet with 2-4 pits; leaves mostly cauline .... **B. insculpta**
      4. Lower glume not as above; leaves mostly confined to base .... **B. pertusa**

Bothriochloa bladhii (Retz.) S. T. Blake, Proc. Roy. Soc. Queensland 80: 62. 1969; *Andropogon bladhii* Retz., Observ. Bot. 2:27. 1781; *Bothriochloa glabra* (Roxb.) A. Camus, Ann. Soc. Linn. Lyon n.s., 76:164. 1931; *Andropogon glaber* Roxb., Fl. Ind. 1:271. 1820; *B. intermedia* (R. Br.) A. Camus, Ann. Soc. Linn. Lyon n.s.,76:164. 1931; *Andropogon intermedius* R. Br. Prodr. Fl. Nov. Holl. 202. 1810; *B. odorata* (Lisboa) A. Camus, Ann. Soc. Linn. Lyon. n.s., 86:165. 1931; *Andropogon odoratus* Lisboa, J. Bombay Nat. Hist. Soc. 4:118. 1889; Frank Gould in Rev. Handbk. Fl. Ceylon, (ed.) W.D. Clayton *et al.,* 8: 82. 1994.

Perennials; culms terete, 5-150 cm tall, erect, nodes bearded; leaf sheath terete, glabrous, ligule membranous; leaf blade linear-ovate, 10-60 cm, glabrous, apex acuminate; inflorescence of 9-10 cm long racemes; sessile spikelets narrowly ovate-elliptic, 3.2-3.5 mm, awned, callus bearded; lower glume membranous, dorsal side hairy in lower half, 9-11-nerved, apex truncate; upper glume membranous, narrowly ovate, margins inflexed, 3-nerved, 1-keeled, apex acute; lower lemma hyaline, narrowly ovate, 2.8-3 mm, nerveless, apex obtuse; palea absent; upper lemma hyaline, bearing geniculate awn at apex, 1-nerved; palea hyaline, short, apex acute; stamens 3; pedicelled spikelet narrowly elliptic; lower glume membranous, narrowly ovate, margins inflexed, 9-11 -nerved, apex obtuse; upper glume membranous, narrowly elliptic, margins inflexed, 3-5-nerved, apex acute; lower lemma hyaline, obtuse, margins ciliate, nerveless, deeply cleft at apex; palea absent; stamens 3; caryopsis 1.8-2 mm.

Fl.-Fr. : Sept. -Dec.

Burhanpur and Khandwa.

Bothriochloa jainii Deshp. & Hemadri, Indian Forester 97:593.1971; *Dichanthium jainii* (Deshp. & Hemadri) Deshp. 1979, *nom. invalid*; Sharma *et al.,* Fl. Maharashtra 3:456. 1996.

Large tufted perennials, about 1.5 m tall; nodes 5-8, bearded; leaves 30-65 × 0.8-1.5 cm, apex acuminate; margins spinulose, bulbous hairy on both surfaces; inflorescence a panicle of racemes on short main axis; racemes 3-15, 5-10 cm long, greenish to purple; spikelets in pairs, both sessile and pedicellate spikelets dissimilar.

*Fl.-Fr.:*Oct. -Jan.

Rare, Burhanpur, Khandwa.

Bothriochloa insculpta (A. Rich.) A. Camus, Ann. Soc. Linn. Lyon n.s.,. 76:165. 1931; *Andropogon insculptus* Hochst. ex A. Rich., Tent. Fl. Abyss. 2:458. 1851; Bor, Grass. Burm. Ceyl. Ind. Pak. 107. 1960; Samvatsar, Fl. West. Trib. Madhya Pradesh 331. 1996.

Perennials, culmms 80-100 cm tall, ascending, nodes bearded; leaves linear, 15-45 × 0.3-0.5 cm, acute or acuminate, ligule membranous, 2 mm long; sheaths 4-10 cm long; racemes 9-14, each 3-9 cm long, corymbosely fasciculate, silky; joints and pedicels 3-4 mm long; sessile spikelets 4-5 mm long, pitted; pedicellate spikelets up to 5.5 mm long; lower glume 2-4 pitted; upper glume equalling the lower; floral lemma hyaline, empty, epaleate; upper lemma reduced to its awn; caryopsis oblong.

Fl.-Fr.: July-Jan.

In open grass land, Alirajpur, Dhar.

Bothriochloa kuntzeana (Hack.) Henr., Blumea 3: 456. 1940; *Andropogon kuntzeanus* Hack., Monogr. Phan. 6: 478. 1889; Bor, Grass. Burm. Ceyl. Ind. Pak. 108. 1960; *Dichanthium kuntzeanum* (Hack.) S.K. Jain & Deshp., Bull. Bot. Surv. India 20: 134. (1978) 1979; Samvatsar, Fl. West. Trib. Madhya Pradesh 331. 1996; Roy in N.P. Singh *et al.,* Fl. Madhya Pradesh 3: 377.2001.

Perennials; culms tufted, 30-90 cm tall; rhizome short; leaves linear, acuminate, hairy at base, margins scabrid; sheath glabrous; ligule a shallow membrane; inflorescence a panicle, lower branches shorter; spikelets in pairs, sessile and pedicelled; pedicel and rachis joints flattened and longitudinally grooved, long ciliate; sessile spikelet 2-3 mm long; callus short, hairy; lower glume 7-nerved, pitted on back, 3.2 mm long; upper glume 3-nerved 3.5 long; lower lemma neuter; upper lemma hyaline, 1-nerved with scaberulous awn, keeled; awn, 1.2 cm long; anthers 2, yellow, 1.2-1.4 mm long; pedicelled spikelet neuter; caryopsis *ca* 2 mm long, without callus.

Fl.-Fr. : Sept. -Dec.

Occasional on rocky hill slopes, Burhanpur, Khandwa.

Bothriochloa pertusa (L.) A. Camus, Ann. Soc. Linn. Lyon n.s., 76:164. 1931; *Holcus pertusus* L.,Mant. PL. 2:301. 1771; *Andropogon pertusus* (L.) Willd., Sp. Pl. 4:922. 1806; Bor, Grass. Burm, Ceyl. Ind. Pak. 109. 1960; Samvatsar, Fl. West. Trib. Madhya Pradesh 331. 1996; Roy in N.P. Singh *et al.*, Fl. Madhya Pradesh 3: 378. 2001. *Phulera*

Perennial, 30-45 cm tall herbs; nodes bearded; leaves up to 30 cm long, narrowly linear, glabrous or pubscent; racemes 3-15, 2.5-4.5 cm long, silky hairy; sessile spikelets 0.3-0.4 cm long, oblong-lanceoate; pedicelled spikelets narrower than sessile ones; lower glume elliptic-lanceolate, hairy below the middle, with 1 or 2 pits; upper glume boat -shaped, equalling the lower, faintly 3-nerved; lower lemma empty, hyaline, lanceolate; palea absent; upper lemma reduced to its awn; palea absent; caryopsis oblong.

Fl.-Fr. : Aug. -Jan.

Common in open grass-lands, along field borders, in all districts.

BRACHIARIA (Trin.) Griseb. W. 125; M P 7; SA 7

Key to Species

1. Panicle linear, lower glume minute, scale-like, 1-nerved .... **B. eruciformis**
1. Panicle oblong, lanceolate or spreading; lower glume well developed, nerves more than 1
    2. Spikelets up to 2.5 mm long
        3. Plants and spikelets glabrous; pedicels with long white bristles; spikelets 1.2-2 mm long; lower glume truncate .... **B. reptans**
        3. Plants and spikelets hairy; pedicels without long hairs; spikelets 2-2.5 mm long; lower glume acute .... **B. villosa**
    2. Spikelets longer than 2.5 mm; lower glume 5 or more nerved, acute or obtuse
        4. Spikelets turgid, paired
            5. Spikelets crowded on the rachis, 2.5-4.5 mm long; leaves 2-3 cm broad . ...**B. ramosa**
            5. Spikelets loosely scattered on the rachis, 2.5 mm long; 0.6-1.6 cm broad ....**B. deflexa**
        4. Spikelets not turgid, paired or arranged singly, 3-3.5 mm long
            6. Branches of panicle sub-verticillate; plants robust .... **B. semiverticillata**
            6. Branches of panicle alternate; plants weak; spikelets remotely placed on the rachis ....**B. remota**

Brachiaria deflexa (Schumach.) C.E. Hubb. & Robyns, Bull. Jard. Bot. Etat 9:181. 1932; *Panicum deflexum* Schumach., Beskr. Guin. Pl. 63. 1827; *Urochloa deflexa* (Schumach.) H. Scholz, Bull. Mus. Natl. Hist. Nat., Adansonia 11:443. 1989; Samvatsar, Fl. West. Trib. Madhya Pradesh 334. 1996.

Annual herbs; culms 30-60 cm tall; nodes pubescent; leaves linear to oblong-lanceolate, softly hairy, scaberulous on margins; spikelets loosely scattered, elliptic-obovate; lower glume hyaline, ovate, pubescent, 5-7 nerved; upper glume membranous, ovate-elliptic, pubescent, 7-nerved; palea hyaline; upper lemma crustaceous, transversely rugose on the back; palea similar and enclosed in its lemma; caryopsis 2-2.5 mm long.

*Fl.-Fr.:*July-Nov

Common in hedges and bushes, Alirajpur, Dhar.

Brachiaria eruciformis (Sm.) Griseb., Fl. Ross. 4: 469. 1853; *Panicum eruciforme* Sm., Fl. Graec. 1: 44. 1806; *P. isachne* Roth Syst. Veg 2: 458. 1817; Hook.f., Fl. Brit. India 7: 28. 1896; Bor, Grass. Burm. Ceyl. Ind. Pak. 283. 1960; *Moorochloa eruciformis* (Sm.) Veldkamp, Reinwardtia 12: 139-140.2004; Samvatsar, Fl. West. Trib. Madhya Pradesh 334, 1996; Roy N.P. Singh *et al.,* Fl. Madhya Pradesh 3: 379.2001. *Sanyawara*

Annuals; culms 20-60 cm long, spreading; nodes pilose; leaves flat; sheaths glabrous; ligules hairy; racemes 2-8, erect, appressed to the rachis, 1.0-2.5 cm long; spikelets solitary, linear-lanceolate, 2.0-2-5 mm long, pubescent; lower glume scale- like, minute; upperglume 1.5-2.0 mm long, membranous, pubescent, 5-nerved; lower lemma empty; upper lemma elliptic or lanceolate, 1.5-2.0 mm long, muticous, faintly nerved; anthers 3, yellow, 1 mm long; caryopsis ellipsoid, 1.0-1.5 mm long.

*Fl.-Fr. :*Throughout the year.

Common weed of cultivated fields, in all districts.

Brachiaria ramosa (L.) Stapf, Fl. Trop. Africa 9:542. 1919; *Panicum ramosum* L., Mant. Pl. 1: 29. 1767; Hook.f., Fl. Brit. India 7: 36. 1896; Bor, Grass. Burm. Ceyl. Ind. Pak. 284. 1960; *Urochloa ramosa* (L.) Nguyen, Syst. Pl. Vasc; Acad. Sci. URSS; 1966; Samvatsar, Fl. West. Trib. Madhya Pradesh 334. 1996; Roy in N.P. Singh *et al.,* Fl. Madhya Pradesh 3:380.2001. *Popti*

Annuals; culms 30-45 cm tall, decumbent-ascending, rooting from the lower nodes; leaves 7-8 × 0.8-0.9 cm, lanceolate, cordate at base, glabrous or finely pubescent, ligule a ring of hairs; racemes 5- many, 3-6 cm long, rachis scabrid; spikelets usually in pairs, 3 mm long, ovoid or obovoid, apiculate; upper glume 7-nerved, pubescent; lower lemma 5-nerved, glabrous, epaleate, empty; upper lemma 2.0-2.25 mm long, chartaceous, transversely rugose, apiculate; palea equal to lemma; stamens 3; caryopsis ovoid, apiculate, brown.

*Fl.-Fr.:*Aug. -Dec.

Common in hedges and waste places in all districts.

Brachiaria remota (Retz.) Haines, Bot. Bihar & Orissa 3: 1052. 1961,(rep. ed. BSI); *Panicum remotum* Retz., Observ. Bot. 4: 17. 1786; Hook.f., Fl. Brit. India 7:38. 1896; Bor, Grass. Burm. Ceyl. Ind. Pak. 285. 1960; *Urochloa remota* (Retz.) Ashalatha & V.J. Nair, Bull. Bot. Surv. India 35 (144): 29. 1993; Samvatsar, Fl. West. Trib. Madhya Pradesh 334. 1996.

Annual, *ca* 50 cm tall herbs, branched from the base; leaves 7.5 × 0.25 cm, linear-lanceolate, acuminate glabrescent; spikes *ca* 10 cm long; splikelets distant, ellipsoid; lower glume 1-1.25 mm long; upper glume ovate apiculate; lower lemma similar to upper glume, 5-nerved; palea elliptic, mucronate; upper lemma and its palea transversely rugulose; stamens 3; caryopsis elliptic-ovate, 1 mm long.

Fl.-Fr.: Sept. -Dec.

Frequently found as forest undergrowth, spreading over adjacent vegetation, in all districts.

Brachiaria reptans (L.) C.A. Gardner & C.E. Hubb., Hooker's, Icon Pl. 34 t. 3363. 1938; *Panicum reptans* L. Syst. Nat. ed. 10 2:870.1759; *Urochloa reptans* (L.) Stapf, Fl. Trop. Afr.

9: 601. 1920; Bor, Grass. Burm. Ceyl. India Pak. 285. 1960; Samvatsar, Fl. West. Trib. Madhya Pradesh 336. 1996; Roy in N.P. Singh *et al.,* Fl. Madhya Pradesh 3: 380.2001. *Chiman chara*

Annuals; culms slender, decumbent, 20-30 cm tall, nodes pubescent, internodes 2.5-7.0 cm long; leaves ovate-lanceolate, cordate at base, sheaths 1-2 cm long, margins ciliate, ligule hairy; Inflorescence of 5-8 racemes, alternate, 1.5-3.0 cm long; spikelets in pairs, secund, 2 mm long, pedicels with bulbous-based bristles; lower glume hyaline, minute; upper glume cymbiform, 9-nerved, boat-shaped, pointed, 2 mm long; lower floret neuter or male; upper floret hermaphrodite; lemma indurate, 1.6 mm long, golden brown, apiculate, palea clasped by lemma; anthers 3, yellow, 0.3 mm long; caryopsis flattened-oval, 1-2 mm long.

*Fl.-Fr.:*July-Dec.

Common along water courses and in shade of trees, in all districts.

Brachiaria semiverticllata (Rottler) Alston, Handb. Fl. Ceylon 4:(Suppl.)318. 1931; *Panicum semiverticillatum* Rottler, Mat. Med. Hindt. 219. 1813; Bor, Grass. Burm. Ceyl. Ind. Pak. 286. 1960; Samvatsar, Fl. West. Trib. Madhya Pradesh 336. 1996.

Annuals, culms, robust, 150-200 cm tall, erect, branched; nodes hairy; leaves linear-lanceolate, acuminate, base rounded; panicle up to 10-20 cm long; spikelets elliptic, up to 3.5 mm long; lower glume, 5-nerved, up to 2 mm long; upper glume, longer, up to 3 mm long, 7-nerved; lower lemma equalling upper glume, 5-nerved; palea hyaline; upper lemma rugulose; caryopsis 2 mm long.

Fl.-Fr.: Sept. -Oct.

In wet and shady places, only **Alirajpur**, Chatliya pani, Kattiwada range.

Brachiaria villosa (Lam.) A. Camus, Fl. Indo-chine 7:433. 1922; *Panicum villosum* Lam., Tabl. Encycl. 1:173. 1791; Hook.f., Fl. Brit. India 7:34. 1896; Bor, Grass. Burm. Ceyl. Ind. Pak. 286. 1960; Samvatsar, F. West. Trib. Madhya Pradesh 336. 1996.

Annual, 10-50 cm tall grasses, creeping below, nodes pubescent; leaves ovate-oblong, acute or obtuse, densely pubescent on both surfaces; racemes second, pubescent; spikelets up to 2.5 mm long, pubescent; lower glume very short, nerveless; upper glume ovate, acute 7-nerved; lower lemma ovate-oblong; palea obovate, apiculate; upper lemma crustaceous, transeversely rugose on back; palea similar; caryopsis 2 mm long.

Fl.-Fr.: Sept. -Oct.

Rare, **Jhabua** only, Ranapur, near streams.

CAPILLIPEDIUM Stapf W. 17; M P 4; SA 4

Key to Species

1. Annuals; culms weak; sessile spikelets 4-4.5 mm long .... **C. filiculme**
1. Perennials
    2. Spikelets paired; sessile spikelets 3-3.5 mm long; pedicellate spikelets 3-4 mm long............ **C. huegelii**
    2. Spikelets in triads or 2-6

3. Spikelets in triads; primary branches of panicle whorled; sessile spikelets 1.8-2.5 mm; pedicellate spikelets 1.5 mm, represented by pedicel only .... **C. parviflorum**
3. Spikelets 2-6; sessile spikelets 3-4 mm; pedicellate spikelet 3-4.5 mm .... **C. assimile**

Capillipedium assimile (Steud.) A. Camus, Fl. Indo-Chine 7: 314. 1922; *Andropogon assimilis* Steud., Syn. Pl. Glumac. 1:397. 1854; Hook.f., Fl. Brit. India 7: 179. 1896; Bor, Grass. Burm. Ceyl. Ind. Pak. 110.1960; Roy in N.P. Singh *et al.,* Fl. Madhya Pradesh 3:382.2001.

Perennials; culms 30-75 cm tall, hard, geniculate; leaves linear-lanceolate, ± 13 cm × 3 mm; midrib hirsute, tubercle hairs on basal part; ligule membranous; inflorescence open panicle, branches capillary; spikelets paired, pale green, dimorphic, upper pedicelled; pedicels longitudinally grooved, margins ciliate; lower spikelet sessile, fertile, 2.5 mm long; lower glume subcoriaceous, lanceolate, laterally 2-grooved, ciliate on back; upper glume broadly lanceolate, 2-keeled, ciliate, 5-nerved; upper lemma 1- nerved, awned, awns 1.0-1.5 cm long; anthers 3, yellow; caryopsis brownish.

Fl.-Fr.: Sept. -Dec.

Barwani, Burhanpur, Khargone.

Capillipedium filiculme (Hook.f.) Stapf, in Hooker's Icon. Pl. 31:t. 3085. 1922; *Andropogon filiculme* Hook.f., Fl. Brit. India 7:181. 1896; Bor, Grass. Burm. Ceyl. Ind. Pak. 111. 1960; Samvatsar, Fl. West. Trib. Madhya Pradesh 340.1996; Roy in N.P. Singh *et al.,* Fl. Madhya Pradesh 382.2001.

Annuals; culms tufted, 50-100 cm tall, erect or geniculately ascending, slender, weak, simple or sparingly branched, glabrous, nodes bearded; leaf sheath glabrous; ligule membranous; panicles decompound, oblong, 5-10 cm long, aromatic, rachis filiform, branches capillary, bearded in axils, joints capillary, with a translucent groove in center, margins ciliate; sessile spikelets 4-4.5 mm long, elliptic, awned, callus short, bearded; lower glume coriaceous, elliptic, 2-keeled, 7-11nerved, apex subacute; upper glume coriaceous, narrowly elliptic, 1-keeled, 3-nerved, apex acute; lower lemma empty, hyaline, narrowly ovate, apex obtuse; upper lemma hyaline to a base of awn, 1-nerved, awn geniculate, 15-18 mm long; stamens 3; pedicels capillary, 1.8-2 mm long, with a translucent groove in center, margins ciliate; pedicelled spikelet narrowly elliptic, 3.5-4 mm long, unawned; lower glume subcoriaceous, narrowly elliptic, 10-11-nerved, margins inflexed, sparsely ciliate, apex acute; upper glume membranous, narrowly elliptic, 5-7-nerved, margins ciliate apex acute.;lower lemma hyaline, apex acute; palea absent; stamens 3; caryopsis elliptic, 1 mm long.

Fl.-Fr. : Sept. -Dec.

Common on slopes of hills, amidst bushes and hedges, in all districts. Endemic to India.

Capillipedium huegelii (Hack.) Stapf, Rev. Bot. Appl. Agric. Colon 1:306. 1921; *Andropogon huegelii* Hack., Monogr. Phan. 6:492. 1889; Bor, Grass. Burm. Ceyl. Ind. Pak. 111. 1960; Hook.f., Fl. Brit. India 7: 180.1896; Samvatsar, Fl. West. Trib. Madhya Pradesh 340.1996; Roy in N.P. Singh *et al.,* Fl. Madhya Pradesh 3: 382.2001. Endemic to India.

Perennials; culms tufted 30-80 cm tall, rooting at lower nodes, erect or branched, shortly creeping, often red; nodes hairy; leaves 10-25 cm, linear, margins scabrid; sheaths compressed; ligule lacerate (membranous ciliate); panicle branches capillary with bearded axils; sessile spikelets ± 3 mm long, pale green-purplish, callus hairy; lower glume ovate, truncate, villous; upper glume glabrous, apiculate; lower lemma ovate, glabrous; upper lemma represented by an awn, about 2 cm long; anthers 3, yellow; caryopsis elliptic.

Fl.-Fr. : Sept. -Nov.

Alirajpur, Dhar, Indore, Khandwa. Endemic to India.

Capillipedium parviflorum (R. Br.) Stapf, Fl. Trop. Afr. 9: 169. 1917; *Holcus parviflorum* R. Br., Prodr. Fl. Nov. Pl. Holl. 199. 1810; Bor, Grass. Burm. Ceyl. Ind. Pak. 112. 1960; Roy in N.P. *et al.,* Fl. Madhya Pradesh 3:383.2001. *Morpakhadia*

Perennials; culms 1-15 cm tall, usually erect, tufted, branched, inter nodes channeled, glabrous; leaves up to 30 cm long, glabrous or pubescent; sheath pubescent-hirsute; ligule membranous; inflorescence compound panicle; racemes reduced to 1 sessile and 2 pedicelled spikelets, joints and pedicels 1.8 mm long, shortly ciliate; sessile spikelets 3-5 mm long, olive brown; lower glume truncate, shallowly concave on back, 6-nerved, scabrid; upper glume similar to lower; lemma 2-2.5 mm long, awn 1-2 cm long; pedicelled spikelet usually male, similar to sessile; anthers 3, yellow, 1.5-2 mm long; caryopsis not seen.

Fl.-Fr. : Nov. -March

Rare in irrigated fields and forest edges, Indore, Khandwa.

CENCHRUS L. W. 23; M P 3; SA 3

Key to Species

1. Bristles or spines of the involucres retrorsely scabrid .... **C. biflorus**
1. Bristles or spines of the involucres antrorsely scabrid
    2. Bristles connate at the base only, all not erect .... **C. ciliaris**
    2. Bristles connate into a cup, all erect .... **C. setiger**

Cenchrus biflorus Roxb. Fl. Ind. 1:238. 1820; Hook., f., Fl. Brit. India 7:90.1896; Bor, Grass. Burm. Cey. Ind. Pak. 287. 1960; Samvatsar, Fl. West. Trib. Madhya Pradesh 340.1996; Roy in N.P. Singh *et al.,* Fl. Madhya Pradesh 3:384.2001. *Dhaman*

Annuals; culms, prostrate, 10-45 cm tall, rooting at lower nodes; leaf sheath glabrous, terete, 2-8 cm long; ligule hairy, 1-1.5 mm long; leaf blade 2-10 cm long, linear ovate, glabrous; racemes 3-15 cm long, usually more or less enclosed in uppermost leaf sheath; spikelets geminate, 4-6 mm long, shorter than inner bristles; bristles hard, retrorsely bearded, sharpy pungent; outermost short, inner ciliate hairy at base, dorsally grooved; lower glume membranous, ovate, glabrous, 1- nerved, apex acute; upper glume membranous, ovate, glabrous, 3-nerved, apex acuminate; lower lemma membranous, ovate, 3. 8-4 mm, glabrous, 5-nerved, apex acuminate; palea membranous, narrowly ovate, margins inflexed, 2-nerved, apex acute; upper lemma coriaceous, ovate, glabrous, 5-nerved, margins inflexed, apex acuminate; palea coriaceous, ovate, glabrous, margin inflexed, 2-nerved, apex acuminate; lodicules absent; stamens 3; caryopsis broadly elliptic.

Fl.-Fr.: Aug. -Dec.

Burhanpur, Dhar, Khandwa, in open gravelly habitat.

Cenchrus ciliaris L., Mant. Pl. 2: 302. 1771; Hook., f., Fl. Brit. India 7:88. 1896; Bor, Grass. Burm. Cey. Ind. Pak. 287. 1960; Samvatsar, Fl. West. Trib. Madhya Pradesh 340.1996; Roy in N.P. Singh *et al.,* Fl. Madhya Pradesh 3:384.2001. *Charwa*

Perennials; culms tufted, 20-60 cm tall, erect or decumbent, nodes glabrous; ligule hairy; leaf blades linear-ovate, 10-50 cm , hairy on both sides, tapering to a fine point; racemes dense, cylindric, 5-15 cm long, pinkish, rachis finely scaberulous; involucels subsessile, bristles antrorsely scabrid, outer bristles slender, short, not ciliate, inner bristles dilated and connate into a ring at the base; spikelets 3-4 in an involucre, narrowly ovate, 4-5 mm long; lower glume membranous, ovate, 1-nerved, apex acute; upper glume membranous, ovate, 3-nerved, apex acute; lower lemma membranous, narrowly ovate, glabrous, 5-nerved, apex acute; palea membranous, narrowly elliptic-ovate, margins inflexed, 2-nerved, apex acute; upper lemma coriaceous, narrowly ovate, glabrous, margins inflexed, 5-nerved, apex acuminate; palea subcoriaceous, narrowly ovate, margins inflexed, 2-nerved, apex acuminate; lodicules absent; stamens 3; caryopsis truncate, dorsally compressed.

*Fl.-Fr. :*Sept. -Dec.

Common in all districts, in rocky, dry localities.

Cenchrus setiger Vahl, Enum. Pl. Obs. 2: 395. 1806; Hook., f., Fl. Brit. India 7:89. 1896; Bor, Grass. Burm. Cey. Ind. Pak. 290.1960; Samvatsar, Fl. West. Trib. Madhya Pradesh 340. 1996; Roy in N.P. Singh *et al.,* Fl. Madhya Pradesh 3:385. 2001. *Kala dhaman*

Perennials; culms 20-60 cm tall, tufted, base woolly, extravaginal sheaths woody, nodes glabrous; leaf sheath terete, glabrous, hairy on margins, ligule membranous, hairy, 1 mm long; leaf blade 5-30 cm long, linear-ovate, glabrous; racemes 2-10 cm long, solitary; involucres usually 2-3 flowered; bristles thick, connate into a cup: outer ones 1.5-2.5 mm long; spikelets in clusters of 2-4, sessile, usually 4 mm long; lower lemma barren or male, 5-7 nerved; upper lemma bisexual, 4-4.5 mm long; palea *ca* 3 mm long; lodicules absent; anthers 3; caryopsis oblong.

Fl.-Fr.: Aug. -Dec.

Common on rocky, soils of open grasslands, Airajpur, Barwani, Burhanpur, Dhar, Indore, Khargone.

CHIONACHNE R. Br.W. 9; M P 1; SA 1

Chionachne gigantea (J. Koenig) Veldkamp, Blumea 47:559.2002; *Coix gigantea* J. Koenig, Naturforscher (Halle) 23:211. 1788; *Coix koenigii* Spreng., Syst. Veg. 1:228. 1824; *Chionachne koenigii* (Spreng.) Thwaites, Enum. Pl. Zeyl. 357. 1864; *Polytoca barbata* (Roxb.) Stapf, Fl. Brit. India 7:102. 1896; Bor, Grass. Burm. Cey. Ind. Pak. 164. 1960; Samvatsar, Fl. West. Trib. Madhya Pradesh 341. 1996; Roy in N.P. Singh *et al.,* Fl. Madhya Pradesh 3:394.2001. *Barru*

Annuals or perennials; culms terete, 50-300 cm tall; nodes glabrous; ligule membranous, 1.5-2 mm long; leaf blade linear-ovate, 10-120 cm long, covered with stiff bulbous-based hairs on upper side, lower side glabrous, margins scabrous, apex acuminate; racemes spiciform, panicled; female spikelets basal, while male spikelets terminal; spathes herbaceous, many nerved; male spikelets 10-11 mm long, solitary or binate; lower glume 9-11 nerved, margins inflexed; upper glume 7-nerved, 2-keeled, margins broadly inflexed; lower lemma hyaline, 3-nerved, 2-keeled; palea hyaline, 2-keeled; upper lemma 1-nerved; palea nerveless; female spikelets sessile, 4-6 mm long; lower glume obscurely nerved, embracing the upper glume and lemmas; upper glume membranous; lower lemma empty, 3-nerved with inflexed margins; palea nerveless; caryopsis enclosed in the hardened lower glume, white, shining, glabrous, oblong, rounded above.

Fl.-Fr. : Sept. - Jan.

Common along streams and river beds all over the area.

CHLORIS Swartz W. 63; M P 10; SA 4

Key to Species

1. Empty lemmas always one above the fertile floret, with, 4.5-5 mm long awn ....**C. virgata**
1. Empty lemmas two or more above the fertile floret
   2. Culms hirsute above the iflorescence; empty lemmas 3-4 .... **C. quinquesetica**
   2. Culms puberulous or glabrous below the inflorescence
      3. Culms puberulous below; empty lemmas 3-4 ....**C. montana**
      3. Culms glabrous below; empty lemmas 2 .... **C. barbata**

Chloris barbata Sw., Fl. Ind. Occid. 1: 200.1797; Hook.f., Fl. Brit. India 7: 292. 1896; Bor, Grass. Burm. Ceyl. Ind. Pak. 465. 1960; Samvatsar, Fl. West. Trib. Madhya Pradesh 341. 1996; Roy in N.P. Singh *et al.,* Fl. Madhya Pradesh 3:387.2001. *Kali sukli.*

Perennials; culms 25-75 cm tall, geniculate at base; leaves flat, glabrous; sheaths compressed, keeled, glabrous; ligule a narrow membrane; inflorescence of digitate, 4-17 spikes, rachis scabrous; spikelets 2.5-3.0 mm long, 2-siriate, imbricate, subsessile; lower floret solitary, perfect; upper two florets imperfect; rachilla a disarticulates above glumes with a tuft of hairs at base; lower glume 1 mm long, 1-nerved, short-awned; upper glume long, with awn; lower lemma broadly elliptic, 2.5-3 mm long, membranous, 3- nerved, often tinged with purple; keel ciliate; awn bristle like; empty lemmas 2, smaller, truncate, with short awns; anthers 3, yellow; caryopsis fusiform.

Fl. & Fr. : Aug. -Dec.

Common around fields, waste lands and on old walls, throughout the area.

Chloris montana Roxb., Fl. Ind. 1:331. 1820; Hook.f., Fl. Brit. India 7:292. 1896; Bor, Grass. Burm. Ceyl. Ind. Pak. 466. 1960; Samvatsar, Fl. West. Trib. Madhya Pradesh 336. 1996; Roy in N.P. Singh *et al.,* Fl. Madhya Pradesh 3:388. 2001.

Perennial, 10-50 cm tall grasses; leaves lanceolate, flat or folded, sometimes pilose along the margins; sheaths compressed glabrous, mouth long-hairy; ligule membranous; panicle of 3-6 digitate spikes; rachis scabrid and pilose; spikelets imbricate, 2-seriate, ovate *ca* 3 mm long; lowest floret perfect and 3-4 imperfect florets above it; rachilla bearded at the base; lower glume 2-2.5 mm long, acute, keeled; upper glume 2.5 -3 mm long; lower lemma elliptic, 2.5-3 mm long; awns 6.5-7 mm long; upper lemma gradually smaller with smaller awns; palea scabrid on keel; caryopsis brownish, 2-2.5 mm long.

Fl.-Fr.: Aug. -Jan.

Weed of fields on bunds, **Jhabua** at Ranapur.

Chloris quinquesetica Bhide, J. Proc. Asiat. Soc. Bengal 8: 311. 1912; Bor, Grass. Burm. Ceyl. Ind. Pak. 466. 1960; Sharma *et al.,* Fl. Maharashtra 3:427. 1996.

Perennials; culms terete, 10-60 cm tall, prostrate, stoloniferous; nodes glabrous, swollen, rooting; leaves 2-15 × 0.25-0.5 cm, flat, apex acuminate, base truncate; ligule membranous; Spikes few to many, fascicled, 2-5 cm long; spikelets 4-5 flowered, awns 5; lower glume membranous, 1-nerved, apex acuminate; upper glume membranous; lower lemma coriaceous, obovate, keel with white hars on both sides; palea membranous, obovate, 2-nerved, apex acute; stamens 3; caryopsis trigonous or biconvex.

Fl.-Fr. : Aug. -Jan.

Khandwa only, in open, moist situation, at Kalibhit range.

Chloris virgata Sw., Fl. Ind. Occid. 1: 203. 1797; Hook., f., Fl. Brit. India 7: 291. 1896; Bor, Grass. Burm. Ceyl. Ind. Pak. 468. 1960; Samvatsar, Fl. West. Trib. Madhya Pradesh 342. 1996; Roy in N.P. Singh *et al.,* Fl. Madhya Pradesh 3:388. 2001;

Perennials; culms 15-50 cm tall, erect or decumbent, usually flattened and profusely branched; leaves flat, glabrous; sheaths loose, glabrous, keeled; ligules membranous; inflorescence of 4-9, digitate, 2.5-3.5 cm long spikes; rachis capillary, scabrid; spikelets 2-seriate, 3-4 mm long, subsessile; lower glume 2 mm long, 1-nerved; upper glume lanceolate, scabrid on keel, with short awn; lower lemma oblong-lanceolate, 2- fid at tip; awn *ca* 7 mm long; empty lemma oblanceolate, truncate, with 5 mm long awn; anthers 3, yellow; caryopsis trigonous, shining.

*Fl.-Fr. :*Aug. -Dec.

Common weed of wastelands, on old walls, throughout the area.

CHRYSOPOGON Trin. W. 48; M P 5; SA 5

Key to Species

1. Inflorescence compact; primary branches not whorled
   2. Spikelets ternate, 4-7 mm long, callus bearded, acicular; upper glume entire .... **C. aciculatus**
   2. Spikelets binate, 4-5 mm long, callus white to rusty-red; upper glume bifid .... **C. polyphyllus**
1. Inflorescence open; primary branches whorled
   3. Spikelets ternate, 7-8 mm long; upper glume 2-toothed, not awned .... **C. fulvus**
   3. Spikelets binate
      4. Sessile spikelet 6 mm long; lower lemma 6 mm long; upper glume awned, awn 10 mm long; upper lemma 4-4.5 mm long, apex bifid, awn 2-3 cm long . ... **C. lawsonii**
      4. Sessile spikelet 3.5-5 mm long; lower lemma 3.5-5 mm long; upper glume not awned; upper lemma 2-3 mm long unawned .... **C. zizanoides**

Chrysopogon aciculatus (Retz.) Trin., Fund. Agrost. 188. 1820; *Andropogon aciculatus* Retz., Observ. Bot. 5:22. 1788; Hook.f., Brit. India 7:188. 1896; Bor, Grass. Burm. Ceyl. Ind. Pak. 115. 1960; Roy in N.P. Singh *et al.,* Fl. Madhya Pradesh 3:390.2001. *Gorio*

Perennials; culms 30-50 cm tall; leaves 15-30 × 0.5-0.8 cm long, mostly sub-basal, linear, lanceolate obtuse or acute; ligule a hairy rim; inflorescence an erect, 10-15 cm long panicle; spikelets in groups of 3, one sessile, 2 pedicelled; sessile spikelets hermaphrodite, 3.5-4.0 mm long, callus with golden brown hairs; lower glume lanceolate, bidentate, 3-nerved, its upper part hirsute on the nerves; upper glume 3.5 mm long, narrow, boat-shaped, aristate, ciliate; lower lemma 2.5-2.8 mm long, hyaline, 2-nerved; upper lemma hyaline, 1-nerved; palea oblong-acute hyaline; anthers 1 mm long; pedicelled spikelets: lower glume 4-5 mm long, glabrous, upper glume 3-4 mm long; lower floret empty; lower lemma 3.2-3.3 mm long; upper lemma mucronate; palea small; stamens 3, anthers 2.5 mm long, yellow; callus pungent, as long as the spikelet.

Fl.-Fr. : July-Nov.

Specially on slopes with yellow soils, **Khargone** only, at Sirvel forest.

Chrysopogon fulvus (Spreng.) Chiov., Fl. Somala 1:327. 1929; *Pollinia fulva* Spreng.,Neue Entdeck. Pflanzenk. 2:93. 1821; *Andropogon monticola* Schult. & Schult. f., Mant. 3:665. 1827; Bor, Grass. Burm. Ceyl. Ind. Pak. 116. 1960; Samvatsar, Fl. West. Trib. Madhya Pradesh 342. 1996; Roy in N.P. Singh *et al.,* Fl. Madhya Pradesh 3:390.2001. *Gurgusi*

Perennials; culms 50-70 cm tall, tufted, erect or ascending from a shortly geniculate base; leaves 15-30 × 0.5-0.8 cm long, linear-acuminate, glabrous; ligule a shallow hairy rim; inflorescence a terminal panicle, 5-10 cm long, yellowish or purple; spikeletes in group of 3, one sessile and 2 pedicellate; sessile spikelets 4-7 mm long; lower glume 4-6 mm long, narrowly oblong, 2-4-nerved, hispid or glabrous on the back; upper glume 5 mm long, compressed, 3-nerved, margin hyaline with 3-6 mm long awn; lemma as long as the upper glume, linear, hyaline, ciliate, obtuse; palea absent or small; pedicelled spikelets 4-5 mm long, compressed; lower glume lanceolate, acuminate, 7-nerved; upper glume lanceolate, acuminate, 3-nerved, shortly aristate; lemma lanceolate hyaline; palea similar; stamens 3, yellow or purple; caryopsis yellow.

*Fl.-Fr. :*July-Dec.

Common, grows in open grassland and along the roads in forests, everywhere in the area.

Chrysopogon lawsonii (Hook.f.) Veldkamp, Austrobaileya 5:515. 1999; *Andropogon lawsonii* Hook.f., Fl. Brit. India 7: 187. 1896; *Vetiveria lawsonii* (Hook.f.) Blatt. & McCann., J. Bomby. Nat. Hist. Soc. 32: 409. 1928; Bor, Grass. Burm. Ceyl. Ind. Pak. 258. 1960; Roy in N.P. Singh *et al.,* Fl. Madhya Pradesh 3:544.2001.

Perennials; rhizome woody; culms 25-170 cm tall, erect, base leafy, tufted; leaves linear, up to 50 cm long, glabrous or bristly; sheath compressed, keeled; ligule membranous, minute; inflorescence a whorled panicle made of several racemes; rachis smooth; callus bearded; sessile spikelets 5-6 mm long, linear-lanceolate, polished, awned; lower glume 5-6 mm long, linear, glabrous, muriculate; upper glume equalling the lower, boat-shaped, with 9-10 mm long arista; lower floret empty, upper hermaphrodite; lemma 4.0-4.5 mm long, 2-lobed, awn 25-29 mm long; palea hyaline, nerveless; anthers 3, yellow; caryopsis ± 4 mm long, broader towards base.

Fl.-Fr. : Sept. -Dec.

Occational in open grasslands, Khandwa, Khargone.

Chrysopogon polyphyllus (Hack.) Blatt. & McCann, J. Bombay Nat. Hist. Soc. 32:416. 1928 & Bombay, Grass. 72. 1935; *Andropogon polyphyllus* Hack. ex Hook.f., Fl. Brit. India 7: 194. 1896; Bor, Grass. Burm. Ceyl. Ind. Pak. 118. 1960; Samvatsar, Fl. West. Trib. Madhya Pradesh 343. 1996; Roy in N.P. Singh *et al.,* Fl. Madhya Pradesh 3:391.2001.

Perennials; culms 60-100 cm tall, rootstock creeping, nodes glabrous; ligule hairy; leaf blade linear, 10-40 cm, glabrous; panicle 10-20 cm long; sessile spikelet narrowly ovate, awned; lower glume narrowly elliptic, glabrous, 4-nerved, 1-keeled, apex acute; upper glume narrowly elliptic, margins ciliate, 3-nerved, awned, 1-keeled; lower lemma membranous, narrowly obovate, glabrous, margins ciliate, 1-nerved, apex acute; palea absent; upper lemma membranous, narrowly linear, glabrous, 3-nerved, cleft at apex in to two teeth, awned between teeth, awn geniculate; palea hyaline, linear, apex two toothed; stamens 3; pedicelled spikelet narrowly ovate; lower glume membranous, narrowly ovate, puberulous, margins inflexed, 6-nerved, apex acuminate; upper glume narrowly ovate, margins inflexed, 3-nerved, apex acute; lower lemma narrowly ovate, margins inflexed, ciliate, 2-nerved, apex acute; palea absent; upper lemma membranous, linear, margins ciliate, 3-nerved, apex two toothed; palea hyaline, narrowly oblong, margins ciliate, nerveless, apex obtuse; stamens 3; caryopsis narrowly obovate.

Fl.-Fr. : Sept. - Dec.

Very common in rocky river and stream beds, in all districts. Endemic to India

Chrysopogon zizanioides (L.) Roberty, Bull. Inst. Franc. Afrique Noire, A 22: 106. 1960; *Phalaris zizanoides* L. Mant. Pl. 2: 183. 1771; *Andropogon squarrosus* L.f., Suppl. Pl. 433. 1782 (*nom. conf.*); Hook.f. Fl. Brit. India 7:186. 1896; *Vetiveria zizaniodes* (L.) Nash, Fl. South East U.S. 67. 1903; Bor, Grass. Burm. Ceyl. India Pak. 258. 1960; Roy in N.P. Sigh *et al.,* Fl. Madhya Pradesh 3: 545. 2001. *Khus*

Strong perennials; culms 1.0 – 1.5 m tall; roots fibrous, aromatic; leaves 30-70 × 0.4-1.8 cm, linear-acute, margins scabrid; ligule a minute membranous line; inflorescence a contracted panicle, 20 – 30 cm long; spikelets in pairs; sessile spikelets lanceolate, acute, yellowish; lower glume 3. 8 mm long, lanceolate, subacute, muricate on the back, 5-nerved; upper glume 4 mm long, concave, spinulosely muricate on the keel at the top, 1-nerved; lower lemma 3.6 mm long, empty, lanceolate-acute, retrorsely ciliate; upper lemma 3-nerved, mucronate; palea 2.5 mm long; stamens 3, orange; pedicelled spikelets more slender, similarly muriculate and scabrid; upper flor*et al.,*ways male; caryopsis *ca* 2 mm long, broader at base.

Fl.-Fr. : -Aug. -Nov.

Khandwa only, on the banks of rivers and streams, Nagchun and Daryapur.

COIX L. W. 4; M P 3; SA 2

Key to Species

1. Lower glume of male spikelets narrowly winged; spikelets 8-12 mm long .... **C. lacryma-jobi**
1. Lower glume of male spikelet not winged; spikelets 8-9 mm long .... **C. aquatica**

Coix aquatica Roxb. Fl. Ind. ed. 1832 3: 571. 1832; Hook., *f.,* Fl. Brit. India 7: 100.1896; Bor, Grass. Burm. Ceyl. Ind. Pak. 264. 1960; Roy in N.P. Singh *et al.,* Fl. Madhya Pradesh 3:393.2001. *Ran jondha*

Perennials; culms erect, 100-200 m tall, nodes glabrous; leaves 10-30 × 2-3 cm, linear-lanceolate, cordate at base, ciliate on margins, acuminate at apex; sheath compressed, glabrous, ligule 1 mm long, membranous; inflorescence erect or pendent, racemes axillary, about 6 cm long; male spikelets terminal; pistillate basal, enclosed by a bead-like hard bract; spikelets unisexual; male spikelets in 2-3 nate, imbricate, elliptic-lanceolate, 10-12 mm long, pedicelled or all sessile; lower glumes 10-11 mm long, elliptic-lanceolate, many-nerved, concave, keels 2, winged; upper glume elliptic-lanceolate, 7-9-nerved; lower lemma long, membranous, hyaline, 3-5-nerved; anthers 3; female spikelets 10-11 mm long; lower glume oblong-lanceolate, acuminate, beaked, margins inrolled; upper glume hyaline, muticous, deeply 2-grooved, keeled; lower lemma empty, oblong-lanceolate, acuminate, faintly nerved; caryopsis ellipsoid.

Fl.-Fr. :Sept. - Dec.

Commonly grows on marshy areas near ponds and river banks, Indore, Khandwa, Khargone.

Coix lacryma-jobi L., Sp. Pl. 972. 1753; Hook.f., Fl. Brit. Ind. 7: 100.1896; Bor, Grass. Burm. Ceyl. Ind. Pak. 264. 1960; Samvatsar, Fl. West. Trib. Madhya Pradesh 343. 1996; Roy in N.P. Singh *et al.,* Fl. Madhya Pradesh 3:394.2001. *Gania*

Perennials; culms erect, 0.8-1.5 m tall, rooting at lower nodes; leaves 10-50 × 2-4 cm, linear-lanceolate, cordate at base, acuminate at the tip, ligule 1 mm long, truncate, lacerate; inflorescence erect or pendent; racemes axillary, about 8 cm long; staminate spikelets terminal; pistillate spikelets basal, enclosed by a bead-like hard involucres; spikelets unisexual; staminate spikelets in pairs, one pedicelled, the other sessile; lower glume 8 mm long, ovate, many-nerved, margins scabrid, winged; upper glume thin, oblong-lanceolate, apex acuminate, 9-nerved; lower lemma hyaline, oblong-lanceolate, acuminate, feebly 5-nerved; palea 6 mm long, hyaline, 2-keeled; upper lemma oblong-lanceolate, acute, 5-nerved; stamens 3; female spikelet basal, ovoid, surrounded by the hardened bract forming a bead-like bony structure, 6-11 mm long; lower glume ovate-oblong, acute; upper glume thinner; lemma equal to spikelet; caryopsis 6-8 mm long, ovoid or globose, white.

Fl.-Fr. :Sept. - Nov.

Common, along margins of ponds and river beds in all districts.

CRYPSIS Aiton W. 10; M P 1; SA 1

Crypsis schoenoides (L.) Lam., Tab. Encycl. 1: 166. 1791; *Phleum schoenoides* L., Sp. Pl. 60.1753; Hook.f., Fl. Brit. India 7:235. 1896; *Heleochloa schoenoides* (L.) Host, Icon. Descr. Gram. Austriac. 1:23. 1801; Bor, Grass. Burm. Ceyl. Ind. Pak. 622. 1960; Samvatsar, Fl. West. Trib. Madhya Pradesh 343. 1996; Roy in N.P. Singh *et al.,* Fl. Madhya Pradesh 3:395. 2001. *Gangridio*

Annuals; culms 5-15 cm, prostrate, glabrous; leaves narrowly lanceolate, 2-3 cm long, acuminate at apex; sheaths deeply furrowed, glabrous or ciliate on margins. ligules hairy; inflorescence a compressed, spiciform panicle, major portion coverd by the sheath; spikelets 3-3.2 mm long; lower glume linear-lanceolate, folded, glabrous; keels ciliate, 1 nerved; upper glume ovate-lanceolate, 2 mm long, acute, 1-nerved, ciliate; lemma broadly ovate, acute, membranous, 1-nerved; palea oblong, glabrous; anthers 2-3, very small, yellow, filaments long; caryopsis oblong or ellipsoid.

Fl.-Fr. :Dec. -May

Common in drying ponds, moist grass, in all districts. Not easy to locate because of its small size and mixed with other grasses.

CYMBOPOGON Spreng. W. 52; M P 4 ;SA 2

Key to Species

1. Lowermost pedicel on the rhachis fused with adjacent internode .... **C. schoenanthus**
1. Lowermost pedicel on the rhachis free from adjacent internode .... **C. martini**

Cymbopogon martini (Roxb.) W. Watsson, Gaz. N. W. India 392. 1882; *Andropogon martini* Roxb., Fl. Ind. 1:280.1820; Bor, Grass. Burm. Ceyl. Ind. Pak. 129. 1960; Samvatsar, Fl. West. Trib. Madhya Pradesh 344. 1996; Roy in N.P. Singh *et al.,* Fl. Madhya Pradesh 3: 397.2001. *Tikhadi* **(Plate 10, Fig. 4)**

Perennials; culms erect, 1.5-2.0 m tall, smooth, glabrous, lower nodes often swollen; leaves 30-40 × 2.0-2.5 cm, lanceolate or linear-lanceolate, acuminate, with a semi-amplexicaul or rounded-cordate base, glabrous above, pubescent beneath, ligule membranous, oblong; inflorescence a decompound panicle, 3-6 cm long; spatheole 2 cm long, consisting of 3-6 spikelets; sessile spikelet ovate-oblong; lower glume ovate-oblong, obtuse, often emarginate or notched, 2-nerved towards the apex; upper glume lanceolate, acute with a winged keel which is serrulate above the middle; lower lemma empty, nerveless, ciliate awned; palea absent; upper lemma, bisexual, 2-lobed, awn geniculate, from the sinus, 16-20 mm long; palea often absent; pedicels of pedicelled spikelets male, up to 1 cm long; lower glume lanceolate-oblong, obtuse; upper glume 3-nerved; lemma hyaline; palea absent; stamens 3; caryopsis oblong.

*Fl.-Fr. :*Sept. -Nov.

Common in open grasslands, along dry streams, freshly cut ground, in all districts.

Cymbopogon schoenanthus (L.) Spreng., Pl. Min. Cogn. Pug. 2:15. 1815; *Andropogon schoenanthus* L., Sp. Pl. 1046. 1753; Hook.f., Fl. Brit. India 7:204. 1896; Bor,Grasses of Burma, Ceylon, India and Pak. 131. 1960; Samvatsar, Fl. West. Trib. Madhya Pradesh 344. 1996; Roy in N.P. Singh *et al.,* Fl. Madhya Pradesh 3:398. 2001.

Perennials; culms 100-200 cm tall, erect; leaves linear, glabrous, glaucous; ligule membranous, notched; panicles made of spikes enclosed in spathes; joints and pedicels narrowly clavate, margin ciliate; sessile spikelets 4-5 mm long, glabrous; lower glume 4-5 mm long, irregularly toothed; upper glume similar, ciliate; lemma oblong, ciliate; awn 12-15 mm long; anthers 3, 1.5-2 mm long; pedicellate spikelet male, similar to sessile ones.

Fl.-Fr.: Sept. -Nov.

Rare, along streams and shady places, **Dhar** only, at Katnera.

CYNODON Rich. W. 11; M P 3; SA 3

Key to Species

1. Lemmas equaling to or shorter than the upper glume; hairs on lemma clavellate....................... **C. barberi**
1. Lemmas longer than the upper glume; hairs usually not clavellate
   2. Inflorescence of 2-6 spikes, up to 6 cm long; plants rhizomatous; ligule a ring of white hairs .... **C. dactylon**
   2. Inflorescence of 4-several spikes, up to 10 cm long; plants not rhizomatus .... **C. radiatus**

Cynodon barberi Rang & Tandul., J. Bomb. Nat. Hat. Hist. Soc. 24: 846. 1916; Bor, Grass. Burm. Ceyl. Ind. Pak. 469. 1960; Samvatsar, Fl. West. Trib. Madhya Pradesh 345. 1996; Roy in N.P. Singh *et al.,* Fl. Madhya Pradesh 3: 399.2001.

Perennials; culms 25-40 cm tall, creeping, rooting at nodes, slender, often branched; leaves linear- lanceolate, acute, margins scabrid; sheaths compressed, keeled, glabrous, mouth bearded; ligule ciliate; inflorescence of 4-5, whorled spikes, rachis scaberulous, hairy at base; spikelets sessile, 2.0-2.5 mm long, compressed, 2-seriate, imbricate, 1-flowered; lower glume membranous, 1-nerved; upper glume similar, 2.25-2.5 mm long, acute or mucronate at apex; lemma boat-shaped, membranous, keel and margins hispid with clavellate hairs, acute; anthers 3, small; caryopsis oblong.

*Fl.-Fr. :*Aug. -Dec.

Sandy river beds, forest streams and paths, Alirajpur, Dhar, Khandwa. Endemic to India.

Cynodon dactylon (L.) Pers., Syn. Pl. 1:85. 1805; *Panicum dactylon* L., Sp. Pl. 58. 1753; Hook.f., Fl. Brit. India 7: 100.1896; Bor, Grass. Burm. Ceyl. Ind. Pak. 469. 1960; Samvatsar, Fl. West. Trib. Madhya Pradesh 345. 1996; Roy in N.P. Singh *et al.,* Fl. Madhya Pradesh 3: 399.2001. *Dub*

Perennials; culms slender, 5-30 cm tall, stoloniferous with slender underground rhizomes; leaves 2-15 × 0.2-0.3 cm, narrow-linear, flat; inflorescence of 4-6 digitate spikes, 1.5-6.0 cm long; spikelets 1.7-2.8 mm long, 1-flowered, awnless, laterally compressed, sessile, in two rows on one side of the rachis; glumes subequal, distinctly mucronate, 1-nerved, keels smooth or scabrid; lemma oblong to semi-ovate, subobtuse 3-nerved, keels ciliate; palea linear- oblong, 2-keeled, scaberulous; stamens 3; caryopsis oblong.

*Fl.-Fr. :*All round the year in moist places.

Very common all over the area.

Cynodon radiatus Roth, Syst. Veg. 2:411. 1817; *C. arcuatus* J. Presl, Reliq. Haenk. 1: 290.1829; Bor, Grass. Burm. Ceyl. Ind. Pak. 469. 1960; Samvatsar, Fl. West. Trib. Madhya Pradesh 344. 1996; Roy in N.P. Singh *et al.,* Fl. Madhya Pradesh 3:399.2001.

Perennials; culms 25-35 cm tall, slender glabrous, often stoloniferous, not rhizomatous, leafy with slender or geniculately-ascending; leaves 2-10 × 0.4-0.6 cm, linear-lanceolate, acuminate, smooth except on midrib, scaberulous on margins, ligule 0.5 mm long, membranous truncate; inflorescence 5-10 cm long, digitate, spikes 4-6, rarely 8; spikelets 1.5-2.0 mm long, 1-flowered, awnless, narrow, sessile in two rows on the side of the rachis; lower glume 1 mm long, curved, lanceolate, acute or acuminate, 1-nerved; upper glume 1.5 mm long, straight, similar in shape; lemma 2 mm long, obliquely ovate, oblong, chartaceous, 3-nerved, margin and keel hairy; palea linear oblong, 2-nerved, 1-keeled, scaberulous; stamens 3; caryopsis elliptic, brownish.

Fl.-Fr. : July-Sept

Alirajpur, Burhanpur, Khandwa, Khargone.

CYRTOCOCCUM Stapf W. 14; M P 2; SA 2

Key to Species

1. Plants 15-20 cm tall; leaf-sheath pilose with tubercle-based hairs; lamina 1-8 cm long; panicle effuse, 2-15 cm long; fertile spikelets 2-2.5 cm long; upper lemma 7-nerved, ... **C. deccanense**

1. Plants 20-6- cm tall; leaf-sheath smooth; lamina 2-20 cm long; panicle contracted, 5-20 cm long; fertile spikelet1.5-2 mm long, gibbous; upper lemma 5-nerved .... **C. oxyphyllum**

Cyrtococcum deccanense Bor, Kew Bull. 11:255. 1956 & Grass. Burm. Ceyl. Ind. Pak. 291. 1960; *Panicum patens auct. non* L. 1753: Hook.f., Fl. Brit. India 7:57. 1896; Davidse in Clayton *et al.,* Rev. Handb. Fl. Ceylon 8:139. 1994.

Annuals; culms 15-50 cm long, creeping or decumbent; nodes glabrous or sparsely bearded; leaves elliptic or elliptic-lanceolate, 1-8 cm, acuminate, narrowed at base, softly villous or glabrous; sheaths with sparse, tubercle-based hairs in the upper half, ciliate along the margins, ligule membranous, thin; panicles 2-15 cm long, lax, pedicels longer than spikelets; spikelets ovate or obovate, 2-3 mm; lower glume ovate, chartaceous, 3-nerved; upper glume boat-shaped, 2 mm long, chartaceous, 5-nerved; lower floret empty, 5-nerved; palea rudimentary; upper floret bisexual, up to 2 mm long; stamens 3; caryopsis *ca* 1 mm long.

Fl.-Fr. :Oct. -Dec.

Rare, in damp places at Mortakka, **Khandwa** only.

Cyrtococcum oxyphyllum (Steud.) Stapf, in Hooker's Icon. Pl. 31:t. 3096. 1922; *Panicum oxyphyllum* Hochst. ex Steud. Syn. Pl. Glumac. 1: 65. 1853; Bor, Grass. Burm. Ceyl. Ind. Pak. 291-292. 1960; Roy in N.P. Singh *et al.,* Fl. Madhya Pradesh 3:400.2001.

Perennials; culms 20-60 cm tall, creeping, rooting at the nodes; leaves 6-20 cm long, sparsely hairy, margins scabrid; sheaths shorter than internodes, shortly hairy or smooth; ligule membranous; inflorescence a panicle, 5-20 cm long, branches filiform with scattered hairs; spikelets crowded, 1.5-2.0 mm long, glabrous; lower glume 1.0-1.5 mm long, acute, 3-nerved; upper glume 1.5-1.8 mm long, elliptic 3-nerve; lower lemma coreaceous, shining, smooth, a brown scar at top, 5-nerved; palea absnt or rudimentary; upper lemma smooth; anthers 2; caryopsis 2 mm long, grooved along the centre.

Fl.-Fr. : Sept. -Dec.

Rare, moist shady places, in Kalibhit range, **Khandwa** only.

DACTYLOCTENIUM Willd. W. 12; M P 12; SA 3

Key to the Species

1. Stoloniferous perennials ....**D. scindicum**

1. Annuals
   2. Plants 15-50 cm tall; spikes 2-6, 2-4.5 cm long; lower glume 3.5-4 mm long .... **D. aegyptium**
   2. Plants 3-10 cm tall; spike solitary, 1.5-2 cm long; lower glume 2.5-3 mm long .... **D. aristatum**

Dactyloctenium aegyptium (L.) Willd., Enum. Pl. 1029. 1809; *Cynosurus aegyptius* L., Sp. Pl. 72. 1753; *Eleusine aegyptia* (L.) Desf., Fl. Atlant. 1:85. 1798; Hook.f., Brit. India 7: 295. 1896; Bor, Grass. Burm. Ceyl. Ind. Pak. 489. 1960; Samvatsar, Fl. West. Trib. Madhya Pradesh 345. 1996; Roy in N.P. Singh *et al.,* Fl. Madhya Pradesh 3: 401.2001. *Ghud chaba*

Annuals; culms 10-70 cm tall, geniculately ascending, rooting at lower nodes; leaves 10-20 × 0.4-0.5 cm, linear-lanceolate, hairy on both surfaces, papillose-hispid along the margins, ligule 1 mm long, membranous truncate; inflorescence of 2-6 digitate spikes, 1.5-30 × 0.5 cm; spiketets accrescent upwards, 3-4-flowered, broadly ovate, 3.0-3.5 mm long; lower glume 2-3 mm long, acute with a ciliate winged keel; upper glume awned; lemma 3 mm long, 7-nerved, mucronate; palea 1.8-2.0 mm long, 2-keeled, winged and ciliate on the keels; anthers 0.25-0.5 mm long; caryopsis broadly obovate, transversely rugose.

Fl.-Fr. : Aug.-Jan.

Common, in wasteland, along the roads, dry river bed, throughout the area.

Dactyloctenium aristatum Link, Hort. Berol. 1: 59. 1827; Hook.f., Fl. Brit. India 7:296. 1896; Bor, Grass. Burm. Ceyl. Ind. Pak. 489. 1960; Samvatsar, Fl. West. Trib. Madhya Pradesh 345. 1996; Roy in N.P. Singh *et al.,* Fl. Madhya Pradesh 3: 402.2001.

Annuals; culms 5-50 cm tall; leaves linear-lanceolate, sparsly hairy or glabrous; sheath compressed, hairy; ligule very small, membranous; inflorescence usually a solitary spike; rachis slender, hairy at base, apical prolongation 3-4 mm long beyond the spikelets; spikelets 3-flowered, densely crowded, laterally compressed, sessile in two rows; lower glume 1-nerved, keel winged; lower lemma membranous, 1-nerved, similar to upper glume; empty lemma one, similar to fertile; keel of palea hispid; anthers 3; caryopsis globose, rugose.

Fl.-Fr. : July-Oct.

Common on field borders, hill slopes, grasslands, on roofs of houses and old walls, all over the area.

Dactyloctenium scindicum Boiss., Diagn. Pl. Orient. 11, 4:131. 1859; *Eleusine scindica* (Boiss.) Duthie, Fodder Grasses N. India 58. 1888; Bor, Grass. Burm. Ceyl. Ind. Pak. 489. 1960; Roy in N.P. Singh *et al.,* Fl. Madhya Pradesh 3: 402.2001.

Perennials; culms spreading, 5-40 cm long, tomentose with long hairs,rooting at nodes, nodes villous; leaves linear, small, hairy; sheath compressed, hairy; ligule hairy; inflorescence of digitate spikes forming a cross, hairy at base, rachis projection 2mm long; spikelets 3.2 mm long; lower glume 2 mm long, keeled, arista 1 mm long; upper glume 2.7 mm long, ovate, acute, cuspidate, glabrous; anthers 3, yellow, 0.5-0.8 mm long; caryopsis ovoid, rugose.

Fl.-Fr. : Sept. - Jan.

Burhanpur, Khandwa, Khargone.

DENDROCALAMUS Nees W. 53; M P 1; SA 1

Dendrocalamus strictus (Roxb.) Nees, Linnaea 9: 476. 1834; *Bambos stricta* Roxb., Pl. Coromandel 1: 58. 1798; Gamble in Hook., f., Brit. India 7: 404. 1896; Samvatsar, Fl. West. Trib. Madhya Pradesh 346. 1996; Roy in N.P. Singh *et al.,* Fl. Madhya Pradesh 3: 403.2001. *Bans*

Perennial, deciduous densely tufted bamboos, culms fistular or solid, 6-20 m tall, branches spinescent at the nodes, internodes 30-45 cm long; sheaths 7-30 cm long; leaves 2.5-25 cm long, linear-lanceolate with subcordate or rounded base, hairy, 1.3-3.0 cm wide, ligule 1 mm long, membranous truncate; inflorescence a large branching panicle (not flowering every year); spikelets spinulose, terete, in congested spicate heads 1-1.5 cm; glumes mucronate, hairy at the top, several-nerved, 5-6 mm long; lemmas *ca* 13 mm long, keeled, prolonged into a sharp spine; palea 2-keeled, truncate; stamens 6, anthers linear apiculate; lodicules absent; caryopsis brown, shining, ovoid to subglobose, beaked.

Fl.-Fr.: Nov. -Dec. (when flowering occurs).

Very common in all districts, wild as well as intensively planted by the forest department.

DESMOSTACHYA (Stapf) Stapf Monotypic

Desmostachya bipinnata (L.) Stapf, Fl. Cap. 7: 632. 1900; *Briza bipinnata* L., Fl. Palaest. 12. 1756; *Poa cynosuroides* Retz., Observ. Bot. 4:20.1786; *Eragrostis cynosuroides* (Retz.) P. Beauv. Ess. Agrostogr. 71, 162. 1812; Bor, Grass. Burm. Ceyl. Ind. Pak. 491.1960; Samvatsar, Fl. West. Trib. Madhya Pradesh 346. 1996; Roy in N.P. Singh *et al.,* Fl. Madhya Pradesh 3: 404.2001. *Darbha*

Strong perennials; culms 30-90 cm tall, with a thick creeping rhizomes, nodes glabrous; leaves 20-40 × 0.5-0.8 cm, linear,convolute,narrowed into a filiform tip, ligule a ring of hairs; sheath glabrous with a hairy throat; inflorescence a 30-45 cm long panicle; spikes crowded, 1.5-2.0 cm long; spikelets purple brown, 2-seriate, 3-6-flowered; lower glume 0.5-0.7 mm long, keeled, mucronate, 1-nerved; upper glume obtuse; lemma sub acute with scabrid keels, 3-nerved; palea 1.3 mm long; stamens 3; caryopsis obliquely ovoid, compressed.

Fl.-Fr. : Aug. -Nov.

Alirajpur, Barwani, Dhar, Indore, Khargone, in waste places, along roadsides, water courses.

DICHANTHIUM Willemet W. 21; MP 6; SA 6

Key to Species

1. Spikes spatheate, enclosed in a spatheole
  2. Lower glume of sessile spikelet pitted, smooth........ **D. foveolatum**
  2. Lower glume of sessile spikelet not pitted, with strong, tuberculate nerves . ... **D. tuberculatum**
1. Spikes espatheate, not enclosed in a spatheole
  3. Peduncle of the raceme hairy . ... **D. aristatum**
  3. Peduncle of the raceme glabrous
    4. Lower glume of the pedicellate spikelet with bulbous-based hairs . ... **D. maccannii**
    4. Lower glume of pedicellate spikelets without marginal, bulbous-based bristles
      5. Sessile spikelets 3-3.5 mm long, median nerve present; leaf sheath terete .... **D. annulatum**
      5. Sessile spikelets 5-6 mm long, callus not pilose .... **D. caricosum**

Dichanthium annulatum (Forssk.) Stapf, Fl. Trop. Africa 9; 178. 1971; *Andropogon annulatus* Forssk., Fl. Aegypt.-Arab. 173. 1775; Hook.f., Brit. India 7: 196.1896; Bor, Grass. Burm. Ceyl. Ind. Pak. 133. 1960; Samvatsar, Fl. West. Trib. Madhya Pradesh 347.1996; Roy in N.P. Singh *et al.,* Fl. Madhya Pradesh 3: 405. 2001. *Chhoti marvel*

Perennials; culms 0.5-1.0 m tall, erect or geniculate-ascending, simple or sparingly branched; leaves 15-30 × 0.3 × 0.6 cm, narrow linear-acuminate, hairy on both surfaces, margins scabrid; inflorescence subdigitate panicle; sessile spikelets hermaphrodite, 3-5 mm long, ovate-oblong, or oblong-concave, compressed, dorsally awned; lower glume 5 mm long, boat-shaped, 5 nerved, 2-keeled, ciliate; upper glume narrow, lanceolate, base ciliate; lower lemma linear-oblong, obtuse, nerveless, hyaline glabrous; upper lemma hermaphrodite, rather thick, very narrow with 1.6-2.5 cm long awn; stamens 3; pedicelled spikelets male or neuter with 2 mm long pedicel; caryopsis *ca* 2 mm long.

Fl.-Fr. : Throughout the year, if moisture is available.

Dichanthium aristatum (Poir.) C.E. Hubb., Bull. Misc. Inform. Kew 1939. 654. 1940; *Andropogon aristatus* Poir., Encycl. Suppl. 1(2): 585. 1811; Hook.f., Fl. Brit. India 7:196. 1896; Bor, Grass. Burm. Ceyl. Ind. Pak. 134. 1960; Samvatsar, Fl. West. Trib. Madhya Pradesh 347. 1996; Roy in N.P. Singh *et al.,* Fl. Madhya Pradesh 3:406.2001. *Kel*

Perennials; culms 0.7-2 m tall, erect or ascending, leafy, branched, nodes hairy; leaves linear, glabrous; sheath compressed glabrous; ligule short, ciliate; inflorescence of 1-7 racemes; peduncles 3-10 cm long; sessile spikelets 5-6 mm long, compressed, awned; lower glume 5.0-5.5 mm long, 7-9-nerved; upper glume similar, 3-nerved; lower lemma empty, hyaline, 4-4.5 mm long, margins ciliate, epaleate; upper lemma reduced, bifid, awned from sinus, awns 2-3 cm long; pedicelled spikelets similar to sessile except being narrow and rusty; anthers 1.2-1.5 mm long; caryopsis fusiform, 2 mm long, slate coloured.

Fl.-Fr.: Aug. -Jan.

Common grass all over the area.

Dichanthium caricosum (L.) A. Camus, Bull. Mus. Natl. Hist. Nat. 27: 549. 1921; *Andropogon caricosus* L., Sp. Pl. ed. 2, 1480.1763; Hook., f., Brit. India 7: 196. 1896; *Dichanthium pallidum* (Hook.f.) Stapf ex C.E.C. Fischer, Fl. Madras Presi. 1741. 1934; Bor, Grass. Burm. Ceyl. Ind. Pak. 134. 1960; Samvatsar, Fl. West. Trib. Madhya Pradesh 347. 1996; Roy in N.P. Singh *et al.,* Fl. Madhya Pradesh 3: 406.2001. *Motha marvel*

Perennials; culms slender, tufted, often basally creeping, 30-80 cm tall; leaves 15-20 × 0.25-0.5 cm, narrowly linear, rounded at the base; inflorescence sub-digitate; spikelets in pairs, almost distichous, imbricate; sessile spikelets hermaphrodite, 4-5 mm long, obovate-elliptic or obovate-oblong, dorsally compressed; lower glume 4-5 mm long, broadly ovate, truncate, 9-11-nerved, margin inflexed, winged, upper surface ciliate; upper glume similar, 3-nerved; lower floret neuter; lemma 3.5 mm long, ovate truncate, hyaline with twisted awn; palea absent; stamens 3; pedicelled spikelets: lower glume 4.5 mm long, obovate truncate, many-nerved, margin inflexed; upper glume 3 mm long, 3-nerved, hyaline, ciliate; lower lemma 2-6 mm long, ovate-truncate, hyaline, nerveless, tip ciliate; upper lemma male or neuter; caryopsis ovoid, olivaceous.

*Fl.-Fr. :*July-Dec.

Common in all districts, along streams, bunds of fields, open grasslands.

Dichanthium foveolatum (Delile) Roberty, Monogr. Syst. Andropog. du Globe 163. 1960; *Andropogon foveolatus* Delile, Descr. Egypte, Hist. Nat. 2:160.1813; *Eremopogon foveolatus* (Delile) Stapf, Fl. Trop. Afr. 9: 183. 1917; Hook, *f.*, Fl. Brit. Ind. 7: 168. 1896; Bor, Grass. Burm. Ceyl. Ind. Pak. 148. 1960; Samvatsar, Fl. West. Trib. Mdhya Pradesh 360.1996; Roy in N.P. Singh *et al.*, Fl. Madhya Pradesh 3: 439.2001. *Chamari*

Perennials; culms 30-70 cm tall, densely tufted, usually decumbent; nodes bearded; leaves narrow, linear, rigid, scaberulous beneath; sheath shorter than the internodes, basal sheaths silky-villous; ligule ciliate, hyaline, truncate; inflorescence of 3-6 racemes, 2-4 cm long; joints and pedicels with long silky hairs; spikelets purplish, oblong-lanceolate, *ca* 5 mm long, glabrous; callus shortly hairy; lower glume 4-5 mm long, 4-nerved, usually with a pit above the middle; upper glume non pitted, equal to lower, 3-nerved; lower lemma much shorter, glabrous, nerveless, empty; upper lemma reduced to a stipe, passing into a geniculate awn; palea absent; pedicelled spikelets similar to sessile ones, but not or rarely pitted; caryopsis oblong.

Fl.-Fr.: July-Dec.

Common in open grassland, throughout the area.

Dichanthium maccannii Blatt., J. Bombay, Nat. Hist. Soc. 32: 357. 1927; Bor, Grass. Bamb. Ind. Pak. 135: 1960; Deshpande, Fasc. Fl. India 15:19. 1984; Samvatsar, Fl. West. Trib. Madhya Pradesh 347. 1996; Naik, Fl. Marathwada 2:1018. 1998. *Dania marvel*

Annuals; culms tufted, 10-45 cm tall, slender, glabrous, nodes densely bearded; leaf sheath terete, upper side covered with bulbous based hairs, ligule hairy; leaf blade linear ovate, 2-6 mm, covered with bulbous based bristles; racemes 1-2, up to 3.7 cm long, lower 3-4 pairs of spikelets all male, joints and pedicels with ciliate hairs; sessile spikelet narrowly ovate, 3.5-4.2 mm long, awned, callus bearded with short hairs; lower glume membranous, narrowly ovate-oblong, up to 4 mm long, margins inflexed, 7-nerved, apex obtus;. upper glume narrowly ovate, slightly longer than lower, apex acuminate; lower lemma hyaline, narrowly oblong, 3.2-3.5 mm long, nerveless, apex obtuse; palea absent; upper lemma 1-nerved, reduced to hyaline base of geniculate awn, up to 28 mm long; palea absent; pedicelled spikelet narrowly obovate, 4-5 mm long; lower glume membranous, narrowly obovate, covered with bulbous based bristles, 10-nerved, apex obtuse; upper glume narrowly ovate, margins inflexed, ciliate, 3-nerved, apex acute; lower lemma hyaline, narrowly oblong, margins ciliate, apex rounded; palea hyaline, narrowly ovate, apex acute.

Fl.-Fr. : Sept. -Dec.

Bunds of fields and near water courses, Burhanpur, Dhar, Khandwa. Endemic to India.

Dichanthium tuberculatum (Hack.) T. A. Cope, Kew Bull. 35; 703. 1980; *Andropogon tuberculatus* Hack., Mongr. Phan. 6; 404. 1889; Hook.f., Fl. Brit. India 7: 168. 1896; *Eremopogon tuberculatus* (Hack.) A. Camus, Ann. Soc. Linn. Lyon, n.s., 68: 208. 1922; Bor, Grass. Burm. Ceyl. Ind. Pak. 149. 1960; Samvatsar, Fl. West. Trib. Madhya Pradesh 360.1996.

Perennials; culms 1.2-1.5 m tall, stout; leaf sheath glabrous, ligule membranous, auricled; leaf blade 30-60 × 02-0.4 cm, rigid, base narrowed, apex acuminate; racemes solitary, 2-4 cm long; sessile spikelet 4-5 mm long, awned; lower glume not pitted, 4-5 mm long, nerves strongly tuberculate.

Fl.-Fr. : July-Dec.

Rare, Barwani, Khandwa. Endemic to India.

DIGITARIA Haller W. 260; MP 9; SA 7

Key to Species

1. Hairs on spikelets always verrucose; spikelets ternate; racemes 2-3; lower glume absent ..... **D. longiflora**
1. Hairs on spikelets never verrucose, either clavate or not; racemes more than 2
    2. Lower glume well developed; spikelets of each pair heteromorphous .... **D. bicornis**
    2. Lower glume absent or insignificant; spikelets homomorphous
        3. Racemes up to 5; spikelets 1.2-1.7 mm long; rhachis triquetrous; upper glume 1-5 nerved; upper lemma brown at maturity; hairs clavate
            4. Spikelets ternate; plants 15-60 cm tall; spikelets 1.2-1.3 mm long; upper glume 1-3 nerved .... **D. stricta**
            4. Spikelets binate; plants 30-90 cm tall; spikelets 1.3-1.7 mm long; upper glume 3-5 nerved ....**D. abludens**
        3. Racemes 2-12
            5. Rhachis not triquetrous; hairs clavate; racemes 2-9 .... **D. ciliaris**
            5. Rhachis triquetrous
                6. Leaf 15-25 cm long; plants up to 90 cm tall .... **D. griffithii**
                6. Leaf 5-20 cm long; plants 30-100 cm tall . ...**D. setigera**

Digitaria abludens (Roem. & Schult.) Veldkamp, Blumea 21: 53. 1973; *Panicum abludens* Roem. & Schult., Syst. Veg. 2: 457. 1817; *Paspalum granulare* Trin., Neue Entd. 2:47. 1821; *Paspalum pedicellatum* Nees ex Duthie, List Grasses N. W. India 1. 1883; Hook.f., Fl. Brit. India 7:19. 1896; *Digitaria granularis* (Trin.) Henrard, Monogr. **Digitaria** 302. 1950; Bor, Grass. Burm. Ceyl. Ind. Pak. 301. 1960; Roy in N.P. Singh *et al.,* Fl. Madhya Pradesh 3: 409.2001. *Sondriya*

Annuals; culms terete, 10-35 cm tall, erect or geniculately ascending from the base; leaves 6-10 × 0.2-0.3 cm, linear-acuminate, ligule 1.5-2.0 mm long, membranous truncate; inflorescence very slender, of 3-7 racemes 3-5 cm long on a long, glabrous peduncle; rachis filiform, triquetrous, narrowly winged; pedicels unequal, slender with clavate hairs; spikelets 1.2-1.5 mm long, elliptic-lanceolate; lower glume absent; upper glume 1mm long, elliptic, acute, membranous, 3-5-nerved, with clavate -tipped hairs between the nerves; lower floret sterile; lower lemma elliptic-lanceolate, 5-7-nerved, hairy, epaleate; upper lemma elliptic, brownish-black at maturity; palea similar; stamens 3; caryopsis 1 mm long, elliptic.

Fl.-Fr. : Sept. -Dec.

On sandy soil along the banks of river, Barwani, Burhanpur, Khandwa, Khargone.

Digitaria bicornis (Lam.) Roem. & Schult., Syst. Veg. 2:470 1817; *Paspalum bicorne* Lam., Tabl. Encycl. 1:176. 1791; *Digitaria biformis* Willd., Enum. Pl. 1: 92. 1809; Henrard., Monogr. Gen. **Digitaria** 71. 1950; Bor, Grass. Burm. Ceyl. Ind. Pak. 299. 1960; Samvatsar, Fl. West. Trib. Madhya Pradesh 349. 1996; Roy in N.P. Singh *et al.,* Fl. Madhya Pradesh 3: 409.2001. *Harkailo*

Annuals; culms erect, decumbent or ascending, 30-50 cm tall, nodes glabrous; leaves 0.5-0.7 cm broad, scabrid, sparsely soft hairy, ligule 0.2 cm long, membranous, truncate; sheaths ciliate, hirsute in the lower half; inflorescence: racemes 2-3, digitate, up to 12 cm long, on a long, glabrous peduncle; rachis 0.05-0.08 cm broad; spikelets 2-nate, lanceolate, acute, 0.28-0.3 cm long, glabrous or slightly pubescent; lower glume variable, acute, 2-fid; upper glume 3-nerved, patently hairy; sterile lemma equal to spikelets; sessile spikelets 7-nerved, glabrous or slightly pubscent and pectinate; fertile lemma slightly shorter than spikelet; anthers 3, purple; caryopsis yellowish.

Fl.-Fr.: Aug. -Dec.

In moist places, cultivated and fallow fields, Alirajpur, Barwani, Dhar, Indore, Khandwa, Khargone.

Digitaria ciliaris (Retz.) Koeler, Descr. Gramin. 27. 1802; *Panicum ciliare* Retz., Observ. Bot. 4: 16. 1786; *Panicum adscendens* Kunth, Nov. Gen. Sp. 1:97. 1816; *Digitaria adscendens* (Kunth) Henrard., Blumea 1: 92. 1934; Henrard, Monogr. Gen. **Digitaria** 129. 1950; Bor, Grass. Burm. Ceyl. Ind. Pak. 298. 1960; Samvatsar, Fl. West. Trib. Madhya Pradesh 349. 1996; Roy in N.P. Singh *et al.,* Fl. Madhya Pradesh 3: 410.2001. *Kot*

Annuals; culms 20-60 cm tall, erect or decumbent and rooting at the nodes; leaves 8-20 × 0.4-0.6 cm, linear-lanceolate-acuminate, glabrous except bulbous based hairs on the base, ligule membranous truncate; inflorescence of 2-9 racemes, up to 20 cm long, digitately arranged; rachis 0.5-1.0 mm wide, scabrid, flat with a sharp median keel on one side; spikelets in pairs, lanceolate, acute, 2.5-3.0 mm long, one with 2-3 mm long pedicel, other with ± 1 mm long pedicel; lower glume triangular; upper glume narrowly triangular, acute, 3-nerved, silky hairy; lower floret sterile; lower lemma 3 mm long, narrowly ovate-acute, 5-nerved; upper floret hermaphrodite; upper lemma hyaline, 3-nerved, narrowly ovate, acute, glabrous; palea 2.8 mm long, 2-nerved, similar in texture; stamens 3, anthers purple; caryopsis 2 mm long.

Fl.-Fr. : Aug. -Dec.

Common in bushes , hedges and moist ground, all over tha area.

Digitaria griffithii (Hook.f.) Henrard, Blumea 1:100.1934; *Paspalum sanguinale* var. *griffithii* Hook.f., Fl. Brit. India 7:15. 1896; *Paspalum griffithii* Arn. ex Hook.f., *loc. cit;* Samvatsar, Fl. West. Trib. Madhya Pradesh 349. 1996.

A tufted geniculate annual, to 90 cm tall; leaves 15-25 cm long; racemes 5-10, peduncle long, rachis trigonous; spikelets binate, lower glume very small; upper glume silky hairy between nerves; empty lemma hairy; caryopsis oblong.

Fl.-Fr.: Aug. -Dec.

Rare, weed of fields, **Dhar** proper.

Digitaria longiflora (Retz.) Pers., Syn. Pl. 1: 85. 1805; *Paspalum longiflorum* Retz. Observ. Bot. 4:15.1786; Hook.f., Fl. Brit. India 7:17. 1896; Henrard, Monogr. Gen. **Digitaria** 408. 1950; Bor, Grass. Bur. Ceyl. Ind. Pak. 302. 1960; Samvatsar, Fl. West. Trib. Madhya Pradesh 349. 1996; Roy in N.P. Singh *et al.,* Fl. Madhya. Pradesh 3: 411.2001.

Annuals; culms 10-50 cm tall, with short creeping stolons; leaves linear, hairy or glabrous, ligule membranous; inflorescence of 2-3 racemes; rachis flat, winged, glabrous; spikelets ellipsoid, 1.5-2 mm long; lower glume absent; upper glume 1.5-1.7 mm long, membranous, covered with verrucose hairs; lower lemma empty, 5-7-nerved; upper lemma bisexual, elliptic, ± 1.5 mm long; anthers 3, yellow; caryopsis pale, ellipsoid.

Fl.-Fr. : Sept. -Dec.

Common in grassland, moist areas, garden weed, in all districts.

Digitaria setigera Roth, Syst. Veg. 2: 474. 1817; *Cynodon setiger* (Roth) A. Rich. ex Hassk., Cat. Hort. Bot. Bogor. 17. 1844; Bor, Grass. Burm. Ceyl. Ind. Pak. 305. 1960; Hook.,f.,Brit. India 7: 15. 1896; Henr., Monogr. Gen. **Digitaria** 684. 1950; Roy in N.P. Singh *et al.,* Fl. Madhya Pradesh 3: 411.2001.

Annuals; culms, erect 40-60 cm tall, rooting at the nodes; leaves 10-30 × 0.5-1.0 cm, linear-lanceolate, acuminate, upper surface pubescent, lower glabrous, ligule 0.5-2.0 mm long, membranous; inflorescence of 4-15 racemes, digitate or sub-digitate, up to 6 cm long; spikelets elliptic-acute, rachis winged, one pedicellate, other, sessile; lower glume small, hyaline scale-like; upper glume 1.25 mm long, narrow, lanceolate, 3-nerved; lower florets sterile; lower lemma 2.5 mm long, hyaline, 5-7-nerved, palea absent; upper floret hermaphrodite; lemma elliptic-acute, chartaceous, margin turned in and clasping the palea; stamens 3, anthers 1.25 mm long; caryopsis *ca* 1 mm long, elliptic, oblong.

Fl.-Fr. : Aug. -Nov.

Along roads and outer skirts of the forests, Barwani, Khandwa, Khargone.

Digitaria stricta Roth, Syst. Veg. 2: 474. 1817; *Paspalum royleanum* Nees ex Hook.f., Fl. Brit. India 7: 18. 1896; Henrard, Monogr. Gen. **Digitaria** 715. 1950; Bor, Grass. Burm. Ceyl. Ind. Pak. 305. 1960; Samvatsar, F. West. Trib. Madhya Pradesh 349. 1996; Roy in N.P. Singh *et al.,* Fl. Madhya Pradesh 3: 412.2001. *Kankri*

Annuals; culms 15-60 cm tall, decumbent-ascending; leaves 5-15 cm long, scabrid or ciliate at base; sheaths glabrous, with long hairs at the throat, ligule 0.7-2.5 mm long, membranous, truncate; inflorescence of 2-3 racemes; spikelets paired, elliptic or lanceolate, acute; upper floret sterile; lower floret fertile; lemma purple or dark brown, apiculate; palea *ca* 1 mm long; caryopsis deep purple, ovoid, *ca*1 mm long.

Fl.-Fr. : Aug. -Dec.

Common weed of cultivated fields, throughout the area.

DIMERIA R. Br.W. 59; MP 2; SA 2

Key to Species

1. Rhachis of inflorescence angular, narrowly winged; lower glume 3 mm long; upper glume 1-keeled, winged all along on keel; fertile lemma 2.5 mm long; awn 9 mm long .... **D. connivens**
1. Rhachis of inflorescence flat; lower glume 0.8-3mm long; upper glume not keeled; fertile lemma 0.8-1.8 mm long, awn 4-5 mm long .... **D. ornithopoda**

Dimeria connivens Hack., Monogr. Phan. 6: 689. 1889; Hook.f., Fl. Brit. India 7: 104. 1896; Bor, Grass. Burm. Ceyl. Ind. Pak. 140.1960; Sharma *et al.,* in Fl. Maharashtra 3:467. 1996.

Annuals; culms erect or geniculate, 20-50 cm high; nodes sparsely bearded; leaves linear-acuminate or lanceolate, mostly basal, 3-10 × 0.2-0.4 cm, with sparse tubercle-based hairs, especially along the margins; ligule ovate, membranous fimbriate; racemes 2, rarely one, 4-8 cm long, divergent; rachis flat, *ca* 0.75 mm wide, narrowly winged and ciliate along the margins; spikelets oblong or oblanceolate, greenish-yellow, 4-4.5 mm; callus bearded; lower glume linear-lanceolate, chartaceous, keels ciliate; upper glume lanceolate or elliptic-acute, narrowly winged all along, wing corky, ciliate, chartaceous, margins hyaline, ciliolate; lower floret empty; upper floret bisexual; first lemma oblanceolate or oblong, hyaline, faintly 1-nerved, ciliate towards apex; second lemma notched, awned, 2 mm, hyaline, faintly 1-nerved, awn 8-12 mm long, column 3-4 mm, brownish; stamens 2, anthers 2 mm, bright yellow; ovary oblong or elliptic.

Fl.-Fr.: Aug. -Oct.

Rare, on hard ground, Khandwa, Khargone. Endemic to India.

Dimeria ornithopoda Trin., Fund. Agrost. 167. 1820; Hook., f., Fl. Brit. India 7: 104. 1869; Bor, Grass. Ceyl. Ind. Pak. 142. 1960; Roy in N.P. Singh *et al.,* Fl. Madhya Pradesh 3: 413.201. *Suwalio*

Annuals; culms 25-40 cm tall, weak, slender; leaves 10-13 cm long, basal portion with tuberculate hairs; sheath keeled, glabrous, hairy at throat, ligule ciliate; inflorescence usually consists of 2 spikes; rachis glabrous, scabrid on margins; spikelets solitary, 1.5-2.0 mm long, linear; callus hairy; lower glume *ca* 1.5 mm long, coriaceous, keeled, 1-wined, margins ciliate; upper glume equal to spikelet, papery, lanceolate, keeled, fimbriate; lower lemma minute, linear-acute, nerveless, glabrous, 1 mm long; upper lemma fertile, oblong-acute, 2-toothed with 9-10 mm long awn; anthers 3, yellow; caryopsis cylindrical.

Fl.-Fr. : Sept. -Dec.

Rare, in wet grass lands and on rocky soil along river banks, Burhanpur, Khandwa.

DINEBRA Jacq. W. 3; MP 1; SA 1

Dinebra retroflexa (Vahl) Panz., Denkschr. Konigl. Akad. Wissensch. 4:270.1814; *Cynosurus retroflexus* Vahl, Symb. Bot. 2:20.1791; Bor, Grass. Burm. Ceyl. Ind. Pak. 491. 1960; Samvatsar, Fl. West. Trib. Madhya Pradesh 350.1996; Roy in Singh *et al.,* Fl. Madhya Pradesh 3; 413.2001. *Kharia*

Annuals; culms 30-60 cm tall, loosely tufted, erect or decumbent-ascending grass; leaves 4.5-20.0 × 0.4-0.6 cm, linear-acuminate, ligule 1.6-2.0 mm long, lacerate; inflorescence a spike-like raceme, 8-15 cm long, main axis angled, spikes many, 0.5-3.0 cm long, stiff, deciduous at maturity; spikelets 1-3-flowered, 5-6 mm long, rachis narrowly winged; lower glume narrowly elliptic, 1-nerved; upper glume 5.5 mm long, same as lower; lemma narrowly ovate, appressed pilose along the lateral nerves, acute to emarginated; palea 1.7 mm long, appressed pilose on the flaps along the sides of keel; stamens 3; caryopsis trigonous, brownish.

Fl.-Fr.: Aug. -Nov.

Common weed of cultivated fields, in all districts.

ECHINOCHLOA P. Beauv. W. 35; MP 4; SA 4

Key to Species

1. Ligule present; spikelets long awned .... **E. stagnina**
1. Ligule absent; spikelets not long awned (except in *E. crusgalli*)
   2. Plants robust, branched from lower nodes; midrib of leaves prominently white; spikelets densely crowded; lower glume *ca* 2.5 mm long; caryopsis not deciduous .... **E. frumentacea**
   2. Plants and midrib of leaves not as above; spikelets not crowded; lower glume less than 2 mm long; caryopsis deciduous
      3. Spikes distantly arranged; spikelets 2.5-3 mm long; lower lemma acute, not produced into an awn .... **E. colona**
      3. Spikes relatively crowded; spikelets 3-4 mm long (excluding awn); lower lemma usually produced into an awn .... **E. crusgalli**

Echinochloa colona (L.) Link, Hort. Berol. 2: 209.1833; *Panicum colonum* L., Syst. Nat. ed. 10 2:870.1759; Hook., f., Brit. India 7: 32.1896; Grass. Burm. Ceyl. Ind. Pak. 308.1960; Samvatsar, Fl. West. Trib. Madhya Pradesh 350.1996; Roy in N.P. Singh *et al.,* Fl. Madhya Pradesh 3: 416.2001. *Sawa*

Annuals; culms 30-45 cm tall, erect or decumbent, rooting below, nodes glabrous; leaves 5-15 × 0.7-0.8 cm, linear-acuminate, narrow, scaberulous; sheaths compressed glabrous, ligule absent; inflorescence panicle, 5-15 cm long; spikelets 2-3-nate, elliptic, acute, hispid, 2.5-3.0 mm long; lower glume 1.5-1.7 mm long, ovate, acute, 5-nerved; upper glume 2.5-3.0 mm long, acute, 7-nerved; lower lemma cuspidate, 2.0-2.5 mm long, 5-nerved, hispidulous on the back; upper lemma 1.8 mm long; palea 1.6 mm long, hyaline; stamens 3, anthers 0.6 mm long; caryopsis 1.25 mm long, ovoid, white.

Fl.-Fr. : Aug. -Dec.

Common weed of cultivated fields, in all districts.

Echinochloa crus-galli (L.) P. Beauv. Ess. Agrostogr. 53. 1812; *Panicum crus-galli* L., Sp. Pl. 56. 1753; Hook.f., Fl. Brit. India 7: 30.1896; Bor, Grass. Burm. Ceyl. Ind. Pak. 310.1960; Samvatsar, Fl. West. Trib. Madhya Pradesh 350.1996; Roy, in N.P. Singh *et al.,* Fl. Madhya Pradesh 3: 416.2001.

Annuals; culms terete, 30-160 cm tall, erect or ascending, branched from the lower nodes, and rooting at the lower nodes, glabrous, nodes glabrous; leaf sheath compressed, 6.22 cm long, glabrous, ligule absent; leaf blade flat, linear ovate, 6-40 cm, smooth or scaberulus, glabrous, margins scaberulus, keeled, apex acute to acuminate; raceme pyramidal; spikes solitary or 2-nate, 1-5 cm long, upper smaller, appressed to the rachis or sub erect; rhachis stout, with a tuft of hairs at the base, triquetrous, scabrid, bristly near the nodes; pedicels short with discoid tip, scabrid, bristly at the base; spikelets paired, crowded, ovate-elliptic, 3. 8-5 mm, hispidulus, cuspidate or awned; lower glume membranous, broadly ovate, 3-nerved, clasping at base, scaberulous on the nerves, apex cuspidate; upper glume membranous, broadly elliptic, 5-7-nerved, scaberulous on nerves, apex acute to cuspidate; lower lemma broadly ovate, 5-nerved, scaberulous on nerves, apex acute to cuspidate; palea hyaline, elliptic, 2-nerved, 2-keeled, margins inflexed in the lower half; upper lemma chartaceous, broadly elliptic, 2.2-2.6 mm, 5-nerved, apex inflexed with basal flaps, apex acute; stamens 3; caryopsis broadly elliptic to orbicular.

Fl.-Fr. : Sept. -Dec.

Common around ponds, marshy places, waste lands, in all districts.

Echinochloa frumentacea Link, Hort. Berol. 1: 204. 1827; *Panicum frumentaceum* Roxb., 1820, *non* Salisb., 1796, *nom. illeg; P. crusgalli* L. var. *frumentacea* (Link) Trimen, Syst. Cat. Fl. Pl. Ceylon 104. 1885; Hook.f., Fl. Brit. India 7:31. 1896; Bor, Grass. Burm. Ceyl. Ind. Pak. 311. 1960; Samvatsar, Fl. West. Trib. Madhya Pradesh 351. 1996; Roy, in N.P. Singh *et al.,* Fl. Madhya Pradesh 3:417.2001. *Kodo*

Annuals; culms up to 1.5 m tall, very robust, branched from lower swollen nodes; leaves 1-2 cm broad; midrib white; sheaths loose, membranous, glabrous, longer than internodes, ligule absent; inflorescence of racemes, 2-nate or 3-nate; rachis angled, scabrid, villous in the axils; spikelets 3-4 mm long, cuspidate, turgid, awnless; lower glume *ca* 2 mm long, apiculate, minutely hispid, 5-nerved; upper glume similar to spikelet, compressed, cuspidate, 5-7-nerved; lower lemma barren; upper lemma fertile, elliptic, acute, minutely ciliolate, awnless; caryopsis 2.0- 2.5 mm long, ellipsoid.

Fl.-Fr. : Sept. -Dec.

Common in fields as a weed, marshy places, cultivated at some places for the grain.

Echinochloa stagnina (Retz.) P. Beauv., Ess. Agrostogr. 53. 1812; *Panicum stagninum* Retz., Observ. Bot. 4: 17. 1789; *P. crusgalli auct. non* L., 1753; Hook.f., Fl. Brit. India 7: 30.1896; Bor, Grass. Burm. Ceyl. Ind. Pak. 311. 1960; Roy, in N.P. Singh *et al.,* Fl. Madhya Pradesh 3: 417.2001.

Annuals or perennials; culms up to 1 m tall, spongy, glabrous; leaves rigid, glabrous, scabrid on margins; sheaths, glabrous, ciliate on mouth, ligule hairy; inflorescence of racemes, rachis hispid or scabrid; spikelets paired, elliptic, 5-6 mm long, acuminate; lower glume elliptic, 1.5-2.5 mm long, acute-cuspidate, membranous, pubescent, 5-nerved; lower lemma barren; upper lemma fertile, straw-coloured, 3-. 5-4.0 mm long, acuminate-mucronate, polished; anthers 3; caryopsis 2-3 mm long, ellipsoid, yellowish-white.

Fl.-Fr. : Sept. -Dec.

In moist places and cultivated fields, Burhanpur, Indore, Khandwa.

ELEUSINE Gaertn. W. 10; MP 2; SA 1

Eleusine indica (L.) Gaertn., Fruct. Sem. Pl. 1: 8. 1788; *Cynosurus indicus* L., Sp. Pl. 72. 1753; Hook., f., Brit. India 7: 293. 1896; Bor. Grass. Burm. Ceyl. Ind. Pak. 493. 1960; Samvatsar, Fl. West. Trib. Madhya Pradesh 351 . 1996; Roy in N.P. Singh *et al.,* Fl. Madhya Pradesh 3: 419.2001. *Pandhar* **(Plate 10, Fig. 5)**

Annuals or biennials; culms 30-50 cm tall, loosely tufted, nodes glabrous; leaves 10-30 × 0.5-0.7 cm, linear, acute, usually folded, ligule 1 mm long, membranous with a tuft of hairs on the auricles; inflorescence of 4-8 sub-digitate spikes, pubescent at base; spikelets 2- seriate, ovate-elliptic, serrulate at the top; lower glume 1.5-2.5 mm long, lanceolate-acute, 1-nerved, scabrid on the back; upper glume 2-4 mm long, ovate-acute, scabrid on the back; lemma 2.5-4.0 mm long, ovate-acute, scabrid on the back; palea narrowly ovate, margin infolded; stamens 3, anthers yellow; caryopsis rugose, ovoid, sharp pointed at both ends, dark brown.

*Fl.-Fr.:*Aug. -Jan.

Common weed in all districts. (very difficult to uproot).

ELIONURUS Humb. Bonpl. ex Willd. W. 15; MP 1; SA 1

Elionurus royleanus Nees ex. A. Rich., Tent. Fl. Abyss 2: 474.1850; Hook.f., Fl. Brit. India 7:161.1896; Bor, Grass. Burm. Ceyl. Ind. Pak. 145.1960; Samvatsar, Fl. West. Trib. Madhya Pradesh 351.1996; Roy, in N.P. Singh *et al.,* Fl. Madhya Pradesh 3: 420.2001. *Seren*

Herbaceous annuals; culms 10-25 cm tall, tufted, decumbent, branched, nodes glabrous; leaves linear, glabrous or sparsely hairy, acute; sheath compressed, glabrous, mouth bearded, ligule lacerate; inflorescence solitary raceme, often tinged with purple, partly enclosed in a flattened sheath, joints 2-2.5 cm long, densely bearded at apex with long hairs, callus bearded; sessile spikelet *ca* 12 mm long, linear, awned; lower glume 11-12 cm long, 7-9 nerved, coriaceous, linear-lanceolate, apex beaked, 2-keeled, keels hairy, beak 6 mm long, 2-fid, violet coloured, margins below the beak with violet-coloured tubercles each provided with a pencil of whitish-glistending hais; upper glume 5-5.5 mm long, membranous, boat-shaped, 1-nerved, acuminate to aristate; lower lemma empty, palea absent; upper lemma hermaphrodite, obovate, 1-nerved; palea absent; anthers 3; caryopsis about 3 mm long; pedicelled spikelet 8-9 mm long, pedicels 2-2.5 mm long; lower glume herbaceous, obscurely nerved, keels winged; upper glume 6.5-7 mm long, concave, awns scabrid; lemma hyaline.

*Fl.-Fr.:*Aug. -Nov.

Rare, in gravelly soil on hill slopes, Alirajpur, Burhanpur, Khandwa.

ELYTROPHORUS P. Beauv. W. 2; MP 1; SA 1

Elytrophorus spicatus (Willd.) A. Camus, Fl. Indo-Chine 7: 547. 1923; *Dactylis spicata* Willd., Neue Ges. Schriften. Naturf. Freunde Berlin, 3: 416. 1801; *Elytrophorus articulatus* P. Beauv. Ess. Agrostogr. 67. 1812; Hook.f., Fl. Brit. India 7: 306. 1896; Bor, Grass. Burm. Ceyl. Ind. Pak. 493. 1960; Roy in N.P. Singh *et al.,* Fl. Madhya. Pradesh 3: 420.2001.

Annuals; culms 10-35 cm tall, erect or decumbent, slender, tufted, nodes glabrous; leaves mostly basal, linear, glabrous, acute, sheath compressed, glabrous, ligule membranous, truncate; inflorescence of false spikes, narrow, cylindric, interrupted towards base, green or straw coloured, rachis scaberulous; spikelets sessile, laterally compressed, 2-4 mm long, awned; glumes subequal, 2-3 mm long, shortly appressed hairy. 1-nerved, keel scaberulous, awns short; lemma 2.5-3 mm long, 3-keeled, acute, margins ciliate, awns 1-1.2 mm long; palea 2-keeled, broadly winged, apex lacerate; anthers 3, yellow; caryopsis ellipsoid, reddish.

*Fl.-Fr.:*Sept. -March

Rare, in drying ditches and wet lands, Indore, Khandwa, Khargone.

ENTEROPOGON Desv. ex P. Beauv. W. 17; MP 2; SA 1

Enteropogon dolichostachyus (Lag.) Keng, Clav. Gen. Sp. Gram. Prim. Sin. 197. 1957; *Chloris dolichostachya* Lagasca, Gen. Sp. Pl. 5. 1816; Hook.f. Fl. Brit. India 7:290.1896; Bor, Grass. Burm. Ceyl. Ind. Pak. 466. 1960; Samvatsar, Fl. West. Trib. Madhya Pradesh 341. 1996; Roy in N.P. Singh *et al.*, Fl. Madhya Pradesh 3:387.2001. *Bamma*

Annuals; culms, 60-90 cm tall, tufted, erect or decumbent-ascending; leaves 12-30 × 0.7-0.8 cm, narrow-linear, upper surface glabrous, lower pubescent, ligule a rim of dense long hairs, 2-4 mm long; sheath compressed, keeled, scabrous on the margins; inflorescence usually of 4 spikes, occasionally varying up to 10; rachis 10-20 cm long, scabrid; spikelets 4.5-5.0 mm long, one bisexual and one sterile flower on a short glabrous pedicel; lower glume 1.7-2.0 mm long, linear-lanceolate, 1-nerved; upper glume 4.5-5.0 mm long, linear-lanceolate, 1-nerved, scaberulous on the back, aristate; lemma 4 mm long, lanceolate, 3-nerved, glabrous with 8 mm long awn; palea 3 mm long, lanceolate, keeled, 2-toothed; stamens 3, anthers 1-4 mm long; caryopsis 2-6 mm long, brown.

*Fl.-Fr.:*Sept. -Dec.

Common, in river beds and among bushes in grassland, Alirajpur, Barwani, Dhar, Khandwa, Khargone.

ERAGROSTIELLA Bor W. 6; MP 4; SA 3

Key to Species

1. Leaves small, up to 6 cm long; spikelets 7-30 flowered, 6-8 mm long; spikes 8- 20 cm long .... **E. brachyphylla**
1. Leaves longer, up to 22 cm long
    2. Spikelets 10-30 flowered, 6-10 mm long; spikes 10-30 cm long........ **E. bifaria**
    2. Spikelets 11-15 flowered, 8-9 mm long; spikes 15-20 cm long .... **E. nardoides**

Eragrostiella bifaria (Vahl) Bor, Indian Forester 66: 270.1940 & Grass. Burm. Ceyl. Ind. Pak. 494.1960; *Poa bifaria* Vahl., Symb. Bot. 2: 19.1791; *Eragrostis coromandeliana* (J. Koenig ex Rottler) Trin., Mem. Acad. Imp. Sci. Saint- Petersbourg, Ser. 6, Sci. Math. 1:415.1830; *Poa coromandeliana* J. Koenig ex Rottler, Neue Schriften. Ges. Naturf. Freunde Berlin 4:191.1803; *E. bifaria* (Vahl) Wight, Cat. Ind. Pl. 105.1834; Stapf in Hook.f., Fl. Brit. India 7: 325.1896; Samvatsar, Fl. West. Trib. Madhya Pradesh 353.1996; Roy in N.P. Singh *et al.*, Fl. Madhya Pradesh 3:423.2001. *Chiraka*

Perennials; culms 20-50 cm tall, base with fibrous remains of old leaf sheaths; leaves 4-16 × 0.6-2.0 cm, filiform, convolute; ligule 0.4-0.6 mm long, membranous, truncate; inflorescence a raceme, 8-20 cm long; spikelets overlapping by half their length, 6-20 mm long, elliptic-linear, oblong with 9-50-flowers; lower glume 1.7 mm long, lanceolate-narrowly ovate, subacute; upper glume 1.8 mm long, oblong-narrowly ovate, rounded or subacute; lemma 2 mm long, ovate, acute; palea broadly winged along the keels, ciliate along the winged margins; stamens 3; caryopsis broadly ovate, pale brownish.

Fl.-Fr. : Aug.-Dec.

Very common in open grassland, on the bunds of fields, old walls, in all districts.

Eragrostiella brachyphylla (Stapf) Bor, Indian Forester 66: 270.1940 & Grass. Burm. Ceyl. Ind. Pak. 494. 1960; *Eragrostis brachyphylla* Stapf in Hook.f., Fl. Brit. India 7: 327. 1896; Samvatsar, Fl. West. Trib. Madhya Pradesh 353 . 1996; Roy in N.P. Singh *et al.,* Fl. Madhya Pradesh 3:424.2001.

Perennials; culms 25-45 cm tall, erect, slender; leaves nearly radical, 5-10 cm long, ligule 1.5 mm long, hair; inflorescence 7-20 cm long, slender, compressed; spikelets 6-18 mm, linear-oblong, secund, biseriate, about 20-flowered; lower glume 2 mm long, acute, keeled, more wide than upper glume; upper glume 2 mm long, acutely keeled; lemma 2.0-2.5 mm long, 3-nerved, lateral nerves very weak; palea obovate, concave, keel scabrid; stamens 3; caryopsis ellipsoid, obscurely trigonous.

Fl.-Fr.: Aug. -Dec.

Rare, on old walls and rocky places, Alirajpur, Burhanpur, Dhar, Indore, Khandwa.

Eragrostiella nardoides (Trin.) Bor, Indian Forester 66: 270.1940 & Grass. Burm. Ceyl. Ind. Pak. 495. 1960; *Eragrostis nardoides* Trin,. Mem. Acad. Imp. Sci. Saint-Petersbourg Ser. 6, Sci. Math. 1: 415. 1830; Stapf in Hook.f., Fl. Brit. India 7: 326. 1896; Roy in N.P. Singh *et al.,* Fl. Madhya Pradesh 3:425. 2001.

Perennials; culms 25-45 cm tall, erect, tufted; leaves as long as the stem, convolute, sparsely hairy on the lower portion; sheaths glabrous, slightly compressed; ligule a minute membrane; inflorescence slender, 15-20 cm long; rachis terete, smooth; spikelets oblong to linear, 5-10 mm long, 20-30-flowered; glumes subequal, ovate, 1.5-1.7 mm long, subacute, nerves faint; palea equal to lemma, wings shortly ciliate; anthers 3, yellow; caryopsis brown, suboviod or sub globose, embryo end pointed.

Fl.-Fr. : Aug. -Dec.

On rocky areas and old walls, Burhanpur, Khandwa.

ERAGROSTIS Wolf W. 424; MP 24; SA 15

Key to Species

1. Spikelets breaking from above downwards; rhachilla fragile
  2. Keels of palea ciliate
    3. Panicles compact and dense, annuals . ... **E. ciliaris**
    3. Panicles effuse
      4. Panicles glandular, glands present on branches and glumes .... **E. viscosa**
      4. Panicles not glandular .... **E. amabilis**
  2. Keels of palea scabrid or smooth, not ciliate
    5. Branches of panicle whorled, panicle not thyrsiform; lemma less than 1 mm long .... **E. japonica**
    5. Branches of panicle not whorled, panicle thyrsiform; lemma 1.5 mm or more long .... **E. aspera**
1. Spikelets breaking from below upwards; rhachilla tough
  6. Annuals

7. Plants totally devoid of glands; spikelets many flowered
   8. Lemmas over 1.5 mm long
      9. Spikelets more than 2 mm wide, ovate to oblong, pink . ... **E. unioloides**
      9. Spikelets up to 1.8 mm wide, linear, often curved
         10. Panicles 20-30 cm wide; spikelets reddish-purple; pedicels over 5 mm long; lemma ovate; grain ovoid .... **E. tremula**
         10. Panicles up to 5 cm wide; spikelets slatey-grey; pedicels less than 2 mm long; lemma elliptic; grain ellipsoid . ... **E. minor**
   8. Lemmas up to 1.5 mm long; spike;ets 1 mm broad, linear-lanceolate, slatey-grey
      11. Plants 40-75 cm tall; lemmas acute to obtuse .... **E. gangetica**
      11. Plants up to 20 cm tall; lemmas very acute .... **E. multicaulis**
7. Plants with or without glands
   12. Lower branches of panicle whorled, with long, white hairs in the axils .... **E. pilosa**
   12. Lower branches of panicle not whorled, hairs absent; spikelets 2-4 mm wide; leaves and sheaths often with pitted glands; lemmas up to 2.2 mm long .... **E. cilianensis**
6. Perennials
   13. Spikelets 5-7 mm long; lemmas 2 mm long; pedicel with 1-2 glands .... **E. patula**
   13. Spikelets 1-2.5 mm long; lemmas 1-1.8 mm long; pedicel eglandular
      14. Spikelets 1.75-2.5 mm wide; palea deciduous; panicle up to 20 cm × 8-9 cm; lemma 1.5-2 mm long .... **E. atrovirens**
      14. Spikelets 1-1.25 mm wide; palea persistent; panicle up to 12 × 3 cm; lemma 1.5 mm long .... **E. nutans**

Eragrostis amabilis (L.) Wight & Arn., Cat. Indian Pl. 2: 105. 1834; *Poa amabilis* L. Sp. Pl. 68. 1753; *Eragrostis tenella* (L.) P. Beauv. ex Roem. & Schult., Syst. Veg. 2: 576. 1817; Stapf in Hook., f., Fl. Brit. India 7: 315. 1896; Bor, Grass. Burm. Ceyl. Ind. Pak. 513. 1960; Samvatsar, Fl. West. Trib. Madhya Pradesh 357 . 1996; Roy in N.P. Singh *et al.,* Fl. Madhya Pradesh 3: 436.2001. *Charpat*

Erect, tufted, annual grass, up to 46 cm high; leaves linear, flat 2.3-16 cm long, acuminate, scabrid; mouth of sheath naked; panicle purplish oblong, up to 17 cm long, effuse, branches patent; spikelets ovate-oblong, 2.0 × 1.2, 4-6-flowered; pedicels scabrid with glandular pits; glumes lanceolate, sub-equal, 1-nerved; lemma ovate-oblong, 3-nerved, obtuse; palea up to 1 mm long, keeled, keel ciliate; stamens 3; caryopsis ovoid

Fl.-Fr.: July-Oct.

On sandy soil and river bed, Barwani, Burhanpur, Khandwa, Khargone.

Eragrostis aspera (Jacq.) Nees, Fl. Afr. Austral. 408. 1841; *Poa aspera* Jacq. Hort. Bot. Vindob. 3: 32.1776; Stapf in Hook.f., Fl. Brit. India 7: 314. 1896; Bor, Grass. Burm. Ceyl. Ind. Pak. 501.1960; Roy in N.P. Singh *et al.,* Fl. Madhya Pradesh 3: 429.2001.

Annuals; culms terete, 20-60 cm tall, erect, simple or branched, glabrous, nodes glabrous; leaf sheath compressed, 2-10 cm long, glabrous, bearded at mouth; ligule a fringe of long hairs; leaf blade linear, 8-35 cm, glabrous, apex acuminate; panicles lax, 15-40 cm long, branches filiform, pseudo-whorled, base bearded with long hairs, scaberuolus; pedicels capillary, longer than the spikelets, scaberulous; spikelets oblong, 2-6, 5-14 flowered, breaking up from above downwards; lower glume membranous, narrowly ovate, 1-keeled, 1-nerved, apex truncate; upper glume membranous, narrowly ovate, 3-nerved, apex truncate; palea hyaline, 1.3-1.5 mm, 2-keeled, keels scabrid, apex obtuse; stamens 3; caryopsis orbicular.

*Fl.-Fr.:*Sept. -Dec.

Along the bank of streams and rivers, also as weed of cultivated fields, on black cotton soil, Barwani, Burhanpur, Khandwa, Khargone.

Eragrostis atrovirens (Desf.) Trin. ex Steud., Nomencl. Bot. ed. 2, 1: 562.1840; *Poa atrovirens* Desf. Fl. Atlant. 1: 73.1798; Bor, Grass. Burm. Ceyl. Ind. Pak. 503.1960; Samvatsar, Fl. West. Trib. Madhya Pradesh 354.1996; Roy in N.P. Singh *et al.,* Fl. Madhya Pradesh 3:430. 2001.

Strong perennials; culms 50-80 cm; leaves 10-20 cm long, narrow, linear, acute; sheath striate, ligule membranous, truncate; inflorescence panicle, 5-20 cm long, ovate or oblong; spikelets 5-10 mm long, 6-12-flowered, glaucous, grey or purplish; pedicels 1-3 mm long; l,ower glume 1.4 mm long, lanceolate, acute, 1-nerved; upper glume 1.8 mm, boat-shaped, lanceolate acute; lemma 1.4-2.0 mm long, acute, 3-nerved, scabrid along mid nerve; palea equalling lemma, narrowly ovate, scabrid along nerves; anthers 3, brownish- yellow, sometime purple; caryopsis red brown, cylindrical.

*Fl.-Fr. :*Aug. -Nov.

In moist places and fields, Alirajpur, Barwani, Khargone.

Eragrostis cilianensis (All.) Janch., Mitt. Naturwiss. Vereins Univ. Wien 5:110.1907; *Poa cilinensis* All., Fl. Pedem. 2: 246. 1785; *Eragrostis major* Host, 1809, *nom. illeg;* Hook.f., Fl. Brit. India 7: 320.1896; Bor, Grass. Burm. Ceyl. Ind. Pak. 503. 1960; Samvatsar, Fl. West. Trib. Madhya Pradesh 354 . 1996; Roy in N.P. Singh *et al.,* Fl. Madhya Pradesh 3: 430.2001. *Ran poha*

Annuals; culms 20-80 cm tall, loosely tufted, erect or ascending; leaves 10-15 × 0.5-0.8 cm, linear-lanceolate, glandular along the midrib and margins; ligule a ring of hairs; inflorescence an open panicle, 5-20 cm long; spikelets 3-13 mm long, narrowly oblong, 5-40-flowered; glumes long, ovate, boat-shaped, 1-3-nerved; lemma broadly ovate, glandular on the keels; palea 1.2 mm long, oblanceolate, chartaceous, 2-keeled, ciliate on the keels; stamens 3; caryopsis subglobose, rugulose, reddish- brown.

*Fl.-Fr.:*Sept. -Dec.

Common as a weed in cultivated fields on black cotton soil and in moist places, in all districts.

Eragrostis ciliaris (L.) R. Br., Narr. Exped. Zaire 478. 1818; *Poa ciliaris* L. Syst. Nat. ed. 10 2:875 1759; Stapf in Hook.f., Fl. Brit. India 7: 314. 1896; Bor, Grass. Burm. Ceyl. Ind. Pak. 506. 1960; Samvatsar, Fl. West. Trib. Madhya Pradesh 354. 1996; Roy in N.P. Singh *et al.,* Fl. Madhya Pradesh 3: 431.2001. *Murmuri*

Annuals; culms tufted, terete, 10-40 cm tall, slender, erect, ascending or geniculute at base, glabrous, nodes glabrous; leaf sheath terete, 1-3 cm long, glabrous, bearded at mouth; ligule a rim of hairs; leaf blade flat or convolute, linear, 2-8 × 0.2 mm, glabrous, apex acuminate; panicles compact, cylindric, 2-6 cm long, rachis slender, smooth, branches short, smooth, flexuous; pedicels very short; spikelets compressed, crowded, broadly elliptic-ovate, 2.2-2.6 mm, 3-9 flowered, breaking up from above downwards; lower glume membranous, narrowly ovate, 0.5-1 mm, 1-keeled, keel scabrid, 1-nerved, apex acute; upper glume membranous, narrowly ovate, 1-keeled, keel scaberulous, 1-nerved, apex acute; lemma membranous, narrowly oblong, puberulous, 1-keeled, 3-nerved, apex acute; palea hyaline, oblong, 2-keeled, keels with long rigid cilia, apex obtuse; stamens 3; caryopsis elliptic.

Fl.-Fr. : -Aug. -Oct.

Common in cultivated fields, sandy soil, in open grasslands all over.

Eragrostis gangetica (Roxb.) Steud., Syn. Pl. Glumac. 1: 266. 1854; *Poa gangetica* Roxb., Fl. Ind. 1: 341. 1820; *Eragrostis stenophylla* Hochst. ex Miq., Anal. Bot. Ind. 2:27. 1851; Hook.f., Fl. Brit. India 7: 318. 1896. Bor, Grass. Burm. Ceyl. Ind. Pak. 508. 1960; Samvatsar, Fl. West. Trib. Madhya Pradesh 355 1996; Roy in N.P. Singh *et al.,* Fl. Madhya Pradesh 3: 433.2001. *Panjaro*

Annuals or perennials; culms 15-60 cm tall, tufted, erect or decumbent-ascending, brached from the base; leaves 2-20 × 0.1-0.3 cm, linear, flat or rolled, tapering to a fine point; ligule membranous, truncate; inflorescence a narrow panicle, 3-10 cm long; spikelets 4-10 mm long, linear or narrowly oblong, grayish- green; lower glume 0.8 mm long, lanceolate, 1-nerved; upper glume lanceolate, tip sharply acute; lemma acute to sub-obtuse; palea scaberulous, falling soon after the lemma; stamens 2, anthers deep pink, appearing blackish; caryopsis 0.4-0.6 mm long, subglobose, dark brown.

Fl.-Fr. : Aug. -Dec.

Very common along water streams, margin of ponds, moist places in all districts.

Eragrostis japonica (Thunb.) Trin., Mem. Acad. Imp. Sci. Saint-Petersbourg Ser. 6, Sci. Math. 1: 405.1830; *Poa japonica* Thunb., Fl. Jap. 51.1784; *Poa diarrhena* Schult., Syst. Veg. Mant. 2:616.1824; *Eragrostis diarrhena* (Schult.) Steud., Syn. Pl. Glumac. 1:266.1854; *E. diplachnoides* Steud., Syn. Pl. Glumac. 1:268. 1854; Bor, Grass. Burm. Ceyl. Ind. Pak. 509. 1960; *E. namaquensis* Nees var. *diplachnoides* (Steud.) Clayton, Kew Bull. 25:251.1971; Samvatsar, Fl. West. Trib. Madhya Pradesh 355.1996; Roy in N.P. Singh *et al.,* Fl. Madhya Pradesh 3: 433.2001. *Ghodila*

Perennials; culms 15-70 cm tall, tufted, erect or geniculate at the base, slender to stout; leaves 2-20 × 0.1-0.4 cm, linear-acute, glabrous, flat or convolute; ligule 1.0-1.2 mm long, membranous; inflorescence panicle, greenish or purplish, 6-25 cm long; spikelets loosely scattered; pedicels up to 3 mm long, scabrid; lower glume 0.5 mm long, ovate-oblong, obtuse, 3-nerved; palea equal to lemma, 2-keeled, scabrid along the keel; stamens 2, anthers yellow; caryopsis elliptic, polished, reddish- brown.

*Fl.-Fr.:*Sept. -Dec.

Common weed of fields, moist places, open wet ground, in all districts.

Eragrostis minor Host, Gram. Austr. 4: 15. 1809; Stapf in Hook.f., Fl. Fl. Brit. India 7: 327. 1896; *E. pappiana* Chiov. in Ann. Inst. Bot. Roma 8:371.1908; *E. poaeoides* P. Beauv. Ess.

Agrostogr. 162.1812; Bor, Grass. Burm. Ceyl. Ind. Pak. 512. 1960; Samvatsar, Fl. West. Trib. Madhya Pradesh 357.1996; Roy in N.P. Singh *et al.,* Fl. Madhya Pradesh 3:435.2001.

Annuals; culms loosely tufted, 10-60 cm high, erect or geniculately ascending; nodes glabrous; leaves linear, 5-12 × 0.2-0.5 cm, flat or convolute, glabrous or pilose on upper surface; sheaths compressed, ligule a rim of hairs; panicles ovate-oblong or oblong, 5-25 cm long, dense; spikelets linear or narrowly oblong, 5-15-flowered, olive-green or yellowish, pedicels with glands; lower glume ovate-lanceolate, 1-2 × 1 mm, chartaceous, 1-nerved; upper glume ovate-lanceolate, 1.5 mm, chartaceous, 1-3-nerved; lemmas ovate, 1.5-2 mm, chartaceous, 3-nerved; palea elliptic, 1.2 × 1 mm, scabrid on the keels; stamens 3; anthers c.0.3 mm long; grains globose, 0.5-0.7 mm long.

*Fl.-Fr.:*Oct. -Dec.

Weed of cultivated fields, Indore, Khandwa.

Eragrostis multicaulis Steud., Syn. Pl. Glumac. 1:426 . 1854; Bor, Grass. Burm. Ceyl. Ind. Pak. 510.1960; Samvatsar, Fl. West. Trib. Madhya Pradesh 355. 1996; Roy in N.P. Singh *et al.,* Fl. Madhya Pradesh 3:433 2001.

Annuals; culms up to 20 cm tall, erect, tufted; leaves linear to lanceolate, up to 10 cm long, acute; ligule a ring of hairs; panicle terminal; spikelets 6-8 flowered, 3-4 mm long, breaking from below upwards; rachis tough; glumes 1 mm long; lemma 1-1.5 mm long, acute; palea up to 1.5 mm long, deciduous, keel scabrid; anthers 2, dark pink; caryopsis dark brown, *ca* 1 mm long.

*Fl.-Fr.:*Aug. -Oct.

Occasional in rocky habitats, Alirajpur, Indore.

Eragrostis nutans (Retz.) Nees ex Steud. Nomencl. Bot. ed. 2, 1: 563.1840; *Poa nutans* Retz., Observ. Bot. 4: 19. 1786; *Eragrostis elegantula* Nees ex Steud., Syn. Pl. Glumac. 1: 266. 1854; Stapf in Hook.f., Fl. Brit. India 7: 318. 1896; Bor, Grass. Burm. Ceyl. Ind. Pak. 511. 1960; Samvatsar, Fl. West. Trib. Madhya Pradesh 357. 1996; Roy in Singh *et al.,* Fl. Madhya Pradesh 3: 434.2001.

Perennials; culms 20-45 cm tall, erect tufted; leaves 7-12 × 0.3-0.4 cm, narrow, convolute or flat, glabrous; ligule friged; inflorescence a contracted panicle; spikelets crowded, 4-6 × 1 mm long, 20-30-flowered; rachis zigzag; lower glume 0.8 mm long, acute; upper glume 1 mm long, lanceolate-acute; lemma 1.25 mm long, oblong-acute, their tip over lapping base of next lemma, lateral nerves straight; palea 1.25 mm long, linear-oblong, persistent, keels scabrid; anthers 1 mm long; caryopsis ellipsoid-globose, reddish- brown.

Fl.-Fr. : July-Sept.

Common near water courses, open ground, in all districts. Endemic to India.

Eragrostis patula (Kunth) Steud., Nomencl. Bot. ed. 2, 1.564.1840; *Poa patula* Kunth, Nov. Gen. Sp. 1: 558.1816; *Eragrostis tenuifolia* (A. Rich.) Hochst. ex Steud., Syn. Pl. Glumac. 1: 268.1854; *Poa tenuifolia* A. Rich.,Tent. Fl. Abyss. 2:425.1850; Hook.f., Fl. Brit. India 7: 322.1896; Bor, Grass. Burm. Ceyl. Ind. Pak. 514. 1960; Roy in N.P. Singh *et al.,* Fl. Madhya Pradesh 3: 437.2001.

Perennials; culms 10-70 cm tall, tufted, basal shoots laterally compressed; leaves 4-30 × 0.1-0.3 cm flat, filiform; ligule a ring of hairs; inflorescence a narrowly elliptic panicle, 5-15 cm long; pedicels with or without glandular ring; spikelets 4-16 × 1-3 mm, linear, dark green or blackish, 4-16-flowered; lower glume 0.5-1.0 mm long, lanceolate; upper glume reaching the base of the second lemma; lemma oblong-elliptic, bluntly acute to narrowly obtuse, inconspicuously nerved; palea green, scaberulous on the keels; stamens 3, anthers dark brown; caryopsis oblong, strongly laterally flattened, embryo more than half of the seed length.

*Fl.-Fr.:*Aug. -Nov.

Along field bunds and forest boundaries, Indore, Khandwa, Khargone.

Eragrostis pilosa (L.) P. Beauv., Ess. Agrostogr. 71. 1812; *Poa pilosa* L., Sp. Pl. 68. 1753; Stapf in Hook.f., Fl. Brit. India 7: 323. 1896; Bor, Grass. Burm. Ceyl. Ind. Pak. 512. 1960; Samvatsar, Fl. West. Trib. Madhya Pradesh 357 . 1996; Roy in N.P. Singh *et al.,* Fl. Madhya Pradesh 3: 435. 2001. *Phuler*

Annuals; culms 5-70 cm tall, loosely tufted, erect or geniculate; leaves 5-10 × 0.2-0.3 cm, linear-filiform, glabrous; ligule ciliate; inflorescence an open panicle, 2-20 cm long, elliptic to ovate; in mature plant lowest branches sub-whorled with glandular bands and few long white hairs in the axil; spikelets 3-7 mm long, 4-10-flowered, purplish- green; lower glume narrowly-ovate, hyaline, nerveless; lemma broadly ovate, 3-nerved with truncate purple tips; palea 1.2 mm long, oblanceolate, 2-nerved, smooth; stamens 3, anthers deep pink; caryopsis ellipsoid, brownish- yellow.

*Fl.-Fr.:*Aug. -Dec.

Common in moist places, open grassland, waste places in all districts.

Eragrostis tremula Hochst. ex Steud., Syn. Pl. Glumac. 1: 269.1854; *Poa tremula* Lam., Tab. Encycl. 1: 185.1791; *Poa multiflora* Roxb., 1820, *nom. illeg*; Stapf in Hook.f., Fl. Brit. India 7:320.1896; Bor, Grass. Burm. Ceyl. Ind. Pak. 514.1960; Samvatsar, Fl. West. Trib. Madhya Pradesh 357.1996; Roy in N.P. Singh *et al.,* Fl. Madhya Pradesh 3:437.2001. *Chirio*

Annuals; culms 15-70 cm tall, tufted, erect or geniculate; leaves 2-20 × 0.2-0.5 cm, linear-lanceolate, tapering to a fine point, glabrous; ligule membranous; inflorescence a diffuse panicle, branches solitary, filiform; spikelets 8-50 mm long, pallid to purplish; lower glume 1.2 mm long, narrowly ovate; lemma broadly ovate, acute with irregular wavy, hyaline margins; palea persistent, scabrid on the keels; stamens 2, anthers deep pink; caryopsis broadly-ovoid, beaked at the embryo end.

*Fl.-Fr.:*Aug. -Dec.

Common in rocky habitats, near streams, grasslands , all over the area.

Eragrostis unioloides (Retz.) Nees ex Steud., Syn. Pl. Glumac. 1: 264.1854; *Poa unioloides* Retz., Observ. Bot. 5: 19.1789; Stapf in Hook., f., Brit. India 7: 317.1896; Bor, Grass. Burm. Ceyl. Ind. Pak. 515.1960; Samvatsar, Fl. West. Trib. Madhya Pradesh 360.1996; Roy in N.P. Singh *et al.,* Fl. Madhya Pradesh 3: 438.2001. *Ram pona*

Annuals; culms 15-40 cm tall, densely tufted, erect or decumbent; leaves 10-15 × 0.5 cm, linear with rounded or subcordate base; leaf sheath compressed, 3-5 cm long, glabrous; ligule a ring of hairs; inflorescence a contracted panicle, 3-15 cm long; spikelets strongly compressed, ovate to ovate-oblong, straw-coloured, often purple-tinged; lower glume lanceolate-acute, 1-nerved; upper glume 1.3 mm long, ovate-acute; lemma 1.8-2.0 mm long, elliptic acute, glabrous; palea 1.5 mm long, obtuse or bidentate, scabrid on the keels; stamens 2, anthers deep pink; caryopsis *ca* 1.0 mm long, obovoid, dark brown.

*Fl.-Fr.:*Sept. -Nov.

Very common on wet soil, along margins of ponds, all over the area.

Eragrostis viscosa (Retz.) Trin., Mem. Acad. Imp. Sci. Saint-Petersbourg Ser. 6, Sci. Math. 1: 397.1831; *Poa viscosa* Retz., Observ. Bot. 4: 20.1786; *Eragrostis tenella var. viscosa* (Retz.) Stapf in Hook., f., Fl. Brit. India 7: 315. 1896; Bor, Grass. Burm. Ceyl. Ind. Pak. 515. 1960; Roy in Singh *et al.,* Fl. Madhya Pradesh 3: 438. 2001. *Bhumsi*

Sweet scented glandular annuals; culms 10-40 cm tall; inflorescence covering more than half length of the plant; all parts sticky due to viscous glands; ligule base and mouth of sheath with very long cilia; spikelets 1.5-405 mm long; rachis fragile; lemma oblong-elliptic, smooth or scaberulous on keel; palea keel ciliate; anthers 2-3 mm long; caryopsis pale brown, polished.

*Fl.-Fr.:*Sept. -Nov.

In irrigated fields and river beds, Barwani, Burhanpur, Khandwa, Khargone.

ERIOCHLOA Kunth W. 35; MP 2; SA 2

Key to Species

1. Spikelets shortly but distinctly awned; axis and pedicels hairy .... **E. fatmensis**
1. Spikelets not awned; axis and pedicels glabrous .... **E. procera**

Eriochloa fatmensis (Hochst. & Steud.) Clayton, Kew Bull. 30: 108. 1975; *Panicum fatmense* Hochst. & Steud. Exsicc. (Pl. Schimp.) 1837 :806. 1837; Bor, Grass. Burm. Ceyl. Ind. Pak. 312: 1960; Samvatsar, Fl. West. Trib. Madhya Pradesh 361 . 1996.

Perennial herbs; culms 30-60 cm tall, geniculately ascending and rooting at lower nodes; sheaths compressed, 1.5-7.0 cm long; ligule hairy; leaves linear-lanceolate, scaberulous along margins, acute at apex; panicles of simple or branched racemes, 3-12 cm long; spikelets paired or rarely solitary, ellipsoid, 4-4.5 mm long, hispid with appressed hairs, distinctly awned; lower glume reduced; upper glume elliptic, 5-nerved, acuminate or shortly awned; lower lemma empty, faintly 5-nerved, dorsally hairy; palea absent; upper lemma chartaceous, elliptic-oblong, whitish-yellow; palea chartaceous, 2-keeled; caryopsis ellipsoid-oblong.

Fl.-Fr. : -Aug.-Feb.

Not common, along water courses, moist places, Alirajpur, Burhanpur, Khargone.

Eriochloa procera (Retz.) C.E. Hubb., Bull. Misc. Inform. Kew 1930:256.1930; *Agrostis procera* Retz., Observ. Bot. 4:19.1786; Hook.f., Fl. Brit. India 7: 20.1896; Bor, Grass. Burm. Ceyl. Ind. Pak. 312. 1960;Roy in N.P. Singh *et al.,* Fl. Madhya Pradesh 3: 440.2001. *Pelu*

Rhizomatous perennials; culms 20-100 cm tall, terete, erect; leaves flat, linear-lanceolate, glabrous, sheaths compressed, glabrous, ligule hairy; inflorescence a panicle, rachis angular, pedicels short, scabrid; spikelets paired, rarely solitary, 3-3.5 mm long, acute, hispid; lower glume fused with rachis; upper glume 3 mm long, 5-nerved, lemmas ovate-oblong, 3-nerved, obtuse; palea keeled, keel ciliate; stamens 3; caryopsis ellipsoid.

Fl.-Fr.:July-Dec.

Rare, in moist places on sandy- loamy soil, Burhanpur, Khargone.

EULALIA Kunth W. 36; MP 3; SA 3

Key to Species

1. Leaf-sheaths covered with white wool in the basal part; spikes 2-4, 1-2 cm long; spikelets 1-2 mm long .... **E. staintonii**
1. Leaf-sheaths glabrous or with scattered, tubercled-based hairs at base; spikes up to 8, 4-6 cm long
   2. Spikes 3-8, 5-10 cm long; spikelets 4-5 mm long .... **E. trispicata**
   2. Spikes 3-6, 7-10 cm long; spikelets 5-6 mm long .... **E. quadrinervis**

Eulalia quadrinervis (Hack.) Kuntze, Revis. Gen. Pl. 2:775. 1891; *Pollinia quadrinervis* Hack., Monogr. Phan. 6:158. 1889; Hook.f. Fl. Brit. India 7:110.1896; Bor, Grass. Burm. Ceyl. Ind. Pak. 156. 1960; Roy in N.P. Singh *et al.,* Fl. Madhya Pradesh 3:442.2001.

Perennials; culms 50-100 cm tall, slender, leafy up to inflorescence; leaves linear, acuminate at apex, scabrid on margins; sheaths glabrous or with scattered tubercle-based hairs; ligule short, hairy; inflorescence of 3-6, subdigitate, 7-15 cm long, false spikes with golden brown or mauve hairs; spikelets oblong-lanceolate, 5-6 mm long; lower glume ovate-lanceolate, 5-6 mm long; keels hispid, 4-nerved; upper glume lanceolate, equal to lower, keels glabrous, 3-nerved; lower floret empty; upper bisexual; lemma 2.5-3 mm long, 2-lobed, awn 15-20 mm long; stamens 3, *ca* 3 mm long.

Fl.-Fr.: Aug. -Nov.

Very rare, **Indore** only, Dev guradiya hill.

Eulalia staintonii Bor, Kew Bull. 1957:411.1958 & Grasses of Burma, Ceylon India and Pakistan 157.1960; Sharma *et al.,* in Fl. Maharashtra 3:495. 1996; Naik, Fl. Marathwada 2:1044. 1998.

Herbs up to 70 cm tall; nodes glabrous; leaves linear, 15-50 cm long, basal sheaths covered with whitewool, upper glabrous; spikes 2-4, fascicled, 1-2 cm long; spikelets oblong-lanceolate, 1-2 mm long; caryopsis ellipsoid.

Fl.-Fr.:Oct. -Dec.

Rare, **Burhanpur** only, Jainabad.

Eulalia trispicata (Schult). Henrard, Blumea 3: 453. 1940; *Andropogon trispicatus* Schult., Mant. 2: 452.1824; *Eulalia argentea* (Trin.) Brongn.,Voy. Monde Phan. 92. 1831; *Pollinia argentea* (Brongn.) Trin., Bull. Sci. Acad. Imp. Sci. Saint-Petersbourg 1: 71.1836; Hook., f., Fl. Brit. India 7: 111.1896; Bor, Grass. Burm. Ceyl. Ind. Pak. 157.1960; Samvatsar, Fl.

West. Trib. Madhya Pradesh 361.1996; Roy in N.P. Singh *et al.,* Fl. Madhya Pradesh 3: 442.2001.

Perennials; culms 50-100 cm tall, leafy at base, turning reddish-brown with age; leaves linear-lanceolate, glabrous or sparsely hairy, usually flat; sheath terete, glabrous except mouth, ligule lacerate, truncate; inflorescence of 3-7 digitate racemes, densely silky; sessile spikelets 4-5 mm long, lanceolate, awned, obtuse; lower glume papery, 3.5-4 mm long, hairy at base, 2-keeled; upper glume similar, but boat-shaped, 1-nerved; lower lemma empty; upper lemma linear, apex clefted, awned from sinus, awns 9-12 mm long, ciliate; palea absent; anthers 3, yellow, 2 mm long; caryopsis oblong; pedicelled spikelets similar to sessile ones.

Fl.-Fr. : -Sept. -Dec.

Along streams, in bushes, common in all districts.

EULALIOPSIS Honda W. 2; MP 1; SA 1

Eulaliopsis binata (Retz.) C.E. Hubb., Hooker's, Icon. Pl. 33: t. 3262:p. 6 1935; *Andropogon binatus* Retz., Observ. Bot. 5: 21. 1789; *Spodiopogon angustifolius* Trin., Mem. Acad. Imp. Sci. Saint-Petersbourg Ser. 6, Sci. Math. 2:300.1883; *Ischaemum angustifolium* (Trin.) Hack., Monogr. Phan. 6: 24. 1889; Hook.f., Fl. Brit. India 7: 129. 1896; Bor, Grass. Burm. Ceyl. Ind. Pak. 158. 1960; Roy in N.P. Singh *et al.,* Fl. Madhya Pradesh 3: 444.2001. *Bagai*

Perennials; culms 40-100 cm tall, erect with densely tufted rootstock and basal sheaths; leaves 20-45 × 0.3-0.5 cm, with a rounded base, upper surface glabrous, lower hirsute with tubercle-based hairs; ligule hairy; inflorescence of 2-4 racemes, pale or whitish- brown, up to 5 cm long; spikelets 4.5 4. 6 mm long, lanceolate; sessile spikelets 4.5-4. 6 mm long, lanceolate; lower glume 3.3-3.5 mm long, lanceolate, 3-4-toothed, 5-7-nerved, ciliate in the upper half with a tuft of hairs at the margin and on the back of base; upper glume 4 mm long, obtuse ciliate in the upper half, with a tuft of hairs on the keel, 5-nerved; lower lemma, 3.0-3.2 mm long, lanceolate, oblong, obtuse, 1-nerved, male or neuter, paleate or not; upper lemma hermaphrodite, bidentate, ciliate, 3. 4-4.0 mm long, with 5-8 mm long awn; palea 2.5-3.0 mm long, ovate, obtuse, ciliate; pedicelled spikelets similar to sessile spikelets; stamens 3, yellow; caryopsis black, *ca* 1.5 mm long.

Fl.-Fr. : -Sept.-Nov.

Rare, along streams, Barwani, Burhanpur, Khargone.

HACKELOCHLOA Kuntze W. 2; MP 1; SA 1

Hackelochloa granularis (L.) Kuntze, Revis. Gen. Pl. 2: 776.1891; *Cenchrus granularis* L., Mant. Pl. 2:575. 1771; *Manisuris granularis* L., Amoen. Acad. 10:40.1779; Hook.f., Fl. Brit. India 7: 159. 1896; Bor, Grass. Burm. Ceyl. Ind. Pak. 159. 1960; Samvatsar, Fl. West. Trib. Madhya Pradesh 361. 1996; Roy in N.P. Singh *et al.,* Fl. Madhya Pradesh 3: 446.2001.

Annuals; culms 60-90 cm tall, erect, hirsute with tubercle-based hairs; leaves 5-10 cm long, cordate at base, hirsute; sheaths loose, hispid; ligule membranous, hairy at tip; inflorescence a terminal, axillary raceme; peduncles 1.5-2.0 cm long, joints flattened, fused with pedicels; sessile spikelets globose, 1 mm long; lower glume rounded-orbicular, pitted and tubercled; upper glume

1.0- 1.2 mm long, obtuse, membranous; lower lemma empty, 1 mm long; upper lemma obtuse, bisexual; anthers 3, orange yellow; caryopsis ovoid, shining.

Fl.-Fr. : -July-Dec.

Common on hill slopes, in grassland, as a weed in cultivated fields, in all districts.

HEMARTHRIA R. Br.W. 12; MP 1; SA 1

Hemarthria compressa (L.f.) R. Br., Prodr. Fl. Nov. Holl. 207. 1810; *Rottboellia compressa* L.f., Suppl. 114. 1781; Hook.f., Fl. Brit. India 7:153. 1896; Bor, Grass. Burm. Ceyl. Ind. Pak. 161. 1960; Roy in N.P. Singh *et al.,* Fl. Madhya Pradesh 3: 447.2001.

Perennials; culms 40-80 cm tall, erect or decumbent; leaves 9-15 × 0.2-0.5 cm, linear, acuminate with rounded or subcordate base, margins scaberulous; ligule membranous; sheath glabrous, hairy at the throat; inflorescence a solitary raceme or false spike, 6-10 cm long; sessile spikelet 4-6 mm long; lower glume 5 mm long, lanceolate, oblique, obtuse, ciliate, 5-7-nerved, winged; upper glume 4 mm long, boat-shaped, membranous, mucronate; lower lemma 3.0-3.2 mm long, oblong hyaline; palea 3.0-3.0 mm long, similar, 2-nerved; stamens 3, anthers 2.8 mm long, yellow; pedicelled spikelets 4-5 mm long, acute, acuminate, and similar to sessile spikelets; caryopsis brownish, about 1 mm long.

*Fl.-Fr.:*Aug. -Dec.

Around fields in moist places, Barwani, Indore, Khandwa, Khargone.

HETEROPOGON Pers. W. 6; MP 2; SA 5

Key to Species

1. Culms 100-250 cm tall; racemes 8-15 cm long .... **H. triticeus**
1. Culms 25-100 cm tall; racemes up to 8 cm long
    2. Lower glume of pedicellate spikelet with a longitudinal median series of depressed glands .... **H. melanocarpus**
    2. Lower glume of pedicellate spikelet not as above
        3. Glumes of pedicellate spikelets glabrous .... **H. polystachyus**
        3. Glumes of pedicellate spikelets armed with tubercled based hairs
            4. Ligule vert short, truncate, not divided into segments .... **H. contortus**
            4. Ligule long, divided into many acute segments .... **H. ritchiei**

Heteropogon contortus (L.) P. Beauv. ex Roem. & Schult. Sys. Veg. 2: 836. 1817; *Andropogon contortus* L., Sp. Pl. 1045. 1753; Bor, Grass. Burm. Ceyl. Ind. Pak. 163. 1960; Samvatsar, Fl. West. Trib. Madhya Pradesh 363 . 1996; Roy in N.P. Singh *et al.,* Fl. Madhya Pradesh 3: 448. 2001. *Suklya* **(Platge 10, Fig. 6)**

Perennials; culms 25-100 cm tall, erect tufted; leaves 4-20 × 0.3-0.7 cm, linear acute or acuminate, ciliate at bas; ligule membranous, shortly hairy; inflorescence a terminal spike- like raceme, up to 12 cm long; sessile spikelets: lower one male or neuter, awnless; hermaphrodite spikelet: terete, 6.5-7.0 mm long; lower glume 6.5 mm long, chartaceous, oblong-lanceolate, glabrous or hairy, 7-9-nerved; upper glume 3-nerved; lower lemma empty, hyaline epaleate; upper lemma hyaline, cartilaginous at base, passing into a stout, geniculate awn, 7-8 cm long, epaleate; stamens 3; pedicelled spikelets 8-9 mm long; lower glume 7.5-8.5 mm long, linear-lanceolate, hispid with tubercle-based hairs; upper glume 8. 3-9.8 mm long, lanceolate-acuminate, 5-nerved; lower lemma 5.5-6.0 mm long, oblong-lanceolate, hyaline epaleate; stamens 3; caryopsis cylindric, callus very sharp.

Fl.-Fr. : -Aug. -Jan.

Very common in open grassland, in forest, throughout the area (very polymorphic species).

Heteropogon melanocarpus (Elliot) Benth.,J. Linn. Soc. Bot. 19:71.1881; *Andropogon melanocarpus* Elliot, Sketch Bot. S. Carolina 1:146. 1816; Bor, Grass. Burm. Ceyl. Ind. Pak. 165. 1960; Hook.f. Fl. Brit. India. 7: 200.1896.

Erect annuals, up to 2 m high; culm stout, with stilt roots from the lower nodes; leaf sheaths compressed, leaf blades linear, long tapering to a fine point; inflorescence with few to many racemes; sessile spikelets long, linear-oblong; lower glume 8-15 cm long, linear-oblong; upper glume narrower, nearly equal to lower glume; lower lemma long, hyaline, 4.5 mm long; upper lemma very narrow passing to an awn; pedicelled spikelets of all pairs male; lower glume long, herbaceous, keeled on one side, keel serrulate with an inflexed flap while the other side involute; upper glume lanceolate acute; lower lemma oblong-ovate, ciliate; upper lemma very narrow.

Fl.-Fr. : -Sept. -Dec.

Khandwa only, Mortakka, sandy river bed of Narmada.

Heteropogon polystachyus (Roxb.) Schult., Mant. 2: 460.1824; *Andropogon polystachyos* Roxb., Fl. Ind. 1: 265. 1820; Hook.f., Fl. Brit. India 202. 1896; Bor, Grass. Burm. Ceyl. Ind. Pak. 165. 1960;

Annuals; culms erect, tufted, 15-25 cm tall; leaf sheath terete, glabrous or hairy; ligule a short ciliate membrane; leaf blade 5-10 cm, hairy on both sides; raceme solitary, with lower homogamous spikelets in pairs; hermaphrodite sessile spikelet awned, callus sharp, bearded with dark brown hairs; lower glume coriaceous, linear to narrowly oblong, covered with brown hairs on dorsal surface, margins inflexed, 5-nerved, obtuse or rounded at apex; upper glume membranous, narrowly obovate, hairy on the middle nerve, 3-nerved, apex apiculate; lower lemma hyaline, ovate, nerveless, apex acute; palea absent; upper lemma represented by the acute base of the geniculate awn, 4-5 cm long, hairy; pedicelled spikelets 5.5-6 mm; lower glume membranous, narrowly elliptic-ovate, glabrous, 1-keeled, keel winged, 8-10-nerved, apex acute; upper glume membranous, narrowly ovate, glabrous, margins inflexed, ciliate, 3-nerved, apex acuminate; lower lemma membranous, narrowly elliptic-ovate, ciliate on margins, apex rounded or obtuse; stamens 3; caryopsis narrowly obovate-oblong.

Fl.-Fr. : -July-Nov.

Rare, in open grassland at wet places at Burhanpur , Khandwa. Endemic to India.

Heteropogon ritchiei (Hook.f.) Blatt. & McCann., J. Bombay Nat. Hist. Soc. 32: 623. 1928 & Bombay Grass. 108. 1935; *Andropogon ritchiei* Hook.f., Fl. Brit. India 7:201. 1896; Sharma *et al.,* Fl. Maharashtra 3:504. 1996; Bor, Grass. Burm. Ceyl. Ind. Pak. 165. 1960. *Pochati*

Culms 90-120 cm high; leaf sheaths glabrous; blades 15-45 cm linear, acuminate, hairy on both surfaces, margins serrulate; ligule brown, membranous, divided into numerous narrow acute segments; racemes 2.5-5.0 cm long, hispid, lower 1-2 pairs of spikelets homogamous; upper sessile spikelets 4.5-6.0 mm long, oblong; lower glume bluntly acuminate-truncate, hairy on the back; upper glume linear oblong, 3-nerved; lower lemma hyaline, nerveless; awn 3-4 cm long; pedicelled spikelets 10-13 mm long; lower glume lanceolate, keeled, keels with a row of long, tubercle based golden-yellow bristles; upper glume lanceolate; lower lemma oblanceolate, hyaline; upper lemma narrowly linear, acute; stamens 3 mm long.

*Fl.-Fr.:*Sept. -Dec.

Slopes of hills and river banks, Burhanpur, Khandwa. Endemic to India.

Heteropogon triticeus (R.Br.) Stapf ex. Craib, Bull. Misc. Inform. Kew 1912: 432. 1912; *Andropogon triticeus* R. Br., Prodr. Fl. Nov. Holl. 201. 1810; Hook.f., Fl. Brit. India 7: 200.1896; Bor, Grass. Burm. Ceyl. Ind. Pak. 1965. 1960; Samvatsar, Fl. West. Trib. Madhya Pradesh 363. 1996; Roy in Singh *et al.,* Fl. Madhya Pradesh 3: 448. 2001. *Sukal*

Perennials; culms tufted, terete, 100-300 cm tall, erect, simple, glabrous, nodes glabrous; ligule membranous; leaf blade flat, linear-ovate, 5-80 × 1-2 cm, glabrous, or sparsely pilose, margins scaberulous, apex acuminate; raceme solitary, 10-20 cm long, with lower homogamous awnless and upper awned heterogamous spikelets, joints very short; hermaphrodite spikelets narrowly ovate-elliptic, awned, callus pungent, bearded with rufous hairs; lower glume coriaceous, narrowly ovate, 7-9 nerved, glabrous, apex truncate; upper glume coriaceous, narrowly ovate-oblong, 3-5 nerved, margins ciliate, apex obtuse; lower lemma hyaline, narrowly ovate-oblong, nerveless, apex acuminate; palea hyaline, ovate-oblong, 2-2.6 mm, apex ciliate hairy; pedicels slender, glabrous; pedicelled spikelets narrowly ovate, 20-26 mm, awnless; lower glume chartaceous, narrowly ovate, glabrous, 2-keeled, keel winged on one margin 20-25 nerved; stamens 3; caryopsis cylindric.

*Fl.-Fr.:*Aug. -Jan.

Banks of streams and river bed, Alirajpur, Dhar, Khandwa.

HYGRORYZA Nees Monotypic

Hygroryza aristata (Retz.) Nees ex Wight & Arn. Edinburg New Philos. J. 15: 380.1833; *Pharus aristatus* Retz., Observ. Bot. 5: 23. 1788; Hook.f., Brit. India 7: 95. 1896; Bor, Grass. Burm. Ceyl. Ind. Pak. 597. 1960; Samvatsar, Fl. West. Trib. Madhya Pradesh 363 . 1996; Roy in N.P. Singh *et al.,* Fl. Madhya Pradesh 3: 450.2001. **(Plate 10, Fig. 7)**

Aquatic, floating grass with fibrous roots; leaves 2.0-3.5 × 0.8-1.7 cm, ovate to ovate-lanceolate, adaxial surface, papillate; ligule 0.8 mm long, membranous, truncate; inflorescence a globose panicle; spikelets 1-flowered, 7- 18 mm long, including awn, glumes reduced; lemma 7 mm long, chartaceous, hispid, apex acute, 3-veined; stamens 6, anthers 3.5 long, yellowish; caryopsis 4.5 mm long, cylindrical.

*Fl.-Fr.:*Oct. -Dec.

Occasional in ponds in the vicinity of villages, Dhar, Indore, Khargone.

IMPERATA Cirillo W. 11; MP 1; SA 1

Imperata cylindrica (L.) Raeusch., Nomencl. Bot. ed 3:10.1797; *Lagurus cylindricus* L., Syst. Nat. ed. 10 2:878. 1759; Bor, Grass. Burm. Ceyl. Ind. Pak. 170.1960; Samvatsar, Fl. West. Trib. Madhya Pradesh 364. 1996; Roy in N.P. Singh *et al.*, Fl. Madhya Pradesh 3: 451.2001. *Pankunda*

Erect perennial grass, 15-76 cm long, rhizome deep-seated, creeping, giving rise to culms, solid above and fistular at the base; leaves linear, flat, 7.5 - 96 mm × 5-8 mm, acuminate, much scabrid, midrib white, sheath longer than internodes, glabrous; ligule membranous, ciliate, cleft; panicles dense, cylindric, silvery-white, soft silky, 5-18 cm long; spikelets linear-lanceolate 3 mm long; pedicels up to 2 mm long, callus hairs 1.5 cm long, soft, white; lower glume-lanceolate, membranous, acuminate; lower lemma empty, nerveless, oblong, up to 2 mm long, apex ciliate, truncate; upper lemma hermaphrodite, ovate-lanceolate, often glabrous; palea linear; stamens 2, anther orange, stigmas 2, purple, caryopsis ellipsoid.

*Fl.-Fr.:*July-Dec.

Common in open grassland, along watercourses, in all districts.

ISACHNE R. Br.W. 91; MP 1; SA 1

Isachne globosa (Thunb.) Kuntze, Revis. Gen. Pl. 2: 778. 1891; *Milium globosum* Thunb., Fl. Jap. 49. 1784; *Isachne australis* R. Br., Prodr. Fl. Nov. Holl. 196. 1810; Hook.f., Brit. India 7: 24. 1896; Bor, Grass. Burm. Ceyl. Ind. Pak. 1965. 1960; Roy in N.P. Singh *et al.*, Fl. Madhya Pradesh 3:451.2001.

Annuals; culms 20-60 cm tall, erect or ascending; leave 5-10 × 0.4-0.5 cm, linear -lanceolate or ovate; ligule a dense tuft of hairs, 2 mm long; inflorescence a contracted panicle, 4-10 × 3-5 cm, branches capillary, flexuous, scabrid; spikelets 2-flowered, 2 mm long, nearly globose; lower glumes light green, ovate-obtuse, 5-nerved with a few hairs on the back; upper glume broader, 9-nerved, scabrid on the back; lower florets male, 1.8 mm long, glabrous; lemma flat with inflexed edges; palea membranous, similar to lemma; stamens 3, anthers 1 mm long, dark brown; upper florets hermaphrodite or female, 1.5 mm long; lemma thick, edge narrowly inflexed; palea flat of same texture; caryopsis deep pink, *ca* 1 mm long.

Fl. -Frt. : Aug. -Sept.

Rare, in moist places and bunds of fields, **Khandwa** only, at Chandgarh range.

ISCHAEMUM L. W. 81; MP 4; SA 8

Key to Species

1. Annuals
   2. Racemes 1 or 2; culms 80-120 cm; sessile spikelets 4-5 mm long; upper lemma awn 7-20 mm long .... **I. molle**
   2. Racemes 2
      3. Culms 15-60 cm; sessile spikelets 2.5-3 mm long; upper lemma awn 6-16 mm long .... **I. timorense**

3. Culms 10-30 cm; sessile spikelets 6 mm long; upper lemma awn 14 mm long ....**I. tumidum**

1. Annuals or perennials

4. Racemes 2-3; culms 40-150 cm; sessile spikelets 6.5-7.5 mm long; upper lemma awn 7-8 mm long ....**I. afrum**

4. Racemes 2

5. Upper lemma awn absent . ...**I. travancorense**

5. Upper lemma awn developed

6. Sessile spikelets 4-5 mm long; upper lemma awn 5-30 mm long ....**I. thomsonianum**

6. Sessile spikelets 5-7 mm long

7. Culms 40-100 cm; sessile spikelets 5-6 mm long; upper lemma awn 12-15 mm long ....**I. rugosum**

7. Culms 60-120 cm; sessile spikelets 5.5-7 mm long; upper lemma awn 10-15 mm long ....**I. barbatum**

Ischaemum afrum (J.F. Gmelin) Dandy, Pl. Sudan 3: 476.1956; *Andropogon afer* J.F. Gmelin. Syst. Nat. 166. 1791; *Andropogon pilosus* Klein ex Willd., Sp. Pl. 4:920.1806; *Ischaemum pilosum* (Willd.) Wight, Madras J. Lit. Sci. J. 3:138.1835; Hook.f., Fl. Brit. India 7: 130.1896; Bor, Grass. Burma. Ceyl. Ind. Pak. 183.1960; Samvatsar, Fl. West. Trib. Madhya Pradesh 367.1996; Roy in N.P. Singh *et al.,* Fl. Madhya Pradesh 3:454.2001. *Kunda* **(Plate 10, Fig. 8)**

Perennial, decumbent herbs; culms 60-90 cm tall, terete, glabrous; leaves linear, finely acuminate, glabrous; sheaths glabrous; ligule membranous, glabrous; racemes sub-digitate, 2-6, fascicled, 2.5-12 cm long, brown, pilose, compressed, sparsely ciliate; sessile spikelets narrowly lanceolate, *ca* 6 mm long; callus bearded with hairs; lower glume narrow, chartaceous, dorsally hairy, margins incurved, ovate-lanceolate, longer than lower glume, laterally compressed, 5-nerved; upper glume 5-9 nerved, pilose on keel; lower lemma hyaline,elliptic-lanceolate, 2-keeled; palea similar; upper lemma hyaline, 3-nerved, cleft at apex, awn geniculate,7-8 mm long; caryopsis oblong; pedicel of the pedicelled spikelet 3-4 mm long, densely hairy; pedicelled spikelets oblong-lanceolate, hairy, awned, more or less similar to sessile ones.

*Fl.-Fr.:*Sept. -Feb.

Occasional in marshy places, Alirajpur, Burhanpur, Dhar, Indore, Khandwa.

Ischaemum barbatum. Retz., Observ. Bot. 6:35.1791; *Ischaemum mangaluricum* (Hack.) Stapf ex C.E.C. Fisch., Fl. Madras 3: 1193.1956 (BSI reprint); Bor, Grass Burma. Ceyl. Ind. Pak. 182.1960

Perennials; culms up to 80 cm tall, densely tufted, rarely rooting at lower nodes; leaves linear-lanceolate, 3-20 × 0.45-0.7 cm, narrowed at base; ligule ovate, acute; racemes 2, 3-10 cm long, joints clavate, ciliate on one angle; sessile spikelets oblong, 4-6 mm long, awned; callus beared; lower glume crustaceous below, usually with 4-6 side nodules; upper glume boat-shaped, chartaceous, dorsally keeled, keel narrowly winged at apex; lower lemma 3-5-nerved; lower palea oblong-lanceolate; upper lemma notched, awned; awn 1-1.5 cm long; upper palea oblong, with wart- like dots in upper half; lower floret male; upper floret bisexual; stamens 3; ovary ovate-oblong; pedicelled spikelets ovate-oblong, pedicels, densely ciliate; lower glume ovate-oblong, broadly winged on one margin; upper glume, lemma and palea similar to sessile spikelets but smaller; floret male; upper lemma awnless.

*Fl.-Fr.:*Oct. -Feb.

Weed in agricultural fields and other moist places, Burhanpur, Khandwa, Khargone.

Ischmaemum molle Hook.f., Fl. Brit. India 7:128. 1896; Bor, Grass. Burm. Ceyl. Ind. Pak. 183. 1960; Samvatsar, Fl. West. Trib. Madhys Pradesh 366. 1996.

Annuals; culms 80-120 cm long; leaves linear; ligule pubescent or sometimes glabrous; racemes 1-2, axillary and terminal, densely villous with soft, white hairs; spikelets 4-5 mm long; sessile spikelets: lower glume thinly coriaceous, broadly ovate, dorsally villous, many nerved; upper glume chartaceous, keel with narrow ciliolate wing from the middle upwards; lower lemma ovate, 3-nerved, paleate; upper lemma shorter than the lower, cleft in the middle into 2 lanceolate, acuminate ciliolate lobes; palea small, narrow; awn nearly 2 cm long; pedicelled spikelets smaller than the sessile ones.

*Fl.-Fr.:*Sept. -Dec.

Occasional in marshy places, along water courses, Alirajpur, Dhar, Indore.

Ischaemum rugosum Salisb., Icon, Strip. Rar. 1: t. 1 1791; Hook.f., Fl. Brit. India 7: 127. 1896; Bor, Grass. Burm. Ceyl. Ind. Pak. 184. 1960; Samvatsar, Fl. West. Trib. Madhya Pradesh 366 . 1996; Roy in N.P. Singh *et al.,* Fl. Madhya Pradesh 3: 454.2001. *Todi jara* **(Plate 11, Fig. 1)**

Annuals or perennials; clums 50–60 cm tall, erect or ascending; leaves 10-20 cm long, linear-lanceolate, acuminate, hairy on the margins; ligule membranous, truncate; inflorescence of 2, yellowish racemes, up to 10 cm long, joints and pedicel trigonous, ciliate on the dorsal angle; sessile spikelets 5.5–6.0 mm long, ovate-oblong, obtuse, callus bearded; lower glume chartaceous in the lower two thirds, membranous in the upper one; upper glume 5.5 mm long, obtuse or shortly bidentate with a winged, ciliate keel; lower lemma lanceolate, acute; lower floret usually male; upper lemma 4 mm long, oblong, 2-lobed, awn geniculate; palea 3 mm long, linear, acute; stamens 3, anther 2 mm long; caryopsis, ovoid, brown *ca* 2mm long; pedicellate spikelets 4. 1 mm long; lower glume winged along one side; upper glume long without a winged keel; lower lemma hyaline, 3-nerved, margins folded; upper lemma ovate, obtuse; palea hyaline.

*Fl.-Fr.:*Aug. -Jan.

Common in agricultural fields and other moist places in all districts.

Ischaemum thomsonianum Stapf ex C.E. C. Fisch., Fl. Pres. Madras 3:1193. 1956 (BSI reprint); *I. murimum* Hook., f., Fl. Brit. India 7: 135. 1896 non Forst. 1786; Bor, Grass. Burma. Ceyl. Ind. Pak. 185. 1960.

Annuals or perennials; culms 20-100 cm long, often creeping or rooting at nodes; leaves lanceolate, 2-5 × 0.5-1.5 cm, rounded at base, acuminate, densely or sparsely hairy; ligule membranous; racemes 2 rarely 3; joints turbinate, densely long ciliate on one angle with a tooth-like projection on inner side; sessile spikelets elliptic-lanceolate, awned; lower glume ovate-lanceolate, 11-15 nerved, upper glume boat-shaped acuminate, shortly bifid, awned, 5-nerved, dorsally keeled, obscurely winged at apex; lower lemma elliptic, acute, ciliolate towards apex;

lower palea oblong-lanceolate, 2-nerved; upper lemma notched, awned; lower floret male; upper bisexual; stamens 3; ovary oblong; pedicelled spikelets oblong-lanceolate; lower glume ovate-lanceolate, acuminate, aristate, 11-15 nerved, dorsally keeled; upper glume and florets as in sessile spikelets.

*Fl.-Fr.:*Dec. -March

Along muddy fields, moist places roadsides and wastelands, all over the area. Endemic to India.

Ischaemum timorense Kunth, Revis. Gramin. 1: 369. 1830: Hook.f., Fl. Brit. India 7:136. 1896: Bor, Grass. Burm. Ceyl. Ind. Pak. 185. 1960; Sharma *et al.,* Fl. Maharashtra 3:527. 1996.

Perennials; culms 60-120 cm tall, creeping, stoloniferous, rooting at nodes; leaves linear-lanceolate, 1-2 cm, cordate, acuminate; racemes solitary or in pairs, 1-4 cm long, hairy; joints 2-3 mm long, ciliate along one angle; sessile spikelets ovate-lanceolate, 3-4 mm long, shortly bicuspidate, 7-9-nerved; upper glume lanceolate, shortly bifid at apex, aristate in between; lower lemma oblong-lanceolate, 3 mm long, 3-nerved; lower palea lanceolate, 2-nerved; upper lemma notched, awned; awn 1 cm long; lower floret male; upper bisexual; stamens 3, anthers 1 mm long; caryopsis oblong; pedicelled spikelets oblong-lanceolate, 3-4 mm long, pedicels long ciliate on one angle; lower glume ovate-lanceolate, acuminate, shortly aristate, 5-7-nerved; upper glume and floret as in sessile spikelets.

*Fl.-Fr.:*Dec. -March

Along muddy fields, Barwani, Khargone.

Ischmaemum travancorense Stapf ex C.E. C. Fisch., Bull. Misc. Inform. Kew 1933: 353. 1933; *I. aristatum* L. subsp. *rottleri* Hook.f., Fl. Brit. India 7: 127. 1896; Bor, Grass. Burm. Ceyl. Ind. Pak. 186. 1960; Sharma *et al.,* Fl. Maharashtra 3:530.1996.

Perennial straggling herbs; up to 2 m tall, rooting at lower nodes; leaves lanceolate, 5-30 × 0.5-1.5 cm, cordate, acuminate; racemes 2, adpressed together, appears solitary; sessile spikelets oblong-lanceolate, awnless; lower glume oblong with few side nodules connected by shallow ridges, 13-15-nerved, minutely winged at apex; upper glume boat-shaped, 5-7-nerved, dorsally keeled; lower lemma oblong-lanceolate, 3-nerved, margins infolded; lower palea 2-nerved; upper lemma notched with a rudimentary awn in between; lower floret male; upper bisexual; stamens 3; caryopsis oblong; pedicelled spikelets ovate-oblong, awnless; lower glume ovate-oblong, winged on one margin; upper glume and florets as in sessile spikelets, but smaller.

*Fl.-Fr.:*Sept. -March

Along streams, muddy fields and ponds, Barwani, Khandwa. Endemic to India.

Ischaemum tumidum Stapf ex Bor, Kew Bull. 1951: 450.1952 & Grass. Burm. Ceyl. India. Pak. 186. 1960; Sharma *et al.,* Fl. Maharashtra 3:530. 1996.

Annuals; culms 15-30 cm tall, creeping or geniculate; upper nodes bearded; leaves ovate-lanceolate, up to 4 cm long, hairy on both sides; ligule truncate, membranous; racemes clavate, 2-4 cm long, swollen, shining; sessile spikelets oblong, awned; lower glume coriaceous with 2-3 nodules on rounded keels, obscurely transversely ridged; upper glume elliptic, boat-shaped; lower lemma oblong-lanceolate, delicate, 3-nerved, ciliate in upper half; lower palea 3-nerved; upper lemma deeply notched, awned, lobes acute; lower floret male; upper bisexual; stamens 3; pedicelled spikelets lanceolate, awnless, hairy; pedicels clavate, hairy; lower glume oblong-lanceolate, 9-11-nerved, hairy; glume and florets as in sessile spikelets.

*Fl.-Fr.:*Oct. -Dec.

Along dry, open and grassy hill slopes, Burhanpur, Khandwa. Endemic to India.

ISEILEMA Anderss. W. 25; MP 4; SA 4

Key to Species

1. Perennials; spatheoles with tubercled hairs on midrib; involucral spikelets 2.5-4.5 mm long; fertile spikelets 2.5-3 mm long; apex of lower glume 2-fid; upper glume 3-nerved; awn of upper lemma geniculate, 5-12 mm long . ... **I. prostratum**

1. Annuals
   2. Spatheoles glabrous on keels and margins .... **I. laxum**
   2. Spatheoles granulate or with tubercle based hairs on keels or margins
      3. Involucral spikelets 2-3 mm long; fertile spikelets 4.5-5 mm long; upper lemma awn 17-18 mm long . ... **I. holei**
      3. Involucral spikelets 6.5-7 mm long; fertile spikelets 5-6.5 mm long; upper lemma awn 12-15 mm long .... **I. anthephoroides**

Iseilema anthephoroides Hack., Monogr. Phan. 6: 683. 1889; Hook., f., Fl. Brit. India 7: 219. 1896; Bor, Grass. Burm. Ceyl. Ind. Pak. 187. 1960; Samvatsar, Fl. West. Trib. Madhya Pradesh 369. 1996; Roy in N.P. Singh *et al.,* Fl. Madhya Pradesh 3: 455. 2001. *Musel*

Annuals; culms 20-100 cm tall, tufted; leaves flat, glabrous of sparsely hairy, with tubercle based hairs on margins; sheaths compressed, glabrous; ligule lacerate; inflorescence a narrow panicle; spathes with tubercle-based hairs; involucral spikelet 6.5-7 mm long, lanceolate, 3-5 nerved; lemma nil; pedicelled spikelet reduced to a long pedicel; hermaphrodite spikelet 6-6.5 mm long, lanceolate, awned; lower glume 6-6.5 mm long, bifid; upper glume similar but acuminate; lower lemma empty; upper lemma reduced to 12-14 mm long awn; stamens 3, yellow, *ca* 2.5 mm long; caryopsis oblong-ellipsoid, brown.

*Fl.-Fr.:*Aug.-Dec.

Common in open grassland in all districts. Endemic to India.

Iseilema holei Haines., Bot. Bihar Orissa 3:1103. 1961 (BSI Reprint); Bor, Grass. Burm. Ceyl. Ind. Pak. 188. 1960; Roy in N.P. Singh *et al.,* Fl. Madhya Pradesh 3: 456.2001.

Annuals; culms 60-100 cm tall, tufted; leaves linear, glabrous; sheath glabrous; ligule membranous, hyaline; inflorescence of racemes enclosed in spathes; involucral spikelets 5 mm long, 5-nerved; upper glume similar, 3-nerved; anthers 3, yellow, 2 mm long; hermaphrodite spikelets 4.5-5 mm long, glabrous, awned; lower glume 4-4.75 mm long, membranous, bifid; upper glume similar, acute; lower lemma absent; upper lemma reduced to 17-18 mm long awn; palea nerveless; anthers 3, yellow, 2-2.5 mm long; caryopsis ellipsoid.

*Fl.-Fr.:*Oct. - Jan.

Moist places, Indore, Khandwa. Endemic to India.

Iseilema laxum Hack., Monogr. Phan. 6: 682. 1889; Hook.f., Fl. Brit. India 7: 218. 1896; Bor, Grass. Burm. Ceyl. Ind. Pak. 188. 1960; Samvatsar, Fl. West. Trib. Madhya Pradesh 369. 1996; Roy in N.P. Singh *et al.,* Fl. Madhya Pradesh 3: 456.2001. *Mushan*

Perennials; culms 70-90 cm tall, erect, densely tufted; leaves 15-30 cm long, setaceous, rigid, scabrid, sheath glabrous; ligule hairy; inflorescence a solitary raceme, 7-11.5 cm long; spikelets 8.4-9.2 mm long, linear-lanceolate, callus shortly bearded; lower glume 8.4-9.2 mm long, chartaceous, bidentate deeply grooved, keeled, winged; upper glume 4-5 mm boat-shaped, awn 1.5 cm long; upper lemma equal to lower, bisexual, 2-lobed, awn 3.5 cm long; palea ciliate; pedicelled spikelet 9.5-11.0 mm long, glabrous; anthers 3, yellow, 2.5-3.0 mm long, glabrous; caryopsis 3.5 mm long, boat-shaped.

Fl.-Fr.: Aug. -Nov.

Common in moist places, in open grass lands, throughout the area.

Iseilema prostratum (L.) Andersson, Nova Acta Regiae Soc.Sci. Upsal. 3, 2: 251.1856; *Andropogon prostratum* L., Mant. Pl. 2:304.1771; *Iseilema wightii* (Steud.) Andersson, *op cit*; Hook., f., Fl. Brit. India 7: 218. 1896; Bor, Grass. Burm. Ceyl. Ind. Pak. 188. 1960; Samvatsar, Fl. West. Trib. Madhya Pradesh 369. 1996; Roy in N.P. Singh *et al.,* Fl. Madhya Pradesh 3: 457.2001. *Ghania*

Perennials; culms 30-80 cm tall, erect or decumbent; leaves 8-18 × 0.5-0.6 cm, linear, acute; ligule 1 mm long, oblong-lanceolate; lower glume 6 mm long, subacute with five ribs on the back; upper glume 1-nerved, glabrous with broadly inflexed margins; lower lemma absent; upper lemma male; stamens 3, anthers 1.5-2.0 mm long; pedicelled spikelets 3.0-3.5 mm long; sessile spikelets 2.5-3.0 mm long, lanceolate; lower glume 2.5 mm long, acute, 4-nerved; upper glume elliptic-lanceolate, acute or obtuse and mucronate, keel scabrid; lower lemma 2.5 mm long, linear hyaline; upper lemma hermaphrodite; palea absent; stamens 3, anthers *ca* 2 mm long, dark brown.

Fl.-Fr. : -Sept. -Dec.

Along the streams and river banks, in all districts.

LEERSIA Sw. W. 18; 1; SA 1

Leersia hexandra Sw., Prodr. Veg. Ind. Occ. 21. 1788; Hook.f., Fl. Brit. Ind. 7 : 94. 1896; Bor, Grass. Burm. Ceyl. Ind. Pak. 599. 1960; Samvatsar, Fl. West. Trib. Madhya Pradesh 370.1996; Roy in N.P. Singh *et al.,* Fl. Madhya Pradesh 3:458. 2001.

Perennial, aquatic grasses, culms 50-120 cm. long, forming dense masses around the swamps or lakes in shallow water, nodes hairy; leaves flat, rigid, glaucuous, serrulate along margins; sheath smooth or scaberulous; ligule membranous; inflorescence a terminal, contracted panicle, 5-10 cm long; branches 2-3cm long; pedicels very short; spikelets 3-4 mm long, closely imbricate, oblong, compressed; glumes reduced to a rim; lemma 3-4 mm long, cartilaginous, 5 nerved; lateral nerves forming thick border, compressed; keel rigidly ciliate; palea shorter, dimidiate, linear-oblong, 3-nerved; keels ciliate; palea linear-lanceolate; anthers 6, equalling the palea; caryopsis almost equal to lemma, glabrous, hardly enclosed by the lemma and palea.

*Fl.-Fr.:*Sept. -Dec.

In shallow water, Burhanpur, Dhar, Indore, Jhabua, Khargone.

LEPTOCHLOA P. Beauv. W. 36; MP 2; SA 2

Key to Species

1. Spikelets 2-6 mm long, 3-7 flowered .... **L. chinensis**
1. Spikelets 1.5-2 mm long, 2-flowered .... **L. panicea**

Leptochloa chinensis (L.) Nees, Syll. Pl. Nov. 1: 4. 1824; *Poa chinensis* L., Sp. Pl. 69. 1753; Hook.f., Fl. Brit. India 7: 299. 1896; Bor, Grass. Burm. Ceyl. Ind. Pak. 516. 1960; Roy in N.P. Singh *et al.,* Fl. Madhya Pradesh 3: 459.2001.

Annuals; culms 40-90 cm tall, tufted, erect or geniculate, nodes glabrous; leaves linear, glabrous; sheaths compressed, glabrou; ligule lacerate, short; inflorescence a loose panicle, branches alternate; rachis stiff, scaberulose; spikelets 4-6 mm long, 3-7 flowered; lower glume 1.25-1.5 mm long, membranous, 1-nerved, keel scaberulous; upper glume 1.4-1.6 m long, 1-nerved, keel scaberulous; lemma 1.8-2 mm long, 3-nerved, pubescent or glabrous; palea keels ciliate; anthers 3, 0.3-0.4 mm long; caryopsis *ca* 1 mm long, obtusely trigonous, brown.

*Fl.-Fr.:*Sept. -Feb.

Along moist places, Barwani, Indore, Khargone.

Leptochloa panicea (Retz.) Ohwi, Bot. Mag. Tokyo 55:311. 1941; *Poa panicea* Retz. Observ. Bot. 3:11. 1783; Hook.f., Fl. Brit. India 7:298. 1896; Samvatsar, Fl. West. Trib. Madhya Pradesh 370.1996; Roy in N.P. Singh *et al.,* Fl. Madhya Pradesh 3:459.2001.

Annuals; culms 25-60 cm tall, slender, geniculate-ascending; leaves 10 – 25 × 0.5-0.6 cm, linear-lanceolate with tubercled hairs; ligule 1.0-1.5 cm long, lacerate; sheath compressed hairy or glabrous at the top; inflorescence a terminal panicle, elliptic to oblong, often purplish, 10-30 cm long, composed of numerous slender, ascending racemes, 4-11 cm long; spikelets usually 3-flowered, narrowly elliptic, 1.9-2.5 mm long, overlapping; lower glume 0.7-1.5 mm long, lanceolate-acute, chartaceous; upper glume 0.9-1.6 mm long, narrowly oblong, obtuse and mucronate; lemmas 0.8-1.2 mm long, elliptic-oblong, 3-veined, margin slightly inrolled; palea 1 mm long, elliptic chartaceous, 2-keeled; caryopsis 0.5 mm long, broadly elliptic, trigonous in cross section.

*Fl.-Fr.:*Aug. -Dec.

Common in wet and shady places, throughout the area.

LOPHOPOGON Hack. W. 2; MP 1; SA 1

Lophopogon tridentatus (Roxb.) Hack., Monogr. Phan. 6:254. 1889; *Andropogon tridentatus* Roxb. Fl. Ind. 1:261. 1820; Hook.f., Fl. Brit. India 7: 149. 1896; Bor, Grass. Burm . Ceyl. Ind. Pak. 190.1960; Samvatsar, Fl. West. Trib. Madhya Pradesh 370.1996.

Culms to 40 cm, erect; nodes glabrous; leaf-sheaths fibrous at base; ligule stiff, membranous, ciliate, to 1.5 mm; leaf blades linear, 2-6 × 0.15 cm, flat or involute; raceme *ca* 2.5 cm, solitary; branches *ca* 4, closely appressed, enclosed within ventricose sheaths or exserted;

joints cupular at tip; sessile spikelet: lower glume oblong, to 4 mm, 5-7-nerved, with tufts of brownish hairs about the middle, and long bristly hairs in between tufts, margins ciliate, apex truncate with 2 long lateral teeth and 1 or 2 shorter in between; upper glume to 4.5 mm, tufted-pilose outside, 1-keeled, 2-dentate, awn to 6.5 mm; lower lemma to 4 mm, 1-nerved, 1-keeled; palea to 4 mm; stamens 3; upper lemma to 5 mm, base hyaline, apex coriaceous, 2-lobed, lobes ciliate, awn to 1.5 cm; stamens 2 or 3; caryopsis elliptic, to 0.8 mm; pedicelled spikelet similar to sessile, to 6 mm.

*Fl.-Fr.:*Aug. -Nov.

Common in open grass land in rocky habitat, in all districts. Endemic to India.

MELANOCENCHRIS Nees W. 3; MP 1; SA 1

Melanocenchris jacquemontii Jaub. & Spach., Ill. Pl. Orient. 4: 36. 1851; Bor, Grass. Burm. Ceyl. Ind. Pak. 473. 1960; Samvatsar, Fl. West. Trib. Madhya Pradesh 370 . 1996; Roy in N.P. Singh *et al.,* Fl. Madhya Pradesh 3: 462.2001. *Dongri* **(Plate 11, Fig. 2)**

Annuals; culms slender, 7.5-30.0 cm tall, densely tufted; leaves 2.5-5.0 × 0.15 cm, linear-lanceolate, acutely pointed, glabrous, margin ciliate; ligule a fringe of hairs; spikes filliform, 1.3-5.0 cm long, spikelets 3-5 in a cluster; spikelets 6.0 – 8.5 mm long (including awns), green or purple; involucral glumes 4; outer glumes 7.0-7.25 mm long, hairy below, awn retrorsely barbed; inner glume 6.5-7.0 mm long, more wide than lower glume; lower lemma bisexual, chartaceous, ovate- oblong, shortly awned, 3-nerved; palea as long as the glume, 2-toothed, hermaphrodite; inner floral glumes stalked, distinctly 3-toothed at apex, scabrid at back, paleate, male; stamens 3; rachilla ending 2-toothed, produced beyond the palea; caryopis 1.5–2.0 mm long, oblong, polished slaty- brown.

*Fl.-Fr.:*Sept. -Nov.

Common in grassland, hill slopes, in all districts.

MELINIS P. Beauv. W. 22; MP 2 subsp; SA 2 subsp.

Key to Subsp.

1. Ligules hairy; pedicels of the spikelets with a few long hairs; spikelets 4-4.5 mm long . ... subsp. **repens**
1. Ligules obsolete; pedicels of the spikelets without long hairs; spikelets 5-12 mm long .... subsp. **grandiflora**

Melinis repens (Willd.) Zizka, subsp. **repens** Biblioth. Bot. 38: 55.1988; *Saccharum repens* Willd. Sp. Pl. 1: 322.1797; *Rhynchelytrum repens* (Willd.) Hubb., Kew Bull. 1934:110. 1934; Bor, Grass. Burm. Ceyl. Ind. Pak. 355. 1960; Roy in N.P. Singh *et al.,* Fl. Madhya Pradesh 3: 502.2001.

Annuals or perennials; culms 20-50 cm tall, rooting at lower nodes; leaves flat, linear, glabrous or finely pubescent; sheath glabrous or pilose, long hairy at mouth; ligule hairy; inflorescence a lax panicle, white or pink, branches capillary; spikelets solitary or paired, 4.0-4.5 mm long, densely hairy, appessed when young; lower glume 1.0-1.3 mm long, hairy at base, 1-

nerved; upper glume equalling the spikelet, 5-nerved, hairy, longer than glumes, tip bifid, mucronate to awned from sinus; lower lemma empty; palea ciliate on keels; upper lemma *ca* 2.5 mm long, 3-nerved; anthers 3, *ca* 2 mm long; caryopsis 2.5 mm long, elliptic, brownish.

*Fl.-Fr.:*Sept. -Dec.

Rare, on dry, rocky hill slopes, Burhanpur, Khandwa.

Melinis repens subsp. **grandiflora** (Hochst.) Zizka, Biblioth. Bot. 38:60.1988; *Rhynchelytrum grandiflorum* Hochst., Flora 27:249. 1844; *Monachyron villosum* Parl., Niger Fl. 191. 1849; Bor, Grass. Burm. Ceyl. Ind. Pak. 355. 1960; Roy in N.P. Singh *et al.,* Fl. Madhya Pradesh 503.2001.

Annuals or short lived perennials; culms tufted,40-80 cm tall; leaves linear, flat; sheaths glabrous or hairy; ligule obsolete; inflorescence a narrow panicle; branches and pedicels capillary; spikelets variable, 5-12 cm long, callus with long hairs; lower glume *ca* 3 mm long, concave, spreading; upper glume and lower lemma shortly 2-lobed, awned or awnless; upper glume gibbous, tapering to a narrow, glabrous beak; palea ciliate on keels; stamens 3, anthers purplish, *ca* 2 mm long.

Fl.-Fr.: Oct. -Feb.

Rare, Khandwa, Khargone.

MICHROCHLOA R. Br.W. 6; MP 1; SA 1

Microchloa indica (L.f.) P. Beauv. Ess. Agrostgr. Atlas 13. 112. 1812; *Nardus indica* L.f. Suppl. Pl. 105. 1782; Hook.f., Fl. Brit. India 7: 283. 1896; Bor, Grass. Burm. Ceyl. Ind. Pak. 473. 1960; Samvatsar, Fl. West Tribal Madhya Pradesh 371. 1996.

Culms tufted, to 10 cm, slender, nodes glabrous; leaf-sheaths 1-3.5 cm; ligule shortly membranous, 0.2 mm; blades linear, to 2 cm, involute, apex acuminate; spike solitary, to 8 cm, slender, curved; spikelets lanceolate, to 2 mm, sessile, 1-flowered; rachilla disarticulating above the glumes, not produced beyond lemma; glumes lanceolate-oblong, subequal, 1-nerved, keeled, cuspidate or acuminate; lower glume to 3 mm; upper glume to 2.5 mm; lemma ovate, to 1.5 mm, 3-nerved, 2-keeked; palea to 1.5 mm, 2- lobed, 2-keeled; anthers 3; caryopsis oblong-ellipsoid, to 1.2 mm.

*Fl.-Fr.:*Aug. -Nov.

Rocky hill slopes and open grass lands, all over the area, missed because of its small size.

MNESITHEA Kunth W. 2; MP 1; SA 1

Mnesithea laevis (Retz.) Kunth, Revis. Gramin. 1: 154. 1829; *Rottboellia laevis* Retz., Observ. Bot. 3: 11. 1783; *R. perforata* Roxb., Pl. Coromandel 2:43. 1798; Hook.f., Fl. Brit. India 7:158. 1896; Bor, Grass. Burm. Ceyl. Ind. Pak. 197. 1960; Roy in N.P. Singh *et al.,* Fl. Madhya Pradesh 3: 465. 2001. *Bari simpi*

Perennials; culms 30–160 cm tall, erect, tufted; leaves 15 – 30 cm long, linear, glabrous, except hairy base; ligule short; inflorescence a raceme, 8 – 20 cm long; sessile spikelets glabrous, 3.5–4.5 mm long; lower glume 3.5 mm long, asymmetric, obtuse or slightly emarginate, smooth or with vertical rows of minute depressions, faintly 4–5-nerved between the marginal nerves; upper glume 4 mm long, 3-nerved; lower lemma with faint marginal nerves; palea absent; upper lemma 3-nerved; pedicelled spikelets reduced; stamens 3, anthers pinkish; caryopsis 1.5-2 mm long.

*Fl.-Fr.:*Sept. -Dec.

As a weed in cultivated fields and moist places, Burhanpur, Khandwa.

OPHIUROS Gaertn. f. W. 4; MP 3; SA 3

Ophiuros exaltatus (L.) Kuntze, Revis. Gen. Pl. 2: 780.1891; *Aegilops exaltata* L., Mant. Pl. 2: 575.1771; *Ophiuros corymbosus* (L.f.) C.F. Gaertn. f., Suppl. Carp. 2: 4.1805; Hook., f., Brit. India 7: 160.1896; Bor, Grass. Burm. Ceyl. Ind. Pak. 199. 1960; Samvatsar, Fl. West. Trib. Madhya Pradesh 371. 1996; Roy in N.P. Singh *et al.,* Fl. Madhya Pradesh 3: 467.2001. *Lahi*

Strong perennials; culms slender, numerous, glabrous, 60-120 cm tall, bulbous at base; leaves 30-40 × 1.0-1.5 cm, linear-linceolate, glabrous, base rounded, margins with tubercle-based hairs; ligule 0.5-1.0 mm long, membranous truncate; inflorescence spicate, spikes 2-5, 5-12 cm long, slender, fragile; spikelets 2.0-3.5 mm long, equalling the joint; lower glume of sessile spikelets 3 mm long, glabrous with many longitudinal lines of small pits, chartaceous oblong, tip rounded; upper glume 3 mm long, boat-shaped, obtuse; lower lemma 2 mm long, hyaline, 2-nerved or nerveless; upper lemma 2 mm long, hyaline; palea 2 mm long, similar, hyaline; stamens 3, anthers 1.5- 2.0 mm long, yellow; caryopsis dorsally compressed, oblong.

*Fl.-Fr.:*Sept. -Oct.

Along bunds of cultivated fields in black cotton soil, in all districts.

OPLISMENUS P. Beauv. W. 7; MP 2; SA 2

Key to Species

1. Awns barbellate, pale; spikelets more or less 3 mm long; annuals ....**O. burmanni**
1. Awns smooth; spikelets 3. 8-5 mm long; perennials ....**O. compositus**

Oplismenus burmanni (Retz.) P. Beauv., Ess. Agrostogr. 54. 1812; *Panicum burmanni* Retz., Observ. Bot. 3: 10.1783; Hook.f., Fl. Brit. India 7: 68. 1896; Bor, Grass. Burm. Ceyl. Ind. Pak. 317. 1960; Samvatsar, Fl. West. Trib. Madhya Pradesh 371 . 1996; Roy in N.P. Singh *et al.,* Fl. Madhya Pradesh 3: 469.2001. *Kombo jhara*

Annuals; culms prostate or decumbent-ascending, often rooting below, glabrous or hairy with a single line of hairs with tubercled hairs, nodes glabrous, or pubescent; leaves acute or acuminate; ligule a ciliate rim; sheaths compressed, hairy along margins; panicle of 3-7 racemes, up to 10 cm long; spikelets solitary or fascicled, 3-3.5 mm long, elliptic-lanceolate; lower glume ovate-lanceolate, obtuse, ciliate, 0.1-0.12 cm long, hairy on the back, 5-nerved, awn 0.8-1.0 cm long; upper glume similar to the lower, 5-7 nerved, awn 2.5-7 mm long;lower lemma empty, lanceolate, 2-keeled, awn 1-1.5 mm long; palea absent; upper lemma chartaceous, 5-7 nerved; palea sub-hyaline, 2-keeled; caryopsis oblong.

*Fl.-Fr.:*July-Dec.

Very common in shady and moist places all over the area.

Oplismenus compositus (L.) P. Beauv., Ess. Agrostogr. 54.1812; *Panicum compositum* L., Sp. Pl. 57.1753; Hook.f., Fl. Brit. India 7: 66.1896; Bor, Grass. Burm. Ceyl. Ind. Pak. 317.1960; Samvatsar, Fl. West. Trib. Madhya Pradesh 373. 1996; Roy in N.P. Singh *et al.,* Fl. Madhya Pradesh 3: 469.2001. *Bashwa*

Perennials; culms up to 1 m long, decumbent, rooting at base, terete, branched, glabrous or hairy along a line adaxial to the supporting leaf; leaves lanceolate, base rounded, minutely hairy to velvety below; sheaths markedly striate, densely ciliate along the margin; ligule short, ciliate, truncate; panicle of 3-5 racemes, second, alternately arranged on triquetrous rachis; spikelets 4-5 mm long, lanceolate; lower glume ovate-lanceolate, acuminate, 5-nerved, awn 3-3.5 mm long; upper glume subequal to lower, 5-7 nerved, with a short awn; lower lemma 7-9 nerved; palea very narrow; upper lemma shorter than lower, ovate, acute, coriaceous, margins incurved; palea acute with involute margins; caryopsis dorsally compressed, 2.5-3 mm long.

Fl.-Fr.: Aug. -Oct.

Common in shady and moist places as forest undergrowth, in all districts.

OROPETIUM Trin. W. 6; MP 3; SA 3

Key to Species

1. Perennials, leaf-sheaths at base investing base of stem; plants 10-25 cm tall; rhachis fracturing into irregular segments apical floret 1, male, 1.2-2 mm long .... **O. roxburghianum**

1. Annuals; rhachis flexuous

 2. Rhachis equally broad on two faces; spikes much compressed; spikelets much longer and only half immersed in cavities of rhachis, 1.5-2 mm long; lemmas glabrous .... **O. thomaeum**

 2. Rhachis unequal on two faces; spikes slightly compressed; spikelets almost equal to the cavities of rhachis, 2-2.5 mm long; lemmas hairy .... **O. villosulum**

Oropetium roxburghianum (Schult.) S. M. Phillips, Kew Bull. 30: 469. 1975; *Rottboellia roxburghiana* Schult., Mant. 2:430.1824; Hook.f., Fl. Brit. India 7: 365. 1896; *Tripogon roxburghianus* Bhide, J. Proc. Asiat. Soc. Bengal 7:515. 1912; Bor, Grass. Burm. Ceyl. Ind. Pak. 524. 1960; Roy in N.P. Singh *et al.,* Fl. Madhya Pradesh 3: 470.2001. *Triyo*

Annuals; culms 10-18 cm tall, erect, densely tufted; leaves 6-10 × 0.1-0.15 cm, filiform or setaceous with long hairs; ligule minutely ciliate; sheath glabrous winged on both sides, 0.4-0.5 mm long; inflorescence of solitary spikes, 4-5 cm long; spikelets a group of 5-6 flowers, 3 mm long, linear- lanceolate; lower glume 2 mm long, broadly 1-nerved, hyaline; upper glume 2 mm long, very coriaceous and thick, 2-nerved; lemma 1.5 mm long, 3-nerved, mucronate, toothed; palea 1.25 mm long, truncate, 2-keeled, winged; stamens 3; caryopsis 1 mm long, narrow at upper end.

Fl.-Fr.: July-Nov.

Rare, grows on the top of the hills at rocky places, Burhanpur, Khargone.

Oropetium thomaeum (L.f.) Trin., Fund. Agrost. 98. 1820; *Nardus thomaea* L.f., Suppl. Pl. 105. 1782; Hook.f., Fl. Brit. India 7: 366. 1896; Bor, Grass. Burm. Ceyl. Ind. Pak. 474. 1960; Samvatsar, Fl. West. Trib. Madhya Pradesh 373 . 1006; Roy in N.P. Singh *et al.,* Fl. Madhya Pradesh 3: 472.2001.

Annuals; culms ulms 5-10 cm tall, densely tufted, branched; leaves setaceous, acute at apex, glabrous or sparsely long hairy; sheaths compressed, spathaceous, glabrous; ligule lacerate;spikes compressed erect, lower part enclosed in the sheath; rachis flexuous; spikelets sessile, sunken into the hollow of the rachis, 1-flowered, 1.8-2 mm long; lower glume hyaline, oblong, nerveless; upper glume rigid, oblong, closely 3-nerved, acute at apex; lemma hyaline, boat-shaped, acute at apex; palea subequal to lemma, 2-keeled; caryopsis oblong.

*Fl.-Fr.:*July-Nov.

Rare, grows on rocky places in yellow soils, also on roofs of houses, common.

Oropetium villosulum Stapf ex Bor, Kew Bull. 4: 571. (1949) 1950 & Grass, Burm. Ceyl. Ind. Pak. 474. 1960; Samvatsar, Fl. West. Trib. Madhya Pradesh 373. 1996; Roy in N.P. Singh *et al.,* Fl. Madhya Pradesh 3: 472.2001.

Annuals; culms 5-10 cm tall; leaves setaceous, up to 2.5 cm long, scabrid, glabrous; ligule short, membranous; spikes subterete, erect, lower part enclosed in the sheath; rachis flexuous; spikelets sessile, deeply sunk into the hollow of the rachis, 1-flowered, 2-2.5 mm long; lower glume hyaline oblong or ovate-oblong, nerveless, acute or truncate at apex; upper glume oblong, lanceolate, 3-nerved, margins inflexed, hyaline, apex cuspidate; lemma membranous, boat-shaped, 3-nerved, hispid, acute to truncate at apex; caryopsis oblong.

Fl.-Fr.: July-Sept.

Common on wet, exposed rocks , Alirajpur, Burhanpur, Dhar, Indore, Khargone. Endemic to India.

ORYZA L. W. 18; MP 4; SA 1

Oryza rufipogon Griff., Not. Pl. Asiat. 3: 5. 1851; Bor, Grass. Burm, Ceyl. Ind. Pak. 605. 1960; Samvatsar, Fl. West. Trib. Madhya Pradesh 374. 1996; Roy in N.P. Singh *et al.,* Fl. Madhya Pradesh 3: 475. 2001. *Jangli dhan*

Annuals; culms 1-1.5 m tall, lower parts spongy, nodes glabrous; leaves linear, acuminate at apex, scabrid on margins and veins; sheaths terete, loose, glabrous, auricled; ligule up to 1.5 cm long, splitting at tip; inflorescence an exserted, compound panicle; main branches angular and scabrid; spikelets 8-9 mm long, falling with age; first lemma 1-nerved, 2.5 mm long; second lemma 2.4 mm long; third lemma 0.7 mm long, strongly folded above the mid nerves; awn 7 cm long; caryopsis 4-5 mm long, similar to *Oryza sativa.*

*Fl.-Fr.:*Aug. -Dec.

Common, found along the margin of ponds and rivers, in all districts.

PANICUM L. W. 442; MP 9; SA 8

Key to Species

1. Annuals
   2. Panicles embraced at base by subtending leaf, branches ascending or spreading; spikelets 1.5-2 mm long; lower glume 1-3 nerved; upper glume 3-5 nerved; fertile lemma and palea indurate .... **P. humile**
   2. Panicles not embraced, branches drooping or capillary
      3. Panicle branches drooping, stout; leaves with tubercle-based hairs; spikelets 2.8-3.5 mm long, gaping wide; lower glume 5-nerved, cuspidate-acuminate .... **P. curviflorum**
      3. Panicle branches capillary
         4. Nodes black; leaf base not amplexicaul; spikelets 3.5-4 mm long; lower glume 3-5 nerved; upper glume 13-15 nerved; palea 2-keeled .... **P. sumatrense**
         4. Nodes not black; leaf base amplexicaul; spikelets 1.5 mm long, gibbous; rhachilla elongated between glumes; lower glume 3-nerved; fertile lemma indurate .... **P. brevifolium**
1. Perennials
   5. Spikelets always with a purple tinge; panicle 8-15 cm long, sparingly branched; upper lemma 7-nerved .... **P. phoiniclados**
   5. Spikelets not purple tinged
      6. Lower glume one-half to three quarters the length of spikelet or more .... **P. notatum**
      6. Lower glume one quarter to half the length of spikelet; spikelets narrowly lanceolate to lanceolate-oblong
         7. Culms spongy, floating; leaf blades broad, flat .... **P. paludosum**
         7. Culms tough, rhizomatous; leaf blades narrow, often involute .... **P. repens**

Panicum brevifolium L., Sp. Pl. 59. 1753; *Panicum ovalifolium* Poir., Encycl. Suppl. 4: 279. 1816; Hook.f., Fl. Brit. India 7: 44. 1896; Bor, Grass. Burm . Ceyl. Ind. Pak. 324. 1960; Roy in N.P. Singh *et al.,* Fl. Madhya Pradesh 3: 477.2001.

Annuals; culms 20-50 cm tall, decumbent, rooting at nodes; leaves ovate-elliptic, acuminate, 2-8 × 1-3 cm; sheath with long hairs on throat, hairy, rarely glabrous; ligule membranous; inflorescence a panicle, exserted from the sheath; main axis hairy; branchlets glabrous; spikelets slighty gibbous, glabrous to hairy, 1.5-2.0 mm long; lower glume acute, 3-5-nerved, hairy, 1.5 mm long; upper glume 5-nerved, boat-shaped, 1.8-2.0 mm long, acute hairy; lower floret neuter; upper floret hermaphrodite, narrow elliptic; lemma 5-nerved, 1.5 mm long; palea hyaline; anthers 3, yellow; caryopsis ± 1.5 mm long, narrowly elliptic.

*Fl.-Fr.:*Sept. -Feb.

In moist and shady places, Khandwa, Khargone.

Panicum curviflorum Hornem.,Hort. Bot. Hafn. Suppl. :116.1819; *P. trypheron* Schult., Mant. 2: 244.1824; Hook.f., Fl. Brit. India 7: 47.1896; *P. deccanense* Naik & Patunkar, Reinwardtia 9: 405.1980; Sharma *et al.,* Fl. Maharashtra 3:553.1996; Naik, Fl. Marathwada 2:1066. 1998. (*P. trypheron* Schult.). *Bhatur*

Culms tufted, 35-50 cm, glabrous at nodes; leaf-sheaths 3.5-7 cm; ligule membranous, with long pilose tufts, to 4 mm; blades linear to narrowly lanceolate, 4.5-8.5 × 0.4-0.6 cm, hispid; panicles 5-15 cm, open, branches alternate; pedicels 2-7 mm; spikelets ellipsoid, to 3 mm, gaping at maturity, glabrous; glumes ovate, acuminate-cuspidate; lower glume 2.2 mm, 3-5-nerved, scaberulous without; upper glume to 3 mm, 7-9-nerved; lower lemma to 2.2 mm, paleate; upper lemma to 2 mm, glossy, smooth; palea to 2 mm; stamens to 1.5 mm; caryopsis ovoid, to 1 mm.

*Fl.-Fr.:*Sept. -Nov.

In open grass land and edges of ditches, **Khandwa** only, Kalibhit range.

Panicum humile Steud., Syn. Pl. Glumac. 1: 84. 1854; *P. walense* Mez, Bot. Jahrb. Syst. 34:146. 1904; Hook. f, Fl. Brit. India 7:48. 1896; Bor, Grass. Burm. Ceyl. Ind. Pak. 324. 1960; Samvatsar, Fl. West. Trib. Madhya Pradesh 376 . 1996; Roy in N.P. Singh *et al.,* Madhya Pradesh 3: 481.2001.

Annuals; culms 15-45 cm tall, tufted, erect or geniculate, sometimes quite dwarf; leaves 10-15 × 0.4-0.5 cm, linear-lanceolate, narrowly acuminate; ligule 1.0-1.5 mm long, ciliate truncate; inflorescence an open, 5-10 cm long panicle; spikelets 1.5-1.75 mm long, acuminate, greenish or tinged with purple; lower glume 1 mm long, obliquely lanceolate, acuminate, 3-nerved; upper glume 1.75 mm long, 3-nerved, narrowly acuminate; lower floret barren; lemma 1.5 mm long, 5-nerved; palea long, boat-shaped; stamens 3, anthers 1 mm long, purplish; caryopsis whitish, lanceolate- elliptic.

Fl.-Fr.: Aug. -Dec.

Along the roads, in ditches, ponds, paddy fields, Alirajpur, Burhanpur, Khargone.

Panicum notatum Retz. Observ. Bot. 4: 18. 1786; *Panicum montanum* Roxb. Fl. Ind. 1: 313. 1820; Hook.f. Fl. Brit. India 7: 53. 1896; Bor, Grass. Burm. Ceyl. Ind. Pak. 701. 1960; Roy in N.P. Singh *et al.,* Fl. Madhya Pradesh 3: 478. 2001. *Jokar band*

Perennials; culms up to 120 cm tall, base woody; leaves acuminate, linear-lanceolate, 15-18 × 3-3 cm, broad and cordate at base, with bulbous based hairs on both surfaces, margins scabrid; sheath hairy on margins; ligule densely hairy; inflorescence a narrow panicle with slender branches; spikelets numerous, ovoid, acute, 15-2.5 mm long; lower glume ovate-acute, 1-2 mm long, 5-nerved, margins long ciliate; upper glume ovate-acute, 5-nerved, glabrous, 2.0-2.5 mm long; lower floret neuter; lemma 5-nerved, 2.2 mm long, epaleate; upper floret hermaphrodite; lemma1.8 mm long, indurated, boat-shaped, nerves obscure; palea 1.7 mm long; anthers 3, yellow; caryopsis compressed, enclosed by hardened lemma and palea.

Fl.-Fr.: Aug. -Dec.

Open grass land, margins of forest in shady places, Burhanpur, Khandwa, Khargone.

Panicum paludosum Roxb., Fl. Ind. 1: 310.1820; Hook.f., Fl. Brit. India 7: 50.1896; Bor, Grass. Burm. Ceyl. Ind. Pak. 329. 1960; Samvatsar, Fl. West. Trib. Madhya Pradesh 374 . 1996; Roy in N.P. Singh *et al.,* Fl. Madhya Pradesh 3: 478. 2001.

Annuals; culms 90-125 cm tall, decumbent-ascending or creeping; stem rooting from lower nodes; leaves 20-30 × 0.8-1.0 cm, linear-lanceolate, acuminate; ligule a ring of hairs; inflorescence effuse panicle with patent branches, 10-15 × 10-12 cm; spikelets 3.0-3.5 mm long, lanceolate, acute; pedicel scabrid; lower glume 1 mm long, suborbicular, obtuse or truncate, 3-5-nerved; upper glume 3.2 mm long, ovate-lanceolate-acute, 9-11 nerved; lower-floret barren; lemma 2.5 mm long, acute, empty; palea 1.6 mm long, ovate, 2-keeled, pointed, hyaline; upper floret hermaphrodite, 2.0-2.5 mm long narrowly elliptic-acute; lemma 2.2 mm long, shining white, boat-shaped; palea 2-keeled, texture similar; stamens 3, anthers 1 mm long, purple; caryopsis *ca* 1 mm long lanceolate-elliptic.

Fl.-Fr.: Aug. -Nov.

Common on wet lands, along banks of streams throughout the area.

Panicum phoiniclados Naik & Patunkar, Reinwardtia 9 (4):403,f. 1. 1980; Grass. Marathwada 159 f. 49. 1980 & Fl. Marathwada 2:1069. 1998; Samvatsar, Fl. West. Trib. Madhya Pradesh 374 . 1996.

Perennials; culms erect, 90-100 cm tall; leaves flat or convolute, linear-lanceolate; ligule a hairy ring; panicles pyramidal, 8-15 cm long, sparingly branched, pedicels solitary or paired; spikelets ovate-lanceolate; lower glume ovate-lanceolate; upper glume linear-lanceolate, acute; lower lemma similar, hyaline, closely 2-nerved; upper lemma 7-nerved; caryopsis oblong.

Fl.-Fr.: Sept. -Dec.

Occasional along water streams, wet and shady places in the forest, Alirajpur, Dhar, Indore, Khandwa. Endemic to India.

Panicum repens L., Sp. Pl. ed. 2.87. 1762; Hook.f., Fl. Brit. India 7:49. 1896; Bor, Grass. Burm. Ceyl. Ind. Pak. 330.1960; Samvatsar, Fl. West. Trib. Madhya Pradesh 376. 1996; Roy in N.P. Singh *et al.,* Fl. Madhya Pradesh 3: 480.2001.

Perennials; culms 50-100 cm tall; nodes glabrous; leaves linear-lanceolate, hairy; ligule hairy; panicle oblong, 10-15 cm long, rachis scabrid, angular; spikelets 2.5-3 mm long, elliptic-lanceolate, glabrous, apex acute; lower glume 0.5-1 mm long, ovate-orbicular, obscurely 1-5 nerved; upper glume elliptic-lanceolate, 5-9 nerved; lower lemma subequal to upper glume; palea hyaline, oblong; upper lemma chartaceous, 7-nerved; palea similar; caryopsis oblong.

Fl.-Fr.: Aug. -Dec.

Rare, in marshy places, Alirajpur, Indore.

Panicum sumatrense Roth Syst. Veg. 2: 434. 1817; *P. psilopodium* Trin., Gram. Panic. 217. 1826; Hook.f., Fl. Brit. India 7: 46. 1896; Bor, Grass. Burm. Ceyl. Ind. Pak. 329. 1960; Samvatsar, Fl. West. Trib. Madhya Pradesh 376. 1996; Roy in N.P. Singh *et al.,* Fl. Madhya Pradesh 3: 480.2001. *Kutki*

Annuals; culms 25-85 cm tall, erect or geniculate; leaves linear, flat, glabrous or sparsely hairy; sheaths loose, ciliate at mouth; ligule ciliate; inflorescence a contracted panicle; spikelets 2.5-3.5 mm long, elliptic, green or tinged, glabrous; lower glume 1 mm long, acuminate, 3-5-nerved; upper glume 2.0-3.5 mm long, 11-13-nerved, concave; lower floret neuter; upper floret hermaphrodite; lemma 2-3 mm long, cartilaginous; stamens 3, anthers 1 mm long, purple; caryopsis 1.5 mm long, subglobose, cream-coloured.

Fl.-Fr.: July-Nov.

Common on stream banks, other moist, open places, Alirajpur, Barwani, Burhanpur, Dhar, Indore, Khandwa.

PARAHYPARRHENIA A. Camus W. 6; MP 1; SA 1

Parahyparrhenia khannae Tiwari, A.P. & Chorghe in Phytotaxa 446(1): 55-60.2020

Annuals; culms 10-40 cm tall; leaves involute, glabrous; racemes solitary, slender; sessile spikelets 3.5-4 mm long; lower glume equal to spikelet, linear-oblong, 2-toothed, 2-keeled at apex with a deep median groove on the dorsal surface; upper glume oblong-lanceolate; lower lemma ovate-oblong; upper lemma a hyaline base of awn; awn 5-6 cm long; pedicelled spikelet longer than the sessile one; caryopsis oblong-ovate.

Fl.-Fr.: Sept. -Nov.

Shady places in forest, Alirajpur, Burhanpur, Dhar, Khandwa. Endemic to India.

PASPALIDIUM Stapf W. 34; MP 3; SA 3

Key to Species

1. Annuals; spikes shorter than the internodes . ... **P. flavidum**
1. Perennials
   2. Rhachis produced into a sharp point; upper lemma not granular .... **P. punctatum**
   2. Rhachis not produced into a sharp point; upper lemma granular . ... **P. geminatum**

Paspalidium flavidum (Retz.) A. Camus, Fl. Indo-Chine 7: 419. 1922; *Panicum flavidum* Retz., Observ. Bot. 4: 15. 1786; Hook.f., Fl. Brit. India 7:28. 1896; Bor, Grass. Bur Ceyl. Ind. Pak. 333. 1960; Samvatsar, Fl. West. Trib. Madhya Pradesh 379 . 1996; Roy in N.P. Singh *et al.,* Fl. Madhya Pradesh 3: 483.2001. *Londwa*

Annuals or perennials; culms 15 – 100 cm tall, decumbent-ascending, nodes glabrous; leaves 4-15 × 0.3-1.0 cm, linear, apex acuminate, glabrous; ligule membranous; spikes 5-10, remote, combined into a 15-20 cm long, recurved rachis; spikelets 2-nate, subsessile, ovoid, oblique; lower glume 1.2-1.6 mm long, suborbicular, obtuse, 5- nerved; upper glume 2.0-2.2 mm long, ovate, rounded, obtuse, 7-nerved; lower lemma 2.5 mm long, empty, chartaceous, ovate, subacute, 5-nerved; palea 1.8-2.2 mm long, membranous, ovate-oblong; upper floret 2.3 mm long; upper lemma as long as the florets, convex, coriaceous, broadly ovate or suborbicular, mucronate; palea similar to lemma; stamens 3, anthers *ca* 1 mm long, dark yellow; caryopsis ovoid .

Fl.-Fr.: July-Dec.

Common along shady places in the forest, in all districts.

Paspalidium geminatum (Forssk.) Stapf, Fl. Trop. Afr. 9: 583.1920; *Panicum geminatum* Forssk., Fl. Aegypt.-Arab. 18.1775; *P. fluitans* Retz., Observ. Bot. 3: 8.1783; *P. paspaloides* Pers., Syn. Pl. 1: 81.1805; Hook.f., Fl. Brit. India 7: 30.1896; Bor, Grass. Burm. Ceyl. Ind. Pak. 333.1960; Samvatsar, Fl. West. Trib. Madhya Pradesh 379.1996; Roy in N.P. Singh *et al.,* Fl. Madhya Pradesh 3: 483.2001.

Perennials; culms 0.3-1 m high, creeping or floating, stoloniferous or spongy- rhizomatous, rooting at lower nodes; leaves lanceolate or linear-lanceolate, 5-30 cm long, acuminate at apex, cordate at base; ligule a tuft of hairs; inflorescence a 5-30 cm long panicle; spikes several, congested; spikelets solitary, 2.5-3 mm long, second; lower glume 0.5-0.7 mm long, hyaline, nerveless; upper glume 1.8-2 mm long, 5-nerved; lower floret male; upper floret bisexual; caryopsis *ca* 2 mm long, broadly ellipsoid.

*Fl.-Fr.:*July-Dec.

Common, grows along the ponds, ditches and riverside marshy places, in all districts.

Paspalidium punctatum (Burm.f.) A. Camus, Fl. Indo-Chine 7: 419. 1922; *Panicum punctatum* Burm.f., Fl. Ind. 26. 1768; Hook.f., Fl. Brit. India 7:29. 1896; Bor, Grass. Bur. Ceyl. Ind. Pak. 333. 1960; Roy in N.P. Singh *et al.,* Fl. MP 3: 484.2001.

Perennials; culms 0.5 – 1.25 mm tall, basal portion spongy, floating, rooting at nodes; leaves linear, acute, scabrid on margins; sheaths glabrous, ligule a ring of hairs; inflorescence of several false spikes, distant and appressed to the rachis; spikelets imbricate, plano-convex, 1.5–2.5 mm long, glabrous; lower glume 0.5-1.0 mm long, membranous, truncate; upper floret hermaphrodite; lemma indurate, acuminate, cuspidate, finely punctate, transversely rugose; palea similar, fitting into the lemma; stamens 3; caryopsis orbicular, compressed.

Fl.-Fr.: -Aug. -Nov.

Not common, grows along the ponds, ditches and riverside marshy places at Barwani, Burhanpur, Khandwa.

PASPALUM L. W. 329; MP 5; SA 4

Key to Species

1. Annuals; racemes 2-6; spikelets 2-2.5 mm long, orbicular-ovate.................................... **P. scrobiculatum**

1. Perennials; racemes up to 5

  2. Rhizomatous, stout, woody plants; racemes 2; spikelets 3-3.5 mm long . ...**P. notatum**

  2. Stoloniferous plants; rhachis flat on the back, narrowly winged on the axial side

    3. Racemes 3-5, subdigitate; upper glume 5-nerved; lower lemma 5-nerved; upper lemma chartaceous, obscurely 3-nerved . ...**P. distichum**

    3. Racemes 2; both glumes obscure or absent; upper lemma indurate .... **P. vaginatum**

Paspalum disticum L., Syst. Nat. ed. 102:855. 1759; *P. paspalodes* (Michx.) Scribn., Mem. Torrey Bot. Club 5: 29. 1894; *Digitaria paspalodes* Michx., Fl. Bor. -Amer. 1:46. 1803; Hook.f., Fl. Brit. India 7: 12. 1896; Bor, Grass. Burm. Ceyl. Ind. Pak. 338. 1960; Samvatsar, Fl. West. Trib. Madhya Pradesh 380.1996. *Koda*

Perennial, stoloniferous; culms 15-40 cm tall, nodes glabrous; leaves flat, linear-lanceolate, glabrous, acute at apex; ligule membranous; spikes 3-5, rachis flattened on the back, keeled on the axial side, keel narrowly winged and wavy; spikelets solitary, elliptic; lower glume absent; upper glume membranous, elliptic, 5-nerved; lower lemma empty, similar in texture and shape to the upper glume, 5-nerved; palea absent; upper lemma chartaceous, elliptic-oblong, obscurely 3-nerved; palea similar in texture and shape with inflexed margins; caryopsis elliptic-oblong.

*Fl.-Fr.:*Sept. -Dec.

Common along margins of ponds and ditches in open grass lands, all over the area.

Paspalum notatum Fluegge, Gram. Monogr. **Paspalum** 106. 1810; *P. disticum* L. sensu Hook.f., Fl. Brit. India 7:12. 1896; Bor, Grass. Burm. Ceyl. Ind. Pafk. 339. 1960; Samvatsar, Fl. West. Trib. Madhya Pradesh 380.1996.

Perennials; rhizome short, stout, woody; leaves linear-lanceolate, acuminate; ligule truncate, minute; spikes 2; spikelets elliptic,ovate; lower glume minute or absent; upper glume membranous; lower lemma similar to upper glume; upper lemma elliptic, subcoriaceous, smooth; caryopsis elliptic, oblong.

Fl.-Fr.: Oct. -Dec.

Rare, in marshy or wet soil, near water bodies, **Alirajpur** only, Aamkhut forest.

Paspalum scrobiculatum L., Mant. Pl. 1:29. 1767; Hook.f., Fl. Brit. India 7: 10.1896; Bor, Grass. Burm. Ceyl. Ind. Pak. 340.1960; Samvatsar, Fl. West. Trib. Madhya Pradesh 380 . 1996; Roy in N.P. Singh *et al.,* Fl. Madhya Pradesh 3: 487.2001. *Kudrali*

Annuals; culms 25-80 cm tall, erect, ascending, tufted; leaves 6-18 cm long, acuminate at apex, glabrous; ligule 1 mm long, membranous; inflorescence of 2-6 spikes; rachis flattened, 2.0-2.5 mm broad; spikelets *ca* 2.5 mm long, 2-nate, elliptic-ovate; lower glume absent; upper glume 2.5 long, 3-5-nerved, membranous; lower lemma similar to upper glume; upper floret hermaphrodite; lemma rounded, stiff, shining, edges slightly inflexed; stamens 3, cream-coloured; caryopsis 2 mm long, globose, biconvex, pale coloured.

Fl.-Fr.: Aug. -Dec.

Common in moist places along the margins of ditches, all over the area.

Paspalum vaginatum Sw., Prodr. Veg. Ind. Occ. 21. 1788; Hook.f., Fl. Brit. India 7:12 . 1960; Bor, Grass. Burm. Cey. Ind. Pak. 341 1960; Roy in N.P. Singh *et al.,* Fl. Madhya Pradesh 3**:** 489.2001. **(Plate 11, Fig. 4)**

Aquatic perennials; culms 1.0-1.5 mlong, rooting from the lower nodes; leaves numerous, distichous, 5-10 × 0.8-1.0 cm, linear- lanceolate, acuminate; sheaths, compressed; ligule 1 mm long, membranous, truncate; inflorescence of 2-3 spikes; spikes 2.5-5.0 cm long with flattened rachis, 1.0-1.5 mm wide; spikelets 3.5 mm long, solitary or 2-nate, elliptic, oblong, acute, dorsally flattened; lower glume absent; upper glume 3.5 mm long, ovate, acute, 5-nerved; lower lemma 3.5 mm long with a mid-nerve; upper lemma 2.8 mm long, boat- shaped, hairy at the tip; palea 3 mm long, similarly indurated; stamens 3, anthers 1 mm long; caryopsis ellipsoid, 2.5-2.8 mm long.

Fl.-Fr.: Aug. -Sept.

Rare, grows along the bank of streams, Burhanpur, Khandwa, Khargone.

PENNISETUM Rich. ex Pers. W. 83; MP; 6; SA 5

Key to Species

1. Perennials
   2. Spikelets 3 in a cluster, 1 sessile, the others pedicellate, 4.5-6.5 mm long; lower glume well developed, 1-nerved; panicle branches accrescent .... **P. orientale**
   2. Spikelets solitary; lower glume minute
      3. Spikelets 7-8 mm long; upper glume 1-3 nerved; lower lemma 9-11 neved, entire, bristles glabrous, uniform . . .. **P. hohenackeri**
      3. Spikelets 4-5 mm long; upper glume 3-5 nerved; lower lemma 5-nerved, 3-toothed; outer bristles not ciliate, inner ciliate, one stouter than the rest .... **P. polystachion**
1. Annuals; nodes glabrous; palea 2-keeled; upper glume and lower lemma 5- nerved
   4. Bristles densely villous, plumose; spikelets solitary or 2-3 together, 5-6 mm long; lower glume nerveless; both glumes silky hairy on back; lower lemma 3-toothed at apex; upper lemma 5-nerved .... **P. pedicellatum**
   4. Bristles scaberulous, not plumose; spikelets solitary, 3-4 mm long; lower glume absent; both glumes not silky; lower lemma 3-lobed at apex; upper lemma nerveless .... **P. hordeoidedes**

Pennisetum hohenackeri Hochst. ex Steud., Syn. Pl. Glumac., 1:103. 1854; *Pennisetum alopecuros* Nees ex Steud., *op. cit.* 102 *non* Jacq. 1844; Hook.f., Fl. Brit. India 7: 84.1896; Bor, Grass. Burm. Ceyl. Ind. Pak. 344. 1960; Samvatsar, Fl. West. Trib. Madhya Pradesh 380.1996; Roy in N.P. Singh *et al.,* Fl. Madhya Pradesh 3:490.2001. *Mhol*

Perennials; culms densely tufted, compressed, 50-150 cm tall, branched, glabrous, nodes glabrous; leaf sheath compressed, glabrous; ligule a rim of hairs; leaf blade convolute or compressed, linear, apex acuminate; panicles spiciform; spikelets sessile, solitary, narrowly ovate7-8.5 mm, glabrous, acuminate; lower glume absent; upper glume hyaline, ovate, glabrous, 3-nerved, apex acute; lower lemma membranous, linear- ovate, 9-11-nerved, apex acute; palea absent; upper lemma subcoriaceous, linear-ovate, 5-nerved, apex acuminate; stamens 3; caryopsis lanceolate, 1.5 mm long, awned.

*Fl.-Fr.:*Aug. -Feb.

Common along streams, on bunds of fields, river beds, in all districts.

Pennisetum hordeoides (Lam.) Steud.,Syn. Pl. Glumac. 1:103. 1854; *Panicum hordeoides* Lam., Tab. Encycl. Meth. Bot. 1:170.1791; *P. parviflorum* Trin., Gramin. Panic. 65. 1826; Hook.f., Fl. Brit. India 7:86. 1896; Samvatsar, Fl. West. Trib. Madhya Pradesh 382 . 1996.

Annual decumbent herbs, rooting at lower nodes; culms 60-75 cm tall; nodes glabrous; leaves flat, narrowly lanceolate, 10-15 × 0.5-0.8 cm, base cuneate, apex acuminate; sheath 6-8 cm, base and mouth ciliate; ligule ciliate; spikes axillary, tinged with purple, 6-10 × 0.5-0.7 cm, linear; spikelets sessile, linear-oblong, 3-4 mm × 0.75-1 mm, bristles free up to base, longest up to 13-15 mm, not plumose, scaberulous; lower glume absent; upper glume 4 mm long, hyaline, 5-nerved; lower lemma 2.5 mm long, acute, hyaline, 5-nerved, 3-lobed at apex; epaleate; upper lemma 2 mm, chartaceous, acute, nerveless; palea 1.75 mm long, 2-keeled, keels rounded, nerveless; stamens 3; caryopsis linear-oblong, 2 mm long.

Fl.-Fr.: Nov. -Feb.

Rare, in association with *Eragrostis tremula* Hochst. ex Steud., **Alirajpur** only, at Harod.

Pennisetum orientale Rich., Syn. Pl. 1: 72. 1805; Hook.f., Fl. India 7: 86. 1896; Bor, Grass. Burm. Ceyl. Ind. Pak. 345. 1960; Roy in N.P. Singh *et al.,* Fl. Madhya Pradesh 3: 491.2001.

Strong perennials with stout rhizomes; culms 60-100 cm tall, nodes glabrous; leaves linear, flat, glabrous or pilose, margins scabrid; sheath smooth, bearded at mouth; ligule hairy; inflorescence a spike; rachis angular, hispid; involucres unequal, bristly, antrorsely barbed; spikelets 2-3, lanceolate, 6.0-6.5 mm long; lower glume 2.5-3.0 mm long, hyaline; upper glume lanceolate, 4.0-4.5 mm long; lower lemma empty; upper lemma fertile, ovate, 6 mm long, 5-nerved, acuminate; anthers 3, yellow, 2 mm long; caryopsis ellipsoid, 2 mm long.

Fl.-Fr.: Aug. -Nov.

Rare, in dense forest, on hill slopes, **Khandwa** only, Chandgarh range and Mortakka forest.

Pennisetum pedicellatum Trin., Mem. Acad. Imp. Sci. Saint- Petersbourg Ser. 6, Sci. Math. Seconde Pt. Sci. Math. 3(2)**:** 184. 1834; Hook.f., Fl. Brit. India 7: 86. 1896; Bor, Grass. Burm. Ceyl. Ind. Pak. 346. 1960; Samvatsar, Fl. West. Trib. Madhya Pradesh 382. 1996; Roy in N.P. Singh *et al.,* Fl. Madhya Pradesh 3: 491.2001. *Dinanath* **(Plate 11, Fig. 5)**

Annuals; culms 0.3-1.2 m tall; nodes glabrous or hispid, scabrid on margins; sheaths terete, usually glabrous; ligule membranous; inflorescence of several spikes; rhachis angular; involucres sessile; bristles free to base, inner once wooly; spikelets solitary or in groups of 2-3, lanceolate, 5-6 mm long; lower glume lanceolate, 2-3 mm long, hairy; upper glume oblong –lanceolate, 5-6 mm long, cuspidate to acuminate, hairy, 5-nerved; lower lemma 4-5 mm long, pubescent, 3-dentate at apex; palea hyaline, lacerate; upper lemma elliptic, 3.0-3.5 mm long, chartaceous, ciliolate on margins; stamens 3, yellow; caryopsis lanceolate, 2.5-3.0 mm long.

Fl.-Fr.: Aug. -Feb.

Common in open grasslands and hill slopes, all over the area.

Pennisetum polystachion (L.) Schultz., Mant. 2:146. 1824; *Panicum polystachion* L., Syst. Nat. ed. 10 2: 870.1759; Bor, Grass. Burm. Ceyl. Ind. Pak. 346. 1960; Samvatsar, Fl. West. Trib. Madhya Pradesh 382. 1996; Roy in N.P. Singh *et al.,* Fl. Madhya Pradesh 3: 392.2001. *Shiwati*

Perennials; culms up to 2 m tall, glabrous; leaves linear-lanceolate, 3-15 mm wide, hirsute; ligule hairy; sheath glabrous or sparsely hairy; inflorescence a false spike, compact, branches angled, scaberulous; involucres sub-sessile; bristles many, densely ciliate, one bristle stouter than the rest and up to 3 cm long; spikelets 4-5 mm long, solitary, glabrous; lower glume 0.5-1.0 mm long, acute nerveless; upper glume oblong-elliptic, mucronate, 5-7-nerved; lower floret male or barren; upper floret hermaphrodite, 2-3 mm long, ciliate at apex; anthers 3, yellow, *ca* 1.5-2.0 mm long, tip glabrous; caryopsis elliptic-oblong, 2 mm long.

*Fl.-Fr.:*Sept. -Dec.

Alirajpur, Dhar, Khandwa.

PEROTIS Aiton W. 13;MP 1; SA 1

Perotis indica (L.) Kuntze, Revis. Gen. Pl. 2: 787. 1891; *Anthoxanthum indicum* L., Sp. Pl. 28. 1753; Hook.f., Brit. India 7: 98. 1896; Bor, Grass. Burm. Ceyl. Ind. Pak. 611. 1960; Samvatsar, Fl. West. Trib. Madhya Pradesh 382 . 1996; Roy in N.P. Singh *et al.,* Fl. Madhya Pradesh 3: 493.2001. *Nakka toka*

Annuals; culms erect, 40-50 cm tall, loosely tufted; leaves 1.0-3.5 × 0.2-0.6 cm, lanceolate, ovate, base cordate, margin scabrid; ligule 0.5 mm long, membranous, truncate; inflorescence a terminal spike, 2-20 cm long; spikelets shortly hispid, pedicelled, 1.5-2.5 mm long with 2 awns; lower glume 1. 9 mm long, narrowly linear-lanceolate, 1-nerved, hispid over the back, awn scaberulous, 1cm long; upper glume similar to lower but hispid only along the nerves, awn 7.5 mm long; lemma 0.6 mm long, lanceolate hyaline, 1-nerved; palea nerveless; stamens 3; caryopsis 4 mm long, terete, brown.

*Fl.-Fr.:*Aug. -Dec.

Common, grows on sandy soil, in open grasslands and along river banks, Alirajpur, Barwani, Burhanpur, Khandwa, Khargone.

PHALARIS L. W. 19; MP 1; SA 1

Phalaris minor Retz. Observ. Bot. 3: 8. 1783; Hook.f., Fl. Brit. India 7: 221. 1896; Bor, Grass. Burm. Ceyl. Ind. Pak. 616. 1960; Samvatsar, Fl. West. Trib. Madhya Pradesh 383 . 1996; Roy in N.P. Singh *et al.,* Fl. Madhya Pradesh 3: 495. 2001. *Chirio*

Annuals; culms 70-80 cm tall, erect or decumbent; leaves 20-25 × 1.0-1.5 cm, lanceolate, acuminate, glabrous; ligule 5-10 mm long, truncate; inflorescence compact, cylindrical panicle, 7-10 × 1.2-1.5 cm; spikelet laterally compressed, ellipsoid, acute; pedicles 2 mm long; glumes equal, 5-6 mm long, 3-nerved, acute with winged keels; lowest lemma absent; middle lemma reduced; uppermost fertile 2.8-3.0 mm long, ovate-acute, 5 nerved; palea 2.0-2.3 mm long, lanceolate, 2-keeled, ciliate at the top; stamens 3, anthers pale yellow; caryopsis 2 mm long, ovoid, apiculate.

*Fl.-Fr.:*July-Dec.

Occasional, near streams, on wasteland, bunds of fields, Alirajpur, Dhar, Indore, Khargone.

PHRAGMITES Adans. W. 4; MP 1; SA 1

Phragmites karka (Retz.) Trin. ex Steud., Normencl. Bot. ed. 2, 2: 324. 1841; *Arundo karka* Retz., Observ. Bot. 4:21. 1786; Hook.f., Fl. Brit. India 7: 304. 1896; Bor, Grass. Burm. Ceyl. Ind. Pak. 416. 1960; Samvatsar, Fl. West. Trib. Madhya Pradesh 384. 1996; Roy in N.P. Singh *et al.,* Fl. Madhya Pradesh 3: 496.2001. *Narkul*

Perennials; culms fistular, erect, 1.5-2.0 m tall, with creeping rhizomes; leaves 30-40 × 1-3 cm, linear-lanceolate, acuminate, glabrous above, lower surface rough to touch; ligule 2-3 mm long, margin fimbriate; sheaths with a hairy throat; inflorescence a panicle, 30-45 cm long, lower nodes often branched in whorls; spikelets 9-12 mm long; pedicels 1-6 mm long; rachilla hairs white, 4-7 mm long; lower glume 3-4 mm long, chartaceous, lanceolate, 1-nerved; upper glume 4.0-5.5 mm long, elliptic to narrowly elliptic, acute; lower lemma empty, 7.5-12.0 mm long, narrowly elliptic; fertile lemma 8-10 mm long, narrowly lanceolate; palea 2.5-3.0 mm long, 2-keeled, ciliate at apex; stamens 3, anthers 2.0-2.5 mm long; caryopsis purplish, *ca* 1.5 mm long.

Fl.-Fr.: Oct. - Dec.

In shallow water along the river and stream banks, Alirajpur, Burhanpur, Dhar, Indore, Khandwa.

POGONATHERUM P. Beauv. W. 4; MP 2; SA 2

Key to Species

1. Erect perennials; racemes 1.5-5 cm long; lower floret present; upper floret with 2 stamens .... **P. paniceum**
1. Prostrate perennials; racemes 1-3 cm long; lower floret absent or obsolete; upper floret with 1 stamen .... **P. crinitum**

Pogonatherum crinitum (Thunb.) Kunth, Enum. Pl. 1: 478. 1833; *Andropogon crinitus* Thunb., Syst. Veg. ed. 14 903. 1784; Hook.f., Fl. Brit. India 7:141. 1896; Bor, Grass. Burm. Ceyl. Ind. Pak. 200.1960; Roy in N.P. Singh *et al.,* Fl. Madhya Pradesh 3: 497.2001.

Perennials; culm10-50 cm tall, slender; leaves linear or linear-lanceolate; ligule of long hairs; racemes solitary, terminal; sessile spikelets up to 2 mm long, broader above; upper glume longer than lower, glabrous or shortly ciliate; lower lemma absent; lemma of upper floret dentate, awn 15-20 mm long; palea 2/3rd the length of lemma; stamen usually 1, rarely 2.

*Fl.-Fr.:*Sept. -Dec.

Rare, in moist shady places, **Khandwa** only, Punasa and Kalibhit.

Pogonatherum paniceum (Lam.) Hack., Allg. Bot. Z. Syst. 12:178. 1906; *Saccharum paniceum* Lam., Encycl. 1**:** 595. 1785; Hook.f., Fl. Brit. India 7:141. 1896; Bor, Grass. Burm. Ceyl. Ind. Pak. 202. 1960; Roy in N.P. Singh *et al.,* Fl. Madhya Pradesh 3: 497.2001.

Perennials; culms 10-15 cm tall, densely tufted; nodes shining; leaves rounded at base, scabrid; sheaths lax, smooth; ligule reduced, fimbriate; inflorescence a solitary, terminal raceme, 1.5-3.0 cm long; joints and pedicels trigonous, 1-1.5mm long, with white hairs on margins; sessile spikelets linear-oblong, 2.5-3 mm long; lower glume truncate at top, membranous, conve on back, 2-3-nerved, ciliate below the apex; upper glume equalling spikelets, obtuse, membranous; keels hispid or scabrid, hairy at apex; awn 12-14 mm long; lower floret male with 2 stamens, rarely bisexual; upper floret bisexual; lemma hyaline, narrow, almost keeled, dentate, nerveless; anthers 2, rarely 1, yellow; caryopsis linear.

*Fl.-Fr.:*July-Sept.

Occasional in moist, shady places along streams and river side, Burhanpur, Khandwa.

POLYPOGON Desf. W. 21; MP 1; SA 1

Polypogon monspeliensis (L.) Desf., Fl. Atlant. 1: 66. 1798; *Alopecurus monospeliensis* L., Sp. Pl. 61. 1753; Hook.f., Fl. Brit. India 7: 245. 1896; Bor, Grass. Burm. Ceyl. Ind. Pak. 403 1960; Roy in N.P. Singh *et al.,* Fl. Madhya Pradesh 3**:** 498. 2001. *Kaalio*

Annuals; culms 20-60 cm tall, tufted, prostrate or decumbent; leaves 4-10 × 0.2-0.8 cm, linear-lanceolate; ligule 5-7 mm long, lacerate; sheath glabrous; inflorescence cylindrical or spiciform panicle, 10 cm long; spikelets 1.2-2.0 mm long, linear lanceolate; pedicels 0.8 mm long; glumes nearly equal, 1.5-2.0 mm long, lanceolate, 2-toothed with a scabrid awn, 4-7 mm long; lemma oblong, 4-toothed, 5-nerved with 1 mm long awn; palea as long as the lemma, oblong, 2-keeled, 2-toothed; stamens 3; caryopsis 1 mm long, elliptical.

Fl.-Fr.: Dec. –Feb.

Common in moist, shady places and in nursery beds, in all districts.

POLYTRIAS Hack. Monotypic

Polytrias indica (Houtt.) Veldkamp, Blumea 36: 180.1991; *Phleum indicum* Houtt. Nat. Hist. II, 13: 198. 1782; *Ischaemum ciliare* Retz., Observ. Bot. 6:36. 1791; Hook.f., Fl. Brit. India 7:133. 1896; *Ischaemum indicum* (Houtt.) Merrill, J. Arnold. Arbor. 19: 320.1938; Bor, Grass. Burm. Ceyl. Ind. Pak. 180.1960; Samvatsar, Fl. West. Trib. Madhya Pradesh 364. 1996; Roy in N.P. Singh *et al.,* Fl. Madhya Pradesh 3: 453.2001. *Putena*

Perennials; culms 30-60 cm tall, erect or decumbent or sometimes creeping, rooting from the lower nodes; leaves 10-20 cm long, linear, acuminate, pubescent or villous on both surfaces, sheath glabrous; ligule 1.0-1.5 mm long, membranous truncate; inflorescence of 2-3 spikes; sessile spikelets 5-6 mm long, oblong-lanceolate; lower glume 5.2 mm long, sub-obovate, glabrous, terminal wings ciliate on upper surface; upper glume 5mm long, boat-shaped, keel scabrid, awn 2 cm long; lower lemma 4.2 mm long, hyaline boat- shaped, glabrous; palea 4.2 mm long, linear-lanceolate, 2-nerved, scabrid; anthers 3, yellow, 2 mm lng; upper lemma deeply notched, 3-nerved, awns twisted, 1.8 cm long; upper glume 5-8 mm long, 5-nerved; lower lemma 2-nerved, margin ciliate; anthers 3, yellow, 2 mm long; caryopsis 1.5 mm long with persistent style.

Fl.-Fr.: Aug. -Dec.

Common in open grassland, particularly in sandy alluvial soils, in all districts.

PSEUDANTHISTIRIA (Hack.) Hook.f. W. 4; MP 1; SA 1

Pseudanthistiria heteroclita (Roxb.) Hook.f., Fl. Brit. India 7: 219. 1896; *Anthistiria heteroclita* Roxb. Fl. Ind. 1:253. 1820; *Pseudanthistiria hispida* Hook.f., *loc.cit*; Bor, Grass. Burm. Ceyl. Ind. Pak. 203. 1960; Samvatsar, Fl. West. Trib. Madhya Pradesh 384. 1996; Roy in N.P. Singh *et al.,* Fl. Madhya Pradesh 3: 499.2001. *Phuli ghas*

Annuals; culms tufted, terete, 30-120 cm tall, erect or geniculate, simple or branched, slender, glabrous, rooting at lower nodes; nodes glabrous; leaves 7-15 cm, linear, leaf blade flat, linear to linear ovate-elliptic, covered with tubercle based hairs, margin scaberulous, apex acuminate; ligule membranous, truncate; panicles *ca* 12 cm long, with loose cluster of spikelets; spikes 0.6-0.8 cm long, spathaceous; sessile spikelets 0.3-0.4 cm long, linear-oblong; pedicelled spikelets lanceolate; lower glume coriaceous, narrowly ovate-oblong; upper glume membranous, linear, ovate-oblong; lower lemma absent; upper lemma reduced to a hyaline base of 20-25 mm long geniculate awn; stamens 3, anthers 2-2.5 mm long; caryopsis not seen.

Fl.-Fr.: Aug. -Dec.

Open grass lands in black cotton soil, Alirajpur, Burhanpur, Dhar, Indore, Khandwa. Endemic to India.

PSEUDORAPHIS Griff. W. 6; MP 2; SA 1

Pseudoraphis spinescens (R.Br.) Vickery, Proc. Roy. Soc. Queensland 62: 69.1952; *Panicum spinescens* R.Br., Prodr. Fl. Nov. Holl. 193. 1810; *Chamaeraphis spinescens* Poir., Encycl. Suppl. 2: 189.1811; Hook.f., Fl. Brit. India 7: 62. 1896; Bor, Grass. Burm. Ceyl. Ind. Pak. 353. 1960; Samvatsar, Fl. West. Trib. Madhya Pradesh 385.1996; Roy in N.P. Singh *et al.,* Fl. Madhya Pradesh 3: 501.2001.

Perennial; culms 20-100 cm tall, floating, much branched and leafy; leaves linear or lanceolate, flat; sheaths compressed, scabrid along margins; ligule short, hairy; inflorescence a spreading panicle; branches simple; rhachis produced into 8–11 mm long bristles beyond the terminal spikelets; spikelets solitary, lanceolate, 6.0-6.5 mm long; lower glume orbicular, 0.5mm long; upper glume lanceolate, 6.0-6.5 mm long, 9-11-nerved; lower lemma 4.5-4.7 mm long, acuminate, 5-7-nerved; anthers 3, yellow, 1.5 mm long; caryopsis ellipsoid, 2 mm long.

Fl-Fr. : Sept. -Dec.

Rare, in shallow water forming patches in ponds and tanks, Alirajpur, Dhar, Indore, Khandwa.

PSEUDOSORGHUM A. Camus W. 2; MP 1; SA 1

Pseudosorghum fasciculare (Roxb.) A. Camus, Bull. Mus. Natl. Hist. Nat. 26: 662. 1920; *Andropogon fascicularis* Roxb. Fl. Ind. 1: 269.1820; Hook.f., Fl. Brit. India 7: 177. 1896; Bor, Grass. Burm. Ceyl. Ind. Pak. 205. 1960; Roy in N.P. Singh *et al.,* Fl. Madhya Pradesh 3: 501.2001.

Annuals; culms 30-75 cm tall, terete, decumbent; leaves flat, lanceolate; sheath compressed, glabrous or sparsely hairy; ligule membranous, truncate; inflorescence of racemes in pairs, one short and the other long peduncled; joints and callus hairy; sessile spikelets 4 mm long, elliptic, awned; lower glume 4 mm long, lanceolate, 7-nerved, shortly hispid; upper glume equal to lower, boat- shaped, 7-nerved; lower lemma equal to glume, broadly lanceolate, margins ciliate; upper lemma cleft into two acute lobes, with 18-20 mm long awn from sinus; anthers 3, yellow; caryopsis 4 mm long, lanceolate.

*Fl.-Fr.:*Sept. -Dec.

Rare, in marshy places along stream banks in forest, Indore, Khandwa.

ROTTBOELLIA L.f. W. 6; MP 1; SA 1

Rottboellia cochinchinensis (Lour.) Clayton, Kew Bull. 35: 817. 1981; *Stegosia cochinchinensis* Lour., Fl. Cochinch. 1: 51. 1790; *Rottboellia exaltata* L.f., Suppl. 114. 1781, non (L.) L.f., 1779; Hook.f., Fl. Brit. India 7: 156. 1896; Bor, Grass. Burm. Ceyl. Ind. Pak. 206. 1960; Samvatsar, Fl. West. Trib. Madhya Pradesh 385. 1996; Roy in N.P. Singh *et al.,* Fl. Madhya Pradesh 3: 503.2001. *Lahu*

Annuals; culms tufted, erect, 1.0-1.5 m tall; rhizomes elongated; leaves 30-40 cm long, membranous, truncate; ligule lacerate; racemes yellowish- green, narrowed towards the apex into a subulate point, cylindrical 10-15 cm long, joints oblong with rounded ends, 4.5 mm long; sessile spikelets 5-6 mm long, ovate-lanceolate; lower glume 5.5 mm long, ovate lanceolate, subacute or slightly 2-toothed with 9-10 faint nerves; upper glume 5.5 mm long, concave, entirely sunken in the cavity of the rachis; lower lemma lanceolate, acute, 3-nerved, margins slightly folded; upper lemma bisexual, 1.5-2.0 mm long , concave acute, margins folded; palea 2-nerved; stamens 3; pedicelled spikelets 5 mm long, lanceolate acuminate, reduced to empty glumes; caryopsis 3 mm long, embryo end broader.

*Fl.-Fr.:*Aug. -Dec.

Common, along water streams, fields, in all districts.

SACCHARUM L. W. 36; MP 3; SA 2

Key to Species

1. Peduncle hairy below the panicle; panicles silvery white; rhachis very fragile ....**S. spontaneum**
1. Peduncle not hairy; panicles not silvery white; rhachis less fragile ....**S. bengalense**

Saccharum bengalense Retz.,Observ. Bot. 5: 16.1789; *S. munja* Roxb., Fl. Ind. 1:246.1820; Hook.f., Fl. Brit. India 7: 119.1896; Bor, Grass. Burm. Ceyl. Ind. Pak. 211.1960; Samvatsar, Fl. West. Trib. Madhya Pradesh 385.1996; Roy in N.P. Singh *et al.,* Fl. M. adhya Pradesh 3:506.2001. *Munj*

Perennials; culms 2.5-3.5 m tall, erect; internodes solid; leaves basal, up to 1 m long, linear, scabrid on the margin, purplish to cream colured; ligule lacerate; spikelets 4-6 mm long, covered with hairs of callus; sessile spikelets longer than the pedicels of pedicellate spikelets; lower glume *ca* mm long, 2- nerved, long hairy; upper glume equal to lower; lower floret empty; upper floret hermaphrodite; lemma hyaline, elliptic, ciliate, mucronate; stamens 3, anthers 2.0-2.5 mm long, yellow or purplish; caryopsis cymbiform, 2 mm long.

Fl.-Fr.: Oct. -Nov.

Common along the banks of rivers and in rocky river beds, all over the area.

Saccharum spontaneum L., Mant. Pl. 2:183. 1771; Hook.f., Fl. Brit. India 7: 118. 1896; Bor, Grass. Burm. Ceyl. Ind. Pak. 214. 1960; Samvatsar, Fl. West. Trib. Madhya Pradesh 386. 1996; Roy in N.P. Singh *et al.,* Fl. Madhya Pradesh 3: 506.2001. *Kans*

Perennials; culms 1-2 m tall, erect, glabrous, nodes hairy with extensive creeping rhizomes; leaves 20-50 cm long, linear, filiform, glaucous beneath, often reduced to the midrib, scabrid on the margin, ligule a scabrid rim; sheaths auriculate, glabrous except hairy throat; inflorescence a panicle, 20-45 cm long, silvery white; spikelets 3-6 mm long, lanceolate; sessile and pedicelled spikelets similar, 1-flowered, hermaphrodite; sessile spikelet 3.5-4.0 mm long, glabrous or ciliate, linear- lanceolate; lower glume 3.5 mm long, lanceolate, sharply acute, 2-nerved, margin ciliate; upper glume 4 mm long, boat-shaped, acuminate, 1-keeled; lower lemma 3 mm long, oblong, ciliate, empty, epaleate; upper lemma bisexual, 3 mm long , linear-elliptical, pointed; caryopsis 1.5 mm long.

*Fl.-Fr.:*Oct. -Feb.

Very common, grows in variable habitats like muddy places along the banks of rivers, ditches and in sandy to clay soils in open fields, in all districts.

SACCIOLEPIS Nash W. 25; MP 3; SA 2

Key to Species

1. Perennials; inflorescence 4-30 cm long; spikelets 4-4.5 mm long; lower glume 5-7 nerved; lemma of sterile spikelet 9-nerved; lemma of fertile spikelet 3-3.5 mm long; palea cartilaginous .... **S. interrupta**

1. Annuals; inflorescence 1-13 cm long; spikelets 2-3 mm long; lower glume 3-5 nerved; lemma of sterile; spikelet 5-7 nerved; lemma of fertile spikelet 1-2 mm long; palea coriaceous .... **S. indica**

Sacciolepis indica (L.) A. Chase, Proc. Biol. Soc. Wash. 21: 8. 1908; *Aira indica* L., Sp. Pl. 1231. 1753, *Panicum indicum*(L.) L., 1771 *nom. illeg*; Hook.f., Fl. Brit. India 7: 41. 1896; Bor, Grass. Burm. Ceyl . Ind. Pak. 357. 1960; Roy in N.P. Singh *et al.,* Fl. Madhya Pradesh 3: 508. 2001.

Annuals; culms 20-60 cm tall, slender, speading; leaves flat to involute, glabrous; sheaths slightly keeled, shorter than internodes; ligule membranous; inflorescence a spike- like, 1-4 cm long panicle; pedicels discoid-tipped; spikelets mm long, glabrous or pilose near the summit; lower glume ovate, half as long as the spikelets, 3-7-nerved; upper glume 7-11-nerved; lower lemma empty; upper lemma pale yellow, narrowly ovate, acute, shining; stamens 3, anthers purple; caryopsis ellipsoid.

*Fl.-Fr.:*Aug. -Dec.

In swamps or along the bank of ponds and ditches, Burhanpur, Khandwa, Khargone.

Sacciolepis interrupta (Willd.) Stapf, Fl. Trop. Africa 9:757. 1920; *Panicum interruptum* Willd., Sp. Pl. 1: 341. 1798; Hook.f., Fl. Brit. India 7: 40.1896; Bor, Grass. Burm. Ceyl. Ind. Pak. 358. 1960; Roy in N.P. Singh *et al.,* Fl. Madhya Pradesh 3: 508. 2001. **(Plate 11, Fig. 6)**

Perenniais; culms 1.0-1.2 m tall, ascending from the creeping and rooting base; leaves 15-20 cm long, linear, acuminate, with cordate or rounded base, glabrous; ligule 2 mm long, membranous truncate; inflorescence spiciform panicle, 15-30 cm long; spikelets 4-5 mm long, ovoid-lanceolate; lower glume 1.5 mm long, ovate, obtuse, 3-5-nerved; upper glume 3.5-4.0 mm long, ovate, acute, 9-nerved; lower floret empty or male; lower lemma 3.5-4.0 mm long, ovate, 9-nerved; palea absent; upper floret bisexual; lemma 2.2 mm long, ovate-oblong , subacute, thinly coriaceous white; palea similar to lemma; stamens 3, anthers dark purple; caryopsis 1.5 mm long, ovoid, apiculate, polished.

*Fl.-Fr.:*Sept. -Dec.

In swamps or along the banks of ponds and ditches, Burhanpur, Khandwa, Khargone.

SCHIZACHYRIUM Nees W. 59; MP 3; SA 2

Key to Species

1. Plants up to 40 cm tall; spatheole green; sessile spikelets 3-3.5 mm long; awn 9-10 mm long; palea 2-keeled .... **S. brevifolium**
1. Plants up to 100 cm tall; spatheole red; sessile spikelets 5-8 mm long; awn 10-25 mm long; palea absent .... **S. exile**

Schizachyrium brevifolium (Sw.) Buse, Pl. Jungh: 359. 1854; *Andropogon brevifolius* Sw., Prodr. Veg. Ind. Occ. 26. 1788; Hook. f., *f.,* Fl. Brit. India 7: 185. 1896; Bor. Grass. Burm. Ceyl. Ind. Pak. 215. 1960; Samvatsar, Fl. West. Trib. Madhya Pradesh 386. 1996; Roy in Singh *et al.,* Fl. Madhya Pradesh 3:510.2001.

Annuals; culms 40-60 cm tall, decumbent, slender, tufted; leaves linear, acute, margins scabrous; sheath compressed, keeled, glabrous; ligule a shallow membrane; inflorescence of a series of racemes; spikelets in pairs; sessile spikelet linear-lanceolate, ± 3 mm long; lower glume bidentate, scaberulous on back; upper glume boat-shaped, acute, 1-nerved, ciliate; lower represented by an empty lemma; upper floret hermaphrodite, 2-toothed, 1.5 mm long; awns 8-6 mm long; anthers 2, purple; caryopsis tapering upwards, ± 2 mm long.

Fl.-Fr.: Aug. -Dec.

Shady and moist places, Alirajpur, Burhanpur, Dhar, Khandwa, Khargone.

Schizachyrium exile (Hochst.) Pilger, Bot. Jahrb. Syst. 54: 284. 1917; *Andropogon exile* Hochst., Flora. 27:142. 1844; Hook.f., Fl. Brit. India 7: 166. 1896; Bor, Grass, Burm. Ceyl. Ind. Pak. 216. 1960; Roy in N.P. Singh *et al.,* Fl. Madhya Pradesh 3: 510.2001.

Annuals; culms 15-16 cm tall, erect, slightly compressed; leaves linear, glabrous; sheath loose, glabrous; ligule a short membrane; inflorescence of short racemes in the spathe; sessile spikelet up to 8 mm long, with bearded callus; glumes equal; lower glume with hairs, scaberulous on keels; upper glume subulate, tip with long-linear point; lower floret empty; upper floret hermaphrpdite, 2-toothed, ± 4 mm long; awn 12-20 mm long; anthers 2, linear; caryopsis linear, ± 4 mm long.

Fl.-Fr.: Aug. -Nov.

Rare, in open grassland, **Khandwa** only, Chandgarh range.

SCHOENEFELDIA Kunth W. 2; MP 1; SA 1

Schoenefeldia gracilis Kunth, Revis. Gramin. 1: 283. 1830; *S. pallida* Edgew., J. Asiat. Soc. Bengal 21:183. 1853; Bor, Grass. Burm. Ceyl. Ind. Pak. 474. 1960; Samvatsar, Fl. West. Trib. Madhya Pradesh 386. 1996; Roy in N.P. Singh *et al.,* Fl. Madhya Pradesh 3: 512.2001.

Annuals; culms 20-40 cm tall, slender, filiform; leaves very narrow, tip attenuate; sheath glabrous; ligule of few hairs; inflorescence golden yellow,of 3-5 spikes; rachis flattened, serrulate; spikelets one-flowered, one-sided, sub-sessile, 1.8-2.0 mm long; glumes sub-equal, compressed, keeled; upper bifid, with curved awn ± 3 cm long; fertile lemma shorter than glumes, ovate-elliptic, laterally compressed, ciliate on nerves, minutely 2-toothed, with a fine awn between the teeth; anthers 3, enclosed in floret; caryopsis narrow, acute, 2.0-2.5 mm long, free from pericarp.

*Fl-Fr.:*Aug. -Oct.

On rocky soil, common in all districts.

SEHIMA Forssk. W. 5; MP 3; SA 4

Key to Species

1. Annuals; racemes 1.5-4 cm long; lower glume of sessile spikelet 4-nerved .... **S. ischaemoides**
1. Perennials
   2. Raceme 9-15 cm long; lower glume of sessile spikelet 8-nerved, sessile spikelet *ca* 12 mm long; joints ciliate on both margins .... **S. notatum**
   2. Raceme 4-10 cm long; lower glume of sessile spikelet 6-nerved
      3. Joints hairy on both margins; lower glume of sessile spikelet 7-8 mm long ....**S. nervosum**
      3. Joints hairy on one margin; lower glume of sessile spikelet 8.8-9. 3 mm long ....**S. sulcatum**

Sehima ischmaeoides Forssk. Fl. Aegypt. -Arab. 178. 1778; *Andropogon sehima* Steud., Syn. Pl. Glumac. 1, 369. 1854; Hook.f., Fl. Brit. India 7:136. 1896; Bor, Grass. Burm. Ceyl. Ind. Pak. 218. 1960; Samvatsar, Fl. West. Trib. Madhya Pradesh 387. 1996.

Annuals; culms 1-25 cm tall, nodes glabrous; leaves flat or involute, linear, scabrid on margins, apex acuminate; ligule a line of hairs; racemes terminal; sessile spikelets lanceolate, 9-12 mm long, awned; lower glume chartaceous, lanceolate, acuminate, grooved on back, 4-nerved, 2-keeled, winged on the keel, apex 2-toothed; upper glume chartaceous, boat-shaped, 5-nerved, keel apically winged, scabrid, produced into a bristle-like, 10-20 mm long awn; lower lemma hyaline, linear-lanceolate, 2-keeled; upper lemma hyaline, 3-nerved, cleft at apex into 2 lobes, 5-6 mm long, awned from the sinus; palea 2-keeled; caryopsis 3.5 mm long; pedicellate spikelets lanceolate, acuminate, 10-12 mm long.

Fl.-Fr.: Aug. -Dec.

On rocky soil, Alirajpur, Dhar, Indore.

Sehima nervosum (Rottler ex Stapf) Stapf, Fl. Trop. Afr. 9:36. 1917: *Andropogon nervosus* Rottler, Neue Schiften. Ges. Naturf. Freunde, Berlin. 4:218. 1803; *Ischmaemum laxum* R. Br., Prodr. Fl. Nov. Holl. 205. 1810; Hook.f., Fl. Brit. India 7:136. 1896; Bor, Grass. Burm. Ceyl. Ind. Pak. 218. 1960; Samvatsar, Fl. West. Trib. Madhya Pradesh 387. 1996; Roy in N.P. Singh *et al.,* Fl. Madhya Pradesh 3: 513.2001. *Bara ponia*

Perennials; culms tufted, erect, up to 90 cm tall; leaves linear, tip filiform, rigid, scabrous; sheaths glabrous, striate; ligule hairy; inflorescence a solitary raceme; pedicels and joints compressed, ciliate on both edges; sessile spikelets 6-8 mm long, linear-lanceolate; callus shortly bearded; lower glume 6-7 mm long, 2-toothed, deeply grooved on back; upper glume boat-shaped, 5.5-6.5 mm long; arista 1.5 mm long; lower floret male; lemma 6 mm long, lanceolate, ciliate; upper floret hermaphrodite; lemma deeply cleft into two lobes; awn 2.5 cm long; anthers 3, yellow, 3-4 mm long; pedicelled spikelet equal to sessile; caryopsis *ca* 2 mm long, narrowly cylindric.

*Fl.-Fr.:*Sept. -Dec.

Common in open grass land, in all districts.

Sehima notatum (Hack.) A. Camus, Bull. Mus. Natl. Hist. Nat. 27: 373. 1921; *Ischaemum notatum* Hack., Monogr. Phan. 6: 246. 1889; Hook., f., Fl. Brit. India 7: 138. 1896; Bor, Grass. Burm. Ceyl. Ind. Pak. 218. 1960; Sharma *et al.,* Fl. Maharashtra 3:590.1996.

Perennial herbs; culms 50-120 cm tall, erect, simple; sheaths 4-14 cm long, ligule membranous, densely hairy; leaves 10-40 cm long, hispid on both surfaces, acuminate; racemes terminal, 9-15 cm long; sessile spikelets lanceolate-acuminate, 1.2-1.3 cm long, awned; pedicelled spikelets lanceolate-acuminate, 1.25-1.3 cm long, 9-nerved, grains ellipsoid.

Fl.-Fr.: Sept. -Dec.

Rare, on bunds of cultivated fields, Khandwa, Khargone. Endemic to India.

Sehima sulcatum (Hack.) A. Camus, Bull. Mus. Natl. Hist. Nat. 27: 373. 1921; *Ischaemum sulcatum* Hack., Monogr. Phan. 6: 248. 1889; Hook.f., Fl. Brit. India 7: 137. 1896; Bor, Grass. Burm, Ceyl. Ind. Pak. 219. 1960; Samvatsar, Fl. West. Trib. Madhya Pradesh 387. 1996; Roy in N.P. Singh *et al.,* Fl. Madhya Pradesh 3: 515. 2001. *Sainer*

Perennials; culms 20-50 cm tall, slender, tufted; leaves linear, glabrous; sheath terete, glabrous; ligule ciliate; inflorescence a terminal raceme; sessile spikelet 9-10 mm long, linear-lanceolate, awned; lower glume 8-9 mm long, deeply grooved, 6-nerved, 2-toothed; upper glume boat-shaped, 3-5-nerved; awn 12-14 mm long; lower lemma male, 4-5 mm long; upper lemma 6 mm long, 2-toothed; awn from sinus, 2.0-3.5 cm long; anthers 3, yellow, 3-4 mm long; caryopsis oblong, *ca* 3 mm long. pedicelled spikelets equal to sessile, awnless; caryopsis oblong, *ca* 3 mm long.

Fl-Fr.: Aug. -Dec.

Common in open grass lands and bunds of fields and slopes of hills, all over the area. Endemic to India.

SETARIA P. Beauv. W. 104; MP 8; SA 4

Key to Species

1. Leaf blades plicately folded like a fan; annuals; branches of panicle divided from the base, bearing short racemes .... **S. barbata**
1. Leaf blades folded or flat;
    2. Bristles retrorsely barbed .... **S. verticillata**
    2. Bristles antrorsely barbed
        3. Inflorescence a cylindrical false spike; spikelets 2-3 mm long .... **S. pumila**
        3. Inflorescence a narrow panicle; spilelets 1.8- 2 mm long .... **S. intermedia**

Setaria barbata (Lam.) Kunth, Revis. Gramin. 1. :47. 1829; *Panicum barbatum* Lam., Tabl. Encycl. 1:171. 1791; Hook.f., Fl. Brit. India 7:56. 1896; Bor, Grass. Burm. Ceyl. Pak. 360.1960; Roy in N.P. Singh *et al.,* Fl. Madhya Pradesh 3:516.2001.

Decumbent, 1-1.5 m tall annuals; leaves linear-lanceolate; sheath margins hairy; ligule obscure; inflorescence a loose, narrow panicle; bristles 5-10 mm long; spikelets *ca* 2.5 mm long; upper glume up to 2 mm long, 5-7 nerved; lower lemma male; upper lemma bisexual, 0.8-2 mm long, acuminate, transversely rugose; caryopsis *ca* 08 mm long, acuminate.

Fl.-Fr.: July-Oct.

Rare, **Khandwa** only, Mortakka forest area.

Setaria intermedia Roem. & Schult., Syst. Veg. 2.489. 1817; *Panicum intermedium* (Roem. & Schult.) Roth, 1821, *nom. illeg*; *Setaria tomentosa* (Roxb.) Kunth, Revis. Gramin. 1:47. 1829; *P. tomentosum* Roxb., Fl. Ind. 1:103. 1832; Hook.f., Fl. Brit. India 7: 79. 1896; Bor, Grass. Burm. Ceyl. Ind. Pak. 365. 1960; Samvatsar, Fl. West. Trib. Madhya Pradesh 389. 1996; Roy in N.P. Singh *et al.,* Fl. Madhya Pradesh 3: 518. 2001. *Chilata.*

Annuals; culms erect, 75-100 cm tall, nodes glabrous; leaves 15-30 cm long, linear-lanceolate, acuminate, narrowed at base, covered with tubercle-based hairs; ligule 1.0-1.5 mm long , membranous; inflorescence a panicle, 10-15 cm long; rachis scabrid; involucres of bristles 1-3, up to 7 cm long; spikelets solitary or often in pairs, 1.8-2.0 mm long; lower glume 0.7-0.8 mm long, apiculate, 3-nerved, glabrous; upper glume 1.0 -1.2 mm long, ovate-rounded, shortly apiculate, 7-nerved, glabrous; lower lemma 1.5 mm long, ovate-oblong, 5-nerved, empty; upper lemma transversely rugose; palea 2-nerved; stamens 3; caryopsis *ca* 1.5 mm long, brown, rugose, upper portion convex, one side ovate.

*Fl.-Fr.:*July-Dec.

Common, under bushes along the roads and wasteland around villages, in all districts.

Setaria pumila (Poir.) Roem. & Schult., Syst. Veg. 2: 891. 1817; *Panicum pumilum* Poir., Encycl. Suppl. 4 : 273. 1816; *P. palledefuscum* Schumach., Beskr. Guin. Pl. 58. 1827; *Setaria pallide-fusca* (Schumach.) Stapf & C.E. Hubb., Kew Bull. 1930: 259. 1930; Bor, Grass. Burm. Ceyl. Ind. Pak. 363. 1960; Samvatsar, Fl. West. Trib. Madhya Pradesh 389. 1996; Roy in N.P. Singh *et al.,* Fl. Madhya Pradesh 3: 519.2001. *Langat*

Annuals; culms 50 – 60 cm tall, prostrate or decumbent, nodes glabrous; leaves 5-20 long, linear-acuminate to acute, slightly contracted base; ligule hairy; sheaths glabrous; inflorescence a panicle, rachis pubescent; spikelets 2.0-2.5 mm long, elliptic, acute; lower glume 1 mm long, ovate, acute, 3-nerved; upper glume 1.3-1.5 mm long, acute, 5-nerved; lemma empty, 5-nerved, paleate; stamens 3 or rarely barren; palea 2 mm long, hyaline, 2-nerved; upper lemma slightly shorter than the floret, transversely rugose, not keeled; palea 1.8 mm long, indurate; stamens 3; caryopsis *ca* 1.5 mm long, rugose, plano-convex.

*Fl.-Fr.:*Aug. -Dec.

Common, grows in rocky yellow soils of fields and grassy localities everywhere.

Setaria verticillata (L.) P. Beauv., Ess. Agrostogr. 51. 1812; *Panicum verticillatum* L., Sp. Pl. ed. 2, 82. 1762; Hook.f., Fl. Brit. India 7: 80.1896; Bor, Grass. Burm. Ceyl. Ind. Pak. 365. 1960: Samvatsar, Fl. West. Trib. Madhya Pradesh 389. 1996; Roy in N.P. Singh *et al.,* Fl. Madhya Pradsesh 3: 519.2001. *Chiktya*

Annuals; culms tufted, ascending or erect, 30-150 cm tall; leaves 10-30 cm long, linear, acuminate, base rounded, sparsely pilose; ligule 1.0-1.5 mm long, truncate, hairy; inflorescence spiciform panicle, 2.5-12.0 cm long; rachis slender, angular, scabrid; spikelets 2.0-2.5 mm long, broadly ovate; lower glume 0.8-1.0 mm long, broadly ovate or orbicular, 3-nerved; upper glume 2.0-2.2 mm long, ovate, acute, 5-7- nerved, empty; palea absent or 1.5 mm long, hyaline; upper lemma 1.8-2.0 mm long, broadly oblong; stamens 3; caryopsis *ca* 1.5 mm long, finely rugose.

*Fl.-Fr.:*July-Dec.

Common in waste land around villages and on old buildings, in all districts.

SORGHUM Moench W. 31; MP 7; SA 6

Key to Species

1. Annuals; lower branches of panicle verticillate or nearly so; sessile spikelets up to 6 mm long .... **S. arundinaceum**
1. Perennials
   2. Lower panicle branches always alternate; sessile spikelets 7.5-8.5 mm long ...... **S. purpureocericeum**
   2. Lower panicle branches verticillate or nearly so
      3. Pedicellate spikelet rudimentary; nodes villous .... **S. nitidum**
      3. Pedicellate spikelets developed; nodes glabrous
         4. Panicle 22-55 cm long; spikelets 4-5.5 mm long; lower glume 8-12 nerved; upper glume 7-nerved, lower 1-nerved .... **S. miliacem**
         4. Panicle 10-45 cm long; lower lemma 2-nerved
            5. Anthers purplish; lower glume 5.5-7 nerved; upper glume 5-7 nerved .... **S. controversum**
            5. Anthers yellow; lower glume 5-11 nerved; upper glume 2-nerved .... **S. halepense**

Sorghum arundinaceum (Desv.) Stapf, Fl. Trop. Africa 9:114. 1917; *Rhaphis arundinacea* Desv., Mem. Soc. Agric. Angers 1:173. 1831; *Andropogon verticilliflorus* Steud. Syn. Pl. Gumamac. 1:393. 1854; *Andropogon stapfii* Hook. f, Fl. Brit. India 7:184. 1896; Bor, Grass. Burm. Ceyl. Ind. Pak. 223. 1960; Samvatsar, Fl. West. Trib. Madhya Pradesh 391. 1996.

Annual or perennial grasses; culms 150-200 cm tall, stout; nodes glabrous or pubescent; leaves flat, linear-lanceolate, margins scabrid, apex acuminate; ligule a fringe of long, stiff hairs; panicle oblong-lanceolate, 30-40 cm long, primary branches verticillate, bearded at the base; sessile spikelets elliptic-lanceolate, 4.5-5.5 mm long, hairy, awned; lower glume chartaceous, dorsally hairy, 7-9 nerved, 2-keeled at apex; upper glume similar, boat-shaped, 5-nerved; lower lemma empty, hyaline; palea absent; upper lemma hyaline, cleft at apex into 2 lobes; awn geniculate, 12-15 mm long; palea hyaline, 2-keeled; caryopsis elliptic; pedicellate spikelets linear-lanceolate, 5-5.5 mm long.

Fl.-Fr.: Sept. -Dec.

Along streams, Alirajpur, Indore.

Sorghum controversum (Steud.) Snowden, J. Linn. Soc. Bot. 55: 210.1955; *Andropogon controversus* Steud. Syn. Pl. Glumac. 1: 391. 1854; Bor, Grass. Burm. Ceyl. Ind. Pak. 222.

1960; Samvatsar, Fl. West. Trib. Madhya Pradesh 390.1996; Roy in N.P. Singh *et al.,* Fl. Madhya Pradesh 3: 523.2001. *Barru*

Rhizomatous perennials; culms 1-2 m tall, terete, nodes glabrous or pubescent; leaves flat, glabrous, mid-nerve prominent, apex tapering to a fin point; sheath terete, glabrous, glaucous; ligule hairy; inflorescence an effuse panicle; pedicellate spikelets linear-lanceolate, ± 6 mm long, hairy; sessile spikelet elliptic, 5-6 mm long, hairy, awned; lower glume lanceolate, 5.0-5.5 mm long, dorsally hairy, 7-nerved; upper glume equal, boat-shaped, 5-7-nerved; lower lemma empty; upper lemma 4.0-4.5 mm long, ciliate, 1-nerved, cleft at apex; awn 1.5-2.0 cm long; anthers 3, purplish, ± 3 mm long; caryopsis obovate, 3.0-3.5 mm long.

*Fl.-Fr.:*Aug. -Nov.

Along water courses and hill slopes in all districts.

Sorghum halepense (L.) Pers., Syn. Pl. 1 : 101. 1805; *Holcus halepensis* L. Sp. Pl. 1047. 1753; Hook.f., Fl. Brit. India 7: 182. 1896; Bor, Grass. Burm. Ceyl. Ind. Pak. 222. 1960; Samvatsar, Fl. West. Trib. Madhya Pradesh 390.1996; Roy in N.P. Singh *et al.,* Fl. Madhya Pradesh 3: 523.2001. *Boru*

Perennials; culms erect, 2.5-3.0 m tall with extensive rhizomes, rooting at the lower nodes; leaves 30-35 cm long, linear-lanceolate, acuminate, margins scabrid; ligule 2-3 mm long membranous truncate; inflorescence an effuse panicle, 15-25 cm long ,hairy at nodes; sessile spikelets 4.5-5.5 mm long, elliptic, acute; lower glume 4.5 mm long flattened, ovate, acute, 5-11-nerved, ciliate on the back side; upper glume 4.5-5.0 mm long, lanceolate, acuminate, 3-nerved, hairy on the back; lower lemma 4.0 – 4.5 mm long, ovate, subacute; palea 1.5 mm long; caryopsis ± 1.4 mm long; pedicellate spikelets 4.5-5.5 long, lanceolate, acute with 2.0-2.5 mm long pedicel; glumes membranous, glabrous, 7-nerved, ciliate on the keel; lemma empty; stamens 3; anthers yellow; caryopsis *ca* 4.5 mm long.

*Fl.-Fr.:*Sept. -Dec.

Common on the hill slopes, bank of rivers, hedges, forest clearings, in all districts.

Sorghum miliaceum (Roxb.) Snowden, J. Linn. Soc. Bot. 55: 205. 1955; *Andropogon miliaceus* Roxb., Fl. Ind. 1: 276. 1820; Hook.f., Fl. Brit. India 7: 182. 1896; Bor, Grass. Burm. Ceyl. Ind. Pak. 223. 1960; Roy in N.P. Singh *et al.,* Fl. Madhya Pradesh 3: 524.2001.

Perennials; culms erect, 1.5-1.75 cm tall with creeping rootstock, rooting below from the nodes; leaves 30-35 cm long, linear-lanceolate, filiform; ligule membranous; inflorescence an effuse panicle; branches tomentose, up to 50 cm long; sessile spikelets 5 mm long, elliptic-lanceolate, acute; lower glume 5 mm long, coriaceous except membranous tip, 2-toothed, hairy on the back; upper glume 5 mm long, sharply acute, ciliate, keeled; lower lemma 4.5 mm long, lanceolate, acute, ciliate on the nerves, empty, epaleate; upper lemma 3.5 mm long, broadly ovate, obtuse or subacute; palea linear 1.5 mm long, ciliate; pedicellate spikelets 5 mm long, lanceolate acute; pedicel 2.5-2.8 mm long, scabrid, callus bearded; lower glume 5 mm long, membranous, ciliate on margins, inflexed; upper glume 5 mm long, ciliate; upper lemma usually male or empty; stamens 3, anthers ± 1mm long, yellow; caryopsis *ca* 1 mm long.

*Fl.-Fr.:*Aug. -Nov.

Common as weed in agricultural fields and open forest areas, all over the area.

Sorghum nitidum (Vahl) Pers., Syn. Pl. 1: 101. 1805; *Holcus nitidus* Vahl, Symb. Bot, 2: 102. 1791; *Andropogon nitidus* (Vahl) Kunth, Revis. Gramin. 1:166. 1829; Hook.f., Fl. Brit. India 7: 185. 1896; Bor, Grass. Burm. Ceyl. Ind. Pak. 245. 1960; Samvatsar, Fl. West. Trib. Madhya Pradesh 390.1996; Roy in N.P. Singh *et al.,* Fl. Madhya Pradesh 3: 524.2001.

Perennials; culms erect, 1.0-2.5 cm tall, nodes villous; Leaves 35-50 cm long, sparsely hairy, scabrid on margins, midrib white; sheath glabrous with long hairs at mouth; ligule membranous, truncate, dorsally ciliate; inflorescence an elongated panicle; spikelets paired, dissimilar; sessile spikelets 4 mm long, elliptic-acute, ciliate, appressed hairy at top; lower glume dorsally flattened, brown hairy on keels, 5-7-nerved; upper glume similar to the lower but glabrous and shining at the basal portion; lower floret reduced, barren; upper floret 2 mm long, hyaline, ciliate on margins; stamens 3, anthers 2 mm long, yellow; pedicellate spikelets rudimentary, 3 mm long; callus densely hairy; caryopsis 2 mm long , brownish.

*Fl.-Fr.:*Sept. -Nov.

Rare, grows as weed in agricultural fields, Alirajpur, Barwani, Dhar.

Sorghum purpureosereceum (A. Rich.) Schweinf. & Asch., Beitr. Fl. Aethiop. 310.1867; *Andropogon purpureosericeus* Hochst. ex A. Rich. Tent. Fl. Abyss. 2: 469. 1850; *Sorghum deccanense* Stapf ex Raizada, Indian Forester 80:43. 1954; Bor. Grass. Burm. Ceyl. Ind. Pak. 245. 1960; Samvatsar, Fl. West. Trib. Madhya Pradesh 390.1996; Roy in N.P. Singh *et al.,* Fl. Madhya Pradesh 3: 523.2001. *Gandhal*

Annuals; culms 1-2 m tall, erect, terete, nodes bearded; leaves linear-lanceolate, flat, hairy, margins scabrid; sheath terete, glabrous; ligule hairy; inflorescence a contracted panicle; rachis glabrous, swollen at base; pedicellate spikelets slightly shorter than sessile, male; anthers 3; sessile spikelet 6-7 mm long, densely hairy, awned; lower glume *ca* 7 mm long, densely hairy, 7-nerved, keeled, beaked at tip, minutely 2-toothed; upper glume similar, 5-nerved, 2-toothed; lower lemma empty, 2-nerved, ciliate along margins; upper lemma hyaline, cleft at apex; margins ciliate; awn 3 cm long; stamens 3, purple, 3.5-4.0 mm long; caryopsis elliptic, 4 mm long.

*Fl.-Fr.:*Sept. -Dec.

Common along bunds and fields, Alirajpur, Burhanpur, Dhar, Indore, Khandwa. Endemic to India.

SPODIOPOGON Trin. W. 15; MP 2; SA 1

Spodiopogon rhizophorus (Steud.) Pilg. Nat. Pflanzenfam. ed. 2, 14 e :119. 1940; *Andropogon rhizophorus* Steud., Syn. Pl. Glumac. 381. 1854; *Spodiopogon albidus* Benth., J. Linn. Soc. Bot. 19:66. 1881; Hook.f. Fl. Brit. India 7:108. 1896; Bor, Grass. Burm. Ceyl. Ind. Pak. 246. 1960; Samvatsar, Fl. West. Trib. Madhya Pradesh 391. 1996; Roy in N.P. Singh *et al.,* Fl Madhya Pradesh 3: 526.2001. *Math jhara*

Annuals or perennials; culm 20-70 cm tall, decumbent, branched; leaves linear-lanceolate, petioled, glabrous; sheath compressed, margins ciliate; ligule chartaceous, obtuse; inflorescence an oblong panicle, branches hairy at base; racemes silky, joints of rachis fiattened ± 3 mm long; pedicellate spikelets linear-lanceolate, slightly smaller than sessile, awned; sessile spikelets 5.0-5.5 mm long, hairy, awned; lower glume equal to spikelet, pilose, 7-nerved; upper glume equal to lower, boat-shaped, with short arista; lower lemma 3.0-3.25 mm long, nerveless, ovate; anthers 3,

3 mm long; upper lemma 3.5-4.0 mm long, bilobed, with 10- 11 mm long awn; anthers 3, 3 mm long; caryopsis *ca* 3 mm long, fusiform.

*Fl.-Fr.:*Oct. -Jan.

Frequent on hill slopes along shady places, along the road sides in forest, in all districts.

SPOROBOLUS R. Br.W. 184; MP 8; SA 10

Key to Species

1. Perennials
    2. Stamens 2; panicles effuse; lower glume 0.3-0.4 mm long ..... **S. diandrus**
    2. Stamens 3; panicle contracted; lower glume more than 0.4 mm long
        3. Plants stoloniferous; lateral branches lacking; spikelets 2.25-2.50 mm long; leaves not distichous;. ..... **S. maderaspatanus**
        3. Plants not stoloniferous
            4. Spikelets 2.5 mm long; roots very thick, covered with soft felt of root hairs ..... **S. iocladus**
            4. Spikelets under 2.5 mm long; roots not so covered
                5. Glumes minutely fringed; palea plicate in the middle; inflorescence usually interrupted; leaf sheath longer than internodes ..... **S. indicus**
                5. Glumes not fringed; palea not plicate, neither the inflorescence interrupted; leaf sheath compressed, equalling the internodes ..... **S. fertilis**
1. Annuals
    6. Upper glume as long as or little exceeding length of lemma
        7. Panicle contracted; lower branches not truely whorled; spikelets 1.25-1.5 mm long; stamens 2 ..... **S. piliferus**
        7. Panicle effuse; lower branches whorled; spikelets 1.5-2 mm long; stamens 3 ..... **S. capillaris**
    6. Upper glume shorter than lemma
        8. Panicle branches whorled ..... **S. coromandelianus**
        8. Panicle branches alternate . ..... **S. tenuissimus**

Sporobolus capillaris Miq., Anal. Bot. Ind. 2: 25. 1851; Bor, Grass. Burm. Ceyl. Ind. Pak. 627. 1960; Samvatsar, Fl. West. Trib. Madhya Pradesh 393. 1996; Roy in N.P. Singh *et al.,* Fl. Madhya Pradesh 3: 529.2001.

Annual; culms 20-40 cm tall, erect, tufted; leaves linear-lanceolate, flat, hispid, bristly on margins; sheaths compressed; ligule hairy; inflorescence an effuse panicle, branches capillary and scabrous; pedicels longer than spikelets; spikelets 1.5-2.0 mm long, grayish- green; lower glume 1 mm long, 1-nerved; upper glume 2 mm long, 1-nerved, acuminate; lemma ovate, similar and equalling the upper glume; palea scabrid; anthers pinkish; caryopsis 0.8 mm long, discoid.

Fl -Fr. : July-Jan.

Along road sides, weed of gardens, fields, common in all districts.

Sporobolus coromandelianus (Retz.) Kunth, Revis. Gramin. 1: 68. 1829; *Agrostis coromandeliana* Retz., Observ. Bot. 4:19. 1786; Hook.f., Fl. Brit. India 7: 252. 1896; Bor, Grass. Burm. Ceyl. Ind. Pak. 627. 1960; Samvatsar, Fl. West. Trib. Madhya Pradesh 393. 1996; Roy in N.P. Singh *et al.,* Fl. Madhya Pradesh 3: 529.2001. *Chinari*

Annuals; culms 15-40 cm tall, tufted, erect or ascending; leaves lanceolate, glabrous or pilose, margins papillate; sheath compressed glabrous, hairy at mouth; ligule very short, hairy; inflorescence an effuse panicle; branches capillary, whorled; spikelets 1.2 mm long, lanceolate; lower glume nerveless; upper glume 1 mm long, hyaline, 1-nerved, acute; lemma similar, equalling the upper glume; anthers 3, yellow; caryopsis obovate.

Fl.-Fr.: July-Sept.

Common in open grass lands and along road sides, all over the area.

Sporobolus diandrus (Retz.) P. Beauv., Ess. Agrostogr. 26. 1812; *Agrostis diandrus* Retz., Observ. Bot. 5:19. 1789; Hook.f., Fl. Brit. India 7: 247. 1896; Bor, Grass. Burm. Ceyl. Ind. Pak. 629. 1960; Samvatsar, Fl. West. Trib. Madhya Pradesh 393. 1996; Roy in N.P. Singh *et al.,* Fl. Madhya Pradesh 3: 529.2001.

Perennials; culms 30-50 cm tall, densely tufted; leaves 15-25 cm long, linear, narrowed toward apex, inrolled when dry; ligule hairy; sheath glabrous; inflorescence a terminal panicle, 10-20 cm long, purplish or slaty-grey; spikelets 1.2-1.4 mm long, ellipsoid-lanceolate, acute; lower glume 0.3-0.4mm long, ovate, truncate, nerveless; upper glume 0.6-0.8 mm long, elliptic obovate, 1-nerved; lemma 1.5 mm long, lanceolate, subacute, 1-nerved; palea 1.2 mm long, olivaceous, nerveless; stamens 2, anthers purplish; caryopsis 0.5-0.6 mm long, sub truncate, oblong-obovoid.

*Fl.-Fr.:*July-Dec.

Common, throughput the area.

Sporobolus fertilis (Steud.) Clayton, Kew Bull. 19: 291. 1965; *Agrostis fertilis* Steud.,Syn. Pl. Glumac. 1:170.1854; *Sporobolus indicus auct non* (L.) R. Br.,1810; Hook.f., Fl. Brit. India 7: 247. 1896; Bor, Grass. Burm. Ceyl. Ind. Pak. 630.1960; Samvatsar, Fl. West. Trib. Madhya Pradesh 393. 1996; Roy in N.P. Singh *et al.,* Fl. Madhya Pradesh 3:530.2001.

Perennials; culms 40-70 cm tall, tufted, terette; leaves linear, convolute, glabrous, acuminate; sheaths compressed, glabrous; ligule hairy; inflorescence a contracted panicle, branches short; pedicels shorter or longer than spikelets; spikelets 1-flowered, 1.6-2.0 mm long, lanceolate; lower glume 0.5 mm long, nerveless, acute; upper glume ± 1 mm long, 1-nerved; lemma ovate, 1.5-2.0 mm long, 1-nerved; palea subequal to lemma, 2-nerved; anthers 3, ± 0.5 mm long; caryopsis 0.6-0.7 mm long, elliptic-obovate, brownish.

Fl.-Fr.: July- Nov.

Common along margins of ponds and in open grass lands, in all districts.

Sporobolus indicus (L.) R. Br. Prodr. Fl. Nov. Holl. 170.1810; *Agrostis indica* L. Sp. Pl. 63. 1753; Bor, Grass. Burm. Ceyl. Ind. Pak. 630.1960; Samvatsar, Fl. West. Triabal MP 393. 1996; Roy in N.P. Singh *et al.,* Fl. Madhya Pradesh 3: 530.2001. *Machhario*

Perennials; culms 50-120 cm tall, tufted, erect; leaves mostly crowded near the base, tips filiform, glabrous; sheaths longer than internodes, ribbed, glabrous; ligule minute, hairy; inflorescence a contracted panicle, often interrupted, branches very short; spikelets 1.5-2.0 mm long, pale green; pedicels shorter than spikelets; glumes unequal, minutely fringed, hyaline, upper one nerved, 1/3 of spikelet; lemma 1 mm long, broadly ovate, 1-nerved; palea plicate in the middle; anthers 3, purplish; caryopsis red brown, obovoid, adnate to pericarp.

*Fl.-Fr.:*Sept. -Nov.

Common in open grasslands and along road sides, Alirajpur, Burhanpur, Dhar, Indore, Khandwa, Khargone.

Sporobolus ioclados (Trin.) Nees, Fl. Afr. Austral. III. 161. 1841; *Vilfa ioclados* Nees ex Trin., Mem. Acad. Imp. Sci. Saint-Petersbourg, Ser. 6, Sci. Math. Seconde Pt. Sci Nat 5 (2):65. 1840; *Sporobolus marginatus* Hochst. ex A. Rich., Tent. Fl. Abyss. 2:397. 1850; Hook.f., Fl. Brit. India 7:252. 1896; Bor, Grasss. Burm. Ceyl. Ind. Pak. 632. 1960; Samvatsar, Fl. West. Trib. Madhya Pradesh 393. 1996; Roy in N.P. Singh *et al.,* Fl. Madhya Pradesh 3:530.2001.

Perennials; culms 20-60 cm tall; roots very thick covered with soft felt of root hairs; leaves flat or convolute, margin serrulate, ciliate, ligule hairy; panicle branches capillary; spikelets 2.5 mm long; lower glume 0.5-0.8 mm long, ovate, nerveless; upper glume 2.5 mm long, 1-nerved; lemma similar to upper glume; caryopsis sub-globose.

*Fl.-Fr.:*July-Dec.

Rare, Alirajpur, Dhar.

Sporobolus maderaspatanus Bor, Kew Bull. 1957: 234. 1957 & Grass. Burm. Ceyl. Ind. Pak. 632. 1960; *Sporobolus orientalis* (Nees) Kunth, 1883, *nom. illegit;* Hook., f., Fl. Brit. India. 7: 251. 1896; Sharma *et al.,* Fl. Maharashtra 3:614. 1996.

Herbaceous perennials; Culms 15-50 cm tall, stoloniferous; leaves 3-15 cm long, linear to linear-lanceolate, scaberulous on margins; leaf sheath glabrous; ligule a ciliate membrane; panicles pyramidal, 3-10 cm long; skpikelets 2-2.5 mm long; caryopsis 1 mm long, elliptic-obovate, pale brown.

Fl.-Fr.: Aug. -Dec.

In grass lands, not common, Burhanpur, Khandwa.

Sporobolus piliferus (Trin.) Kunth, Enum. Pl. 1:211. 1833; *Vilfa pilifera* Trin., Gram. Unifl. Sesquifl. 157. 1824; Hook.f., Fl. Brit. India 7:251. 1896; Bor, Grass. Burm. Ceyl. Ind. Pak. 632. 1960; Samvatsar, Fl. West. Trib. Madhya Pradesh 394. 1996.

Annuals; culms 10-30 cm tall; nodes glabrous; leaves flat or convolute, margins scabrid or serrulate; ligule a fringe of hairs; panicle linear, 3-10 cm long, ovate-lanceolate, hyaline, nerveless; upper glume 2 mm long, oblong-lanceolate, 1-nerved; palea oblong; caryopsis ellipsoid.

*Fl.-Fr.:*Aug. -Dec.

Occasional, in rocky habitats and near streams, Alirajpur, Dhar.

Sporobolus tenuissimus (Schrank) Kuntze, Revis. Gen. Pl. 3(2): 369.1898; *Panicum tenuissimum* Mart. ex Schrank, Denkschr. Koenigl.-Baier. Bot. Ges. Regensberg, 2: 26.1822; *Sporobolus minutiflorus* (Trin.) Link, Hort. Berol. 1: 88.1827; *Vilfa minutiflora* Trin., Gam. Unifl. Sesquifl. 158.1824; Hook.f., Fl. Brit. India 7: 248.1896; Bor, Grass. Burm. Ceyl. Ind. Pak. 633.1960; Samvatsar, Fl. West. Trib. Madhya Pradesh 394.1996; Roy in N.P. Singh *et al.,* Fl. Madhya Pradesh 3: 531.2001.

Annuals; culms erect, slender, 18 -70 cm tall; leaves 4-20 cm long, narrow, flat, finely acuminate; ligule a ring of hairs; inflorescence a panicle, 10-35 long, narrowly oblong, delicate and diffuse; spikelets 0.8-1.0 mm long, gaping at maturity; lower glume 0.1-0.2 mm long, oblong, truncate; upper glume 0.3-0.5 mm long, ovate-oblong, subacute; lemma ovate, as long as spikelet; palea 0.8 mm long; stamens 3, anthers dark brown; caryopsis 0.5 mm long, truncate.

*Fl.-Fr.:*July-Dec.

Common, grows on old walls, open forests, along streams, Alirajpur, Barwani, Burhanpur, Dhar, Khandwa, Khargone.

Sporobolus tetragonus Bor, Kew Bull. 4:251.1949 & Grass. Burm. Ceyl. Ind. Pak. 633.1960; Hook.f., Fl. Brit. India 7:252.1896; Roy in N.P. Singh *et al.,* Fl. Madhya Pradesh 3:531.2001.

Annuals; culms 20-70 cm tall, slender; leaves narrowly lanceolate; sheaths ciliate at mouth; ligule obscure; inflorescence a narrow panicle with capillary, whorled, branches; spikelets *ca* 1.5 mm long, pedicels short; lower glume lanceolate, *ca* 0.8 mm long, hyaline, *ca* 1.4 mm long; lemma similar to upper glume; anthers 2, yellow; caryopsis reddish-brown, about 1 mm long.

Fl.-Fr.: Aug. -Oct.

Rare, **Indore** only, Mendh forest range.

TETRAPOGON Desf. W. 5; MP 1; SA 1

Tetrapogon tenellus (Roxb.) Chiov., Annulario Reale Ist. Bot. Roma. 8: 352. 1908; *Chloris tenella* J. Koenig ex Roxb., Fl. Ind. 1: 330.1820; Hook.f., Fl. Brit. India 7: 290.1896; Bor, Grass. Burm. Ceyl. Ind. Pak. 475. 1960; Samvatsar, Fl. West. Trib. Madhya Pradesh 394. 1996; Roy in N.P. Singh *et al.,* Fl. Madhya. Pradesh 3: 532.2001.

Annuals; culms 20-50 cm tall, erect or decumbent; leaves linear-lanceolate, glabrous or sparsely hairy; sheath compressed, keeled, glabrous; ligule membranous; inflorescence usually an erect spike; spikelets 5-6 mm long, consisting 3-4 perfect and 1-3 sterile florets above them; lower glume 3 mm long, 1-nerved, keeled; upper glume 4.5-5.0 mm long with short awn; lowest lemma 5.0-5.5 mm long, 3-nerved, hairy; callus crustaceous; awn 7-8 mm long; palea keels ciliate; other fertile lemmas relatively shorter with shorter awns; empty lemmas usually 2, obovate-cuneate, with short mucro or not; anthers 3, yellow, 3 mm long; caryopsis 3.0-3.5 mm long, sub -trigonous.

*Fl.-Fr.:*Aug. -Nov.

Shade of trees, rocky areas, old walls, bunds of fields. Alirajpur, Burhanpur, Dhar, Indore, Khargone.

THELEPOGON Roth W. 2; MP 1; SA 1

Thelepogon elegans Roth, Syst. Veg. 2: 788. 1817; Hook.f., Fl. Brit. India 7: 1896; Bor, Grass. Burm. Ceyl. Ind. Pak. 247. 1960; Samvatsar, Fl. West. Trib. Madhya Pradesh 394. 1996; Roy in N.P. Singh *et al.,* Fl. Madhya Pradesh 3: 534.2001. *Farodi*

Perennials; culms 20-60 cm tall, erect stout; leaves flat, hispid, margins scabrid and wavy; sheaths terete, scabrid; ligule membranous; inflorescence of 2-several spikes; callus hairy; sessile spikelet 6-7 mm long, lanceolate; lower glume 5-6 mm long, transversely rugose, tubercled, apex beaked; upper glume 6.0-6. 7 mm long, boat-shaped, acuminate; lower lemma elliptic, 2-nerved, ciliate; anthers 3, yellow; upper lemma lanceolate, ± 5 mm long bilobed; awn 20-25 mm long; anthers 3, yellow; caryopsis cylindric, 2 mm long.

*Fl.-Fr.:*Aug. -Dec.

Common in open grasslands and cultivated fields, in all districts.

THEMEDA Forssk. W. 27; MP 4; SA 4

Key to Species

1. Annuals; margin of spatheoles with tubercled based hairs; involucral spikelets 4-6 mm long; awn 25-40 mm long .... **T. quadrivalvis**

1. Perennials

 2. Ligule absent; spatheole 2.5 3.5 cm long; involucral spikelets 5-6 mm long; spikelets 4.25 mm long; callus dark brown; awn 30-35 mm long .... **T. laxa**

 2. Ligule present

  3. Spatheole bristly; awn 40-50 mm long; involucral spikelets 6-6.5 mm long; plants perennials or robust annuals .... **T. triandra**

  3. Spatheole not bristly, 1-2 cm long; sessile spikelets 3-3.5 mm long; involucral spikelets 5.5-7.5 mm long; callus red; awn 25-40 mm .... **T. tremula**

Themeda laxa (Andersson) A. Camus, Bull. Mus. Natl. Hist. Nat. 26: 423. 1920; *Anthistiria laxa* Andersson, Nova Acta Regiae Soc. Sci. Upsal. 3, 2: 243. 1856; Hook.f., Fl. Brit. India 7: 213. 1896; Bor, Grass. Burm. Ceyl. Ind. Pak. 251. 1960; Roy in N.P. Singh *et al.,* Madhya Pradesh 3: 535. 2001. *Charha*

Perennials; culms 0.6-1.2 m tall, tufted, branched, leafy glabrous; leave 15 cm long, glabrous, scabrid on margins; sheaths spatheolate; ligule absent; inflorescence of fan-shaped clusters of racemes; spatheoles 2.0-3.5 mm long, acuminate; involucral spikelets male or neuter, 4-5 mm long; callus glabrous; lower glume narrow, spinously acute, tuberculate, hairy, 9-11-nerved; upper glume ± 5 mm long, ciliate; lemma equal to glume; sessile spikelets ± 4 mm long; lower glume truncate, 9-nerved; upper glume narrower; lower lemma ± 3 mm long; upper lemma with awn; pedicellate spikelets ± 2.5 mm long, very narrow, sharply acute; anthers long, yellow; caryopsis 3-4 mm long, brown.

Fl.-Fr.: July-Nov.

Occasional in river beds and open grass land, Khandwa, Khargone.

Themeda quadrivalvis (L.) Kuntze, Revis. Gen. Pl. 2:793. 1891; *Andropogon quadrivalvis* L., Syst. Veg. 13 2:758. 1774; *Anthistiria ciliata* L.f., 1782, nom. illeg; Hook.f., Fl. Brit. India 7:213. 1896; Bor, Grass. Burm. Ceyl. Ind. Pak. 252. 1960; Roy, in N.P. Singh *et al.,* Fl. Madhya Pradesh 3:536.2001. *Guned* **(Plate 11, Fig. 7)**

Annuals; culms 30-90 cm tall; leaves 15-30 cm long, glabrous, sheaths loose; ligules membranous; inflorescence a dense panicle, spatheoles cymbiform, acuminate, 1.7-2.5 cm long; involucral spikelets 4.5-5.5 mm long, acute; lower glume 4-5 mm long, with golden-brown tubercled hairs; upper glume glabrous, equal to the lower, 3 – nerved; lemma hyaline, 1-nerved; palea absent; sessile spikelets bisexual, *ca* 4.5mm long; callus hairy; lower glume *ca* 4 mm long, 6-7 nerved; upper glume similar; upper lemma with a very short awn; anthers 3, yellow; caryopsis brown, subcylindrical, 5-6 mm long, callus hairy.

*Fl.-Fr.:*Sept. -Oct.

Very common in all districts.

Themeda tremula (Nees ex Steud.) Hack., Monogr. Phan. 6: 667. 1889; *Anthistiria tremula* Nees ex Steud. Syn. Pl. Glumac. 1: 401. 1854; Hook.f., Fl. Brit. India. 7: 214. 1896; Bor, Grass. Burm. Ceyl. Ind. Pak. 254. 1960; Sharma *et al.,* in Fl. Maharashtra 3:623. 1996.

Perennial herbs; culms 30-120 cm high, erect or ascending from a creeping rootstock, nodes glabrous; leaves linear, tips setaceous, margins scabrid; ligule narrow, membranous; leaf sheath compressed; leaf blade linear ovate, 5-40 × 0.2-1 cm, sparsely hairy, margins scaberulous, apex acuminate; panicles 30-60 cm long with loose or congested clusters of racemes; involucral spikelets *ca* 7mm long; pedicelled spikelets same as involucral ones, but glabrous; bisexual spikelets *ca* 4 mm long.

*Fl.-Fr.:*Sept. -Dec.

Rare, Barwani, Khargone.

Themeda triandra Forssk., Fl. Aegypt-Arab. 178. 1775; *Anthistiria imberbis* Retz., Observ. Bot. 3: 11. 1783; Hook.f., Fl. Brit. India 7: 211. 1896; Bor, Grass. Burm. Ceyl. Ind. Pak. 254. 1960; Samvatsar, Fl. West. Trib. Madhya Pradesh 396. 1996; Roy in N.P. Singh *et al.,* Fl. Madhya Pradesh 536.2001. *Ghonyad*

Perennials; culms 70-100 cm tall, erect; leaves 15-30 cm long, linear, rounded at the base, acuminate; ligule 2 mm long, membranous, truncate; inflorescence a leafy panicle; spatheoles 3 cm long, lanceolate, boat-shaped, keeled, compressed; involucral spikelets subsessile, linear-lanceolate to acuminate. 6.2-6. 4 mm long, margin scarious; lower glume 6.0-6.2 mm long, oblong-lanceolate, acute, many nerved; upper glume similar to lower, 3-nerved; lemma hyaline, equal to the glume; stamens 3, anthers 2.0 -2.5 mm long; sessile spikelets hermaphrodite; callus bearded with stiff, brown pairs; lower glume 5.5-6.0 mm long, 9-11-nerved; upper glume 5.5-6.0 mm long lanceolate, acute , 3-nerved; lower lemma linear, oblong, acute or obtuse; upper lemma 4.0 -4.5 mm long, awn 3-6 cm long; colum hispid brownish, empty; palea 2- toothed, glabrous; stamens 3, anthers 2 mm long; pedicelled spikelets similar to the involucral spikelets, consisting of 2 glumes and one lemma.

*Fl.-Fr.:*Aug. -Dec.

Common in open grasslands, cultivated fields and bank of streams, in all districts.

THYSANOLAENA Nees Monotypic

Thysanolaena latifolia (Roxb. ex Hornem.) Honda, J. Fac. Sci. Univ. Tokyo, Sect. 3, Bot 3:312. 1930; *Melica latifolia* Roxb. ex Hornem., Hort. Bot. Hafn. Suppl. : 117. 1819; *Thysanolaena maxima* (Roxb.) Kuntze, Revis. Gen. Pl. 2:784. 1891; *Agrostis maxima* Roxb., Fl. Ind. 1:319. 1820; Bor, Grass. Burm. Ceyl. India Pak. 650.1960; Samvatsar, Fl. West. Trib. Madhya Pradesh 396. 1996; Roy in N.P. Singh *et al.,* Fl. Madhya Pradesh 3: 537.2001. *Phul buhari* **(Plate 11, Fig. 8)**

Pernnials; culms 1-3 m tall, erect, solid, nodes glabrous; leaves 30 -40 cm long, lanceolate, narrowed into an acuminate apex; ligule 1.5 mm long, membranous, backed by short stiff hairs; inflorescence an effuse panicle, up to 60 cm long with filiform, glabrous branches; spikelets numerous, 1-2-nate, minute, greenish or purple, 2-flowered; lower glue 0.5 mm long, ovate-acute, obscurely 1- nerved; upper glume 0.7 mm long, oblong-cuneate; lower lemma 1.8 mm long lanceolate-acuminate; upper lemma 0.8 mm long, hyaline; stamens 2; caryopsis 0.5 mm long, ovoid, reddish- brown.

*Fl.-Fr.:*Nov. - May

Confined to hilly areas only, Alirajpur, Burhanpur, Khandwa.

TRAGUS Haller W. 8; MP 1; SA 1

Tragus mongolorum Ohwi, Acta Phytotax. Geobot. 10: 268. 1941; Bor, Grass. Burm. Ceyl. Ind. Pak. 682. 1960; *T. roxburghii* Panigrahi, Kew Bull. 29: 496. 1974; Samvatsar, Fl. West. Trib. Madhya Pradesh 397. 1996; Roy in N.P. Singh *et al.,* Fl. Madhya Pradesh 3: 528. 2001.

Annuals; culms 7-20 cm tall, erect or decumbent; leaves flat, narrowly lanceolate, ciliate on margins; sheath glabrous, with a few hairs at mouth; ligule membranous, truncate; inflorescence a terminal receme, rachis pubescent, ending with spikelet; burs on spikelets deciduous; spikelets 4.0-4.5 mm long sessile; lower glume minute nerveless; upper glume equal to spikelet, 5-ribbed, with hooked spines; lemma hyaline, 2.0- 2.5 mm long, 3- nerved, acute; anthers 3, yellow; caryopsis 4.0-1.5 mm long.

*Fl.-Fr.:*July-Sept.

Common on rocky soils, edges of tar roads, throughout the area.

TRIPOGON Roem. & Schult. W. 46; MP 2; SA 6

Key to Species

1. Rhizomatous plants; lower lemma 2-fid at apex
   2. Rhachis angular; lateral branches present; fertile spikelets 5-7 flowered; lower glume 1-1.3 mm long; upper glume 2- 2.5 mm long; leaves pungent at apex .... **T. pungens**
   2. Rhachis flattened
      3. Spikelets 40-50 flowered; lower glume 3-3.25 mm long; upper glume 4.5 mm long; leaves not pungent .... **T. polyanthus**
      3. Spikelets 5-10 flowered; lower glume 2-3 mm long, upper glume 3-3.5 mm long; leaves acuminate .... **T. lisboe**

1. Non rhizomatous plants; rhachis flattened or angular
   4. Apex of lower lemma 2-lobed; spikelets 8-30 flowered . ... **T. jacqemontii**
   4. Apex of lower lemma 4-lobed
      5. Rhachis angular; lateral branches lacking; spikelets 5-8 flowered; leaves filiform .... **T. filiformis**
      5. Rhachis flattened; lateral branches present; spikelets 5-14 flowered; leaves not filiform .... **T. bromoides**

Tripogon bromoides Roth, Syst. Veg. 2: 600.1817; Hook.f., Fl. Brit. India 7: 287. 1896; Bor, Grass. Burm. Ceyl. Ind. Pak. 521. 1960; Samvatsar, Fl. West. Trib. Madhya Pradesh 397. 1996; Roy in N.P. Singh *et al.,* Fl. Madhya Pradesh 3: 540.2001.

Tufted perennials; culms 20-80 cm tall; leaves mostly basal, glabrous, convolute; sheath glabrous, striate; ligule membranous; inflorescence of racemes; rachis flat, glabrous; spikelets 0.6-3.0 mm long, 10-30-flowered; lower glume 4-5 mm long, membranous, 1-nerved, notched at apex; upper glume similar, except being mucronate; lemma 4.5-5.5 mm long, with awn 1.5-2.0 mm, 3-nerved. 2-lobed; anthers 3, yellow,± 0.5 mm long; caryopsis 1.5-2. o mm long, oblong, brown.

*Fl.-Fr.:*Aug. -Nov.

In open grasslands, on hill slopes and forest margins, Alirajpur, Burhanpur, Dhar, Indore, Khandwa.

Tripogon filiformis Nees ex Steud. Syn. Pl. Glumac. 1: 301. 1854; Hook., f., Fl. Brit. India 7: 288.1896; Bor, Grass. Burm. Ceyl. Ind. Pak. 521. 1960; Sharma *et al.,* Fl. Maharashtra 3:628. 1996.

Perennials; culms tufted, terete, 10-45 cm tall, slender, erect, glabrous, nodes glabrous; leaf sheath terete, glabrous or sparsely hairy; ligule obsolete; leaf blade filiform, 5-17 cm long, glabrous, apex acuminate to setaceous; racemes 5-25 cm long, rachis flattened, slender, glabrous; spikelets linear-oblong, sessile, 3-7.5 mm long; lower glume membranous, ovate-oblong, 1.5-1.8 mm long, 1-nerved, apex notched; upper glume membranous, bearded at base, ovate-oblong, 7-8 mm long, 3-nerved, apex 4-lobed, linear lobes obtuse, outer lobes shortly aristate, median awn nearly twice as long as its lemma or shorter, 4-5 mm long; palea hyaline, elliptic-oblong, 2-2.4 mm long, 2-keeled, keels ciliate, apex obtuse; stamens 3; grain oblong-elliptic.

*Fl.-Fr.:*Aug. -Dec.

Moist rocky soil, Burhanpur, Khandwa, Khargone.

Tripogon jacquemontii Stapf, Bull. Misc. Inform. Kew, 1892. 85.1892; Hook., f., Fl. Brit. India 7: 286.1896, Bor, Grass. Burm. Ceyl. Ind. Pak. 522.1960; Samvatsar, Fl. West. Trib. Madhya Pradesh 397.1996; Roy in N.P. Singh *et al.,* Fl. Madhya Pradesh 3: 540.2001. *Tintagio*

Tufted perennials; culms 20-80 cm tall; leaves mostly basal, glabrous, convolute; sheath glabrous, striate; ligule membranous; racemes spiciform; rachis flat, glabrous; spikelets 0.6-3.0 mm long, 10-30-flowered; lower glume 4-5 mm long, membranous, 1-nerved, notched at apex; upper glume similar, except being mucronate; lemma 4.5-5.5 mm long with awn 1.5-2.0 mm, long, 3-nerved, 2-lobed; anthers 3, yellow; caryopsis 1.5-2.0 mm long, oblong, brown.

*Fl.-Fr.:*Aug. -Dec.

Common in open grassland, on old walls, open places, in all districts.

Tripogon lisboe Stapf, Bull. Misc. Inform. Kew, 1892:84. 1892; Hook.f. Fl. Brit. India 7:286. 1896; Bor, Grass. Burm. Ceyl. Ind. Pak. 522. 1960; Roy, in N.P. Singh *et al.,* Fl. Madhya Pradesh 3:541.2001.

Perennials; culms 50-120 cm tall; nodes glabrous; leaf-sheath terete; ligule membranous; leaves 20-40 cm long, glabrous, apex acuminate; racemes 10-30 cm long, rhachis flattened; spikelets 5-7 mm long, 5-10 flowered; lower glume 1-nerved, upper glume 1-nerved, 2-toothed; lemmas membranous, 3-nerved, apex notched into 2 lobes, awned in the sinus, awn 1-2 mm long; palea 2-keeled, 2-nerved.

*Fl.-Fr.:*Sept. -Dec.

Rare, **Khargone**, hanging from damp rocks, Peepal zopa range.

Tripogon polyanthus Naik & Patunkar, Bull. Bot. Surv. India 15: 1976; Naik, Fl. Marathawada 2: 1114. 1998.

Perennials; culms tufted, 30-80 cm tall; leaf sheath 1.5-3 cm long; ligule membranous; leaf blade 3-7.5 × 0.2-0.3 cm, apex acuminate; racemes spiciform, 15-18 cm long; spikelet sessile, linear, 2-6.5 cm long, 40-50 or more flowered, dense, erect or ascending; lower glume membranous, 1-nerved, caudate-acuminate at apex; upper glume membranous; lemma ovate-lanceolate, apex acutely 2-lobed; awn from the sinus up to 1-1.25 mm long, keels shortly hairy.

*Fl.-Fr.:*Aug. -Dec.

Rare along stream banks, **Khandwa** only, Mortakka forest area. Endemic to India.

Tripogon pungens C.E. C. Fisch., Bull. Misc. Inform. Kew, 1934: 170.1934; Bor, Grass. Burm. Ceyl. Ind. Pak. 522. 1960; Sharma *et al.,* in Fl. Maharashtra 3:629. 1996.

Tufted perennial with short rhizome; culms filiform, 6-10 cm tall, erect, simple, 2-3-noded; leaf sheaths compressed; ligule membranous; leaves convolute or rolled, 1.0-2.5 cm long, glabrous or sparsely pilose, apex obtuse, apiculate; racemes spiciform, 2-5 cm long; spikelets linear, 0.25-0.35 cm long, 5-7-flowered, distant or imbricate; grain cylindric, *ca* 0.05 cm long.

*Fl.-Fr.:*Aug. -Nov.

Rare in rocky soil in grassland, Burhanpur, Khandwa. Endemic to India.

UROCHLOA P. Beauv. W. 17; MP 2; SA 2

Key to Species

1. Perennials, erect; racemes 2-8; rachis filiform triquetrous .... **U. setigera**
1. Annuals, decumbent, trailing; racemes 4-12; rachis flat .... **U. panicoides**

Urochloa panicoides P. Beauv., Ess. Agrostogr. 52. 1812; *Panicum javanicum* Poir., Encycl. Suppl. 4: 274. 1816; Hook.f., Fl. Brit. India 7: 35. 1896; Bor, Grass. Burm. Ceyl. Ind. Pak. 372. 1960; Samvatsar, Fl. West. Trib. Madhya Pradesh 398. 1996; Roy in N.P. Singh *et al.,* Fl. Madhya Pradesh 3: 542.2001. *Chatu latu*

Annuals; culms 30-60 cm tall, tufted; leaves finely hirsute with tubercle-based hairs; sheaths loose, hairy upwards; ligule hairy; inflorescence of 4-10 recemes; rachis ± 1 mm broad; spikelets 2-seriate, ovate-elliptic, 4-5 mm long, acuminate; lower glume 1.5-2.0 mm long, glabrous, 5-nerved; upper glume similar to lower, 9-11-nerved; lower floret male or barren; lemma equal to upper glume, 5-7 nerved; stamens 3, 2 mm long, anthers yellow; caryopsis ellipsoid or oblong, compressed, 2 mm long, yellow.

*Fl.-Fr.:*July-Jan.

Common, in cultivated fields, along river banks, ponds, all over the area.

Urochloa setigera (Retz.) Stapf, Fl. Trop. Afr. 9: 598. 1920; *Panicum setigerum* Retz. Observ. Bor. 4: 75. 1786; Hook.f., Fl. Brit. India 7: 36. 1896; *Brachiaria setigera* (Retz.) C.E. Hubb., Hooker's Icon. Pl. 34: t. 3363. 1938; Bor, Grass. Burm. Ceyl. Ind. Pak. 286. 1960; Samvatsar, Fl. West. Trib. Madhya Pradesh 336. 1996.

Herbs, 60-90 cm long, straggling; nodes bearded, the lower rooting; leaves 3-10 × 1-2 cm, ovate-lanceolate; ligule a tuft of long, woolly hairs; leaf blade linear-ovate, 10-30 cm long, glabrous or sparsely hairy, scaberulous on margins, apex finely acuminate; panicles of 3-10 racemes, 2-5 cm long; spikelets 0.25-0.3 cm long, 2-seriate, ovoid, acute or acuminate; grains 0.24-0.26 cm long, elliptic; lower glume membranous, ovate, 1-nerved; upper glume membranous, broadly ovate, glabrous 5-nerved, apex acute; lower lemma membranous; palea membranous, narrowly ovate, 3 mm, margin inflexed; stamens 3, anthers 1.3 mm long; upper lemma coriaceous; palea coriaceous, elliptic, 2-nerved.

Fl.-Fr.: Aug. -Dec.

Moist places, bunds of fields, in all districts.

ANALYSIS OF THE FLORA

I. STATISTICAL

	POSITION OF FAMILIES, GENERA AND SPECIES AT THE WORLD, INDIA AND STUDY AREA			
	WORLD	INDIA	MP	STUDY AREA
FAMILIES	405	247	166	120
GENERA	14559	2984	930	641
SPECIES	304459	17000	2214	1432

Source : **World**-The Plant List; **India** : Karthikeyan 2000, **MP** Khanna & Anand Kumar 2006, **Study Area**- The Present Work

Note:- P. Singh *et al.,* have given the number of flowering plants, as 18,159, in **New Additions to Indian Flora in 2014,** Phytotaxonomy 16: 1-11.2017.

GROUP

DICOTYLEDONS	FAMILIES	GENERA	SPECIES
POLYPETALAE	51	200	448
GAMOPETALAE	28	221	402
MONOCHLAMYDEAE	17	51	134
SUBTOTAL	96	472	984
MONOCOTYLEDONS	24	169	448
GRAND TOTAL	120	641	1432

RATIOS

GEN.	:	SPP.
641	:	1432
1	:	2.23
FAM.	:	GEN.
120	:	641
1	:	5. 31
FAM.	:	SPP.
120	:	1432
1	:	11. 93

MONOCOTYLEDONS

FAM.	:	DICOTYLEDONS
24	:	96
1	:	4
GEN.	:	DICOT
169	:	472
1	:	2.79
SPP.	:	DICOT
448	:	984
1	:	2.19

This shows that on an average, each family has 5. 31 genera and 11. 93 species, while the genus to species ratio is 1: 2.23. The Plant List, Sept. 2013, version 1. 1 records a total of 304419 accepted species of Angiosperms, 14559 genera and 405 families. But these figures are dynamic as the research is continuing, resulting in change of figures. The living world has no last word.

Another important aspect of this flora is the representation of Monocotyledons, both in respect of genera and species. The bulk of genera and species in this group is due to Cyperaceae and Poaceae (103 genera and 323 species) out of the 169 genera and 448 species in that group. Whereas the very large number of grasses shows a predominance of plain areas and weeds, the sedges indicate moist ground. This latter aspect is further corroborated by families like Campanulaceae, Gentianaceae, Lentibulariaceae, Linderniaceae, Menyanthaceae, Onagraceae etc. and several members of the Monocotyledons.

The comparatively large number of Orchids (12 genera and 18 species), in a predominantly dry-deciduous forest area, shows high percentage of humidity in selected pockets. A perusal of the flora shows that the Eastern two districts of the Satpura, viz., Burhanpur and Khandwa are rich, both in the number of genera and species compared to the Western two districts, viz. Barwani and Khargone, and both in the same range. This is due to more rain in the Eastern Nimar compared to the Western Nimar.

Distrct Alirajpur, especially Aamkhut-Kattiwada range and Mathwad-Shakarja range has a large number of rare and noteworthy floral elements. The same is true for Khandwa, in pockets like Aawlia, Sunder deo, Kalibhit. The Khaknar range in Burhanpur, Sirvel in Khargone, Nahar-jhabua in Indore, Mandu plateau in Dhar are equally rich.

10 DOMINANT FAMILIES

NAME	GENERA	SPECIES
POACEAE	89	239
LEGUMINOSAE	64	191
CYPERACEAE	14	84

COMPOSITAE	48	73
MALVACEAE	23	64
LAMIACEAE	24	54
ACANTHACEAE	21	49
CONVOLVULACEAE	11	39
EUPHORBIACEAE	9	39
APOCYNACEAE	24	30
	327	862

The 862 species within these 10 families constitute 60.27 % of the total species. Similarly, there are 16 genera having 10 or more species within each. The percentage of the generic component of these 10 families 50.78, while that of species is 60.27. In other words, these 10 families contribute more than half of the total flora of this area.

The dominant genera, each having 10 or more species are:

Cyperus L. 34	**Ipomoea** Jacq. 16	**Arthraxon** P. Beauv. 10
Euphorbia L. 22	**Eragrostis** Wolf 15	**Grewia** L. 10
Crotalaria L. 21	**Alysicarpus** Neck. ex Desv. 14	**Justicia** L. 10
Indigofera L. 18	**Leucas** R. Br. 13	**Phyllanthus** L. 10
Acacia Miller 16	**Blumea** DC. 12	**Sporobolus** R. Br. 10
Fimbristylis Vahl 16	**Ficus** L. 13	

The total number of species within these 16 genera is 250 or just 6. 4 %. Except **Acacia, Ficus** and some species within **Euphorbia** and **Phyllanthus emblica,** others are herbaceous.

II. PHYTOGEOGRAPHICAL ANALYSIS

For a better appreciation of the flora, the 1432 species are placed in 15 categories. This data is based upon:

1. T. Cooke The flora of Bombay Presidency (3 vols., B.S.I. reprint, 1958), **2**. K.M. Mathew, The Flora of Tamilnadu Carnatic, 3 vols. 1983, **3**. A Revised Handbook to the Flora of Ceylon, 14 vols., M.D. Dassanayake and F.R. Fosberg and M.D. Dassanayake and W.D. Clayton, 1980-2000.

Category	Number	Percentage
1. India (including Brit. India & present)	485	33.86
2. Indo-Malaya	214	14.94
3. Pantropical	88	06.14
4. Paleotropical	74	05.16

5. Cosmopolitan	30	02.09
6. Mediterranean	42	02.94
7. Trop. America	85	05.89
8. Africa	22	01.54
9. Afro-Asia	34	02.38
10. India-Australia	66	04.63
11. Africa-India-Australia	73	05.12
12. Asia-Australia	13	00.91
13. Africa-India	104	07.19
14. Asia	47	03.29
15. Africa-Malaysia	55	03.85
	1432	99.93

It would be seen from the above figures that the percentage of Indian elements is almost 34 or 1/3 rd of the flora, rest of the elements being from regions as given above. Actually, the number of species shown in each category is the natural, geographical range of that species, meaning thereby that after having evolved, it has spread over those regions. The study area does not offer any formidable challenge for migrating elements as well.

III. LIFE FORMS

Here we have not followed Raunkier (1934) either in categories, criteria or terminology, but relied upon the habit(habitat in case of hydrophytes). The predictive value remains unaltered.

Category	Number	Percentage
1. Herbs	847	59.14
2. Shrubs	152	10.61
3. Trees	142	09.91
4. Climbers	141	09.91
5. Hydrophytes	61	04.25
6. Epiphytes	12	00.83
7. Parasites	15	01.04
8. Geophytes	62	04.32
	1432	100.10

It is clear that almost 60 per cent of the flora is herbaceous. This is due to large tracts of open land and comparatively open canopy in parts of the forest. The presence of shrubs in this open land results in a savannah type of forest. The presence of trees and shrubs provide an ideal support for climbers. Where the forest is dense, e.g., Kalibhit East in Khandwa Dist., the humidity percentage is more, supporting epiphytes and a rich ground cover.

Similar isolated patches of rich forest are present in other districts also.

The richness of habit forms, apart from the herbaceous elements is a good indicator of phytodiversity in the flora.

Vegetation pattern of this area shows a great species diversity and density within each category. Floristically, this area has yielded 1432 wholly wild and naturalised species, which show richness of biodiversity within such a small area. This is mainly due to forest cover, hilly terrain and very little anthropogenic influence. Moreover, there are only 4-5 cities in the whole area and very few industries or factories.

Some species occur commonly with high population size but do not spill over throughout the area although abundantly present in limited niches e. g., *Capparis decidua* or *Salvadora persica* occur in plenty in Dhar and Khargone dists., while they are scattered in the remaining area.

Population size of some herbaceous elements is quite high and their distribution in the area is also even, e.g., *Eragrostis ciliaris, E. gangetica, Tephrosia purpurea* (Shrubby,) *Themeda quadrivalvis, Tridax procumbens, Xanthium strumarium* etc.,

Lantana × *aculeata* is so well and evenly spread all over the study area that the existence of indigenous elements has become endangered.

The population size of grasses like *Apluda mutica, Dichanthium annulatum, D. caricosum, Digitaria bifaria, Eragrostis gangetica, Sehima nervosum, Themeda quadrivalvis* etc., is quite high.

Narmada and its tributaries offer a very favorable habitat for plants like *Hydrilla verticillata, Lagarosiphon alternifolia, Ottelia alismoides, Potamogeton* spp., etc., and this niche has been fully exploited by members of these taxa. *Crinum viviparum* also occurs all along the linear belt but in a scattered way. *Sagittaria triphylla* is another example in the above category.

Phoenix sylvestris occurs in wasteland and along banks of water streams. The number of individuals is also quite high but distribution is uneven.

Vegetation of the past: During the last 30-35 years it has been observed that quite a large number of species have either disappeared or have become “locally extinct”. This data is given district-wise for better appreciation.

Alirajpur: **Alysicarpus roxburghianus, Aspidopterys cordata, Cleome scaposa, Crotalaria filipes** var. **trichophora, Hesperethusa crenulata, Hibiscus beddomei, Vicia peregrina**

Barwani: **Olax scandens, Rothia indica**

Burhanpur: **Asystasia mysorensis, Euphorbia elegans**

Dhar: **Acalypha alnifolia, Aloysia citriodora, Corallocarpus conocarpus, Jatropha nana, Talinum portulacifolium, Vahlia digyna**

Indore: **Cassytha filiformis, Cleome felina, Pueraria phaseoloides, Rhynchosia acutissima, Seseli diffusum**

Jhabua: Nil

Khandwa: **Caldesia oligococca, Cullen plicatum, Desmodium scorpiuros, Holostemma ada-kodien, Orthosiphon thymiflorus, Trifolium fragiferum**

Khargone: **Acilepis divergens, Ardisia solanacea**

*Reasons for disappearance of these species are:*1. Modification of the habitat, either due to urbanization, industrialization, and road-building, 2. Conversion of forest land to agriculture or simply encroachment. This is a point of no return, unless these species are found again and relocated to identical habitats, definitely a tall order.

A number of species are restricted to one district only. The district-wise list is given below. The taxa appear as they do in the Flora. The area from which they are collected is given after the name of species. Authorities for binomials are not given.

ALIRAJPUR

1. **Cleome scaposa**- Bakhat garh range 2. **Hibiscus beddomei**- Phulmal range 3. ***Aspidopterys cordata**- Aamkhut forest 4. * **Hesperethusa crenulata**- on way to Mogarba 5. **Alysicarpus heterophyllus**- Kawant road (Chhaktala) 6. **A. roxburghianus**- Kitty forest 7. **Crotalaria filipes** var. **filipes**-Kadwal, Chhachkuwan 8. **C. filipes** var. **trichophora** Bhabra-Kattiwada road 9. **C. leptostachya**- Shakarja 10. ***Dalbergia volubilis**- Kattiwada range 11. **Indigastrum parviflorum**- Aamkhut range 12. **Indigofera prostrata**- Kattiwada forest 13. **Vicia perigrina**- Shakarja along Narmada bank 14. **Kalanchoe olivacea** Chatliya pani (Kattiwada) 15. **Rotala densiflora**- Aamkhut 16. **R. mexicana**- Kadwal 17. **Ludwigia hyssopifolia** Kattiwada range 18. **L. prostrata**- Chatliya pani (Kattiwada) 19. **Oldenlandia attenuata**- Shakarja 20. **Acmella radicans**- Kattiwada forest nursery 21. **Erigeron trilobus**-Kattiwada range 22. **Tricholepis radicans**- Shakarja 23. **Wahlenbergia erecta**- Chatliya pani (Kattiwada) 24. **Ipomoea tuberculata**- Kattiwada range 25. **Stictocardia tiliifolia**- Phulmal (Chhaktala) 26. **Lindernia anagallis**- Aamkhut forest 27. **Dopatrium nudicaule**- Aamkhut range 28. **Justicia vahlii**- Aamkhut range 29. **Anisomeles malabarica**- Bakhat garh range 30. **Pogostemon paniculatus**- Aamkut range 31. **P. parviflorus**- Aamkhut range 32. **P. purpurascens**- Aamkhut range 33. **Salvia aegyptiaca**- Shakarja 34. **Euphorbia cristata** Bakhat garh 35. **E. fusiformis**- Kattiwada, open ground 36. **Jatropha nana**- Shakarja 37. **Ficus caulocarpa**-Aamkhut range 38. **Peristylus lawii**- Kadya pahar 39. **Commelina suffruticosa**- Chatliya pani (Kattiwada) 40. **Cyanotis arcotensis**- Chatliya pani (Kattiwada) 41. **Eriocaulon truncatum**- Harod, Mordundiya 42. **Arthraxon nudus**-Bhabhra, Shakarja 43. **Brachiaria semiverticillata**- Chatliya pani (Kattiwada) 44. **B. villosa** Bhabra 45. **Paspalum notatum**-Aamkhut range 46. **Pennisetum hordeoides**- Harod

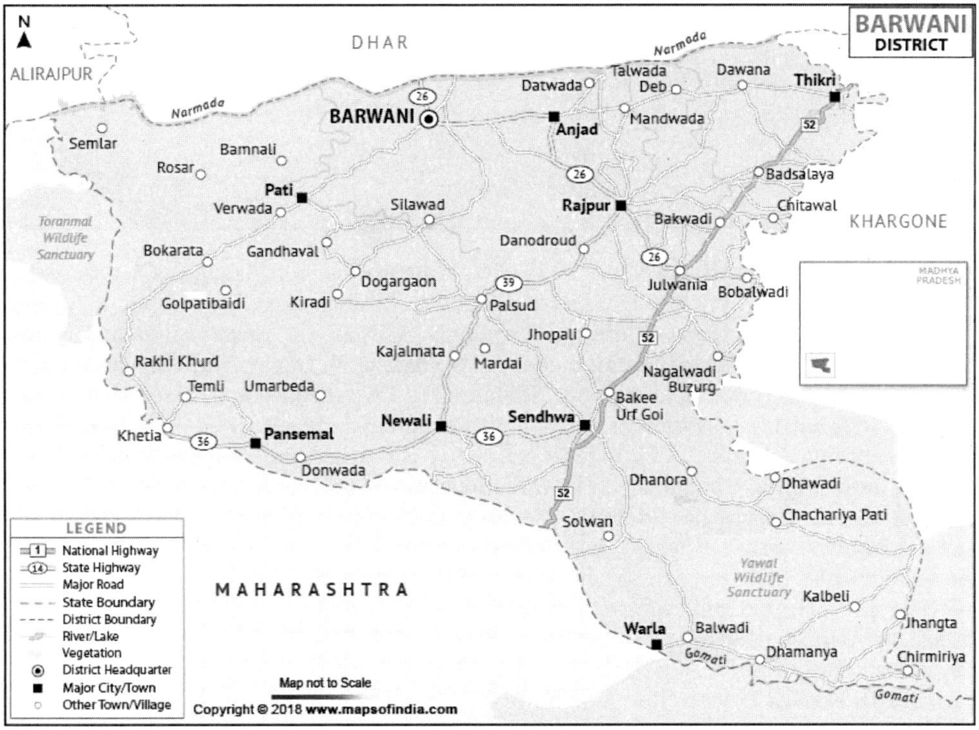

1. **Rothia indica-** Bhatki village 2. **Euphorbia perbracteata**- Weed of fields, Barwani proper 3. **E. serpens**- common in fields, Barwani

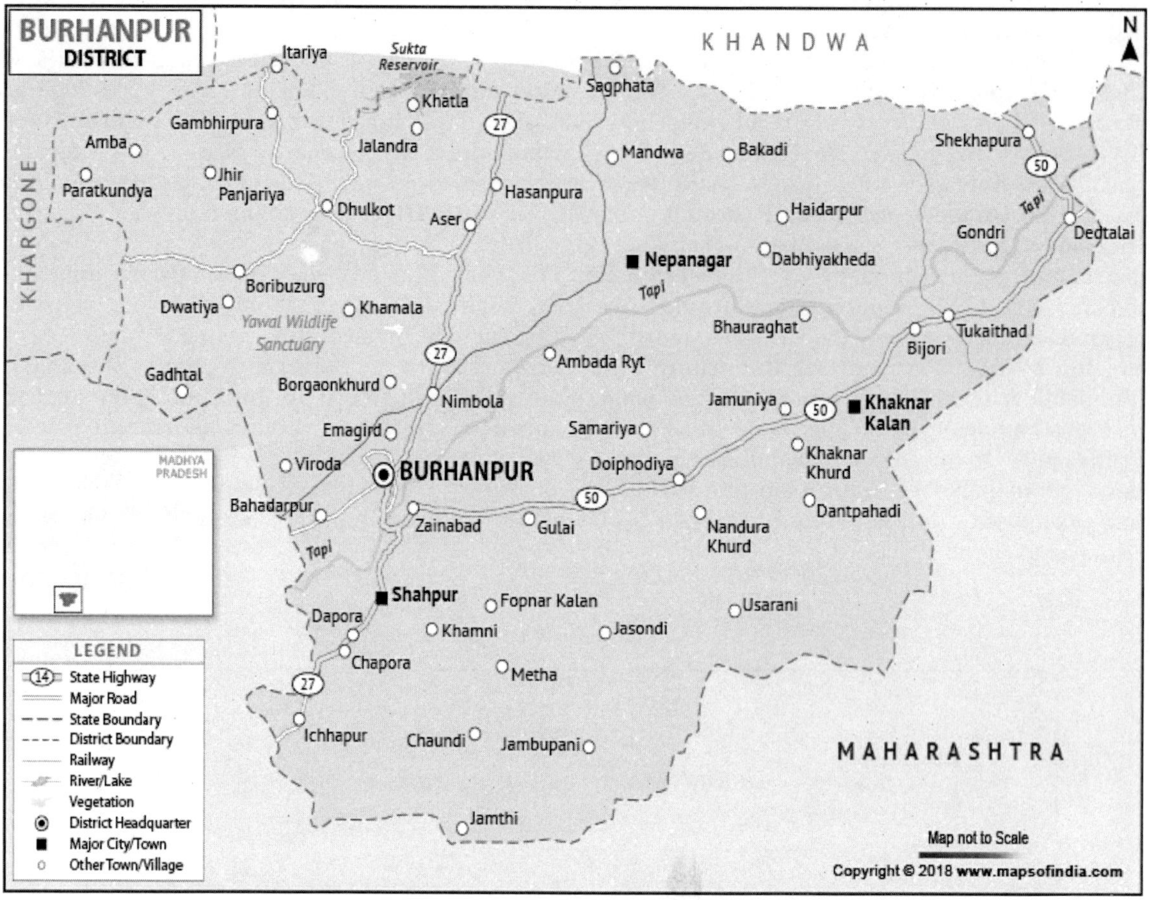

1. **Clematis heynei**-Jambupani, Badal khora 2. **Cocculus pendulus**-Nepanagar 3. **Cleome rutidosperma** var. **burmanni** – Chandni Rly. track 4. **Drymaria cordata** –Mirachia park, Burhanpur 5. ***Grewia tenax**-Dhulkot & Hansapura 6. ***Maytenus rothiana**- Chandni & Asirgad 7. **Cissus vitiginea**- Hedges of fields, Asirgad 8. **Cardiospermum microcarpum**- Jainabad & Shahpur 9. **Desmodium scorpiurus**- Chandni & Asirgad 10. **Tephrosia maxima**- Jambupani & Shahpur 11. **Trigonella corniculata**- Jainabad 12. **Vigna indica**-Asirgad (Addition to the Flora of India) 13. **Ctenolepis cerasiformis**- Lal bag, Burhanpur 14. **Momordica balsamina**- Renuka devi mandir area 15. **Zaleya decandra**-Jainabad & Shahikila 16. **Hydrocotyle sibthorpioides**- Pamakhedi 17. ***Salvadora alii**-Main market & Saifia college (Addition to the Flora of India) 18. **Ceropegia odorata**- Nepanagar 19. **Cressa cretica**-Daryapur & Shahpur 20. **Solanum americanum**- Asirgad range 21. **Evolvulus nummularius**- Hina garden & Shahikila 22. **Alectra parasitica**- Tukai thad 23. **A. parasitica** subsp. **chitrakutensis**- Tukai thad 24. **Barleria acanthoides**- Jainabad & Shahpur 25. **Isodon rugosus**- Jambu pani 26. **Nepeta graciliflora**- Asirgad range 27. **Boerhavia erecta**-Asirgad range 28. **Loranthus cordifolius**- Jambu pani 29. **Phyllanthus scabrifolius**- Asirgad range 30. **Elatostema cuneatum**- Asirgad range 31. ***Ficus rumphii**- Madni, banks of river Chhoti Tawa 32. **Dioscorea belophylla**- Asirgad & Nepanagar 33. **Typhonium inopinatum** Jambu pani 34. **T. flagelliforme** Nepanagar 35. **Cyperus elatus**-Asirgad & Dahinala 36. **Eleocharis retroflexa**- Jaswadi 37. **Kyllinga melanosperma**- Asirgad, Shahi kila 38. **Arthraxon junnarensis**- Asirgad range 39. **Eulalia staintonii**- Jainabad

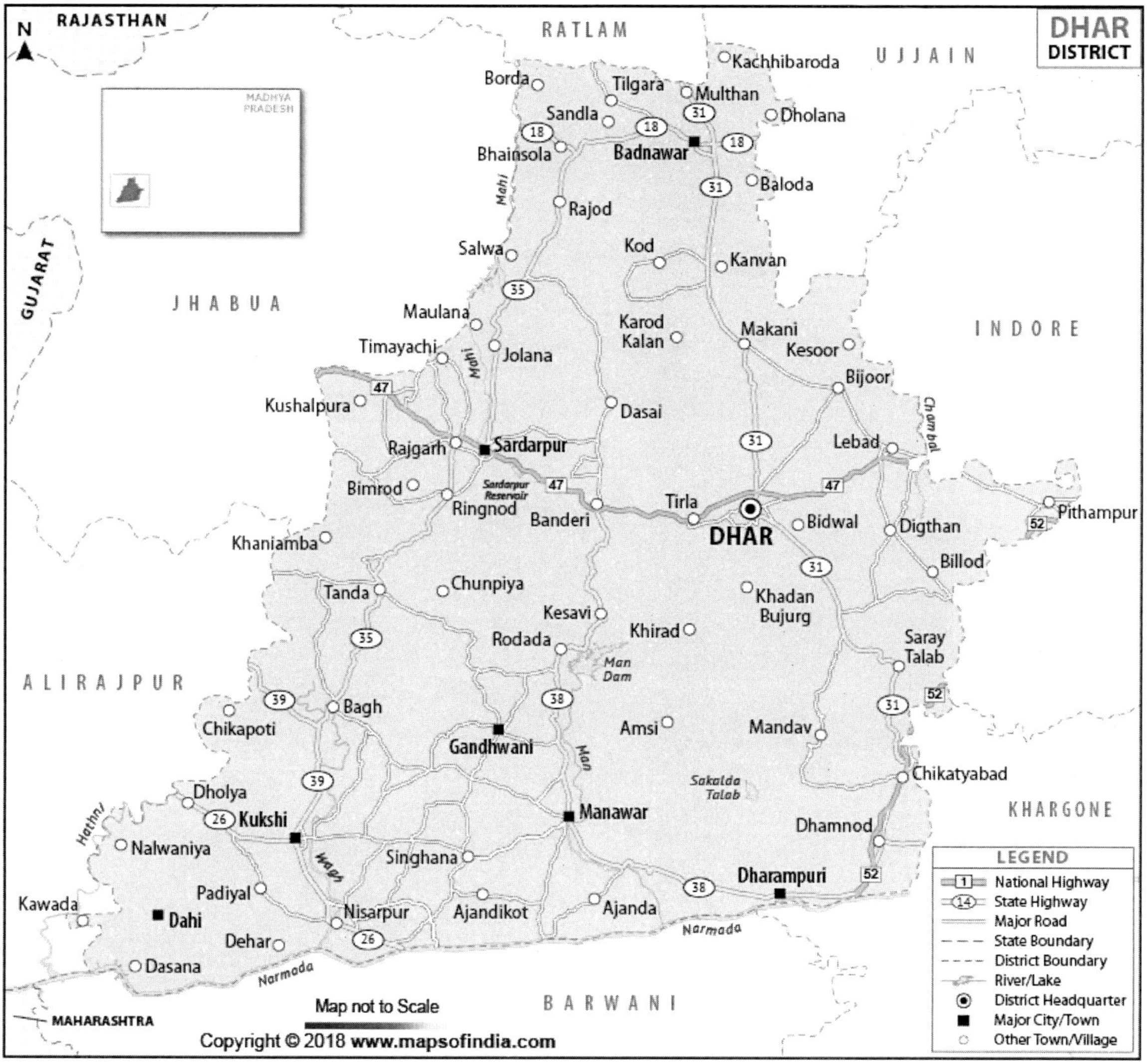

1. **Talinum portulacifolium**- Dharam puri bet 2. **Indigofera trigonelloides**- Jama Masjid area-Mandu plateau 3. **Tephrosia senticosa**- Dharam puri bet 4. **T. tinctoria**- Shri Ram ghat hill 5. **Vahlia digyna**- Katnera 6. **Rotala verticillaris**- Mordundiya (Katnera) 7. **Zaleya pentandra**- Dharam puri & Katnera 8. **Synedrella nodiflora**- Mandu plateau 9. **Justicia betonica**- Dhar proper in hedges of fields 10. **Aloysia citriodora**- Dharam puri bet 11. ***Vitex peduncularis**- Mandu plateau 12. **Habenaria grandifloriformis**- Mandu plateau & Mogarba hills 13. ***Chamaerops humilis**- Kukshi 14. **Cymbopogon schoenanthus**- Katnera 15. **Digitaria griffithii**- weed of fields, Dhar proper

1. **Cleome felina**- Kalakund 2. **Polygala jacobi**-Khurda-Khurdi 3. **Abelmoschus crinitus**- Nahar jhabua 4. **Linum mysorense**- Janapao hills 5. **Melhania futtyporensis**- Tinchha fall & Patal pani 6. **Biophytum umbraculum**- Nahar jhabua 7. **Indigofera angulosa**- Bai ghat 8. **Neptunia oleracea**- Gautampura 9. **Pueraria phaseoloides**- Chhapariya 10. **Rhynchosia acutissima** 11. *****Anogeissus sericea**- Tinchha falls 12. **Luffa echinata**- Manpur hills 13. **Seseli diffusum**-Kshipra & Chambal river banks 14. **Neanotis lancifolia**- Berchha pond, Mhow 15. **Blumea bifoliata**-Gautampura 16. **Chrysanthellum indicum**-Bijasan hill, Indore 17. *****Jasminum arborescens**- Nandlal ghati 18. **J. coarctatum**- Nahar jhabua-19. **J. flexile**- Kalakund-Patal pani track 20. **Ipomoea arachnosperma-** Hatod. 21. **Merremia hederacea**-Hedges of fielda, Yeshwant Sagar pond 22. **Torenia indica**-Kishanpura pond-Chhota Betma 23. **Mimulus strictus**- Yeshwant nagar pond area 24. **Striga densiflora**- Grass land, Harsola area 25. **Utricularia australis**- Banedia pond 26. **U. caerulea**- Banedia pond 27. **U. gibba** subsp. **exoleta**- Lekoda pond 28.

Justicia mollissima- Nehru Park as a weed 29. **Leonurus sibiricus**- Rala Mandal road 30. **Leucas wightiana**- field weed, Mhow 31. **Plectranthus barbatus**-Sitala mata falls area 32. **Amaranthus roxburghianus**- Field weed, Indore 33. **Euphorbia pycnostegia**- Simrol grass land 34. **Eulophia ramentacea**- Berchha pond area 35. **Papilionanthe teres**- over flow area beyond Pipliya Pala 36. **Monochoria hastata**- Puddles on Mhow road 37. **Arisaema tortuosum**- Kala kund 38. **Zanichellia palustris**- Yeshwant nagar pond 39. **Cyperus bulbosus**-Graasland, Hirniya khedi 40. **C. kyllingiella**- Pond near Machal 41. **C. pilosus**- Banks of Sirpur pond 42. **Fimbristylis albo-viridis**- Wet soil, Banedia pond 43. **F. cymosa**-Choral 44. **Schoenoplectiella juncoides**- Simrol 45. **Scleria stocksiana**- Army grass land, Simrol 46. **Aristida cynantha** Fields at Rala mandal 47. **Aundinella setosa**- Kampel 48. **Eulalia quadrinervis**- Dev Guradiya fielda 49. **Sporobolus tetragonus**- Mendh

JHABUA

1. Rungia elegans

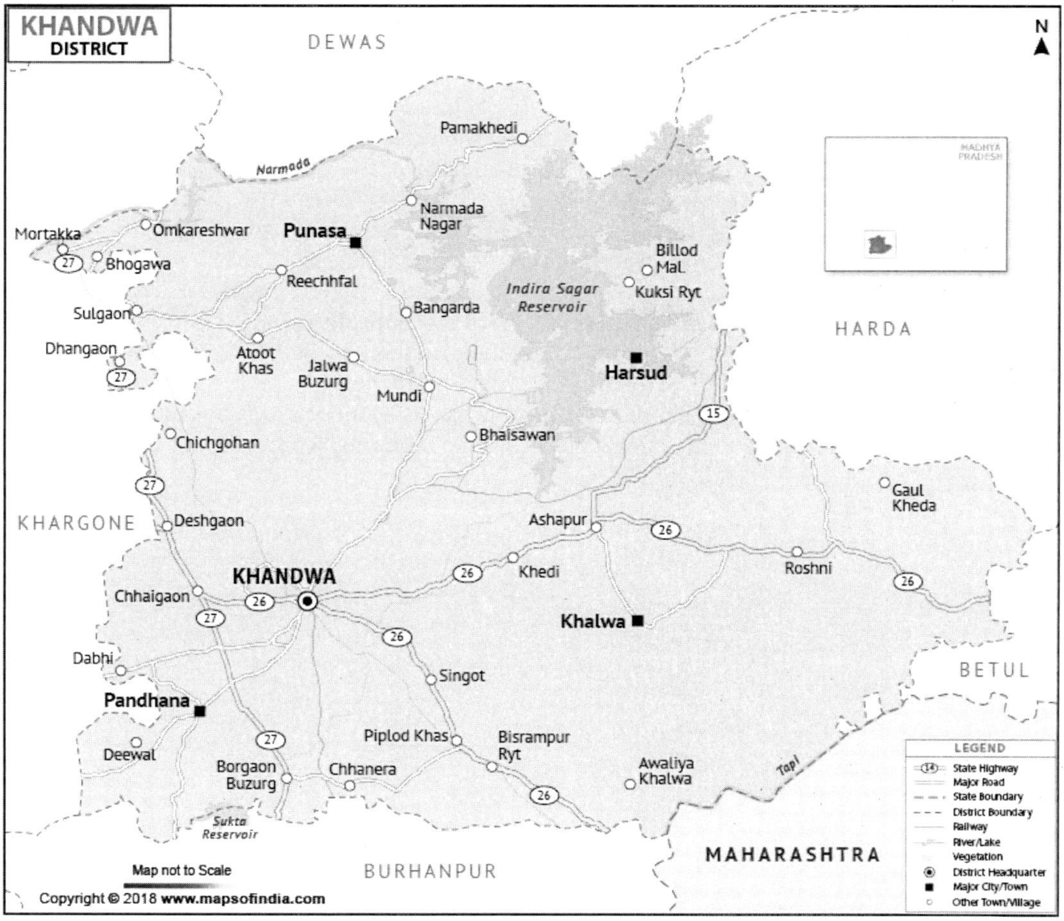

Khandwa: 1. **Tinospora maqsoodiana** Babangaon, Kalibhit 2. **Impatiens dasysperma** Forest undergrowth, Kalibhit 3. * **Acacia senegal** Khandwa proper 4. **Crotalaria pallida** Open groud, Mortakka 5. **Cullen plicatum** Khandwa proper 6. **Flemingia macrophylla** Klibhit & Sunder dev 7. **Trifolium fragiferum** Field weed, Mortakka 8. **Uraria rufescens** Mahlu area, Bhagpura 9. **Vigna vexillata** Battu barga area, Kalibhit 10. **Rotala illecebroides** Chand garh range, Kalibhit 11. **Heracleum nepalense** Mortakka, Narmada river bed 12. **Dentella repens** var. **serpyllifolia** Moist places along railway tracks, *Khandwa 13. **Oldenlandia ovatifolia** Kalibhit range 14. **Centipeda minima** Chand garh range, Kalibhit 15. **Laggera crispata** Weed of wheat fields, Khandwa 16. **Phyllocephalum indicum** Kalibhit range 16. **Holostemma ada-kodien** Kalibhit & Sunder dev 17. **Telosma cordata** Pandhana 18. **Exacum tetragonum** Mortakka 19. ***Cordia monoica** Kalibhit range20. **Heliotropium zeylanicum** Chand garh range 21. **Ipomoea pes-caprae** Banks of Narmada, Mortakka 22. **Merremia dissecta** Tapal chal & Awlia forest 23. **Dopatrium lobelioides** Moist areas, civil lines, Khandwa 24. **Glossostigma diandrum** Chira khadan aquatic habitat 25. **Mazus pumilus** Wet ground, Nimar Nursery 26. **Nelsonia canescens** Kalibhit forest 27. **Lippia javanica** Mandhana area 28. **Isodon coetsa** Amulla khurda, Kalibhit & Nalwadiya 29. **Orthosiphon thymiflorus** Singot & Khalwa 30. **Achyranthes aspera** var. **porphyristachya** Singot-Khandwa road 31. **Persicaria dichotoma** Golia area 32. **P. pulchra** Wet ground, Mortakka 33. ***Girardinia diversifolia** Kalibhit range 34. ***Neodistemon indicum** Forest undergrowth, Kalibhit *35. **Ficus drupacea** Chand garh range **36. Eulophia epidendraea** Sunder dev 37. **Habenaria digitata** Kalibhit range 38. **H. plantaginea** Banks of streams, Bhagpura 39. **Pecteilis susannae** Kalibhit forest 40. **Rhynchostylis retusa** Bhagpura & Battu-Bagra forest 41. **Vanda testacea** Kali ghodi Mata mandir forest 42. **Curcuma amada** Forest undergrowth, Chand garh range 43. **C. pseudomontana** Kalibhit range 44. **Pancratium triflorum** Sunder dev (Khalwa) 45. **Chlorophytum arundinaceum** Sunder dev & Golai 46. **C. diwanji** Along forest streams at Dhama village 47. ***Smilax zeylanica** Battu-Bagra 48. **Commelina rufipes** var. **glabrata** Mundi, Singhana 49. **Cyanotis tuberosa** Kalibhit range 50. **Amorphophallus konkanensis** Chand garh range 51. **A. longiconnectivus** Sunder dev & Golai 52. **Theriophonum dalzellii** Kalibhit range 53. **Eriocaulon heterolepis** Kalibhit range 54. **E. oryzoterum** Kalibhit range 55. **Cyperus imbricatus** Mortakka, Narmada river bank 56. **Eleocharis khandwensis** Village Bamangaon 57. **Lipocarpha kernii** Jaswadi 58. **Pycreus polystachyos** Mortakka, Narmada river bank 59. **Andropogon fastigiatus** 60. **Bambusa bambos** Extensively Planted in forest nursery and escape 61. **Chrysopogon zizanoides** Along streams, Nagchun & Daryapur 62. **Cyrtococcum deccanense** Damp places, Mortakka 63. **C. oxyphyllum** Kalibhit range 64. **Heteropogon melanocarpus** Mortakka 65. **Isachne globosa** Chand garh range 66. **Panicum curviflorum** Open grassland, edges of ditches, Kalibhit 69. **Pennisetum orientale** Chand garh & Mortakka 70. **Pogonatherum crinitum** Punasa & Kalibhit 71. **Schizachyrium exile** Chand garh range 72. **Setaria barbata** Mortakka forest area 73. **Tripogon polyanthus** Mortakka forest area

1. **Chrysopogon aciculatus** 2. **Tripogon lisboe**

This shows that in the yester years there were a number of species, now not seen. The reasons for their disappearance or rarity are the same as explained above. Species marked with * are woody elements, numbering 15.

The two districts of eastern Nimar, viz. Burhanpur and Khandwa are richer in vegetation, compared to the two districts of west Nimar, viz. Barwani and Khargone. All these 4 districts are in the Satpura range, but the eastern side receives more rainfall and is hillier. The floristic richness of Dhar is due to the hilly Mandu plateau and that of Indore is due good forest in Nahar jhabua, Choral, Khurda khurdi and Manpur hills. In Khargone district Sirvel area is alo rich, as it is hilly and just next to Pal Wild Life Sanctuary in Maharashtra.

Compared to these districts, Barwani and Jhabua are poor, as they are devoid of hills.

Aliens naturalized in the area: 1. **Acanthospermum** Schrank 2. **Acmella** A. Rich. 3. **Adansonia** L. 4. **Adenostemma** J. and G. Forst. 5. **Ageratum** L. 6. **Aloysia** Ortega ex Pers. 7. **Anagallis** L. 8. **Annona** L. 9. **Argemone** L. 10. **Bidens** L. 11. **Caesulia** Roxb. 12. **Campanula** L. 13. **Capsella** Medik. 14. L. 15. **Chamaecrista** Moench 16. **Cichorium** L. 17. **Conyza** L. 18. **Croton** L. 19. **Cryptostegia** R. Br. 20. **Drymaria** Willd. ex Roem. & Schult. 21. **Echinops** L. 22. **Eichhornia** Kunth 23. **Enicostemma** Blume 24. **Erigeron** L. 25. **Evolvulus** L. 26. **Flaveria** A. Juss. 27. **Galinsoga** Ruiz & Pavon28. **Gnomophalium** Greuter 29. **Gomphrena** L. 30. **Hyptis** Jacq. 31. **Jatropha** L. 32. **Lagascea** Cav. 33. **Lantana** L. 34. **Lavandula** L. 35. **Lawsonia** L. 36. **Lepidium** L. 37. **Lobelia** L. 38. **Malachra** L. 39. **Malvastrum** A. Gray 40. **Martynia** L. 41. **Mecardonia** Ruiz & Pavon 42. **Medicago** L. 43. **Melilotus** Miller 44. **Neanotis** W. H. Lewis 45. **Nopalea** Salm-Dyck 46. **Opuntia** Miller 47. **Oxalis** L. 48. **Parkinsonia** L. 49. **Parthenium** L. 50. **Passiflora** L. 51. **Phyla** Lour. 52. **Pithecellobium** Mart. 53. **Portulaca** L. 54. **Potentilla** L. 55. **Scoparia** L. 56. **Senna** Miller 57. **Stachytarpheta** Vahl 58. **Sonchus** L. 59. **Synedrella** Gaertn. 60. **Tamarindus** L. 61. **Tridax** L. 62. **Trifolium** L. 63. **Trigonella** L. 64. **Verbascum** L. 65. **Veronica** L. 66. **Vicia** L. 67. **Wahlenbergia** Schrad. ex Roth 68. **Waltheria** L. 69. **Xanthium** L. 70. **Zinnia** L.

Majority of them are from Trop. America

Monotypic Genera: 1. **Apluda** L. 2. **Caesulia** L. 3. **Chamaerops** L. 4. **Chloroxylon** DC. 5. **Coldenia** L. 6. **Colebrookia** J. E. Smith 7. **Desmostachya** (Stapf) Stapf 8. **Digera** Forssk. 9. **Endosamara** R. Geesink 10. **Gnomophalium** Greuter 11. **Goniocaulon** Cass. 12. **Haldina** Ridsdale 13. **Hesperethusa** M. Roem. 14. **Holoptelea** Planch. 15. **Holostemma** R. Br. 16. **Hydrilla** L. C. Rich. 17. **Hygroryza** Nees 18. **Kavalama** Rafin. 19. **Limonia** L. 20. **Microcarpaea** R. Br. 21. **Mukia** Arnott 22. **Neodistemon** Babu & A.N. Henry 23. **Nothosaerva** Wight 24. **Oroxylum** Venten. 25. **Parasopubia** H. -P. Hoff. & Ed. Fisch. 26. **Pistia** L. 27. **Polytrias** Hackel 28. **Pongamia** Venten. 29. **Rivea** Choisy 30. **Rotula** Lour. 31. **Soymida** A. Juss. 32. **Spermadictyon** Roxb. 33. **Synedrella** Gaertn. 34. **Tamarindus** L. 35. **Tamilnadia** Tirveng. & Sastre 36. **Thysanolaena** Nees 37. **Trichuriella** Bennet

Source: The Plant List, version 1. 1, Sept, 2013.

Khanna and Anand Kumar (2006) have deleted plants from Chhattisgarh area from the Flora of MP and the new figures are : 166 families, 930 genera and 2214 species, out of which the share of Monocotyledons is : 34 families, 219 genera and 562 species and that of Dicotyledons is : 132

families, 711 genera and 1652 species. **The percentage of 1432 species from 8 districts of this flora is 64. 67, compared to the total 2214 species in M.P.**

They have given a list of species which are common to **Western Himalayas**. Out of that list, the present flora has the following species :

1. **Bauhinia vahlii** Wight & Arn. 2. **Bothriochloa kuntzeana** (Hack.) Henr. 3. **Celastrus paniculatus** Willd. 4. **Colebrookia oppositifolia** Sm. 5. **Crotalaria albida** Roth 6. **Cynoglossum lanceolatum** Forssk. 7. **Dioscorea pentaphylla** L. 8. **Erigeron trilobus** (Decne.) Boiss. 9. **Ficus arnottiana** (Miq.) Miq. 10. **Garuga pinnata** Roxb. 11. **Plumbago zeylanica** L. 12. **Polygala arvensis** Willd. 13. **Pueraria tuberosa** (Willd.) DC. 14. **Sterculia villosa** Roxb. 15. **Trema orientalis** (L.) Blume16. **Vallaris solanacea** (Roth) Kuntze 17. **Zornia gibbosa** Span.

Species common to Eastern Himalayas:

1. **Acacia catechu** (L.f.) Willd. 2. **Capillipedium assimile** (Steud.) Clayton 3. **Cayratia trifolia** (L.) Domin 4. **Cyrtococcum oxyphyllum** (Steud.) Stapf 5. Geodorum densiflorum (Lam.) Schlechter 6. **Haldina cordifolia** Ridsdale 7. **Oroxylum indicum** (L.) Venten. 8. **Pogostemon benghalensis** (Burm.f.) Kuntze 9. **Pseudosorghum fasciculare** (Roxb.) A. Camus 10. **Rhynchostylis retusa** (L.) Blume 11. **Spermadictyon suaveolens** Roxb. 12. **Themeda quadrivalvis** (L.) Kuntze

Species common to Peninsular India:

1. **Cochlospermum religiosum** L. 2. **Diospyros montana** Roxb3. **Grewia abtulifolia** Venten. ex Juss. 4. **Habenaria plantaginea** Lindl. 5. **Helicteres isora** L. 6. **Justicia betonica** L. 7. **Mimulus strictus** Benth. 8. **Moringa concanensis** Nimmo 9. **Oldenlandia ovatifolia** (Cav.) DC. 10. **Persicaria pulchra** (Blume) Sojak11. **Schizachyrium bervifolium** (Sw.) Buse 12. **Schleichera oleosa** (Lour.) Oken13. **Strobilanthes callosa** Nees

Khanna and Anand Kumar (2006) have revised the flora of MP after deleting the species occurring in Chhatis garh. In spite of this, a number of species have been recorded in M.P. This can be considered as addition to the flora of M.P. The names of districts where these species are found is given below. These species are reported by Ph. D. scholars (unpublished thesis) and some are new additions after the publication of Khanna & Anand Kumar (2006). This is but natural. State floras are not revised so frequently and in the meantime, new explorations reveal new taxa.

This list is not alphabetical, but according to the system followed in this flora.

1. **Clematis heynei** M.A. Rau *et al.,* Burhanpur 2. **Drymaria cordata** (L.) Willd., Burhanpur 3. **Polycarpaea aurea** Wight & Arn., Barwani, Khandwa 4. **Malachra capitata** (L.) L., Burhanpur, Indore, Khandwa 5. **Corchorus tridens** L., Alirajpur, Dhar, Indore 6. **Oxalis debilis** Kunth, Barwani, Burhanpur, Khargone 7. **Ventilago maderaspatana** Gaertn., Burhanpur, Dhar, Khandwa 8. **Leea indica** (Burm.f.) Merr., Alirajpur, Barwani, Burhanpur, Dhar 9. **Cajanus platycarpus**(Benth.) Maesen, Alirajpur, Burhanpur, Indore 10. **Rhynchosia acutissima** Thw., Indore 11. **R. aurea** (Willd.) DC., Common in all 8 districts 12. **Tadehagi triquetrum** (L.) Ohashi, Burhanpur 13. **Tephrosia tinctoria** (L.) Pers., Dhar14. **T. uniflora** Pers. subsp. **petrosa** (Blatt. & Hallb.) Gillet & Ali, Burhanpur 15. **Trifolium fragiferum** L., Khandwa 16. **Vahlia**

digyna (Retz.) Kuntze, Dhar 17. **Vigna indica** Dixit *et al.,* Burhanpur 18. **Nesaea lanceolata** Koehne, Alirajpur, Dhar 19. **Rotala verticillaris** L., Dhar 20. **Momordica balsamina** L., Burhanpur 21. **Nopalea cochenellifera** (L.) Salm-Dyck, all 8 districts 22. **Mollugo nudicaulis** Lam., all 8 districts 23. **Heracleum nepalense** D. Don, Khandwa 24. **Seseli diffusum** (Roxb. ex Sm.) Santapau & Wagh, Indore 25. **Blumea laciniata** (Wall. ex Roxb.) DC., Alirajpur, Burhanpur, Dhar, Indore, Khandwa 26. **Tricholepis amplexicaulis** C.B. Clarke, Burhanpur, Khandwa 27. **Merremia dissecta** (Jacq.) Hall.f., Khandwa 28. **Solanum americanum** Mill., Burhanpur 29. **Lippia javanica** (Burm.f.) Spreng., Khandwa 30. **Callicarpa macrophylla** Vahl, Barwani, Khargone 31. **Orthosiphon thymiflorus** (Roth) Sleesen, Khandwa 32. **Persicaria pulchra** (Blume) Sojak, Khandwa 33. **Euphorbia caducifolia** Haines, Burhanpur, Indore, Jhabua 34. **E. perbracteata** Gage, Barwani 35. **Neodistemon indicum** (Wedd.) Babu & A.N. Henry, Khandwa 36. **Ficus rumphii** Blume, Burhanpur 37. **Eulophia epidendraea** (J. Koenig ex Retz.) C.E. C. Fisch., Khandwa 38. **Pancratium triflorum** Roxb., Khandwa 39. **Cyperus javanicus** Houtt., Alirajpur, Dhar 40. **C. tenuiculmis** Boeck., Alirajpur, Dhar 41. **Eleocharis retroflexa** (Poir.) Urban, Burhanpur 42. **Arundinella tuberculata** Munro ex Lisboa, Dhar, Indore 43. **Bothriochloa jainii** Desh. & Hemadri, Burhanpur, Khandwa 44. **Brachiaria deflexa** (Schumach.) C.E. Hubb. & Robyns, Alirajpur, Dhar 45. **B. remota** (Retz.) Haines, Common in all 8 districts 46. **B. semiverticillata** (Rottl.) Alston, Alirajpur 47. **Dimeria connivens** Hack., Khandwa, Khargone 48. **Ischaemum molle** Hook.f., Alirajpur, Dhar, Indore 49. **I. thomsonianum** Stapf ex C.E. C. Fisch., Common in all 8 districts 50. **I. timorense** Kunth, Barwani, Khargone 51. **I. travancorense** Stapf ex C.E. C. Fisch., Barwani, Khandwa 52. **I. tumidum** Stapf ex Bor, Burhanpur, Khandwa

Flora of Maharashtra (Vol. 1:76-102.2000) has given a list of 694 species endemic to that state. In our study area, 103 species are found and are listed below. This list extends the distribution of these species to MP. The names of species are not alphabetical, but starts from Ranunculaceae and ends in Poaceae. Authorities for binomials are not given. The names are according to the The Plant List.

Name of the species	Family
1. **Clematis heynei**	Ranunculaceae
2. **Cleome felina**	Cleomaceae
3. **C. simplicifolia**	Cleomaceae
4. **Aspidopterys cordata**	Malpighiaceae
5. **Impatiens dasysperma**	Balsaminaceae
6. **Alysicarpus pubescens** var. **vasavadae**	Leguminosae
7. **A. roxburghianus**	Leguminosae
8. **Clitoria biflora**	Leguminosae
9. **Crotalaria filipes** var. **filipes**	Leguminosae
10. **C. filipes** var. **trichophora**	Leguminosae
11. **C. leptostachya**	Leguminosae
12. **C. notonii**	Leguminosae
12. **C. pusilla**	Leguminosae
13. **C. vestita**	Leguminosae
14. **Hardwickia binata**	Leguminosae
15. **Indigofera angulosa**	Leguminosae
16. **I. deccanensis**	Leguminosae
17. **I. glandulosa** var. **sykesii**	Leguminosae
18. **I. prostrata**	Leguminosae
19. **I. trifoliata** var. **duthiei**	Leguminosae
20. **Pterocarpus marsupium** subsp. **acuminatus**	Leguminosae
21. **Kalanchoe olivacea**	Crassulaceae
22. **Corallocarpus conocarpus**	Cucurbitaceae
23. **Seseli diffusm**	Apiaceae
24. **Trachyspermum roxburghianum**	Apiaceae
25. **Neanotis lancifolia**	Rubiaceae
26. **N. montholonii**	Rubiaceae

Species	Family
27. **N. subtilis**	Rubiaceae
28. **Blumea belangeriana**	Compositae
29. **B. eriantha**	Compositae
30. **Tricholepis amplexicaulis**	Compositae
31. **T. radicans**	Compositae
32. **Ceropegia odorata**	Apocynaceae
33. **Argyreia sericea**	Convolvulaceae
34. **Torenia indica**	Linderniaceae
35. **Andrographis longipedunculata**	Acanthaceae
36. **Asystasia mysorensis**	Acanthaceae
37. **Barleria gibsonii**	Acanthaceae
38. **Eranthemum roseum**	Acanthaceae
39. **Hemigraphis urens**	Acanthaceae
40. **Haplanthus tentaculatus**	Acanthaceaee
41. **H. verticillatus**	
42. **Strobilanthes callosus**	Acanthaceae
43. **Pogostemon purpurascens**	Lamiaceae
44. **Jatropha nana**	Euphorbiaceae
45. **Phyllanthus scabrifolius**	Phyllanthaceae
46. **Aerides maculosa**	Orchidaceae
47. **Eulophia dabia**	Orchidaceae
48. **Habenaria grandifloriformis**	Orchidaceae
49. **Curcuma amada**	Zingiberaceae
50. **C. decipiens**	Zingiberaceae
51. **C. pseudomontana**	Zingiberaceae
52. **Zingiber cernuum**	Zingiberaceae
53. **Ensete superbum**	Musaceae
54. **Dioscorea belophylla**	Dioscoreaceae
55. **Chlorophytum borivilianum**	Asparagaceae
56. **Iphigenia pallida**	Colchicaceae
57. **Cyanotis arcotensis**	Commelinaceae
58. **C. cucullata**	Commelinaceae
59. **Amorphophallus commutataus**	Araceae
60. **A. konkanensis**	Araceae
61. **Theriophonum dalzellii**	Araceae
62. **Eriocaulon duthiei**	Eriocaulacaeae
63. **E. elenorae**	Eriocaulacaeae
64. **E. heterolepis**	Eriocaulacaeae

Species	Family
65. **Aristida redacta**	Poaceae
66. **A. stocksii**	Poaceae
67. **Arthraxon jubatus**	Poaceae
68. **A. junnarensis**	Poaceae
69. **A. meeboldii**	Poaceae
70. **A. santapaui**	Poaceae
71. **Arundinella tuberculata**	Poaceae
72. **Bothriochloa jainii**	Poaceae
73. **Capillipedium filiculme**	Poaceae
74. **C. huegelii**	Poaceae
75. **Chloris quinquesetica**	Poaceae
76. **Chrysopogon lawsonii**	Poaceae
77. **C. polyphyllus**	Poaceae
78. **Cynodon barberi**	Poaceae
79. **Dichanthium maccannii**	Poaceae
80. **Dimeria connivens**	Poaceae
81. **Eragrostiella brachyphylla**	Poaceae
82. **Heteropogon polystachyos**	Poaceae
83. **H. ritchiei**	Poaceae
84. **Ischaemum afrum**	Poaceae
85. **I. molle**	Poaceae
86. **I. thomsonianum**	Poaceae
87. **I. travancorense**	Poaceae
88. **I. tumidum**	Poaceae
89. **Iseilema anthephoroides**	Poaceae
90. **I. holei**	Poaceae
91. **Lophopogon tridentatus**	Poaceae
92. **Oropetium roxburghianum**	Poaceae
93. **O. villosulum**	Poaceae
94. **Panicum phoiniclados**	Poaceae
95. **Pseudanthistiria heteroclita**	Poaceae
96. **Sehima sulcatum**	Poaceae
97. **Sorghum purpureosericeum**	Poaceae
98. **Spodiopogon rhizophorus**	Poaceae
99. **Tragus mongolorum**	Poaceae
100. **Tripogon jacquemontii**	Poaceae
101. **T. lisboe**	Poaceae
102. **T. polyanthus**	Poaceae
103. **T. pungens**	Poaceae

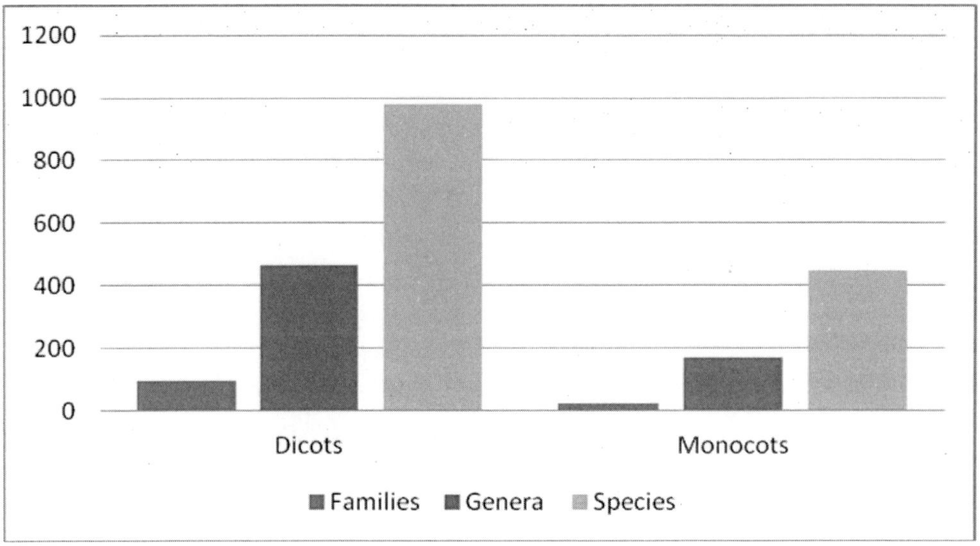

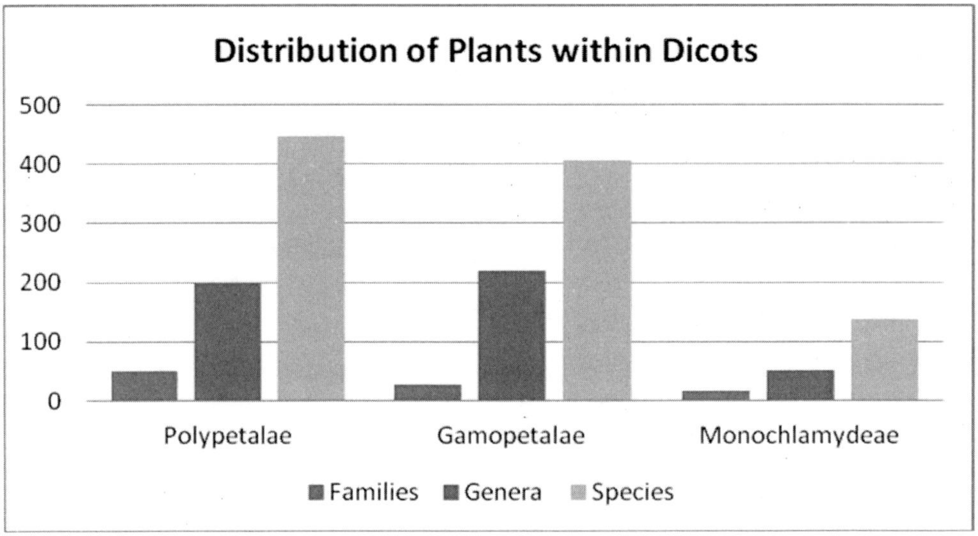

Verma (1996) has published a list of rare plants of MP along with the names where they occur in M.P. and outside M.P. To this list, the following taxa are added. Authorities of binomials omitted.

Name of the species	Place of occurrence	Outside M.P.	Study area
Alysicarpus pubescens var. *vasavadae*	West Nimar	Maharashtra	Khandwa, E. Nimar
Amorphophallus sylvaticus	Raipur	Deccan Peninsular, Sri Lanka	Alirajpur, Burhanpur, Khandwa

Name of the species	Place of occurrence	Outside M.P.	Study area
Andropogon pumilus	Indore	W. India	All 8 districts
Arthraxon echinatus	Rewa	Deccan Peninsula	East Nimar
Aspidopterys cordata	Chhindwara, Seoni	Deccan Peninsula	Alirajpur
Crotalaria notonii	Raipur	Nilgiri, Pulney hills, Maharashtra	Alirajpur, Dhar, Indore
Crotalaria triquetra	Hoshangabad, Indore	Assam, S. India, Konkan	Alirajpur, Dhar
Eragrostis minor (*E. pappiana*)	Bilaspur, Raipur	N.W. India, S. Africa	Indore, Khandwa
Euphorbia caducifolia	Raipur (endemic)	-	Burhanpur, Indore, Jhabua
E. perbracteata	Banda (endemic)	-	Barwani
Indigofera angulosa	Rajnandgaon	Bundelkhand, Rajsthan	Indore
Iseilema anthephoroides	Indore	Deccan Peninsula	All 8 districts
Maytenus rothiana	Shivpuri	Maharashtra	Burhanpur
Oropetium villosulum	Raipur, Bhopal	Maharashtra	Alirajpur, Burhanpur, Dhar, Indore, Khargone
Pulicaria wightiana	Hoshangabad	Odisha, S. Indian hills	All 8 districts
Scleria stocksiana	Balaghat	Maharashtra	Indore
S. tessellata	Guna, Indore	W. India, Africa	Common, all districts
Sorghum purpureosericeum (*S. deccanense*)	Indore	Deccan Peninsula	Alirajpur, Burhanpur, Dhar, Khandwa

CHAPTER - VI

CONSERVATION STRATEGIES

The history of humankind is one of the modifications of forested environment, which by the degradation and fragmentation of forest ecosystems, their alteration through the harvesting of forest products, use of fire, or more general environmental alteration and by the introduction of pests, pathogens, or exotic species resulted in the erosion of biodiversity. We cannot expect to achieve the long-term goal of an ecological sustainability without monitoring and modification of our forest management through time. Hence, we should treat forestry as adaptive management which must be accompanied with monitoring programmes to accumulate key information. In this context, the monitoring of forest genetic resources is one important component. But meaningful and operational feasible genetic monitoring is a far from simple task. Because complete genetic inventories are neither possible nor practical, we are forced to use surrogate measures, each of which is adequate in its own right. As biodiversity monitoring is a long-term process it only rarely provides instant results of direct value for political decisions makers. Hence, funding is often lacking.

Any activity to conserve forest biodiversity must be based on reliable information about the current status of resources, their likely development and of course the conservation objective. Inventories and monitoring programmes can play a pivotal role but in order to optimize efforts it must be first decided which biodiversity level (genes, species), geographical scale, and time frame is aimed at. Secondly, availability of adequate resources and sound understanding of ecological dynamics in forests are cornerstones for an accurate evaluation and thirdly, standardized protocols wherever possible make inventories and monitoring comparable and allow later the validation and calibration of findings. Therefore, a statistical framework is essential and should be planned ahead of any field work.

A first important step towards forest genetic conservation in this region would be the data collection on forest tree species abundance and distribution. Generally, information on the natural range and the potential natural vegetation also would come in handy to identify and declare *in situ* conservation areas.

Genetic conservation of biodiversity can both be *ex situ* and *in situ*. While *ex situ* conservation involves the sampling, transfer and storage of target taxa to target areas, the *in situ* strategy involves the designation, management and monitoring or target taxa where they are encountered *In situ* species conservation measures usually emhasize the protection and management of ecosystems and communities to avoid the loss of resident species. The key characteristic of *in situ* conservation is its dynamic nature allowing for continuous evolution. In forestry *in situ* conservation is commonly restricted to naturally regenerated forests of certain genetic value.

The situation in the study area needs to be described accepting the blame of repetition, as most of the rare species identified for both types of conservation, are restricted to small areas and do not occur elsewhere. Some of these are enumerated below.

S. No.	Botanical Name	Vern. / Local Name	Occurrence
1	**Acacia lenticularis** Benth.	Kanta siris, Kirangi	Dharampuri Bet, **Dhar**
2	**Acacia polyacantha** Willd.	Safed Khair, Papariya Khair	**Indore, Khandwa**
3	**Acacia senegal** (L.) Willd.	Kumpta	**Khandwa**
4	**Anogeissus sericea** Brandis	Indu koda, Kala Dhawda	**Indore**
5	**Careya arborea** Roxb.	Kumbhi	**Alirajpur, Barwani, Burhanpur, Dhar, Khandwa, Khargone**
6	**Ceriscoides turgida** (Roxb.) Tirveng.	Safed Phetra, Kharhar	**Alirajpur, Barwani, Burhanpur, Dhar, Khandwa**
7	**Chamaerops humilis** L.	Chhind	Kukshi **Dhar**
7	**Cochlospermum religiosum** (L.) Alsto	Galgal, Gabdi	**Alirajpur, Barwani, Burhanpur, Dhar, Indore, Khargone**
8	**Commiphora wightii** (Arn.) Bhandari	Guggal	**Alirajpur, Burhanpur, Dhar, Indore, Khandwa**
9	**Cordia monoica** Roxb.	Gondi Prajati, Lasoda Prajati	**Khandwa**
10	**Crateva adansonii** subsp. **odora** (Buch. -Ham.) Jacobs	Barna, Varun	**Burhanpur, Dhar**
11	**Crateva religiosa** G. Forst.	Barna, Varun	**Alirajpur, Dhar**
12	**Dalbergia volubilis** Roxb.	Dhoban bel	**Alirajpur**
13	**Dillenia pentagyna** Roxb.	Karkat, Shalyakarni	Aamkhut,**Alirajpur**, Sirvel, **Khargone**
14	**Diospyros chloroxylon** Roxb.	Kala Tendu	**Barwani, Indore, Khargone**
15	**Embelia basaal** (Roem. & Schult.) DC.	Waividang	**Barwani, Dhar, Khandwa, Khargone**
16	**Endosamara racemosa** (Roxb.) R. Geesink	Ardhaga	Neelkanth - Mandu **Dhar**
17	**Ficus caulocarpa** (Miq.) Miq.	Anjir Prajati	Aamkhut, **Alirajpur**

S. No.	Botanical Name	Vern. / Local Name	Occurrence
18.	**Ficus drupacea** Thunb.	Peepal Prajati	**Khandwa**
19.	**Ficus rumphii** Blume	Gajna, Jangli Peepal	**Burhanpur**
20.	**Firmiana colorata** (Roxb.) R. Br.	Kuwarin	Mathwad- **Alirajpur**, Janapao-**Indore**
21.	**Grewia eriocarpa** A. L. Juss.	Phalsa Prajati	Mandu plateau, **Dhar**
22.	**Hesperethusa crenulata** (Roxb.) M. Roem.	Bhelsai	Mogarba – **Alirajpur**
23.	**Hymenodictyon orixense** (Roxb.) Mabb.	Bhorsal, Bhudkut	**Alirajpur, Barwani, Dhar, Indore, Khargone**
24.	**Jatropha nana** Dalz. & Gibs.	Ratan jyote prajati	Mathwad-Shakarja, **Alirajpur**
25.	**Maytenus rothiana** Lobr. -Callen	Baikal Prajati	**Burhanpur**
26.	**Oroxylum indicum** (L.) Kurz	Tetu, Shyonak	**Alirajpur, Burhanpur, Khargone**
27.	**Phoenix acaulis** Roxb.	Khajuri	Bagh-Kukshi, **Dhar**
28.	**Pterocarpus marsupium** Roxb. subsp. **acuminatus** (Prain) Thoth.	Bija, Bijasal	**All districts**, but population very small
29.	**Radermachera xylocarpa** (Roxb.) K. Schum.	Garud vriksha, Kharsingha	**Alirajpur, Dhar, Khargone**
30.	**Schrebera swietenioides** Roxb.	Mokha	**Alirajpur, Barwani, Dhar, Khandwa, Khargone**
31.	**Semecarpus annacardium** L.f.	Bhilawa	**Alirajpur, Barwani, Dhar, Khandwa, Khargone**
32	**Soymida febrifuga** (Roxb.) A. Juss.	Rohan, Rohina	**Alirajpur, Barwani, Burhanpur, Dhar, Khandwa**
31	**Sterculia villosa** Roxb. ex DC.	Udal	Rani sati gate - Mandu Dhar, Asirgad-Burhanpur
32	**Stereospermum chelonoides** (L.f.) DC.	Padar	Harod Alirajpur
33	**Stereospermum tetragonum** DC.	Chhota pader	Mathwad Alirajpur
34	**Tamilnadia uliginosa** (Retz.) Tirveng. & Sastre	Kala phetra	Aamkhut - Alirajpur, Neelkanth - Mandu Dhar
35	**Terminalia crenulata** Roth	Sadad, Saj	Way to Mandu - Dhar, Kadwal - Alirajpur
36	**Ventilago maderaspatena** Gaertn.	Keoti	Jhabua
37	**Wrightia arborea** (Denstt.) Mabb.	Indra nav	Shakarja – Alirajpur

Protected Areas: A tool of Biodiversity Conservation

To begin with, the most effective and efficient way of conserving biodiversity is to prevent the conversion or degradation of habitat. The measures must be complemented by a wide array of techniques to conserve individual species, populations and genes. Maintenance of plant and animal genetic material in the wild (*in situ*) and outside their natural habitats, e. g. in plantations, seed storages, gene or pollen banks (*ex situ*) is essential of managing the human use of genetic diversity. *In situ* conservation maintains not only the genetic diversity of a population but also the evolutionary interactions that allow it to adapt continually to shifting environmental conditions, such as changes in pest populations or climate. Of all the various categories for conservation of forest areas, 'Genetic Reserve Forests', 'National Parks' , and 'Strict Forest Reserves' provide a high status for *in situ* conservation of forest genetic resources. Both strict protected forests and commercial forests have potential to maintain biodiversity. In both cases, minimum size of forest areas, natural stand structures and adequate population sizes are required. The focus of genetic conservation is to maintain or enhance genetic diversity in all taxa.

Protected Forest Areas

At the Fourth 'World Congress on National Parks and Protected Areas', held in Caracas, Venezuela, in 1992, protected areas were defined as 'areas of land or sea especially dedicated to the protection and maintenance of biological diversity, and of natural and associated cultural resources, and managed through legal or other effective means' (IUCN, 1994). Protected areas are important reservoirs for biodiversity and ensure that other benefits, such as soil and watershed protection, research and education, are secured. The long-term importance of protected areas is dependent on how they are chosen and managed for biodiversity conservation.

In situ conservation methods

Basic principles of conservation of genetic variation are essentially the same for all living organisms. The methods, however, may vary according to the specific objectives of conservation and the distribution and biological nature of the material to be conserved (FAO 1989). The term 'method' is often used to denote different concepts such as *in situ* conservation, *ex situ* conservation, ecosystem conservation, species conservation, static conservation, dynamic conservation and others. Here species, ecosystems, populations and individuals are considered objects of conservation and the term 'method' will be used to distinguish between basically different ways to conserve genetic resources like *ex situ* or *in situ*, dynamic or static, active or passive conservation.

Definition of *in situ* conservation is not very clear and the term has been used in different ways. The lack of clarity is in part due to *in situ* conservation being applied to wild species on the one hand and to domesticated species on the other hand. *In situ* conservation implies that a given population is maintained within the community of which it forms a part, in the environment in which it has developed. The term is frequently associated with wild, naturally regenerating populations in protected areas. However, *in situ* conservation has also been integrated into managed and multiple-use forests. In its essence, *in situ* conservation focuses on conserving a genetic resource in its original ecosystem, irrespective of whether such ecosystems have been subject to human interference. It simply means that the germplasm is conserved in the locality where it is currently found, either where it is naturally located or where it has developed distinctive traits under cultivation. *In situ* genetic conservation thus involves the saving of appropriate populations over generations, in order to maintain their potential for future evolution given by the adaptively or randomly developed genetic structures within the species.

The *in situ* method has several advantages. It is dynamic methods of conservation, allowing for natural selection processes, i.e. further evolutionary potential of the gene pool and the adaptive capacity of the population. This is important since no species is static but is continually interacting with its physical environment and is competing with other species in the ecosystem. For a species to be viable in the future it must be able to compete and it will only maintain its competitive ability if the evolutionary process is allowed to continue. This aspect is of particular importance today in the light of the worldwide climate changes which are taking place as a result or global warming. In addition, the *in situ* conservation method has the general advantage of conserving the functions of an ecosystem rather than a population or a species which means that it normally includes a great number of associated animal and plant species including all interactions among them. Finally, another advantage of *in situ* conservation, which is most important to the evolutionary development of a species, is that it is much easier, more secure and financially more efficient to conserve a viable population of a species in its natural habitat than in a *ex situ* situation. This is particularly true for tree species, since they require a lot of area to conserve thousands of individuals.

Ex situ conservation methods

When populations, individuals or reproductive material are maintained outside their original growth environment, it is *ex situ* conservation.

Ex situ populations or collections with forest trees are established and maintained for a number of objectives. Often *ex situ* conservation will be used as a complement to, or substitute for, *in situ* conservation of unique populations that are threatened in their natural habitat. *Ex situ* conservation becomes very useful when it is desirable to establish a well-defined seed source for commercial plantations, without doing expensive breeding operations, and at the same time assure that the reproductive materials produced are well adapted to the ecological conditions in the forest. This will be an evolutionary or dynamic approach to genetic conservation, if both natural processes and human management will deterrnine the genetic structure of the next generation. Seed orchards and done archives are examples of static *ex situ* conservation units as no change occurs in their genetic structure.

PROTECTED AREAS IN MADHYA PRADESH

S. NO.	NATIONAL PARKS	DATE OF ESTABLISHMENT	AREA Sq. Kms.
1	Bandhavgarh	11/05/1982	448.85
2	Fossil	17/06/1983	0.27
3	Kanha	10/10/1953	940.00
4	Madhav	01/01/1959	375. 22
5	Panna	19/07/1973	542.67

S. NO.	NATIONAL PARKS	DATE OF ESTABLISHMENT	AREA Sq. Kms.
6	Pench (Priyadarshini)	22/11/1975	295. 85
7	Sanjay	23/09/1981	466.88
8	Satpura	13/10/1981	585. 17
9	Van Vihar	27/07/1979	04. 45
WLF			
1	Bagdara	15/02/1978	478.00
2	Bori	01/08/1977	485. 72
3	Gandhi Sagar	18/04/1981	368. 62
4	Ghatigaon	12/06/1981	511.00
5	Karera	21/05/1981	202.21
6	Ken Gharial	21/10/1981	45. 20
7	Kheoni	04/12/1982	122.70
8	Narsingarh	21/09/1978	59. 18
9	National Chambal	22/12/1978	435.00
10	Nauradehi	01/02/1984	1194. 67
11	Orchha	22/09/1994	44. 91
12	Pachmarhi	01/06/1977	417. 78
13	Palpur Kuno	16/01/1981	344. 68
14	Panna (Gangau)	16/04/1979	64. 14
15	Panpatha	04/06/1983	245. 84
16	Pench	22/11/1975	118. 47
17	Phen	10/03/1983	110.74
18	Ralamandal	09/02/1989	02.74
19	Ratapani	02/07/1978	823. 84
20	Sailana	04/06/1983	12. 96
21	Sanjay Dubri	30/08/1975	364.59
22	Sardarpur	10/02/1983	348. 12
23	Singhori	02/07/1976	287. 91
24	Son Gharial	23/09/1981	41.80
25	Veerangana Durgavati	06/01/1987	23. 97
	Total		
	Area		**10814. 75**

Source- Floristic Diversity and Conservtion Strategies in India, Vol. V Ed. -N.P. Singh and K. P. Singh Botanical Survey of India 2002

BIBLIOGRAPHY

Anand Kumar *et al.,* **2014** Rediscovery of **Typhonium inopinatum** (Araceae) from India, with notes on the identity of **T. khandwense** Rheedea 24(2):120-123

Anonymus 2017 State of Forest Report-Forest and Tree Resources of States and Union territories, MoEF,Govt. of India

Bailey, L. H. 1954 (second printing) Manual of Cultivated Plants. The Macmillan Co. New York

Bhandari, M. M. 1995 Flora of the Indian Desert MPS Repros, 39, BGKT Extn., New Pali Road, Jodhpur

Bor, N. L. 1960 The Grasses of Burma, Ceylon, India and Pakistan (excluding Bambusae) Pergamon Press, London

Champion, H. G. and S.K. Seth 1968 A Revised Survey of Forest Types of India, New Delhi

Cooke, T. 1901-1908 Flora of the Presidency of Bombay, vols. 1-3 (BSI reprint) 1958, Calcutta

Dassanayake, M.D. and F.R. Fosberg, and Dassanayake,M.D. and W.D. Clayton 1980-2000 A Revised Handbook to the Flora of Ceylon vols. I-XIV. Oxford IBH New Delhi

Dixit, T.M., *et al.,* **2011 Vigna indica** A new species in the genus **Vigna** Savi Rheedea 21(1) : 1-7

Gamble, J.S. and C.E.C. Fisch. 1915 Flora of the Presidency of Madras, vols I-III (BSI Reprint) 1957, Calcutta

G. Gnansekaran, K.V. *et al.,* **2017** Taxonomy and distribution of **Andrographis longipedunculata** (Acanthaceae) Rheedea 27(2):159-164

Haines, H. H. 1921-1924 The Botany of Bihar and Orissa vols. I-III, (BSI Reprint) 1961, Calcutta

Hooker, J. D. 1872-1897 The Flora of British India vol. I-VII. Reeve and Co. London

Jain, S.K. and R. R. Rao 1977 A Handbook of Field and Herbarium Methods. Today & Tomorrow's Printers & Pub. New Delhi

Jalil, V.A., *et al.,* Revision of **Amorphophallus** Blume ex Decne. Sect. **Rhapidophyllus** (Schott) Engl. (Araceae) in India Bangladesh J. Pl. Taxon. 18(1): 1-26.2011.

Karthikeyan, S. A. 2000 Statistical Analysis of Flowering Plants of India, in Singh *et al.,* (eds.) Flora of India, Introductory Vol. (part II) 201-217 B.S.I. New Delhi

Khanna, K. K. *et al.,* **2001** Supplement to the Flora of Madhya Pradesh B.S.I. Calcutta

Khanna, K. K. & Anand Kumar 2006 Floristic Analysis of Angiosperms from MP. Bionature 26(1):13-20

Mathew, K.M. 1983 The Flora of Tamilnadu Carnatic vol. I-III Parts 1 and 2. Rapinat Herbarium, Tiruchirapalli

Mudgal, V., *et al.,* **1997** Flora of MP vol. II B.S.I. Calcutta

Mujaffar Shaikh *et al.,* **2017** A new species of **Chlorophytum** (Asparagaceae) from MP, India Phytotaxa 291(2):159-162

Naik, V. N. and Associates 1998 Flora of Marathwada Vol. I and II Amrut Prakashan, Aurangabad

Potdar, G. G., C.B. Salunkhe & S.R. Yadav 2012 Grasses of Maharashtra Shivaji University, Kolhapur

Rao, A. S. & A. S.K. Sastry 1964 An account of the Flowering Plants of Indore Dist. MP. Bull. Bot. Surv. India 6: 267-286

Raunkier, C. 1934 Life Forms of Plants and Statistical Plant Geography. Oxford, London

Saldanha, C. J. & D. H. Nicolson 1976 Flora of Hassan District, Karnataka, India, Amerind Publ. Co., Bombay

Samvatsar, S. 1996 The Flora of West. Trib. Madhya Pradesh. Scientific Publishers, Jodhpur

Samvatsar, S. and V.B. Diwanji 2007 Floral Diversity Assesment & Conservation with Reference to Sardar Sarovar Project Bishen Singh Mahendra Pal Singh Dehra Dun

Sanjappa, M. 1992 Legumes of India. Bishen Singh Mahendra Pal Singh, Dehra Dun

Sengupta,G. 1977 A resume of Botanical Explorations and Floristic Studies in the Central Indian State of M.P. Bull. Bot. Surv. India 19:71-88

Shaikh Mujaffar 2011 Studies on the Flora of East Nimar region of M.P. Ph. D. dissertation (unpublished) DAVV, Indore

Shaikh Mujaffar *et al.,* **2012 Amorphophallus konkanensis** Hett.,*et al.,* A new record from M.P. Sci. Res. Reporter 2(3): 298-301

Shaikh Mujaffar *et al.,* **2012 Salvadora alii** (Salvadoraceae) A new record for India Rheedea 22(2):80-82

Shaikh Mujaffar *et al.,* **2014 Eleocharis khandwensis** sp. nov. (Cyperaceae) from Madhya Pradcsh, India Nordic J. Bot. 32: 710-712.2014.

Shaikh Mujaffar 2015 Ceropegia odorata Nimmo ex J. Graham Addition to flora of Central India J. Threatened Taxa 7(4):6810-12

Singh, N.P. ***et al.,*** **2001** Flora of MP vol. III B.S.I. Calcutta

Singh, N.P. & S.P. Singh 2002 Floristic Diversity and Conservation Strategies in India vol. V Bot. Surv. India

Singh, P. *et al.,* New Additions to Indian Flora in 2014 Phytotaxonomy 16:1-11.2016

Sinha, B.K. and B.K. Shukla 2007 Synoptic flora of Khargone Dist., M.P. J. Econ. Taxon. Bot. 31(2):687-695

Sinha, B.K. and B.K. Shukla 2009 Ibid J. Econ. Taxon. Bot. 33(1): 120-147

Sisodiya, S. 2012. Study of Floristic Composition and Phytoresources of Barwani District. Ph.D. Dissertation (unpublished) DAVV, Indore

Sivadasan, M. & Abdul Jaleel 1998 Rediscovery of **Amorphophallus longiconnectivus** Bogner, a little known rare endemic Species of Araceae Rheedea 8(2): 243-247

Solanki,C. M. 1983 Flora of Indore Dist. Ph.D. Dissertation (unpublished) DAVV Indore

Subramanyam, K. 1962 Aquatic Angiosperms Bot. Monog. C.S.I.R. New Delhi

Surage, Kiran 2015 Diversity, Distribution and Ethnobotanical Studies on the Grasses of Nimar Region, M.P. Ph. D. Dissertation, (Unpublished) DAVV Indore

The Plant List 2013 (version 1.1) Accessed December, 2017

Tiwari, A.P., *et al.,* A new species of *Parahyparrhenia* (Poaceae : Andropogoneae) from India. Phytotaxa 446(1): 055-060. 2020

Tiwari, A.P., S. Gavade & Shaikh Mujaffar 2016 Lectotypification of *Eriocaulon duthiei* (Eriocaulaceae) Phytotaxa ISSN 1179-3155

Verma, D.M., *et al.,* **1993** Flora of MP vol. I, B.S.I. Calcutta

Verma, D. M & V. Mudgal 1999 in Mudgal, V. & P. K. Hajra (eds.) Floristic Diversity and Conservation Strategies in India 1999 Vol III BSI Calcutta.

ADDENDA

The following 2 species were collected when the mss had gone to the publisher.

Macrotyloma ciliatum (Willd.) Verdc., Kew Bull. 24(3):404. 1970; *Dolichos ciliatus* Klein ex Willd., Sp. Pl. 3:1049. 1802; Baker in Hook.f., Fl. Brit. India 2:210.1876; Khanna *et al.,* in Suppl. Fl. Madhya Pradesh 64.2001.

Perennial twiners with tuberous roots; leaves 3-foliolate; leaflets oblong, up to 7 × 2 cm, margin ciliate, appressed hairy on both surfaces; flowers yellow, 3-6, in subsessile cymes; calyx teeth setaceous; corolla exserted; pods linear-oblong, up to 4 × 1.2 cm, with a persistent, hooked base of style;seeds 3-4.

Fl.-Fr.: Sept. -Dec

Dhar only in hedges of fields.

Solanum violaceum Ortega, Nov. Pl. Deser. Dec. 56. 1798; Nicolson *et al.,* in Regnum Veg. 119: 252. 1988; *S. anguivi auct non.* Lam., 1794; Hara *et al.,* Pl. Nepal 3: 111. 1982; *S. indicum auct non* L., 1753; C.B. Clarke in Hook.f., Fl. Brit. India 4: 234. 1883; Srivastava & Dixit in Mudgal *et al.* Fl. Madhya Pradesh 2: 194. 1997. *Barhanta*

Erect, tomentose undershrubs with hooked prickles; leaves 6-15 × 5-10 cm, ovate-oblong, apex obtuse or subacture, base subcordate or truncate, margin sinuate-pinnatifid; flowers in corymbose cymes on *ca* 1.5 cm long pedicels; calyx 5 mm long, pubescent outside; corolla violet, *ca* 2.5 cm acrss, stellate-pubescent; berries globose; seeds smooth.

Fl.-Fr.: Throughtout the year.

Rare, Khandwa only (Kalibhit area).

INDEX TO BOTANICAL NAMES

Q

INDEX TO VERNACULAR NAMES